Lecture Notes in Networks and Systems

Volume 215

The series "Lecture Notes in Networks and Systems" publishes the latest developments in Networks and Systems—quickly, informally and with high quality. Original research reported in proceedings and post-proceedings represents the core of LNNS.

Volumes published in LNNS embrace all aspects and subfields of, as well as new challenges in, Networks and Systems.

The series contains proceedings and edited volumes in systems and networks, spanning the areas of Cyber-Physical Systems, Autonomous Systems, Sensor Networks, Control Systems, Energy Systems, Automotive Systems, Biological Systems, Vehicular Networking and Connected Vehicles, Aerospace Systems, Automation, Manufacturing, Smart Grids, Nonlinear Systems, Power Systems, Robotics, Social Systems, Economic Systems and other. Of particular value to both the contributors and the readership are the short publication timeframe and the world-wide distribution and exposure which enable both a wide and rapid dissemination of research output.

The series covers the theory, applications, and perspectives on the state of the art and future developments relevant to systems and networks, decision making, control, complex processes and related areas, as embedded in the fields of interdisciplinary and applied sciences, engineering, computer science, physics, economics, social, and life sciences, as well as the paradigms and methodologies behind them.

Indexed by SCOPUS, INSPEC, WTI Frankfurt eG, zbMATH, SCImago.

All books published in the series are submitted for consideration in Web of Science.

More information about this series at https://link.springer.com/bookseries/15179

S. Jyothi · D. M. Mamatha · Yu-Dong Zhang · K. Srujan Raju
Editors

Proceedings of the 2nd International Conference on Computational and Bio Engineering

CBE 2020

Springer

Editors
S. Jyothi
Department of Computer Science
Sri Padmavati Mahila University
Tirupati, India

D. M. Mamatha
Department of Bioscience and Sericulture
Sri Padmavati Mahila Univesity
Tirupati, India

Yu-Dong Zhang
School of Informatics
University of Leicester
Leicester, UK

K. Srujan Raju
Department of Computer Science
Engineering
CMR Technical Campus (CMRG)
Secunderabad, India

ISSN 2367-3370 ISSN 2367-3389 (electronic)
Lecture Notes in Networks and Systems
ISBN 978-981-16-1940-3 ISBN 978-981-16-1941-0 (eBook)
https://doi.org/10.1007/978-981-16-1941-0

This Springer imprint is published by the registered company Springer Nature Singapore Pte Ltd.
The registered company address is: 152 Beach Road, #21-01/04 Gateway East, Singapore 189721, Singapore

Preface

The objective of this conference is to provide a forum to experts, researchers and post-graduate students from different disciplines to present problems in several fields of Science, Technology and Engineering such as theoretical and computational aspects in Informatics, Physics, Chemistry, Mechanics, Biology, Economics and other sciences from the entire world in order to discuss high-level scientific questions and exchange solid knowledge of pure and applied sciences and investigate diverse backgrounds theoretically and practically.

The international conference on **Computational and Bio Engineering (CBE'20)** was held in Tirupati, Andhra Pradesh, India, from 4 to 5 December 2020 at Sri Padmavati Mahila Visvavidyalaym (SPMVV) which is a women's university. Tirupati City is situated in southeastern Andhra Pradesh State under Rayalaseema Region in Chittoor District, where Tirupati is considered as a major holy city for Hinduism and receives a significant amount of pilgrimage traffic. It is referred to as the "Spiritual Capital of Andhra Pradesh". Many attractions here are geared towards religious tourisms and another one is Education Hub. And one of the educational universities is SPMVV which was started in 1983 by Sri N. T. Rama Rao, the Chief Minister of Andhra Pradesh, to train the women students as better builders of the nation and to inculcate skills of leadership in all aspects of life.

Computational Engineering (CE) is a relatively new discipline that deals with the development and application of computational models and simulations often coupled with high-performance computing to solve complex physical problems arising in engineering analysis and design in natural phenomena.

Bio Engineering (BE) is a significant task of computational biology, and computational bio systems aim to develop and use efficient algorithms, data structures, visualization and communication tools with the goal of computer modelling of biological system. It involves the computer simulations of biological systems which includes cellular subsystems to analyse and visualize the complex connections of these cellular processes. In addition, computational models help investigators to systematically analyse system perturbations, develop hypotheses to guide the design of new experimental tests and ultimately assess the suitability of specific molecules as novel therapeutic targets. Today, engineering approaches are essential for biologists, enabling them to analyse complex physiological processes as well as for the pharmaceutical industry as a supporting drug discovery and development programme.

The tradition and root of this conference is to bring researchers from the international scientific community to discuss the advancements and the future perspectives in Data Science, Big Data Analytics, Cloud Computing, Bio Engineering, Computational Biology, Biomaterials, Bioinformatics and Biomedical fields.

Nowadays cloud and big data are new platforms to manage and analyse the biological databases. To address these aspects, scientists with computer science, biotechnology background, academicians, decision-makers, policy-maker's stakeholders, students, government agencies, industries and public consumers have shared their research finding and thoughts on this international conference.

The authors submitted more than 200 papers from across the global. This certainly attests to the widespread, international importance of the theme of the conference. Each paper was reviewed on the basis of originality, novelty and rigorousness. After the reviews, 90 papers were accepted for presentation, out of which 76 papers are finally being published in the proceedings.

The event was a 2-day programme comprised of 10 paper presentation sections. The themes are contributed, and scientific sessions range from theories to applications, reflecting a combination of computational and bio engineering. We are very gratified to have an exciting line up of featured speakers who are among the leaders in changing the view of computational bio engineering.

Anantha S. Babbili (Ph.D., The University of Iowa) is former Provost and Vice President for Academic Affairs of Texas A&M University-Corpus Christi (2007–2010) and former Dean of the College of Mass Communication (2002–2007) at Middle Tennessee State University. Invited as a chief guest has given gratitude for meeting researchers from around the world, widens professional contact to create new opportunities including establishing new collaborations. **Samuel jones** has been invited as Keynote Speaker, who is working as a **Sr. Principal IT Architect, USA** and has given a view on the role of bioengineering and the potential of computational methods for ground breaking healthcare products, therapies and services. **Prof. El-Sayed A. El-Sheikh, Zagazig University, Egypt** has been invited as Guest of Honour, and has encouraged the delegates to study the biology relating with computational engineering to be developed and the applications of computational models are simulations to solve complex physical problems for analysing and designing.

We give our gratitude to **Prof. D. Jamuna, Vice Chancellor of SPMVV** who has given valuable thoughts.

The conference would truly not be successful without the contributions and support received from authors, participants, keynote speakers, programme committee members, session chairs, organizing committee members, steering committee members and others in their various roles. Their valuable support, suggestions, dedicated commitment and hard work have made the **CBE'20** successful.

It has been a great honour to serve as programme chairs for the **CBE'20** and to work with the conference team. We believe this event will certainly help further disseminate new ideas and inspire more collaborative work.

Tirupati, India — Prof. S. Jyothi
Tirupati, India — Prof. D. M. Mamatha
Leicester, UK — Yu-Dong Zhang
Secunderabad, India — K. Srujan Raju

Contents

Implementation of Healthcare System Using Android Application Storing Health Records in Cloud 1
P. Divyaja and M. Usha Rani

Threshold and Segmentation-Based X-ray Imaging Analysis for Covid-19 Detection 11
K. Himabindu, K. Sekar, and K. Vandana Rao

Feature-Based Squids Classification Using Fuzzy Logic 17
K. Himabindu, S. Jyothi, and D. M. Mamatha

In Silico Platforms for Systems Toxicology 25
M. Lalasa, S. Nithya, K. Nagalakshmamma, A. Suvarnalatha, and P. Nageshwar Rao

Comparative Analysis on Factors Affecting Accuracy of Seizure Detection Using EEG Signal Processing 33
M. Zaheer Ahamed and S. Aruna Mastani

Identification of Coronavirus (COVID-19) Long Non-coding RNAs (lncRNAs) Using Machine and Deep Learning Approaches 43
A. Revathi, S. Jyothi, and P. Swathi

Predictive Analysis of Damage Occurred Due to Natural Disasters Using Whale-Optimization Algorithm-Based Hybrid Computation 51
K. Hemalatha, K. Hema, and V. Deepika

Remote Access for Advanced Mobile Communication Systems 59
Swetha Rani Karanam and Suseela Kallepalli

A Statistical Study on Coronary Artery Disease Using Discriminant Analysis 75
A. Vani, T. Suneetha, K. Sukanya, and T. Sukeerthi

A Study to Assess the Knowledge and Opinion of Mid-Level Health Providers of Chittoor District Regarding m-Health 85
B. G. Bhavani

Machine Learning Techniques for Identifying Diabetes and Its Complications Based on Long Non-coding RNAs 93
P. Swathi, S. Jyothi, and A. Revathi

Computational Method on Breast Cancer Survival Data Using Binary Classification Models 107
T. Sukeerthi, K. Sukanya, and K. Vandana Rao

Percentage Nucleotide Concentration and Classification of SARS–Corona Viruses 115
F. Amul Mary and S. Jyothi

The Role of Law and Regulation on the Intersection of Bioprospecting and Bio Piracy 129
Samykya Mukku, M. Lalasa, K. Nagalakshmamma, G. Savitri, P. Sujata, and K. S. Shanthi Sree

Robust Face Recognition Using Multi-scale Feature Pattern Sparse Representation with Extreme Learning Machine 135
Pallavaram Venkateswar Lal, Uppalapati Srilakshmi, and D. Venkateswarlu

A Review of Diabetic Retinopathy Screening Using Machine Learning 151
S. Nalini Durga and K. Usha Rani

A Novel Framework for Modeling Medical-Sensitive Big Data Using Document-Based Database 163
Sai Jyothi Bolla and S. Jyothi

Studies on Effect of Leaf Roller (*Diaphania pulverulentalis*) Infestation on the Mineral Composition of Mulberry (Morus Sp.) Varieties 177
C. T. Bhagyamma and N. Vijaya Kumari

Performance Comparison of Semantic Web Data Using Virtual and Cloud Services 187
C. Lakshmi and K. Usha Rani

Analysis of Bio Vanillin Production from Waste Material by HPTLC ... 195
M. Sujatha and R. Jaya Madhuri

Phytochemical Studies and In Vitro Biological Activity of Citrus and Pomegranate Peel Extracts 209
K. Gnaneswari, M. Vani, and B. Lakshmi Thanuja

Present and Future Prospectives of Microbial Fibrinolytic Enzyme Production and Its Applications 219
K. Gowthami and R. Jaya Madhuri

Molecular Docking of Bacterial Laccases for the Decolorization of Textile Reactive Azo Dyes 229
Uma Maheswari Devi Palemaplli and Vani Mathakala

Classification and Detection of Brain Tumor Using Segmentation Techniques on MRI Images 239
Gira Aparna, Gollapalli Sumana, and Gade Anitha Mary

Enhancing the Structure of Cyber Risks in Cloud Environment Using Cloud Forensics Technique 251
Vadetay Saraswathi Bai and T. Sudha

Facial Mask Detection Using Deep Learning and Machine Learning Approaches 259
P. Venkateswara Rao, K. Sai Susheel, and V. Sucharita

Application of Artificial Intelligence in Public Health Care in India 267
Sathvik Yerragolla, Sai Vennela Bandi, and Vijaya Lakshmi Peddiboyana

Comparative Study of Enzyme Kinetics in Biotransformation of Loratadine by Selected Fungi 279
Keerthana Morusu and Vidyavathi Maravajhala

A Study Report on Introducing Sanskrit Language Based on the Technologies Related to Current Trends 291
G. Nagalakshmi and T. Sarath

Celecoxib Topical Nanoemulgel: Formulation, *Ex-Vivo*, Pharmacodynamic, and Pharmacokinetic Studies 299
Y. Indira Muzib, Y. Sarah Sujitha, and Y. R. Ambedkar

Cloud-Based SAAS Applications with Better Solution Using Fiori 311
M. Usha Rani and Narasimhappagari Chandana Priya

A Modified BAT Optimization Algorithm to Segment MRIs of Brain Subregions for Early Detection of Alzheimer's Disease 323
Sai Sindhuri Nasina and A. Rama Mohan Reddy

Morphological and Molecular Identification of l-Asparaginase-Producing Marine Fungus Sarocladium kiliense 333
M. Bhargavi and R. Jaya Madhuri

Elicitation of Induced Systemic Resistance in Groundnut (*Arachis Hypogaea* L.) Plants by *Trichoderma* Spp. Against *Sclerotium Rolfsii* 343
Saraswathi Maddu and Jaya Madhuri Ravuri

Applications of Anomaly Detection in Manufacturing 355
P. Balasubramanyam and K. Sreenivasa Murthy

Pollution Monitoring System Using Raspberry Pi and Php Web Server 365
T. Monica and A. Sai Suneel

Genetic Diversity of Rice Cultivars (*Oryza Sativa L.*) Assessed Through Simple Sequence Repeat Marker 381
R. Usha and Ch Indhravathi

Classification of Diabetic Retinopathy Using PSO Clustering and Raspberry Pi 395
Bhimavarapu Usharani, Raju Anitha, and Ravi Kumar Tata

A Study on the Sericulture Farm Automation System Using the Internet of Things (IoT) 403
S. Jyothi, D. M. Mamatha, P. J. Raju, Sufia Sultana, and J. Seetharamulu

Secured Data Transfer Between Fog Nodes Using Blockchain 417
R. Priyadarshini and N. Malarvizhi

Computational Screening of Potential Phenolic Phytochemicals Against Targets of SARS-CoV-2 Main Protease (Mpro), S-Protein and RdRp by Molecular Docking 423
Sarath Nalla, Ramesh Eluri, and Vijetha Pendyala

The Research Importance and Possible Problem Domains for NoSQL Databases in Big Data Analysis 433
G. Somasekhar, Raj Kumar Patra, and K. Srujan Raju

Classification of Normal/Abnormal Heart Sound Recording Through Convolution Neural Network Through the Integration of Baseline and AdaBoost Classifier 441
P. Jyothi and G. Pradeepini

Integration of Artificial Intelligence and the Internet of Things with Blockchain Technology 449
K. Saritha, Muralidhar Kurni, K. Madhavi, and D. Nagadevi

Prominent Features in Sleep Disorder Disease in Bioinformatics Using Relevant Data Sets 459
P. Naga Deepthi, Naga Madhavi Latha Kakarla, and K. Swathi

High-Performance Encoded-Driven LFSR for Improved Fault Coverage 467
Avvaru Sai Mounika, Chilakapati Sripriya, and V. Sarada

Novel Hadamard Transform-Based Companding Technique for PAPR Mitigation in OFDM IDMA 477
Korada Lavanya and Recharla Divya

Automatic Vehicle Number Plate Recognition System (AVNPR) Using OpenCV Python ... 487
K. Suneetha and K. Mounika Raj

A Hybrid Framework for Prediction of Heart Disease Using Rough Set and Fuzzy Set Approach ... 497
Srikanth Meda and Raveendra Babu Bhogapathi

Coastal Indian Prawn Species Recognition System Through Molecular Identification and Neural Networks ... 507
D. M. Mamatha, S. Jyothi, K. Hari Priya, and S. Sharmila

A Substantial Approach to Predict Air Quality Using LVQ Neural Network ... 523
Sumaya Sanober and K. Usha Rani

An Automatic Identification of Diabetic Macular Edema Using Transfer Learning ... 533
Y. Nagendra Prasad, C. Shoba Bindu, E. Sudheer Kumar, and P. Dileep Kumar Reddy

Pharmacological Evaluation and Molecular Docking Studies of *Sorghum bicolor* for Diuretic Activity ... 543
S. Shobha, A. Sreedevi, and K. Sai Sruthi

A Review on Bio-computational Measures for the Control of Invasive Pest: *Tuta absoluta* (Tomato Leaf Miner) ... 553
Kalpana Sriramadasu and D. M. Mamatha

In Silico Studies on Juvenile Hormone Epoxide Hydrolase in *Spilarctia Obliqua* to Elucidate JH Analogs as Green Biopesticides ... 565
Swetha kumari Koduru and D. M. Mamatha

Optimisation of Forecast Error Through Combining the Forecast Results ... 577
M. Chandrasekhar Reddy and P. Balasubramanyam

A Comparative Performance Analysis of AOMDV and PAR Algorithms for Mobile Ad Hoc Networks ... 585
Salini Suresh, M. Manjunath, K. Ravikumar, and M. Hanumanthappa

Effect of Chemoradiation on Haematological Parameters in Cervical Cancer ... 593
B. Sai Lalitha, M. Malini, and Mahendran Botlagunta

A Novel Approach for Heart Disease Prediction Based on Risk Factors Using Machine Learning ... 603
K. Santhi and A. Rama Mohan Reddy

Big Data-Based Smart Grid Data Management Model for Improved Monitoring of Power Consumption 615
Doosetty Padmaja Kalala and P. Bhargavi

Metaheuristic Optimization Using Hybrid Algorithm in Cloud-Based Big Data Analytics 625
Amitkumar Manekar and G. Pradeepini

An Enhanced Autism Spectrum Disorder Detection Model Using Convolutional Neural Networks and Machine Learning Algorithms 631
T. Lakshmi Praveena and N. V. Muthu Lakshmi

Building an Efficient Feature Selection for Intrusion Detection System on UNSW-NB15 641
P. G. V. Suresh Kumar and Shaheda Akthar

An Adaptive Channel Assignment Strategy for Multi-radio Multi-channel Wireless Mesh Networks with Mobile Nodes Based on Internet Traffic 651
V. Sujatha, C. Shoba Bindu, G. Vijay Kumar, and P. Dileep Kumar Reddy

Multi-standard Schema-Based Classification of Geospatial Metadata in Spatial Data Infrastructures Using Feature Weight Induced Probabilistic Learning Scheme 661
Gangothri Rajaram and K. R. Manjula

Solar Irradiation Prediction Using Weather Forecast Data 679
S. Gayathri Devi, K. R. Manjula, P. P. Sasisharan, V. S. Rohith, and N. Ramya

EEG-Based Brain-Electric Activity Detection During Meditation Using Spectral Estimation Techniques 687
Padmavathi Kora, A. Rajani, M. C. Chinnaiah, K. R. Madhavi, K. Swaraja, and K. Meenakshi

Design and Implementation of Opportunistic Routing in WSN Using Multi-hop River Formation for IOT-Based Indoor Patient Smart Monitoring 695
Venkateswari Pichaimani and K. R. Manjula

A Cross-Cultural Assessment of the Influence of Knowledge Management Practices and Acceptance of Massive Open Online Courses (Moocs) Among Technical Students in India 709
Preethika Basava, Chetan Reddy Bayana, Priyanka Battula, and K. R. Manjula

Review Paper on Anomaly Detection in Data Streams 721
G. Sandhya Madhuri, Yamuna, and M. Usha Rani

Significance of Primary Baseline Databases for Conservation of Biodiversity 729
Suvarna Latha Anchapakula, Bharathi Depuru, Lakshmi Padmavathi Pydipati, and Hemavathi Buddareddy

Effectiveness of Plant-Based Biomaterials in the Field of Biomedical Engineering 735
Lakshmi Padmavathi Pydipati, Bharathi Depuru, Suvarna Latha Anchapakula, and Hemavathi Buddareddy

Real-Time Video Tracking Framework With Moving Object Segmentation in Stream Data 745
K. R. Manjula and S. Anand Kumar Varma

Short Circuit Fault Analysis of Three Phase Transmission System Using Matlab 759
D. Srilatha and R. V. S. Lakshmi Kumari

Text Summarization of Multiple Documents Using Binary Fruit Fly Optimization Algorithm 769
Kishore Kumar Mamidala and Suresh Kumar Sanampudi

Correction to: Classification of Normal/Abnormal Heart Sound Recording Through Convolution Neural Network Through the Integration of Baseline and AdaBoost Classifier C1
P. Jyothi and G. Pradeepini

Author Index 779

Editors and Contributors

About the Editors

Dr. S. Jyothi is Senior Professor and Head of the Department in Computer Science at Sri Padmavathi Women's University, Tirupati, Andhra Pradesh, India. She is also Secretary of SPMVV Society for Innovation Incubation Entrepreneurship (SSIIE) which has TBI & WBIF, Coordinator of CURIE-AI, National Innovation Startup Policy (NISP), Coordinator and Vice-President of Institute's Innovation Council (IIC). She handled many administrative positions in the University. She is BOS Member of several Universities. She is NAAC peer team Member. Seventeen Ph.D.s and seven M.Phil.s were awarded and nine Ph.D. scholars are being guided under her supervision. More than 250 papers published and presented in international journals and conferences. 12 book chapters are published and 12 books are authored and edited by her. She has filed 2 patents and one international patent publication in her credit. She has conducted 3 international and 4 national conferences, 10 hands-on workshops & faculty development programmes and acted as Keynote Speaker, Invited Speaker and Session Chair of international and national conferences.

She has completed UGC, DBT major Projects and ISRO-NRSC consultancy projects and handling DBT major project. She is honoured and awarded Marquis Marque Lifetime achievement award, Venus Lifetime achievement award, Best Research award, AP State Best Teacher Award, Rashtriya Gaurav Award and Best Citizens of India Award. She acted as Editorial Board Member of international journals, Technical Committee Member of international conferences and Reviewer of many research papers for international journals and international conferences. She is Senior Member of IEEE, ACM and IACSIT, Fellow of ISCA, IETE & SSARSC, Member of IAENG & Gyancity Research Labs and Life Member of CSI, ISTE, ISCA, IFERP, IUPRAI and ISRS. Her research interest is in image processing, artificial intelligence, machine learning, soft computing, data mining, big data analytics, bioinformatics and hyperspectra image analysis.

Dr. D. M. Mamatha is Fulbright Scholar and Professor from the Department of Bioscience & Sericulture at Sri Padmavathi Women's University, Tirupati, Andhra

Pradesh, India. She is the current Registrar of Sri Padmavati Women's University. She obtained six major collaborative multidisciplinary research projects from DBT, DST and UGC to the tune of Rs. 250.56 lakhs (approximately) and completed three international research fellowships like Indo-US & Indo-Egypt and Fulbright Fellowship during the period 2010 to 2018. Her area of research is wide ranging on cloning, expression, rDNA technology, DNA barcoding and silk biomaterials. She had been to several countries, viz. Singapore, Malaysia, Thailand, USA, Hong Kong, Canada, Egypt, Nepal and USA on various academic, research and international collaborative assignments. Her relentless efforts made her to publish several international publications in Scopus journals with high impact factor and a high citation index of 1505. Her works speak about her scientific calibre, which had been appreciated and posted in UCDAVIS website by her mentor Prof. Bruce Hammock, Distinguished Professor and Well-Renowned Scientist upon her visit to UCDAVIS.

Apart from the 'Fulbright award', Dr. Mamatha received several significant awards like 'Achievement award' from WHO-UNESCO, 'DBT-Young scientist Associateship' award from Government of India, Twice received INSA-NASI-IAS 'National science academies Summer research fellowship awards' for the years 2009 and 2011, 'Best Teacher award' from The Lion's club, 'Best Research award' from Indus foundation, and several 'Best research paper' and presentations awards which demonstrate her academic and research capacities. Her multidisciplinary research on automated identification of prawn species using DNA coding and neural networks has gained research team a national and international patent for the technology generated. Another thrust area is the 'Development of Silk fibroin Biomaterials' on the offing, where India has a large potential to hit in the coming years. She is currently guiding postdocs, Ph.D.s and M.Phil. scholars. One postdoc, two Ph.D.s, and one M.Phil. and several master's course dissertations have been awarded under her guidance. She published 5 books and contributed 5 chapters to different books from reputed publishers.

Dr. Yu-Dong Zhang (Eugene) is Professor and Chair in Knowledge Discovery and Machine Learning. He has Memberships in Associate Fellow of Higher Education Academy, IEEE Senior Member, ACM Senior Member and Fellow of IET. He has more than 400 publications with 12654 citations in Google Scholar with h index 62 and 8336 citations in Scopus with h index 52. His research interest includes deep learning, convolutional neural network, neural networks, biomedical image analysis, bio-inspired computing, pattern recognition, transfer learning, biomedical image analysis, etc. Dr. Yu-Dong Zhang is Honorary Professor of Henan Polytechnic University and King Abdulaziz University. He won MDPI Top 10 Most Cited Papers 2015 from MDPI, Switzerland, Emerald Citation of Excellence 2017, Emerald, UK, 'Top Scientist' in Guide2Research, Most Cited Chinese Researchers (Computer Science) by Elsevier, Highly Cited Researcher 2019 by Web of Science and Springer Nature CNY2020 Collection by Springer Nature. Dr. Yu-Dong Zhang (Eugene) served as (Associate) Editor of Neural Networks, Scientific Reports, Journal of

Alzheimer's Disease, IEEE Transactions on Circuits and Systems for Video Technology, IEEE Access, Neurocomputing, etc. He served as Leading Guest Editor of Information Fusion, Neural Networks, IEEE Access, etc.

He served as Guest Editor of IEEE Transactions on Intelligent Transportation Systems, International Journal of Information Management and other peer-reviewed journals. He served as General Chair/ Co-Chair/TPC Chair/ Workshop Chair/ Session Chair of IEEE/Springer conferences. He has delivered academic talk in Doctoral School Teaching and invited talks in reputed Universities. He served as Senior Committee Members in number of conferences. He has research team which consists of 3 staffs, 8 Ph.D. students and 12 visiting students/scholars. Dr. Yu-Dong Zhang has several research projects in various disciplines from various funding agencies.

Dr. K. Srujan Raju is currently working as Dean of Student Welfare and Heading Department of Computer Science & Engineering and Information Technology at CMR Technical Campus. He obtained his Doctorate in Computer Science in the area of Network Security. He has more than 20 years of experience in academics and research. His research interest areas include computer networks, information security, data mining, cognitive radio networks, image processing and other programming languages. Dr. Raju is presently working on 2 projects funded by Government of India under CSRI & NSTMIS; he has also filed 7 patents and 1 copyright at Indian Patent Office, edited more than 14 books published by Springer Book Proceedings of AISC series, LAIS series and other which are indexed by Scopus, also authored books in C Programming & Data Structure, Exploring to the Internet and Hacking Secrets, contributed chapters in various books and published more than 40 papers in reputed peer-reviewed journals and international conferences.

Dr. Raju was invited as Session Chair, Keynote Speaker, Technical Programme Committee Member, Track Manager and Reviewer for many national and international conferences and also appointed as Subject Expert by CEPTAM DRDO – Delhi & CDAC. He has undergone specific training conducted by Wipro Mission 10X & NITTTR, Chennai, which helped his involvement with students and is very conducive for solving their day to day problems. He has guided various student clubs for activities ranging from photography to Hackathon. He mentored more than 100 students for incubating cutting-edge solutions. He has organized many conferences, FDPs, workshops and symposiums. He has established the Centre of Excellence in IoT, Data Analytics. Dr. Raju is Member of various professional bodies and received a Significant Contributor Award and Active Young Member award from Computer Society of India and also served as Management Committee Member, State Student Coordinator & Secretary of CSI – Hyderabad Chapter.

Contributors

Shaheda Akthar Department of Computer Science, Government College for Women, Guntur, AP, India

Y. R. Ambedkar Sri Venkateswar Veterinary University, Tirupati, India

F. Amul Mary Department of Computer Science, SPMVV, Tirupathi, India

S. Anand Kumar Varma E.G.S. Pillay College, Namakkal, Nagapattinam, India

Suvarna Latha Anchapakula Faculty, Department of Biosciences and Sericulture, Sri Padmavati Mahila Visvavidyalayam (Women's University), Tirupati, Andhra Pradesh, India

Raju Anitha Department of Computer Science and Engineering, Koneru Lakshmaiaha Education Foundation, Vaddeswaram, Andhra Pradesh, India

Gira Aparna ECE Department, University College of Engineering, Hyderabad, Telangana, India

S. Aruna Mastani ECE, JNTUA, Anantapuram, India

Vadetay Saraswathi Bai Department Computer Science and Engineering, School of Engineering and Technology, SPMVV, Tirupati, India

P. Balasubramanyam Goa Institute of Management, Goa, India

Sai Vennela Bandi Department of Computer Science, SRM University, Chennai, India

Preethika Basava Department of Computer Science Engineering, SASTRA University, Thanjavur, India

Priyanka Battula Department of Computer Science Engineering, SASTRA University, Thanjavur, India

Chetan Reddy Bayana Department of Computer Science Engineering, SASTRA University, Thanjavur, India

C. T. Bhagyamma Department of Biosciences and Sericulture, Sri Padmavati Mahila Visvavidyalayam (Women's University), Tirupati, Andhra Pradesh, India

M. Bhargavi Department of Applied Microbiology, Sri Padmavati Mahila Visvavidyalayam, Tirupati, India

P. Bhargavi Department of Computer Science, Sri Padmavati Mahila University, Tirupati, AP, India

B. G. Bhavani College of Nursing, SPMVV, Tirupati, India

Raveendra Babu Bhogapathi Chalapathi Institute of Engineering and Technology, Guntur, India

Sai Jyothi Bolla Department of Information Technology, Vasireddy Venkatadri Institute of Technology, Guntur, India

Mahendran Botlagunta Department of Indo-American Cancer Research Foundation, Basavatarakam Indo-American Cancer Hospital and Research Centre, Hyderabad, Telangana, India

Hemavathi Buddareddy Faculty, Department of Biosciences and Sericulture, Sri Padmavati Mahila Visvavidyalayam (Women's University), Tirupati, Andhra Pradesh, India

M. Chandrasekhar Reddy Department of Statistics, College of Natural Sciences, Arba Minch University, Arba Minch, Ethiopia

M. C. Chinnaiah BVRIT, Hyderabad, Telangana, India

V. Deepika Sri Venkateswara College of Engineering, Tirupati, India

Bharathi Depuru Faculty, Department of Biosciences and Sericulture, Sri Padmavati Mahila Visvavidyalayam (Women's University), Tirupati, AP, India

P. Dileep Kumar Reddy Department of CSE, SV College of Engineering, Tirupathi, Andhra Pradesh, India

Recharla Divya Department of ECE, Sri Padmavathi Mahila University, Tirupati, AP, India

P. Divyaja Department of Computer Science, SPMVV, Tirupati, India

Ramesh Eluri Department of Computer Science and Engineering, RVR & JC College of Engineering, Guntur, AP, India

S. Gayathri Devi Computer Science & Engineering, School of Computing, SASTRA Deemed to be University, Thanjavur, India

K. Gnaneswari Department of Applied Microbiology & Biochemistry, Sri Padmavati Mahila Visvavidyalayam, Tirupati, AP, India

K. Gowthami Department of Applied Microbiology, Sri Padmavathi Mahila Visvavidhyalayam, Tirupati, India

M. Hanumanthappa Bangalore University, Bangalore, India

K. Hari Priya Department of Biosciences and Sericulture, Sri Padmavati Women's University, Tirupati, AP, India

K. Hema Sri Venkateswara College of Engineering, Tirupati, India

K. Hemalatha Sri Venkateswara College of Engineering, Tirupati, India

K. Himabindu Department of MCA, SVCE, Tirupati, India

Ch Indhravathi Department of Biotechnology, Tirupati, Andhra Pradesh, India

R. Jaya Madhuri Department of Applied Microbiology, Sri Padmavati Mahila Visvavidyalayam, Tirupati, Andhra Pradesh, India

P. Jyothi Department of CSE, Koneru Lakshmaiah Education Foundation, Vaddeswaram, India

S. Jyothi Department of Computer Science, SPMVV, Tirupati, India;
Department of Computer Science, Sri Padmavati Women's University, Tirupati, AP, India

Naga Madhavi Latha Kakarla Department Of Computer Science and Engineering, Sir C R Reddy College of Engineering, Vatluru, AP, India

Doosetty Padmaja Kalala Department of Computer Science and Engineering, Sri Padmavati Mahila University, Tirupati, AP, India

Suseela Kallepalli Department of ECE, Sri Padmavathi Mahila University, Tirupati, AP, India

Swetha Rani Karanam Department of ECE, Sri Padmavathi Mahila University, Tirupati, AP, India

Swetha kumari Koduru Molecular Cloning Lab, Department of Biosciences and Sericulture, Tirupati, India

Padmavathi Kora GRIET, Hyderabad, Telangana, India

Muralidhar Kurni JNTUA, Anantapuramu, AP, India

R. V. S. Lakshmi Kumari Department of EEE, Gayatri Vidya Parishad College of Engineering for Women, Vishakapatnam, AP, India

T. Lakshmi Praveena Sri Padmavathi Mahila Visvavidyalayam, Tirupati, AP, India

B. Lakshmi Thanuja Department of Applied Microbiology & Biochemistry, Sri Padmavati Mahila Visvavidyalayam, Tirupati, AP, India

C. Lakshmi Department of Computer Science, Sri Padmavati Mahila Visvavidyalayam, Tirupati, AP, India

M. Lalasa Department of Computer Science, Swetha Institute of Technology and Science, JNTU, Tirupati, AP, India

Korada Lavanya Department of ECE, Sri Padmavathi Mahila University, Tirupati, AP, India

Saraswathi Maddu Department of Applied Microbiology, Sri Padmavati Mahila Visvavidyalayam, Tirupati, A.P., India

K. Madhavi Department of CSE, JNTUACEA, Anantapuramu, AP, India

K. R. Madhavi Sree Vaidyanathan Engineering College, Tirupati, AP, India

R. Jaya Madhuri Department of Applied Microbiology, Sri Padmavati Mahila Visvavidyalayam, Tirupati, India

N. Malarvizhi Veltech Rangarajan Dr. Sagunthala R&D Institute of Science and Technology, Chennai, India

M. Malini Department of Biomedical Engineering, University College of Engineering, Osmania University, Hyderabad, India

D. M. Mamatha Department of Biosciences and Sericulture, Sri Padmavati Women's University, Tirupati, AP, India;
Molecular Cloning Lab, Department of Biosciences and Sericulture, Tirupati, India

Kishore Kumar Mamidala Department of Computer Science & Engineering, Vivekananda Institute of Technology & Science, Karimnagar, India

Amitkumar Manekar Department of CSE, Koneru Lakshmaiah Education Foundation, Vaddeswaram, Andhra Pradesh, India

K. R. Manjula Jain University, School of Computing, SASTRA Deemed University, Thanjavur, Tamil Nadu, India;
Computer Science & Engineering, School of Computing, SASTRA Deemed to be University, Thanjavur, India

M. Manjunath R V College of Engineering, Bangalore, India

Vidyavathi Maravajhala Institute of Pharmaceutical Technology, Sri Padmavati Mahila Visvavidyalayam, Tirupati, AP, India

Gade Anitha Mary MCA Department, Loyola College, Hyderabad, Telangana, India

Vani Mathakala Department of Applied Microbiology & Biochemistry, Sri Padmavati Mahila Visvavidyalayam, Tirupati, AP, India

Srikanth Meda Acharya Nagarjuna University, Guntur, India

K. Meenakshi GRIET, Hyderabad, Telangana, India

T. Monica Department of Electronics and Communication Engineering, School of Engineering and Technology, Sri Padmavati Mahila Visvavidyalayam, Tirupati, India

Keerthana Morusu Institute of Pharmaceutical Technology, Sri Padmavati Mahila Visvavidyalayam, Tirupati, AP, India

K. Mounika Raj Sree Vidyanikethan Engineering College, Rangampet, AP, India

Samykya Mukku New Delhi, India

N. V. Muthu Lakshmi Sri Padmavathi Mahila Visvavidyalayam, Tirupati, AP, India

Y. Indira Muzib Sri Padmavati Mahila ViswaVidyalayam, Tirupati, India

P. Naga Deepthi Department Of Computer Science and Engineering, Koneru Lakshmaiah Education Foundation, Vaddeswaram, AP, India

D. Nagadevi Department of ECE, CBIT, Hyderabad, Telangana, India

K. Nagalakshmamma Department of Biosciences and Sericulture, Sri Padmavati Mahila Vishwavidyalayam, Tirupati, AP, India

G. Nagalakshmi Department of CS, Rashtriya Sanskrit Vidyapeetha, Tirupati, India

Y. Nagendra Prasad Department of CSE, JNTUA, Ananthapuramu, Andhra Pradesh, India

P. Nageshwar Rao Department of Computer Science, Rayalaseema University, Kurnool, India

S. Nalini Durga Department of Computer Science, Sri Padmavati Mahila Viswavidyalayam, Tirupati, Andhra Pradesh, India

Sarath Nalla Department of Pharmaceutical Chemistry, University College of Pharmaceutical Sciences, Andhra University, Visakhapatnam, AP, India

Sai Sindhuri Nasina Computer Science and Engineering, Sri Venkateswara University College of Engineering, Tirupati, India

S. Nithya Institite of Pharmaceutical Technology, Sri Padmavati Mahila Visvavidyalayam, Tirupati, India

Uma Maheswari Devi Palemaplli Department of Applied Microbiology & Biochemistry, Sri Padmavati Mahila Visvavidyalayam, Tirupati, AP, India

Raj Kumar Patra Department of Computer Science and Engineering, CMR Technical Campus, Hyderabad, Telangana, India

Vijaya Lakshmi Peddiboyana Department of Communication and Journalism, Sri Padmavati Mahila Visvavidyalayam, Tirupati, India

Vijetha Pendyala Department of Pharmacognosy& Phytochemistry, Chebrolu Hanumaiah Institute of Pharmaceutical Sciences, Guntur, AP, India

Venkateswari Pichaimani CSE, SRC, SASTRA Deemed University, Thanjavur, TN, India

G. Pradeepini Department of CSE, Koneru Lakshmaiah Education Foundation, Vaddeswaram, India

R. Priyadarshini Veltech Rangarajan Dr. Sagunthala R&D Institute of Science and Technology, Chennai, India

Lakshmi Padmavathi Pydipati Faculty, Department of Biosciences and Sericulture, Sri Padmavati Mahila Visvavidyalayam (Women's University), Tirupati, AP, India

A. Rajani JNTUH, Hyderabad, Telangana, India

Gangothri Rajaram Jain University, School of Computing, SASTRA Deemed University, Thanjavur, Tamil Nadu, India

P. J. Raju A. P. State Sericulture Research and Development Institute, Hindupur, India

A. Rama Mohan Reddy CSE Department, SV University College of Engineering, Tirupati, India

N. Ramya Computer Science & Engineering, School of Computing, SASTRA Deemed to be University, Thanjavur, India

M. Usha Rani Department of Computer Science, SPMVV, Tirupati, India

K. Ravikumar Kalinga Institute of Industrial Management, Bhuvaneshwar, India

Jaya Madhuri Ravuri Department of Applied Microbiology, Sri Padmavati Mahila Visvavidyalayam, Tirupati, A.P., India

A. Rama Mohan Reddy Computer Science and Engineering, Sri Venkateswara University College of Engineering, Tirupati, India

A. Revathi Department of Computer Science, SPMVV, Tirupati, India

V. S. Rohith Computer Science & Engineering, School of Computing, SASTRA Deemed to be University, Thanjavur, India

B. Sai Lalitha Department of Biomedical Engineering, University College of Engineering, Osmania University, Hyderabad, India

Avvaru Sai Mounika Department of ECE, Sri Padmavathi Mahila University, Tirupati, Andhra Pradesh, India

K. Sai Sruthi Institute of Pharmaceutical Technology, Tirupati, Andhra Pradesh, India

K. Sai Susheel Department of Computer Science and Engineering, Narayana Engineering College, Gudur, AP, India

Suresh Kumar Sanampudi Department of Information Technology, JNTUH College of Engineering Jagtial, Telangana, India

G. Sandhya Madhuri Department of Computer Science, SPMVV, Tirupati, India

Sumaya Sanober SPMVV, Tirupati, India

K. Santhi CSE Department, SV University College of Engineering, Tirupati, India

V. Sarada Department of ECE, SRMIST, Chennai, Tamil Nadu, India

T. Sarath Department CSE, Siddhartha Institute of Science and Technology, Puttur, India

K. Saritha Sri Venkateswara Degree and PG College, Anantapuramu, AP, India

P. P. Sasisharan Computer Science & Engineering, School of Computing, SASTRA Deemed to be University, Thanjavur, India

G. Savitri Department of Biosciences and Sericulture, Sri Padmavati Mahila Vishwavidyalayam, Tirupati, AP, India

J. Seetharamulu A. P. State Sericulture Research and Development Institute, Hindupur, India

K. Sekar Department of MCA, SVCE, Tirupati, India

K. S. Shanthi Sree Department of Biosciences and Sericulture, Sri Padmavati Mahila Vishwavidyalayam, Tirupati, AP, India

S. Sharmila Department of Biosciences and Sericulture, Sri Padmavati Women's University, Tirupati, AP, India

C. Shoba Bindu Department of CSE, JNTUA, Ananthapuramu, Andhra Pradesh, India

S. Shobha Institute of Pharmaceutical Technology, Tirupati, Andhra Pradesh, India

G. Somasekhar Department of Computer Science and Engineering, CMR Technical Campus, Hyderabad, Telangana, India

A. Sreedevi Institute of Pharmaceutical Technology, Tirupati, Andhra Pradesh, India

K. Sreenivasa Murthy Department of Management, Central University of Tamil Nadu, Thiruvarur, India

Uppalapati Srilakshmi Department of CSE, VFSTR Deemed to be University, Vadlamudi, Guntur, Andhra Pradesh, India

D. Srilatha Department of EEE, Vasireddy Venkatadri Institute of Technology, Nambur, AP, India

Chilakapati Sripriya Department of ECE, Sri Padmavathi Mahila University, Tirupati, Andhra Pradesh, India

Kalpana Sriramadasu Department of Biosciences & Sericulture, Sri Padmavati Women's University, Tirupati, AP, India

K. Srujan Raju Department of Computer Science and Engineering, CMR Technical Campus, Hyderabad, Telangana, India

V. Sucharita Department of Computer Science and Engineering, Narayana Engineering College, Gudur, AP, India

T. Sudha Department Computer Science, SPMVV, Tirupati, India

E. Sudheer Kumar Department of CSE, JNTUA, Ananthapuramu, Andhra Pradesh, India

P. Sujata Department of Biosciences and Sericulture, Sri Padmavati Mahila Vishwavidyalayam, Tirupati, AP, India

M. Sujatha Department of Applied Microbiology, Sri Padmavati Mahila Visvavidyalayam, Andhra Pradesh, Tirupati, India

V. Sujatha Department of CSE, JNTUA, Ananthapuramu, Andhra Pradesh, India

Y. Sarah Sujitha Sri Padmavati Mahila ViswaVidyalayam, Tirupati, India

K. Sukanya Department of Statistics, Sri Padmavati Mahila Visvavidyalayam, Tirupati, India

T. Sukeerthi Department of Statistics, Sri Padmavati Mahila Visvavidyalayam, Tirupati, India

Sufia Sultana JRF-DBT, Department of Biosciences and Sericulture, Sri Padmavati Women's University, Tirupati, India

Gollapalli Sumana Command Control Center, Sri Padmavati Mahila Visvavidyalayam, Tirupati, AP, India

A. Sai Suneel Department of Electronics and Communication Engineering, School of Engineering and Technology, Sri Padmavati Mahila Visvavidyalayam, Tirupati, India

K. Suneetha School of CS & IT, Jain (Deemed-to-be) Univesity, Bangalore, Karnataka, India

T. Suneetha Department of Statistics, Sri Padmavati Mahila Visvavidyalayam, Tirupati, India

P. G. V. Suresh Kumar Department of CSE, Acharya Nagarjuna University, Guntur, AP, India

Salini Suresh Dayananda Sagar College, Bangalore, India

A. Suvarnalatha Department of Biosciences and Sericulture, Sri Padmavati Mahila Visvavidyalayam, Tirupati, India

K. Swaraja GRIET, Hyderabad, Telangana, India

K. Swathi Department Of Computer Science and Engineering, Koneru Lakshmaiah Education Foundation, Vaddeswaram, AP, India

P. Swathi Department of Computer Science, SPMVV, Tirupati, India

Ravi Kumar Tata Department of Computer Science and Engineering, Koneru Lakshmaiaha Education Foundation, Vaddeswaram, Andhra Pradesh, India

K. Usha Rani Department of Computer Science, Sri Padmavati Mahila Viswavidyalayam, Tirupati, Andhra Pradesh, India

M. Usha Rani Department of Computer Science, SPMVV, Tirupati, India

R. Usha Department of Biotechnology, Tirupati, Andhra Pradesh, India

Bhimavarapu Usharani Department of Computer Science and Engineering, Koneru Lakshmaiaha Education Foundation, Vaddeswaram, Andhra Pradesh, India

K. Vandana Rao Department of Computer Science, SPMVV, Tirupati, India

A. Vani Department of Statistics, Sri Padmavati Mahila Visvavidyalayam, Tirupati, India

M. Vani Department of Applied Microbiology & Biochemistry, Sri Padmavati Mahila Visvavidyalayam, Tirupati, AP, India

P. Venkateswara Rao Department of Computer Science and Engineering, Narayana Engineering College, Gudur, AP, India

D. Venkateswarlu Department of CSE, VFSTR Deemed to be University, Vadlamudi, Guntur, Andhra Pradesh, India

G. Vijay Kumar Department of CSE, G.Pulla Reddy Engineering College, Kurnool, Andhra Pradesh, India

N. Vijaya Kumari Department of Biosciences and Sericulture, Sri Padmavati Mahila Visvavidyalayam (Women's University), Tirupati, Andhra Pradesh, India

Yamuna Department of Computer Science, SPMVV, Tirupati, India

Sathvik Yerragolla CIDSE, Arizona State University, Phoenix, USA

M. Zaheer Ahamed ECE, JNTUA, Anantapuram, India

Implementation of Healthcare System Using Android Application Storing Health Records in Cloud

P. Divyaja and M. Usha Rani

Abstract Nowadays, one-day rapid growth in technology has been seen. Within the context of recent healthcare, the utilization of IoT technologies brings their convenience to physicians and patients as they are applied to varied medical areas. The development of new technology in healthcare applications without regard to safety makes patient privacy vulnerable. In this study, we highlighted key security requirements when storing patient information in the cloud. When we consider the hospitality facilities in the villages, they face a lot of problems with the fact that the doctors are not able to visit the rural clinics often; they always want to take all the medical history, medicines, and records with them; and they have to stand in front of the hospitals. Accordingly, providing an honest quality of life for people has become a big social challenge at that point. The rapid proliferation of knowledge and communication technologies has led to the approval of innovative trend solutions and tools that show commitment to meeting challenges. There has been a long struggle to maintain access to quality health services with consulting physicians in rural areas. When we consider the major challenges facing rural public hospitality, such as lack of infrastructure, poor manpower, cost, quality of health services, and time required to provide services to patients, the treatment recommendations of the IoT doctor and patient updated treatment information will be stored directly in the [1] cloud [2]. In order to be effective in storing information on the cloud, we must provide secure communication protocols. We propose a secure IoT based on the sensors used in modern healthcare.

Keywords Cloud · Data privacy · Data integrity · Data authentication · Data refresh and IOT

P. Divyaja (✉) · M. Usha Rani (✉)
Department of Computer Science, SPMVV, Tirupati, India
e-mail: pdivyajahime@gmail.com

M. Usha Rani
e-mail: musha_rohan@yahoo.com

S. Jyothi et al. (eds.), *Proceedings of the 2nd International Conference on Computational and Bio Engineering*, Lecture Notes in Networks and Systems 215,
https://doi.org/10.1007/978-981-16-1941-0_1

1 Introduction

1.1 Cloud Computing

The definition of cloud computing is "It is the way of data faraway from the PC, Desktop and Laptop output of the results computing will be main location i.e., central location that can be depend on the good Internet Connectivity". Today, clouds are smart and automated. In healthcare cloud computing is used for storing patient information in a remote server. Cloud maintains patient data collections, healthcare history, results, quality of service, effective and efficient care.

1.2 Healthcare Using IoT

The Internet of things is a platform where regular devices are connected to the NET so that can interact, collaborate, and exchange data with each other.

The IoT device in the health system gathers and converts patient's health data (PHD), such as blood sugar levels, oxygen, blood pressure (BP), weight, and electrocardiogram (ECG) [3]. These data can be stored in the cloud and shared with an authorized person, who may be a physician or patient, and allow the collected data to be viewed regardless of their location, time, or device.

An IoT system consists of [4] sensors or devices which "talk" to the cloud through some kind of technology. Once the data gets into the cloud, the software processes it and then might decide to show the required page.

1.3 Security in Healthcare System

Security causes a major impact on the cloud storage system using [7] Android application. The two most important security issues in the healthcare system are:

1. As per security part, patient's healthcare organizations (PHO) failed to update their sensitive and secure information quickly.
2. There are many vulnerabilities in finding personal patient information especially with criminals.

2 Existing System

People have a completely different perspective on security, so it is outlined in many ways. In general, security can be a framework equivalent to the security of an overall system. Now, communication in sensor network applications is generally wireless in nature. This can lead to different security threats to those systems. This part unit security issues cause serious problems for cloud wireless detection devices. During this section, we tend to describe the key security-tie requirements in the IoT-based focus system [5].

2.1 Disadvantages

- People in the village suffer to get doctors visiting the hospitals frequently.
- In hospitals patients have to queue up.
- Even though the health problem is minor, illness needs a physical check-up.

3 Proposed System

A system is proposed to protect the stored information about patient health details in the cloud. In rural hospitals, IoT-based smart bed is used [6]. The IoT device part gathers and modifies health information such as blood sugar levels, oxygen, blood pressure (BP), weight, and electrocardiogram (ECG). This information is updated in the main cloud data center and can be shared with an approved individual; it may be a physician or patient, thus allowing the collected data to be viewed irrespective of their location, time, or device. The perceived data in a computer or mobile or tab is retrieved by an IoT device [8]. So, if the disease cannot be cured by the nearest nurse or doctor, then the doctors can monitor the health of the people and guide them to cure the disease accordingly.

Patient data will be updated each time. The updated data can be accessed from anywhere in the world. In urban areas, basic tests can be performed with connected devices and results can be sent directly to their doctors for quick response and record keeping.

3.1 Advantages

- Easy monitoring of patients with more care
- Facilitates remote place consultation of patients
- Facilitates effective treatment
- Gives health forecasting
- Easily access data whenever it is needed.

4 Doctor Application

In this application, first the doctor needs to be authenticated by their reference ID which is already there in the Firebase (cloud).

First, the doctor will login with their E-mail id and Password into the application.

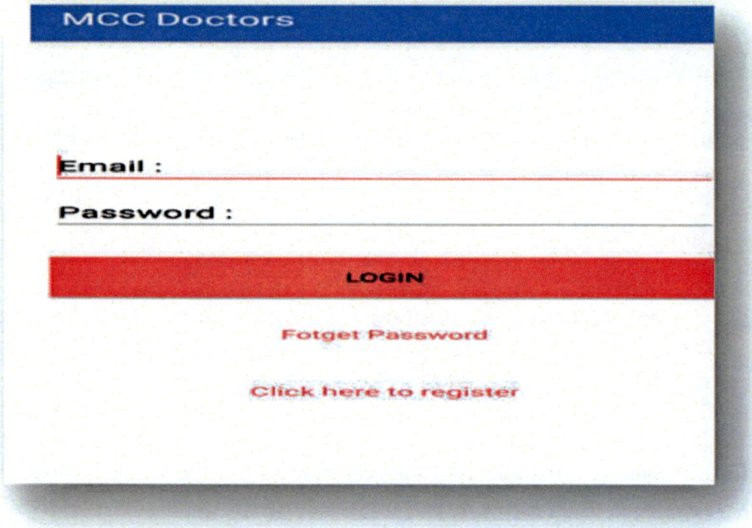

4.1 Doctor Registration Page

On this page the doctor can register with their name, E-mail id, Aadhar number, phone number, password before given access to the firebase to access the details of the patient. Here, the password must be matched with an already entered password or existing password. PIN or password should be uppercase and lower-case characters with the following digits (0–9) and special characters. Then click on the button "click to register". If the person is already registered it shows already exists go-to login, else it shows you are registered successfully. Now the doctor can login to his page and can access the health details of the patient.

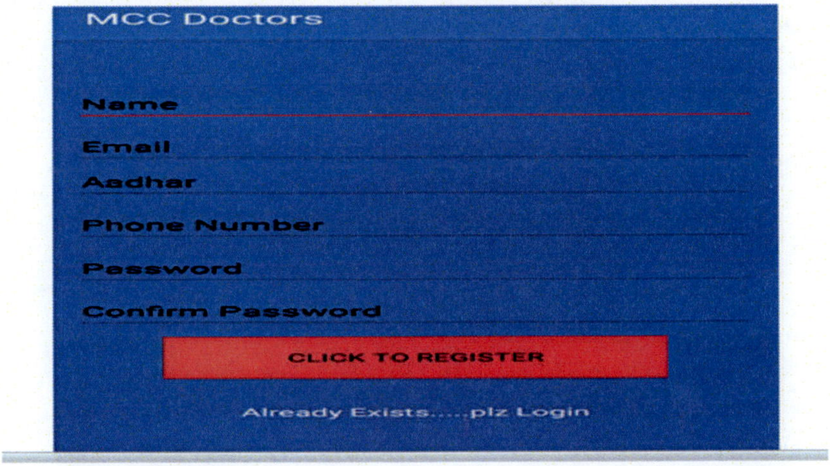

4.2 *Second Page*

On this page, the doctor can access each patient's data individually by entering their Aadhar ID.

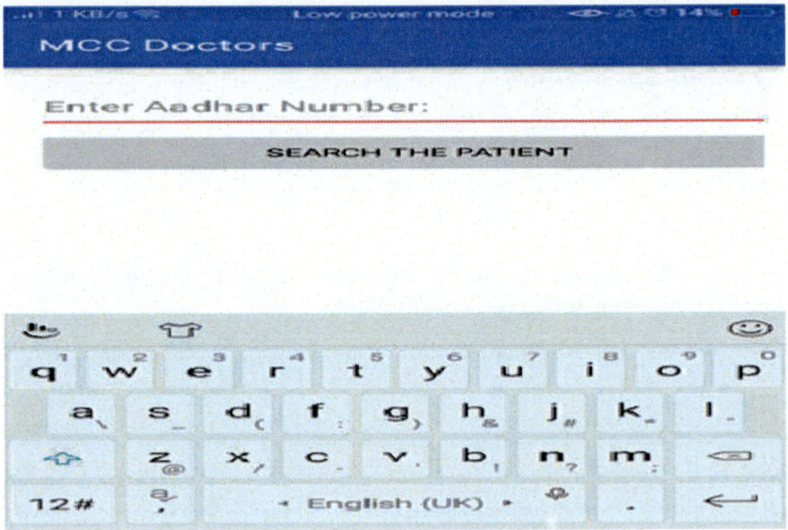

Without an Aadhar number doctor cannot access the data from firebase

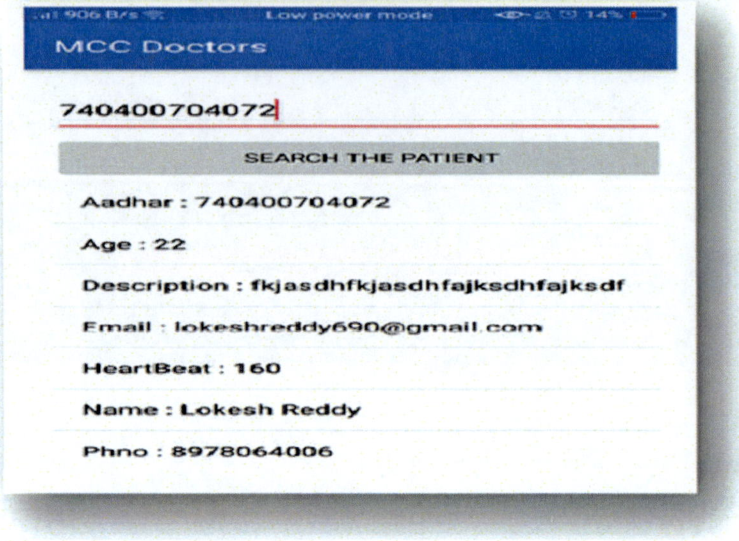

By entering the specific patient's Aadhar ID, each patient's details can be accessed, as shown above.

4.3 *Patient Application*

In this application, patient is also given permission to get the data stored in the firebase.

First page

On this page, the patients can access their health details by entering their phone number. OTP is sent to their registered number.

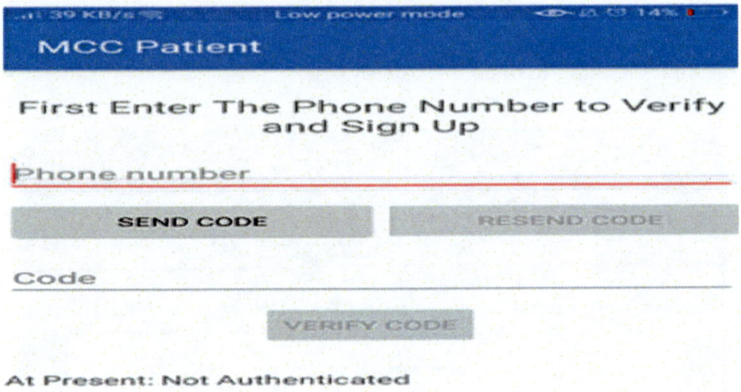

Second page

By entering the OTP number and click on "verify code".

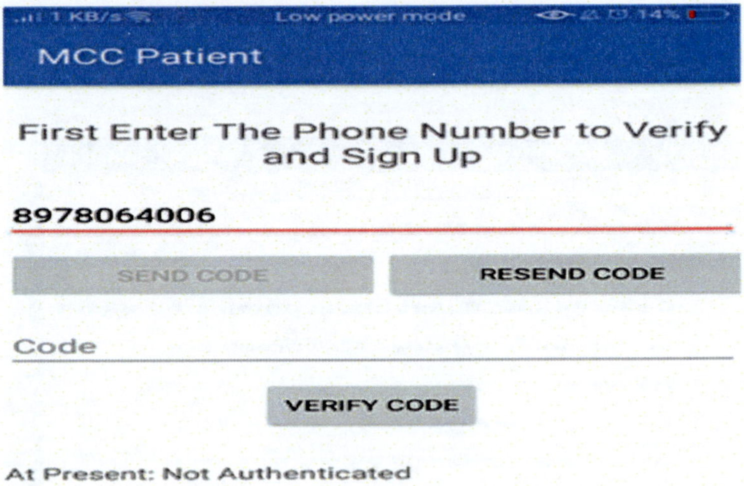

Page loading

Here page starts loading to open the patient page.

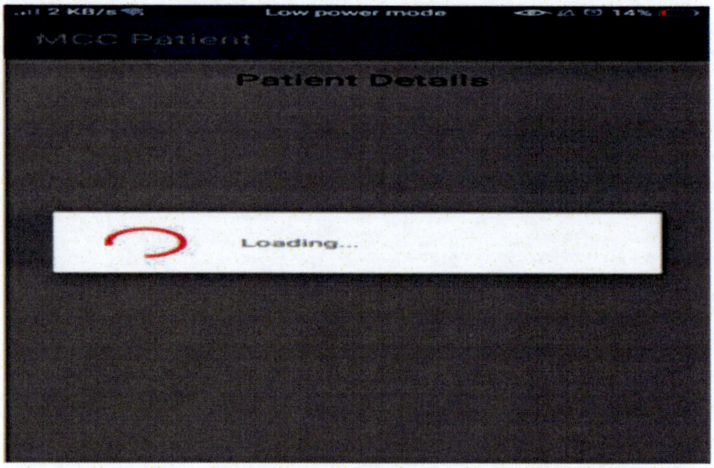

Third page

After entering the Aadhar ID patient health details can be shown as follows.

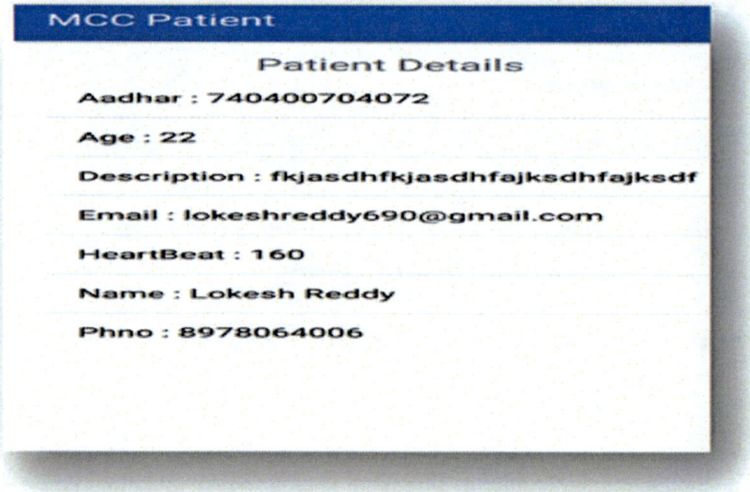

5 Conclusion

Finally, the planned cloud health monitoring smart bed victimization IOT is proficient and safeguards all requirements from all angles. These frameworks result in developing patient's healthcare monitoring epitome system for the benefit of the rural individuals and build the Asian country much stronger within the vital alarms in cloud healthcare monitoring system (CHMS) and to resolve the security issues in cloud healthcare monitoring system.

Here, for the purposes of secured access, OTP is used; it will be implemented by using another way of security access data from firebase in the future.

References

1. Cloud computing in health care to reach $5.4 billion by 2017. http://www.eweek.com/c/a/Health-Care-IT/CloudComputing-in-Heal
2. Mell P, Grance T (2011) NIST Definition of cloud computing. Presented at National Institute of Standards and Technology, US Department of Commerce
3. Harini R, Rama Murthy B, Alam KT (2017) Development of ECG monitoring system using android app. In: International conference on emerging trends in engineering, science and management
4. Khan SF (2017) Health care monitoring system in internet of things (IoT) by using RFID. In: The 6th international conference on industrial technology and management
5. Zaitseva E (2010) Reliability analysis methods for healthcare system. In: 3rd international conference on human system interaction, pp 211–216
6. NiraghatamB, Ramanamurthy MV (2017) Heart beat monitoring system and security using android. Int J Adv Res Comput Sci 8(7)
7. DeshmukhD, Shinde UB, Zanwar SR (2017) Android based health care monitoring system. Int J Adv Sci Res Eng Trends 2(7)
8. Kumar R, Maheshwari R, Aggarwal A, Shanmugasundaram M, Sundar S (2017) IoT based health monitoring system using android app. ARPN J Eng Appl Sci 12(19)

Threshold and Segmentation-Based X-ray Imaging Analysis for Covid-19 Detection

K. Himabindu, K. Sekar, and K. Vandana Rao

Abstract Diagnosing the novel Covid-19 disease is the best way to precluding the loss of human deaths. This research mainly concentrates on visually observable symptoms that can be seen on the lung X-rays of humans. Novel Covid-19 monitoring of health and diagnosing the disease in humans is very critical for sustainability to medical. Nowadays, it is difficult to detect the Covid-19 positive cases because of limited equipments, and also it needs the presence of medical experts in the identification of disease. Moreover, excessive processing time is required. For diagnosing the disease machine learning approaches play a very important role in the normal or abnormal state of Covid-19. For detection various steps are involved, such as acquisition of images, preprocessing, and segmentation of images. For automatic detection of Covid-19 disease, X-ray of lungs plays an important role. Hence we first segment it using various segmentation techniques in artificial intelligence.

Keywords Covid-19 · X-rays · Image analysis · Enhancement · Segmentation

1 Introduction

Covid-19 (a novel coronavirus) is a large family of acute resolved viruses. This disease causes severe illness to humans. It is a global challenge with a fatality rate of 2%. The death rates of Covid-19 patients increase day-by-day. Early diagnosis of novel coronavirus is not possible due to lack or non-availability of equipment. Early diagnosis may be beneficial to humans by immediate and timely isolation of

K. Himabindu (✉) · K. Sekar
Department of MCA, SVCE, Tirupati, India
e-mail: himabindu.karamala@svcolleges.edu.in

K. Sekar
e-mail: sekar.k@gmail.com

K. Vandana Rao
Department of Computer Science, SPMVV, Tirupati, India
e-mail: Vandana.kalapati@gmail.com

S. Jyothi et al. (eds.), *Proceedings of the 2nd International Conference on Computational and Bio Engineering*, Lecture Notes in Networks and Systems 215,
https://doi.org/10.1007/978-981-16-1941-0_2

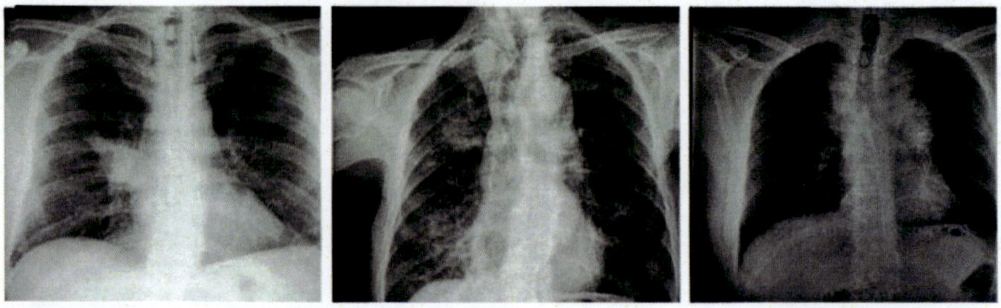

Fig. 1 Sample images from Covid-19 X-ray dataset

patients, quick intubation of severe cases in specific hospitals, and decreases the chances of infection of a healthy population [1]. People are also facing financial issues for the diagnosis of Covid-19, particularly in countries with specific health systems, or limited access to health systems due to excessive prices. Therefore, a society for better action against the Covid-19 disease needed epidemiological data for real-time analysis. For real-time analysis, X-ray or CT scans are the routine tools for diagnosing the disease. Based on these medical images, we can classifying the images according to extracting features [2, 3]. Mild coughing and fever is treated as a mild severity. When the hypoxia is >50% and lung involvement on imaging, the disease is considered severe. When multiorgans fail, it can be considered as critical as per the radiology analysis [4]. Hence medical X-rays are important to find out the disease severity, possibility, or most likely Covid-19 prediction of involvement (Fig. 1).

Machine learning (ML) algorithms provide high performance to image classification using different image processing techniques, like image analysis, feature segmentation, feature extraction, and selection and classification task using different classifiers. Based on the X-ray image the patient class, i.e., positive or negative, is diagnosed. Hence, we have taken X-ray images from Kaggle database for the detection of COVID-19 disease [5].

Image Preprocessing: In this section by using the image preprocessing we have improved the quality of images through filtering and enhancement techniques.

2 Image Enhancement

Image enhancement has to improve the perception of information on images for human visibility. This is completely a subjective process. Enhancement sharpens the image features like edges and boundaries to make a graphic display good for analysis viewing. To improve the boundary for segmentation, edge-aware local-contrast method was used. It enhances the X-ray grayscale image local contrast; even strong edges are unchanged by increasing or smoothing details of image. The

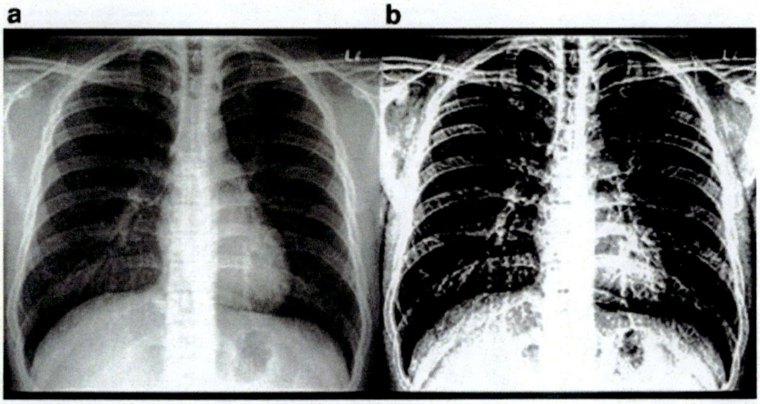

Fig. 2 **a** Original image, **b** enhanced image

minimum intensity amplitude of strong edges mainly defines the edge threshold [6] (Fig. 2).

This enhancement algorithm is very much useful in identifying the opacities in chest X-ray and scan images. Here ground glass opacities represent when lungs are diseased and area of lung attenuation in the view of radiologist [7, 8]. Opacities refer to vessels and bronchial walls. The presence of opacities indicates COVID disease in a person, and this data is very useful for medical experts [9, 10]. To identify certain opacities on X-rays, an enhanced X-ray is needed.

The purpose of X-ray image enhancement is to fix some threshold value by using edge-aware processing technique, and after fixing the threshold value diffrentiate the large-scale edges with small-scale detail images. To get fair edge smoothening and enhancement, image filters are used. Finally, the resultant image shows the contrast-enhanced and smoothening of edge formation [11].

3 Segmentation of Lungs from X-ray

In this paper, segmentation plays an important role in processing and recognizing a particular object in Covid-19 diseased image. Medical experts can diagnose the disease accurately and successfully with this approach [12]. A combination of data is useful in identifying the area of disease from the image to get more exact results [13, 14] (Fig. 3).

A big problem with detecting Covid-19 in X-ray images is all the useless extra information in the images, which includes the diaphragm, bones, trachea, and other regions. These unneeded regions would just make the extraction of the opacities much more difficult, so we decided to simply pull out the lung regions [15, 16]. We did this through the removal of any small white objects with lesser than 1000 pixels

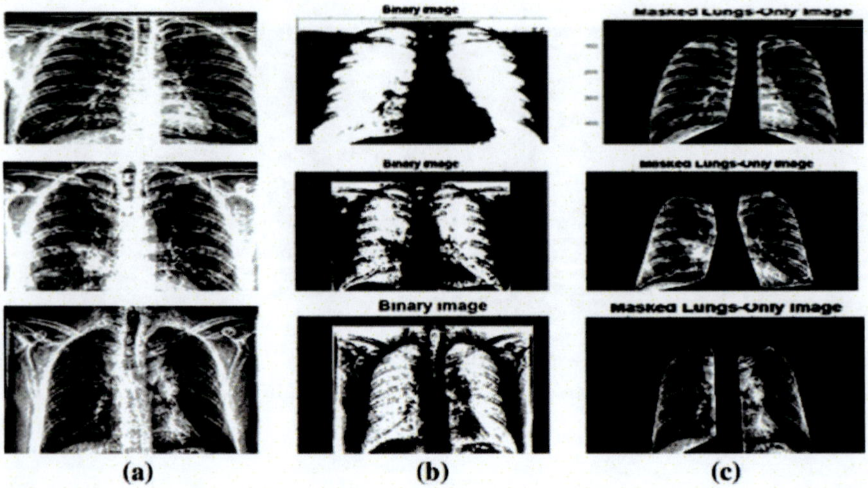

Fig. 3 **a** Thresh image, **b** binarised image, **c** segmented masked lungs

from the image and masking them so that we can pull out the original image. Using the histogram we can find out the needed pixel of the region and zero out these bins so we can see the other bins [17, 18]. By indicating the amount of regions we wanted to extract and finding the areas that fit the lungs, we can detect the regions once the correct threshold was set. The threshold for this image was best at a value of 120. Then extract only the two largest blobs, and make blobs convex and smooth. This will also fill holes in the blobs to make them solid [19, 20]. The regions were then used with the gray image to make anything that was not the lung areas to black, so what is left is just the masked lungs from the gray image.

4 Conclusion

In this paper, a novel Covid-19 lung X-ray infection segmentation method is used, which can find out the area of opacities in lungs. The explicit part of the lungs is not needed to improve the accuracy of identification of infected regions by mainly segmenting the lungs. Moreover, to enhance the image we applied edge-aware contrast method, and segmentation helps to separate the left and right lungs from an image. This approach was very much useful for diagnosing the novel Covid-19 disease and the infected regions.

References

1. Stoecklin SB, Rolland P et al (2020) First cases of coronavirus disease 2019 (COVID-19) in France: surveillance, investigations and control measures, January 2020. Eurosurveillance 25(6)
2. Kong W, Agarwal PP (2020) Chest imaging appearance of COVID-19 infection. Radiol Cardiothorac Imaging 2(1):e200028
3. Elmousalami HH, Hassanien AE (2020) Day level forecasting for coronavirus disease (COVID-19) spread: analysis, modeling and recommendations. Arxiv Preprint 2003.07778
4. Chouhan V, Singh SK, Khamparia A, Gupta D, Tiwari P, Moreira C et al (2020) A novel transfer learning-based approach for pneumonia detection in chest X-ray images. Appl Sci 10:559
5. https://github.com/ieee8023/covid-chestxray-dataset
6. He J, Chen M (2016) Edge-aware local laplacian filters for medical X-ray image enhancement. In: Conference: international conference on health information science. https://doi.org/10.1007/978-3-319-48335-1_11
7. Nishiura H, Linton NM, Akhmetzhanov AR (2020) Serial interval of novel coronavirus (COVID-19) infections. Int J Infect Dis https://doi.org/10.1016/j.ijid.2020.02.060
8. Amya A, Modzelewski R, Ruan S (2020) Multi-task deep learning based ct imaging analysis for COVID-19: classification and segmentation. https://doi.org/10.1101/2020.04.16.20064709
9. https://www.itnonline.com/content/how-does-covid-19-appealungs#:~:text=The%20pair%20discovered%20that%20while,be%20present%20in%20certain%20patients
10. Perlman S, Jasone N (2009) Coronaviruses post-SARS: update on replication and pathogenesis. Nat Rev Microbiol 7(6):439–450
11. Na'am J, Harlan J, Madenda S, Santony J, Suharinto C (2018) Detection of proximal caries at the molar teeth using edge enhancement algorithm. Int J Electr Comput Eng (IJECE) 8(5):3259–3266
12. Fan DP, Zhou T, Ji GP, Zhou Y, Chen G, Fu H, Shen J, Shao L (2020) Inf-Net: automatic COVID-19 lung infection segmentation from CT scans
13. Chijindu VC, Udeze CC, Ahaneku MA, Anoliefo EC (2018) Detection of prostate cancer using radial/axial scanning of 2D trans-rectal ultrasound images. Bull Electr Eng Inform 7(2):222–229. https://doi.org/10.11591/eei.v7i2.727
14. Zhu HQ, Xie QY (2018) A multiphase level set formulation for image segmentation using a MRF-based nonsymmetric Student's-t mixture model. Signal Image Video Process 18:1577–1585
15. Kamble B, Sahu SP, Doriya R (2020) A review on lung and nodule segmentation techniques. In: Advances in data and information science (ADIS). Springer, pp 555–565
16. Gordaliza PM, Barrutia AM, Monica A et al (2018) Unsupervised CT lung image segmentation of a mycobacterium tuberculosis infection model. Sci Rep 8:1–10
17. Keshani M, Azimifar Z, Tajeripour F et al (2013) Lung nodule segmentation and recognition using SVM classifier and active contour modeling: a complete intelligent system. Comput Biol Med 43(4):287–300
18. Shen S, Bui AA, Cong J et al (2015) An automated lung segmentation approach using bidirectional chain codes to improve nodule detection accuracy. Comput Biol Med 57:139–149
19. Mahdy LN, Ezzat KA, Elmousalami HH et al (2020) Automatic X-ray COVID-19 lung image classification system based on multi-level thresholding and support vector machine. https://doi.org/10.1101/2020.03.30.20047787
20. Voulodimos A, Protopapadakis E, Katsamenis I (2020) Deep learning models for COVID-19 infected area segmentation in ct images. https://doi.org/10.1101/2020.05.08.20094664

Feature-Based Squids Classification Using Fuzzy Logic

K. Himabindu, S. Jyothi, and D. M. Mamatha

Abstract Squid is the most diversified member of marine cephalopod mollusk and it is a highly priced crustacean. The authenticity of species is a big issue for food authorities, while given wrong species labels representing the commercial fraud to the consumers, hence this implies substitutiion of one species by another of cheap commercial value. Each squid has got its own features to accurately classify the species. Color, shape, and texture features are very important in classifying the squid species. The most dramatic differences were observed in the shape variation of squid species, which are mantle length, mantle width, head length, head width, fin width, and total length. Not only the shape features are sufficient to identify the squid species but the color and textural features are also considered for classification. The color and texture features of squids are correlation, homogeneity, entropy, R mean, standard deviation, G mean, standard deviation, skewness, B mean standard deviation, and skewness of B. While considering the color, shape, and texture features of squid images, some uncertainty can be found on these features. To avoid uncertainty soft computing techniques have become more popular. In different soft computing paradigms, more flexible and uncertainty-avoiding method is fuzzy logic. Hence, in this work, we developed a feature-based fuzzy inference system design to classify squids into 15 classes. The classifier in this case operates as a Mamdani type fuzzy inference system used for the classification of squid species.

Keywords Squid species · Shape · Color · Texture · Classification · Fuzzy inference system

K. Himabindu
Department of MCA, SVCE, Tirupati, India
e-mail: himabindu.karamala@svcolleges.edu.in

S. Jyothi (✉) · D. M. Mamatha
Department of Computer Science, SPMVV, Tirupati, India
e-mail: jyothi.spmvv@gmail.com

D. M. Mamatha
e-mail: drdmmamatha@gmail.com

S. Jyothi et al. (eds.), *Proceedings of the 2nd International Conference on Computational and Bio Engineering*, Lecture Notes in Networks and Systems 215,
https://doi.org/10.1007/978-981-16-1941-0_3

1 Introduction

As we know the species that has evolved today from some of the earliest known mollusks on earth is squid. Most people know less about squid than most other forms of aquatic life. Squid is one of the cephalopod orders. In this order cephalo indicates head and pod as foot, so the head and foot are dominant in cephalopods. These are swift-moving carnivores, and their bodies are quite streamlined. Squid belongs to the order of Teuthoidea, which comprises over 300 species [1]. It is commercially referred to as calamari. Squid species identification is an important and challenging problem in the real world. Importing and exporting the inclusive range of squids has been a great trend for a long time. At the time of importing and exporting trained taxonomist presence is very essential to assign labeling to species since incorrect labeling represents commercial fraud. Therefore, such skilled experts are always in shortage [2]. Wrong labeling may not give positive sanitary implications such as allergytic symtoms or any toxic syndromes. This can create interest in developing the automatic process for identification. The external anatomy of squid species is shown in Fig. 1.

With this background, this paper is aimed to develop soft computing algorithms for the recognition of species with the use of shape, color, and texture. Soft computing consists of fuzzy logic (FL), neural networks, and evolutionary algorithms. In our problem, while considering shape measurements, uncertainty in color and texture of squid species arouse. For avoiding uncertainty, fuzzy logic is used. Fuzzy logic provides exact conclusion based on ambiguous, vague, and imprecise information [3]. Fuzzy accepted truth value ranges are 0 and 1. Substitution of yes or no, truth value, or membership in fuzzy logic explains a matter of degree. The input is mapped onto fuzzy logic, that is shape parameters, color parameters, and texture parameters to an output space such as squid species family type. For doing this, a list of if–then statements called rules is evaluated in parallel. In this paper, we focused to develop a feature-based fuzzy inference system (FFIS) with 20 input parameters for squid classification into 15 species. The squid species class types are mentioned in Table 1.

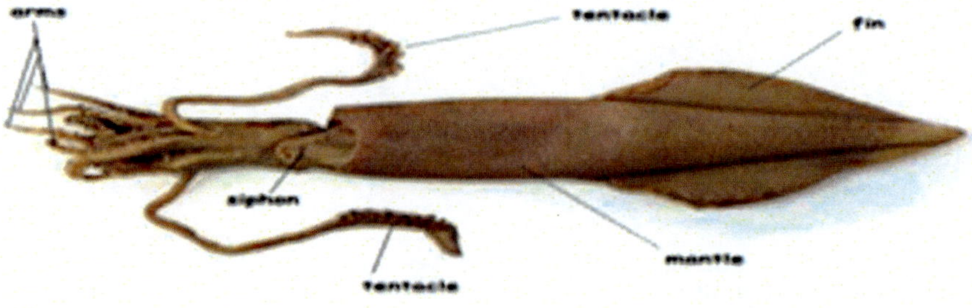

Fig. 1 External anatomy of squid

Table 1 Squids class type

S. no	Species name
1	Todarodes pacificus
2	Todarodes sagittatus
3	Backlight
4	Collosal
5	Dosidicus gigus
6	Gonatopsis makko
7	Gonatus
8	Gonatus antarcticus
9	Illex argentinus
10	Illex illecebrosus
11	Illex steenstrup
12	Loligo Duvauceli
13	Loligo forbessi
14	Loligo opalescens
15	Neon flying

2 Image Acquisition

Before classification, image acquisition plays a very important role. The squid images were taken with white background and top view angle for better analysis. In this study, we used DSLR camera for capturing the species. In the initial stage of image acquisition system, first we collected the squid species from various sources, like harbors, ports, and aqua centers.

Also, we captured images at different angles for better analysis. The acquired squid images are shown in Fig. 2 (Fig. 3).

The output is split into 15 equal intervals that represent 15 squid species. The fuzzy MFs of output are shown in Fig. 4.

For designing a fuzzy inference system, mainly three steps have to be followed [4]:

1. Twenty (species color, shape, and texture features) input features are considered for designing their membership functions.
2. Fifteen (squid species classes) output variables design their membership functions.
3. Rules are defined to connect 20 inputs and their outputs.

Table 2 Rules for shape, color, and texture squid parameters

SC	FL	FW	HL	HW	ML	MW	TL	Con.	Corr.	Hom	Etr.	RM	Rskew.	Rstd.	GM	Gskew.	Gstd.	BM	Bskew.	Bstd.
1	S	M	S	M	L	M	S	S	M	L	S	M	S	M	S	S	M	S	L	L
2	M	S	S	M	S	L	L	S	M	S	M	M	S	S	M	S	S	S	M	S
3	S	L	M	M	S	M	S	M	L	M	S	S	M	S	M	L	M	S	M	L
4	M	M	S	L	S	S	M	S	M	L	M	S	M	L	S	L	L	S	S	S
5	S	S	S	L	S	M	S	S	L	S	M	S	L	S	M	S	L	S	M	M
6	M	M	S	M	M	L	S	M	S	M	S	M	S	S	M	S	M	L	M	S
7	L	S	S	M	S	S	M	S	M	S	M	L	M	S	S	M	L	L	M	S
8	S	M	L	M	M	S	M	L	L	S	S	S	M	S	M	L	M	L	S	L
9	S	S	L	S	M	S	S	S	S	M	S	L	S	L	S	M	M	S	L	S
10	L	M	S	M	M	S	M	M	S	M	M	M	S	S	M	L	S	L	S	M
11	S	S	M	M	L	S	S	M	L	M	S	S	S	M	L	S	L	S	S	M
12	S	S	L	L	S	S	S	L	M	S	S	M	S	L	S	S	L	S	M	L
13	S	M	S	L	S	M	S	S	S	L	M	S	M	M	L	S	M	S	M	S
14	L	S	S	M	M	S	M	S	L	M	S	L	S	L	S	M	M	M	M	S
15	L	S	M	S	S	M	M	S	S	S	M	L	S	M	S	L	S	L	S	M

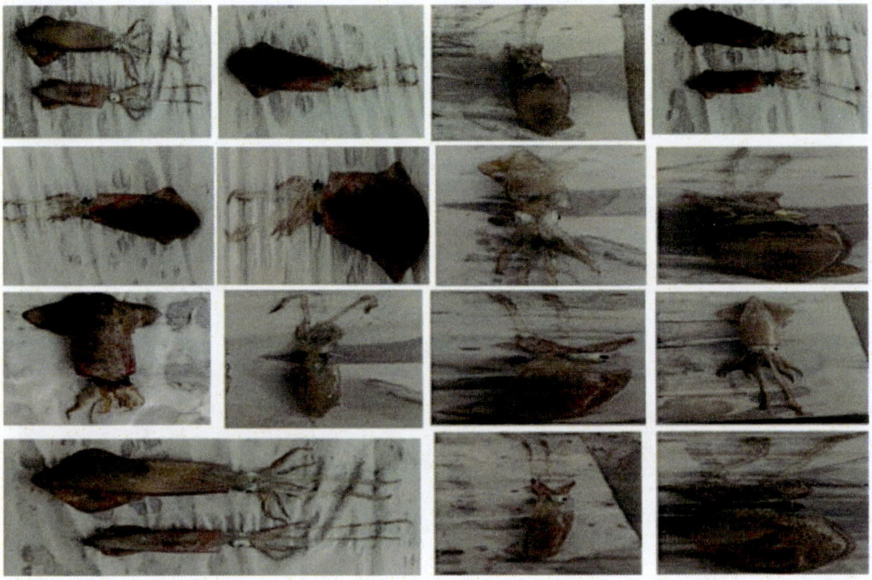

Fig. 2 Squid images collection

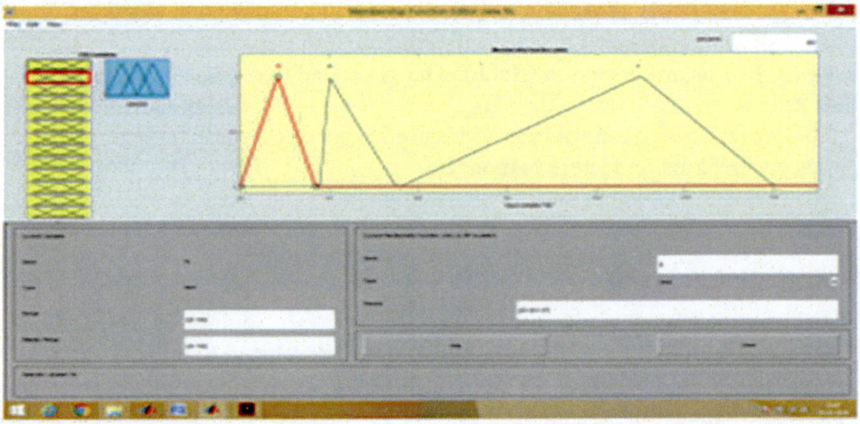

Fig. 3 Sample fuzzy MFs of inputs

3 Fuzzy Rule-Based System

Fuzzy inference system is based on the fuzzy rule-based system; it contains number of fuzzy if–then rules [5–7]. Based on the descriptions of the input (squid color, shape, and texture features) and output variables (squid classes), the rules are generated in the rule editor. The rules are generated by using squids shape parameters, such as

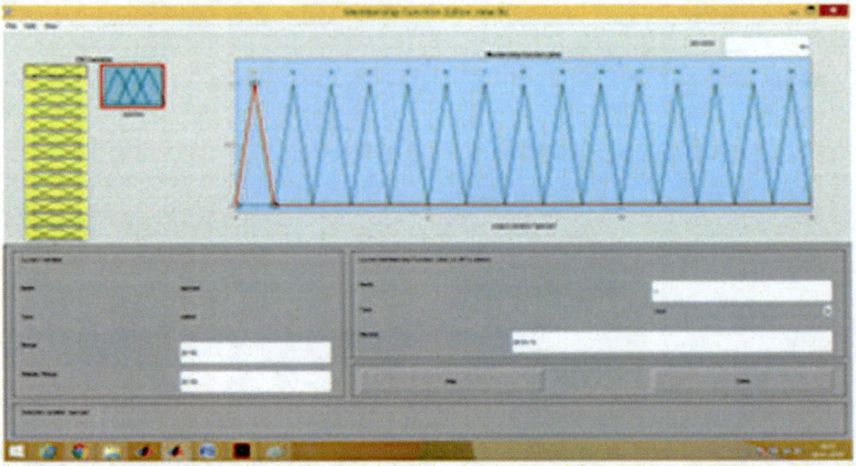

Fig. 4 Fuzzy MFs of outputs

fin length (FL), fin width (FW), head length (HL), head width (HW), mantle length (ML), mantle width (MW), total length (TL), and texture parameters such as contrast (Con.), correlation (Corr.), homogeneity (Hom.), entropy (Etr.), and color parameters such as R mean (RM), R skewness (Rskew.), R standard deviation (Rstd.), G mean (GM), G skewness (Gskew.), G standard deviation (Gstd.), B mean (BM), B skewness (Bskew.), B standard deviation (Bstd.). The generated fuzzy rules are described in Table 2.

For defuzzification, centroid method is used to get an outcome for segregating the species group based on feature parameters.

4 Result Analysis

In this study feature-based FIS with 20 features is developed to classify the squids into 15 classes by evaluating the developed feature-based fuzzy inference system (FFIS). The outputs for 50 squid samples are generated by evaluating the developed FFIS. Confusion matrix is used to calculate the performance measures, like accuracy, recall, and precision using the following formulas [8–10].

$$\text{Accuracy} = (\text{TP} + \text{TN})/(\text{TP} + \text{FP} + \text{TN} + \text{FN}) \tag{1}$$

$$\text{Recall} = \text{TP}/(\text{TP} + \text{FN}) \tag{2}$$

$$\text{Precision} = \text{TP}/(\text{TP} + \text{FP}) \tag{3}$$

Table 3 Performance measures of FFIS

FFIS with 20 features	
Accuracy (%)	85.7
Recall (%)	79.64
Precision (%)	81.25

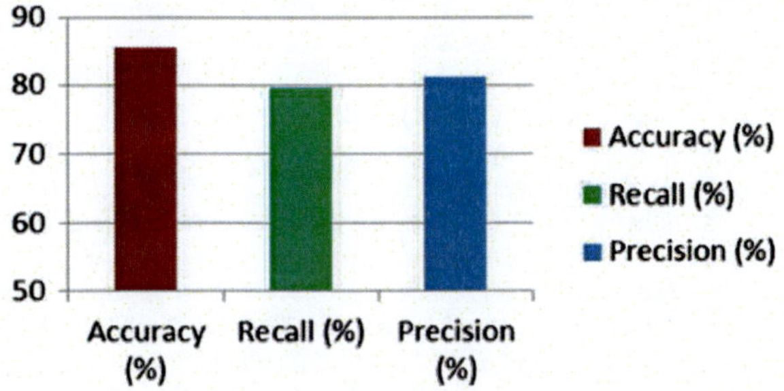

Fig. 5 Performance measures of FFIS

where

TP refers to true positives;
TN refers to true negative;
FP refers to false positive;
FN refers to false negative.

The calculated performance measures of FFIS are tabulated in Table 3.

From Table 3, it can be observed that the developed FFIS classified the squid samples with an accuracy of 85.7%, recall of 79.64%, and precision of 81.25%. The same is represented in graphical form in Fig. 5.

5 Conclusion

In this study, FFIS is designed for squid species classification with 20 features. Totally, 15 fuzzy rules are constructed for classifying the 15 squid species. The developed FFIS successfully classifies the class of squids. The efficiency of this squid species classification can be improved by manipulating the input parameters of the system through MATLAB fuzzy logic toolbox.

Acknowledgements This Research work is done under DBT project, New Delhi.

References

1. Muchlisin ZA, Zulkarnaini B, Purnawan S (2014) Morphometric variations of three species of harvested cephalopods found in northern sea of aceh province, Indonesia. Biodiversitas J Biol Divers 15(2):142–146. ISSN: 1412-033X. https://doi.org/10.13057/biodiv/d150205.
2. Lockley AK, Bardley RG (2000) DNA-based methods for food authentication. Trends Food Sci Technol 11:67–77
3. Kurhe AB, Satonkar SS, Khanale PB (2011) Soft computing and its applications. BIOINFO Soft Comput 1(1):05–07
4. Malik V, Gautam A, Sahai A (2013) Satellite image classification using fuzzy logic. Int J Recent Technol Eng (IJRTE) 2(2). ISSN: 2277-3878
5. Moallem P, Somayeh Mousavi B (2013) Gender classification by fuzzy inference system. Int J Adv Robotic Syst 10(89)
6. Kuncheva LI (2008) Fuzzy classifiers. Scholarpedia 3(1):2925. https://doi.org/10.4249/scholarpedia.2925
7. Ingole K, Katole K, Shinde A (2013) Crop prediction and detection using fuzzy logic in matlab. Int J Adv Eng Technol 6(5):2006–2012. ISSN: 2231-1963
8. Borkar AD, Atulkar M (2013) Fuzzy inference system for image processing. Int J Adv Res Comput Eng Technol (IJARCET) 2(3). ISSN: 2278-1323
9. Shashikanth CC, Kulkarni P (2013) Region based image similarity using fuzzy based sift matching. Int J Comput Appl 67(3). ISSN: 0975-8887
10. Bai Y, Wang D (2006) Fundamentals of fuzzy logic control–fuzzy sets, fuzzy rules and defuzzifications

In Silico Platforms for Systems Toxicology

M. Lalasa, S. Nithya, K. Nagalakshmamma, A. Suvarnalatha, and P. Nageshwar Rao

Abstract Computational toxicology provides scientific insights into the detrimental effects of various chemical compounds at multiple levels, including molecular and functional changes in biological systems. There is a great demand to closely monitor and assess the risks with high precision. Developing approaches with high accuracy requires a good understanding of the mechanism of toxicity. The computational toxicology provides analogous mechanistic acquaintance by amalgamating the modern analytical techniques and computing tools. It also identifies and uses structural alerts and biomarkers to improve the risk assessment and decodes the sequence of events occurring in a biological system upon exposure to a chemical moiety. Statistical and arithmetic models quantitatively describe toxicity. The integrated data explains how the biological system is affected upon exposure and helps in developing arithmetic or statistical models to predict the toxicity process. This perception combines modern bio-analytical techniques and computational tools in potential risk assessment.

Keywords Computational toxicology · Computing tools · Biomarkers · Arithmetic models

M. Lalasa (✉)
Department of Computer Science, Swetha Institute of Technology and Science, JNTU, Tirupati, India
e-mail: mlalasa2020@gmail.com

S. Nithya
Institite of Pharmaceutical Technology, Sri Padmavati Mahila Visvavidyalayam, Tirupati, India

K. Nagalakshmamma · A. Suvarnalatha
Department of Biosciences and Sericulture, Sri Padmavati Mahila Visvavidyalayam, Tirupati, India

P. Nageshwar Rao
Department of Computer Science, Rayalaseema University, Kurnool, India

S. Jyothi et al. (eds.), *Proceedings of the 2nd International Conference on Computational and Bio Engineering*, Lecture Notes in Networks and Systems 215,
https://doi.org/10.1007/978-981-16-1941-0_4

1 Introduction

Toxicology is defined as a study of detrimental effects of chemical compounds or any other toxic substances on living organisms. Substances that are toxic are known as toxins. A toxicologist is the one who resolves detrimental effects of toxicants that are responsible for the modified cellular, biochemical, and molecular mechanism in an organism. Toxins are of two types, viz., organ toxins and systemic toxins. A "systemic toxin" is the compound that targets the entire cell system of the organism and it is not restricted to any particular organ; for example, "potassium cyanide". While the "organ toxin" affects only the particular target organ, the central nervous system, kidneys, and hematopoietic system are the target organs for lead [1].

2 Mechanisms of Toxicity

The knowledge on the toxicity pathways provides basic information to interpret the descriptive toxicity data. The dynamics of the manifestation of toxicities simulate a series of changes, beginning with the exposure of the organism to the toxicant and the multitude of their interactions that culminates in a toxic effect. Greater volumes of potential toxicants and a vast number of biological compounds and their interaction are prone to impairment and manifest in all possible pathways leading to toxicity (Fig. 1).

The potential stages in the development of toxicity after chemical exposure include the delivery of toxicant to its target (Step 1), interaction with the target molecules (Step 2a), toxicant altering the environment (Step 2b), triggering the perturbance of cell function and/or structure (Step 3), and initiation of repair mechanisms at the tissue, cellular, and molecular levels (Step 4).

3 Systems or Predictive Toxicology

Systems/predictive toxicology is the combination of systems biology, chemistry, toxicology, mathematics, statistics, physics, and computer science. It is the translation between the experimental systems of the dry lab and wet lab. The main objective is to reveal the ultimate effect of the toxic substances on the living organisms about the period of exposure and also the varied concentrations of the toxic substance. It estimates the interaction between the adverse effects and the constituents of the biological system. Scientific integration of systems toxicology is shown in Fig. 2 [2].

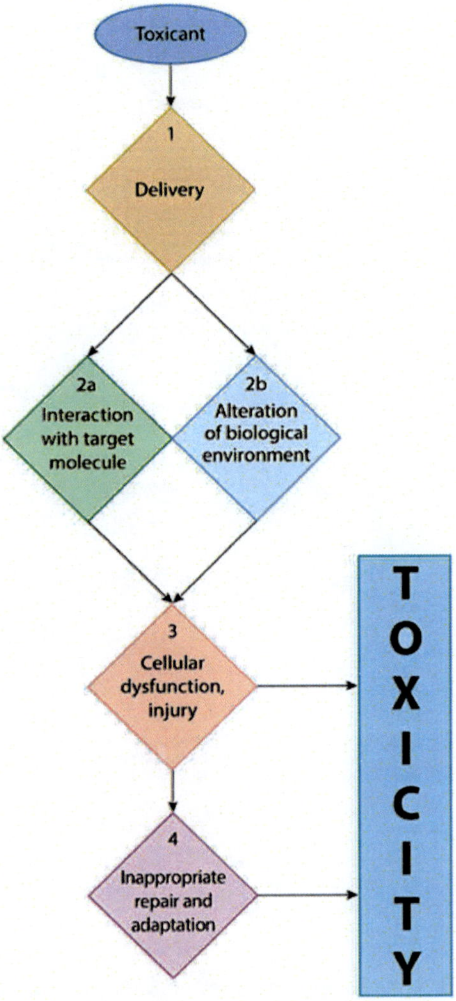

Fig. 1 Toxicity pathways (*Source* Kiaassen CD, Watkins JB & Doull's Essentials of toxicology, 2nd Edition: http://www.accesspharmacy.com)

3.1 The Phenomenon of Systems/Predictive Toxicology

The phenomenon of systems toxicology is the measurement of exposure of the organism and the level of concentration of the toxic substance and/or the polluted environment. The toxic substance may be an abiotic or biotic component. Systems toxicology evaluates the altered biological mechanisms, like lipid peroxidation, oxidative stress, and "omic" mechanisms. The biomarkers are utilized as a reflection of the synergy between the toxicant and organism exposure. Systems biology and systems toxicology provide a pathway for identifying diseases, observation of their succession levels, and remedies [3, 4].

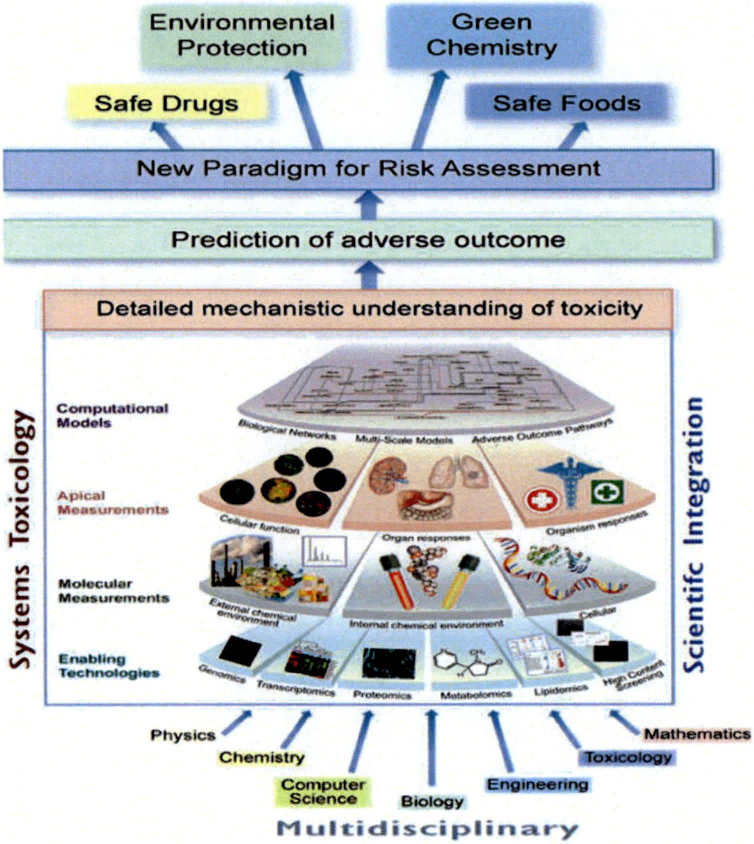

Fig. 2 Systems toxicology and scientific integration (*Chem. Res. Toxicol.* 2014, 27, 3, 314–329 https://doi.org/10.1021/tx400410s)

3.2 Molecular-Level Assessment of Adverse Effects

This is the vital step in systems toxicology/predictive toxicology. It influences the structural, functional change in an organism as an outcome of the toxicant effect. The major developing molecular tools that are crucial supporters of predictive toxicology are transcriptomics, metabolomics, lipidomics, and high-content screening [5, 6].

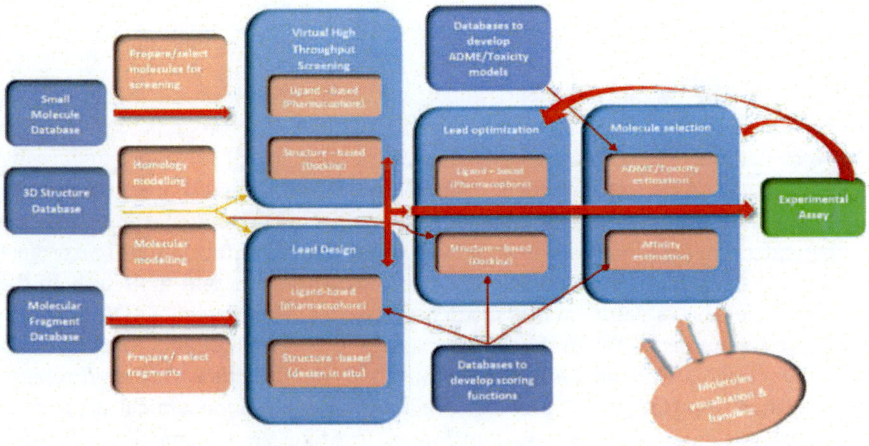

Fig. 3 In silico drug design pipeline, by click2drug

4 Computational Platforms for Systems Toxicology

4.1 CADD

SWISS Institute of Bioinformatics developed the "Directory of tools" by name "Click2Drug" [7] in order to enable drug design with the aid of computers. It consists of a comprehensive list of CADD databases, software, and web services. "Directory of tools" are categorized following their field of application and encapsulating the whole drug design pipeline (Fig. 3).

The directory of tools displayed on the left side of the home page of CADD is a database listed for "Chemical structure representation, Molecular modelling, Homology modelling, Binding site prediction, Docking, Screening, Target prediction, Ligand design, Binding free energy estimation, QSAR, and ADME Toxicity". A left-click on any of these tools directs to the details of the respective tool [8].

4.2 CRDD

It is another web portal of "Computational Resources for Drug Discovery" that hosts the provision of computational resources on a single platform pertinent to drug discovery. It is a very good platform for those researchers who have limited resources. It is an important in silico module of OSDD (Open Source Drug Discovery). The tools of CRDD are framed into three major groups, namely the drug target tools, screening tools, and drug design tools [9].

4.3 Current Computational Toxicological Works

In the USA, five ongoing computational toxicological projects are being carried out by "the Environmental Protection Agency" (EPA). The projects are "Adverse Outcome Pathways" (AOPs). AOPs outline the mechanisms of substance toxicity by creating models based on historical data. These computational models predict the toxicity without the need for animal testing. The project *DA* (Defined Approach) uses the input data for data interpretation and to provide toxicity information. The project "Integrated Chemical Environment (ICE)" provides the data and tools to help the development and evaluation of new testing approaches. The project "IVIV" (In Vitro to In Vivo Extrapolation) is for the development of predictive/computational methods for conducting IVIVE analyses. The project "Open Structure–Activity/Property Relationship App (OPERA)" provides predictions on physicochemical properties, environmental fate, and toxicity. The Indian Institute for Chemical Technology (CSIT), Government of India, launched "OpenTox Community Development", which spotlighted the theme of advances in "predictive toxicology".

5 Limitations

Systems toxicology has numerous setbacks. Resolving these setbacks directs to determine how best this domain continues to integrate with toxicology research by developing accurate biological models for computing. The computational models often fail to replicate the biological mechanism of the organism which drastically changes at every microsecond or even shorter than the microsecond, based on the toxicant interactions. Another drawback from the perspective of toxicology is the synergism of synthetic compounds at whole body/organ level. Hence, it is necessary to explore better alternate approaches to quantify the toxicity of manufactured and natural chemicals at whole body systems and the organ level.

6 Conclusion

Systems toxicology is a new field evolving to benefit toxicology research by designing computational models to study the interaction of the toxicants and their adverse effects in "in silico" studies. The in silico studies minimize the use of animal models and also significantly cut down the expenditure that incurs in pharmaceutical research. Evolving of appropriate and efficient computational models that replicate the live biological mechanism of an organism is required to enhance the reliability and widen the scope for systems toxicology.

References

1. Hodgson E (2010) A textbook of modern toxicology. Wiley, p 10. ISBN: 978-0-470-46206-5
2. Livingart Veterinary, Medical and Scientific Illustration International Company. www.livingart.uk.org
3. Escher BI, Hermens JLM (2004) Internal exposure: linking bioavailability to effects. Environ Sci Technol 38:455A-462A
4. Pogribny IP, Rusyn I (2013) Environmental toxicants, epigenetics, and cancer. Adv Exp Med Biol 754:215–232
5. Fiehn O (2001) Combining genomics, metabolome analysis, and biochemical modelling to understand metabolic networks. Comp Funct Genomics 2:155–168
6. James LP (2013) Metabolomics: integration of a new "omics" with clinical pharmacology. Clin Pharmacol Ther 94:547–551
7. Directory of Computer-Aided Drug Design. https://www.click2drug.org
8. Directory of Computer-Aided Drug Design Mobile Directory. click2drug.org/directory_Mobile.html
9. Computational Resources for Drug Discovery. http://crdd.osdd.net/ga.php

Comparative Analysis on Factors Affecting Accuracy of Seizure Detection Using EEG Signal Processing

M. Zaheer Ahamed and S. Aruna Mastani

Abstract Epilepsy or seizure is a serious disorder that can severely affect and cause damage to a person's health, which can even lead to death. Hence an early detection and prediction system is very much essential for controlling the damage. As the monitoring system requires continuous monitoring of EEG data, the system needs to be a low-power, high-speed, portable system. The design of these systems is severely affected by certain factors, such as the features used, classifier employed, and the platform used for implementation. In this review paper, we have presented the study of the factors that affect the seizure detection system.

Keywords Seizure · Classifiers · EEG database · Feature selection

1 Introduction

Epilepsy is a neurological disorder characterized by a sudden failure of the brain which causes what is called a seizure. It is a dangerous disease that affects many people every year. Generally, seizures occur due to a sudden change in brain functioning. As of today, electroencephalography (EEG) is considered to be one of the most effective methods generally used for the treatment of epilepsy. The system used for detecting the occurrence of seizures is given as follows: pre-processing of EEG signals, computation of features, and finally seizure state classification. Here, once the EEG signals are acquired from electrodes and pre-processed, the seizure detection system then has two main steps. In this paper, a brief review of the following aspects is presented:

1. Analysis of various EEG datasets available
2. Seizure detection and prediction features in various domains

M. Zaheer Ahamed (✉) · S. Aruna Mastani (✉)
ECE, JNTUA, Anantapuram, India
e-mail: zaheer.mtech@gmail.com

S. Aruna Mastani
e-mail: aruna_mastani@yahoo.com

S. Jyothi et al. (eds.), *Proceedings of the 2nd International Conference on Computational and Bio Engineering*, Lecture Notes in Networks and Systems 215,
https://doi.org/10.1007/978-981-16-1941-0_5

3. Classification methods—comparative analysis study.

Our review paper is next organized into the following sections: In Sect. 2, we have the analysis of various datasets of EEG signals and is presented. In Sect. 3, the features used for seizure detection in various domains are presented and compared. In Sect. 4, the conclusion is presented.

2 Review of EEG Datasets Available

Since the recording of EEG signals directly from patients is costly and time-consuming, hence the research for seizure detection and analysis and the researchers generally use the readily available various sources. Various datasets are available.

2.1 CHB–MIT Dataset [1]

The CHB–MIT dataset [1] is given by the collaborative project of Boston Children's Hospital and MIT. This extensive dataset features 24 patients with overall 980 h of EEG recordings. The data is divided into 24 group sets; each group set comprises EEG signals extracted from each patient for more than 12 h continuously.

2.2 Dataset of Bonn University, Germany [2, 3]

This dataset consists of five groups of one-channel EEG denoted by A, B, C, D, and E. The duration of each sample is 23.6 s. The sampling frequency used to convert into digital is 173.6 Hz. The EEG signals are then applied to band pass filter (BPF) with filter cut-off frequencies ranging between 0.53 and 40 Hz. Each data group set has 100 EEG records. EEG signals of A and B set groups are taken from five patients in healthy conditions in two states—one with eyes wide open and the other with eyes fully closed conditions. The signals of set groups C and D are taken from patients having epilepsy but during seizure-less intervals. Finally, the group set E has signals of patients during the seizure occurrence.

2.3 Freiburg Hospital

The Freiburg hospital data group set includes intracranial EEG (iEEG) records of 21 persons with this epilepsy. The signals are captured using a signal frequency of 256 Hz. In this dataset, there are six capturing channels from six selected contacts.

2.4 *Temple University Dataset*

The Temple University database [4] is one of the largest database of clinical EEG signals available for public usage. It consists of 28,000 EEG records taken with 15,000 patients for hours.

2.5 *Karunya University [5]*

Data of 174 patients (95 males and 79 females) of varying ages from 1 to 84 years are available in this database. We have 18 channels, in which there are 16 channels taken from the scalp and two channels are from the periocular. A 256 Hz frequency is used for sampling and pre-processing is carried out by passband filter having a range of 0.01–100 Hz for the data.

2.6 *EPILEPSIAE*

This database is a huge database containing data from 275 persons [4]. It has 30 patient recordings done using scalp EEG with 261 seizures and 2881.4 h. A sampling rate of 256 Hz is used and data are collected by 19 electrodes.

Dataset name	No. of patients	Length of each channel	No. of channels	Sampling rate (Hz)
CHB–MIT	24	12 h	128	256
Frieburg	21	24 h	128	256
Temple University	15,000	1.86 h		
Bonn University	100	23.63 s	128	173.67
EPILEPSIAE	275	10.7 h	19	256
Karunya	174	10 s	16	256

3 Seizure Detection Features: Various Domains

There have been a lot of methods and ways for the determination of seizures. Because the qualities of the features we use have a direct impact on the overall performance, the choice of an appropriate way to extract and organize the valuable features from the original EEG signals is of huge prominence. In general, the methods used for seizure detection features can be classified into four characteristic domains.

3.1 Features in Time Domain

The time-domain features are the ones calculated on the original EEG signals or bandpass filtered signals. The list of time-domain features is given as below.

a. Energy and Average Power
These are the most common and effective features available. The energy of a signal of discrete nature is given by

$$E = \sum_{n=-\infty}^{\infty} x(n)^2 s$$

The average power of a discrete signal is given by

$$P = \lim_{N \to \infty} \left(\frac{1}{2N+1} \sum_{n=-N}^{N} |x(n)|^2 \right)$$

b. The line length of a discrete signal
The line length or curve length of the discrete signal can be defined as the total vertical length of the signal. It is generally computed using the formula

$$\text{LL(X)} = \sum_{i=1}^{N-1} \text{abs}[x(i) - x(i-1)]$$

c. Energy (NE) of a discrete signal in a nonlinear domain
The nonlinear energy of discrete signal can be defined as

$$\text{Non L.E. (x)} = \sum_{i=1}^{N-2} x[i]^2 - (x[i+1] * x[i-1])$$

d. Hurst exponent of discrete-time signal (HE)
It indicates the degree of a time series discrete signal. The hurst exponent of a signal can be calculated as follows:
At first, we compute the cumulative derivative series, or also known as partial cumulative time series

$$z(t) = \sum_{j=0}^{t} (x[j]x')$$

Here $x' = (1/m) \sum_{i=o}^{m-1} x[i]$ is the average of the partial time series.

Next, we compute cumulative range $R(m)$ as:
$R(m) = \max z[t] - \min z[t]$, and $S(m)$ as the standard deviation of the partial time series.
Then, the HE is defined as the slope of the straight line that fits log $(R(m)/S(m))$ as a function of log m by the least-square method.

$$\log(R(m)/S(m)) = \mathrm{HE} \cdot \log m$$

e. Hjorth parameters (activity, mobility, and complexity)
Hjorth activity can be defined as the variance of a signal and is high during seizure events in terms of a high amplitude variation from its mean.
The Hjorth mobility of a signal represents the mean frequency or the proportion of standard deviation of the power spectrum. It is given by

$$\text{Mobility} = \sqrt{\frac{var(\frac{dy(t)}{dt})}{var(y(t))}}$$

Hjorth complexity represents the difference between the signal and a pure sine wave

$$\text{Complexity} = \frac{\text{mobility}\left(\frac{dy(t)}{dt}\right)}{\text{mobility}(y(t))}$$

f. Number of zero-crossings.
g. Number of local extrema.

3.2 *Features in Frequency Domain: Comparison*

Here at first the raw EEG signals are converted into the frequency domain using one of the transforms such as Fourier transform. Then on the frequency components, we extract certain features that help in classification. The most widely used set of features in the frequency domain are:

a. The energy of signal at different frequencies can be computed as parameters for distinguishing normal to epileptic patients.
b. Intensity weighted mean frequency (iwmf) or mean frequency [6].
It tells the mean value of the frequency distribution using the normalized power spectral density and can be defined by the formula

$$\text{iwmf(X)} = \sum_n x[k]f[k]$$

Here $x[k]$ is defined as the normalized power spectral density of the signal $x[n]$ at the frequency $f[n]$.

c. Intensity weighted bandwidth (IWBW), or the SD frequency [6]
This typically measures power spectral density width in terms of standard deviation. It is calculated using the below formula

$$\text{iwbw(X)} = \sqrt{\sum_k x[k](f[k] * iwmf(X))2}$$

where $x[k]$ is the normalized PSD.

d. Spectral edge frequency component (SEF)
Generally, this is defined as the frequency where the entire power of the normalized power spectral density, and is equal to α percent. We thus define the SEFα as the frequency $f[k_{\text{SEF}}]$ such that

$$\sum_{k=0}^{K_{\text{sef}}} x[k] = 0.01$$

e. Spectral entropy (SE) [6]
This is a measure of the random process uncertainty from the frequency distribution [6]. It is measured in a manner similar to that for the Shannon entropy, but here we replace the probability distribution with power spectral density. The formula for this is given by

$$\text{SE (X)} = -\sum_k x[k] \log x[k]$$

f. Peak frequency [6]
Normally, this is the frequency where the power spectral density of the largest average power in its full-width-half-max (FWHM) [6] band has the highest magnitude.

Using the frequency domain analysis, we can extract the information related to the spectral components and the frequency where the seizures occur. But here, the time domain-related data like the corresponding duration of the seizure events or the time transformations between various activities are not covered. Thus frequency domain-based seizure analysis does not give satisfactory accuracy along with an improvement in complexity and size of the detection system.

3.3 Time–Frequency Features

In this domain, we compute the features from both time and frequency domain, i.e., we extract data present in both time components and frequency components. Thus using the time–frequency domain features gives better accuracy when compared to other independent domains. In general, the time–frequency analysis is done by applying one of the following techniques:

a. Wavelet transform
b. WVD
c. STFT
d. HHT (Hilbert–Huang transform).

3.4 Statistical Features

The nonlinearity and the dynamic nature of the EEG signals can be evaluated efficiently using statistical analysis-based features such as mean and entropy. In general, the statistical parameters used are:

(i) Mean
(ii) Standard deviation of signal
(iii) Zero crossing rate of the signal
(iv) the RMS value of signal
(v) Computation of various entropies like approximate entropy (APEN), sample entropy (SPEN), phase entropy (PEN) [7]
(vi) Chaotic parameters such as largest Lyapunov exponent LLE, correlation dimension (CD), fractal dimension (FD), and brain rhythms [7].

4 Classification Techniques

Of late, there is an exponential increase in the domain of artificial intelligence and machine learning domains. There are so many applications where these can be applied to extract effective answers to the problems which were not possible earlier. These can also be utilized to execute the task of automated seizure detection. The general techniques available are:

A. Machine Learning Classifiers
   There are so many different variants of ML classifiers in the literature. In all of these the most widely employed ones are

   1. Support vector machines (SVM)—linear and polynomial
   2. K-nearest neighbor
   3. Random forest

4. Naive Bayes

B. Neural Networks

It is shown in the literature [8–12] that the detection accuracy of the detection system can be increased further by leaps with the help of artificial neural networks such as CNN and deep learning techniques.

5 Conclusion

The paper discusses the key factors affecting the accuracy of seizure detection. From the available databases, an appropriate database is to be selected based on constraints such as memory, speed, and application. The time domain features are simple to calculate but have the lowest accuracy, whereas the wavelett domain-based features give the highest accuracy for preictal and ictal periods for seizure. But the implementation of wavelett domain is complex and power-hungry. Hence for designing, an optimized system with high accuracy can be obtained by considering a power-efficient ASIC with wavelett domain features.

References

1. Alickovic E, Kevric J, Subasi A (2018) Performance evaluation of empirical mode decomposition, discrete wavelet transform, and wavelet packed decomposition for automated epileptic seizure detection and prediction. Biomed Signal Process Control 39:94–102
2. Selvaraj TG, Ramasamy B, Jeyaraj SJ, Suviseshamuthu ES (2014) EEG database of seizure disorders for experts and application developers. Clin EEG Neurosci 45(4):304–309
3. Andrzejak RG, Schindler K, Rummel C (2012) Non randomness, nonlinear dependence, and non stationarity of electroencephalographic recordings from epilepsy patients. Phys Rev E 86(4):046206
4. Devroye L, Györfi L, Lugosi G (2013) A probabilistic theory of pattern recognition, vol 31. Springer Science & Business Media
5. Anand S, Jaiswal S, Ghosh PK (2017) Automatic focal epileptic seizure detection in EEG signals. In: Proceedings of the 2017 IEEE international WIE conference on electrical and computer engineering, pp 103–107
6. Li JW, Barma S, Mak PU, Pun SH, Vai MI (2019) Brain rhythm sequencing using EEG signals: a case study on seizure detection. IEEE Access
7. Sarala B, Zaheer Ahamed M, Sree Hari S, Bhagya sree V (2020) Voice and image BER analysis of the OFDM system with MECCT and MLNST companding techniques over mobile radio channels. In: Saini H, Srinivas T, Kumar DV, Mauryan KC (eds) Innovations in electrical and electronics engineering. Lecture notes in electrical engineering, vol 626. Springer, Singapore. https://doi.org/10.1007/978-981-15-2256-7_72
8. Aarabi A, Wallois F, Grebe R (2006) Automated neonatal seizure detection: a multistage classification system through feature selection based on relevance and redundancy analysis. Clin Neurophysiol 117(2):328–340

9. Boashash B, Ouelha S (2018) Designing high-resolution time-frequency and time-scale distributions for the analysis and classification of non-stationary signals: a tutorial review with a comparison of features performance. Digit Signal Process 77:120–152
10. Bandt C, Pompe B (2002) Permutation entropy: a natural complexity measure for time series. Phys Rev Lett 88(17):174102
11. Bryce RM, Sprague KB (2012) Revisiting detrended actuation analysis. Sci Rep 2:315–320
12. Bandarabadi M, Teixeira CA, Rasekhi J, Dourado A (2015) Epileptic seizure prediction using relative spectral power features. Clin Neurophysiol 126(2):237–248

Identification of Coronavirus (COVID-19) Long Non-coding RNAs (lncRNAs) Using Machine and Deep Learning Approaches

A. Revathi, S. Jyothi, and P. Swathi

Abstract Long non-coding RNAs (lncRNAs) are the molecules of RNA which are greater than 200 nucleotides. They play a major role in pathological processes and the main role in cellular transcripts. In many viral infections, the lncRNAs are involved in viral gene expression enhancement, viral transformation, viral replication and so on. They actually cause many human diseases due to their alterations. The recent research of COVID-19 says that lncRNAs play important role in the replication cycle of "severe acute respiratory syndrome coronavirus 2 (SARS-CoV-2)" by up-regulated and down-regulated protein interactions. The life activities and the disease analysis at the molecular level can be easily analysed by identifying the lncRNAs of a particular disease. The new dimension to the architecture of molecular is added by the lncRNA, which gives a new pathway of opportunities for treatment and also help to know the actual cause of death by knowing the lncRNAs biological functions in the development of the disease. The main problem is identifying the lncRNAs from the huge number of RNA sequences accurately and quickly with sequence technologies is very difficult. Hence the computational identification of COVID-19 lncRNAs-protein interaction is urgently required. The COVID-19 lncRNAs can be easily identified by the recent advancement in computational prediction technology, i.e. machine learning and deep learning approaches which are branches of artificial intelligence (AI). This is a powerful tool to fight against the COVID-19 pandemic. This paper describes how the machine learning and deep learning methods can be used to identify the lncRNAs of COVID-19.

Keywords COVID-19 · Coronavirus · lncRNAs · Viral infections · Respiratory syndrome · Machine learning · Deep learning · Artificial intelligence

---

A. Revathi (✉) · S. Jyothi (✉) · P. Swathi (✉)
Department of Computer Science, SPMVV, Tirupati, India
e-mail: arevathi20@gmail.com

S. Jyothi
e-mail: jyothi.spmvv@gmail.com

P. Swathi
e-mail: swathivinubaby@gmail.com

S. Jyothi et al. (eds.), *Proceedings of the 2nd International Conference on Computational and Bio Engineering*, Lecture Notes in Networks and Systems 215,
https://doi.org/10.1007/978-981-16-1941-0_6

1 Introduction

The novel coronavirus is the virus disease named COVID-19 which is a global disease according to the World Health Organization (WHO). It has affected many people and is spreading over many countries [1]. At the end of 2019, it started at Wuhan of China [2]. The disease symptoms are fever, headache, cough at the early stage; later the person gets respiratory problems which should be treated with ventilation in the intensive care unit [2]. The deaths and infected cases are increasing, which is an uncontrolled situation [1]. It is very dangerous and easily spreads by human-to-human transmission [1]. COVID-19 is the first documented case of severe lung disease in 2019, which is also known as severe acute respiratory syndrome coronavirus 2 (SARS-Cov-2) [3]. It is a positive-strand RNA virus, like other coronaviruses, and is a novel member of the coronavirus (Cov) family caused by the SARS-Cov2 virus [4, 3].

In 2020, the COVID-19 pandemic has affected human society with a longer impact. Still, research is going on for vaccine and treatment solutions [3]. Understanding the mechanisms of molecular in the replication of COVID-19 is needed [3]. Currently, the vaccines for COVID-19 are under development around the world. For approval and distribution, it takes 12–18 months [5]. Generally, to develop a vaccine it takes several years and even decades. To develop a vaccine for smallpox, it nearly took 25 years, 15 years for rotavirus and 28 years for influenza vaccine [5]. The vaccine development of COVID-19 is being allotted a very shorter time, so there is a more chance of failure in clinical testing. In addition to the development of the vaccine, other activities need to be concentrated to come out of this pandemic [5].

The four structural proteins of CoV genomes code are used for major drug targets [4]. There are many variable numbers of open reading frames (ORFs) present in genomes which are accessory proteins that play important role in pathogenic, but not responsible for the replication of the virus [4].

This paper's purpose is to show the machine learning algorithm's usage in finding the COVID-19 lncRNA functionalities that help in finding medicine to overcome the COVID-19 pandemic [1]. This work is organized as follows: Sect. 2 presents about lncRNA. Next, Sect. 3 analyses lncRNA-associated diseases using machine and deep learning approaches and in Sect. 4, the COVID-19-associated lncRNAs identification using machine and deep learning approaches. Finally, Sect. 5 concludes this paper by specifying the issues to be resolved for the identification of Covid-19-associated lncRNAs.

2 About Long Non-coding RNA

Long non-coding RNAs (lncRNAs) transcripts are longer than 200 nucleotides which are annotated very poorly [6]. Recent studies identified the lncRNAs effects in many biological and pathological processes. The functionalities of lncRNAs can help to monitor the progress of diseases of targeted therapies [6]. Recently functional studies of non-coding RNAs (lncRNAs) have shown that it is not a "junk" material, as it plays an important role in regulating gene expression in both physiological and pathological conditions and evidence also has shown that ncRNAs may serve as diagnostic markers for many disorders, and a predictor for drugs response which helps to find new therapeutic methods [7].

Computational approaches are used to overcome the difficulty in identifying lncRNAs from a huge amount of biological data [8]. The various disease development and occurrence involves the role of lncRNA [8]. A total of 150 lncRNAs-associated human diseases are found by lncRNA disease database. Some of them are Alzheimer's, colon cancer, breast cancer, prostate cancer and leukaemia [8]. The evidences show that for drug discovery and clinical treatment, the lncRNA is a potential biomarker [8]. So it is needed urgently to find the lncRNA association with diseases [8].

2.1 COVID-19-Associated lncRNAs Analysis

The understanding of lncRNAs functions in COVID-19 is crucial to the development of new therapeutic targeted approaches [7]. Artificial intelligence and the advancement of its applications can be used to overcome the COVID-19 pandemic [1]. The lncRNAs are involved during viral life cycles such as viral transformation, viral latency maintenance promoting viral replication, and viral gene expression enhancement. Additionally, the immune responses are also hosted by lncRNAs [8]. Based on the molecular classification of COVID-19, identifying the lncRNAs with altered expression helps to identify many lncRNAs of COVID-19 [9].

Figure 1 shows the collection of patient's samples for detection of lncRNA, and the detected RNA can be analysed by real-time PCR, next-generation RNA sequencing, microarray or analyses of computational methods which result in many lncRNA [10]. The new drugs development and other clinical-required applications can be done by studying the functionalities of selected lncRNA [10].

3 The Analysis of lncRNA-Associated Diseases Using Machine and Deep Learning Approaches

Next-generation sequencing is used for the identification of lncRNA, but it is a very time-consuming process, so a machine learning approach is used for lncRNA identification [11]. Recently, machine and deep learning approaches are used for the identification of lncRNAs of a particular disease and have shown promising results [11]. It is necessary for the understanding of the functionalities of non-coding RNAs of particular diseases for effective treatments [11]. Deep learning is one of the best

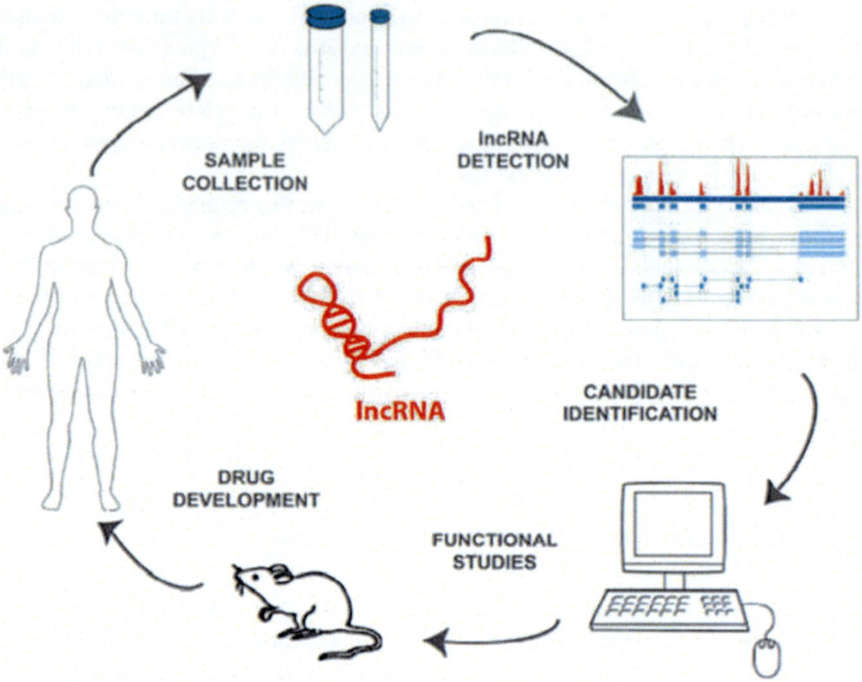

Fig. 1 The process of detection of long non-coding RNA (lncRNA) for clinical-required application analysis [10]

models of machine learning approaches for prediction and classification. It is having multiple hidden layers; it is a neural network-based method [11]. Deep learning is applied to many real-world problems such as language translation, speech recognition and image recognition. Recently, it is used in bioinformatics and medicine [11].

There are many RNAs that do not produce protein products and are identified in our cells by using RNA sequencing. For the past many years, the functional and biochemical characterization of non-coding RNAs is analysed with the help of RNA-seq [12]. The non-coding RNAs are identified as a potential disease modifier, which can be used as biomarkers for therapeutic targets [12]. There are thousands of lncRNAs in human cells, which are one of the groups of non-coding RNAs that are analysed to know the physiological role of it [12]. In the past years, many studies showed that misregulation of lncRNAs leads to the development and progression of the disease [12, 14] which are given in Table 1. To analyse the COVID-19 associated lncRNAs, the samples of patient and profiles expression of different cells are very important for analysis [12].

To develop new biomarkers for the therapeutic target, it is necessary to understand the molecular mechanisms of a particular disease by identifying the disease-associated lncRNAs [15]. The functionalities of many lncRNAs and its implications

Table 1 Some of the identified lncRNAs of different diseases

Name of the lncRNA	Associated disease
MUC5B-AS1	It promotes metastasis in lung cancer [9]
LINC00885	Associated with bladder cancer [9]
DSCAM1-AS1	The most significant upregulated lncRNA in breast cancer [9]
LOC101927136	The most significant downregulated lncRNA in breast cancer [12]
GEPIA2, ZNF667-AS1, ELFN1-AS1, LINC00885 and CELF2-AS1	Some other lncRNAs which are related to breast cancer [12]
lncRNA PCA3	For prostate cancer lncRNA PCA3 treated as a biomarker [14]
lncRNA BC200	When compared to normal tissues, lncRNA BC200 expressively higher in Alzheimer's disease [14]

in disease are unknown [15]. It is a challenging task to extract features of lncRNA since they are mostly unstructured [13].

4 The COVID-19-Associated lncRNAs Identification Using Machine and Deep Learning Approaches

COVID-19 infected many people around the world, which led to many deaths. It is a worldwide health crisis. To prevent the infection, there is an urgent need for vaccines and drugs [16]. Artificial intelligence is used by researchers and companies around the world to find all the possibilities to develop a vaccine, drug discovery and plasma therapy to come out of this COVID-19 pandemic [17]. Machine learning is a sub-field of AI used for fast and accurate results in the present situation [17].

Artificial intelligence (AI) is used in all the fields around the world in which machine learning is part of it. The algorithms of machine learning (ML) helps to solve complex problems, as it consists of many classification and prediction algorithms [2]. In the medical field, machine learning algorithms are used for the prediction of diseases such as cancer early-stage prediction, and even many types of research are conducted for predicting COVID-19 [2]. To study the reasons for the spread of COVID-19, many classifications and prediction algorithms of ML are used and also analysed [2]. Some studies identified the COVID-19 new symptoms such as having loss of taste and smell and some others identified the disease transmission process [2].

A model was developed for COVID-19 by using ANN-based adaptive online techniques of incremental learning, where parameters are updated continuously to adapt intelligently to new data sets [18]. The models of deep learning techniques eliminate the difficulty of retraining from scratch whenever new data sets are given [18]. Recently, a machine learning (ML) model is developed to predict the antibodies of coronavirus by training the model with collected virus antibody sequences of patients [19]. Eight stable antibodies of COVID-19 that potentially inhibit the virus were found by using thousands of hypothetical antibody sequences with the help of ML methods [19]. The advancement in machine learning helped COVID-19 pandemic in many ways, such as clinical and molecular prediction of the protein structure of COVID-19 which helped for treatment [20].

The researchers used easier ways to diagnose the disease with the help of AI. It decreased the transmission by identifying the COVID-19 at the initial stage by monitoring and detecting the infected persons [21]. The viral infections are visualized by a medical imaging machine, which is an advantage to use many different methods of machine and deep learning techniques [21]. The best way to diagnose tumours and infections of various diseases is by machine and deep learning methods, which gave accurate results than radiologists [21].

The performance of the machine learning models of COVID-19 can be improved by giving deep features. The extraction of deep features is easy when compared with traditional learning, as machine learning models take very less time for prediction [14]. So it is better to use the combination of deep feature extraction with machine learning methods for high accurate results in less time [14].

The AI is used in the different fields of analysis of COVID-19 such as dashboards, early prediction, diagnosis and prognosis, treatments and drug discovery. To predict and diagnose COVID-19 machine learning big data-based models are required [22]. The COVID-19 has various issues, such as lack of historical, problems of big data, noisy and outlier data. To resolve these problems careful balancing of data is required [22].

5 Conclusion

It is very difficult to identify lncRNAs because of characteristics sequence similar to the characteristic of protein-coding transcripts. So advanced computational techniques have to be used to distinguish it, and also there are very few experiments done in both biological and computational approaches to find the lncRNA disease associations with other diseases. No experiment was conducted for finding lncRNA disease association of COVID-19. There is an urgent requirement for the use of the machine and deep learning approaches to reveal potential lncRNA disease association of COVID-19 and other diseases for effective treatment and drug development. There is a need for the development of COVID-19 lncRNAs databases. Machine and deep learning models have to be developed to analyse the functionality of COVID-19 lncRNAs. Further investigation is required to know the roles and functions of lncRNAs in viral infection and COVID-19. An intensive study is to be done on lncRNAs of key regulators of COVID-19. The major problem is to be resolved, i.e. lack of standardization is needed to explore the efficient existing lncRNAs data.

References

1. Pham Q-V, Nguyen DC, Huynh-The T, Hwang WJ, Pathirana PN (2020) Artificial intelligence (AI) and big data for coronavirus (COVID-19) pandemic: a survey on the state-of-the-arts. IEEE Trans Artif Intell RG.2.2.23518.38727
2. Theerthagiri P, Jeena Jacob I, Usha Ruby A, Vamsidhar Y (2020) An investigation of machine learning algorithms on covid-19 dataset. https://doi.org/10.21203/rs.3.rs-70985/v1
3. Miladi M, Fuchs J, Maier W, Weigang S, Pedrosa ND, Weiss L, Lother A, Nekrutenko A, Ruzsics Z, Panning M, Kochs G, Gilsbach R, Gruning B (2020) The landscape of SARS-CoV-2 RNA modifications. https://doi.org/10.1101/2020.07.18.204362
4. Michel CJ, Mayer C, Poch O, Thompson JD, Michel et al (2020) Characterization of accessory genes in coronavirus genomes. Virol J 17(131). https://doi.org/10.1186/s12985-020-01402-1
5. Ng WH, Liu X, Mahalingam S (2020) Development of vaccines for SARS-CoV-2. https://doi.org/10.12688/f1000research.25998.1
6. Wang Z, Zhao Y, Zhang Y (2017) Viral lncRNA: a regulatory molecule for controlling virus life cycle. ISSN: 2468-0540. https://doi.org/10.1016/j.ncrna.2017.03.002
7. Vicentinia C, Galuppinib F, Corboa V, Fassanb M (2019) Current role of non-coding RNAs in the clinical setting. https://doi.org/10.1016/j.ncrna.2019.09.001
8. Revathi A, Jyothi S, Swathi P (2020) The role of long non-coding RNA (lncRNA) in health care using big data analytics. Chapter 35 Springer Science and Business Media LLC. https://doi.org/10.1007/978-3-030-46939-9_35
9. Zaheed O, Samson J, Dean K (2020) A bioinformatics approach to identify novel long, non-coding RNAs in breast cancer cell lines from an existing RNA-sequencing dataset. https://doi.org/10.1016/j.ncrna.2020.02.004
10. Sanchez Y, Huarte M (2013) Long non-coding RNAs: challenges for diagnosis and therapies. NuclC Acid Ther 23
11. Amin N, McGrath A, Chen Y-PP (2019) Evaluation of deep learning in non-coding RNA classification. Nat Mach Intell 1:246–256
12. Gambhir E, Jain R, Gupta A (2020) Regression analysis of COVID-19 using machine learning algorithms. In: (ICOSEC), IEEE Xplore part number: CFP20V90-ART; ISBN: 978-1-7281-5461-9
13. Wang C, Wei L, Guo M, Zou Q (2013) Computational approaches in detecting non- coding RNA. Curr Genomics 14:371-377
14. Zenga M, Lua C, Zhanga F, Lia Y, Wub F-X, Lic Y, Li M (2020) SDLDA: lncRNA-disease association prediction based on singular value decomposition and deep learning. https://doi.org/10.1016/j.ymeth.2020.05.002
15. Yan C, Zhang Z, Bao S, Hou P, Zhou M, Xu C, Sun J (2020) Computational methods and applications for identifying disease-associated lncRNAs as potential biomarkers and therapeutic targets. Mol Ther Nuclc Acids
16. Syeda HB, Syed M, Sexton K, Syed S, Begum S, Farhan F, Yu F (2020) The role of machine learning techniques to tackle COVID19 crisis: a systematic review. https://doi.org/10.1101/2020.08.23.20180158
17. Kabra R, Singh S (2020) Evolutionary artificial intelligence-based peptide discoveries for effective covid-19 therapeutics. https://doi.org/10.1016/j.bbadis.2020.165978
18. Farooq J, Bazaz MA (2020) A deep learning algorithm for modeling and forecasting of COVID-19 in five worst affected states of India. ISSN: 1110-0168. https://doi.org/10.1016/j.aej.2020.09.037
19. Magar R, Yadav P, Farimani AB (2020) Potential neutralizing antibodies discovered for novel corona virus using machine learning. arXiv:2003.08447
20. Punn NS, Sonbhadra SK, Agarwal S (2020) COVID-19 epidemic analysis using machine learning and deep learning algorithms. https://doi.org/10.1101/2020.04.08.20057679
21. Rahimzadeh M, Attar A, Sakhaei SM (2020) A fully automated deep learning-based network for detecting COVID-19 from a new and large lung CT scan dataset. https://doi.org/10.20944/preprints202006.0031.v1

22. Baek J, Lee B, Kwon S, Yoon S (2018) LncRNAnet: long non-coding RNA identification using deep learning. https://doi.org/10.1093/bioinformatics/bty418

Predictive Analysis of Damage Occurred Due to Natural Disasters Using Whale-Optimization Algorithm-Based Hybrid Computation

K. Hemalatha, K. Hema, and V. Deepika

Abstract Bio-inspired computing algorithms are inspired by nature that can provide flexible, dynamic, and effective solutions to hard and complex real-life problems. Many bio-inspired algorithms such as ant colony algorithm, firefly algorithm, whale optimization algorithm, grey-wolf algorithm, etc. proved their efficiency in solving various nonlinear, high-dimensional, and complex problems. On the other hand, artificial intelligence techniques can help to build intelligent models to implement the bio-inspired computing algorithms successfully and provide solutions to real-life problems that occurred due to natural disasters. Natural disasters like hurricanes, earthquakes, floods, tsunami, landslides, forest fires, etc. are the inspiration for solving hard and complex problems in computer science since they exhibit robust phenomena. It is necessary for the government agencies to know the accurate measurements of the damage that occurred due to natural disasters. However, these disasters are also called "ACTS OF GOD" because they cannot be found in fixed locations and time. So, here the predictive analysis is necessary to save the economy as well as human and animal lives. The combination of bio-inspired computing algorithms and artificial intelligence techniques will provide flexible solutions in those situations to predict the damage that occurred due to natural disasters. In this study, we proposed a bio-inspired computing-based hybrid approach, which is a combination of the modified version of popular bio-inspired algorithm whale optimization algorithm (MWOA) with significant artificial intelligent technique multi-layer perceptron (MLP) neural network to predict the damage of buildings that occurred due to the natural disasters using the satellite images.

K. Hemalatha (✉) · K. Hema · V. Deepika
Sri Venkateswara College of Engineering, Tirupati, India
e-mail: hemalathakulala@gmail.com

K. Hema
e-mail: goldenhema@gmail.com

V. Deepika
e-mail: deepika.v1@svcolleges.edu.in

S. Jyothi et al. (eds.), *Proceedings of the 2nd International Conference on Computational and Bio Engineering*, Lecture Notes in Networks and Systems 215,
https://doi.org/10.1007/978-981-16-1941-0_7

Keywords Whale optimization · Predictive analysis · Multi-layer perceptron · Hybrid approach · Satellite images

1 Introduction

Natural disasters are the consequences of events triggered by natural hazards. These natural hazards seriously affect the social and economic development of a region. Natural disasters are of various types, such as volcanic eruptions, tsunamis, floods, drought, landslides, earthquakes, hurricanes, etc. There is no human control over the above-mentioned natural disasters. Even though scientists do claim to find ways to predict these natural disasters, most of the time the founded ways have failed to predict natural disasters that cause serious destruction. However, instead of making attempts to stop natural disasters, it is better to be with necessary precautions whenever these disasters occur. These precautions will minimize the damage caused due to natural disasters. Therefore, the prediction of damage that occurred due to natural disasters is necessary. By predicting the damage that can occur after the natural disasters will help us to take necessary precautions to minimize the damage that occurred to lives, building, nature, etc. In this study, we concentrated to build a whale optimization-based hybrid computation model to predict the damage that occurred to the buildings due to hurricanes. This bio-inspired computing-based prediction model will help to estimate the damage that will occur in the future based on the input parameters.

2 Materials and Methods

2.1 Whale Optimization Algorithm (WOA)

WOA is the latest optimization technique for solving optimization problems that simulate the inspiration and foraging behavior of humpback whales. This algorithm is a population-based method proposed by Mirjalili and Lewis [1], which is simulated by the bubble-net hunting strategy [2]. Whale optimization algorithm (WOA) is the latest and popular bio-inspired optimization algorithm. WOA consists of three phases. The first phase is encircling prey; the second one is the exploitation phase: the bubble-net attacking method; and the final phase is the exploration phase: search for prey [3]. The parameters such as population, parameter "a", coefficient vectors A and C, and a random value p are initialized. To maintain exploration and exploitation, the parameter, "a" value lies between 2 and 0. For each iteration, a search agent is chosen randomly when the coefficient vector A is greater than 1. The position of the current search agent will be updated when the coefficient vector A is less than 1. Finally, the algorithm terminates when the termination criteria are satisfied.

2.2 Selection of Mathematical Model for Searching Path

Humpback whales use the bubble-net predation method to hunt the prey. They can recognize the distance between the whales and the prey. Based on the distance, they encircle the prey using a search path to hunt. This unique characteristic of whales is mimicked by the latest bio-inspired algorithm WOA. In this section, few searching paths are discussed. The selection of a searching path is significant in the performance of WOA.

The Logarithmic Spiral Curve. The logarithmic spiral curve can also be called a growth spiral and equiangular spiral. The logarithmic spiral searching path can be represented parametrically as described in the below equation. This curve is constructed from equally spaced rays which start from a point along with one ray and perpendicular to a neighboring ray.

Fermat Spiral Curve. The Fermat spiral curve is a special case of a parabolic spiral. It is a close curve without double points dividing the plane into two connected regions. The thread pitch is fixed in this curve.

2.3 Dataset

Satellite images of damaged buildings on the post-hurricane dataset from the IEEE data port [4] are considered for experimentation in this study. The dataset consists of two classes: (i) damaged and (ii) no damage. A total of 955 samples of satellite images of damaged buildings and 1002 samples of satellite images of no damaged buildings are there in the dataset. The sample images of each class are given below in Figs. 1 and 2.

Fig. 1 Sample images of damaged buildings

Fig. 2 Sample images of no damaged buildings

3 Modified Whale Optimization Algorithm (MWOA)

The standard WOA search mechanism has a limitation. The standard WOA uses logarithmic spiral curves as the search path, in which the thread pitch is not fixed. The logarithmic spiral curve does not search the path completely. This leads to incomplete searches in the search space. Because of the incomplete search, the algorithm may fall into local optimum issues [5, 6]. Therefore, proper selection of the search path is significant. Hence in this study, the standard WOA is modified by modifying the search path. The logarithmic spiral curve search path of WOA is replaced by the Fermat spiral curve path. The workflow of the proposed MWOA-based predict model is presented in Fig. 3.

4 Proposed Hybrid Prediction Model (MWOA-MLP)

In this study, a modified whale optimization algorithm (MWOA)-based hybrid prediction model is proposed to estimate the damage that occurred during post hurricanes. Satellite images of damaged buildings on post hurricanes are considered in this study to test the performance of the proposed model. A total of 1957 such images are preprocessed and the following significant texture features, contrast, correlation, energy, and homogeneity, are extracted. Few values of extracted texture features are presented in Table 1.

MATLAB R2015a is used to do the experiments in this study. The extracted texture features are collected as a dataset. Multi-layer perceptron (MLP) neural network is trained with the texture features dataset to classify the data into damaged and no damaged labels. To enhance the classification rate of MLP, MWOA-based hybrid prediction model is proposed. It is used to optimize the weights and biases of the MLP network to enhance the classification rate. The MLP network is simulated with the optimal set of weights and biases obtained by using MWOA. The network is trained with 764 samples of damaged and 801 samples of no damaged data. The trained network is tested with 191 samples of damaged and 201 samples of no damaged data. Finally, the classification rate is calculated.

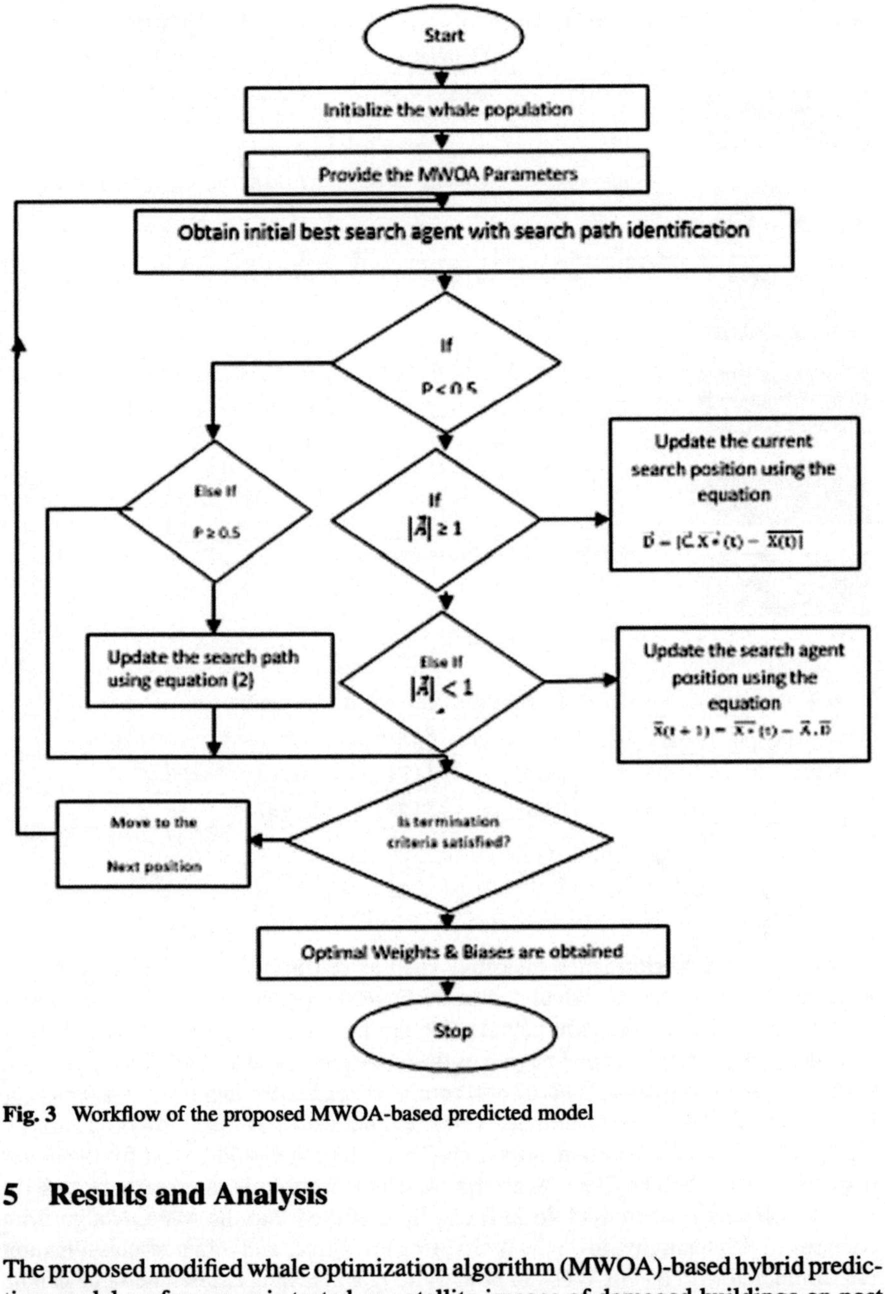

Fig. 3 Workflow of the proposed MWOA-based predicted model

5 Results and Analysis

The proposed modified whale optimization algorithm (MWOA)-based hybrid prediction model performance is tested on satellite images of damaged buildings on post hurricane. The experimental results using MWOA have been compared with that of standard WOA also. The results obtained are given in Table 2.

Table 1 Texture features of satellite images of damaged buildings on post hurricanes

Images	Contrast	Correlation	Energy	Homogeneity
	0.03895527	0.985530367	0.448609858	0.980522365
	0.395849865	0.669064653	0.014665184	0.735906814
	0.870446315	0.655299134	0.034423439	0.281132336
	0.893923438	0.868997022	0.01870931	0.732619196
	0.040656506	0.647881725	0.878719313	0.492876836
	0.052638176	0.141542782	0.674602222	0.963544185

Table 2 Performance analysis of the proposed hybrid prediction model

Training/testing	Algorithm	Sensitivity	Specificity	Error rate	Classification rate
Training	WOA	86.36	43.01	0.265	73.48
	MWOA	**95.83**	**54.79**	**0.137**	**86.26**
Testing	WOA	79.34	**66.67**	0.255	74.49
	MWOA	**97.22**	31.25	**0.082**	**91.84**

MLP network performance measures such as sensitivity, specificity, error rate, and classification rate are calculated to evaluate the performance of the proposed model. Sensitivity measures the actual positives that are correctly identified, whereas specificity measures the actual negatives that are correctly identified. Sensitivity and specificity are inversely related to each other. If sensitivity increases, on the other hand, specificity decreases, and vice versa. Higher sensitivity and lower specificity are useful for increased classification rate. The error rate should be less for the better performance of the classifier. A higher classification rate is necessary to classify the data effectively. From Table 2, it can be observed that the MWOA algorithm contains high sensitivity, low specificity, less error rate, and a higher classification rate compared with that of WOA in both training and testing cases. The same can be observed from Figs. 4 and 5.

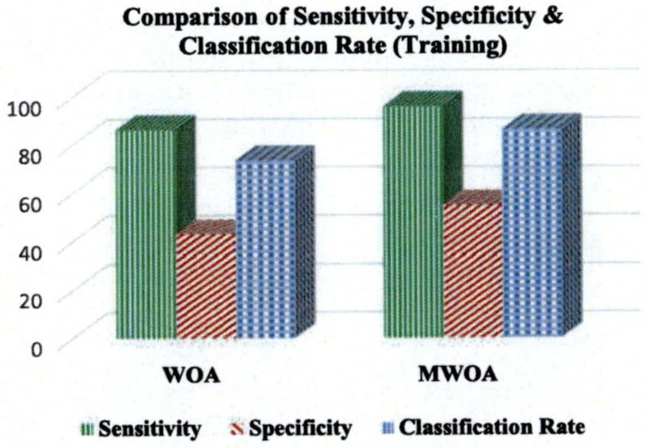

Fig. 4 Comparison of training sensitivity, specificity, and classification rate of the proposed model

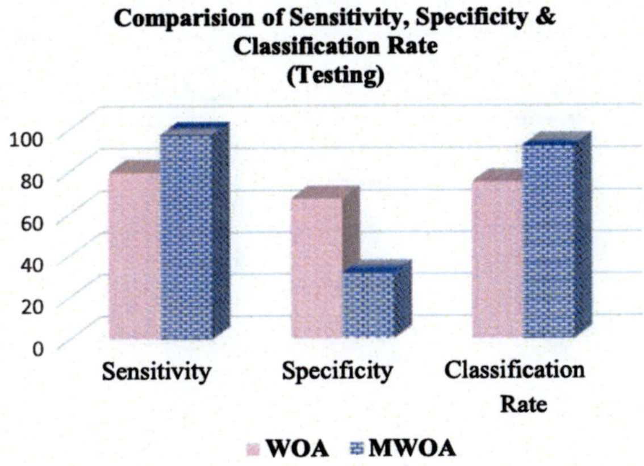

Fig. 5 Comparison of sensitivity, specification, and classification rate of the proposed model (testing)

6 Conclusion

Natural disasters can cause massive destruction to nature and they cannot be stopped by humans. Instead of trying to stop these disasters, necessary precautions are to be taken to minimize the damage that occurred. For this purpose, forecasting damage that will occur during natural disasters is essential. Data scientists are engaged in the development of efficient prediction models to estimate the effect of the damage that occurs due to natural disasters. Hurricanes are one of the most commonly occurred natural disasters. Hence in this study, we concentrated on the prediction of damage that will occur due to hurricanes.

Bio-inspired algorithms are efficient to predict the damage that occurred due to hurricanes. Therefore, in this study, we proposed a bio-inspired-based prediction model to identify the damage that occurred on hurricanes. The proposed model is a combination of the modified whale optimization algorithm (MWOA) and multi-layer perceptron network (MLP). In MWOA, the search path of the standard

WOA is replaced to enhance the searching of the standard WOA to get improved results. MWOA performance is compared with the standard WOA. The proposed hybrid prediction model classified the satellite images of damaged buildings in post hurricanes with a higher classification rate and lower error rate.

References

1. Mirjalili S, Lewis A (2016) The whale optimization algorithm. Adv Eng Softw 9:51–67
2. Aljarah I, Faris H, Mirjalili S (2018) Optimizing connection weights in neural networks using the whale optimization algorithm. Soft Comput 20:1–15
3. https://en.wikiversity.org/wiki/Whale_Optimization_Algorithm. Accessed 04 May 2020
4. Cao QD, Choe Y (2018) Detecting damaged buildings on post-hurricane satellite imagery based on customized convolutional neural networks. IEEE Data port
5. Elghamrawy SM, Hassanien AE (2018) GWOA: a hybrid genetic whale optimization algorithm for combating attacks in cognitive radio network. J Ambient Intell Humlzed Comput 10:4345–4360
6. Canayaz M, Ozdag R (2018) Training ANFIS using the whale optimization algorithm. In: International conference on advanced technologies computer engineering and science

Remote Access for Advanced Mobile Communication Systems

Swetha Rani Karanam and Suseela Kallepalli

Abstract Advanced mobile communication systems have a number of nodes to be intercommunicated in different places across the universe. For better maintenance and to reduce downtime, preventive measures are to be taken. In the same aspect, local maintenance terminals and remote monitoring tools help in monitoring the various network entities of wireless communication systems even in the corona pandemic scenario. The design is done using the framework in the frontend and the database is viewed using Microsoft access which is the backend of the complete tool designing. This tool is implemented in the logical model via Visual Studio and C# language. The data and its associated history are stored in servers, and the logs generated show the interruptions and the cause of failures in the system using characteristics as configuration, performance, alarm management, backup and restoration procedures. This improves system reliability and reduces cost.

Keywords GSM · Mobile Communications · GSM network architecture · UMTS · UTRAN · RBS · Radio Base Station Hardware Unit

1 Introduction

1.1 General

In wireless communication systems there will be a huge number of nodes intercommunicated. To maintain the network, we have to physically visit the sites, identify the faults and rectify the faults accordingly. But it is very difficult to visit a number of sites physically as it demands more manpower. Hence more investments result in the reduction of profit margin. To resolve this, remote access of equipment, the privilege to modify the equipment or unequipment the card in the concerned equipment, to configure the equipment according to the existing demand, etc. can be performed.

S. R. Karanam (✉) · S. Kallepalli
Department of ECE, Sri Padmavathi Mahila University, Tirupati, AP, India

S. Jyothi et al. (eds.), *Proceedings of the 2nd International Conference on Computational and Bio Engineering*, Lecture Notes in Networks and Systems 215,
https://doi.org/10.1007/978-981-16-1941-0_8

It is a radio communication system where the components that have been traditionally executed in hardware are instead executed by means of software on a personal computer [1].

One of the innovations in the field of technology is a system of monitoring and controlling of radio base stations (RBS), i.e., both BTS (base transceiver station) in GSM and NodeB in UMTS [2]. An operator that charges for monitoring and controlling the RBS can roll in hay wherever he is. With the presence of this system monitor becomes easy and efficient, including the efficiency of time. An operator can monitor the condition of the RBS just by opening the appliance via a personal computer (PC). The system is also useful with respect to time efficiency because an operator does not need to visit the RBS but only has to see the condition of the connected devices in the RBS. This tool for RBS acts as a protective shield for telecommunications infrastructure or base stations to prevent damage and hazards to the RBS [3].

The invention relates to the saved program-controlled telecommunications switching systems, and more particularly, to the design and diagnostic tool for such systems. The end result must be accurate and the timely processing of numerous signals affects the control of various hardware components and set up the communication connections. The term network monitoring defines the utilization of a system that continually monitors a network for slow or failing components and notifies the network administration just in case of outages via email, pager or other alarms. It is a subset of the function's associates in network management. There exists a wide collection of software and hardware products that help network system administrations manage a network. Network management covers a wide area, including:

Security: Ensuring that network is shielded from unauthorized users.
Performance: Eliminating bottlenecks in the network.
Reliability: Ensuring the network is out there to users and responding to hardware and software malfunctions.

1.2 Objective and Scope of Work

For designing such graphical system design, we are using the C# language, i.e., .NET. A programming infrastructure created by Microsoft for building, deploying and running applications and services that use .NET technologies, like desktop applications and web services .NET framework is meant to be employed by most new applications created for the Windows platform. Microsoft also produces an integrated development environment extensively for.NET software called Visual Studio. The design is done using this framework in the frontend and the database is viewed using Microsoft access which is the backend of the complete tool designing [5].

The radio system infrastructure for second-generation (2G) GSM, third-generation (3G) UMTS / HSPA and fourth-generation (4G) LTE is an advanced wireless platform that has been specified and deployed based on evolving architectural and technological ideas. Here, we focus on the access system and its monitoring tool where the complete radio base station can be operated remotely for 2G, 3G and 4G technologies [4]. Depending upon the specifications of 2G, 3G and 4G, the complete tool is designed where we can switch between two users. This tool is limited only to these generations.

2 Global Systems for Mobile Communications

In mobile communications, the connectivity of the user (mobile subscriber) with mobile network access is through radio signals and from the network to the user equipment. Although mobile telephony can be seen in a broad sense as the wireless communication and wireless series can be offered through various technologies like GSM, UMTS/WCDMA, CDMA etc., here we will talk about GSM (global system for mobile communication) and UMTS/WCDMA as a wireless communication technology.

The basic unit of a GSM network is the cell, which is identified by BTS. A BTS transmits and receives the radio signals and provides coverage in an area. The third component of the GSM network, OSS, is to support the O & M team during the operation and maintenance of the GSM system. The UMTS network is a network that overcomes the hurdles in GSM and GPRS related to data speed and global access. The changes are related to the access part of the network.

2.1 GSM Network Architecture

GSM system is often looked at as consisting of two main blocks, namely:

- The radio equipment, the one which provides access to mobile subscribers.
- Core equipment, which is responsible for call handling, billing, routing and other important functions.

Core network also has databases like HLR and VLR which keep a record of subscriber data. A GSM network includes various subsystems as described below:

I. **Mobile Station (MS):**

- Comprises mobile handset and SIM (subscriber identity module).

II. **Base Station Subsystem (BSS):**

- BTS—Base transceiver station
- BSC—Base station controller.

III. **Network Switching Subsystem (NSS):**

- MSC—Mobile switching center
- VLR—Visitor location register
- HLR—Home location register
- AuC—Authentication center
- EIR—Equipment identity register.

IV. **OSS**—Operation and support subsystem.

V. **Important Interfaces**: Interface interfacing nodes, air—MS and BTS, Abis—BTS and BSC, Ai—BSC and MSC (Fig. 1).

2.1.1 Mobile Station (MS)

MS is a combination of a GSM handset (IMEI) and a SIM card (this number is called International Mobile Subscriber Identity (IMSI)). MS carries out the signal processing function of digitizing, encoding, error protecting, encrypting and modulating the transmitted signals.

2.1.2 Base Station Subsystem (BSS)

The BSS is a set of base station equipment, namely BTS and BSC, responsible for communicating with MS in a certain area. A BTS comprises radio transmission and reception devices, including the antennas, and also all the signal processing specific to the radio interface. BTSs have number of GSM radio frequencies.

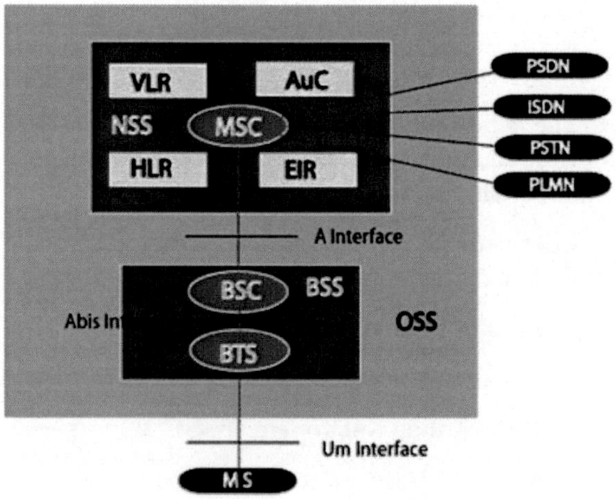

Fig. 1 GSM network architecture

2.1.3 Network Switching Subsystem (NSS)

The NSS consists of the main switch (Exchange) also as some databases, needed for subscriber data and mobility management. As NSS includes some databases to keep the data of subscribers and their locations. The home location register (HLR) is the main database of the GSM system and a temporary database of the subscribers is also maintained in the form of VLR (visitor location register). AuC (authentication centre) is another database of NSS to be used to authenticate a SIM into the network.

2.1.4 Operation and Support Subsystem (OSS)

The OSS may be a server to monitor the network performance. The OSS is connected to all the BSCs and MSCs in a LAN/WAN network. An OMC (operation and maintenance center) is assembled through OSS which monitors the network round the clock. As the OSS is connected to all the network elements, it is also used to control / configure the network entities.

2.2 Basics of Network Performance and Optimization

The basics of network performance are the many parameters to assess the performance of a GSM network and therefore the principle of the assessment procedure. The three items most vital to watch the network performance are coverage, blocking and dropping rate. The different actions in optimization processes that we want to solve in the network problems are summarized as below:

I. Antenna Optimization: This includes changing antenna tilts, orientations and positions. Sometimes the antenna may also be changed.

II. Frequency Changes: Frequency changes help us to regulate the interference within the network and hence security. A careful study of cell coverage area and server area provides in making those changes.

III. Addition of new cell sites: Sometimes to unravel whole coverage problems, we would like to feature more sites. While adding a new site (BTS) in the network, the definition of neighbors must be taken care of to avoid the problems of handover failure and call drop due to this.

3 Operation Support Subsystem

3.1 Network Management System

Network management is the act of initializing, monitoring and modifying the configuration of network entities. NMS could also be used to monitor both software and hardware components during a network. It usually records data from the network's remote points to hold out central reporting to a supervisor. Network management system components assist in:

- **Network device monitoring**: Monitoring at the device level to work out the health of network components and therefore the extent to which their performance matches with the capacity plans and intra-enterprise service-level agreements (SLAs).
- **Network performance analysis**: Tracing performance indicators like bandwidth utilization, packet loss, latency, availability and uptime of routers, switches and other simple network management protocol (SNMP)-enabled devices.
- **Intelligent notifications**: Configurable alerts which will answer specific network scenarios by paging, emailing, calling or texting a network administrator (Fig. 2).

3.2 Network Management Architecture

The network management architecture consists of the following:

I. **Network management system (NMS)**: A system that implements applications that monitor and control managed devices. NMS provides the majority of processing and memory resources that are required for network management.

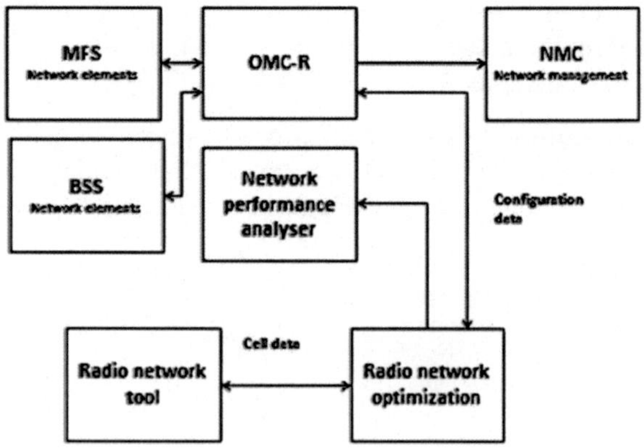

Fig. 2 Network management system environment

II. **Network management protocol**: A protocol that facilitates the exchange of management information medially with the NMS and managed devices, including SNMP, MIB.
III. **Managed devices**: A device (such as a router) administered by an NMS.
IV. **Management agents**: Software, on managed devices, that gathers and stores management information, including SNMP agents and RMON agents.
V. **Management information**: Data that is of interest to a devices' management, usually stored in MIBs (Fig. 3).

A variety of network management applications are often used on a network management system. The selection depends on the network platform (such as the hardware or operating system). The management information abides on network devices management agents that abide on the device to collect and store data in a standardized data definition structure known as the **MIB**.

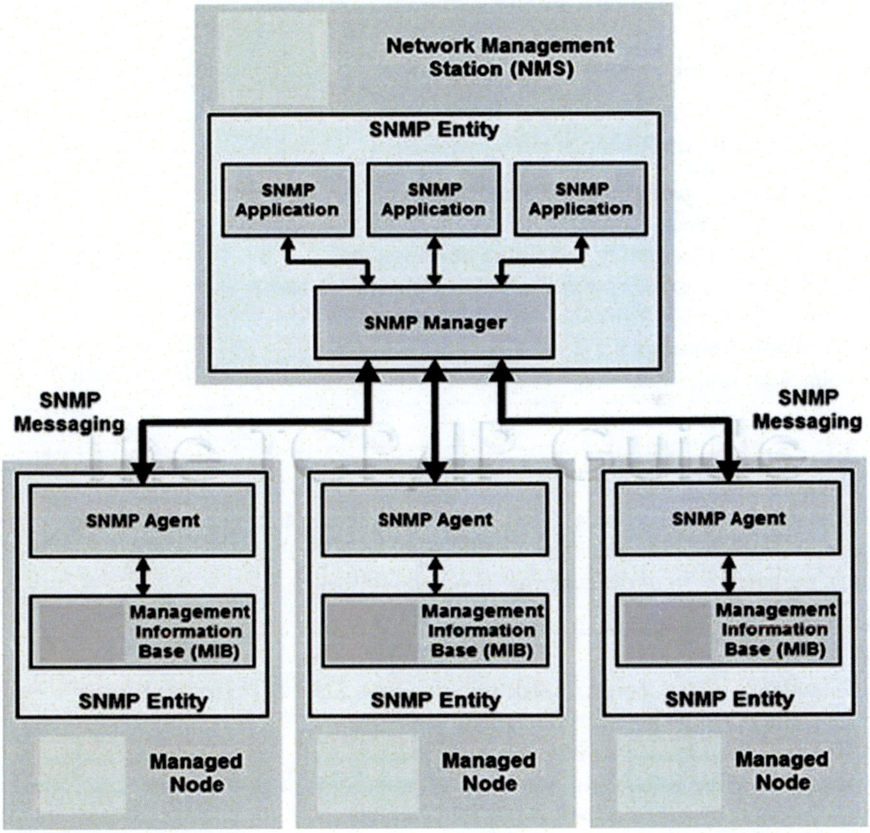

Fig. 3 NMS architecture

The network management application uses **SNMP** (simplest network management protocol) or other network management protocols to retrieve the data that the management agents collect. The retrieved data is usually processed and ready for display with a GUI, which allows the operator to use a graphical representation of the network to regulate managed devices and program the network management application.

3.3 Network Planning

Radio network planning RNP is a complex project that depends on different processes to be defined before the commencement of the actual planning.

RNP is a process made of few phases:

- Preplanning (creating the initial plan for the next phase)
- Planning
- Verification and acceptance
- Optimization.

Factors influence RAN planning:

- Targeted user groups (services and usage)
- Capacity for each service
- Coverage for beginning and future development
- Quality of services (how end-users experience network)
- Area topography
- Available frequency band
- Financial limitation.

3.3.1 RNP Pre Planning

Preplanning is an entry process in RAN planning which aims to get information from the customer about his plans for the future network. In this process, we agree about input parameters from the customer about the following:

- Budget for the investment
- Coverage and capacity
- Frequency plan (operator needs to buy rights, licenses, to transmit signals over specific bands of the electromagnetic spectrum)
- Frequencies for use in a radio access network.
- Number of users (existing, forecast and probably growth)
- Services that will be provided (GSM, UMTS).

3.3.2 Planning

After this input information is collected, the preliminary planning can be done in more detail. This preliminary planning will provide us a number of network elements to fulfill the quality of services set by an operator, the minimum number of RBS needed for coverage and capacity.

Detailed planning process covers:

- Frequencies planning (coverage, capacity and interference problems)
- Parameter planning (system features and services).

A limited number of frequencies forces us to reuse the same frequencies. Frequency reuse implies that in a given coverage area, there are several cells in close range that use the same set of frequencies and lead to the problem called same channel interference (co-channel).

3.3.3 Verification and Acceptance

It is a process that happens before launching the network. The target of this process is to check the quality of installation work and to correct installation mistakes before launching the network.

3.3.4 Optimization

It is a process that is done in an operational environment of the network to reduce coverage holes and to provide better signal strength in the area as decided by the service provider.

3.4 RBS Operation and Maintenance

RBS is locally managed by Ericsson software named "OMT" (Operational and Maintenance Terminal). OMT is usually called a local maintenance terminal (LMT). A special cable is required to attach to the interface of the RBS with the LMT. Operation maintenance center is affianced to maintain and watch the performance of every mobile station (MS), base station (BS), base station controller (BSC) and mobile switching center (MSC) within a GSM system and in the same way as in UMTS system (Fig. 4).

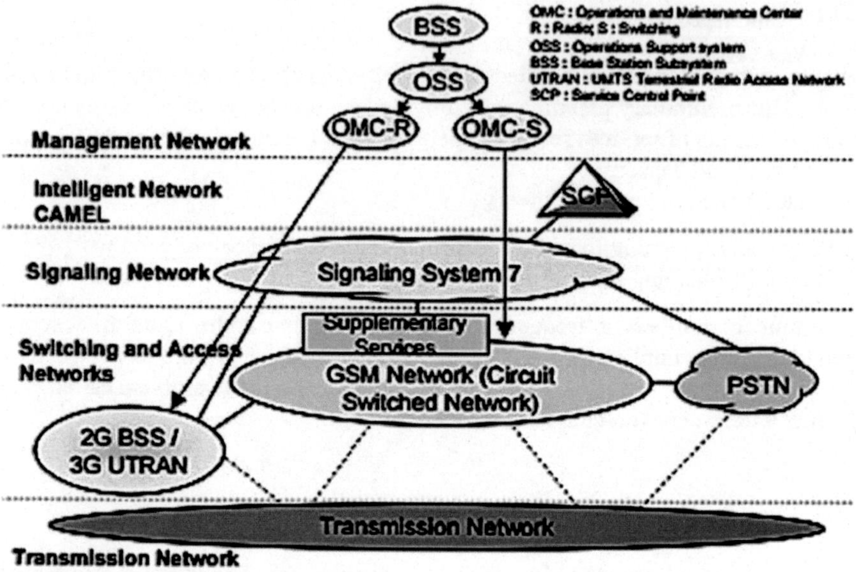

Fig. 4 OMCR and OMCS functionality

3.5 OMCR Functions

The purpose of OMCR is to control the O & M functions of the BSS and RAN. It contains O & M application programs for this purpose. The OMCR connects to the BSS and the MFS over an IP network. Table 1 provides a summary of the main tasks.

4 Radio Base Station Hardware Unit

RBS provides the radio interface to the mobile (radio) stations in a radio cell. RBS consists of one or more radio transceivers. The name radio base station is about cabinet topology which consists of radiofrequency unit, digital signal processing unit, baseband unit, rectifier and battery backup, and transmission module. In most cases all of this equipment was in the same cabinet but as technology develops many RBS sites use remote radio unit (RRU) which are mounted near the antenna and connected with the baseband unit through fiber optic cable.

Today, in the world, there are few vendors of RBS:

- HUAWEI
- ERICSSON

Table 1 OMC functions

O & M function	Task description
Configuration management	The configuration management tasks are: Manage GSM/UMTS/LTE radio resources (cells, carriers, channels …) Manage transmission equipment (configuration of the transmission path for different types of data) at various levels Support the network for O&M communications in GSM/UMTS/LTE Manage the SS7 connection in the system Manage different support activities like alarm management, logs Manage BSS/MFS and RNC database backup and restore Manage logical configuration Manage hardware configuration (add, delete, extend, reduce, display, setup and reconfigure the network) Manage the control stations and MFS database Generate reports about system activity
Fault management	The fault management tasks are: Handle BSS/RAN equipment and resources (monitor states and execute operator commands) Manage BSS/RAN tests (diagnostic tests and audits) Handle BSS/RAN alarms (states and history). Generate alarm reports Performs miscellaneous alarm maintenance functions from the IMT Provide files in formats compatible with optimization tools
Performance management	The performance management tasks are: Manage row counters (QoS alarms) Define and supervise counter thresholds Manage call and IMSI traces

- ALCATEL
- NSN (Nokia Siemens)
- ZTE.

4.1 RBS Functions

RBS functionality is often divided into the subsequent areas:

I. **Radio resources**: The radio base station's main function is to supply a reference to the mobile stations over the radio interface.

   - Radio transmission: To transmit several frequencies using an equivalent antenna, a combiner or a group of combiners are needed. Transmission power is controlled by the BSC and RNC.

- Radio reception: Additionally, for the reception of traffic on the physical channels, a primary RBS function is needed for the detection of channel requests from MSs (e.g., when a call is being made).

II. **Signal Processing**: An RBS is liable for the processing of signals before transmission and after reception.

III. **Signaling Link Management**: An RBS manages the signaling link between the BSC / RNC and MS, applying the appropriate protocols to the information being sent.

IV. **Local Maintenance Handling**: RBS enables operation and maintenance functions to be carried out locally at the RBS site, without BSC / RNC connection. In this way, field technicians can preserve RBS equipment and software on site.

4.2 RBS Implementation

All types of RBS have the following characteristics:

- Support for user flexibility by handing modular hardware and software designs.
- Transceiver-oriented design, which stresses using as little common equipment probably ensuring dependable performance.
- Design and use are aimed toward keeping system life cycle costs low.

The RBS series is predicated on standardized hardware units called replaceable units (RU). The major RUs usually have a combination or subset of these hardware cards:

- Power Supply Unit (PSU)
- Distribution Switch Unit (DXU)
- Double Transceiver Unit (DTRU)
- Combining & Distribution Unit (CDU)
- Alternate Current or Direct Current Connection Unit (ACCU/DCCU)
- Internal Distribution Module (IDM)
- Configuration Switch Unit (CXU)
- Fan Control Unit (FCU)
- DC filter
- Tower Mounted Antenna (TMA).

4.2.1 Power Supply Unit (PSU)

The PSU rectifies the facility supply voltage to the +24 V DC necessary for RBS operation. When employing a battery backup, an additional PSU is suggested for the recharging of batteries. If RBS is provided with a redundant PSU, no extra PSU is required for battery charging.

4.2.2 Distribution Switch Unit (DXU)

The DXU may be a CPU, which acts as an interface between the transmission network and therefore the transceivers. The DXU21 transmission interface has long haul capability and may be run on both 1.544 Mbit/s (T1) and a couple of 0.048 Mbit/s (E1) PCM links.

5 Remote Access Tool

5.1 General

Microsoft Visual Studio is an integrated development environment from Microsoft. It is used to develop computer programs for Microsoft Windows, also as websites, web apps and services and mobile apps. Microsoft development platforms like Microsoft Silverlight, Windows API, Windows Forms, Windows Store and Windows Presentation Foundation are used by Visual Studio. It can produce both native codes and managed code.

A code editor supporting Intel Sense is included in Visual Studio also as code refactoring. The integrated debugger works as a source-level debugger and also a machine-level debugger. Built-in tools like code profiler, web designer, building GUI applications, class designer and database schema designer.

A total of 36 other programming languages are supported by Visual Studio and permit the code editor and debugger to support mostly any programming language, provided a language-specific service exists. It contains built-in languages which include C, C++ and C++, VB.NET (via Visual Basic.NET) and C# (via Visual C#). It also supports XML, HTML, JavaScript and CSS. Java was supported in the past.

5.2 Designers

Visual Studio includes various visual designers to help within the event of applications. These tools include:

I. **Windows Forms Designer**: The Windows Forms designer is used to make GUI applications using Windows Forms. Controls that display data (like text box, list box and grid view) are often bound to data sources like databases or queries.

II. **Class designer**: The class designer is employed to author and edit the classes (including its members and their access) using UML modeling. The category designer can generate C# and VB.NET code outlines for the classes and methods. It can also produce class diagrams from handwritten classes.

III. **Data designer**: Data designer is often used to graphically edit database schemas, including typed tables, primary and foreign keys and constraints. It also can be used to design queries from the graphical view.

5.3 Microsoft Visual C#

Microsoft Visual C#, Microsoft's implementation of the C# language, targets the .NET Framework. In conjunction with the language's services are an area of Visual Studio and a compiler is out there separately as an area of the.NET Framework. The Visual C# compiler support versions 3.0, 4.0 and 5.0 of C# language specifications. Visual C# supports the Visual Studio class designer and data designer among others. VB.NET is an object-oriented programming language and may also be a multiparadigm, implemented on the .NET Framework. In 2002, Microsoft launched VB.NET as the successor to its original Visual Basic language which is used by Microsoft's integrated development environment. Most Visual Studio editions are commercial. The only assurance is Visual Studio Express and Visual Studio Community, which are freeware. Additionally, vbc.exe. Mono is a command-line compiler which is also included in.NET Framework SDK and also includes command-line VB.NET compiler.

- **Application Development in Visual Studio**: Visual Studio provides tools to style, develop, debug and deploy applications.
- **.NET Framework Guide**: The .NET Framework could also be a development platform for building apps for web, Windows, Windows Phone, Windows Server, and Microsoft Azure. It consists of the common language runtime (CLR) and the.NET Framework class library, which includes a broad range of functionalities and support for several industry standards. The.NET Framework provides many services, comprising memory management, type and memory safety, security, networking, and application deployment. We will use a selection of programming languages with the .NET Framework, including C# and Visual Basic.

5.4 Installation

NET Framework includes Windows, enabling to run .NET Framework applications.

I. Getting Started: Connects to additional resources and provides a comprehensive overview of the .NET Framework.

II. Migration Guide: Provides resources and a listing of changes you would wish to believe if you are migrating your application to a replacement version of the .NET Framework.

III. Development Guide: It provides any key technology areas and tasks, including creating, application development, configuring, securing, debugging and deploying your application, and interoperability, knowledge about dynamic programming, extensibility, memory management and threading.

Tools: It distinguishes the tools that assist to develop, configure and deploy applications by using .NET Framework technologies .NET Framework class library provides syntax, code examples, and related information for each class contained within the .NET Framework namespaces. It provides language interoperability (each language can use codes written in other languages) across several programming languages and carries an outsized class library named Framework Class Library (FCL).

Framework Class Library consists of network communications, interface, web application development, data access, cryptography, numeric algorithms and database connectivity. The software is produced by programmers by combining their ASCII document with .NET Framework and other libraries. The Framework is supposed to be used by most new applications created for Windows platform. Microsoft also generates an integrated development (ID) environment widely for .NET software called Visual Studio.

6 Conclusion

The hardware and software of RBS are often prepared for function and capacity. New configurations could even be applied without affecting cell's operational or call ongoing. Future enhancements can in many cases be implemented without affecting installed hardware. The RBS capacity is extended by adding activation keys for baseband resources, carriers and output power. The event in online monitoring tool with this 2G, 3G and 4G access gives the user access to voice data with high speed using remote sensing from all stations which are accessed.

References

1. Telecommunications network and service architectures wiki
2. PPT on BSS. http://www.authorstream.com/presentation/dipendra.ericsson-1330272-bss-training-full/
3. RBS radio base station description. http://tel3pedia.blogspot.in/2011/11/radio-base-station-rbs.html
4. Generations in telecommunications (2G,3G,4G). http://www.zseries.in/telecom%20lab/telecom%20generations/#.WP-I3NKGPIU
5. .NET framework and architecture. http://www.c-sharpcorner.com/UploadFile/puranindia/net-framework-and-architecture/

A Statistical Study on Coronary Artery Disease Using Discriminant Analysis

A. Vani, T. Suneetha, K. Sukanya, and T. Sukeerthi

Abstract Coronary artery disease is a type of disease that develops in the heart, as a result of which thousands of people have died. Because of this most predictable chronic disease, in the Western world, a model is analysed to help in assessing the risk of cardiovascular disease in patients. The study aims to avoid the incidence or lack of heart disease in patients with the disease following treatment and to establish a relation between the disease state (dependent variable) and a set of separate variables.

Keywords Coronary artery · Chest pain · Thal · Old peak · Fasting blood sugar · Serum cholesterol · Discriminant analysis

1 Introduction to Heart Disease

Cardiovascular disorders (CAD), also known as heart disease, refer to as coronary artery disease (CHD), cerebrovascular disease, peripheral artery disease (PAD), and aortic atherosclerosis. CAD results from a lower myocardial infusion causing ischemia and myocardial infarction (MI) and heart insufficiency. CAD is the cause of angina [1]. One of our body's most essential organs is the heart. The heart, essentially a pump, is a muscle consisting of four chambers, split into two halves and divided by values. Each half comprises one chamber called an atrium; the other is known as a ventricle. Blood is stored by the atria and the ventricles contract to release blood from the heart. Oxygen-poor blood is pumped to the lungs by the right half of the heart, which is the blood with a low amount of oxygen, and blood cells can absorb more oxygen. The blood flows from the lungs into the left atrium. The fresh oxygen-rich blood is pumped into the organ and tissues of the body through the left ventricle. This oxygen provides our body with energy and is essential to protect our body.

A. Vani (✉) · T. Suneetha · K. Sukanya · T. Sukeerthi
Department of Statistics, Sri Padmavati Mahila Visvavidyalayam, Tirupati, India
e-mail: akkyamvani1@gmail.com

S. Jyothi et al. (eds.), *Proceedings of the 2nd International Conference on Computational and Bio Engineering*, Lecture Notes in Networks and Systems 215,
https://doi.org/10.1007/978-981-16-1941-0_9

2 Overview

Heart defects are the general term used to address heart failure, often a cardiac condition ("cardiac" is a Latin word meaning the heart). Although there are various forms of heart disease, our review focuses on the common: heart attacks and cardiac insufficiency.

One should learn about cardiac attacks and cardiac failures:

- What leads to this disease?
- What are the forms of such diseases?
- What can be done to treat those diseases?

Since these issues can be somewhat prevented, we can also discuss preventative measures to minimise the risk of dealing with cardiovascular disease or to alleviate the impact of existing cardiac diseases. There are many factors that can increase the risk of heart disease. Some of these variables are not under our influence, but several can be avoided by choosing a healthy lifestyle [2]. The controllable (modifiable) risk factors are: High BP; high levels of blood cholesterol; smoking; diabetes; overweight or obedience; lack of physical activity; poor diets, and stress [2, 3]. Coronary artery disease develops if your major blood vessels are damaged or ill. Cholesterols (plaques) are generally responsible for coronary artery disease in your coronary arteries and inflammation [3].

Purpose

The purpose of the data analysis is to estimate values for the occurrence or absence of cardiovascular disease in patients with a disease following treatment.

Objective:

- Assessment of the sensitivity of patients during treatment or absence of heart disease.
- Determine the correct classification percentage.

3 Methodology

Discriminant Analysis

The discrimination analytical technique developed by R. Fisher A. (1936) is one multivariate group differential study method [4, 5]. The focus of Discriminant Analysis is the association of several separate variables to a categorized variable by forming an independent composite. The range of any composite variables that distinguish between two or more existing groups of subjects can be determined in this kind of multivariate analysis, and a classification model can also be developed for predicting the group members of new observations [5]. What is the discriminatory characteristics study?

- Constructs a community membership prediction model.
- The model contains a biased equation centred on a linear predictive variable mix.
- The best discrimination between groups is expressed in these variables.

Summary

There are several different variables but we are interested in one categorical variable. We would like to clarify the aspects of these relationships. A classification based on the independent variables in use can be predicted or the distinction decided the classes in the grouping of independent variables.

This is like regression analysis. Based on the two-fold outcome the weighted combination was independent variables.

$\mathrm{Di} = a + b_1X_1 + b_2X_2 + \cdots + b_nX_n$; the score for Di is expected (discriminant score)

X is the measure, and *b* is the discriminatory coefficient.

The estimated factor for a separate variable is multiplied for each case by the case score in this variable. These products are summed and added to the constant, and the result is a composite score, that is, the discriminant score for that case [5]. To assign a case from a certain cut-off point, we use the technique of maximum probability to a group.

Cut-off means that the score of the size of the group does not meet the weighted means to determine the cut-off.

- Click ANALYZE → Classify → Discriminant.
- Enter the values in dialogue box for Output.
- Then continue and click on OK button.
- We see our Results.

3.1 Data Collection

- The data is carried out from UCI datasets UCI Machine Learning Repository: Statlog (Heart) Data Set in the google.
- The dataset includes 270 instances. The instances are described by 13 attributes; some of which are linear and some are nominal.

3.2 About the Software

Introduction:

The 'Statistical Package for Social Sciences' is a package that is widely used in social and behavioral sciences for manipulating, analyzing, and presenting data [6]. SPSS is a multifaceted statistical software traditionally common only with applications in social sciences. It is based on Windows and shares all the facilities with other Windows applications. It is an open and structured platform for scalable data analysis and data processing solutions. PLC 16.0 and 20.0. PLC 16.0 is the new version in use.

SPSS is a computer program that enables surveys, data mining, text analysis, collaboration and deployment [7]. It has all the main research methods for managing a large volume of data and complex multivariate analyses [7]. Working data is a must in one of the statistical packages like SPSS for contemporary analysis.

3.3 Uses and Advantages

SPSS is very useful to the consumer in the use of various statistical instruments: Telecoms, banking, finance, health insurance, manufacturing, retail, government and consumer packaging products, higher education and market research. SPSS provides a user-friendly way of using various packages. This is popular and while it certainly does not excuse the selection of a statistical package, it is easy to load several data sets and various programs can be easily imported into SPSS files.

Statistical Tools Used Under Study

Discriminant research assumptions close to linear regression assumptions:

1. Variances equality, normality, linearity, multilinearity
2. Violations of many of these assumptions are very robust. However, they are highly vulnerable to outliers.

Aim of the discriminant function for analysis:

1. To completely distinguish the classes
2. To find the most sterile way to split classes
3. To delete variables that have no association with cluster differences.

The key variables are:

Dependent variable: whether the patient is diseased or not disease (1 = absence, 2 = presence).

Possible explanatory variables: Age, sex (0 = female, 1 = male), chest pain, blood sugar resting, serum cholesterol, fasting blood sugar (0 = absence, 1 = presence), ECG, heart rate limit, exercise (0 = absence, 1 = presence), oldpeak, slopepeak, number of major vessels, thal (3 = normal, 6 = fixed defect, 7 = reversible defect).

Usage ANALYZE-Classify-Discriminant in SPSS.

4 Empirical Work

Significance test:

1. For both groups, the null hypothesis is that the means of distinguishing between centroids are similar for both groups.
2. Centroids is the mean discriminating score for each group.
3. Wilks' lambda is used to find a significant distinction among groups.
4. Wilks' lambda is from zero to one. It tells us about the difference.

The non-discriminatory variable is dependent (Table 1).

1. As eigen is greater, the more the variance in the dependent variable describes the function.
2. Only one distinctive feature is present in two types of dependence.
3. The canonical correlation is the approximation of the relation to the discriminating function of the variable.
4. The percentage of variance specified in the dependent variable is the canonical coefficient square.

It is the total share of discrimination. The proportions of discriminatory ability add up to one for each study. So, one of the last entries in the total column is also considered (Table 2).

Wilks' lambda is the test of the classification of cases.

The values of the little Wilks' lambda reflect the device's greater difference.

The related chi-square statistical test, the equality of means between functional groups defined the small value and implies that the discriminating function is better than a chance of group breakup (Tables 3 and 4).

Pooled associations between distinguishing factors and standardised canonical discriminant functions within classes.

Variables are ordered in the function for optimum correlation size:

1. The structure matrix table shows how each variable is associated with each function.
2. In this analysis, there is only one discriminatory function.
3. Then the associations are the factor analytical burdens; in other words, by identifying the greatest absolute relationships associated with any discriminating function, the investigators can learn how to call each characteristic.

Table 1 Canonical discriminant functions description. Eigen values

Function	Eigen value	% of variance	Cumulative %	Canonical correlation
1	1.144[a]	100.0	100.0	731

[a]In the analysis, one canonical discriminant function was first used

Table 2 Wilks' Lambda

Test of function(s)	Wilks' Lambda	Chi-square	Df	Sig
1	0.466	201.381	8	0.000

Table 3 Standardized canonical discriminant function coefficients

	Function
	1
Sex	0.237
Chest pain	0.328
Resting electrocardiographic results	0.195
Maximum heart rate achieved	−0.22
Exercise induced angina	0.216
Old peak	0.270
Number of major vessels	0.431
Thal	0.433

Table 4 Structure matrix

	Function
	1
Thal	0.577
Number of major vessels	0.478
Exercise induced angina	0.432
Maximum heart rate achieved	−0.431
Chest pain	0.429
Old peak	0.414
Sex	0.292
Slope peak[a]	0.273
Age[a]	0.217
Resting electrocardiographic results	0.173
Fasting blood sugar[a]	0.071
Resting blood pressure[a]	0.025
Serum cholesterol[a]	−0.002

[a] is pooled associations between distinguishing factors and standardised canonical discriminant functions with in the classes

Summary of Discriminant Canonical Functions:

1. The discriminatory function is an independent linear combination created as a latent variable.
2. Differentiating variables are distinct variables.
3. Pearson associations between predictors and discriminant standardised in the Table 5 are shown as canonical functions.
4. The loading prototype <0.30 can be omitted.

The canonical connection in the table is best since the relationship between Pearson's discriminating scores and Pearson is similar between two classes—unnormal coefficients. This table contains the coefficients of the unnormalized racist

Table 5 Canonical discriminant function coefficients structure matrix

	Function
	1
Sex	0.529
Chest pain resting	0.38
Electrocardiographic results	0.199
Maximum heart rate achieved	−0.010
Exercise induced angina	0.505
old peak	0.245
Number of major vessels	0.512
Thal	0.262
(Constant)	−2.199

equation. This is used to construct the actual equation of prediction. In certain regressions, new cases are defined as non-standard (record) variables (Table 6).

Discriminant task: Our methodology should be like this:

Di = −2.199 + 0.529 sex + 0.380 chest pain + 0.199 Resbp − 0.10MaxHR + 0.505 Exercise + 0.245 old peak + 0.512 no. of major vessels + 0.262 thal.

Unstandardized group-assessed canonical discriminant functions imply.

Centroids are calculated based on the function:

Di = −2.199 + 0.529 sex + 0.380 chest pain + 0.199 Resbp − 0.10MaxHR + 0.505 Exercise + 0.245 old peak + 0.512 no. of major vessels + 0.262 thal.

Centroids for each group are a discriminant score when the vector means are introduced into the feature (rather than individual values for each case) (Tables 7 and 8).

Previous probabilities are used in classification. Designs refer to the group sizes found. Measure the probabilities of previous membership of the analytical groups,

Table 6 Functions at group centroids

Disease state	Function
	1
Absence	−0.953
Presence	1.192

Table 7 Classification statistics

Classification processing summary		
Processed		270
Excluded	Missing or out-of-range group codes	0
	At least one missing discriminating variable	0
Used in Output		270

Table 8 Prior probabilities for groups

Disease state	Prior	Cases used in analysis	
		Unweighted	Weighted
Absence	0.500	150	150.000
Presence	0.500	120	120.000
Total	1.000	270	270.000

Table 9 Classification function coefficients

	Disease state	
	Absence	Presence
Sex	1.577	2.712
Chest pain	5.253	6.068
Resting electrocardiographic results	0.984	1.410
Maximum heart rate achieved	0.432	0.410
Exercise-induced angina	2.590	3.672
Old peak	1.838	2.365
Number of major vessels	1.009	2.107
Thal	1.173	1.735
(Constant)	−46.289	−51.261

and this is important if the size is different. If the size of the group is equal, all groups can specify the same probabilities (Table 9).

Fisher's linear discriminant functions.

Two sets of unnormalized coefficients of linear discrimination (one for each group dependent) are calculated and can be used for case differentiation. This is the classical classification method, even though it is less used today (Table 10).

Separate-Groups Graphs

The model understands the presence and absence of heart disease patients. However, the classification of people with heart disease is an incredibly successful task. It does. To identify these individuals, you need not find any other predictor. We have

Table 10 Classification results[a]

		Disease state	Predicted group membership		Total
			Absence	Presence	
Original	Count	Absence	133	17	150
		Presence	21	99	120
	%	Absence	88.7	11.3	100.0
		Presence	17.5	82.5	100.0

[a] 85.9% of original grouped cases correctly classified

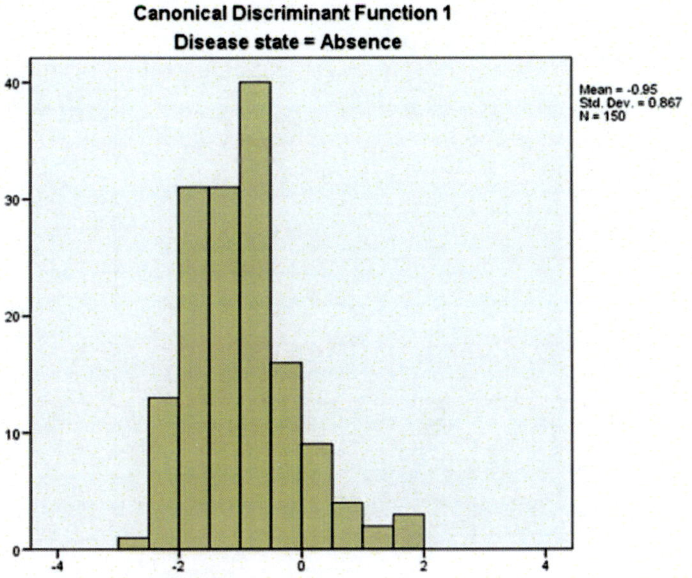

Graph 1 Disease state = Absence

developed a discriminatory model that categorises variables based on each person's demographic data into two predefined groups. The most useful variable has been described to segment the base structure by means of a structural matrix. Finally, findings from classification suggest that the prevalence and lack of individuals are not represented incorrectly in the model. Thereby, if the concern is not the identifier of the sick person, the model may be reasonably accurate (Graphs 1 and 2).

5 Conclusion

- To better understand the impacts of heart disease and its treatment on quality of life, evaluation of heart disease in different age groups and which factors influence the quality of life due to the age is required.
- The square of the canonical correlation coefficient is the percentage of variance explained in the dependent variable.
- Centroids from the Centroids table is a medium discriminating score for any group.
- For two groups the null hypothesis is the means to discriminate between two groups that are the same as the centroids.
- 85.9% of cases classified correctly under the classification table have been predicted to be right.

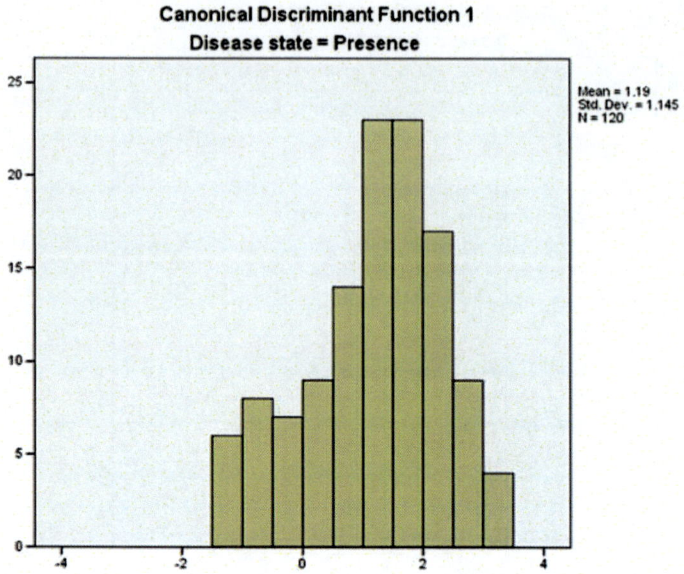

Graph 2 Disease state = Presence

References

1. Lopez EO, Ballard BD, Jan A (2020) Cardiovascular disease
2. Hajar R (2017) Risk factors for coronary artery disease: Historical perspectives. https://www.ncbi.nlm.nih.gov/pmc/articles/PMC5686931/
3. Mayo clinic Disease Refrence. Sourced from Mayo clinic, these articles include detailed information about medical conditions and diseases. https://www.mayoclinic.org/diseases-conditions/coronary-artery-disease/symptoms-causes/syc-20350613
4. Betz NE (1987) Use of discriminant analysis in counseling psychology research
5. Antonogeorgos G, Panagiotakos DB, Priftis KN, Tzonou A (2009) Logistic regression and linear discriminant analyses in evaluating factors associated with asthma prevalence among 10- to 12-years-old children: Divergence and similarity of the two statistical methods
6. Landau S, Everitt BS (2003) A handbook of statistical analyses using SPSS
7. Sarma KVS (2010) Statistics made simple, do it yourself on PC. Prentice Hall of India

A Study to Assess the Knowledge and Opinion of Mid-Level Health Providers of Chittoor District Regarding m-Health

B. G. Bhavani

Abstract Seventy percentage of Indian population live in rural areas. To cater to the health needs of the rural Indian population, the Central Government has set the goal of health for all through National Rural Health Mission. Mid-level health providers are the B.Sc nursing graduates who have been given six months of special training to serve in the primary health centers in rural India. **Objectives**: The aim is to know the MLHP student's knowledge and opinion regarding m-health. m-Health is the component of e-health used widely in advanced countries to provide health care at the grass-roots level of the rural community. **Study Approach**: Survey is conducted on MLHP course trainees of Chittoor district on field. **Data analysis**: 40% of the MLHP students know applications of the m-health system, and 47% of the students know about the m-health programs. The majority of both male and female students want to learn and use m-health to serve the rural population at the grass-root level. 86% of the students agreed that m-health is a supportive, not a substitute for nursing care.

Keywords Mid-level health providers · National rural health mission · Grass-root level · Community · m-Health · e-Health · Knowledge · Opinion

1 Introduction

India is a developing country with a huge geographical area and cultural diversity. 70% of the population lives in villages and inaccessible areas without health care facilities, especially in tribal and hilly regions. Latest technological advancements have considerably increased connectivity to the remote areas of the country through mobile phones and the internet for communication purposes. The health care industry too started using electronic media and mobile phones to serve the health care needs of the community at grass-root levels through e-health and m-health approach, which became very popular in advanced countries.

B. G. Bhavani (✉)
College of Nursing, SPMVV, Tirupati, India
e-mail: mscspmvv@gmail.com

S. Jyothi et al. (eds.), *Proceedings of the 2nd International Conference on Computational and Bio Engineering*, Lecture Notes in Networks and Systems 215,
https://doi.org/10.1007/978-981-16-1941-0_10

According to Wikipedia "**eHealth** (also written **e-health**) is a relatively recent **health care** practice supported by **electronic** processes and communication, dating back to at least 1999. ... It can also include **health** applications and links on mobile phones, referred to as mHealth or m-**Health**." [1].

(eHealth – Wikipedia. https://en.wikipedia.org/wiki/EHealth).

2 Need for the Study

Digital India has revolutionized the communication system at all levels, including health care systems. The health indices and health census are the key elements that health care providers must be trained in apart from training in direct patient care. MLHP students after their training have to go and serve the community at grass-root level in different geographical areas with or without communicational facilities. It is essential for MLHP students to know the different approaches to reach the community at grass-root level to provide health care.

3 Objectives: To Assess

- The knowledge of MLHP trainees regarding e-health.
- The knowledge of MLHP trainees regarding m-health.
- The opinion of MLHP trainees regarding m-health.

4 Review of Literature

According to Ganapathy K, m-health uses applications for emerging mobile communications and network technologies for health care systems. Cell phones are used in computing, medical sensors, and communications technologies for health care. m-health technologies are successful in sending information in a correct form, at right time, to the right place [2].

As per the World Health Organization, mobile phone technologies coupled with infrastructure for telecommunications had been growing rapidly and expanding globally (WHO 2011) [3].

The uses of m-health are to support clinical diagnosis and/or decision-making; improve clinical outcomes from established treatment pathways through behavior change and enhancement of patient adherence and compliance with treatment; act as standalone digital therapeutics; and primarily deliver disease-related education [4].

Thyra de Jongh and others are of the view that **in developing and advanced countries** long-term illnesses affect a significant proportion of the population. Short messages service (SMS) and multimedia message service (MMS) of mobile phones present convenient, cost-effective ways of self-management and improving patients' self-efficacy skills, for example, medication reminders, therapy adjustments etc. [5].

5 Study Methodology

The study is conducted on MLHP trainees of different training centers when they are attending field training at the community health, primary health and subcenters in Chittoor district. The total number of samples is 105 (female 100 and male 5).

All MLHP trainees will be posted at rural health care centers after training. Written informed consent is taken from all the participants.

6 Data Analysis

Knowledge scores' analysis

1. Knowledge regarding the applications of m-health is 43% in female and 40% in male students.
2. Knowledge regarding categories of health programs: 44% in female and 40% in male students.
3. Knowledge regarding promotion of public health through m-health approach: 56% in female and 60% in male students.
4. Knowledge regarding uses of mobile phones in preventing and controlling COVID-19 is 55% in female and 60% in male students.
5. Knowledge regarding the use of mobile phones in tracing, testing and treating corona patients is 57% in female and 60% in male students (Table 1 and Graph 1).

Opinion Scores' analysis

1. 45% of female students are confident of m-health mode of communications at the grass-root level; 50% of the students are somewhat confident whereas 5% are not confident.
2. 85% of female students are ready to learn and use m-health applications for the public under their care, 15% are somewhat agreed to learn and use m-health applications. No one is against it.
3. 63% of female students are of the view that e-health is the solution to reach community's health care needs; 36% of the students are somewhat agreed, whereas one student disagreed with it.
4. 43% of the female students agreed that communicating with mobile phones with all types of population in the community is not possible, whereas 55% have somewhat agreed and 2% disagreed.

Table 1 N = 105 (female 100, male 5)

S. no	Description	Female frequency	Female percentage	Male frequency	Male percentage	Total percentage
1	Knowledge regarding the applications of m-health	43	43	2	40	41.5
2	Knowledge regarding e-health programs in India	44	44	2	40	42.0
3	Knowledge regarding of m-health programs in India	56	56	3	60	58.0
4	Use of mobile phones in preventing and controlling COVID-19	55	55	3	60	57.5
5	Uses of mobile phones in tracing, testing and treating the corona patients	57	57	3	60	58.5

5. 87% of female and 80% of male have agreed that electronic gadgets are only supplementary and not a substitute for human care. 20% of male and 13% of female somewhat agreed to this opinion (Table 2 and Graph 2).

There is no significant difference in the knowledge and opinion scores between male and female students based on gender.

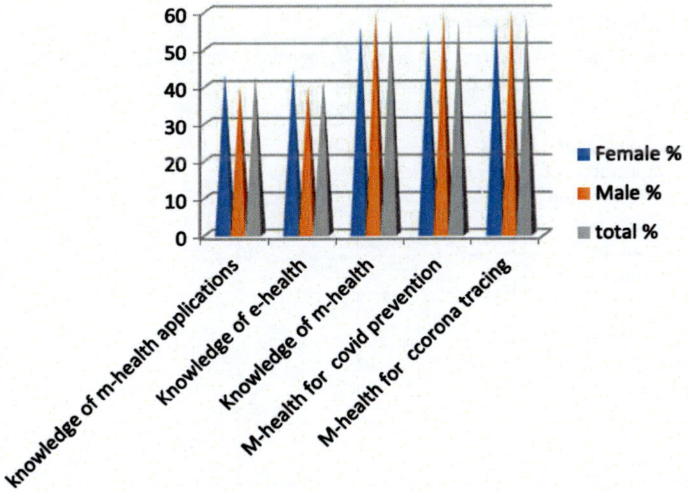

Graph 1 Knowledge scores in percentage

7 Discussion and Conclusion

MLHP trainees' knowledge data analysis shows that there is inadequate knowledge regarding m-health applications and e-health programs in India (42–45%) and the knowledge scores of m-health programs. The use of m-health approach in the prevention, tracing, testing and treating corona is fair (55–60%).

Analysis of opinion scores is good. The students are ready to learn e-health applications, implement m-health programs and ready to serve the community at grass-root levels.

In conclusion, it is recommended that the e-health and m-health modules need to be included in the community health and public health curriculum.

Table 2 N = 105 (female 100, male 5)

S. no	Description	Agree female	Agree male	Somewhat agree female	Somewhat agree male	Disagree female	Disagree male
1	I am confident with m-mode of communication at grass-root level in the community	45	3	50	2	0	0
2	I can learn and use m-health applications as per the needs of people under my care in the community	85	5	15	0	0	0
3	e-health is the solution to reach community at grass-root level	63	5	36	0	1	0
4	Communicating health programs by mobile phone with all types of population in the community is not possible	43	4	55	1	2	0
5	Electronic gadgets are only supplements not substitutes for health care by human beings	87	4	13	1	0	0

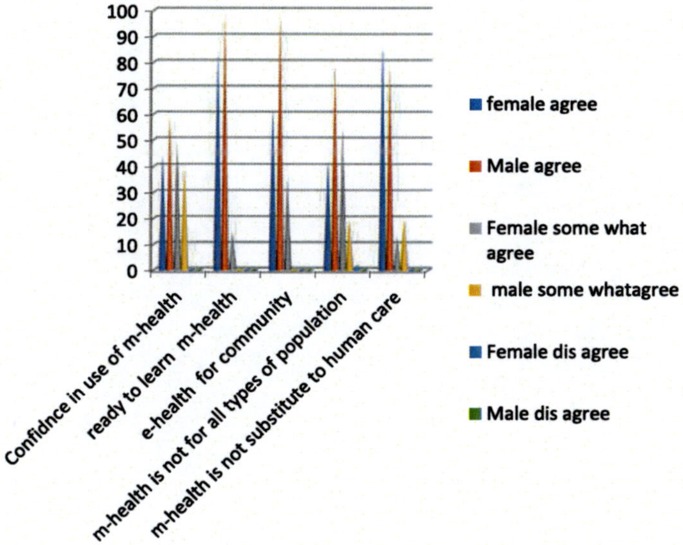

Graph 2 Opinion scores in percentage

References

1. eHealth – Wikipedia. https://en.wikipedia.org/wiki/EHealth
2. Ganapathy K, Ravindra A. m-Health: a potential tool for health care delivery in India. http://citeseerx.ist.psu.edu/viewdoc/download?doi=10.1.1.511.7287&rep=rep1&type=pdf, pg no. 1
3. WHO (2011) http://ncbi.nlm.nih.gov/pmc/articles/pmc4232479/#R39
4. Rowland SP, Edward Fitzgerald J, Holme T, Powell J, McGregor A (2020) NPJ digital medicine what is the clinical value of m health for patients? vol 3. Article number: 4, p 1. http://www.nature.com/npjdigitalmedicine, pg no. 1
5. de Jongh T, Gurol-Urganci I, Vodopivec-Jamsek V et al. Mobile phone messaging for facilitating self-management of long-term illnesses, p 1. PMC6486189. https://doi.org/10.1002/14651858.CD007459.pub2

Machine Learning Techniques for Identifying Diabetes and Its Complications Based on Long Non-coding RNAs

P. Swathi, S. Jyothi, and A. Revathi

Abstract Earlier long non-coding RNA (lncRNA) in gene transcription was termed as "noises". But later in the researchers' development of ENCODE, it was clear that lncRNA can regulate the genomic expression in the system of RNAs. But later in the researchers development of ENCODE, it was clear that lncRNA can regulate the expression genomic in the system of RNAs in transcription, post transcription and epigenetic, those are involved and substantial influences on precautions and occurrences of human diseases. By the deep study in biological research, it has been proven that the involvement and importance of lncRNAs have increased in the whole life cycle of a cell in many critical situations with their various mechanisms. So, there is no surprise that the dysregulation and mutations of lncRNAs would involve and contribute to the various complex disease developments in humans. Diabetes is a very popular human disease, and there is an epidemic growth that has an impact duly on the environment (e.g. diet and lifestyle choices) and genetics. The state of high glycemic defines diabetes and it also stimulates the macrovascular and microvascular diseases which bring diabetes complications and contribute to patient mortality. This paper introduces the long non-coding RNAs functions, types of diabetes and diabetic complications based on the previous researches and also provides the machine learning methods that have been implemented to identify diabetes in humans at the early stages.

Keywords lncRNA · Gene transcription · Diabetes · Diabetic complications · Micro–macrovascular · Machine learning

P. Swathi (✉) · S. Jyothi · A. Revathi
Department of Computer Science, SPMVV, Tirupati, India
e-mail: swathivinubaby@gmail.com

S. Jyothi
e-mail: Jyothi.spmvv@gmail.com

A. Revathi
e-mail: arevathi20@gmail.com

S. Jyothi et al. (eds.), *Proceedings of the 2nd International Conference on Computational and Bio Engineering*, Lecture Notes in Networks and Systems 215,
https://doi.org/10.1007/978-981-16-1941-0_11

1 Introduction

The transcripts length of more than 200 nucleotides and without an extended open reading frame (ORF) are defined as long non-coding RNAs (lncRNAs). The flow of genetic information into the biological system was explained and the central precept of molecular biology by Francis Crick in the year 1985 was proposed. The complete various biological cell processes were defined by central dogma, where RNA acts as an arbitrator among DNA and carries the proteins and genetic information. The Encyclopedia of DNA Elements (ENCODE), the large international consortium, has publicized that a genome is transcribed only 80%, whereas its protein-coding sequences are 1.5% [1]. With this, it has been proven that non-coding RNA is considered more abundant than previous analyses. lncRNAs are broadly classified into two groups based on their transcript length. The primary one contains pretty defined classes such as miRNAs, piRNAs, tRNAs, snoRNAs and so on. Whereas, the secondary one comes with very poorly, functionally annotated classes that include rRNAs and lncRNAs.

Diabetes is well known as a life-altering autoimmune disease affecting the growing percentage of the population each year [2]. The latest edition of the International Diabetes Federation Atlas (IDFA) [2] says that worldwide 425 million people are suffering from diabetes, and in India, 73 million people have diabetes [3].

Diabetes is a group of metabolic diseases, or a chronic one, where humans suffer to an extent level of the blood glucose in the body. The cells in the body respond less to the insulin or there would be no proper production of insulin. The continues hyperglycemia of diabetes will harm humans and lead to brokenness, long-haul harm, failure of various organs, effectively on eyes, nerves, kidney, veins and heart.

Machine learning is a technique that has been growing rapidly with its tremendous support for training the system by making predictions. Machine learning is all about knowing the structures from provided data. In the medical domain, the machine learning is becoming a reliable, evolving and supporting tool in recent times. Keeping in mind the time constraint to interact with patients, detection of disease and taking care in the medical domain leads to fetching great interest in automatic learning.

In DM research, applying ML methods is a vital approach for using a large amount of data that relates to diabetes for knowledge extraction [4]. When it comes to diabetic-related aspects such as diagnosis, management and other clinical administrative, machine learning approaches are of great concern.

2 Diabetes Mellitus

DM [5] is a ceaseless ailment, chronic and a metabolic disorder, where the sugar levels are high and circulate in the blood cells. It occurs because when there is no proper working of the pancreatic β cells. This leads to damage to the body parts, which include pancreas glitch, hypertension, pancreatic issues, heart ailment, kidney disappointment, foot issues, nerve harm, visual influences, ketoacidosis, and other eye problems like glaucoma and so on.

There are many reasons that basically make diabetes a high risk, like threatening lifestyle of a man, sustenance propensities, lack of activeness, overweight (obese), high cholesterol (hyperlipidemia), smoking, high blood pressure (hyperglycemia) etc. There is no age restriction for its influence, which includes kids, youngsters to middle age and old people. An organ is positioned in the abdomen of a body known as the pancreas [6]. Exocrine capacity and endocrine capacity are the two fundamental capacities available in the pancreas. In pancreas, the exocrine segment keeps the sugar levels up in the circulatory structure and endocrine assistants for assimilation. Each time the sugar or glucose levels are high in the circulatory system, then the pancreas β cells release the insulin to the circulatory system, to stop the excess sugar by blood to not reach the liver, then it changes to vitality frame. The other job of pancreas is when the glucose levels are low, the production of insulin activates glucagon that gets generated by alpha cells and starts to maintain the glucose or sugar levels high in the blood flow [7] (Fig. 1).

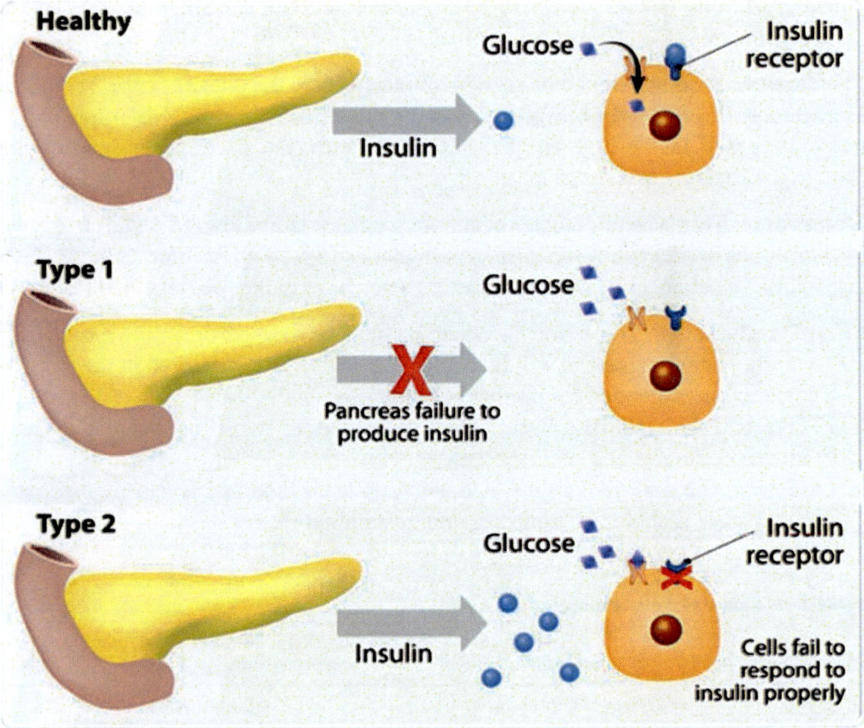

Fig. 1 Diabetes main types. It may be that pancreas is not producing insulin sufficiently or the body cells not absorbing the insulin which is produced [7]

2.1 Types of Diabetes

Diabetes [8, 9] is an issue that has hazardous factors and expands passing rates. It is classified into four kinds: type 1 [10], type 2 [11], prediabetes [12], and gestational diabetes [13].

Type 1: Type 1 diabetes mellitus (T1DM) is a syndrome when there is no proper stability of glucose by an autoimmune decrease of insulin production by pancreatic β cells which leads to deficiency of insulin and results in hyperglycemia. When it is not treated, the insulin deficiency makes metabolic disorder, with worsening ketoacidosis, hyperglycemia, starvation and death. In order to maintain and restore euglycemia, medication to simulate the native β cell by exogenously replacing insulin and also frequent sugar or blood glucose levels monitoring are essential.

Type 2: Type 2 diabetes is the most regular type of diabetes, and it occurs mostly in grownups. Type 2 diabetes occurs when the insulin is not properly used by the cells in the body. It is formally called non-insulin-dependent or adult-onset diabetes. This happens due to some realities, which are metabolic components, hereditary, physical dormancy, family history, heftiness, eating habits, alcohol and smoking, that lead to the danger levels of diabetes [14].

Prediabetes: Prediabetes can be seen in humans before the phase of type 2 diabetes. The glucose levels will be higher typically but not to the levels of type 2. The individuals will face a high risk of getting diabetes, but it can be normal if they change the lifestyle and focus on it.

Gestational: This type of diabetes occurs in pregnant ladies [14, 15]. The changes in the hormones in pregnancy increase insulin that can speed up the blood glucose level very high. The chances of getting diabetes in newborn babies are less [13] (Table 1).

- A1C tests/tested: The blood of a person will be tested for a few months (3–4 months) and the average will be taken. The values will be stored in the records. They are recommended for prediabetes and diabetes.
- FPG tests/tested: Here the plasma glucose test is done in fasting; the result is used to identify diabetes and prediabetes.
- OGT test: The oral glucose test is done; they test blood and levels are estimated for analyzing diabetes, prediabetes and gestational diabetes.

Table 1 Diagnosis levels of diabetes and ranges

	A1C (%)	Fasting plasma glucose (mg/dL)	Oral glucose tolerance test (mg/dL)
Diabetes	6.5 or above	126 or above	200 or above
Prediabetes	5.7 to 6.4	100–125	140–199
Normal	About 5	99 or below	139 or below

3 lncRNA

The transcripts which are longer than 200 nucleotides are broadly classified as long non-coding RNAs (lncRNAs). In various cellular processes, human diseases have been implicated by the lncRNA-mediated biology.

3.1 lncRNAs Classifications and Functions

lncRNAs can be divided on the basis of sequence orientation to protein-coding genes and genomic location into five categories, which can be described with features displayed as "three more", called more patterns, more quality and more types (Fig. 2) [16].

- Sense lncRNAs: The transcription is done from the sense strand of protein-coding genes.
- Antisense lncRNAs: The antisense strand of protein-coding genes is transcribed.
- Bidirectional lncRNAs: The transcription will start from sense direction and antisense.
- Intronic lncRNAs: These were done from the introns of protein-coding genes.
- Intergenic lncRNAs: This will be transcribed from intergenic regions.

The functions executed by most of the lncRNAs are decoys, signals, scaffolds or guides [17]. Disparate to structural features of mRNA, the lncRNAs primary sequence was reported to exhibit less conservation among those species, while the secondary as well as tertiary structures express high conservation and play an important role in gene expression as a regulatory function [18].

- Signals: lncRNAs will be acting as transcriptional activity signals in conjunction, which indicates gene regulation with specific chromatin regulatory complexes or transcription factors.
- Decoys: When lncRNAs are served as decoys for inhibiting, the transcriptional regulation is done via making transcription factors away from their specific location.

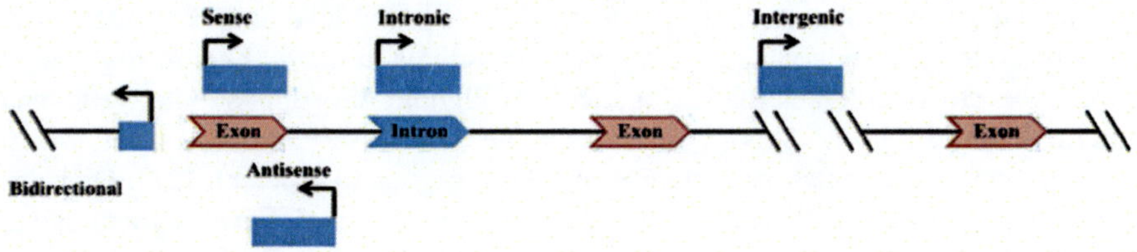

Fig. 2 The lncRNAs of genomic location (blue color) [16]

- Guides: Transcription factors to the gene promoter were done when lncRNAs act as guides to stimulate transcriptional regulation.
- Scaffolds: Through the act of lncRNAs, the scaffold is interacting and recruiting with various proteins to regulate transcription.

3.2 lncRNAs in Diabetes Complications

The human body parts are influenced by lncRNAs in diabetes. To evaluate the role of the functions of lncRNAs in diabetes and its complications, the research study has started in recent years. The study shows that single nucleotide polymorphisms (SNPs) of humans are associated with both types of diabetes (type 1 and type 2), which are integrated within lncRNA loci [19], and indicates the role of lncRNAs involved in diabetes and its complications. When diabetes is not diagnosed accurately, then diabetes complications may arise which intimidates health and endanger life. These complications are trivial contributors to costs, poor quality of life and morality.

Diabetic Neuropathy: Kidney neuropathy [20] or chronic kidney infection occurs when the sugar levels are high in blood, then the kidney vessels are harmed. The kidney's function is to dispose of the waste and excess water in the blood. The sugar level and hypertension in kidneys if goes-on to have a burden to clean the blood, this may lead to prompt kidney malfunctioning, and successive dialysis of blood is required. The treatments like transplantation of kidneys and pancreas may be incorporated.

Diabetic Retinopathy: Loss of vision and retinopathy is a situation when the optic nerve, retina, the focal point is affected. This may lead to vitreous hemorrhage, retinal detachment, glaucoma and blindness (loss of vision). A diabetic individual should have to deal with vision by taking some tests and pharmaceutical checkups regularly from the early stage of diabetes [21]. The treatment incorporates tonometry, student enlargement, visual sharpness testing and optic intelligibility tomography (OCT). Different medications and procedures are available, like focal macular laser procedure, corticosteroid and anti-VEGF infusion therapy.

Cardiovascular: The person with diabetes may lead to heart problems also. American heart affiliation bestowed 68% of humans will experience issues in the heart that lead to heart stroke, driving even to death. Stress and load on heart, atherosclerosis or solidifying of routes supply drive individuals to death. Due to high sugar levels, the blood becomes thicker, it sticks throughout the supply route, veins and put more strain to flow. Accordingly, it damages the vessels and the nerves, which malfunctions the circulatory system or even the organs in a person [22]. Diabetes creates cardiovascular disease by incorporating unusual cholesterol, hypertension, corpulence, high triglycerides and lack of physical activities. The parameters with different clinical effects like low glycemic control and insulin differences of diabetes may lead to major heart issues [23].

Liver problem: Type 2 diabetes is more dangerous to the liver. The liver does an incredible job in fine-tuning the blood glucose level by starch digestion with the models like glycogenesis and neoglucogenesis [24]. The liver tumor will occur when there is a fatty liver. The incorporated difficulties are modified metabolism, malnutrition, renal debilitation, insulin opposition and hyperglycemia, and individuals will be affected by experiencing different antitoxin drugs [25] and the liver admits the other treatment [16] which leads to pharmacological treatment, life alteration, α-glycosidase inhibitors, weight to decrease and TZDs.

3.3 Future of lncRNAs in Diabetic Complications

lncRNAs play an important role among RNAs that can function where a lot of cellular processes are involved in the condition of normal or disease state. In recent studies, lncRNAs in diabetes have been projected as a class; they are dysregulated throughout the process of differencing the pancreatic β-cell and also hyperglycemic and other factors. lncRNAs can also contribute to complications of diabetes by changing the fibrosis, inflammation, oxidant stress and ER stress. Undoubtedly, we have learned more regarding transcriptomes of diabetic issues, and now the turn is to study more regarding the complications in diabetes through the functions of lncRNAs. It is specifically important to know that lncRNAs are individuals. It is better to know the imperviousness of diabetic complications of an individual by understanding the specific lncRNAs expression and it will be useful.

4 Machine Learning

The scientific area that deals with different means wherein the machine learns from the experience is called machine learning. According to scientists, the word "Machine Learning" is the same as the word "Artificial Intelligence"; it is given the main characteristics of knowledge possibilities of any entity known as intelligence in the sense of the universe. The main aim is that machine learning has to construct systems that can easily learn and adapt from its own experience [24].

4.1 Machine Learning Categories

ML algorithms are used to predict diabetes. The aim for selecting a machine learning algorithm embraces the matching of features in data depending on the existing approaches. The machine learning algorithms' catalog is discussed in Fig. 3. Machine learning algorithms are classified into three types: supervised learning, unsupervised learning and reinforcement learning.

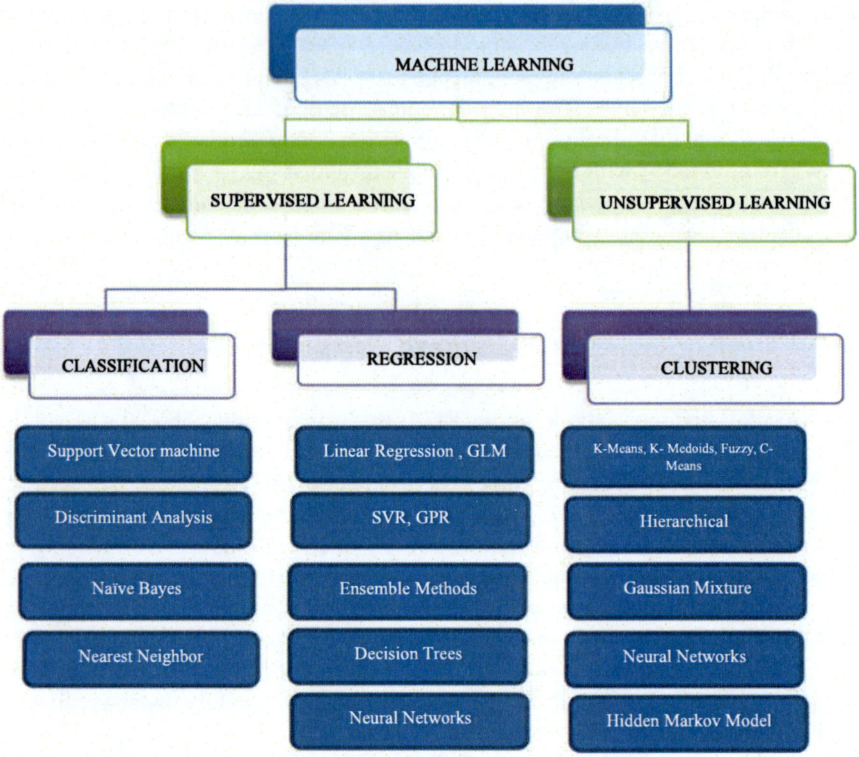

Fig. 3 Classification of machine learning algorithms for diabetes prediction

Supervised learning: All predictive models are constructed by using the supervised learning. All the missing values are predicted by a predictive model which is presented in the database. To build a model for realistic predictions, supervised learning sets up input and output data for the response to new dataset. Supervised learning includes Bayesian method, decision tree, ensemble method, artificial neural network and instance-based learning. These are emerging techniques in machine learning [25].

Unsupervised Learning: It is also known as descriptive models. Unsupervised learning methods are used to implement descriptive models. Here we only know the input data but not the output data. Transactional data is mostly used by unsupervised learning. Unsupervised learning consists of clustering algorithms such as k-means and k-medians clustering [25].

Semi-supervised Learning: In semi-supervised learning we use both the unlabeled and labeled data on the dataset for training. This includes regression and classification techniques. Linear regression and logistic regression are the best examples of regression techniques [25].

Reinforcement Learning: The family of techniques has been allotted a general term known as reinforcement learning, where the system tries to learn itself by interacting directly with the environment by increasing some cumulative reward notions. It is compulsory to know that the system does not have knowledge about the environment behavior, it just finds with the help of trial and error (try and fail). Due to the independence of autonomous systems, the reinforcement learning is applied [26].

There are some guidelines for selecting among supervised and unsupervised machine learning algorithms.

- Select supervised learning depending on the need, to train a model for making a prediction; for example, in data the future value of the continuous variable, such as weather prediction, stock prices.
- Select unsupervised learning if needed for exploring the data and if wanted to train a model to find a perfect internal representation, when data want to be split into clusters.

4.2 *Literature Review*

- To date, many researchers tried to analyze and predict the occurrence of diabetes by using data mining techniques [27–29]. KNN algorithm, expectation–maximization algorithm, amalgam KNN algorithm, K-means algorithm and ANFIS algorithm were used by Vijayan et al. [30] to predict and diagnose diabetes. The UCI dataset was used by them which contains blood test and demographic variables, and they have shown that the output of the expectation–maximization algorithm gives least classification accuracy, while ANFIS and amalgam KNN algorithms give classification accuracy as better as 80 and 80%, respectively.
- Saravananathan et al. [31] conducted a study using the popular classification algorithms, which includes support vector machine (SVM) regression and classification, J48, k-nearest neighbor (KNN) and tree CART for diabetic data. For performance analysis they used indicators like specificity, accuracy, error rate, precision and sensitivity. According to their output, they found that J48 technique's outcome was undoubtedly superior when compared with other techniques for diabetes classification.
- Three data mining models, ANN, logistic regression and decision tree were compared by Meng et al. [28] for predicting prediabetes and diabetes mellitus based on the risk factors. They have collected information for their study about family diabetes history, demographic characteristics, lifestyle risk and anthropometric measurements. They achieved the best classification performance with decision tree model (C5.0) with a sensitivity of 80.68%, accuracy of 77.87% and specificity of 75.13%.
- The logistic regression was used by Saeidi et al. [32] for their study, to evaluate the diagnostic values of HCG27_201 and LY86-AS1 as biomarkers for type 2

diabetes. The result showed a specificity of 79.8% and sensitivity of 64.6%. Here they have not compared any other algorithm to give the best and least performance.

- A study by Zahra Mansoori et al. [33] applied logistic regression with two other lncRNAs which include LINC00994 and LINC00523 expressions for the calculation of potential diagnostic value for type 2 diabetes and got the result as specificity of 61.11% and sensitivity of 81.44% at their best.
- A study of comparison of four classification models was done by Kazerouni et al. [34] for the first time on six lncRNAs (LINC00523, HCG27_201, LINC00994, LY86-AS1, DKFZP and TPT1-AS1) as target genes and their variables. The model includes algorithms like support vector machine (SVM), k-nearest neighbor (KNN), artificial neural networks (ANN) and logistic regression for diagnosing type 2 diabetes. They have done a comparison of the diagnostic power of all algorithms. The KNN algorithm has given the best result with 96% sensitivity compared with other algorithms.

4.3 LncRNA Identification Tools by Machine Learning Algorithms

According to the research done earlier in genomic, there are some algorithms used for identifying the lncRNAs related to health care. Many diseases are associated with lncRNAs, and experiments done with the traditional approaches are difficult in detecting the human disease based on lncRNAs due to the huge amount of biological data. By the computational methods, we can find the significant values which will help to predict and diagnose the disease at an early stage.

SVM algorithm: Vapnik introduced support vector machine (SVM) algorithm for the first time; it is one type of supervised learning that is used maximum by the medical fraternities. The diagnosis was done through regression and classification techniques. SVM simultaneously maximizes the geometric margin and minimizes the practical classification error. The nonlinear classification called kernel trick was efficiently performed by support vector machine, and it also maps its inputs implicitly into high-dimensional feature spaces. Recently, the research community is getting attracted by SVM in machine learning, and many studies reported that the performance delivered was very high in the aspects of classification accuracy than other algorithms of classification.

There are some lncRNA identification tools developed by support vector machine.

- CPC: Coding potential calculator (2007) [35].
- CNCI: Coding–non-coding index (2013) [36].
- PLEK: Predictor of long non-coding RNAs and messenger RNAs based on an improved k-mer scheme (2014) [37].
- CPC2: Coding potential calculator 2 (2017) [38].

Logistic Regression

Logistic regression in statistics is a regression model where the variable is binary dependent and takes only two values as input, which are "1" and "0". This represents the results as win/lose, true/false, pass/fail, dead/alive, and healthy/unhealthy. There are a number of fields where we use logistic regression, namely machine learning, life science and social science, mostly in the medical fields for accurate solutions.

The following are some of the tools developed using logistic regression for the identification of lncRNAs, they are as follows:

- CPAT: Coding Potential Assessment Tool (2013) [39].

Random Forest Algorithm

Random forest is one of the supervised learning techniques and also a very popular machine learning algorithm. Both regression and classification can use random forest to solve the problems in machine learning. It belongs to the concept of ensemble learning, where we can solve a complex problem by combining multiple classifiers and can also improve the model's performance.

The random forest algorithm as a classifier works on multiple decision trees to predict the relevant subsets of given dataset. It doesn't depend on one decision tree. It predicts from different trees depending on the majority predicted votes and gets the final result.

- LncRNApred: Long non-coding RNAs and protein-coding transcripts (2016) [40].
- LncRNA-ID: Long non-coding RNA identification (2015) [41].
- CRlncRC: Cancer-related long non-coding RNA identification (2008) [42].

5 Conclusions

To summarize, increased studies and the interest of researchers show that there is a good connection of lncRNAs with DM progression and its associated complication, like insulin resistance, a disorder in pancreatic β-cells, epigenetic regulation, etc. As a startup regulatory biomarker, lncRNAs participate in various phases of diabetes and its complications by regulating the related genes of expression at the transcriptional, post transcriptional and epigenetic levels. We can say that the knowledge known today about lncRNAs is just a tip of an iceberg. New technologies and novel methods will definitely lift up the veil covered by lncRNAs, by providing new methodologies for early diagnosis, preventing and treatment of DM with the help of machine learning technologies, for the identification of lncRNAs and their role in diabetes and its complications. At the end of the study, the knowledge can be improved by giving importance to the lncRNAs as a biomarker and their role in diabetes and its complications. In future, the knowledge should be improved and the tools for detection and diagnosis of diabetes at an early stage should be developed to avoid the occurrences of diabetes complications.

References

1. Consortium EP (2012) An integrated encyclopedia of DNA elements in the human genome. Nature 489(7414):57–74
2. Cornall RJ, Prins J-B, Todd JA, Pressey A, DeLarato NH, Wicker LS, Peterson LB (1991) Type 1 diabetes in mice is linked to the interleukin-1 receptor and Lsh/Ity/Bcg genes on chromosome 1. Nature 353(6341):262–265
3. Sathishkumar C, Prabu P, Mohan V, Balasubramanyam M (2018) Linking a role of lncRNAs (long non-coding RNAs) with insulin resistance, accelerated senescence, and inflammation in patients with type 2 diabetes. Hum Genomics 12:41
4. Kavakiotis I, Tsave O et al (2017) Machine learning and data mining methods in diabetes research. Comput Struct Biotechnol J 15(2017):104–116
5. Alberti KG, Zimmet PZ (1998) Defnition, diagnosis and classifcation of diabetes mellitus and its complications. Part 1: diagnosis and classifcation of diabetes mellitus provisional report of a WHO consultation. Diabet Med 15(7):539–553
6. Kaddis JS, Olack BJ, Sowinski J, Cravens J, Contreras JL, Niland JC (2009) Human pancreatic islets and diabetes research. J Am Med Assoc 301(15):1580–1587. https://doi.org/10.1001/jama.2009.482
7. World Health Organization (2015) Guideline: sugars intake for adults and children. World Health Organization. http://www.who.int/iris/handle/10665/149782
8. Avogaro P, Crepaldi G, Enzi G, Tiengo A (1967) Associazione di iperlipidemia, diabetemellito e obesità di mediogrado. Acta Diabetol Lat 4:36–41
9. American Diabetes Association (2012) Diagnosis and classifcation of diabetes mellitus. Diabetes Care 35(1):S64-71. https://doi.org/10.2337/dc12-S064
10. Raha O, Chowdhury S, Dasgupta S et al (2009) Approaches in type 1 diabetes research: a status report. Int J Diabetes Dev Ctries 29(2):85–101. https://doi.org/10.4103/0973-3930.53126
11. Bellamy L, Casas JP, Hingorani AD, Williams D (2009) Type 2 diabetes mellitus after gestational diabetes: a systematic review and meta-analysis. Lancet 373:1773–1779
12. Meigs JB, D'Agostino RB Sr, Wilson PW, Cupples LA, Nathan DM, Singer DE (1997) Risk variable clustering in the insulin resistance syndrome: the Framingham Offspring Study. Diabetes 46:1594–1600
13. Anna V, van der Ploeg HP, Cheung NW, Huxley RR, Bauman AE (2008) Socio-demographic correlates of the increasing trend in prevalence of gestational diabetes mellitus in a large population of women between 1995 and 2005. Diabetes Care 31(12):2288–2293
14. Barengo NC, Katoh S, Moltchanov V, Tajima N, Tuomilehto J (2008) The diabetes-cardiovascular risk paradox: results from a Finnish population-based prospective study. Eur Heart J 29(15):1889–1895
15. World Health Organization (2013) Diagnostic criteria and classification of hyperglycemia first detected in pregnancy. World Health Organization. http://www.who.int/iris/handle/10665/85975
16. Chen Y, He Y, Zhou H (2020) The potential role of lncRNAs in diabetes and diabetic microvascular complications. Jpn Endocr Soc 67(7):659–668
17. Li X, Wu Z, Fu X, Han W (2014) lncRNAs: insights into their function and mechanics in underlying disorders. Mutat Res Rev Mutat Res 762:1–21
18. Li R, Zhu H, Luo Y (2016) Understanding the functions of long non-coding RNAs through their higher-order structures. Int J Mol Sci 17:702
19. Wallace C, Smyth DJ, Maisuria-Armer M, Walker NM, Todd JA, Clayton DG (2010) The imprinted DLK1-MEG3 gene region on chromosome 14q32.2 alters susceptibility to type 1 diabetes. Nat Genet 42:68–71
20. Long J, Danesh FR (2018) Values and limitations of targeting lncRNAs in diabetic nephropathy diabetes 67:552–553. https://doi.org/10.2337/dbi17-0052
21. Oliver F, Rajendra AU, Ng EY, KwanHoong N, Jasjit SS (2012) Algorithms for the automated detection of diabetic retinopathy using digital fundus images: a review. J Med Syst 36(1):145–157. https://doi.org/10.1007/s10916-010-9454-7

22. Scott MG, Ivor JB, Gregory LB, Alan C, Robert HE, Barbara VH, William M, Sidney CS, James RS (1999) Diabetes and cardiovascular disease a statement for healthcare professionals from the American Heart Association. Circulation 100(10):1134–1146
23. de Mattos Matheus AS, Tannus LR, Cobas RA, Sousa Palma CC, Negrato CA, de Brito GM (2013) Impact of diabetes on cardiovascular disease: an update. Int J Hypertens 65:15. https://doi.org/10.1155/2013/653789
24. Wilson RA, Keil FC (1999) The MIT encyclopaedia of the cognitive sciences. MIT Press, Cambridge
25. Nithya B, Ilango V (2017) Predictive analytics in health care using machine learning tools and techniques. In: International conference on intelligent computing and control systems. ISBN 978-1-5386-2745-7
26. Joshi TN, Chawan PM (2018) Diabetes prediction using machine learning techniques (S. Dewangan et al.). Int J Eng Res Appl 8(1):09–13 (Part-II). ISSN 2248-9622
27. Daghistani T, Alshammari R (2016) Diagnosis of diabetes by applying data mining classification techniques. Int J Adv Comput Sci Appl 7(7):329–332
28. Meng X-H, Huang Y-X, Rao D-P, Zhang Q, Liu Q (2013) Comparison of three data mining models for predicting diabetes or prediabetes by risk factors. Kaohsiung J Med Sci 29(2):93–99
29. Wu H, Yang S, Huang Z, He J, Wang X (2018) Type 2 diabetes mellitus prediction model based on data mining. Inform Med Unlocked 10:100–107
30. Vijayan V, Ravikumar A (2014) Study of data mining algorithms for prediction and diagnosis of diabetes mellitus. Int J Comput Appl 95(17)
31. Saravananathan K, Velmurugan T (2016) Analyzing diabetic data using classification algorithms in data mining. Indian J Sci Technol 9(43):196–201
32. Saeidi L, Ghaedi H, Sadatamini M, Vahabpour R, Rahimipour A, Shanaki M, Mansoori Z, Kazerouni F (2018) Long non-coding RNA LY86-AS1 and HCG27_201 expression in type 2 diabetes mellitus. Mol Biol Rep 45(6):2601–2608
33. Mansoori Z, Ghaedi H, Sadatamini M, Vahabpour R, Rahimipour A, Shanaki M, Kazerouni F (2018) Downregulation of long noncoding RNAs LINC00523 and LINC00994 in type 2 diabetes in an Iranian cohort. Mol Biol Rep 45(5):1227–1233
34. Kazerouni F, Bayani A, Asadi F, Saeidi L, Parvizi N, Mansoori Z (2020) Type2 diabetes mellitus prediction using data mining algorithms based on the longnoncoding RNAs expression: a comparison of four data mining approaches. BMC Bioinform 21:372. https://doi.org/10.1186/s12859-020-03719-8
35. Kong L, Zhang Y, Ye ZQ et al (2007) CPC: assess the protein-coding potential of transcripts using sequence features and support vector machine. Nucleic Acids Res 35(Suppl 2):W345–W349
36. Sun L, Luo H, Bu D et al (2013) Utilizing sequence intrinsic composition to classify protein-coding and long non-coding transcripts. Nucleic Acids Res 41(17):e166
37. Li A, Zhang J, Zhou Z et al (2014) PLEK: a tool for predicting long non-coding RNAs and messenger RNAs based on an improved kmer scheme. BMC Bioinform 15:311
38. Kang YJ, Yang DC, Kong L et al (2017) CPC2: a fast and accurate coding potential calculator based on sequence intrinsic features. Nucleic Acids Res 45(W1):W12–W16
39. Wang L, Park HJ, Dasari S, et al (2013) CPAT: coding-potential assessment tool using an alignment-free logistic regression model. Nucleic Acids Res 41(6):e74
40. Pian C, Zhang G, Chen Z, Chen Y, Zhang J, Yang T, Zhang L (2016) LncRNApred: classification of long non-coding RNAs and protein-coding transcripts by the ensemble algorithm with a new hybrid feature 11(5):e0154567
41. Achawanantakun R, Chen J, Sun Y, Zhang Y (2015) LncRNA-ID: long non-coding RNA identification using balanced random forests 31(24):3897–3905
42. Hang X, Wang J, Li J, Chen W, Liu C (2018) CRlncRC: a machine learning-based method for cancer-related long noncoding RNA identification using integrated features. BMC Med Genomics 11(Suppl 6):120. https://doi.org/10.1186/s12920-018-0436-9

Computational Method on Breast Cancer Survival Data Using Binary Classification Models

T. Sukeerthi, K. Sukanya, and K. Vandana Rao

Abstract Overall around the world, mammary cancer is one of the most common tumors among women and it almost causes death. By distinction, breast cancer is an increasingly critical problem among all countries around the globe. This paper study reports an analysis of data to predict the values of the patient's survived or may die within 5 years after mammary cancer treatment regarding the survival of breast cancer patients who have undergone surgery. Logistic regression and decision tree were binary classification models performed to determine the correct percent of breast cancer survival rates and death rates. In this experimental study, data analysis was examined in Statistical Package for the Social Sciences (SPSS) software to guide statistical students and researchers about performing statistical tools easily on a computational basis. An attempt has been made to enhance the model to predict and describe the best fitting correlation between the dichotomous and explanatory values implementing through binary classification models.

Keywords Mammary cancer · SPSS software · Logistic regression · Decision tree

1 Introduction

When cells in the breast begin to grow out of control it leads to mammary cancer. Most of them begin in the ducts that carry milk to the mammary glands and some of the glands lead to lobular cancers. If the cells grown in surrounding tissues spread to distant areas of the body, it is called a tumor and it is felt as a lump that can be displayed

T. Sukeerthi (✉) · K. Sukanya
Department of Statistics, SPMVV, Tirupati, India
e-mail: tptsukeerthi2309@gmail.com

K. Sukanya
e-mail: sukanya.sudha27@gmail.com

K. Vandana Rao
Department of Computer Science, SPMVV, Tirupati, India
e-mail: vandana.kalapati@gmail.com

S. Jyothi et al. (eds.), *Proceedings of the 2nd International Conference on Computational and Bio Engineering*, Lecture Notes in Networks and Systems 215,
https://doi.org/10.1007/978-981-16-1941-0_12

on an X-ray. A new lump or hard mass with irregular edges, swelling of breast parts, redness or solidify nipple or breast skin and nipple discharge are signs of mammary cancer. Treatments performed are lumpectomy, mastectomy, chemotherapy, radiation therapy, hormonal therapy, oncoplastic surgery, targeted therapy drugs, and immunotherapy.

Cancer cells get going to grow in lymph nodes through lymph vessels. The majority of the lymph vessels of the breast drain into lymph nodes under the auxiliary nodes, around the collar bone and inside the chest near the internal mammary lymph nodes. Mammary cancer occurs particularly in women, but men obtain it too. Cancers can be detected at an early stage through mammograms. The second cause of death in women is breast cancer. The survival rates of breast cancer by stages are discussed in the section.

2 Data Collection

Data is collected from the UCI data sets: https://archive.ics.uci.edu/ml/datasets in Google. The case study of data set between 1958 and 1970 at the University of Chicago's Billings Hospital is performed on the survival of patients who had undergone surgery for mammary cancer. https://archive.ics.uci.edu/ml/datasets/Haberman%27s+Survival. This data set comprises 305 members of patient's information and four attributes such as patient's age at treatment processing time, year, number of positive auxiliary nodes and survival significance (i.e. patient lived within 5 years or prolong and may die within 5 years).

3 About Software

SPSS is powerful statistical software, which was earlier known for its applications in social sciences only. It is a comprehensive integrated system for workable statistical data examination and data processing [1, 3, 5]. This program is used for survey authoring, data extracting, text mining, collaboration and categorization. It has all major analytical tools for handling a large volume of data as well as complicated multivariate analyses. The current versions in use are IBMSPSS 16.0 and 20.0 [4].

4 Statistical Tools Used Under Study

4.1 Logistic Regression

Logistic regression is a statistical method for analyzing a data set in which there are one or more independent variables that determine an outcome [2]. The results are measured with binary variables and lead to a model which can be used to predict the chance of an occurrence happening for an individual. The aim of the tool is to find the best fitting version to describe the correlation between the binary and predicted values.

$$l = \log_b \frac{p}{1-p} = \beta_0 + \beta_1 X_1 + \beta_2 X_2 + \cdots + \beta_n X_n$$

where l arise log-difference, b exists as base logarithm and β_n are parameters of the model.

4.2 Logistic Regression in SPSS

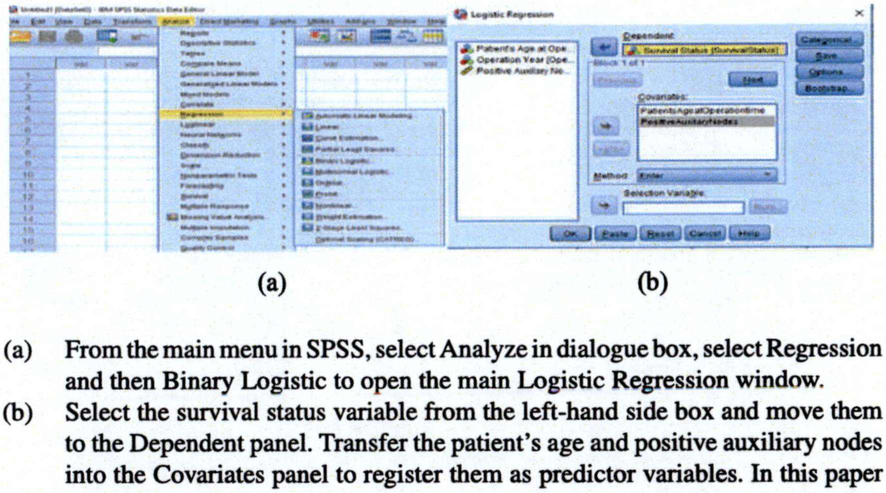

(a) (b)

(a) From the main menu in SPSS, select Analyze in dialogue box, select Regression and then Binary Logistic to open the main Logistic Regression window.
(b) Select the survival status variable from the left-hand side box and move them to the Dependent panel. Transfer the patient's age and positive auxiliary nodes into the Covariates panel to register them as predictor variables. In this paper retain the default method of enter to perform a standard logistic regression analysis [5].

When variables are moved into the Covariates panel, selecting the categorical push button opens the Define Categorical variables window. We move patient's age at operation time into the Categorical Covariates panel. The second step is the specification of the References Category. Males and females are arbitrary codes represented as 0 and 1, respectively. In this paper data is about females and it is coded as the last category (the highest numeric code).

In the Save window, choose for the Probabilities and Group membership for predicted values. In this paper select Unstandardized residuals and Leverages values in Influence box and get through on Continue button.

In the Options window, call for the Classification plots, Hosmer-Lemeshow goodness of fit, correlation of estimates as well as 95% confidence level for the odds ratio.

4.3 Decision Tree Analysis Method

This method provides an exploratory, confirmatory, classification analysis procedure and tree-based categorization process. This is a tool to display an algorithm that only prefers conditional control statements including chance event outcomes, possible consequences and resource cost which support the decision. It divides cases into groups or predict values of a dependent variable based on the values of independent variables [1]. In this paper, the CRT method is used and the other methods are chi-squared automatic interaction detection (CHAID), quick, unbiased, efficient statistical tree (QUEST) and exhaustive CHAID to estimate data models.

4.4 Decision Tree in SPSS

From the main menu in SPSS, choose, analyze and classify and then go to Tree to open the Decision Tree window.

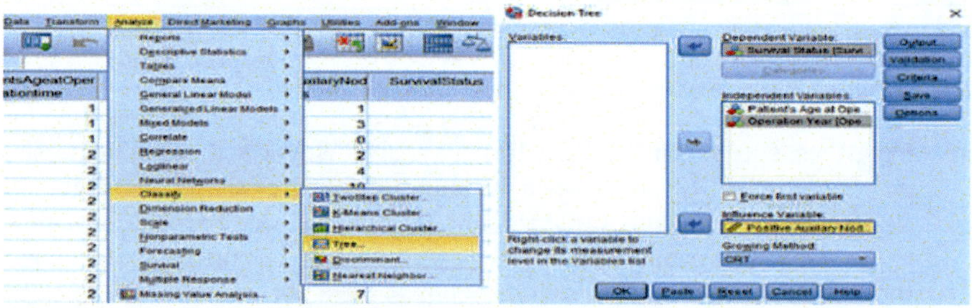

Select the survival status variable from the left hand side box and move them to the Dependent panel. Transfer patient's age and operation year to the Independent panel and then positive auxiliary nodes into influence variables.

By clicking on Growing Method, three types of trees appear on the screen, i.e. CRT, CHAID, exhaustive CHAID and QUEST. In this analysis, the CRT to split the data into sets that are homogeneous is performed with respect to the dichotomous variable.

When variables are moved into Influence panel, the Output push button becomes available. Select it and opens the Output window having Trees, statistics, plots and rules in the dialogue box. Output produces the Trees presentation with orientation, size and the nodes are displayed with statistics.

In **Statistics** window, click on Summary, risk and classification table. Click on Continue. By selecting Rules window, click on Generate classification rules in that default syntax, type and nodes and then click Continue. Especially in large data sets validation separates the sample data set into training and testing samples. Here, cross-validation divides the sample into a number of subsamples and produces a single or final tree model.

On selecting the Criteria push button, Growth limits, CRT and Pruning appeared on the screen. The Growth limits tab allows you to limit the number of levels in the tree and control the minimum number of cases for parent and child nodes. The CRT growing method attempts to maximize within-node homogeneity. Clicking on these options default values are taken and then press Continue.

In the **Save** window, ask for the Terminal node number, predicted value and predict probabilities and then click on Continue button.

5 Experimental Results

5.1 Logistic Regression

Variables in the equation

		B	S.E.	Wald	df	Sig.	Exp(B)	95% C.I. for EXP(B)	
								Lower	Upper
Step 1[a]	Patients age at operation time			0.586	2	0.746			
	Patients age at operation time (1)	−20.071	23179.188	0.000	1	0.999	0.000	0.000	
	Patient's age at operation time (2)	−0.239	0.312	0.586	1	0.444	0.787	0.427	1.452
	Operation year	−0.139	0.186	0.564	1	0.453	0.870	0.604	1.252

(continued)

(continued)

Variables in the equation

		B	S.E.	Wald	df	Sig.	Exp(B)	95% C.I. for EXP(B)	
								Lower	Upper
	Positive auxiliary nodes	0.087	0.020	19.591	1	0.000	1.091	1.050	1.134
	Constant	−0.974	0.464	4.410	1	0.036	0.378		

[a]Variable(s) entered on Step 1: Patients age at operation time, Operation year, Positive auxiliary nodes

Results of the analysis are presented in the variables in the equation. It presents for each predictor the raw score partial regression coefficient and its standard error (S.E.). The sig. column represents the p-value for testing whether a predictor is significantly associated with a patient's survival for the other predictors.

$$\text{Survival} = -0.974 + (-20.071)\,(\text{Patients age at operation time}(1)) + (-0.239)\,(\text{Patients age at operation time}(2)) + (-0.139)\,(\text{Operation year}) + 0.87\,(\text{Positive auxiliary nodes}).$$

The Exp(B) column provides the odds ratios associated with each predictor with the 95% confidence interval associated with each provided in the final two columns. The adjusted odds ratio for Patient's age at operation time is 0.787 with a 95% confidence interval 0.427–1.452, Operation year of the patient is 0.870 with a 95% confidence interval 0.604–1.252 and Positive auxiliary nodes is 1.091 with a 95% confidence interval 1.050–1.134.

Classification table

	Observed		Predicted		
			Survival status		Percentage correct
			Patient survived 5 years or longer	Patient died within 5 years	
Step 1	Survival status	Patient survived 5 years or longer	217	8	96.4
		Patient died within 5 years	67	13	16.3
	Overall percentage				75.4

[a]The cut-off value is 0.500

The classification table shows the overall predictive accuracy as 75.4% of predictors in the model, although the model predicted better for patients survived (96.4%) than for patients died (16.3%).

5.2 Decision Tree Output

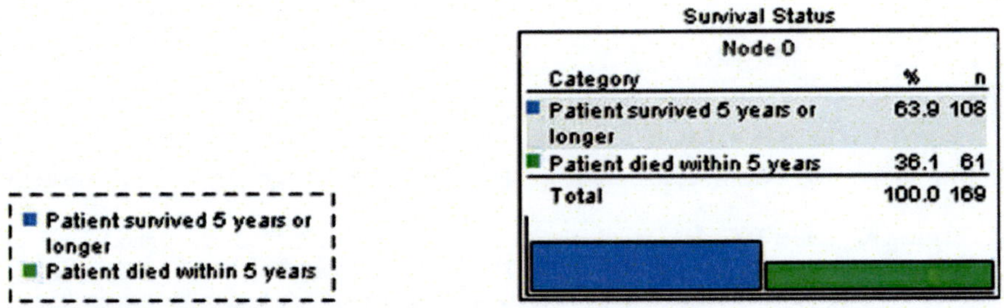

By default, the tree diagram is displayed in the Viewer and the importance of bar chart is to display an independent variable which is predictable. Bar chart can only visualize by using CRT growing method.

Classification			
Observed	Predicted		
	Patient survived 5 years or longer	Patient died within 5 years	Percent Correct (%)
Patient survived 5 years or longer	108	0	100.0
Patient died within 5 years	61	0	0.0
Overall percentage	100.0%	0.0%	63.9

Growing method: CRT
Dependent variable: survival status

Here, the classification table of decision tree shows the overall predictive accuracy is 63.9%, although the model predicted better for patients survived (108%) than for patients died (61%). Thus, predictors certainly contributed to successful prediction.

6 Conclusion

Here, two types of binary classification models are performed to predict the breast cancer survival data.

- Overall predicted percentage for breast cancer survival data values through logistic regression analysis is 75.4%.
- Overall predicted percentage value of decision tree analysis is 63.9%.

By comparing these two percentage values of binary classification models, this paper concludes that logistic regression analysis is the best predicted model than decision tree analysis.

References

1. Argyrous G (2005) Statistics for research: with a guide to SPSS. SAGE, London. ISBN 1-4129-1948-7
2. Bryman A, Cramer D (2011) Quantitative data analysis with IBM SPSS 17, 18 and 19: a guide for social scientists. Routledge, New York. ISBN 978-0-415-57918-6
3. Foster JJ (2001) Data analysis using SPSS for Windows 8.0–10.0, a beginner's guide
4. Meyers LS, Gamst GC, Guarino AJ (2013) Performing data analysis using IBM SPSS
5. Sarma KVS (2010) Statistics made simple, do it yourself on PC. Prentice Hall of India

Percentage Nucleotide Concentration and Classification of SARS–Corona Viruses

F. Amul Mary and S. Jyothi

Abstract One of the most advantages that we have today to fight against corona virus which was not available when SARS outbreak in the year 2003 is big data and high-level technologies. China used big data and machine learning techniques and other digital tools as the virus spread through the nation in order to control the spread of the virus. Most of the viruses are too small which is difficult to see through a light microscope, and sequencing is one of the main tools in virology to identify and study the virus. Viral genomes can be based on DNA or RNA. RNA viruses are more time-sensitive for genome sequencing, as they degrade faster in clinical samples. There are more than 2.3 million unique viral sequences in GenBank. In this paper, Genome Data of eight SARS–corona type viruses are taken for analysis, which are archived in National Center for Biotechnology (NCBI) for analyzing and finding common features among the genomes of adenine, thymine, guanine and cytosine. Percentage concentration of nucleotides pA, pT, pG and pC is evaluated for each genome, the differences in percentage nucleotide concentration of one genome with respect to all other genomes are identified and the classification is done. It is identified that the genetic features are approximately the same for some genomes. There is a considerable structural relationship between the genomes that are identified in this analysis. These relationships can be further understood when spectral-domain characterization is carried out for these eight genomes.

Keywords SARS–corona viruses · Big data analysis · Machine learning techniques

F. Amul Mary (✉) · S. Jyothi (✉)
Department of Computer Science, SPMVV, Tirupathi, India
e-mail: amulmary@gmail.com

S. Jyothi
e-mail: Jyothi.spmvv@gmail.com

S. Jyothi et al. (eds.), *Proceedings of the 2nd International Conference on Computational and Bio Engineering*, Lecture Notes in Networks and Systems 215,
https://doi.org/10.1007/978-981-16-1941-0_13

1 Introduction

Genomics data have two-sided aspects. The first one is monolithic sequencing data and the second is ordered lists of nucleotides. In human genomics, generally, they are mapped to the genome and are used to generate coverage or variation data [1]. The canonical structure of DNA has four bases. They are adenine (A), thymine (T), guanine (G) and cytosine (C). DNA sequencing is the determination of the physical order of the bases in a molecule of DNA. DNA was first discovered and isolated by Friedrich Miescher in 1869. RNA sequencing was one of the earliest forms of nucleotide sequencing. It is the sequence of the first complete gene identified and published by Walter Fiers [2]. Big data is offering ongoing opportunities to genetic research and its associated Sciences. Big data is also affecting the business the way we do either from startups to enterprises regardless of the size of data, data analysis and understanding of data. New biomedical techniques such as next-generation genome sequencing are generating large volumes of data [3]. The practical application of these data in genetic research is really a challenge and concern.

2 Data Used for Analysis

Twelve genomes of viruses are taken from NCBI website for data analysis. These genomes are considered to be genuine and analysis is carried out on them.

2.1 Severe Acute Respiratory Syndrome Corona Virus 2 2019-nCoV_PH_nCOV_20_026 N Gene for Nucleocapsid Phosphoprotein, Partial Cds

GenBank: **LC523807.1**
Locus: LC523807 357bp RNA linear VRL 13-FEB-2020.
Definition: Severe acute respiratory syndrome coronavirus 2 2019-nCoV_PH_nCOV_20_026 N gene for nucleocapsid phosphoprotein, partial cds.

Accession: LC523807
Version: LC523807.1
Source: Severe acute respiratory syndrome coronavirus 2 (SARS-CoV2).
Organism: Severe acute respiratory syndrome coronavirus 2 viruses; Riboviria; Nidovirales; Cornidovirineae; Coronaviridae; Orthocoronavirinae; Betacoronavirus; Sarbecovirus.
Reference: 1
Authors: Mercado, E. S., Manalo, J. I., Nicolasora, A. D., Medado, I. P., Tujan, M. A., Onza, O. T., Cruz, K. M. and Polotan, F. M.
Title: Molecular Detection of 2019 Novel Coronavirus (2019-nCoV) in Philippines.
Journal: Unpublished.
Reference: 2 (bases 1–357).
Authors: Manalo, J. I. and Nicolasora, A. D.
Title: Direct Submission.
Journal: Submitted (11-Feb-2020) Contact: Joanna Ina Manalo Research Institute for Tropical Medicine, Molecular Biology Laboratory; 9002 Research Dr., Alabang, Muntinlupa 1781, Philippines.
Features: Location/Qualifiers.

2.2 *Wuhan Seafood Market Pneumonia Virus Genome Assembly, Chromosome: Whole_Genome*

GenBank: **LR757995.1**
Locus: LR757995 - 29872 bp RNA linear VRL 06-MAR-2020.
Definition: Wuhan seafood market pneumonia virus genome assembly, chromosome: whole_genome.
Accession: LR757995
Version: LR757995.1
DBlink: BioProject: PRJEB36487, BioSample: SAMEA6507893.
Source: Severe acute respiratory syndrome coronavirus 2 (SARS-CoV2).
Organism: Severe acute respiratory syndrome coronavirus 2 viruses; Riboviria; Nidovirales; Cornidovirineae; Coronaviridae; Orthocoronavirinae; Betacoronavirus; Sarbecovirus.
Reference: 1
Authors: Hunter, C. and Wei, X.
Title: Direct Submission.
Journal: Submitted (30-Jan-2020) BGI, GigaScience, 16 Dai Fu Street, Tai Po Industrial Estate, N.T., Hong Kong.
Comment: Corona virus sequence assembly isolated from 2019/2020 Wuhan outbreak patient.
Features: Location/Qualifiers

Likewise, eight complete genomes are considered for analysis from NCBI website. The genome data and its Numerical Sequence of LC523807.1 are shown here in this paper. Since this data has less number of nucleotides, in order to reduce the number of pages for publication.

Genome Data for LC523807.1

aactggcagt aaccagaatg gagaacgcag tggggcgcga tcaaaacaac gtcggcccca

aggtttaccc aataatactg cgtcttggtt caccgctctc actcaacatg gcaaggaaga

ccttaaattc cctcgaggac aaggcgttcc aattaacacc aatagcagtc cagatgacca

aattggctac taccgaagag ctaccagacg aattcgtggt ggtgacggta aaatgaaaga

tctcagtcca agatggtatt tctactacct aggaactggg ccagaagctg gacttcccta

tggtgctaac aaagacggca tcatatgggt tgcaactgag ggagccttga atacacc.

Genome data of LC523807.1 Numerical Sequence:

1142334132 1144131123 3131143413 2333343431 2411114114 3243344441

1332221444 1121121423 4324223322 4144342424 1424114123 3411331131

4422111224 4424313314 1133432244 1122114144 1121341324 4131231441

1122334214 2144311313 4214413143 1122432332 3323143321 1112311131

2424132441 1312332122 2421421442 1331142333 4413113423 3142244421

2332342114 1113143341 2412123332 2341142313 3313442231 1214144.

Total Number of Nucleotides in each Genome:

Total number of nucleotides in each genome is the total number of adenine, thymine, guanine and cytosine (Table 1).

Table 1 Total number of nucleotides in each genome

S. no	Gene Bank ID	No. of adenine	No. of thymine	No. of guanine	No. of cytosine	Total no. of nucleotides
1	LC523807.1	110	73	86	88	357
2	LC523809.1	108	73	86	88	355
3	LR757995.1	8932	9589	5861	5490	29,872
4	LR757996.1	8933	9586	5861	5488	29,868
5	LR757997.1	5738	6101	3804	3620	19,263
6	LR757998.1	8937	9584	5859	5486	29,866
7	MN970003.1	79	99	57	55	290
8	MN970004.1	79	99	57	55	290

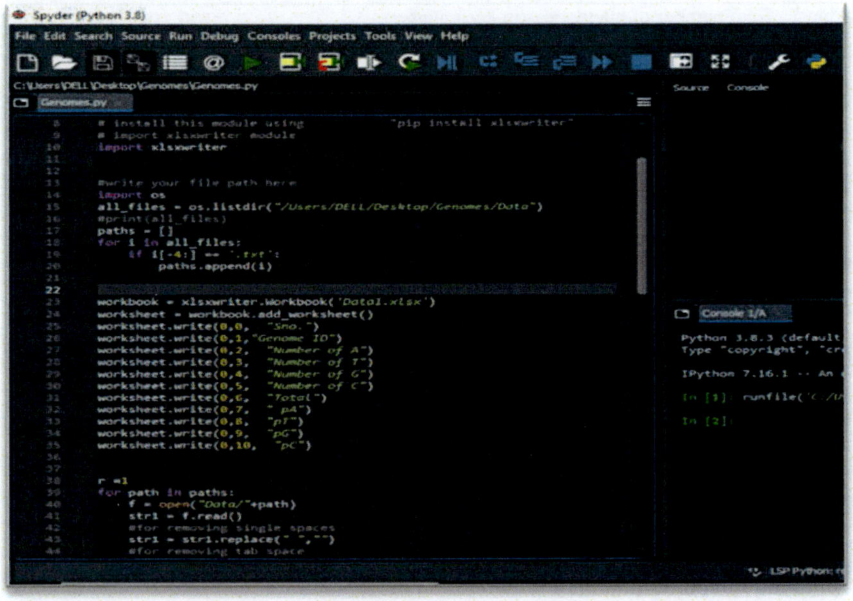

Fig. 1 Coding to calculate number of adenine, thymine, guanine, cytosine and pA, pT, pG, pC

By executing the following code, numbers of adenine, thymine, guanine and cytosine are calculated (Fig. 1).

3 Percentage Nucleotide Concentration

Similar to the pH value of a solution, the genome sequence values of pA, pT, pG and pC are calculated. These values are like the proportion pA:pT:pG:pC used to classify and characterize genome sets. Percentage nucleotide concentration is calculated by the formula: The total number of adenine divided by total number of nucleotides multiplied by 100. Similar to thymine, guanine and cytosine also pT, pG, and pC are calculated which are displayed in Table 2.

The graphical representation for the above percentage nucleotide concentration is given in Fig. 2.

The color coding is given as blue for adenine, yellow for thymine, green for guanine and red for cytosine in red. From the above line graph and data, it is identified that the percentage concentration of nucleotides is similar for most of the genomes.

Table 2 Total number of nucleotides in each genome

S. No	Gene Bank ID	pA	pT	pG	pC
1	LC523807.1	30.812	20.448	24.090	24.650
2	LC523809.1	30.423	20.563	24.225	24.789
3	LR757995.1	29.901	32.100	19.620	18.378
4	LR757996.1	29.908	32.095	19.623	18.374
5	LR757997.1	29.788	31.672	19.748	18.793
6	LR757998.1	29.924	32.090	19.618	18.369
7	MN970003.1	27.241	34.138	19.655	18.966
8	MN970004.1	27.241	34.138	19.655	18.966

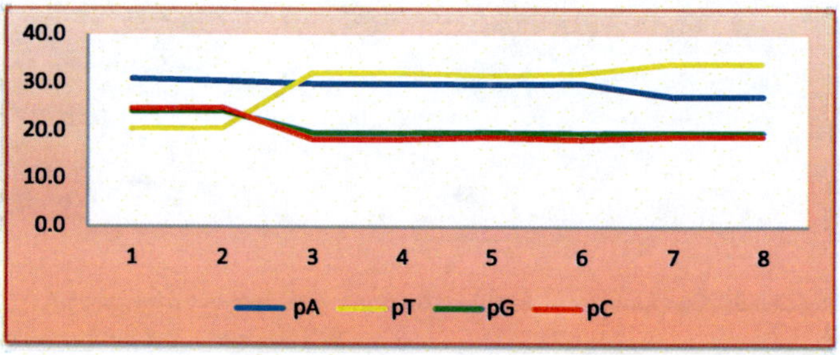

Fig. 2 Graphical representation of pA, pT, pG, **pC**

3.1 Some Important Observations from pA, pT, pG and pC

- LC523807.1 and LC523809.1 virus genomes form one class
- LR757995.1, LR757996.1 and LR757998.1 virus genomes form one class.
- MN970003.1 and MN970004.1 virus genomes form one class.
- LR757997.1 forms a separate class.
- There is no direct relationship between the length of the virus genome sequence and their nucleotide concentration.

4 Changes in Percentage Nucleotide Concentration

The change in percentage nucleotide concentration gives a statistical overview of how the percentage distribution of nucleotide changes from one genome to another. A positive value depicts an increase in concentration from Genome X to Genome Y. A negative value depicts a decrease in the concentration from Genome X to Genome Y. **Differences in percentage nucleotide concentration with respect to Genome No: 1, that is Genome ID: LC523807.1** (Table 3).

Table 3 Differences of % concentration with respect to Genome ID: LC523807.1

Genome no	Gene Bank ID	Genome (X, Y)	$p(A_X) - p(A_Y)$	$p(T_X) - p(T_Y)$	$p(G_X) - p(G_Y)$	$p(C_X) - p(C_Y)$
1	LC523807.1	(1, 1)	0.000	0.000	0.000	0.000
2	LC523809.1	(1, 2)	0.390	−0.115	−0.136	−0.139
3	LR757995.1	(1, 3)	0.911	−11.652	4.469	6.271
4	LR757996.1	(1, 4)	0.904	−11.646	4.467	6.276
5	LR757997.1	(1, 5)	1.025	−11.224	4.342	5.857
6	LR757998.1	(1, 6)	0.889	−11.642	4.472	6.281
7	MN970003.1	(1, 7)	3.571	−13.690	4.434	5.684
8	MN970004.1	(1, 8)	3.571	−13.690	4.434	5.684

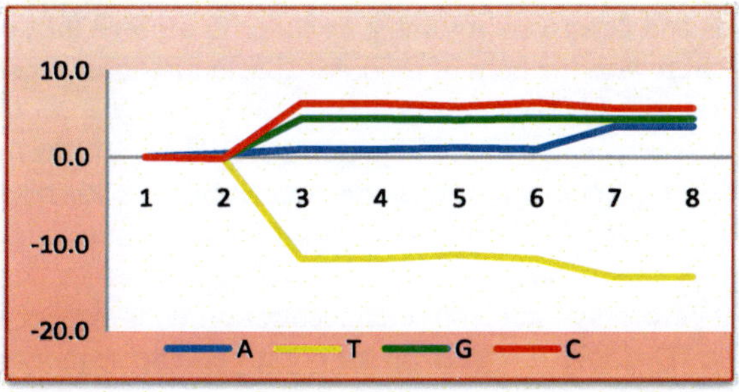

Fig. 3 Graphical representation of % concentration with respect to Genome ID: LC523807.1

Figure 3 shows a change in nucleotide percentage (%) concentration values for Genome1, that is Genome ID: LC523807.1, with all 12 genome sequences. These values are used to decide the evolutionary nature of virus genomes. Most of the virus genomes are active and they do not change their structure; in the sense, their nucleotide percentage concentration will not change drastically. But their functional behavior depends on the distribution of the nucleotides. If the nucleotide percentage concentrations and their distribution patterns remain unchanged, then the viruses are passive in nature or a vaccine would indeed deactivate them. But, if the nucleotide distribution patterns change in spite of their nucleotide percentage concentrations remaining the same, they are classified as active, which means they are evolutions from earlier ones. If the evolution pattern obeys a rule, then it is easy to predict future evolutions. If the evolution pattern is random, it is difficult to predict future evolutions. If the random evolution is natural, then it is almost not possible to control the pandemic nature of the virus. If the random evolution is empirical, that is the virus is modeled in a laboratory, it is easy to predict and decrypt the evolutions by treating the evolutions as a random process. One can create in the laboratory an antidote, that is a vaccine, also termed as antivirus, which will be acting as a standard genetic vaccine. This antivirus also would change its evolution, combine with any type of

Table 4 Differences of % concentration with respect to Genome ID: LC523809.1

Genome no	Gene Bank ID	Genome (X, Y)	$p(A_X) - p(A_Y)$	$p(T_X) - p(T_Y)$	$p(G_X) - p(G_Y)$	$p(C_X) - p(C_Y)$
1	LC523807.1	(2, 1)	−0.390	0.115	0.136	0.139
2	LC523809.1	(2, 2)	0.000	0.000	0.000	0.000
3	LR757995.1	(2, 3)	0.522	−11.537	4.605	6.410
4	LR757996.1	(2, 4)	0.514	−11.531	4.602	6.415
5	LR757997.1	(2, 5)	0.635	−11.109	4.478	5.996
6	LR757998.1	(2, 6)	0.499	−11.527	4.608	6.420
7	MN970003.1	(2, 7)	3.181	−13.575	4.570	5.823
8	MN970004.1	(2, 8)	3.181	−13.575	4.570	5.823

evolutionary virus and deactivate it jointly. In order to achieve this, one has to carry out involved research with the help of theoretical scientists and genetic experts who work in wet laboratories [4, 5].

The same process is repeated for all 12 genome sequences and the results presented below. Differences in percentage nucleotide concentration with respect to Genome No: 2, that is Genome ID: LC523809.1 with all eight genome sequences (Table 4 and Fig. 4).

Differences in percentage nucleotide concentration with respect to Genome No: 3, that is Genome ID: LR757995.1 with all eight genome sequences (Table 5 and Fig. 5).

Differences in percentage nucleotide concentration with respect to Genome No: 4, that is Genome ID: LR757996.1, with all eight genome sequences (Table 6 and Fig. 6).

Differences in percentage nucleotide concentration with respect to Genome No: 5, that is Genome ID: LR757997.1, with all eight genome sequences (Table 7 and Fig. 7).

Differences in percentage nucleotide concentration with respect to Genome No: 6, that is Genome ID: LR757998.1, with all eight genome sequences (Table 8 and Fig. 8).

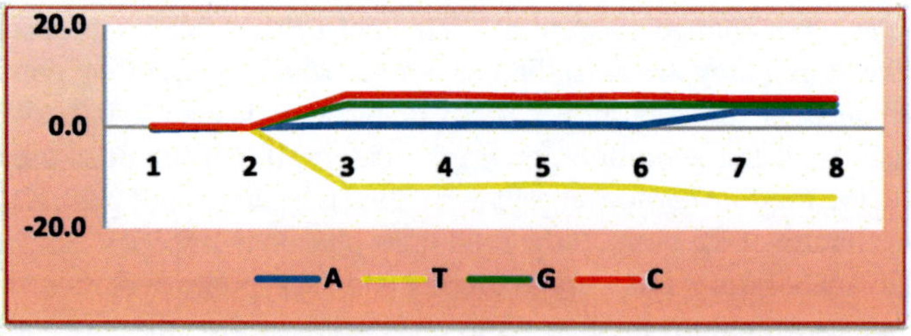

Fig. 4 Graphical representation of % concentration with respect to Genome ID: LC523809.1

Table 5 Differences of % concentration with respect to Genome ID: LR757995.1

Genome no	Gene Bank ID	Genome (X, Y)	$p(A_X) - p(A_Y)$	$p(T_X) - p(T_Y)$	$p(G_X) - p(G_Y)$	$p(C_X) - p(C_Y)$
1	LC523807.1	(3, 1)	−0.911	11.652	−4.469	−6.271
2	LC523809.1	(3, 2)	−0.522	11.537	−4.605	−6.410
3	LR757995.1	(3, 3)	0.000	0.000	0.000	0.000
4	LR757996.1	(3, 4)	−0.007	0.006	−0.003	0.004
5	LR757997.1	(3, 5)	0.113	0.428	−0.127	−0.414
6	LR757998.1	(3, 6)	−0.023	0.010	0.003	0.010
7	MN970003.1	(3, 7)	2.660	−2.038	−0.035	−0.587
8	MN970004.1	(3, 8)	2.660	−2.038	−0.035	−0.587

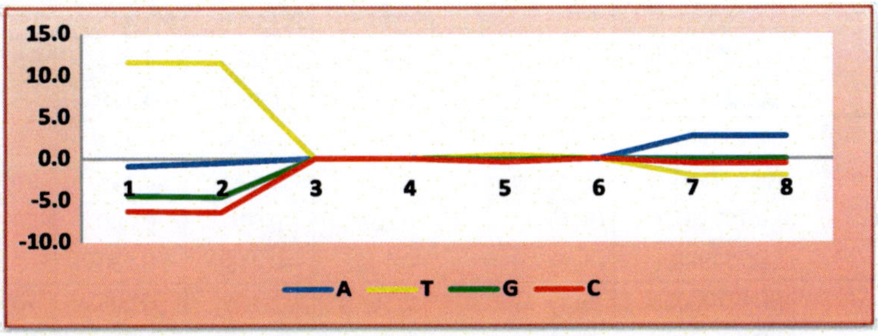

Fig. 5 Graphical representation of % concentration with respect to Genome ID: LR757995.1

Table 6 Differences of % concentration with respect to Genome ID: LR757996.1

Genome no	Gene Bank ID	Genome (X, Y)	$p(A_X) - p(A_Y)$	$p(T_X) - p(T_Y)$	$p(G_X) - p(G_Y)$	$p(C_X) - p(C_Y)$
1	LC523807.1	(4, 1)	−0.904	11.646	−4.467	−6.276
2	LC523809.1	(4, 2)	−0.514	11.531	−4.602	−6.415
3	LR757995.1	(4, 3)	0.007	−0.006	0.003	−0.004
4	LR757996.1	(4, 4)	0.000	0.000	0.000	0.000
5	LR757997.1	(4, 5)	0.121	0.422	−0.125	−0.418
6	LR757998.1	(4, 6)	−0.015	0.005	0.005	0.005
7	MN970003.1	(4, 7)	2.667	−2.043	−0.032	−0.591
8	MN970004.1	(4, 8)	2.667	−2.043	−0.032	−0.591

Differences in percentage nucleotide concentration with respect to Genome No: 7, that is, Genome ID: MN970003.1, with all eight genome sequences (Table 9 and Fig. 9).

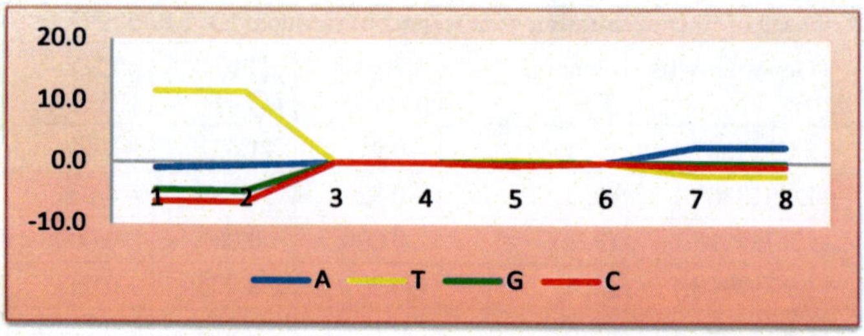

Fig. 6 Graphical representation of % concentration with respect to Genome ID: LR757996.1

Table 7 Differences of % concentration with respect to Genome ID: LR757997.1

Genome no	Gene Bank ID	Genome (X, Y)	$p(A_X) - p(A_Y)$	$p(T_X) - p(T_Y)$	$p(G_X) - p(G_Y)$	$p(C_X) - p(C_Y)$
1	LC523807.1	(5, 1)	−1.025	11.224	−4.342	−5.857
2	LC523809.1	(5, 2)	−0.635	11.109	−4.478	−5.996
3	LR757995.1	(5, 3)	−0.113	−0.428	0.127	0.414
4	LR757996.1	(5, 4)	−0.121	−0.422	0.125	0.418
5	LR757997.1	(5, 5)	0.000	0.000	0.000	0.000
6	LR757998.1	(5, 6)	−0.136	−0.418	0.130	0.424
7	MN970003.1	(5, 7)	2.546	−2.466	0.093	−0.173
8	MN970004.1	(5, 8)	2.546	−2.466	0.093	−0.173

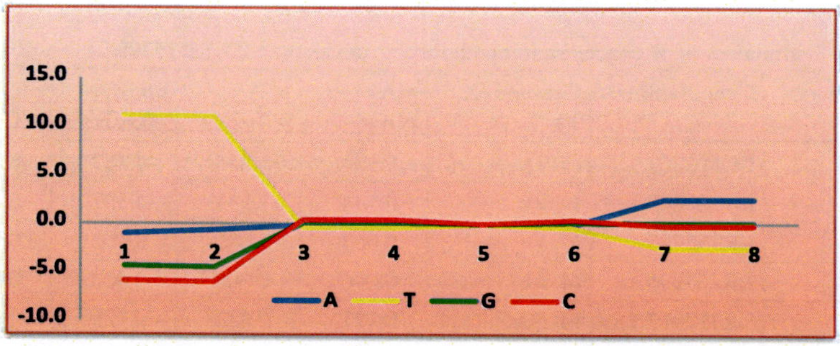

Fig. 7 Graphical representation of % concentration with respect to Genome ID: LR757997.1

Differences in percentage nucleotide concentration with respect to Genome No: 8, that is, Genome ID: MN970004.1, with all eight genome sequences (Table 10 and Fig. 10).

Table 8 Differences of % concentration with respect to Genome ID: LR757998.1

Genome no	Gene Bank ID	Genome (X, Y)	$p(A_X) - p(A_Y)$	$p(T_X) - p(T_Y)$	$p(G_X) - p(G_Y)$	$p(C_X) - p(C_Y)$
1	LC523807.1	(6, 1)	−0.889	11.642	−4.472	−6.281
2	LC523809.1	(6, 2)	−0.499	11.527	−4.608	−6.420
3	LR757995.1	(6, 3)	0.023	−0.010	−0.003	−0.010
4	LR757996.1	(6, 4)	0.015	−0.005	−0.005	−0.005
5	LR757997.1	(6, 5)	0.136	0.418	−0.130	−0.424
6	LR757998.1	(6, 6)	0.000	0.000	0.000	0.000
7	MN970003.1	(6, 7)	2.682	−2.048	−0.038	−0.597
8	MN970004.1	(6, 8)	2.682	−2.048	−0.038	−0.597

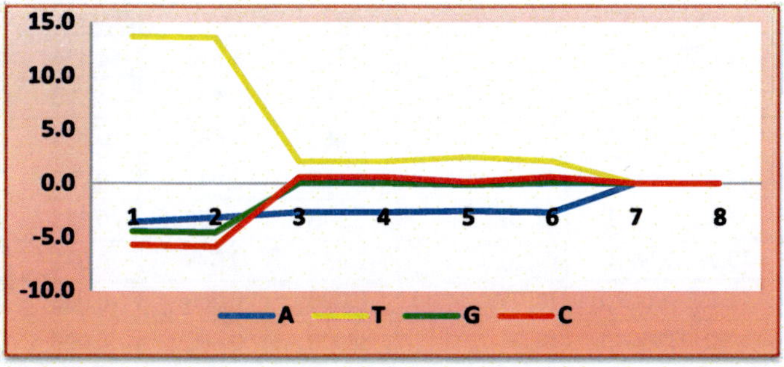

Fig. 8 Graphical representation of % concentration with respect to Genome ID: LR757998.1

Table 9 Differences of % concentration with respect to Genome ID: MN970003.1

Genome no	Gene Bank ID	Genome (X, Y)	$p(A_X) - p(A_Y)$	$p(T_X) - p(T_Y)$	$p(G_X) - p(G_Y)$	$p(C_X) - p(C_Y)$
1	LC523807.1	(7, 1)	−3.571	13.690	−4.434	−5.684
2	LC523809.1	(7, 2)	−3.181	13.575	−4.570	−5.823
3	LR757995.1	(7, 3)	−2.660	2.038	0.035	0.587
4	LR757996.1	(7, 4)	−2.667	2.043	0.032	0.591
5	LR757997.1	(7, 5)	−2.546	2.466	−0.093	0.173
6	LR757998.1	(7, 6)	−2.682	2.048	0.038	0.597
7	MN970003.1	(7, 7)	0.000	0.000	0.000	0.000
8	MN970004.1	(7, 8)	0.000	0.000	0.000	0.000

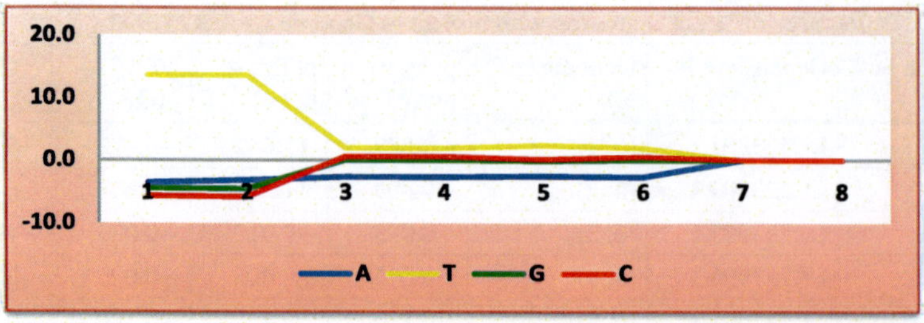

Fig. 9 Graphical representation of % concentration with respect to Genome ID: MN970003.1

Table 10 Differences of % concentration with respect to Genome ID: MN970004.1

Genome no	Gene Bank ID	Genome (X, Y)	$p(A_X) - p(A_Y)$	$p(T_X) - p(T_Y)$	$p(G_X) - p(G_Y)$	$p(C_X) - p(C_Y)$
1	LC523807.1	(8, 1)	−3.571	13.690	−4.434	−5.684
2	LC523809.1	(8, 2)	−3.181	13.575	−4.570	−5.823
3	LR757995.1	(8, 3)	−2.660	2.038	0.035	0.587
4	LR757996.1	(8, 4)	−2.667	2.043	0.032	0.591
5	LR757997.1	(8, 5)	−2.546	2.466	−0.093	0.173
6	LR757998.1	(8, 6)	−2.682	2.048	0.038	0.597
7	MN970003.1	(8, 7)	0.000	0.000	0.000	0.000
8	MN970004.1	(8, 8)	0.000	0.000	0.000	0.000

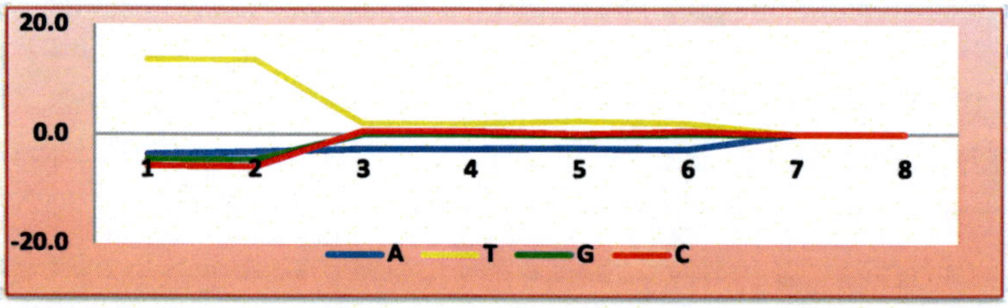

Fig. 10 Graphical representation of % concentration with respect to Genome ID: MN970004.1

5 Conclusions

The eight SARS–corona virus data taken from NCBI website for analysis arrive at certain findings. The findings show that the percentage nucleotide concentration varies within the tolerable limits. The differences or the changes in nucleotide concentration exhibit a relationship between the viruses. A positive value depicts an increase in concentration from Genome X to Genome Y. A negative value depicts a decrease in the concentration from Genome X to Genome Y. It is identified that 1 and 2 virus genomes form one class; 3, 4, 6 virus genomes form one class; 7 and 8 virus genomes for one class and virus genome no: 5 forms another separate class. Basically, they all are the same and they evolve from one generic form. There is a significant structural relationship between genomes which is identified in this analysis. Such relationships can be further understood when spectral-domain characterization is carried out for these eight genomes.

References

1. https://genomebiology.biomedcentral.com/articles/10.1186/s13059-019-1724-1
2. https://en.wikipedia.org/wiki/DNA_sequencing
3. https://www.getsmarter.com/blog/market-trends/how-big-data-is-changing-genetic-research/
4. Rajan EG et al (2020) Pairwise spatial correlation of SARS-Corona viruses. Lond J Res Comput Sci Technol (London Journals Press) 20(1). ISSN: 2514-8648
5. https://www.forbes.com/sites/bernardmarr/2020/04/09/the-vital-role-of-big-data-in-the-fight-against-coronavirus/#2735b8853806

The Role of Law and Regulation on the Intersection of Bioprospecting and Bio Piracy

Samykya Mukku, M. Lalasa, K. Nagalakshmamma, G. Savitri, P. Sujata, and K. S. Shanthi Sree

Abstract "Bioprospecting" is the "exploration of biodiversity for commercially valuable genetic resources and bio chemicals". There is an abundance of variety in the biosphere which manifests through a plethora of species of flora, fauna, and microorganisms. Biodiversity maintains ecological balance and evolutionary processes and has several uses and applications in agriculture, health, and food industries. Bio piracy is the exploitative misappropriation of traditional knowledge by parties motivated by commercial interests and also includes the pursuit of creating new compounds from existing biological resources. There is an inequity that is inherent between the exploring entities, often backed by commercial actors, and the indigenous people who possess traditional knowledge. This inequity, fueled by modern patent laws, has led to the conception of "bio piracy". This paper examines the role of environmental law and regulation in shaping the application of emerging technologies in bioprospecting.

Keywords Bioprospecting · Bio piracy · Biotechnology · Patent · Intellectual property rights · Environmental law

---

S. Mukku
New Delhi, India

M. Lalasa
Department of Computer Science, Swetha Institute of Technology and Science, JNTU, Tirupati, AP, India

K. Nagalakshmamma (✉) · G. Savitri · P. Sujata · K. S. Shanthi Sree
Department of Biosciences and Sericulture, Sri Padmavati Mahila Vishwavidyalayam, Tirupati, AP, India
e-mail: nala2020@gmail.com

S. Jyothi et al. (eds.), *Proceedings of the 2nd International Conference on Computational and Bio Engineering*, Lecture Notes in Networks and Systems 215,
https://doi.org/10.1007/978-981-16-1941-0_14

1 Introduction

"Bioprospecting" is the "exploration of biodiversity for commercially valuable genetic resources and bio chemicals" [1]. There is an abundance of variety in the biosphere which manifests through a plethora of species of flora, fauna, and microorganisms. Each of these species possesses unique characteristics that create a unique value to them. Post-colonial advancements in science and technology have led to the exploration of indigenous biodiversity for the medicinal other commercial value attached to these species. Biodiversity maintains ecological balance and evolutionary processes and has several uses and applications in agriculture, health, and food industries [2].

Indigenous farming and tribal communities possess valuable knowledge that is inherited through generations which allows them to live sustainably and effectively use the biological and genetic natural resources that are present within their ecological environment. This knowledge equips them with valuable insight into the ecological processes and the ability to extract products from the local habitat for sustainable use. There is an inequity that is inherent between the exploring entities, often backed by commercial actors, and the indigenous people who possess traditional knowledge. This inequity, fueled by modern patent laws, has led to the conception of "bio piracy".

Bio piracy is the exploitative misappropriation of traditional knowledge by parties motivated by commercial interests and also includes the pursuit of creating new compounds from existing biological resources [3]. Bio piracy is the result of research teams appropriating biological resources and indigenous knowledge without the official sanction, by exercising their access to superior financial resources over marginalized people, often belonging to less affluent countries [3]. Bio piracy, although extensively rampant in drug-related research, is also present in other agricultural and industrial activities. Recent activity and cases suggest that bioprospecting has been rendered as a front for research organizations that engage in bio piracy.

Bioprospecting, in its ideal form that is unique from bio piracy, involves an ethically and morally correct process. Bioprospecting requires obtaining prior informed consent, agreements that allow access and benefit-sharing, and transfer agreements before initiating research. Post research, the gains arising from the study or exploration should be shared proportionately with the indigenous people for their contribution. Further, funds arising from commercial products developed through the research should be utilized for local conservation effects and the development of local infrastructure [2].

2 Colonial Origins of Bio Piracy

Although bio piracy within a country is not unheard of, it is a practice that is inherently colonially motivated. Historically, colonial times have witnessed the origins of bio piracy resources belonging to the colonized countries that were forcefully removed to sustain colonial trade interests. Commodities such as sugar, opium, rubber, and coffee have a colonial past of bio piracy [4].

An analysis of these colonial origins reflects the issues of ownership that are inherent in the practice of bio piracy. Law related to patents and trademarks has been extensively debated and lobbied for at international and domestic levels to facilitate easier access to ownership over biological and genetic property. Multinational corporations and transnational trade companies have been actively lobbying for the assignment of ownership to an individual as against a community of indigenous people [4].

3 Indigenous Methods and Bioprospecting

Bio piracy has the potential to impoverish the biological diversity of a nation. Indigenous communities that possess traditional knowledge tend to follow the inherited practices that are ecologically friendly and are designed to facilitate sustainable use of resources [5]. In contrast, corporations are profit-driven and engage in monoculture to maximize the yield. They also extensively use artificial and chemical fertilizers to accelerate crop production. Because of this extensive use of chemicals, the self-replenishing quality of the soil diminishes every season, rendering it uncultivable. Indigenous communities depend on the resources that are naturally occurring in the biosphere and engage in sustainable practices that ensure the ecosystem is not harmed [5] (Fig. 1).

Multinational corporations have been attempting to appropriate traditional knowledge by way of patenting. One such particular instance is the case of the *Indian Neem* [5]. In 1994, W. R. Grace and the United States Department of Agriculture patented a method for combating fungi affecting plants by using neem oil. The grant of the aforesaid patent was challenged on the basis that the medicinal and healing properties of the neem tree have been identified and employed by indigenous and even urban population in India for centuries, which renders the purported innovation attached to the patent nonexistent. Consequently, the European Patent Office revoked the patent. This revocation was further upheld in an appeal preferred by W. R. Grace. The neem case is only one amongst many instances of bio piracy through patenting, such as the cases of *haldi*, *karela*, *jamun*, and *basmati* rice [5].

4 Regulation of Bioprospecting

As is the case with most environmental concerns, the opposing considerations involved in the debate around bio piracy are trade against the protection of natural resources. Regulation of bioprospecting involves a myriad of provisions, declarations, legislations, and conventions under International Law and domestic law. It also

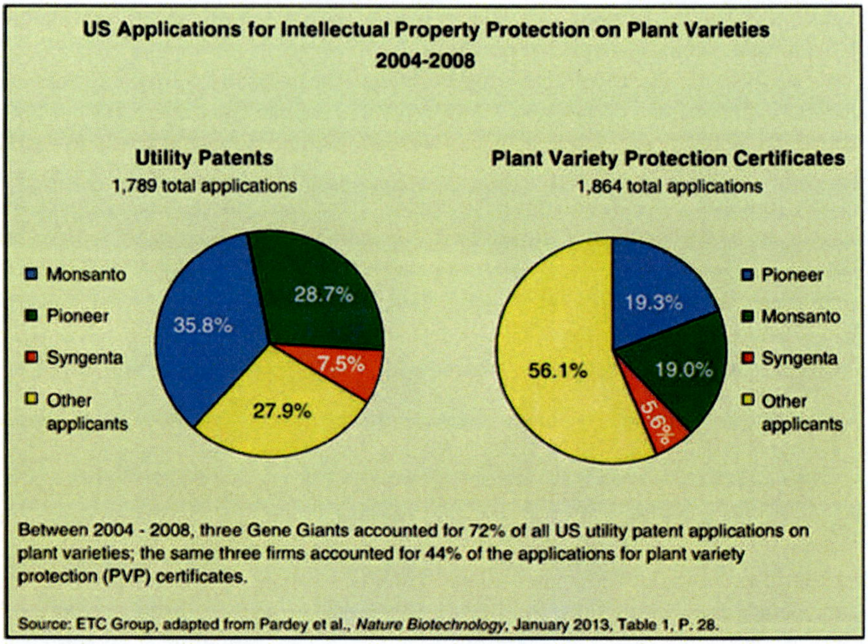

Fig. 1 Patent applications relating to plant varieties in the United States of America (*Source* ETC Group, adapted from Nature Biotechnology, January 2013)

involves various branches of law including intellectual property rights, environmental legislations, and trade-related laws.

Together, these provisions related to bio piracy collectively form confusing assimilation that often contradicts each other. Laws relating to IPR and international trade, which are commercially motivated, attempt to maximize benefit from natural resources. In contrast, environmentally inclined provisions seek to protect biodiversity and the rights of indigenous people. These conflicting aims often lead to outcomes that led to a conflict of interests (Fig. 2).

5 Global and National Efforts to Combat Bio Piracy

The "Convention on Biological Diversity" is a consolidation of environmental concerns and espouses the need to conserve biodiversity through principles such as sustainable development and intergenerational equity [6]. It recognizes the inherent inequity between developing and developed nations and highlights the need to ensure that the latter does not exploit the former. It also advocates for benefit sharing and informed consent. It reinforces state sovereignty over natural resources [6].

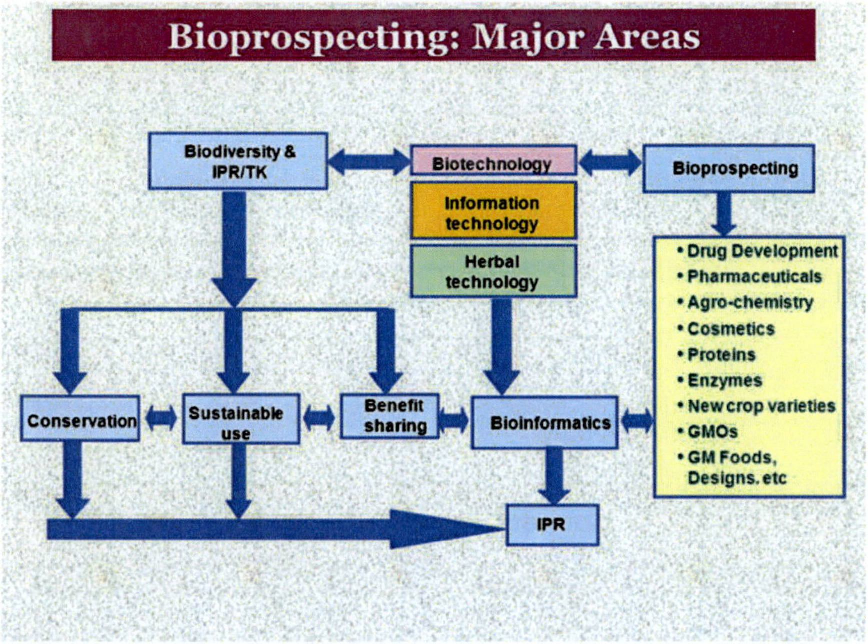

Fig. 2 Interface of bioprospecting, technology and intellectual property rights law (*Source* Issues in Bioprospecting: Lessons from the field)

The Nagoya Protocol of 2010 provides the means for enforcing the Convention on Biological Diversity. It extensively promotes the "Access and Benefit Sharing" system. It provides for better and more efficient regulation of access to genetic resources by compelling the states to form an agency to regulate the licenses for research. States are also required to establish a well-functioning mechanism to execute and enforce the ABS system [7].

The Protection of Plant Varieties and Farmers Rights Act aims to protect farmers' rights by recognizing the role of farmers and their contribution to the biodiversity of the country. The Act establishes a National Gene Fund to promote the sustainable use of agro biodiversity and plant varieties. It also establishes the Farmers' Right Authority as a regulator to protect plant varieties.

The Indian Patent Act, 1970 prohibits grant of patents over plants, traditional knowledge, or animals. Further, the Patent Act requires disclosure of the source and geographical origin of the underlying organic material used for an invention.

India ratified the Convention on Biodiversity by enacting "The Biological Diversity Act" in 2002. In consonance with the Convention, the Act governs the accessing, collecting, and utilizing biological and genetic resources and traditional knowledge. It also provides for sharing the benefits arising from the commercial use of such resources. The Act establishes the "National Biodiversity Authority" and mandates that the prior permission of the National Biodiversity Authority is a prerequisite in respect of intellectual property rights being claimed for any invention based on resources or knowledge originating in India. The National Biodiversity Authority is empowered to determine that a component of the revenue generated from the commercial use of such intellectual property rights is shared with the indigenous community.

Indian Government also launched the Traditional Knowledge Digital Library (“**TKDL**”) in order to catalog traditional knowledge and protect it from bio piracy. The TKDL forms a database that catalogs indigenous traditional knowledge to make it available to the public at large. This database aids patent examiners at both national and global levels to check against patent applications.

6 Conclusion

The prevalent measures and legislation enacted by the Government of India make it evident that India is on the right path to protecting its biological and genetic resources from bio piracy. It is crucial that access is regulated and benefit-sharing is strictly enforced in order to ensure that the biological resources of the nation are not depleted. Commercialization and commodification of biodiversity through bio piracy, masquerading as bioprospecting will inevitably impoverish the biological diversity of a nation. It is paramount that combating such incidences should be given the highest priority.

References

1. Reid WV, Laird SA, Gámez R, Sittenfeld A, Janzen DH, Gollin MA, Juma C (1993) A new lease on life. In: Biodiversity prospecting: using genetic resources for sustainable development
2. Shiva V (2007) Bioprospecting as sophisticated bio piracy. Signs 32(2):307–313 (Winter 2007)
3. Cluis C (2013) Bioprospecting: a new western blockbuster, after the gold rush, the gene rush. Sci Creat Q (Univ B C)
4. Shand H (2000) BIOPIRACY, BIODIVERSITY and PEOPLE: the right to say “no” to monopoly patents exploiting the South’s resources and knowledge. Race Poverty Environ 7(2):13–14 (a PLACE at the TABLE: Food and Environmental Justice (WINTER 2000))
5. Shuler P (2004) Biopiracy and commercialization of ethnobotanical knowledge. In: Michael Finger J, Schuler P (eds) Poor people’s knowledge: promoting intellectual property in developing countries, pp 159–181. World Bank and Oxford University Press, Washington, DC
6. Convention on Biological Diversity (1992), https://www.cbd.int/youth/0003.shtml
7. Nagoya Protocol (2010), https://www.cbd.int/abs/

Robust Face Recognition Using Multi-scale Feature Pattern Sparse Representation with Extreme Learning Machine

Pallavaram Venkateswar Lal, Uppalapati Srilakshmi, and D. Venkateswarlu

Abstract Robust vision-assisted facial recognition is the top stimulating job in automation machines. Most of the face recognition techniques introduced till now considered sparse coding as an effective method, whereas the sparse coding prototypes are being hindered through the tremendously expensive evaluation cost in terms of execution. However, these sparse coding models could tolerate robust face recognition having high dimensionality. Therefore, it is essential to develop new face recognition approaches with reduced complexity and less computational time. Hence, novel robust face recognition approaches are presented in this paper. The recognition of robust facial images is performed through ELM classifier which is priory preprocessed by extracting the multi-scale sparse coding feature patterns from the sample images. Then finally the feature patterns dictionaries are obtained for every sample image that is employed with the primary function to work with the SLFN neural network. The experimental outcomes revealed that the suggested methodology has higher recognition rates under all three circumstances compared to other algorithms such as SRC, RSC, and MPRSC-ELM classification algorithms and has less recognition time in seconds.

Keywords Robust face recognition · Sparse code representation · Extreme learning machine · Classification · SLFN · Occluded facial images

1 Introduction

Intellectual science is the interdisciplinary knowledge of cognizance and intellect, artificial intelligence, psychology, philosophy, neuroscience, semantics, anthropology, and brain principles [1]. From the previous years, there are numerous intellectually motivated evaluations for image processing [2–4].

P. V. Lal (✉)
VFSTR Deemed to be University, Guntur, India

U. Srilakshmi · D. Venkateswarlu
Department of CSE, VFSTR Deemed to be University, Vadlamudi, Guntur, Andhra Pradesh, India

S. Jyothi et al. (eds.), *Proceedings of the 2nd International Conference on Computational and Bio Engineering*, Lecture Notes in Networks and Systems 215,
https://doi.org/10.1007/978-981-16-1941-0_15

Facial detection stands amongst the whole exciting investigation area in the realms of cognition computation field, computer vision, and pattern recognition. The foremost causes comprise extremely overlying intra- and inter-individual disseminations pertaining to the posture, age, communication, obstruction, and exterior imaging features like disparities in brightness. Considering face recognition as one of the utmost popular techniques in biometrics, it could be employed in collective robotics to accomplish the individual's recognition task in a usual and non-interaction approach. Face recognition applications can also prospect in public security, digital authentication, video monitoring, and other fields, and after a decade of development, the research on face recognition has made numerous successes.

In general, the traditional FR methods have the following difficulties. Firstly, pertaining to the vast dimensions of gallery images, the processes, like brute force nearest neighbor, do not scale good enough. Secondly, whenever new pictures are added to the gallery, the model in the gallery needs to be re-trained. Hence the consequences could increase the computational complexity of the methods. In recent times, sparse coding [5] and deep learning [6–8] are being extensively functioned on face recognition approaches. Even though deep learning has exhibited favorable precisions, it further has certain drawbacks like the need for vast quantities of training examples and explicitly higher estimating equipment, deficiency of robust hypothetical investigation, and precise prototypes for facial recognition having numerous occlusions. A simplified methodology employing sparse coding to stable facial detection is sparse representation-aided classifier (SRC) [5] that has given a promising outcome in face recognition with obstruction.

However, most of the previous sparse coding depictions-based face recognition approaches are more usual to work with strong recognition and to tolerate distributed image pixels such as occluded pixels specifically for the high dimensional sample images. Furthermore, in numerous real-world experiments, obstruction in the experimental instances is in its local space. These local occlusions prompt noises in the training vocabulary along with the sparse representation of the innovative image for face recognition that consumes higher evaluation time and minimized complexity. To address these issues, this paper suggested a new robust face recognition approach along with the extreme learning machine (ELM) that has fascinated growing concentration in machine learning through more oversimplification improvement at a much quicker speed. Before applying the classification model, the sample images are preprocessed by identifying the un-occluded pixels and partitioning those blocks with different scale variants. Then finally the feature patterns dictionaries are obtained for every sample image.

A brief introduction of robust face recognition and its application in various domains, along with the motivation for the suggested approach, is given in this section. The existing methods of the sparse coding representation-based classification approach and extreme learning machine-based face recognition techniques are provided in Sect. 2. A brief explanation of the given robust face recognition algorithm through multi-scale feature pattern dictionaries is given in Sect. 3. The experimental outcome and its exploration are specified in Sect. 4 followed by the conclusions and references specified in Sects. 5 and 6, respectively.

2 Literature Survey

Face recognition (FR) is being tremendously investigated from the previous two decades [9] and several representative methodologies, like eigenfaces [10], Fisher-faces [11], LBP [12] are suggested. A comprehensive Weber-face [13] for radiance-based robust facial recognition representation (CESR) for facial string recognition is given in [14], which presented a Gaussian kernel-based dependability method to normalize coding residuals, and Gabor feature is presented in traditional SRC to improve its differentiation [15]. Formerly, a global half-quadratic standard that scientifically amalgamates error correction and detection was suggested in [16], which is efficient for robust facial recognition. To work with further natural facial occlusion, in [17] regularized robust coding (RRC) approach is modeled through planning a strong representation term.

The other instance is the robust sparse coding (RSC) approach presented in [18] where sparse coding is modeled like sparsity controlled healthy regression issue. RSC was presented to outstrip general SRC and GSRC, along with being further efficient in the functioning of facial occlusions. Thus, RSC is computationally costly whenever associated with numerous other SRC methodologies. In [19] the advantage of a systematic vocabulary is explored, where every element is accompanied by a class label. With the help of Fisher discrimination strategy [20], a group of class-predefined sub-vocabularies is studied, where every type has a smaller interior group scatter and a huge in between-group scatter.

The performance analysis of different facial recognition approaches, such as principal component analysis (PCA) and linear discriminant analysis (LDA), is suggested in [21]. This approach was accepted on several present PCA and LDA-aided face recognition approaches by means of standard public data samples. A methodology that employed multiple dimensional scaling to change characteristics from reduced qualitative inquiry images to the more qualitative gallery images instantaneously was proposed in [22]. A computer system [23] was developed that recognizes an individual by matching the features of the face to those of recognized individuals in the storage data samples. In [24] significant phases of face recognition are defined, which are very beneficial in numerous appliances such as facial recognition, facial appearance recognition, face tracking, facial feature mining, gender categorization, credential system, and several other applications. The face recognition approach of colored face images was presented in [25], which is different from the background and adequate illumination circumstances. A threshold level is defined to discard the non-human facial images and the unidentified human facial images that do not exist in the input data samples.

In ELM [26], whenever the input weights and the unknown layer prejudices are arbitrarily allotted, the output weights could be evaluated through widespread converse of hidden layer output matrix [27]. There are boundless several ELM dissimilarities that are suggested, comprising the arbitrary hidden layer feature mapping-aided ELM [28], the kernel-aided ELM [29], the completely multi-faceted ELM [30], the incremental ELM (IELM) [31], the online sequential ELM (OS-ELM) [32,

33], the pruning ELM (P-ELM, OP-ELM) [34, 35], the circular-ELM (C-ELM) [36], ELM groups [37], etc., which tends to the existing outputs in numerous appliances for the regression and pattern recognition issues [38–40].

3 Proposed Multi-scale Feature Pattern-Based ELM

In general, various algorithms showed efficient performances in face recognition. Nevertheless, these algorithms do not continuously attain acceptable outcomes due to the difficulties of real-world facial occlusion. Hence, this new robust face recognition algorithm is advisable in this area which is implemented in three stages.

- Extraction of multi-scale feature patterns
- ELM learning
- ELM classification

a. The Building of Multi-Scale Feature Pattern Vocabulary

The feature patterns on the images characterize the organized at a such as spatial consistency and previous understanding, such as the unchanging location of facial disguise of the images in addition to the feature weights so as to fully exploit the facial images. As facial occlusion can have many looks, feature pattern is obtained for current information official occlusion excluding the look. It is considered that the structured data and previous information are inserted into the pixels themselves but, on the other hand, not the intensities of the image. The image pixels in a similar occlusion region can have fairly dissimilar values that exhibit tiny organized data and previous knowledge. Nevertheless, the feature pattern has fairly three-dimensional consistency and is likewise unchanged for some occlusion. In further usual conditions of real-world face detection, facial occlusion typically has specific preceding knowledge, like they conquer local sections of facial portions, and facial disguises had constant positions even though these masks can have random values to theatrically alter the presence of facial images. Despite the fact that these facial occlusions have different appearances, the feature patterns could be designed without difficulties, i.e., the pattern value is set to 1 for un-occluded pixels and 0 for occluded pixels.

In addition to the training data sample, an added group of occluded face images is gathered through a facial disguise. This disguise frequently has a relatively constant location and limited classes. For example, sunglasses, headscarves, cap, and breathing masks are general facial disguises, and their positions are likewise comparatively constant like suspects would shield the area of eyes and breathing mask would conceal nose and mouth. The obstructed feature pattern is mined physically or through employing image segmenting devices depending on occluded facial images. For these feature patterns, the complete image is merely divided into numerous chunks with pixel single chunks values as 1 s in every feature pattern. If the whole image is segregated into $1 \times 1, 2 \times 2, 3 \times 3, 4 \times 4, 6 \times 6 and 8 \times 8$ chunks, the feature pattern vocabulary would have 130 patterns. Symbolize the feature pattern of the queried image having a weight vector (i.e., w). For a facial image, w designates the support of every pixel to query image illustration on predetermined known vocabulary.

One of the crucial issues emerging when analyzing these images begins from the way that face structures show up in various manners, depending upon the scales of observations. To begin with, local facial structures are shown by varying the level of scales, extending from textures of skin at fine scales, through mouth and eyes represented at median scales, to the shape of face contours at large scales. Then in the second step, the characteristic of a local structure is dependent on the scale at which the structure is modeled. Finally, it is still not known in the early stage what the proper scales are to describe various local structures of strange face images. In order to overcome these difficulties, a tool that incorporates the notion of scale is required and crucial. In multi-scale local features are extracted from a huge set of multi-scale image patches that fetch great generalization ability for face recognition. Nevertheless, a multi-scale representation by itself contains no data regarding which image structures are important and what scales are best suitable to explain them. Therefore, for automatic scale selection, it is always primary to complement a multi-scale process by precise mechanisms.

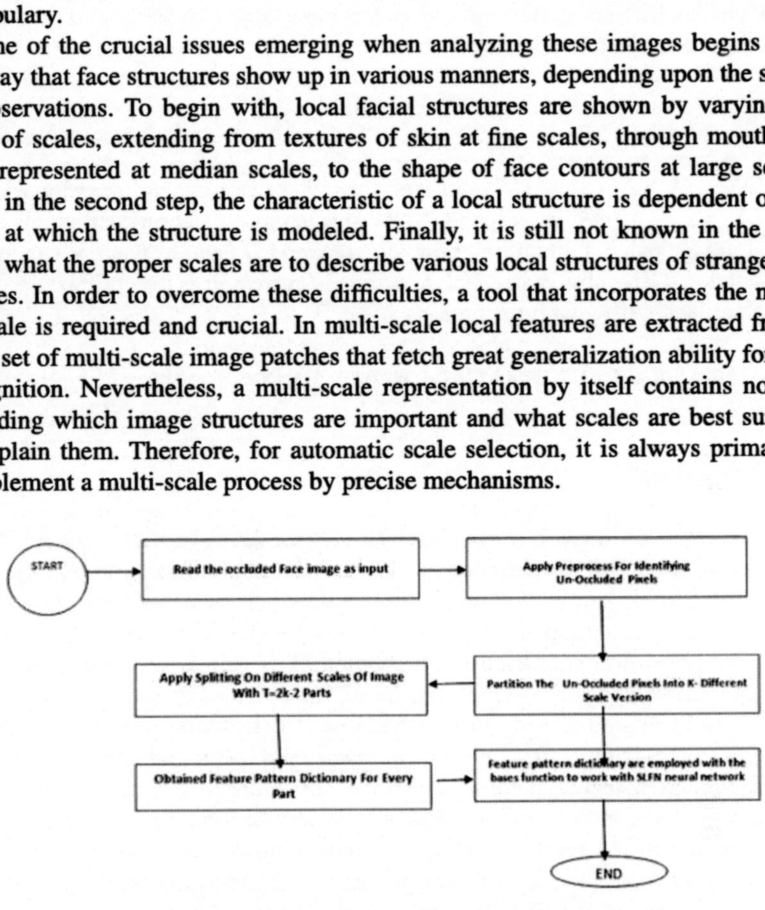

All the sample training and training images with the pattern values of 1 s with the un-occluded pixels partitioned into K different scale version using the square roots such as $\left\{1, \frac{1}{2}, \ldots, \frac{1}{2^K}\right\}$ of the original face image. Then images with the different scales are split to $2K, 2k-1, \ldots, 4, 2$ portions correspondingly. Thus, there are $T = 2K - 2$ parts for a single facial image. After division, the feature pattern dictionary is constructed for every part. The matrix of collected feature patterns $F = \left[F_1^t, F_2^t, F_3^t, \ldots, F_K^t\right] = \left[p_{1,1}^t, p_{1,2}^t, p_{1,3}^t, \ldots, p_{k,n}^t\right]$, for $t = 1, 2, \ldots, T$ signifies the vocabulary that comprises the entire tth portion from the complete training instances, whereas $p_{i,jn}^t$ signifies the tth portion from the jth instance connected with the ith class.

b. **Extreme Learning Machine**

Olshausen and Field [27, 41] first coined sparse coding which helps in finding the sparse linear representation of a specific image to an overcomplete dictionary and supply much efficient sparse coding like the primary visual cortex in the human visual system. The standard model of the sparse coding can be termed as a sparse linear superposition of the basic functions

$$X = DA$$

where $X = [x1, x\,2, \ldots, x^d]^1 \varepsilon R^d$ represents the image patch in a d-dimensional space, $A = [a1, a2, \ldots, a^t]^1 \varepsilon R^t$ corresponds to the sparse representation coefficients of the original image patch X in at-dimensional space, and D denotes the $d \times t$ matrix of base function. Generally, sparse feature coding is expressed as the optimization issue through reducing the cost function E as

$$\min_{A,w,\beta} \|diag(w)(X - DA)\|_2^2 + \lambda\|A\|_{l_p} + \gamma\left(\|w - F\beta\|_2^2 + k\|\beta\|_2^2\right),$$

show that $\sum_j \beta_j = 1$

where the sparse coding will be obtained by $\|A\|_{l_p}$; s:t: $\|diag(w)(X - DA)\|_2^2$, the L2 norm constraint (*X-DA*) on the columns of D can take out the scaling ambiguity, and λ is a positive Lagrange multiplier that influences the significances of the second term to the first. Sparse coding can be solved effectively by imposing L2 norm regularization on the representation coefficients. The bottleneck is primarily the computational speed as sparse coding should encode a broad collection of image patches.

Before officially mimicking the base function D in the sparse coding of worldwide images, preprocessing is first finished by means of constructing multi-scale vocabularies. Numerous feature patterns are mined from arbitrarily composed obstructed images after being attuned through single nonlinear conversion, and the obstructed image patches X and fundamental sparse illustration A would be as simulated in the reiteration procedure at a similar period. The entrenching base function D is implemented through the ELM procedure in the learning method giving the interest to make an attempt to attain the sparse coding vectors A from the image patches X for the face images once the SLFN for simulation is recognized. Then the scattered coding representation S of every image will be given to the next SLFN in the test process. The decision $k*$ will be made in an ensemble, by the output Pk and the weight Xk for the individual SLFN of the kth person. The base function D would be reset using the Gaussian arbitrary variables. The cost function of the sparse feature coding considers the total of every coefficient's action to encounter the principle and selection of nonlinear function among activity circumstances having the least number of non-zero constants. The novel base function D would formally come into existence from the conforming sparse feature representation A for the subsequent generation

$$D = D + \frac{\eta}{N}\sum_{i}^{N}(x_i - Da_i)a_i'$$

where η signifies the learning rate, and a_i is the utmost feasible fundamental indication specified in the image patch observation x_i and the existing base function. Subsequently, to every update, the values are standardized to unit distance to accelerate the learning procedure.

Whenever the number of the iteration steps equals the input that we have allocated before the final basis function D, in the end, the image patches $X0$, and the respective sparse representation $A0$ will be obtained from the natural images, where $X0 = X$ is a collection of the N natural image patches and $A0 = A$ is the sparse representation of all the extracted image patches in the final iteration step.

During the learning process, the base function would be obtained from usual images for feature pattern mining. Instantaneously, after considering a group of the repetitive phrases, the newest obstructed image patches X and the consistent sparse coding representations A would likewise be attained that could be represented like the obstructed image package $P = \{X, A\}$. Presume that there are N random dissimilar training examples $P_0 = \{x_{0,i}, a_{0,i}\}_{i=1}^{N}$ with the input $x_{0,i}$ and the expected output $a_{0,i}$. The occluded image patches X is taken as the inputs for single hidden layer feed forward network (SLFN) to forecast the sparse feature coding constants of the facial image patches with fundamental general feature policy accompanied by ELM as:

$$H\beta = A$$

$$H(\omega_1, \ldots, \omega_L, b_1, \ldots, b_L, x_{0,1}, \ldots, x_{0,N})$$

$$= \begin{bmatrix} g(\omega_1 \times x_{0,1} + b_L) & \cdots & g(\omega_L \times x_{0,1} + b_L \\ \vdots & g(\omega_i \times x_{0,j} + b_i) & \vdots \\ g(\omega_1 \times x_{0,N} + b_1) & \cdots & g(\omega_L \times x_{0,N} + b_N) \end{bmatrix}_{N \times L}$$

$$\beta = [\beta_1, \beta_2, \ldots, \beta_L]_{p^2 \times L}, A = [a_{0,1}, \ldots, a_{0,N}]_{p^2 \times N}$$

Here each input $x_{0,j}$ comprised obstructed image patch and every predicted outcome is the sparse feature coding illustration $a_{0,j}$. The SLFN is developed and set to study the sparse feature coding procedure from the images. Whenever the SLFN is given with the facial image patches X, the real outcome would be deliberated as the approximation of consistent sparse feature pattern coding constants:

$$a_j = \sum_{i=1}^{L}\beta_i g(\omega_i \times x_j + b_i), j = 1, 2, \ldots, N'$$

The model choice of ELM design for face detection issues can be estimated through the simplification error as

$$E = \lim_{N' \to \infty} \sum_{j=1}^{N'} \left(I\left(x_j\right) - a_j\right)^2 / N',$$

$$j = 1, 2, \ldots, N'$$

Here, I is the input–output function of the ELM learning, x_j is facial image patch, $I\left(x_j\right) = \widehat{a_j}$ is the definite cross-verification that can be a good selection for the approach in SLFN that is essentially a distinct instance of k folding cross-verification in the situation where $k = N'$. The whole facial image patches are separated into N' portions for training samples. In everyone, there is a correct single instance that has remained for analysis, and the former approximation of the simplification error attains.

$$E = \sum_{j=1}^{N'} \left(I\left(x_j, -j\right) - a_j\right)^2 / N',$$

$$j = 1, 2, \ldots, N'$$

where $I\left(x_j, -j\right)$ signifies the outcome of the jth training group deprived of the jth sample.

c. **ELM-Based Classification**

Depending on the previous information attained from the ELM learning, the classification of the test image is performed by the decomposition of the occluded facial image through multi-scale feature patterns. These are fed to the subsequent SLFN to accomplish face detection by the ELM classification approach. The number of entire facial images are $Q^1 = K \times Q$ the dimension of every image is $M^1 \times M^1$; the trained group comprised Q_1^1 categorized images from the identified individuals; and the remaining Q_2^1 uncategorized face images comprise the test group.

$$Q_1^1 + Q_2^1 = Q^1$$

$$H\beta = P$$

$$H\left(\omega_1, \ldots, \omega_L, b_1, \ldots, b_L, SF_1, \ldots, SF_{Q_1'}\right) =$$

$$\begin{bmatrix} g(\omega_1 \times SF_1 + b_1) & \cdots & g(\omega_L \times SF_1 + b_L \\ \vdots & g(\omega_i \times SF_j + b_i) & \vdots \\ g\left(\omega_1 \times SF_{Q_1'} + b_1\right) & \cdots & g\left(\omega_L \times SF_{Q_1'} + b_L\right) \end{bmatrix}_{Q_1' \times L}$$

$$\beta = [\beta_1, \beta_2, \ldots, \beta_L]'_{p^2 \times L},$$

$$P = [P_1, \ldots, P_k, \ldots, P_K]'_{K \times Q_1'}$$

where SF_j signifies the sparse coding illustration that comprised the constants $\left\{a_1, a_2, \ldots, a_{N'/Q'}\right\}$ for image patches resulting from the specified jth face image in the training group. The detection approach is to identify the relationship of facial image to be detected through plotting pertaining to ELM learning:

$$P_j = \sum_{i=1}^{L} \beta_i g\left(\omega_i \times SF_j + b_i\right), j = 1, 2, \ldots, Q_2'$$

where SF_j resembles sparse coding illustration of the facial image in test group, and $P_j = \left[P_{j1}, \ldots, P_{jk}, \ldots, P_{jK}\right]'$ is the outcome of the SLFN from the input SFj. ELM ensemble can be set up further here to better generalization performance, with each person an individual SLFN. The important thing here is to make more variations in the recognition process for the same task which when united will better the interpretation and give more stabilized predictions about the face images. Hansen and Salamon already proved this, which open the gates for stable prediction.

The learning procedure in the ELM group is to consider a linear merging policy of every individual outcome. The assessment would be done by $K^* = \text{argmax}_{k=1}^{K} \omega_k P_k$ where K is the number of individuals, P_k is the outcome of kth SLFN, and ω_k resembles significant weight, $0 < \omega_k < 1$. In this situation, the outcome of every individual could be trimmed to $P_k \in R$ for generalization.

The ELM classification prevents problems like local minima, indecent learning rate, and overfitting that are usually faced by iterative learning methods, and finishes training quickly. The evaluation results on multi-scale classification performance of ELM on two data sets (different), namely the Extended Yale B data set and the AR data set are as follows. The performance analysis for the proposed approach is performed on Extended Yale B and AR facial data sets under varying percentages of random pixel occlusion, for varying levels of random block occlusion and real objects disguise.

The experimental outcomes revealed that the suggested approach has higher recognition rates under all three circumstances compared to other algorithms such as SRC, RSC, and MPRSC-ELM classification algorithms and has less recognition time in seconds on these data sets.

We introduced proficient sparse coding here to use on the natural image patch vectors [31]. When the standard generative model assumes that $X - DA$ is distributed as a zero-mean Gaussian distribution, the algorithm is as follows:

Algorithm 1 Basic sparse coding

1. Input all the image patches and set the parameter k.
2. Initialize the basis function matrix D, the sparse coding matrix A with a Gaussian random matrix
3. For $t = 1{:}\, T$

Random select N image patches $X = (x1, x2, x3, \ldots, xn)$ from all image patches

For $t = 1{:}\, N$

Apply the feature-sign search algorithm

Compute the sparse coding coefficient

$$a_i = \arg\min \|x_i - Da_i\|_2^2 + \lambda \|a_i\|_1$$

for each image patch xi

End

Update the basis function matrix $DT = (a_i a_i T + A) - 1\,(x_i\, a_i T)\, T$ where $A = \mathrm{diag}(\lambda)$

Get the sparse coding matrix A ¼ ð$a1$; $a2$; …; aNÞ

End

4. Set $X0 = X$, $A0 = A$, D as the outputs

4 Experimental Results and Analysis

The experimental outcomes for the suggested robust face recognition approach are performed with the help of two different benchmark image data samples. They are Extended Yale B database and AR database. The multi-scale feature is extracted from the data samples by means of pattern representation approach. The suggested methodology is matched with the prevailing face recognition approaches, such as sparse representation-aided classifier (SRC) [5], robust sparse representation-aided classifier (RSC) [19], and improved sparse representation coding classifier (MRCS-ELM) [42]. The proposed approach is an experiment performed using the MATLAB 16a version. The robustness of the proposed hybrid face recognition classifier is authenticated through three different methods, like arbitrary pixel exploitation, arbitrary block obstruction, and actual face disguise through two objects, i.e. sunglasses and scarf.

4.1 Data Set Description

Extended Yale B database: The data samples comprise 5760 solitary one light source images of 10 issues, each perceived underneath 576 observing situations with 9 postures and 64 illumination circumstances. For each topic in a specific posture, an image with ambient illumination was seized. The data set is aimlessly fragmented into two splits where one portion is used for training the classification technique and the other portion for testing.

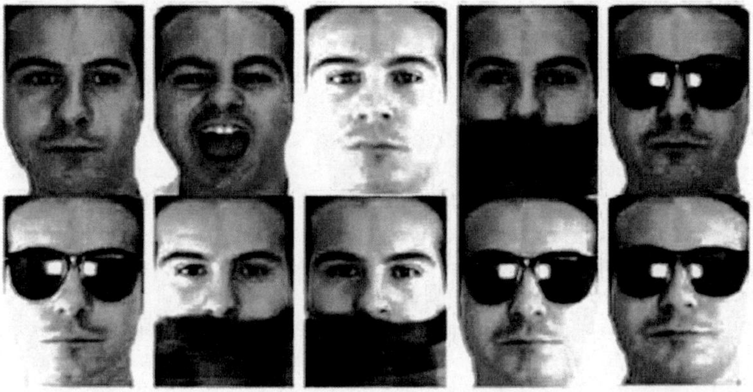

This data set (AR face database) was made by Robert Benavente and Aleix Martinez in CVC at U.A.B. It has more than 4,000 color images with 126 individual faces (56 female and 70 male). Images have various facial expressions, illumination circumstances, and obstructions. The database is arbitrarily fragmented into two splits, where one portion is employed for training the classification approach and another portion for testing.

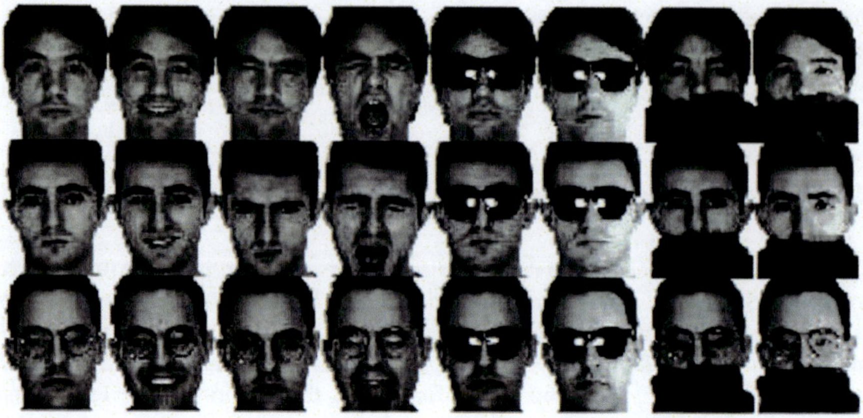

4.2 Result Analysis

Face Detection using Arbitrary Pixel Occlusion: For the testing purpose, specific percentages of pixels are replaced by evenly distributing arbitrary values between [0, 255]. The degraded pixels are aimlessly selected for every test image and positions are unidentified to the procedure. The images are sized to 96 × 84 pixels. The analysis is performed on different percentages varying from 0 to 90% of the random pixel occlusions on the two different databases.

From Fig. 1, it can be clearly inferred that the proposed multi-scale feature pattern-based extreme learning machine algorithm (MSFP-ELM) has a higher recognition rate when compared with the existing classification algorithms for the different percentages levels of pixels values occlusions. It can be seen that 10, 30, and 50%

Fig. 1 Recognition rates under different percentages of random pixel occlusion

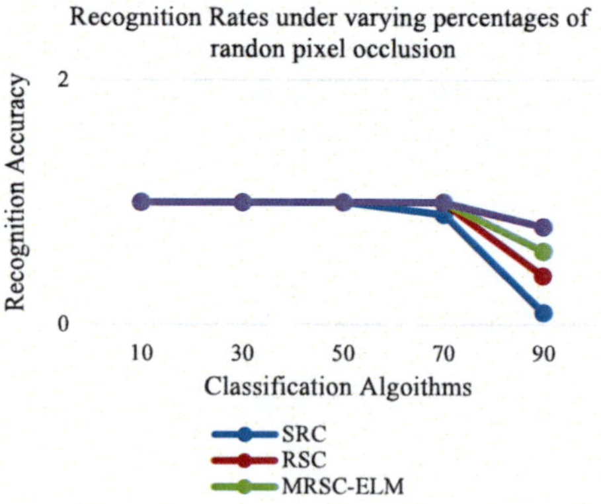

Fig. 2 Recognition rates under varying levels of random block occlusion

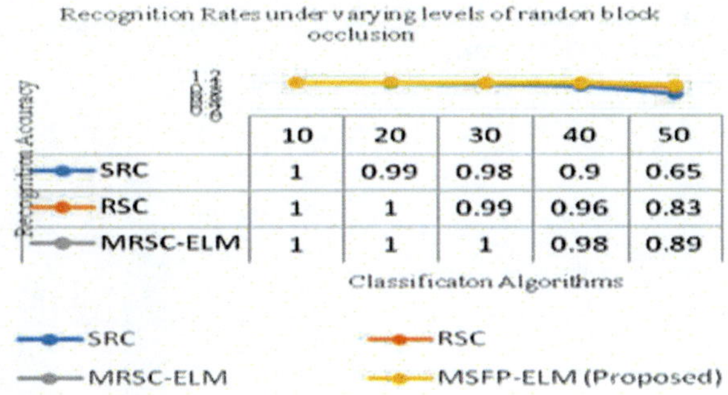

of occluded pixel values have identical recognition rates, whereas from 70 to 90%, it gradually rises for the four different classification algorithms.

Face detection with arbitrary block occlusion: The robustness of the suggested approach is tested for the data sets against the block occlusions. Similar to random pixel occlusion, some of the blocks are occluded and examined where the blocks of size 8×8 are considered randomly and replaced by the values of either 1 or 0. The images were resized to 96×84 pixels. The analysis is performed at different levels varying from 10 to 50% of blocks occluded on two different data sets.

From Fig. 2, it can be clearly inferred that the proposed multi-scale feature pattern-based extreme learning machine algorithm (MSFP-ELM) has a higher recognition rate when compared with the existing classification algorithms for the different levels of random block occlusions. For 10, 20, and 30% of occluded blocks identical recognition rates can be observed, whereas from 30 to 5%, it gradually rises for the four different classification algorithms.

Face Recognition with Real Disguise: Instead of pixel or block occlusion, in this case, particular real objects are employed to occlude the facial images such as sunglasses and scarf. A subset of AR face data sets consisting of sunglasses and scarf is considered for experimentation. The database comprises 600 images (6 non-obstructed front view examples per group, 3 from Session 1 and 3 from Session 2) from 100 subjects (50 males and 50 females) for training and 200 images (2 examples per group, with sunglasses or scarf) from 100 subjects for analysis. Figure 3 exhibits six training instances having facial expression variations and two testing instances having unbiased expression from the initial issues in AR data set. Figures 4 and 5

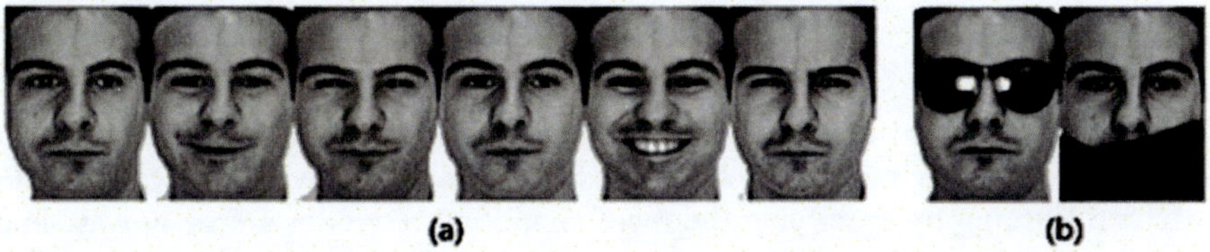

Fig. 3 **a** Training samples with facial expressions. **b** Testing samples with sunglasses and scarf

Fig. 4 Recognition rates of real disguise facial images

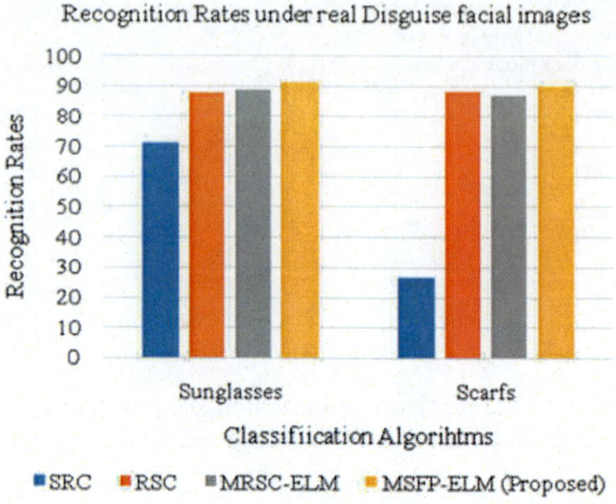

Fig. 5 Average recognition time of real disguise facial images

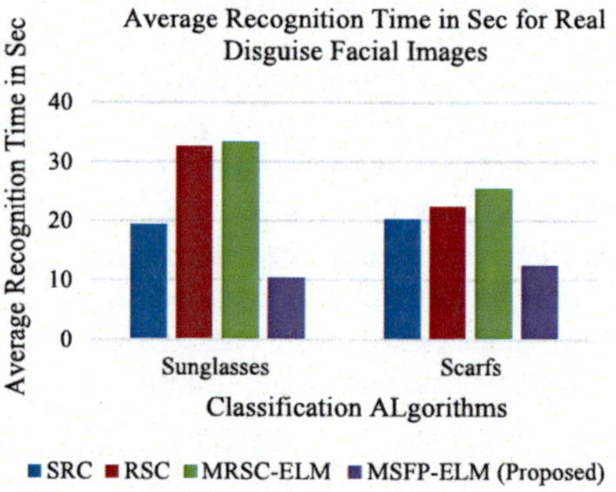

represent the detection rates and average recognition times in seconds of real disguise objects like sunglasses and scarf. It could be inferred that the detection rate is high for the suggested approach when compared to the other approaches where the average recognition time is less for the proposed approach when compared with the other three classification approaches.

5 Conclusions

This paper suggested a novel robust facial recognition methodology along with the extreme learning machine (ELM) that has fascinated growing concentration in machine learning by specifying the more simplified results in a speedy way than machine learning. Before applying the classification model, the sample images are preprocessed by identifying the un-occluded pixels and partitioning those blocks with different scale variants. Then, finally the feature patterns dictionaries are obtained for every sample image that is employed with the basic function to work with the SLFN neural network. The performance analysis for the proposed approach is performed on Extended Yale B, and AR facial data sets under varying percentages of random pixel occlusion, varying levels of random block occlusion, and for real objects disguises. The experimental outcomes revealed that the suggested approach has higher recognition rates under all three circumstances compared to other algorithms such as SRC, RSC, and MPRSC-ELM classification algorithms and has less recognition time in seconds.

References

1. Clark A (2001) Mindware: an introduction to the philosophy of cognitive science. Oxford University Press, New York
2. Underwood G (2009) Cognitive processes in eye guidance: algorithms for attention in image processing. Cogn Comput 1:64–76
3. Cambria E, Hussain A (2012) Sentic album: content-, concept, and context-based online personal photo management system. Cogn Comput 4(4):477–496
4. Nian R, Ji GR, Zhao WC, Feng C (2007) Probabilistic 3D object recognition from 2D invariant view sequence based on similarity. Neurocomputing 70(6):785–793
5. Wright J, Yang AY, Ganesh A, Sastry SS, Ma Y (2009) Robust face recognition via sparse representation. IEEE Trans Pattern Anal Mach Intell 31(2):210–227
6. Sun Y, Chen Y, Wang X, Tang X (2014) Deep learning face representation by joint identification-verification. In: Proceedings of NIPS
7. Zhu Z, Luo P, Wang X, Tang X (2014) Multi-view perceptron: a deep model for learning face identity and view representations. In: Proceedings of NIPS
8. Kan N, Shan SG, Chang H, Chen XL (2014) Stacked progressive auto-encoder (SPAE) for face recognition across poses. In: Proceedings of CVPR
9. Zhao W, Chellappa R, Phillips PJ, Rosenfeld A (2003) Face recognition: a literature survey. ACM Comput Surv 35(4):399–458
10. Turk M, Pentland A (1991) Eigenfaces for recognition. J Cogn Neurosci 3(1):71–86
11. Belhumeur PN, Hespanha JP, Kriengman DJ (1997) Eigenfaces vs. Fisherfaces: recognition using class specific linear projection. IEEE Trans Pattern Anal Mach Intell 19(7):711–720
12. Ahonen T, Hadid A, Pietikainen M (2006) Face description with local binary patterns: application to face recognition. IEEE Trans Pattern Anal Mach Intell 28(12):2037–2041
13. Wu Y, Jiang YY, Zhou YC, Li WF, Lu ZQ, Liao QM (2014) Generalized Weber-face for illumination-robust face recognition. Neurocomputing 136:262–267
14. He R, Zheng W, Hu B (2011) Maximum correntropy criterion for robust face recognition. IEEE Trans Pattern Anal Mach Intell 33(8):1561–1576
15. Yang M, Zhang L (2010) Gabor feature based sparse representation for face recognition with Gabor occlusion dictionary. In: Proceedings of ECCV
16. Hua G, Yang M-H, Learned-Miller E, Ma Y, Turk M, Kriegman DJ, Huang TS (2011) Introduction to the special section on real-world face recognition. IEEE Trans Pattern Anal Mach Intell 33(10):1921–1924
17. Yang M, Zhang L, Yang J, Zhang D (2013) Regularized robust coding for face recognition. IEEE Trans Image Process 22(5):1753–1766
18. Yang M, Zhang L, Yang J, Zhang D (2011) Robust sparse coding for face recognition. In: IEEE conference on computer vision and pattern recognition, pp 625–632
19. Yang M, Zhang L, Feng X, Zhang D (2011) Fisher discrimination dictionary learning for sparse representation. In: IEEE international conference on computer vision, pp 543–550

20. Fisher RA (1936) The use of multiple measurements in taxonomic problems. Ann Eugen 7:179–188
21. Fernandes S, Bala J (2013) Performance analysis of PCA-based and LDA-based algorithms for face recognition. Int J Signal Process Syst 1(1):1–6
22. Biswas S, Aggarwal G, Flynn PJ, Bowyer KW (2013) Pose-robust recognition of low-resolution face images. IEEE Trans Pattern Anal Mach Intell 35(12):3037–3049
23. Javed (2013) Face recognition based on principal component analysis. Int J Image Graph Signal Process 5(2):38
24. Roy MS, Podder MS (2013) Face detection and its applications. Int J Res Eng Adv Technol 1(2)
25. Jindal N, Kumar V (2014) Enhanced face recognition algorithm using PCA with artificial neural networks. Int J Adv Res Comput Sci Softw Eng 3(6)
26. Obaidullah Md Sk, Bose A, Mukherjee H, Santosh KC, Das N, Roy K (2018) Extreme learning machine for handwritten Indic script identification in multi-script documents. J Electron Imaging 27(5):051214
27. Huang GB, Zhu Q, Siew CK (2006) Extreme learning machine: theory and applications. Neurocomputing 70:489–501
28. Huang GB, Zhou H, Ding X, Zhang R (2012) Extreme learning machine for regression and multi-class classification. IEEE Trans Syst Man Cybern 42(2):513–529
29. Frenay B, Verleysen M (2011) Parameter-insensitive kernel in extreme learning for non-linear support vector regression. Neurocomputing 74(16):2526–2531
30. Huang GB, Li M, Chen L, Siew C-K (2008) Incremental extreme learning machine with fully complex hidden nodes. Neurocomputing 71:576–583
31. Huang GB, Chen L (2011) Enhanced random search based incremental extreme learning machine. Neurocomputing 71:3460–3468
32. Rong HJ, Huang GB, Sundararajan N, Saratchandran P (2009) Online sequential fuzzy extreme learning machine for function approximation and classification problems. IEEE Trans Syst Man Cybern 39(4):1067–1072
33. Sun Y, Yuan Y, Wang G (2011) An OS-ELM based distributed ensemble classification framework in P2P networks. Neurocomputing 74:2438–2443
34. Rong H, Ong YS, Tan AH, Zhu Z (2008) A fast pruned extreme learning machine for the classification problem. Neurocomputing 72:359–366
35. Miche Y, Sorjamaa A, Bas P, Simula O, Jutten C, Lendasse A (2010) OP-ELM: optimally-pruned extreme learning machine. IEEE Trans Neural Netw 21(1):158–62
36. Decherchi S, Gastaldo P, Zunino R, Cambria E, Redi J (2013) Circular ELM for the reduced-reference assessment of perceived image quality. Neurocomputing 102:78–89
37. van Heeswijk M, Miche Y, Oja E, Lendasse A (2011) GPU accelerated and parallelized ELM ensembles for large-scale regression. Neurocomputing 74:2430–2437
38. Man ZH, Lee K, Wang DH, Cao ZW, Miao CY (2011) A new robust training algorithm for a class of single-hidden layer feed-forward neural networks. Neurocomputing 74:2491–2501
39. Man ZH, Lee K, Wang DH, Cao ZW, Miao CY (2011) A modified ELM algorithm for single-hidden layer feedforward neural networks with linear nodes. In: 6th IEEE conference on industrial electronics and applications, pp 2524–2529
40. Nian R, He B, Lendasse A (2012) 3D object recognition based on a geometrical topology model and extreme learning machine. Neural Comput Appl 22(3–4):427–433
41. Huang GB, Wang DH, Lan Y (2011) Extreme learning machines: a survey. Int J Mach Learn Cybern 2:107–122
42. Jun-Kai Z, Xiao-Ya G (2015) Face recognition based on improved robust sparse coding algorithm. Int J Signal Process Image Process Pattern Recogn 8(9):339–346

A Review of Diabetic Retinopathy Screening Using Machine Learning

S. Nalini Durga and K. Usha Rani

Abstract Nowadays, the majority of research with machine learning is concentrated on healthcare applications. One of the major health issues that is being observed in the majority of people is diabetes. Diabetes is a chronic disease that can affect other organs, like eyes, kidney, and liver in the human body. Diabetic retinopathy is a kind of abnormal disruption which can occur due to high sugar levels in the blood for a long period of time. This in turn can cause blood leaks and damage the retina, which relentlessly can become worse and would be one of the major reasons for vision falling and blindness. Experienced lab people are required to diagnose this disease based on small and significant features of color fundus images which is a difficult and long time-taking process. With recent advancements in biomedical research, we are getting enormous amounts of data that are helpful in extracting meaningful knowledge with the help of artificial intelligence. Many researchers proposed several machine learning and deep learning techniques to identify the presence of diabetic retinopathy in fundus images and accurately classify its severity. This paper aims to review these learning techniques which diagnose diabetic retinopathy.

Keywords Diabetic retinopathy · Retinal fundus images · Retinal imaging · Machine learning · Deep learning · Image classification · Classifiers

1 Introduction

When surveyed about the five senses of humans—sight, hearing, taste, smell, and touch—among all of these most of the people report that their sight is the mode of perception they value most. Everyone cares about their eyes. Various factors or health issues, like uncontrolled diabetes, can affect eye health that can result in diseases like

S. Nalini Durga (✉) · K. Usha Rani
Department of Computer Science, Sri Padmavati Mahila Viswavidyalayam, Tirupati, Andhra Pradesh, India
e-mail: nalini.sst@gmail.com

K. Usha Rani
e-mail: usharanikuruba@yahoo.co.in

S. Jyothi et al. (eds.), *Proceedings of the 2nd International Conference on Computational and Bio Engineering*, Lecture Notes in Networks and Systems 215,
https://doi.org/10.1007/978-981-16-1941-0_16

diabetic retinopathy (DR), diabetic macular edema (DME), cataracts, and glaucoma. All these diseases have their own unique way of confronting and have to be treated carefully. Choosing improper treatment methods can lead to severe vision loss or may lead to complete blindness.

DR is one of the most common eye diseases caused among the above four possible disorders. It causes damage to the blood vessels and swelling in the macula that is present in the retina, thus resulting in another auxiliary disease called DME. Clouding of eyes (or) cataracts is another kind of disease commonly seen in humans, either due to the age factor or diabetes. While glaucoma can impact the optic nerve of the eye by increasing the blood vessel size in the eye due to diabetes, it can cause vision loss.

Diabetes is the most common disease found among the majority of the population, and it shows various kinds of impacts on almost all the organs of the patient [1]. It can be categorized into Type 1 and Type 2. We are focusing on Type 2 diabetes, which includes the disease with complications like diabetic retinopathy, heart disease (HD), neuropathy, and nephropathy.

DR is an adamant and widely spread eye disease, primarily a consequence of Type 2 diabetes. DR occurs when the blood vessels get damaged inside the retina, leaking blood and fluids into the neighboring tissue resulting in tiny swellings or microaneurysms (μA), hemorrhages (H), hard exudates (HE), neovascularization (NV), and cotton wool spots (CWS) [2, 3]. Figure 1 shows the difference between the unaffected and diabetic retinopathy affected eye.

DR can be sub-categorized into non-proliferative diabetic retinopathy (NPDR) and proliferative diabetic retinopathy (PDR) which are caused by the abnormal growth in blood vessels of retina. This can lead to blindness if not treated in time properly [4]. Between the both, NPDR is the most commonly occurring, which is almost 80% among all cases. Figure 2 shows the retinal comparison of normal and diabetic with all possible signs.

The NPDR can be further classified into mild, moderate, and severe [3]. The grading of DR diagnoses based on various parameters given by most of the medical experts is given in Table 1.

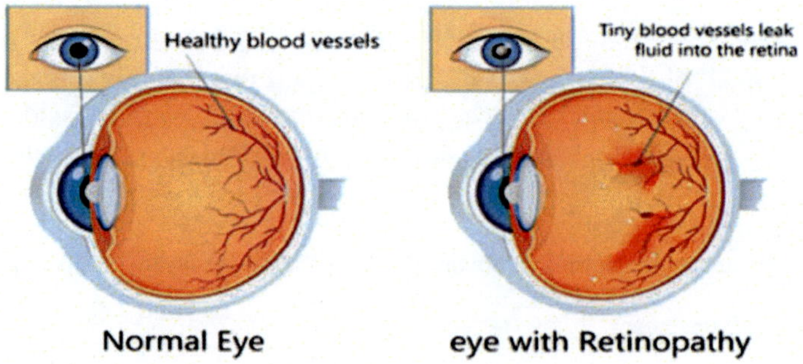

Fig. 1 Comparison of normal and retinopathy eye structure

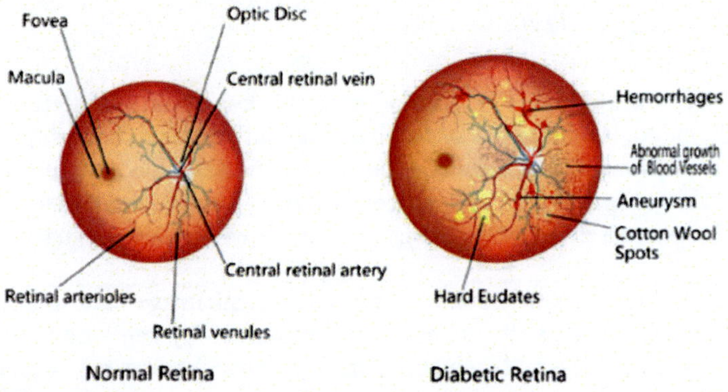

Fig. 2 Retinal comparison of normal and diabetic with all possible signs

Table 1 Retinopathy abnormality grading chart

Value	Type of DR	Abnormality grading range
0	Normal	$\mu A = 0$ and $H = 0$
1	Mild NPDR	$0 < \mu A \leq 5$ and $H = 0$
2	Moderate NPDR	$5 < \mu A < 15$ and $0 < H < 5$ and $NV = 0$
3	Severe NPDR	$\mu A \geq 15$ and $H \geq 5$ and $NV = 1$

DR is so reticent disease, by the time the patients can recognize it; the changes in the retina will be at such an advanced level that in some cases the treatment might be difficult or impossible. The rear part of the eye image, known as fundus image, which is acquired from the patient eyes will undergo retinography image analysis for the diagnosis of DR. In this diagnosis process the crucial step is the extraction of retinal blood vessels which can help in identifying the disease and can be treated preventing the eventual vision loss.

With the advancement in technology to acquire color fundus images as input, various DR screen models were proposed which help ophthalmologists in DR screening and recuperate difference if found any between manual and automated DR screening procedures. The screening models proposed so far use image classification, pattern recognition, and machine learning. Several research works have proposed different frameworks in detecting diabetic retinopathy. The main objective of our work is to propose a universal architecture for DR detection by analyzing the performance of previous work in classifying color fundus images based on the most important geometric features, like segmenting the blood vessels, detection of exudates, microaneurysm, and cotton wool spots.

2 Literature Review

In recent days, the awareness in terms of protecting eye vision has increased. Generally, ophthalmologists make use of features of blood vessels in fundus images, such as retinal vessel length, area, diameter, perimeter, and variation in tortuosity to diagnose the DR [5]. Many researchers propose various advanced models to improve the accuracy in the prediction of DR. In this section, we will walk through some of those models.

Islam et al. [6] proposed a framework that begins with image acquisition, followed by image enhancements such as contrast improvement by applying preprocessing techniques. After successful completion, other features like area, mean, and standard deviation of pixels are extracted from that image. Now the training set of images are classified into various categories using nonlinear classifiers like binary decision tree (BDT), k-nearest neighbor (kNN), and two support vector machines (both radial basis and polynomial function). This research finally shows that the given test data will be classified into two categories (normal or DR) with a sensitivity of 98% and specificity of more than 94%.

Yun et al. [7] have proposed a framework for auto-identifying the different stages (mild, moderate, and severe) of NPDR and PDR using a neural network. Using morphological operations on preprocessed images they extracted features such as a layer of perimeter and a layer of area in all three RGB channels. They have achieved an accuracy of 80% and sensitivity of 90% in the diagnosis process. However, by concentrating on better features they could have improved the efficiency.

Streeter [8] designed a prototype for the reliable detection of μA using non-mydriatic color fundus images. They used a technique called linear discriminant analysis (LDA) which improves the possibility of detecting μA with no false detections in diseased images. This method has proven better when compared with other methods, with some limitations like poor image quality capturing due to the noise and poor detection of small μA which are of size single-digit pixels.

Sujith Kumar [9] emphasized an algorithm to auto-detect μAs in poor quality fundus images. The steps followed are preprocessing of RGB images, feature extraction, followed by classification. In the preprocessing stage, the images were converted into grayscale after normalizing uneven illumination and contrast adjustment. Adaptive histogram equalization is applied for contrast enhancement. This helped in achieving an average sensitivity of 94.44% and specificity of 87.50%. In this model, the rate of false negative and false positive detection is a little higher because of either too small μA or too blurred μA.

Jiang et al. [10] proposed a method to accurately detect the blood vessels in retinal images by using verified multi-threshold probing. It is hard to segment the retinal blood vessels in the images due to the gradients present in the background. This issue was overcome by applying different values of threshold to extract all binary components in the image. They have applied three pruning operations. With this method on the selected dataset, they achieved almost the same rate of true positives as they achieved in the hand-labeled method. This method is best suitable for detecting μAs in digital fundus images.

Spencer et al. [11] by applying digital image-processing on quantification of retinal pathology have proposed an accurate and highly repeatable technique for early DR detection. In this method, the brighter regions of the images are corrected and applied a transformation to eliminate the dark regions. Then images are analyzed for their subsequent characteristics, such as the location, size, shape, brightness, etc., for the detection of μA. The algorithm attained 82% sensitivity.

Bethanney Janney et al. [12] presented work on auto exudates detection using fuzzy "C" means (FCM) algorithm and classified normal and exudates contained in retinal images using KNN. All the necessary preprocessing steps like resizing, removal of noise, and contrast adjustment in the images are carried out, alongside canny edge detection is used to differentiate blood vessels, exudates, and the rest of the components in the retinal image. It is followed by feature extraction using the gray-level co-occurrence matrix (GLCM). Using the FCM segmentation the segmented areas of the input images and the exudates in the retinal images are being found out. The system achieved accuracy, sensitivity, and specificity of 96%, 90%, and 94%, respectively, with KNN classifier.

Jalan [13] has focused on identifying exudates and classifying lesions in color fundus images to detect the DR. The detection of exudates lesion was identified by concentrating on the features such as shape, texture, and color and morphological features. The obtained features are fed to both KNN and SVM algorithms. The severity of the disease was decided by collaborating the output of KNN and SVM classifiers. This helped in achieving more accuracy in screening DR in lesser time.

Radha [14] proposed a method for the identification of exudates using k-means and classified the retinal images. By using thresholding and morphology operations, they identified the ridges of vessels in retinal images. Before going to segmentation of images they applied a probabilistic neural network for training the model with the obtained feature and CLAHE algorithm for contrast enhancement. The segmented images are reconstructed by preserving the thin edges of vessels, and the false edges were eliminated. Improvements in segmentation and thresholding algorithms can improve the accuracy of deceased detection.

Gandhi and Dhanasekaran [15] proposed a method to distinguish between exudates, optic disc, and blood vessels. In preprocessing stage along with enhancement operation, the exudates similar intensity level objects, like optic disc and blood vessels, were removed. Morphological operations like erosion followed by dilation were applied for the successful detection and identification of the location of exudates. After that, GLCM was applied to extract the features, followed by segmentation and classification that was performed using SVM to obtain the images based on the severity grade which helps in confirming the disease more accurately.

2.1 Research Challenges

Most of the research work done so far is facing some challenges in terms of

i. Handling of large volumes of datasets
ii. Similarity in the optic disc and exudates brightness, size, and color
iii. Difficulty in illuminating the retinal vessels
iv. Difficult in differentiating blood vessels with μA
v. Noise while capturing the fundus images
vi. Proposed methods that are limited to specific datasets.

By observing these challenges, the research of DR diagnostics is in preschooler stage and needs to be addressed to improve the accuracy and assist the ophthalmologist to detect the early sign of DR.

2.2 Performance Evaluation Metrics

The performance of the algorithm is evaluated based on its sensitivity, specificity, accuracy, and F-score. These metrics are calculated using four parameters, namely true positive (TP), true negative (TN), false positive (FP), and false negative (FN).

Sensitivity is the proportion of abnormal fundus's categorized as abnormal by the process.

$$\text{Sensitivity (SE)} = \frac{\text{TP}}{\text{TP} + \text{FN}}$$

Specificity is the proportion of normal fundus's categorized as normal by the procedure.

$$\text{Specificity (SP)} = \frac{\text{TN}}{\text{TN} + \text{FP}}$$

Two more evaluating metrics are as follows:

$$\text{Accuracy (A)} = \frac{\text{TP} + \text{TN}}{\text{TP} + \text{TN} + \text{FP} + \text{FN}} \qquad \text{F-Score } (F) = \frac{2\text{TP}}{2\text{TP} + \text{FN} + \text{FP}}$$

3 Generalized System Architecture

Based on research work proposed by many authors, the generalized system architecture for DR screening process includes the stages like image acquisition, preprocessing, segmentation, feature engineering, and classification. Figure 3 shows a generalized framework for automatic DR screening.

3.1 Image Acquisition

For research purposes, color fundus images can be collected from the datasets available publicly from famous repositories, like Messidor, Digital Retinal Images for Vessel Extraction (DRIVE), e-Ophtha-MA, e-Ophtha-EX, DIARETDB, STARE, GOLD STANDARD, and Kaggle. Other datasets like "Standard Diabetic Retinopathy Database" created by Machine Vision and Pattern Recognition Research Group, Finland are also available. Among the available color fundus images, few were unaffected and the remaining were affected by diabetic retinopathy.

3.2 Preprocessing

Preprocessing is an important stage in DR screening. The difficulties aroused due to the blurred image or non-clarity images are addressed here. The main objective of the

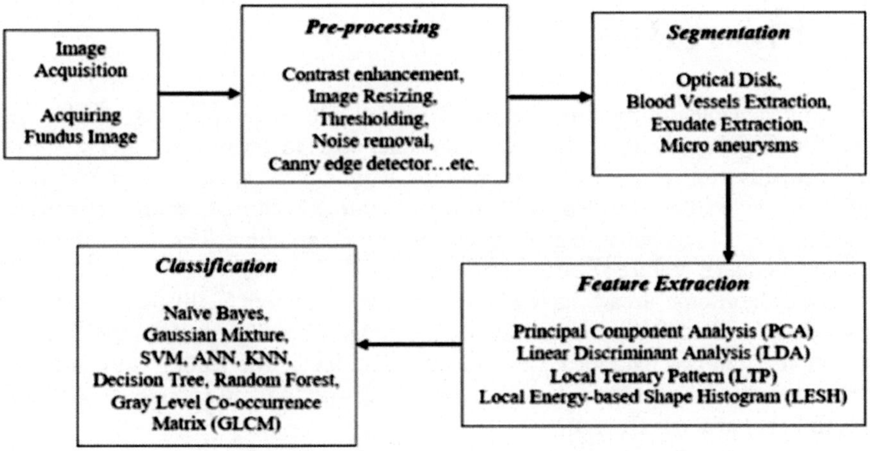

Fig. 3 Generalized framework for the model

preprocessing is to produce a reliable structural representation of the retina by minimizing the intensity difference among the objects and background. The doctors face challenges in diagnosing the retinopathy in the fundus images due to the intra-retinal occlusions that get obscured with less contrast and contain few details regarding the lesions such as blood vessels, μA, etc. So, we need an effective method to improve the texture and features of the retinal structure by applying image enhancement.

To attain image enhancement, the color fundus images are converted into grayscale to cut out the noise which will improve the visual appearance of lesions. An adaptive median filter is applied to remove the salt and pepper noises, which were developed during the image acquisition, to smoothen the image, and to reduce the distortion caused due to the improper thickness of image boundaries. Then finally histogram equalization is applied to improve the contrast of the image.

3.3 Segmentation

Segmentation is the next phase after the preprocessing. In this module, the digital image is segmented into distinctive layers, where each layer comprises harmonized elements/features that are useful in classification, i.e. optic disc, abnormal blood vessels, exudates, and microaneurysms. Image segmentation can be done by using various methods like simple threshold, K-means, and fuzzy C-means. We are limiting the research focus on the segmentation phase to only four features, namely optical disk localization, extraction of blood vessel, exudate extraction, and microaneurysm extraction.

3.4 Feature Extraction

Building valuable information (features) from raw data by reformatting, combining, transforming the primary features into new ones so as to develop a new dataset that still contains most of the useful information will help ML models to achieve their goals. However, there are no possible ways in creating features. Creating meaningful features needs knowledge, creativeness, perception, and time. This arises the need for an automated feature selection process.

Some algorithms already have a built-in feature extraction facility that can extract minimum features. Deep learning is one such algorithm that extracts useful representations from raw data through a hidden neural layer. The feature extraction can be unsupervised (i.e. principal component analysis (PCA)) or supervised (i.e. linear discriminant analysis (LDA)).

3.5 Feature Selection

The process of selecting important features from input data that can contribute to achieving more accuracy in predicting the output that we are interested in is called feature selection. We consider applying feature selection on the input data when we observe the presence of irrelevant or redundant data because they will affect the accuracy of the model by adding noise. Sometimes feature selection is applied on relevant and non-redundant data to condense the number of features so that there will be a possibility of interpretability and computing feasibility or to avoid imbalance in selected features and sample ratio.

The features such as microaneurysm, fovea centralis, optic distance, blood vessel, dot and blot hemorrhages, edema, exudates area, and bifurcation are some of the most important to be considered while diagnosing the DR.

3.6 Classification

Finally, the images are identified and classified into different sets of categories based on abnormalities observed by using the classifiers. As stated in our previous work, S Nalini et al., in order to do classification, several machine learning algorithms are available. The key challenge lies in choosing the optimal algorithm for solving the problem with the key parameters, like speed, forecast accuracy, and training time [16]. The classifiers such as support vector machine (SVM), K-nearest neighbors (KNN), decision tree (DT), naive Bayes (NB), random forest (RF), adaptive boosting (AB), quadratic discriminant analysis (QDA), Gaussian process, and multi-layer perceptron (MLP) neural network Bayesian, and neural network are well suited for classifying the abnormalities of DR, such as μA, hemorrhages, and exudates.

For the training and testing phases of the classifiers, the available data can be selected randomly as 30% and 70% for the testing and training set, respectively.

The more accurate methods are SVM and ANN, which produce more accurate results in medical image processing. The classification results include various parameters, like accuracy, specificity, and sensitivity which can be measured. Table 2 summarizes the various classification techniques available and the datasets are used for implementing the results with their performances. This analysis helps to produce more accurate results in future implementation methods.

4 Conclusion

In this paper, a brief review of the different concepts relating to how the DR disease could be detected at an early stage was presented in a nutshell, which gave a glance on the use of the preprocessing, segmentation, feature extraction, and classification

Table 2 Summary of various classification techniques available with their performances

Type of classifier	Detected abnormality	Disadvantage	Performance measures (%)	Dataset	Authors
NB	μA	Limited datasets	–	–	Lazar et al. [17]
NB	ME	Limited features are used	A: 96.7	–	Chowriappa [18]
NB	HE	OD location performance can be improved	A: 97	–	Köse et al. [19]
NB	μA	Issue of improving the sensitivity	SE: 85.68 SP: 99.99 A: 99.99 Precision 83.34	–	Sopharak et al. [20]
Gaussian mixture	Blood vessel	Glaucoma detection can be included	A >90	DRIVE, DIARETDB0, DIARETDB1, CHASE_DB1, MESSIDOR	Roychowdhury [21]
Gaussian mixture	ME	Normal pixels also detected as candidates	SE: 100	MESSIDOR	Deepak et al. [22]
Gaussian mixture model with SVM	ME	Severity of risk is not computed	SE: 97.3 SP: 95.9 A: 96.8	MESSIDOR	Akram [23]
Gaussian Bayes	Fovea	Limited datasets are used	SE >95 SP >97 A >98	–	Ahmad Fadzil et al. [24]
SVM	Blood vessels, exudates	Limited database is used	–	–	Gandhi et al. [25]
SVM	Blood vessels	Vessel pixels is not included	A: 95.72	AFIO, DRIVE, STARE	Awan et al. [26]
SVM	–	Can develop in e-health	A: 98.10	DRIVE, DIARETDB1	Gurudath et al. [27]
SVM	Hemorrhages	Limited testing images	TP rate 93.2	–	Zhang et al. [28]

schemes. Finally, in this paper, a small review about the general framework to be followed by the researcher in developing systems to diagnose diabetic retinopathy disease. We are also looking forward to proposing a model which can be suitable for all sizes of datasets by addressing the overfitting issue caused by the noise appearing in the form of human annotation errors [29]. Thus, this study gives a brief insight into our future research work on the detection of the second-largest disease in the world, i.e., the diabetic retinopathy disease in the human eyes which could be detected using novel preprocessing, segmentation, and classification techniques [30].

References

1. Shirbahadurkar SD et al (2017) A modern screening approach for detection of DR. In: 2nd international conference of man and machine interfacing
2. Maher R, Kayte S, Dhopeshwarkar DM (2015) Review of automated detection for DR using fundus images. IJARCSE 5(3):1129–1136
3. Browning DJ (2010) Diabetic retinopathy: evidence-based management. Springer Science & Business Media
4. Wilfred Franklin S, Edward Rajan S (2014) Computerized screening of diabetic retinopathy employing blood vessel segmentation in retinal images. Biocybern Biomed Eng 34:117–124
5. Leontidis G (2017) A new unified framework for the early detection of the progression to DR from fundus images. Comput Biol Med
6. Islam S, Hasan MdM, Abdullah S (2018). Deep learning based early detection and grading of DR using retinal fundus images
7. Yun W, Acharya et al (2008) Identification of different stages of DR using retinal optical images. Inf Sci 178:106–121. https://doi.org/10.1016/j.ins.2007.07.020
8. Streeter L, Cree MJ (2003) Microaneurysm detection in colour fundus images. Image Vis Comput NZ (Palmerston North) 11(5):280–285
9. SujithKumar SB, Singh V (2012) Automatic detection of DR in non-dilated RGB retinal fundus images. IJCA 47(19)
10. Jiang X, Mojon D (2003) Adaptive local thresholding by verification-based multithreshold probing with application to vessel detection in retinal images. IEEE Trans Pattern Anal Mach Intell 25(1):131–137. https://doi.org/10.1109/tpami.2003.1159954
11. Spencer T, Olson JA, McHardy KC, Sharp PF, Forrester JV (1996) An image-processing strategy for the segmentation and quantification of microaneurysms in fluorescein angiograms of the ocular fundus. Comput Biomed Res 29(4):284–302. https://doi.org/10.1006/cbmr.1996.0021
12. Bethanney Janney J, Divakaran S, Abraham S, Meera G, UmaShankar G (2015) Detection and classification of exudates in retinal image using image processing techniques. JCPS 8(3)
13. Jalan S, Tayade AA (2015) Review paper on diagnosis of DR using KNN and SVM algorithms. IJARCSMS 3(1)
14. Radha R, Lakshman B (2013) Retinal image analysis using morphological process and clustering technique. Signal Image Process Int J 4:55–69
15. Gandhi M, Raghavan D (2013) Diagnosis of DR using morphological process and SVM classifier. In: ICCSP 2013—proceedings, pp 873–877. https://doi.org/10.1109/iccsp.2013.6577181
16. Nalini Durga S, Usha Rani K (2020) A perspective overview on machine learning algorithms. In: Jyothi S, Mamatha D, Satapathy S, Raju K, Favorskaya M (eds) Advances in computational and bio-engineering. CBE 2019. Learning and analytics in intelligent systems, vol 15. Springer, Cham. https://doi.org/10.1007/978-3-030-46939-9_30

17. Lazar I, Hajdu A (2013) Retinal microaneurysm detection through local rotating cross-section profile analysis. IEEE Trans Med Imaging 32(2):400–407
18. Chowriappa P, Dua S, Acharya UR, Rama MM (2013) Ensemble selection for feature-based classification of diabetic maculopathy images. Comput Biol Med 43(12):2156–2162
19. Kose C (2011) Simple methods for segmentation and measurement of diabetic retinopathy lesions in retinal fundus images, pp 274–293
20. Sopharak A, Uyyanonvara B, Barman S (2013) Computerized medical imaging and graphics simple hybrid method for fine microaneurysm detection from non-dilated diabetic retinopathy retinal images. Comput Med Imaging Graph 37(5–6):394–402
21. Roychowdhury S, Koozekanani DD, Kuchinka SN, Keshab K (2015) Optic disc boundary and vessel origin segmentation of fundus images 2194(c):1–12
22. Deepak KS, Sivaswamy J (2012) Automatic assessment of macular edema from color retinal images. IEEE Trans Med Imaging 31(3):766–776
23. Akram MU, Tariq A, Khan SA, Javed MY (2014) Automated detection of exudates and macula for grading of diabetic macular edema. Comput Methods Programs Biomed 114(2):141–152
24. Ahmad Fadzil MH, Izhar LI, Nugroho HA (2010) Determination of foveal avascular zone in diabetic retinopathy digital fundus images. Comput Biol Med 40(7):657–664
25. Gandhi M, Dhanasekaran R (2013) Diagnosis of diabetic retinopathy using morphological process and SVM classifier, pp 873–877
26. Awan AW (2015) A robust algorithm for segmentation of blood vessels in the presence of lesions in retinal fundus images, pp 3–8
27. Gurudath N, Celenk M, Riley HB (2014) Machine learning identification of diabetic retinopathy from fundus images, pp 1–7
28. Zhang X, Chutatape O (2005) A SVM approach for detection of hemorrhages in background diabetic retinopathy, pp 2435–2440
29. Ramakrishna M, Rama Satish A, Siva Krishna PSS (2021) Design and development of an efficient malware detection Using ML. In: Chaki N, Pejas J, Devarakonda N, Rao Kovvur RM (eds) Proceedings of international conference on computational intelligence and data engineering. Lecture notes on data engineering and communications technologies, vol 56. Springer, Singapore. https://doi.org/10.1007/978-981-15-8767-2_35
30. Kumaran Y et al (2018) A brief review of the detection of DR in human eyes using pre-processing & segmentation techniques. IJRTE 7(4S2). ISSN: 2277-3878

A Novel Framework for Modeling Medical-Sensitive Big Data Using Document-Based Database

Sai Jyothi Bolla and S. Jyothi

Abstract Modeling big data is the most pervasive in the current evolution of big data which cannot be quantified. This arrests the attention of most of the researchers to discover and invent the relationships among data in big data. One emerging big data is medical data. Medical big data consists of data of various types, which is multi-lingual and sensitive. This paper proposes a data model for modeling medical big data into a flexible format that can be used for performing analytics in order to take decisions for better healthcare systems.

Keywords Big data · Medical data · Sensitive data · Data modeling

1 Introduction

The very idea of "big data" though not very novel, the definition of the very same kept changing from time to time. Efforts were made by different people in defining big data as per their perception and the same resulted in characterizing big data as a pool of data whose distinctive features like their extent, speed, type, and/or complexity intrigue many to explore, adopt, and discover novel hardware and software entities so as to store, analyze, and visualize the data in an effective way [1–3]. In the healthcare domain, velocity (swiftness of generation of data), diversity, and volume [4] play a pivotal role with regard to the data it produces. Indeed, the data from healthcare systems, the health insurance sector, researchers, government entities, and the like spreads across various languages and resources. Much of the data produced from these sources contain no global transparency. Nevertheless, another V, veracity apart

S. J. Bolla
Department of Information Technology, Vasireddy Venkatadri Institute of Technology, Guntur, India
e-mail: saidilipyerram@gmail.com

S. Jyothi (✉)
Department of Computer Science, Sri Padmavathi Mahila Visvavidyalayam, Tirupati, India
e-mail: jyothi.spmvv@gmail.com

S. Jyothi et al. (eds.), *Proceedings of the 2nd International Conference on Computational and Bio Engineering*, Lecture Notes in Networks and Systems 215,
https://doi.org/10.1007/978-981-16-1941-0_17

from the above said three Vs is critical for the healthcare data to add meaning to use big data in developing another vertical, translational research.

The finest solutions so far designed in medical big data overcome numerous complexities and hidden relations. McKinsey Global Institute's report suggests that there is more than $300 billion market every year still to be explored if at all the US healthcare systems were to practice big data ingeniously and efficiently. Two-thirds of the presumed value would be in terms of reducing the expenditure accrued on healthcare in the USA [5]. Classical approaches used earlier are confined to the study of changes in physiology as a singular modality of data in the investigation of disease [6]. Research to comprehend the disease in this way muted the elements of variation and interconnectedness for study which are true basic medical mechanisms for study though [7]. However, the recent advancements in technology claim to have the potential to pool massive volumes of information pertaining to individual patients across a wide timescale. As a result, the data captured from various individuals remained redundant.

Countless number of various structured and unstructured data of various modalities is needed for understanding the behavior of a disease. This provoked the notion of big data analytics in medical data. Therefore, big data analysts aggravated their research in understanding the behavior and relationships of such medical big data.

While the sphere of data is growing massive day-in and day-out, their management techniques on the other hand are not evolving at the same velocity, ensuing a considerable demand for an aptly competent data analyzing system. Having originated in 1970s and 1980s, the current database management systems underwent significant revamping [8] including the query structure.

Years of research in big data analytics resulted in a decent number of big data models, however, the data proliferation throws open high scope for more models so as to meet the new challenges. Lockdown across the globe opened more online operations apart from new diseases which opened a big challenge in data pooling and thus required a highly competent model in computational needs. Internet is the buzzword in the current times and online tutorials are a new normal. Netizen's number went up enormously, specifically in social media, resulting in massive and complex data production. The data accumulation over some time, especially during the lockdown, became enormous and the data so accumulated demands more effectual retrieving of accurate information and analyzing in a short span of time; also demanding optimal big data modeling for extracting such results. The data accumulation demands strengthening of the current infra to a massive, effective, and ascendable infra.

Recent developments in big data research have altered the way we have been managing, analyzing, and taking leverage of the data in every sector. Among all the data resources, healthcare stands out with more scope for challenging and interesting contexts. Big data analytics in healthcare claim to have the competency to reduce treatment costs, predict epidemics outbreaks, and can even avert avertible diseases and help enhance the overall quality of human life. The professionals of healthcare are just like data analysts, analyzing healthcare data for understanding which helps in extracting hidden relations. Over some time, the healthcare data which includes capturing and managing data is mounting at a quick pace, demanding more professionals into doing the analyses. As a result, there emerged a necessity for data analytics in healthcare systems than ever before.

In this paper, a generalized approach for modeling medical-sensitive data is recommended. This recommendation is based on priority-based feature document modeling. It also includes why this approach is more flexible than any other data modeling like column-and-row-based modeling.

2 Literature Review

In this section, an attempt is made to take a look at all the existing six data models which are active in the arena of big data (petabyte size) and are considered the most reliable, effective, and leading data models that are able to deal with the current data challenges. However, there are, of course, various data models still relevant for big data. Nevertheless, the six designated models signify a variety with regard to mainstream approaches. Table 1 reveals the summary of these six data models practically which are in use in the context of big data.

These models have been opted as best owing to their competence to critically resolve the bottlenecks encountered while converting the data into big data.

From among the models shown in Table 1, CrowdDB debatably cannot stand as a big data model [9]. The same is discussed here deliberately and strategically in comparison to the others because of its capacity of crowdsourcing in resolving many complex tasks that are far-reaching for the conventional database management systems. The presence of crowdsource tags in CrowdSQL pays well in contributing to the enhancement of language as well as help in dealing effectively with the scaled-up data. Besides, MapReduce, different from other conventional big data models, basically resembles a software programming model well acknowledged for its effective parallel and distributed computing competency. As of late, in the arena of big data, MapReduce programming model has been doing rounds with its interweaving proximity along with the Apache Hadoop framework and attracted the consideration besides admiration, invariably.

Inarguably, in databases the significance of indexing [10] has been proved, predominantly when data extent is enormously large as it enhances retrieving the required data in massive tables by swiftly tracing the data evading the traversal of every one of the rows. Indexes help in effective execution of recurrent queries, particularly for reading operations. In fact, an index contains a column or more than that and occasionally the extent of an index might surpass the size of the table it is formed for; nonetheless, in any case, the index provides speedy admittance to the data. This spells convenience for the overhead of having indexes.

The element of indexing in big data turns out to be a bottleneck for various reasons: basically, the big data hosts a large chunk and velocity of the data; a generated index, on the other hand, must be trivial (a portion of the actual data) and fast as against the data because searching includes scanning billions or more than that of data values in

Table 1 Summary of data models for handling big data

Features	Model					
	BigTable (2004)	Cassandra (2007)	HBase (2007)	MongoDB (2007)	CouchDB (2005)	CrowdDB (2006)
Type	Key-value pair	Hybrid of Key-value pair and tabular	Column-oriented (key-value pair)	Document-oriented (JSON format)	Document-oriented (JSON format)	Tabular
Relational nature	No	Yes	No	No	No	Yes
Developer	Google Inc.	Facebook Inc. (Avinash Lakshman, Prashant Malik)	Microsoft (Powerset)	10gen	IBM (Damien Katz)	Jeff Howe
Language of development	C, C++	Java	–	C++	Erlang	Html, JavaScript, Java
Query language	APIs in C++	CQL	1. Pig latin 2. HQL	MongoDB ad-hoc query language	JavaScript	CrowdSQL
Query nature (SQL based)	No	Yes	1. No 2. Yes	No	No	Yes
High availability	Yes	Yes	Yes	Yes	Yes	No
High scalability	Yes	Yes	Yes	Yes	Yes	No
Single point of failure	N/A	No	Yes	No	No	Yes
Open source	No	Yes	Yes	Yes	Yes	No
Versioning	Yes (built-in timestamp)	Yes (external timestamp at query level, but no built-in)	Yes (built-in timestamp)	Yes (different database for audit trials, not native)	Yes (MVCC)	No (except basic versioning of RDBMS)
Indexing	Based on map, supported by timestamps	Basic primary and secondary	Basic (secondary not by default)	Advance and wide variety (includes spatial indexing)	Basic associated map and reduce functions	Basic as in relational data models

(continued)

Table 1 (continued)

Features	Model					
	BigTable (2004)	Cassandra (2007)	HBase (2007)	MongoDB (2007)	CouchDB (2005)	CrowdDB (2006)
Data processing nature	Batch processing	Streaming and atomic batches	Batch processing	Batch processing and event streaming	Batch processing	Batch processing

a fraction of seconds. All the models of big data allow complicated as well as multi-variable queries which prompt us to consider indexing to deliver quicker access for all search results and even to deliver collective response at large. As regards, indexing models must be capable enough to generate tiny indexes as a part of granularity reduction. So formed indexes must be able to be portioned into such units if there come up opportunities for parallel processing. Owing to the increase in the rate of data generation, it is equally important to generate an index to make hassle-free data processing. The paper thus focuses on indexing issues pertaining to big data models.

In MongoDB, B-tree data structure generates indexes [10] at the collection level (tantamount to the table in the relational databases). To extend backing for a lot of different data as well as queries, varied indexes exist: (1) Default _id—the existing one on the _id collections' field and acts like an exclusive id; (2) Single field—this, however, is a custom demarcated index in a single field of the document; (3) Compound Index—this is yet another variety created by a user over manifold fields; (4) Multi-key Index—this is utilized to index the field that contains values in a selection; (5) Geospatial Index—this is a distinctive index which is a MongoDB (distinct from other models) hosts to help geospatial queries with built-in two types: (i) 2D spatial index pertaining to queries in plane geometry; (ii) 2sphere spatial index refers to queries of spherical geometry; (6) Text Index—yet another type used in searching a string/text within a collection; (7) Hashed Index—supports indexing with hash in the search field.

In Cassandra, the very first index is, nonetheless, the principal key that confirms the exclusivity of every record much similar to that of a relational database. However, the primary index, attached to a column family, behaves like an index for the corresponding row keys and is sustained by every node aimed at the corresponding associated data [11]. Cassandra too offers secondary indexes over the column values that support like a supplementary filter for the resulting set. MongoDB thus gives a way for a wide choice of indexing competence over Cassandra.

In contrast, relational model CouchDB has no feature of indexing. In comparison, the practice of indexing in CouchDB is aided by functions of Map and Reduce pertaining to presumed MapReduce framework; distinctions if at all in the overall data of the document are well captured by these functions encouraging parallelism besides autonomous computation in indexing. The map functions that create indexes try to limit the targeted data's window, while the reduced queries are permitted to seek optimum efficiency. It is thus important to build an index while writing the map function. The union-reduce function and a map in CouchDB are labeled as a view.

HBase does not host a strong indexing ability similar to CouchDB [12]. Though secondary indexes are not built-in by default, it is possible, however, to generate the other tables that may require updating from time to time by MapReduce. The other option is while writing to the data table, it can also be written to the index table. Each of the approaches although has their own merits and demerits, nevertheless, HBase supports indexing but not aggressively.

BigTable data models like HBase are not good agents in giving indexes [13]. Technically speaking, it was already said that BigTable is but a sparse map, multi-dimensional, widespread, sorted, and persistent indexed by a row key, column key, along with a timestamp though in the map. CrowdDB, on the other hand, partially (apart from the human involvement aspect) appears similar to a conventional relational database, hosting all the rudimentary indexing methods like that of any relational database. Thus, MongoDB hosts flexibility of both representing the data and at the same time retrieving the data.

Another significant feature of databases is versioning [14], which is also called version control or revision control. It is an obvious fact that changes in any database are usual and are frequent sometimes and so a mechanism to identify the changes is really worth it and the same is versioning. New alterations/changes are allocated a unique number, also called revision or version. The feature of giving versions lets evaluating the newest or a specific version in the database tables considerably easing the data audit trailing in the systems. However, in general, some basic information like who is involved in editing the data, the changes that have been made, the time of occurrence of change, etc. are required for audit trailing.

Though it is not a built-in feature, MongoDB gives a version number to the document at the application layer. However, a way out to attain efficacy is to own a couple of databases; one that is application data store and the other the audit trail. CouchDB, on the other hand, employs the multi-versioning concurrency control (MVCC) [15, 16] system while addressing issues in versioning. Despite this, CouchDB does not offer any support for default document versioning. However, it implements MVCC form, very specially to prevent the strict requisites pertaining to exclusive data locking files during the write operation. Nevertheless, any collisions in versions are fixed by conflicted data merging into making one file at the application level, succeeded by the abolition of the reeking part.

In HBase, timestamps do the version management of the data. A change at any particular instant is updated by giving a new timestamp earmarking it as the latest version in the database. In HBase, all of the columns stand entitled to have n variety of versions, with data access well facilitated at a specific time or a range of time, though the method involves uncomplimentary issues. Much in a similar way, Google's BigTable using the timestamps, on the other hand, addresses versioning. Each and every cell of a BigTable is competent in accepting the same data with manifold versions, with timestamp tagging (64-bit integers); the versions are organized in descending order of the timestamp, highlighting the latest version at the top of the line for ease of reading. On assigning a tag by BigTable, it shows the real time up to milliseconds, whereas in other database models it must be done by the applications themselves by generating timestamps that are unique, thus avoiding any conflict.

In the Cassandra model, a default mechanism to assign data version is absent; nevertheless, timestamps are utilized to award versions at the query level. In this connection, the Cassandra Query Language (CQL) adopts the semantic versioning regulations strictly [17] in version assigning, tagging the version in the format of X.Y.Z, with X, Y, and Z being integers, representing major, minor, and patch level values, respectively. Except for the basic versioning of CrowdDB at the RDBMS level, the entire version management of CrowdDB throws a challenge giving way for addressing from the scientific community.

Style of data processing is yet another significant element that the present paper also discusses. Indeed, batch processing [18] and real-time data processing (streaming) [19] do possess their own advantages and disadvantages. The former is known to be efficient while processing huge volumes of data with the transactions taken in clusters and analyzed in a time interval. Data is collected, processed, and the results are released in batches. On the other hand, the latter is extra effective while processing when there is a continuous input and output, with the data handled over a short time (near real-time). However, stream processing is effective when the input data is to be processed without storing it.

In comparison, Google's BigTable has a built-in element for batch processing. Much in a similar way, the HBase original does the processing of data in batches. Later, over some time, numerous frameworks came to light that claimed streaming of the data integrated with HBase. However, Cassandra endorses streaming of data while allowing atomic batches. Using batches approach offers an integrity during transactions in the circumstance if the coordinator yields to mid-batch. However, an atomic batch boasts high rate of success among all the batches if at all some part of the batch completes successfully.

In contrast, MongoDB hosts a powerful batch processing capacity when it comes to bulk inputs, updates as well event streaming. Similarly, a CouchDB too performs the processing in batches while importing, updating, or even deleting huge amounts of data. It even implements the processing of data in batches as crowdsourcing is done in batches [20].

When comprehended, these models form the core of the discussions with regard to their usage in the research, academia, and even in the industry while handling any encounters in big data. On the other hand, while discussing about a small entity of the current big data models, equal importance is given to weight and rationale for comparison. None of the models stated above include a unified modeling approach for big data.

3 Modeling Medical-Sensitive Big Data

One of the major impediments to using big data across the healthcare sector is to know how medical data spread across many avenues is handled by various government organizations, hospitals, and private organizations. Novel approaches including new infra requirements and data providers who work together in the data sources specified above are the need of the hour. Besides, it is equally important to implement new data analysis tools and strategies. Nevertheless, the need for healthcare analytics has already advanced from normal regression-based methodologies to more advanced analytics, like prognostic, machine learning, and so on, to keep pace with the industry.

Medical data consists of data of various types, like text specifying about the investigations of the patient, statistical values of various tests (biochemical, pathological, and microbiological), medical images, and their respective impressions. All these data can be obtained from various sources. A few are available in the form of electronic health records; some are available as paper records and a few are available as images. To have effective analytics in healthcare, data must be electronic. A brief flow of making data as electronic is shown in Fig. 1 at a single station.

An abstract form of sample document record of an entity is shown in Fig. 2.

Included are various parameters in association with a physician and few parameters that take part in the document store.

At a station, a document of entity is generated new if the entity is new. Else if the entity is an old one, the existing document is revised and a new document with an extension is linked to the existing document. All documents are made available

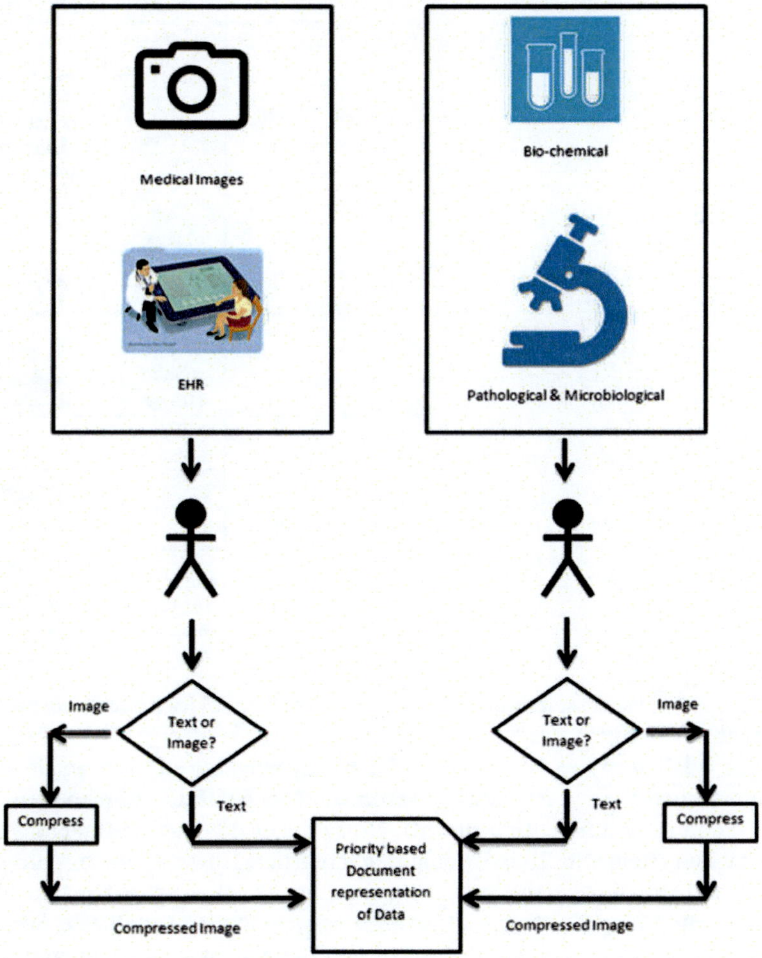

Fig. 1 Flow of making data as electronic

Fig. 2 Document structure

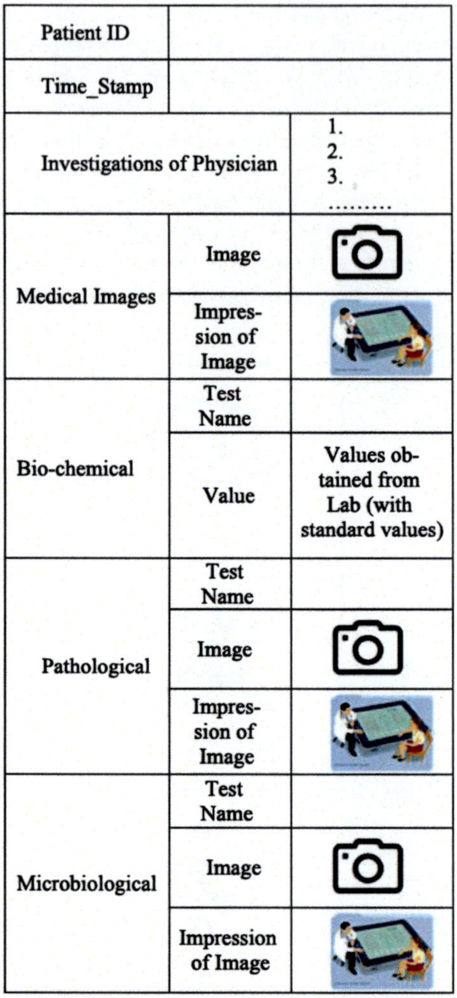

globally but to be accessed by only the authorized personnel for performing analytics. This is shown in Fig. 3.

This modeling approach uses MongoDB for representing an entity and its respective information scraped from various resources. MongoDB is a storehouse of documents comprising different documents. Every document signifies an entity in the ambit of the system with its individual characteristics. However, various properties, content, and size of the entity vary. All entities in a selection are either alike or relative in a way. MongoDB shuns enforcing a schematic plan for all selections alike but uses a schema that is dynamic in nature. These semantics are appropriate for echo systems with variable properties. They reflect non-null properties, however, they ignore properties with nulls; a thin data representation. On the other hand,

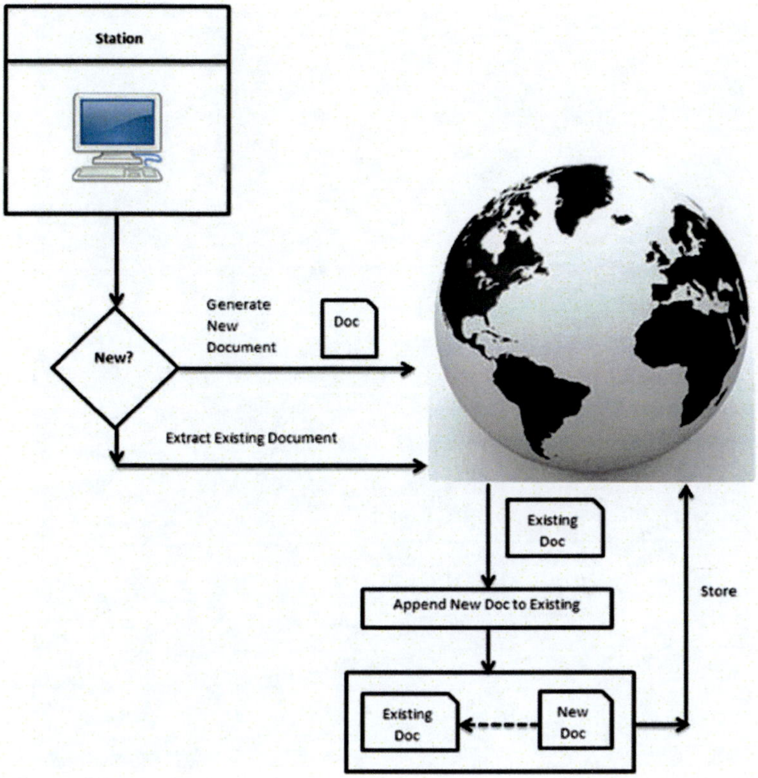

Fig. 3 Documents globally

properties with multiple values are deemed as manifold values as per one mutual property, whereas in other circumstances multiple values pertaining to a property are developed into a solo blended value that turns analyzing into an inheritance.

For implementing this model, we have gathered patient data from local hospitals. While developing documents each test report of a patient is taken along timestamp. Figure 4 shows a part of the document store.

A. *Advantages*

Representing big data of any context using this model is at most flexible. Its approach adds more advantage to medical big data. If medical data is modeled using this approach and all hospitals have global access to these documents, then patients take benefit in terms of proper and adequate care. Patients can also be given advance medication by a remotely located physician.

Values for an entity that are NULL (unknown) are not taken into account. As medical data is sensitive, values that are null must not be replaced with a prediction

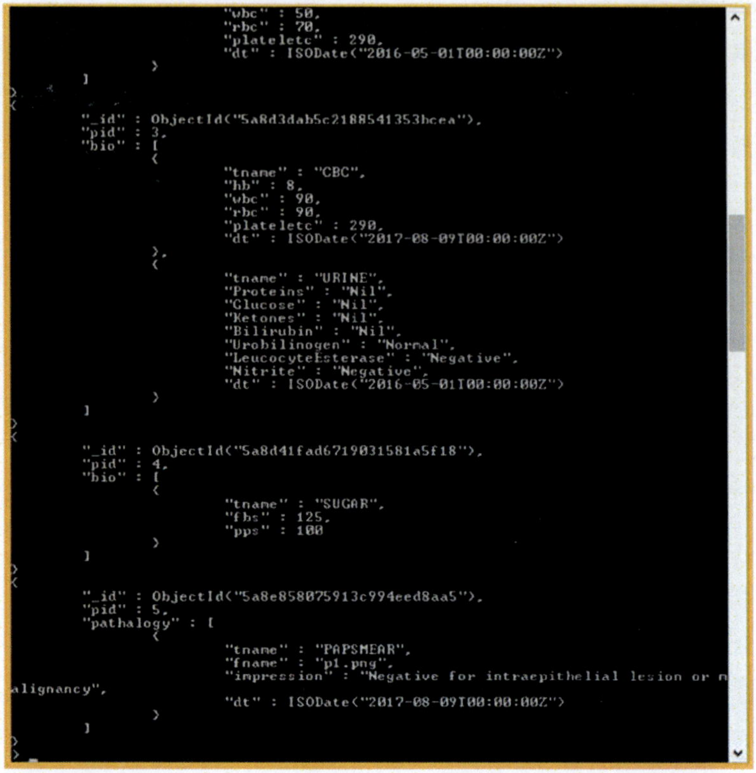

Fig. 4 Sample document

method. Replacing nulls always dilutes the performance of analytics over medical data which is not ethical. Preprocessing of unknown data is not required.

4 Conclusion

A comparison is made among HBase, MySQL, and MongoDB in extracting records and is shown in Fig. 5.

The above graph reveals that MongoDB is flexible in representing medical big data. Global representation of entity documents is the biggest and toughest challenge which demands a unified modeling approach for medical data.

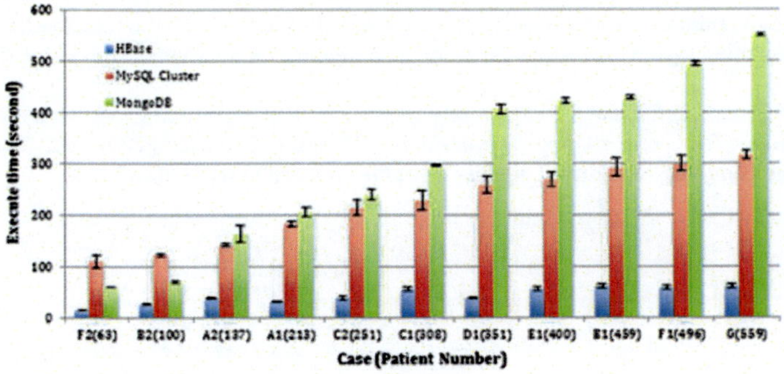

Fig. 5 Comparison of MongoDB with HBase and MySQL

References

1. Jacobs A (2009) The pathologies of big data. Commun ACM 52(8):36–44
2. McAfee A, Brynjolfsson E, Davenport TH, Patil DJ, Barton D (2012) Big data: the management revolution. Harv Bus Rev 90(10):60–68
3. Lynch C (2008) Big data: how do your data grow? Nature 455(7209):28–29
4. Zikopoulos P, Eaton C, deRoos D, Deutsch T, Lapis G (2011) Understanding big data: analytics for enterprise class Hadoop and streaming data. McGraw-Hill Osborne Media
5. Manyika J, Chui M, Brown B et al (2011) Big data: the next frontier for innovation, competition, and productivity. McKinsey Global Institute
6. Borckardt JJ, Nash MR, Murphy MD, Moore M, Shaw D, O'Neil P (2008) Clinical practice as natural laboratory for psychotherapy research: a guide to case-based time-series analysis. Am Psychol 63(2):77–95
7. Celi LA, Mark RG, Stone DJ, Montgomery RA (2013) 'Big data' in the intensive care unit: closing the data loop. Am J Respir Crit Care Med 187(11):1157–1160
8. Carino F, Sterling WM (1998) Method and apparatus for extending existing database management system for new data types. U.S. Patent No. 5,794,250
9. Franklin MJ (2012) CrowdDB: query processing with people and machines. New England Database Society, HP/Vertica Computer Science Lounge, Brandeis University, USA
10. Bertino E, Beng CO, Ron SD, Kian LT, Justin Z, Boris S, Daniele A (2012) Indexing techniques for advanced database systems. Springer Publishing Company, Incorporated
11. Lakshman A, Malik P (2011) The Apache Cassandra project. World Wide Web: http://cassandra.apache.org/. Accessed 31 Oct 2014
12. Dimiduk N, Khurana A (2012) HBase in action, 1st edn. Manning Publications, p 350. ISBN 978-1617290527
13. Jin A, Cheng C, Ren F, Song S (2011) An index model of global subdivision in cloud computing environment. In: 19th international conference on geoinformatics, pp 1–5
14. Shaughnessy ST (2012) Database system providing high performance database versioning. U.S. Patent No. 8,117,174. U.S. Patent and Trademark Office, Washington, DC
15. Cattell R (2011) Scalable SQL and NoSQL data stores. ACM SIGMOD Rec 39(4):12–27
16. Suryavanshi R, Yadav D (2012) Modeling of multiversion concurrency control system using Event-B. In: Proceedings of the federated conference on computer science and information systems, Warsaw, Poland, pp 1397–1401
17. OSGi Alliance (2010) Semantic versioning. Technical Whitepaper, Revision 1.0. World Wide Web: http://www.osgi.org/wiki/uploads/Links/SemanticVersioning.pdf. Accessed 31 Oct 2014
18. Damodaran P, Vélez-Gallego MC (2012) A simulated annealing algorithm to minimize makespan of parallel batch processing machines with unequal job ready times. Expert Syst Appl 39(1):1451–1458

19. Gulisano V, Jimenez-Peris R, Patino-Martinez M, Valduriez P (2010) Streamcloud: a large scale data streaming system. In: 30th international conference on distributed computing systems, pp 126–137
20. Ramesh S (2011) CrowdDB—answering queries with crowdsourcing. Master's thesis. http://www.kutter-fonds.ethz.ch/App_Themes/default/datalinks/RameshMT2011.pdf, https://www.datapine.com/blog/big-data-examples-in-healthcare/. Accessed 4 May 2014

Studies on Effect of Leaf Roller (*Diaphania pulverulentalis*) Infestation on the Mineral Composition of Mulberry (Morus Sp.) Varieties

C. T. Bhagyamma and N. Vijaya Kumari

Abstract Mulberry foliage is the food for the silkworm, *Bombyx mori* L. The healthy growth of silkworm and its larval, cocoon parameters are known to be influenced highly by the nutritional levels of the leaves. The mineral elements are known to a play major role in the production of mulberry quality. Macro and micronutrients were analyzed in three mulberry popular varieties, V_1, S_{36} and Mysore local infested with leaf roller *Diaphania pulverulentalis*. Feeding such infested leaves to the silkworms, mainly young age mulberry plants, will affect the growth, decrease the quantity as well as the quality of raw silk production.

Keywords Micro and macronutrients · Silkworm · Mulberry · Growth · Leaf roller

1 Introduction

Mulberry *Morus* species forms the food plant of silkworm; it depends on the quantity and quality of mulberry foliage highly significant in order to release silkworm rearing that is good profitable. The process of mulberry leaf production is inflated due to the interference caused by insect pests. Since 1995, a new pest called the leaf roller, *Diaphania pulverulentalis* has been recorded to infest mulberry predominantly in the southern silk-producing states of India, viz., Karnataka, Andhra Pradesh and Tamil Nadu [44] (Rajadurai et al. 1999; Siddegowda et al. 1995).

The main target site of leaf roller infestation is the mulberry apical tip part and unopened tender bud of mulberry shoot. The infested leaves become rolled because of silky salivary secretion produced by the larva. Generally, a single leaf is rolled into folded shape with the web from the larva, which remains inside. Sometimes two leaves are rolled into folded shape with the web from the larva. Hence, the pest is commonly called as **"leaf roller"**.

C. T. Bhagyamma (✉) · N. Vijaya Kumari
Department of Biosciences and Sericulture, Sri Padmavati Mahila Visvavidyalayam (Women's University), Chittoor (dt), Tirupati 517502, Andhra Pradesh, India
e-mail: chittibhagyasree@gmail.com

S. Jyothi et al. (eds.), *Proceedings of the 2nd International Conference on Computational and Bio Engineering*, Lecture Notes in Networks and Systems 215,
https://doi.org/10.1007/978-981-16-1941-0_18

Recently, the pest lepidopteron *Diaphania pulverulentalis* highly impacts the quality of mulberry leaves infestation. Leaf roller is one such serious pest that causes a foliage loss up to 12.18%. The average incidence of this pest on mulberry was 27.85% in Karnataka, 20.98% in Andhra Pradesh and 16.48% in Tamil Nadu (Rajadurai et al. 1999). *Diaphania pulverulentalis* is one of the major serious pests known to cause a foliage loss up to 78.05%. The present study is the average incidence of leaf roller on mulberry up to 76.12%. The high activity of leaf roller on mulberry was also observed in the present study in Ananthapuramu district in Andhra Pradesh, India.

It damages the soft green leaf tissues and later damages the apical shoot part, thus leading to drying up of leaf part; hence the plant growth is stunted. The pest leaf roller not only reduces the leaf yield but also reduces the nutritional levels of the foliage. The present study is mainly aimed to investigate the effect of the nutritional composition on various varieties, viz., V_1, S_{36}, Mysore local which are under cultivation in different sericulture villages of Ananthapur district.

2 Materials and Methodology

The present study is done under natural conditions by collecting leaf roller infested as well as healthy leaves from different sericulture gardens of 6–8 years old in Ananthapur district during 2018–2020. For the study mulberry leaves are collected from the following villages consisting of 10 mandals, namely Agali, Amarapuram, Rolla, Gudibanda, Madakaṣira, Parigi, Hindupur, Penukonda, Somandepalli and Roddam. The healthy and infested mulberry leaves were collected in thin cotton green bags for selected mulberry gardens. The leaves were washed thoroughly in tap water; after that washed with double-distilled water, then dry it naturally for 6 h. After that leaves are dried in an electric oven at an optimum temperature between 40 and 45 °C for 36 h.

2.1 Estimation of Nitrogen in Mulberry Leaves

Nitrogen estimation was done in mulberry leaves by **Micro-Kjeldahl's** method as suggested by **Jackson** (1973) and modified by **Iswaran** and **Marwaha**[1] (1980). Nitrogen content was calculated using the following formula:

$$\text{Nitrogen content} = \frac{14.01 \times (\text{titration value} - \text{blank value}) \times \text{molarity of the acid}}{10 \times \text{weight of the sample}}$$

where 14.01 = standard volume, 1 N = molarity of acid.

2.2 *Estimation of Phosphorous in Mulberry Leaves*

The estimation of phosphorous in leaves by the **Vanado-molybdate yellow color method** was followed by **Jackson**[3] (1973).

A standard graph was prepared, and calculations were done using the following formula:

$$\textbf{Phosphorus content} = \frac{\textbf{Graph ppm} \times \textbf{value of digestive sample} \times \textbf{volume made up} \times 100}{10^6 \times \textbf{weight of sample} \times \textbf{aliquot taken}}$$

2.3 *Estimation of Potassium in Mulberry Leaves*

The triple acid digest was used for the estimation of potassium in the sample, and the made-up extract was directly read in the flame photometer by changing the filter for potassium. The percentage of potassium in the leaf tissue is calculated from the photometer readings[3].

$$\textbf{Potassium content} = \frac{\textbf{Graph ppm} \times \textbf{value of digestive sample} \times \textbf{volume made up} \times 100}{10^6 \times \textbf{weight of sample} \times \textbf{aliquot taken}}$$

The mineral analysis is done with one gram of dried mulberry healthy and infested leaf powder taken separately and added in 15 ml of nitric acid, then 10 ml of perchloric acid. This was digested over a sand bath until a clear solution was obtained. It was cooled and the volume was made up to 100 ml with deionized H_2O separately. It was filtered through Whatman no. 1 filter paper.

From the above solution, 25 ml solution was taken and macro and micronutrients are estimated by using **"Elicol CL 360 Flame photometer"** and **"Atomic Absorption Spectrophotometer"** (GBC 932 Plus) Martin[2] et al. (1987) (Figs. 1 and 2).

3 Results and Discussion

Mineral composition of mulberry found to reduce in the leaf roller infested mulberry leaves of three mulberry popular varieties of V_1, S_{36} and Mysore local compared to their healthy leaves.

Fig. 1 Healthy mulberry gardens and different varieties of mulberry plots

Fig. 2 Infested mulberry gardens (leaf roller) and collection of healthy and infested

3.1 Macronutrients

Macronutrients play a very important role in plant growth and development. Their main functions are structural and redox-sensitive agents. Nitrogen values were decreased in leaf roller foliage in the V_1 variety. The reduction was in the range of 2.97% (Mysore local) to **18.04% (V_1)**. It was low (2.97%) and relatively high (3.26%) in the S_{36} variety.

Nitrogen is necessary for the production of chlorophyll and is an important component for proteins and amino acids synthesis. Plants deficient in nitrogen plant appear pale yellow color as a result of reduced chlorophyll production. Nitrogen is the major component of chlorophylls and the most necessary pigment needed for photosynthesis. Therefore, plants may also be stunted. In the present study, it was observed that nitrogen content was decreased. Nitrogen is the main role of all living cells of mulberry.

Phosphorous also falls down by V_1, S_{36} and Mysore local varieties in the range of 5.55% (Mysore local) to **17.39% (V_1)**, respectively. It is involved in main activities like photosynthesis, the transport of energy (ATP) through the plant and lowering and also seed production. In this study plants show weaker stems, less growth and reduction in cell activity.

The leaf roller V_1, S_{36} and Mysore local varieties show decreased **potassium** content range of 3.75 (V_1) to **10.32% (S_{36})**. Mysore local was 4.04% decreased. **Calcium** (Ca) content decreased in three varieties. It was minimum of 2.27% (Mysore local), 4.70% V_1 and S_{36} was 3.61%. **Magnesium (Mg)** and **sulfur (S)** content showed a decrease in leaf roller infestation. Mg highly decreased to 9.87% in $V_{1,}$ followed by 5.17% $S_{36,}$ and Mysore local was 4.83%. Sulfur also decreased and it was 5.88% (Mysore local) to 6.66% (S_{36}), and V_1 was 2.77%.

3.2 Micronutrients

The presence of micronutrients in healthy and infested mulberry leaves was compared. The **zinc** content showed decreased range in Mysore local 1.35 to 3.28% in S_{36}, and V_1 was 2.15%. The **copper** (Cu) content was reduced in three varieties also. The range of Cu content decrease from 0.49 (S36) to 2.47% in Mysore local, and for V_1 it was 1.01%. The **iron** (Fe) content decreased in the range from 1.21 (Mysore local) to 4.66% S_{36} variety, and for V_1 it was 2.28%. The **boron** (B) content was also decreased in the range of 0.12% in S_{36} to 3.55% in V_1 variety, and for Mysore local it was 0.73%. **Molybdenum** (Mo) content also recorded a reduction, and it was in the range from 3.70 (V_1) to 5.88% in Mysore local.

The **chloride** (Cl) content decreased from 2.16 (S_{36}) to 5.08% (Mysore local) and for V_1 it was 2.23%. The **manganese** (Ma) content also recorded a reduction in the range from 1.64 (Mysore local) to 3.52 (V_1) variety, and for S_{36} it was 3.03%. Leaf roller pest-infested plants indicate typically symptoms of nutrient deficiency in

Table 1 Macronutrients changes in leaf roller (%) infested mulberry foliage in Ananthapuramu district Andhra Pradesh during the year 2018–2020

S. no	Macronutrients		Mulberry varieties		
			V_1	S_{36}	Mysore local
1	Nitrogen (%)	Healthy	4.49	3.37	2.35
	% of decrease over control	Infested	3.68	3.26	2.28
			18.04	**3.26**	**2.97**
2	Phosphorus (%)	Healthy	0.23	0.18	0.18
	% of decrease over control	Infested	0.17	0.19	0.17
			17.39	**11.10**	**5.55**
3	Potassium (%)	Healthy	2.93	1.84	2.72
	% of decrease over control	Infested	2.82	1.65	2.61
			3.75	**10.32**	**4.04**
4	Calcium (%)	Healthy	2.34	2.21	2.64
	% of decrease over control	Infested	2.23	2.13	2.58
			4.70	**3.61**	**2.27**
5	Magnesium (%)	Healthy	0.81	0.58	0.62
	% of decrease over control	Infested	0.73	0.55	0.57
			9.87	**5.17**	**4.83**
6	Sulfur (%)	Healthy	0.36	0.15	0.34
	% of decrease over control	Infested	0.35	0.14	0.32
			2.77	**6.66**	**5.88**

particularly nitrogen, phosphorous, micronutrients, manganese, which are important for plant growth and development. The nutrient transport could be affected by a decrease in water and mineral transport uptake of plants. The pest-infested mulberry leaves are quantitatively and qualitatively very poor (Narayanaswamy et al. 1999) (Tables 1 and 2, Figs. 3 and 4).

4 Conclusion

The leaf roller *Diaphania pulverulentalis* infested mulberry gardens have shown significant depletion in nutritive value. Silkworm's growth and its development mainly depend upon the quality of mulberry leaves. If the mulberry leaf quality is decreased, it affects silkworm rearing. The influence of macro and micronutrients, their excess and also deficiencies both show an impact on the mulberry quality. Rearing of silkworm's quality and quantity depends on the mineral nutrition of mulberry leaves. The mineral nutrition decreased due to leaf roller infestation and causes physiological disorders in mulberry.

In this study, mulberry nutrition is not balanced because of leaf roller pest damage. Therefore, the sericulture farmers must avoid feeding silkworms with infested leaf roller. The results concluded that the leaf roller on mulberry altered nutritive parameters causing yield and quality reduction in mulberry leaves and which affects on

Table 2 Micronutrients changes in leaf roller-infested mulberry foliage in Ananthapuramu district Andhra Pradesh during 2018–2020

S. no	Micronutrients		Mulberry varieties		
			V_1	S_{36}	Mysore local
1	Zinc (ppm) **% of decrease over control**	Healthy Infested	139 136 **2.15**	152 147 **3.28**	74 73 **1.35**
2	Iron (ppm) **% of decrease over control**	Healthy Infested	263 257 **2.28**	193 184 **4.66**	246 243 **1.21**
3	Copper (ppm) **% of decrease over control**	Healthy Infested	7.89 7.83 **1.01**	8.16 8.12 **0.49**	6.86 6.65 **2.47**
4	Boron (ppm) **% of decrease over control**	Healthy Infested	27.01 26.05 **3.55**	24.03 23.06 **0.12**	39.36 39.07 **0.73**
5	Manganese (ppm) **% of decrease over control**	Healthy Infested	142 137 **3.52**	132 121 **3.03**	59.01 58.04 **1.64**
6	Molybdenum (ppm) **% of decrease over control**	Healthy Infested	0.27 0.25 **3.70**	0.22 0.21 **4.54**	0.11 0.10 **5.88**
7	Chloride(%) **% of decrease over control**	Healthy Infested	1.34 1.31 **2.23**	1.85 1.81 **2.16**	0.59 0.56 **5.08**

ppm = parts per million

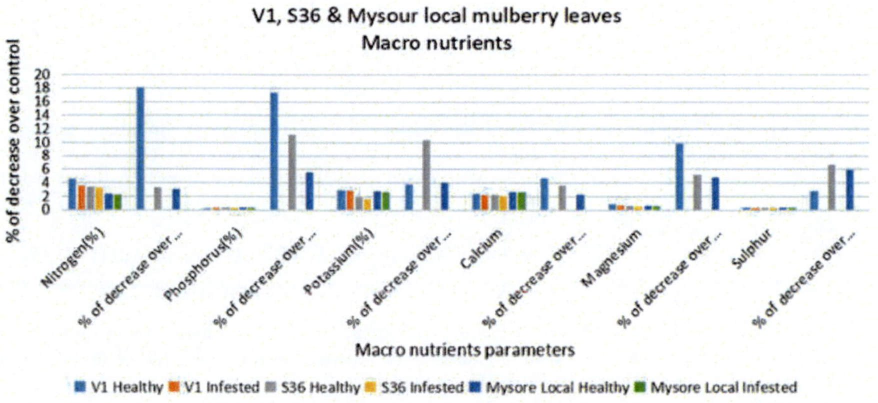

Fig. 3 Effect on leaf roller-infested macronutrients status of different mulberry varieties V_1, S_{36} and Mysore local during the year August 2018–2020 January

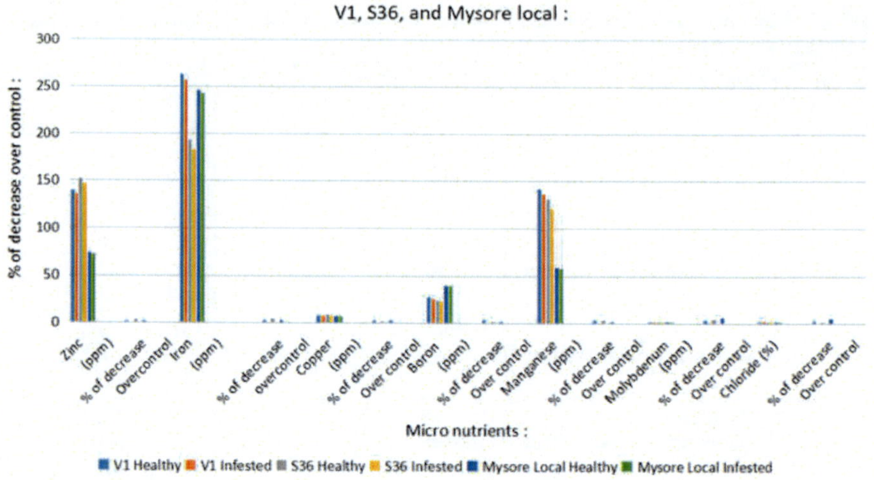

Fig. 4 Effect on leaf roller *Diaphania pulverulentalis* micronutrients of different mulberry varieties V_1, S_{36} and Mysore local during the year August 2018–2020 January

young age, adult silkworm health and quality of cocoon production, and finally leads to an economical loss in sericulture.

Acknowledgements The authors sincerely thank the staff of DST-CURIE, Sri Padmavathi Mahila Visvavidyalayam, Tirupati.

References

1. Consortium EP (2012) An integrated encyclopedia of DNA elements in the human genome. Nature 489(7414):57–74
2. Kavakiotis I, Tsave O, Salifoglou A, Maglaveras N, Vlahavas I, Chouvarda I, Kavakiotis I et al (2017) Machine learning and data mining methods in diabetes research. Comput Struct Biotechnol J 15: 104–116
3. Cornall RJ, Prins J-B, Todd JA, Pressey A, DeLarato NH, Wicker LS, Peterson LB (1991) Type 1 diabetes in mice is linked to the interleukin-1 receptor and lsh/Ity/bcg genes on chromosome 1. Nature 353(6341):262–265
4. Sathishkumar C, Prabu P, Mohan V, Balasubramanyam M (2018) Linking a role of lnc RNAs (long non-coding RNAs) with insulin resistance, accelerated senescence, and inflammation in patients with type 2 diabetes. Human Genomics 12, 41
5. Alberti KG, Zimmet PZ (1998) Defnition, diagnosis and classifcation of diabetes mellitus and its complications. Part 1: diagnosis and classifcation of diabetes mellitus provisional report of a WHO consultation. Diabet Med 15(7): 539–553
6. Kaddis JS, Olack BJ, Sowinski J, Cravens J, Contreras JL, Niland JC (2009) Human pancreatic islets and diabetes research. JAMA J Am Med Assoc 301(15):1580–1587. https://doi.org/10.1001/jama.2009.482
7. World Health Organization (2015) Guideline: sugars intake for adults and children. World Health Organization. http://www.who.int/iris/handle/10665/149782

8. Avogaro P, Crepaldi G, Enzi G, Tiengo A (1967) Associazione di iperlipidemia, diabetemellito e obesita' di mediogrado. Acta DiabetolLat 4:36–41
9. American Diabetes Association (2012) Diagnosis and classifcation of diabetes mellitus. Diab Care 35(1):S64-71. https://doi.org/10.2337/dc12-S064
10. Raha O, Chowdhury S, Dasgupta S et al (2009) Approaches in type 1 diabetes research: a status report. Int J Diab Dev Ctries 29(2):85–101. https://doi.org/10.4103/0973-3930.53126
11. Bellamy L, Casas JP, Hingorani AD, Williams D (2009) Type 2 diabetes mellitus after gestational diabetes: a systematic review and meta-analysis. Lancet 373:1773–1779
12. Meigs JB, D'Agostino RB Sr, Wilson PW, Cupples LA, Nathan DM, Singer DE (1997) Risk variable clustering in the insulin resistance syndrome: the Framingham Ofspring Study. Diabetes 46:1594–1600
13. Anna V, van der Ploeg HP, Cheung NW, Huxley RR, Bauman AE (2008) Socio-demographic correlates of the increasing trend in prevalence of gestational diabetes mellitus in a large population of women between 1995 and 2005. Diab Care 31(12):2288–2293
14. https://www.news-medical.net/health/Diabetes-Mellitus-Subtypes.aspx
15. Barengo NC, Katoh S, Moltchanov V, Tajima N, Tuomilehto J (2008) The diabetes-cardiovascular risk paradox: results from a finnish population-based prospective study. Eur Heart J 29(15):1889–1895
16. World Health Organization (2013) Diagnostic criteria and classification of hyperglycemia first detected in pregnancy. World Health Organization. http://www.who.int/iris/handle/10665/85975
17. Long J, Danesh FR (2018) Values and limitations of targeting lncRNAs in diabetic nephropathy. Diabetes 67: 552–553.l https://doi.org/10.2337/dbi17-0052
18. Oliver F, Rajendra AU, Ng EY, KwanHoong N, Jasjit SS (2012) Algorithms for the automated detection of diabetic retinopathy using digital fundus images: a review. J Med Syst 36(1):145–157. https://doi.org/10.1007/s10916-010-9454-7
19. Scott MG, Ivor JB, Gregory LB, Alan C, Robert HE, Barbara VH, William M, Sidney CS, James RS. Diabetes and cardiovascular disease a statement for healthcare professionals from the American Heart Association. Circulation. 1999;100(10):1134–46.
20. de Mattos Matheus AS, Tannus LR, Cobas RA, Sousa Palma CC, Negrato CA, de Brito Gomes M (2013) Impact of diabetes on cardiovascular disease: an update. Int J Hyperten 65:15. https://doi.org/10.1155/2013/653789
21. Wilson RA, Keil FC (1999) The MIT encyclopaedia of the cognitive sciences. MIT Press
22. Nithya B, Ilango V (2017) Predictive analytics in health care using machine learning tools and techniques. In: International conference on intelligent computing and control systems. 978–1–5386–2745–7
23. Chen Y, He Y, Zhou H (2020) The potential role of lncRNAs in diabetes and diabetic microvascular complications. The Japan Endocr Soc 67(7): 659–668
24. Li X, Wu Z, Fu X, Han W (2014) lncRNAs: insights into their function and mechanics in underlying disorders. Mutat Res Rev Mutat Res 762:1–21
25. Li R, Zhu H, Luo Y (2016) Understanding the functions of long non-coding RNAs through their higher-order structures. Int J Mol Sci 17:702
26. Wallace C, Smyth DJ, Maisuria-Armer M, Walker NM, Todd JA, Clayton DG (2010) The imprinted DLK1-MEG3 gene region on chromosome 14q32.2 alters susceptibility to type 1 diabetes. Nat Genet 42: 68–71
27. Joshi TN, Chawan PM (2018) Diabetes prediction using machine learning techniques. Int J Eng Research Appl 8(1, Part -II): 09–13. ISSN: 2248–9622
28. Daghistani T, Alshammari R (2016) Diagnosis of diabetes by applying data mining classification techniques. Int J Adv Comput Sci Appl (IJACSA) 7(7):329–332
29. Meng X-H, Huang Y-X, Rao D-P, Zhang Q, Liu Q (2013) Comparison of three data mining models for predicting diabetes or prediabetes by risk factors. Kaohsiung J Med Sci 29(2):93–99
30. Wu H, Yang S, Huang Z, He J, Wang X (2018) Type 2 diabetes mellitus prediction model based on data mining. Inform Med Unlocked 10:100–107

31. Vijayan V, Ravikumar A (2014) Study of data mining algorithms for prediction and diagnosis of diabetes mellitus. Int J Comput Appl 95(17)
32. Saravananathan K, Velmurugan T (2016) Analyzing diabetic data using classification algorithms in data mining. Indian J Sci Technol 9(43):196–201
33. Saeidi L, Ghaedi H, Sadatamini M, Vahabpour R, Rahimipour A, Shanaki M, Mansoori Z, Kazerouni F (2018) Long non-coding RNA LY86-AS1 and HCG27_201 expression in type 2 diabetes mellitus. Mol Biol Rep 45(6):2601–2608
34. Mansoori Z, Ghaedi H, Sadatamini M, Vahabpour R, Rahimipour A, Shanaki M, Kazerouni F (2018) Downregulation of long noncoding RNAs LINC00523 and LINC00994 in type 2 diabetes in an Iranian cohort. Mol Biol Rep 45(5):1227–1233
35. Kazerouni F, Bayani A, Asadi F, Saeidi L, Parvizi N, Mansoori Z (2020) Type2 diabetes mellitus prediction using data mining algorithms based on the longnoncoding RNAs expression: a comparison of four data mining approaches. BMC Bioinformatics. 21: 372. https://doi.org/10.1186/s12859-020-03719-8
36. Kong L, Zhang Y, Ye ZQ et al (2007) CPC: assess the protein-coding potential of transcripts using sequence features and support vector machine. Nucleic Acids Res 35(Suppl 2):W345–W349
37. Wang L, Park HJ, Dasari S et al. (2013) CPAT: coding-potential assessment tool using an alignment-free logistic regression model. Nucleic Acids Res 41(6): e74
38. Sun L, Luo H, Bu D et al. (2013) Utilizing sequence intrinsic composition to classify protein-coding and long non-coding transcripts. Nucleic Acids Res 41(17): e166
39. Li A, Zhang J, Zhou Z et al. (2014) PLEK: a tool for predicting long non-coding RNAs and messenger RNAs based on an improved kmer scheme. BMC Bioinformatics 15: 311
40. Kang YJ, Yang DC, Kong L et al. (2017) CPC2: a fast and accurate coding potential calculator based on sequence intrinsic features. Nucleic Acids Res 45(W1): W12–6
41. Pian C, Zhang G, Chen Z, Chen Y, Zhang J, Yang T, Zhang L (2016) LncRNApred: classification of long non-coding RNAs and protein-coding transcripts by the ensemble algorithm with a new hybrid feature 11(5): e0154567
42. Achawanantakun R, Chen J, Sun Y, Zhang Y (2015) LncRNA-ID: long non-coding RNA identification using balanced random forests 31(24): 3897–3905
43. Zhang X, Wang J, Li J, Chen W, Liu C (2018) CRlncRC: a machine learning-based method for cancer-related long noncoding RNA identification using integrated features. BMC Med Genomics 11(Suppl 6): 120. https://doi.org/10.1186/s12920-018-0436-9
44. Geetha bai M, Marimadaiah, B, Narayanaswamy, KC, Rajgopal D (1997) An outbreak of leaf roller pest, Diaphania (Margaronia) pulverulentalis (Hampson) on mulberry in Karnataka. Geobios New Rep 16(2):73–79

Performance Comparison of Semantic Web Data Using Virtual and Cloud Services

C. Lakshmi and K. Usha Rani

Abstract Big Data techniques describe a new generation of technologies and architectures designed to economically extract value from an enormous volume of a wide variety of data. Hadoop does a massively parallel processing technique for processing data distributed on a commodity cluster rather than sequential. The mapper and reducer phase in Hadoop MapReduce Model helps to overcome the processing of large amounts of data in a distributed environment, which will affect the performance. Data can be divided into smaller chunks and stores each part of the data on a separate node within the cluster. Nodes are arranged using the Amazon web service and the virtual environment with different data loading sizes. Compared with both frameworks, Hadoop configured in a cloud plays a vital role in data warehousing and clustering of the data to get the best performance than virtual environment. Hence in this study, we segregate the semantic web data into pre-processing and clustering to enhance the performance improvement.

Keywords Hadoop · MapReduce · Spark · Hive · Data warehousing · Clustering

1 Introduction

Hadoop is an open-source technology that started a Big Data wave. It allows distributed data storage and computing using low-cost hardware. It can scale petabytes of data and can run on clusters with hundreds of nodes. Hadoop has two components: the Hadoop Distributed File System (HDFS) that provides data storage and Map-Reduce is a programming model and implementation that provides distributed computing capabilities with data stored in HDFS. HDFS still helps in keeping the large quantities of data at a minimal cost. It includes scaling and cost

C. Lakshmi · K. Usha Rani (✉)
Department of Computer Science, Sri Padmavati Mahila Visvavidyalayam, Tirupati, AP, India
e-mail: usharanikuruba@yahoo.co.in

C. Lakshmi
e-mail: lakshmibhavya@gmail.com

S. Jyothi et al. (eds.), *Proceedings of the 2nd International Conference on Computational and Bio Engineering*, Lecture Notes in Networks and Systems 215,
https://doi.org/10.1007/978-981-16-1941-0_19

benefits that help in its adoption. It is most suitable for enterprises with in-house data centers to host the data within internal networks; cloud options like AWS S3, Google Cloud Storage, and Azure Blob. They are also becoming increasingly popular. MapReduce, on the other hand, is becoming old. Simultaneously, it scales horizontally over hundreds of compute nodes; it is prolonged and uses disk storage for intermediate caching instead of memory. Newer technologies like Apache Spark and Apache Flink have emerged to execute the same processing at much faster rates. It supports other capabilities and a growing library of connectors, making them a better choice than Map-Reduce. Apache Spark is also an open-source technology that started a more efficient alternative to Hadoop Map-Reduce. It is a large-scale distributed data processing engine; Spark stores its data primarily in memory to speed up computations. It has capabilities in batch processing, real-time streaming, machine learning, and graphs. To process the large amount of the Big Data using the traditional systems becomes ineffective in clustering the big datasets; even k-means provide the fastest clustering algorithm. Day-to-day updations on the repositories cannot afford in clustering on massive datasets using traditional k-means. Hadoop provides the distributed interconnected machines to complete in processing large data sets using node management algorithms. Apache Hadoop is an open-source designed architecture having two main components used in the distributed Map-Reduce model.

Processing part: It recognizes and executes the programming model [1], which is designed using Map-Reduce Paradigm. Hence the k-means clustering algorithm is applied by converting the Map-Reduce model in Hadoop.

Storage part: Using the storage management system in blocks, Hadoop Distributed File System (HDFS) [1] is designed.

2 Overview of Map-Reduce

Map-Reduce 1.0 has several limits inherent in the API. It allows only batch processing. Coding Map-Reduce jobs are tricky and miss enterprise features, like security and high availability essential for enterprise editions. Map-Reduce 2.0 enters with YARN (Yet Another Resource Negotiator). It adds an abstraction layer between Map-Reduce and HDFS and drives Hadoop's adaption more than anything else. This allows the Hadoop ecosystem to be used for not only the traditional batch processing but also provides for other processing to occur on top of the HDFS File system made enjoyable. When the volume of the data increasing exponentially, the business wants to have the cheap storage of HDFS [1] with fast processing, so; YARN allows a lot of possibilities according to the business requirements. There have been a plethora of new libraries coming out of the open-source community.

Mapper stage:

As shown in Fig. 1, Map-Reduce is a massively parallel processing technique for processing data distributed on a commodity cluster rather than processing serially,

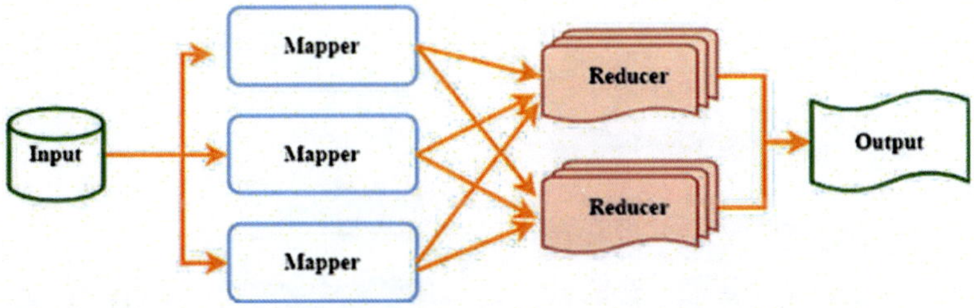

Fig. 1 Map-Reduce data processing model

saving time. Data Flow in Map-Reduce Framework can be considered by taking input consisting of multiple different records. Those records will be split by the input split into numerous logical models that can be handled by the mapper. Record reader collects the data from splitter and converts it into <Key, Value> pairs because mappers can only read and hold key-value pairs. Then the output will be made available to the mapper. A developer can build the logic in mapper according to the business requirement [2]. The mapper will produce an intermediate product [3], and these outputs will not be stored onto HDFS but will be stored on the local HDD then. Shuffling and sorting can be done in this phase and release the output to the reducer stage.

Reducer Stage:

Reducer task optimization in Fig. 1 usually falls under four categories: [2] subdividing the task (changing the jobs), logging/debugging helps in knowing the processing node's status, set thresholds in the configurations. The reducer stage accepts the input from the output of the Mapper stage.

OutputCollector.collect() method

This method writes the output to the reduce phase to the File System. The <key – Values> received from the Mapper stage will be in Text and Int Writable form. The reducers' performance varies for different algorithms for making clusters and finding the maximum temperature from the clusters formed. Here Fig. 2 [4] shows how parallel processing occurs when a Hadoop job is executed. There are two different stages in the Map-reduce execution flow

(1) Map Phase
(2) Reduce Phase

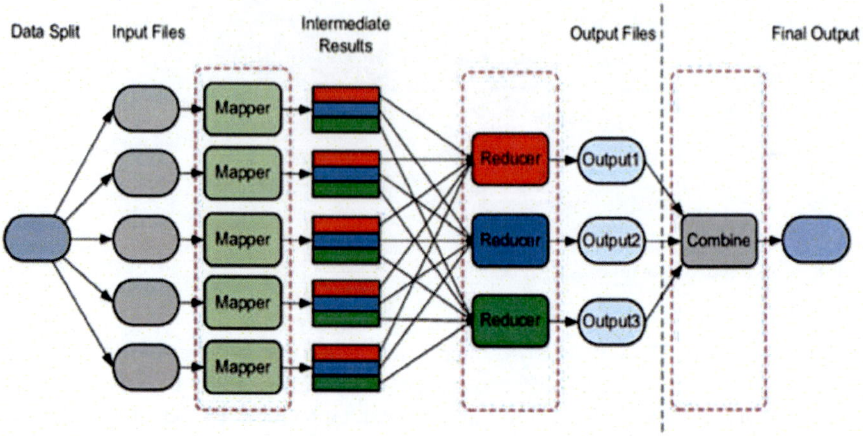

Fig. 2 Map-Reduce execution flow [4]

3 Literature Survey

Lakshmi [5] proposed integration between Big Data processing techniques like Spark and Hive for Querying the RDF semantic Web data and provided efficient query processing difference between Spark Engine and Hive Component.

SAHARA [6] OpenStack and Apache Hadoop represent the largest open source communities. Their integration will benefit users of both of these communities. Considering this, there have been some efforts in this area, the most notable of them is the project. This project is a collaboration of RedHat, Mirantis, and Hortonworks.

Lakshmi [7] proposed an integration between Spark in-memory processing tool and PIG Latin and shown the efficient results than Hive and Spark.

In web data applications, triple stores are playing a vital role, and triples processing is very complicated for those applications. As a consequence, when running in real-time applications, represent the strength and weaknesses. According to Morsey et al [8], creating the SPARQL benchmark procedures have applied to the DBpedia knowledge base and comparing them with relational and triple stores, performance measurement against a relational database was converted into RDF.

3.1 Node Management

Hadoop provides the best scalability in the data storage part. However, it lacks dynamicity in node management [9]. The reason behind this is because all the nodes in Hadoop are tightly coupled from each other. Using this management system, we can manage the Hadoop cluster's node mechanism with a concrete implementation. Nodes can be removed and add from the cluster whenever necessary and available

[10]. Before implementing the system in Hadoop, we need to analyze the node management mechanism. Three states in Hadoop define the status of the nodes.

Normal: It states that the node is in service. In order to remove the node whenever required, Hadoop sets a property for that.

dfs.hosts. exclude

Decommission_Inprogress: When a user changes a hostname and requests the refreshing, then the node's status says Decommission in progress.

Decommissioned: After refreshing is done, the status is transited to decommissioned and sets ready to remove the node safely from the cluster. Every node sends a heartbeat status to the node management system to the name node and gets the information [9]. When the node fails to get the node's heartbeat message for a configured time, dead nodes are found, and the recovery process is executed by the name node.

4 Experimental Work

Tables 2 and 3 are the configuration details installed on top of the Virtual and Cloud Framework, with the same data sets ingested into hadoop components from the Table 1.

Table 1 DBpedia data sets info

S. no.	Name of the file	Number of triples	Size of data
1	infoboxes-fixed.nt	15,472,624	2.1 GB
2	geocoordinates-fixed.nt	447,517	64 MB
3	homepages-fixed.nt	200,036	24 MB

Table 2 Virtual Box configuration setup

Properties	Values
Number of nodes	4 (1 master, 3 slaves)
Virtual Box guest machine	Ubuntu 14.06, 4 GB RAM, 100 GB Hard Disk
Virtual Box host machine	Windows 2007, 8 GB RAM, 1000 GB Hard Disk
Replication factor	3

Table 3 Amazon cloud service EMR configuration setup

Properties	Values
Number of nodes	4 (1 master, 3 slaves)
Virtual Box guest machine	Ubuntu 14.06, 4 GB RAM, 100 GB Hard Disk
Virtual Box host machine	Windows 2007, 8 GB RAM, 1000 GB Hard Disk
Replication factor	3

4.1 Virtual Node Management

Virtual Framework [11] is popular for Hadoop, which will help in processing cost-effective unstructured data analysis [12]. Companies are interested in deploying the Hadoop Cluster in a physical server environment or public clouds. The private Cloud's security and the granular control lead to dominating for medium to large enterprises [3]. It provides unique benefits compared with other remote cloud models.

- Can set up clusters immediately
- Flexible to use a variety of hardware (DAS, SAN, NAS)
- Cost-effective
- High availability and fault tolerance

Virtualization refers to using a hypervisor and virtual machines to isolate and allocate resources in a private cloud environment. Establishing a private cloud environment is not limited to virtualization [3]. Aggregation became a useful alternative for cloud computing, significantly as applications like Hadoop grew in size. With reference to Paper 1 and Paper 2 [5, 7], the performance comparison between the Hadoop Components like SPARK, PIG, and Hive has been done and concluded the adequate tool to execute and process the RDF data sets using Benchmark Queries [13]. The virtual environments setup [11] is done using 4 Node cluster (1 Master and 3 Slave) nodes. When the number of nodes increases according to the data, the execution time is reduced and gives the best results as per the business requirement.

4.2 Cloud Node Management

Virtualization Hadoop created the infrastructure required to run Hadoop in the cloud platforms. Many of the major players offer Hadoop web service. Amazon Web Services (AWS) started beta testing in their Elastic Map-Reduce (EMR) service as early [14]. Even so, public cloud formation is not the point of this review; it is worth noting that it can be useful for batch processing or ad hoc. Mainly, the dataset would be stored in the Cloud already. For a constant live cluster, a company might find

Table 4 Result comparison table

Benchmark	Virtual-HIVE execution time (s)	Cloud-HIVE execution time (s)
1st query	41	34
2nd query	127	86
3rd query	48	29

Table 5 Result comparison table

Benchmark	Virtual-PIG execution time (s)	Cloud-PIG execution time (s)
1st query	34	18
2nd query	33	16
3rd query	29	13

that building its private Cloud is more cost-effective. Additionally, regulated organizations may prefer the security of a remote hosting facility [15]. Cloud services required to frame the Hadoop Environment are:

(1) EMR (Elastic Map Reduce)
(2) EC2
(3) S3

Using Amazon Web Services, the Elastic Map Reduce model has mainly been designed for ingesting Big Data and applying Hadoop tools to process and visualize the data for the best results. With reference to the previous future work, [5, 7] cloud framework is designed with 4 Node Cluster (1 Master, 3 Slave) nodes. Both the environments with the same configurations are configured and executed the Benchmark Queries using the same DBpedia datasets.

Result Analysis:

Table 4 concerning the hive processing time is compared with the Cloud-based results [6]. The Cloud shows extreme differences from virtual execution time. Table 5 showcases the PIG-Latin execution time in virtual and Cloud frameworks. Not only with the Hive component, PIG-Latin even shows the rigorous difference between both frameworks [8].

5 Conclusion

Cloud services are getting updated day by day with more cost-effective models. Software as a service (Saas) has become a standard service for the OpenStack cloud platform, as well as this Hadoop Virtual framework is also extensively used for extensive data processing in the industries according to the business requirement. This paper mainly focuses on integrating the Hadoop Framework with the 4 Node Cluster

Management Cloud Model, executed the benchmark query using Hadoop components, and compared the Hive and Pig virtual and Cloud performance [6] results between both frameworks; Cloud integrated with Hadoop components is shown huge variation and shown the best fit framework with a cost-effective model. Further increasing the number of nodes can facilitate less execution time and performance improvement in future work.

References

1. Anusha K, Usha Rani K (2017) Big data techniques for efficient storage and processing of weather data. Int J Res Appl Sci Eng Technol (IJRASET) 5(7). ISSN 2321-9653
2. Li F et al (2018) Distributed data management using MapReduce. ACM Comput Surv 46(3):1–42
3. Usha Rani K, Anusha K (2019) Data pre-processing on Cassandra data through Spark SQL. Int J Res Eng Appl Manag (IJREAM) 05(04). ISSN: 2454-9150 (accepted for publication)
4. https://tutorials.freshersnow.com/map-reduce-tutorial/hadoop-mapreduce-job-execution-flow-chart/
5. Usha Rani K, Lakshmi C (2020) Effective query processing for web-scale RDF data using Hadoop components. TEST Eng Manag 83:5764–5769. ISSN: 0193-4120 (accepted for publication)
6. Anusha K, Usha Rani K (2020) Performance evaluation of Spark SQL for batch processing. Advances in intelligent systems and computing (accepted for publication in Springer series)
7. Anusha K, Usha Rani K, Lakshmi C (2016) A survey on big data techniques. Spec Issue Comput Sci Math Biol IJCSME-SCSMB. ISSN: 2349-8439
8. Morsey M et al (2017) DBpedia SPARQL benchmark—performance assessment with real queries on real data. In: ISWC 2017, Part I, LNCS, vol 7031, pp 454–469
9. Ryu W (2017) Flexible management of data nodes for Hadoop distributed file system. In: The 3rd international conference on big data, small data, linked data and open data (ALLDATA 2017), IARIA, pp 1–2, Apr 2017
10. Anusha K, Usha Rani K (2016) Survey on big data techniques. In: International conference on computational science, mathematics and biology (IJCSME-16). Organized by Department of Computer Science, SPMVV, Tirupati, 16 Mar 2016
11. Anusha K, Usha Rani K (2018) Comparative evaluation of big data frameworks on batch processing. Int J Pure Appl Math (IJPAM) 119(16). ISSN: 1314-3395 (Scopus indexed journal)
12. Xu G, Xu F, Ma H (2012) Deploying and researching Hadoop in virtual machines. In: Proceedings of the international conference on automation and logistics (ICAL), Zhengzhou, China, pp 395–399
13. A benchmarking case study of virtualized Hadoop performance on VMware framework vSphere5. http://www.vmware.com/files/pdf/VMW-Hadoop-Performance-vSphere5.pdf. Accessed 18 July 2014
14. Anusha K, Usha Rani K (2019) Performance evaluation of Spark SQL for batch processing. In: 2nd international conference on computing communications and data engineering (CCODE 2019). Organized by Department of Computer Science, SPMVV, Tirupati, 1–2 Feb 2019
15. Anusha K, Usha Rani K (2018) Batch data processing with Apache Cassandra and Apache Spark SQL. In: International conference on science, engineering & technology (ICSET-18). Organized by South Asia Institute for Research and Publications (SAIRAP), Guntur, 17 Nov 2018

Analysis of Bio Vanillin Production from Waste Material by HPTLC

M. Sujatha and R. Jaya Madhuri

Abstract Vanillin (4-hydroxy 3-methoxy benzaldehyde) is one of the main flavour existences utilized around the world. Because of its utilization in food, feed beautifiers, and drug businesses, right now, the universe's need for vanillin utilization is around 12,000 tons/year. Normally 50 ton of vanillin is delivered each year and rest of the vanillin is created artificially from lignin-containing waste and oil-based crude material in particular wheat grain, rice bran, corncob, groundnut cake. A basic, quick, explicit, and exact superior slender layer chromatography strategy has been created for the gauge of vanillin in food items. Aluminium plates that pre-coated with silica gel 60 GF 254 were utilized as fixed stage and a combination of methanol–water–chilly acidic corrosive in the proportion of 20:05:02 (v/v) as the versatile stage. The evaluation was done by examining the created spots utilizing a densitometer in absorbance mode at 275 nm. The RF estimation of vanillin was 0.84. The after-effects of the examination have been approved factually and by the recuperation considerations. Linearity was seen in the focus scope of 150–500 μg.

Keywords Vanillin · Waste raw material · Ferulic acid · HPTLC

1 Introduction

Vanillin or 4-hydroxy-3-methoxy-benzaldehyde is the world's chief enhancing compound, which is utilized generally in the food business, perfumery, and drink other than the presence applied in the drug business. Vanillin can be created artificially or normally from vanilla beans [1]. Its fundamental part, vanillin is the key compound adding to vanilla trademark smell and is the second biggest fragrance delivered on the planet. As of now, mainstream (≥80%) of vanillin provided to the global market is synthetically incorporated from an oil-based crude.

M. Sujatha (✉) · R. Jaya Madhuri
Department of Applied Microbiology, Sri Padmavati Mahila Visvavidyalayam, Andhra Pradesh Tirupati, India
e-mail: indlasujatha89@gmail.com

S. Jyothi et al. (eds.), *Proceedings of the 2nd International Conference on Computational and Bio Engineering*, Lecture Notes in Networks and Systems 215,
https://doi.org/10.1007/978-981-16-1941-0_20

Material "guaiacol" with under 12–14% created from lignin-containing waste delivered by the sulphite pulping measure.

The creation of vanillin by the microbial or enzymatic transformation of common forerunners, for example, ferulic corrosive, vanillic corrosive, glucose, and eugenol has been examined. It is an enormously searched after flavour compound and is broadly utilized in the food, dessert shop, perfumery, beauty care products, and drug enterprises. Popularity for characteristic flavours from purchasers has set off examination in common vanillin creation from regular assets through microbial biotransformation [2].

Biotechnology can possibly create through biotransformation completed by microorganisms and their enzymatic systems [3]. Ferulic corrosive, a generally realized characteristic phenolic compound from lignin debasement by organisms and microbes, was profoundly determined as vanillin antecedent. Basic horticultural waste build-ups, for example, grain wheat, rice–wheat, sugar beet mash, and wheat straw are rich wellsprings of ferulic corrosive (FA), which is a significant forerunner of microbial orchestrated vanillin and can be misused as crude material for change to vanillin by microorganisms [4].

The current examination was made to assess the chance of delivering vanillin from ferulic acid recuperated from wheat grain hydrolyzates by utilizing cells of *Bacillus subtilis*

2 Materials and Methods

2.1 Collection and Milling of Raw Materials

Raw materials that are used for the production of vanillin, such as wheat bran, corn cob, rice bran, and groundnut cake, were collected from Tirupati. The collected raw materials were air-dried and small particles.

2.2 Extraction of Ferulic Acid from Wheat Bran [5]

See Fig. 1.

Fig. 1 Wheat bran

2.3 *Ferulic Acid Was Extracted from Wheat Bran by Solvent Method*

About 10 g of wheat bran was refluxed with petroleum ether for 30 min. The solution was filtered and treated with 1m NaOH with a ratio of 1:10 (g/ml) for 4 h at 40 °C and neutralized to pH 6.0 with 6M HCL. Three volumes of ethanol were added to precipitate hemicellulose. The solution was filtered and evaporated the ethanol. 0.1M H_2SO_4 was added to remove the lignin as a precipitate at pH 1.5. Then the filtrate was extracted with three volumes of chloroform and methanol. The ferulic acid-containing methanol layer was dried and obtained.

2.4 *Bioconversion of Ferulic Acid to Vanillin [6]*

The positive strain was transferred to 100 ml of LB broth and incubated 24–48 h at 37 °C. Colonies of positive strain were transferred to the fermentation broth (pH—5.6)) consisting of 10 g of raw materials (wheat bran, rice bran, groundnut cake, corn cob) and 0.05% of ferulic acid as an inducer for biotransformation and incubated at 37 °C for 48–72 h at 180 rpm under shaking conditions. After 72 h of biotransformation, vanillin was estimated from the fermentation medium by HPTLC Analysis.

2.5 *Separation of Vanillin from Fermentation Medium*

After biotransformation, the fermented biomass was centrifuged to remove the particles. Then the supernatant was acidified to pH 2 using 6N HCL. The phenolic compounds were extracted three times with the same volumes of ethyl acetate. Then supernatant was used for the estimation of vanillin by HPTLC.

2.6 High-Performance Thin Layer Chromatography Analysis [7]

Chromatography was implemented on an aluminium upheld silica gel TLC plates washing with methanol (0.2 mm thickness). Standard arrangements of vanillin were set up by moving the standard arrangement (1 mg/ml) weakened with methanol with the end goal that the focuses are 100, 250, and 500 μg/ml and sonicated for 15 min. The norms and four distinctive example arrangements were applied to the silica gel plates as 8.0 mm groups with 9.0 mm space between two groups utilizing a Camag Linnomat V example utensil. The plates were created with a versatile period of glacial acidic acid: Methanol: Water: (02:20:05) in a TLC chamber. After development, the plates were dehydrated for 5 min at 60 °C and evaluated the results by the guidelines and tests were achieved by methods for a Camag TLC scanner III constrained by Win CATS 1.4.3 form programming at 275 nm. The measure of vanillin in the example arrangements was registered from the adjustment design.

3 Results

3.1 Collection of Raw Materials

See Fig. 2.

Fig. 2 Rice bran

Fig. 3 Groundnut cake

Fig. 4 Corn cob

3.2 Extraction of Ferulic Acid from Wheat Bran

See Figs. 3 and 4.

3.3 Ferulic Acid Was Extracted from Wheat Bran by Solvent Method

See Figs. 5 and 6.

3.4 Bioconversion of Ferulic Acid into Vanillin

In the present work, the vanillin was extracted from ferulic acid recovered by wheat bran hydrolyzates by positive strain (Figs. 7, 8, 9 and 10).

Fig. 5 Ferulic acid

Fig. 6 *Bacillus subtilis*

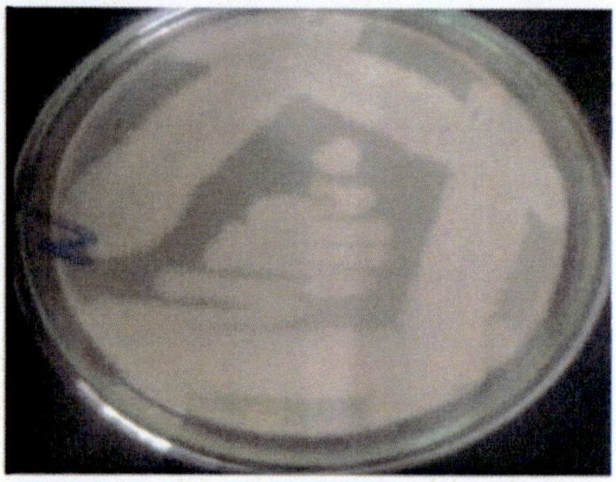

Fig. 7 Wheat bran

Fig. 8 Corn cob

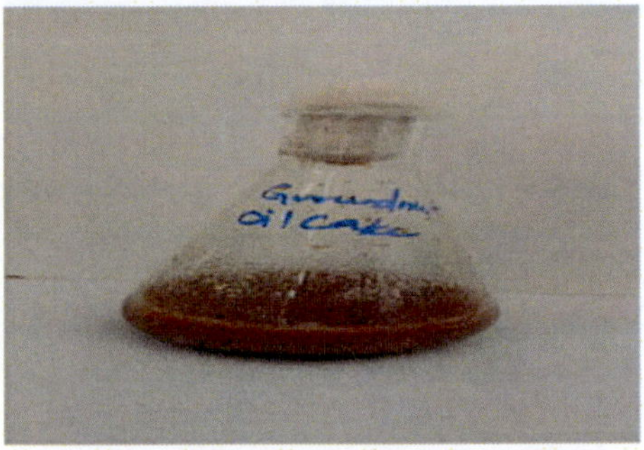

Fig. 9 Groundnut cake

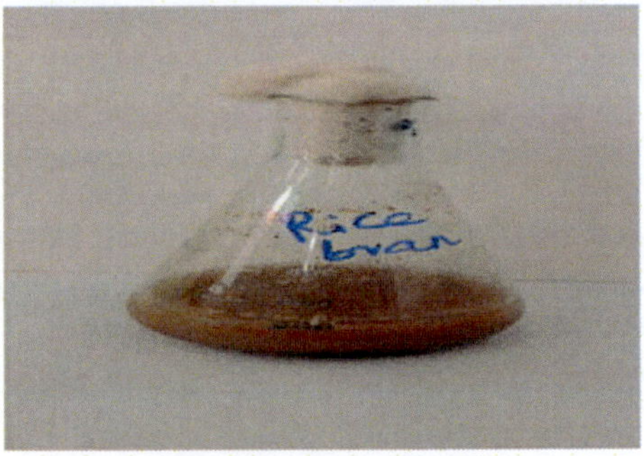

Fig. 10 Rice bran

3.5 *Biotransformation of Ferulic Acid by Positive Strain*

See Figs. 11 and 12.

Fig. 11 Crude vanillin

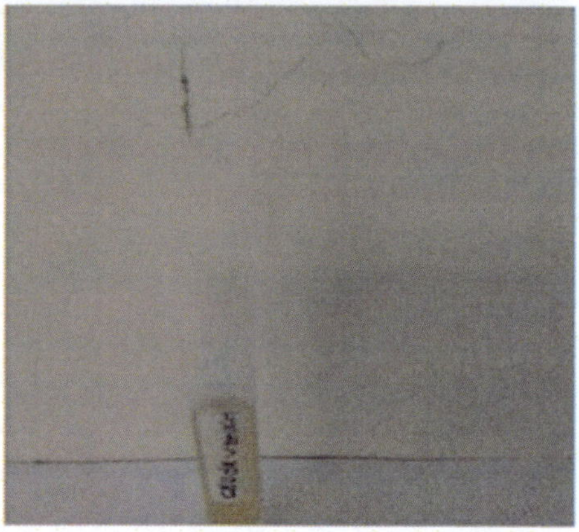

Fig. 12 Crude vanillin

3.6 *High-Performance Thin Layer Chromatography Analysis*

The chose versatile period of glacial acidic acid: Methanol: Water: [02:20:05] settled vanillin proficiently and is appeared in the outcome. The RF estimation of vanillin was discovered to be 0.848. Vanillin arrangement provided an absorbance highest at 275 nm and was chosen for recognition. The strategy was utilized to decide vanillin at ease in the waste material, which is chosen for investigation (Fig. 13 and Tables 1, 2, 3, 4 and 5).

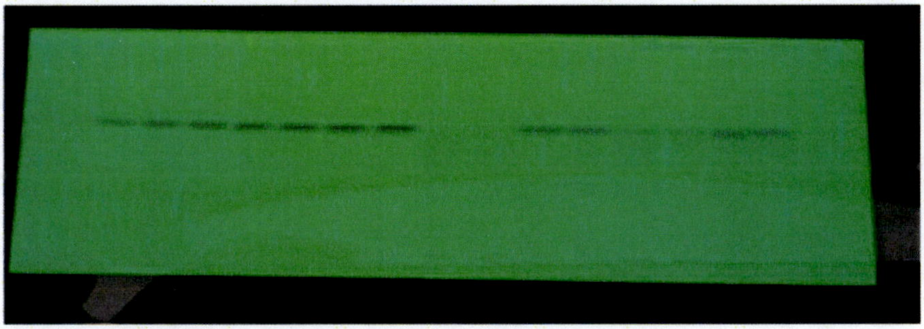

Fig. 13 Image of TLC plate of crude sample and vanillin standard

Table 1 HPTLC chromatogram of vanillin standard

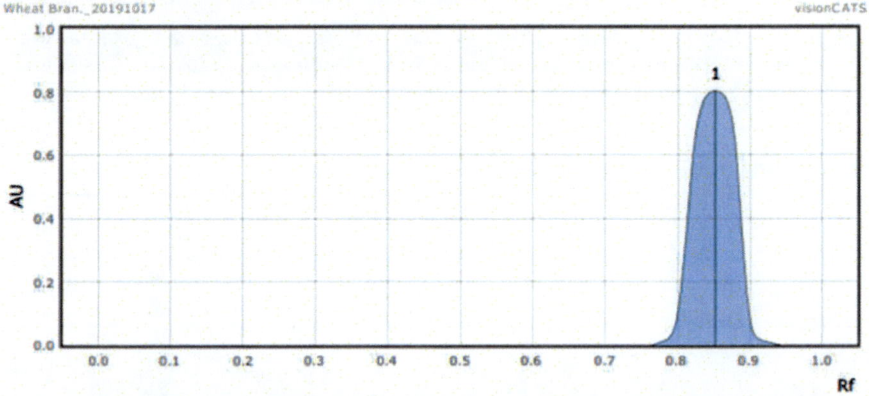

Peak #	Start		Max			End		Area		Manual peak	Substance Name
	Rf	H	Rf	H	%	Rf	H	A	%		
1	0.765	0.0000	0.852	0.8014	100.00	0.942	0.0001	0.05934	100.00	No	vanillin

Table 2 HPTLC chromatogram of groundnut cake

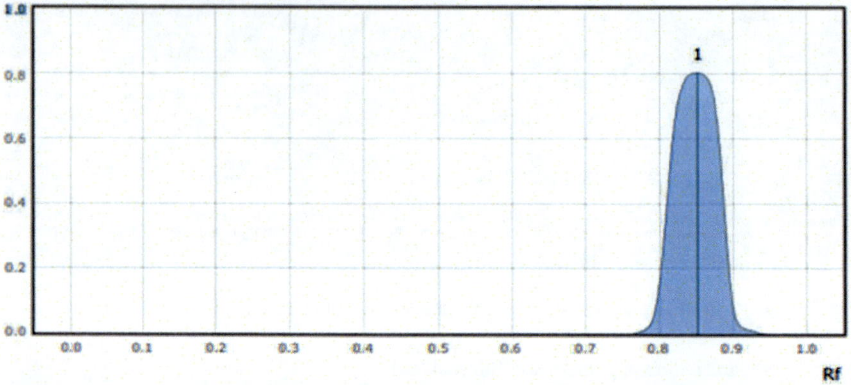

eak #	Start Rf	Start H	Max Rf	Max H	Max %	End Rf	End H	Area A	Area %	Manual peak	Substance Name
1	0.765	0.0000	0.852	0.8042	100.00	0.942	0.0001	0.06188	100.00	No	vanillin

Table 3 HPTLC chromatogram of corn cob

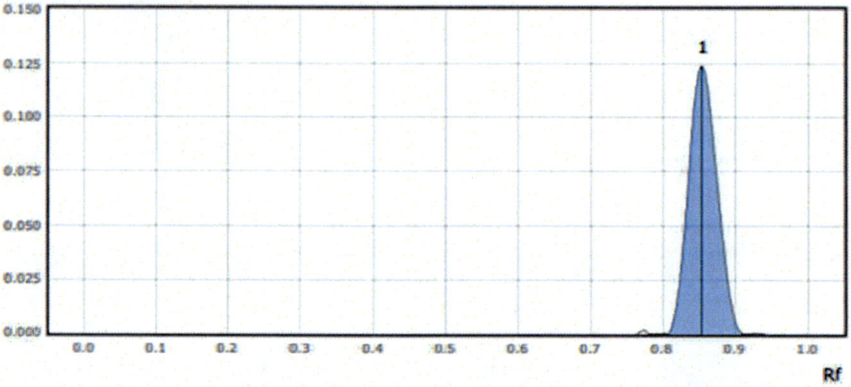

eak #	Start Rf	Start H	Max Rf	Max H	Max %	End Rf	End H	Area A	Area %	Manual peak	Substance Name
1	0.806	0.0000	0.853	0.1237	100.00	0.916	0.0000	0.00567	100.00	No	vanillin

Table 4 HPTLC chromatogram of rice born

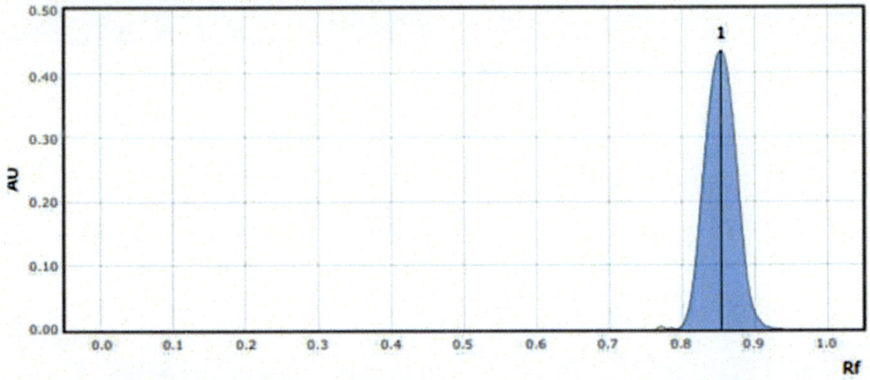

Peak #	Start Rf	Start H	Max Rf	Max H	Max %	End Rf	End H	Area A	Area %	Manual peak	Substance Name
1	0.794	0.0000	0.855	0.4342	100.00	0.935	0.0000	0.02261	100.00	No	vanillin

Table 5 HPTLC chromatogram of wheat bran

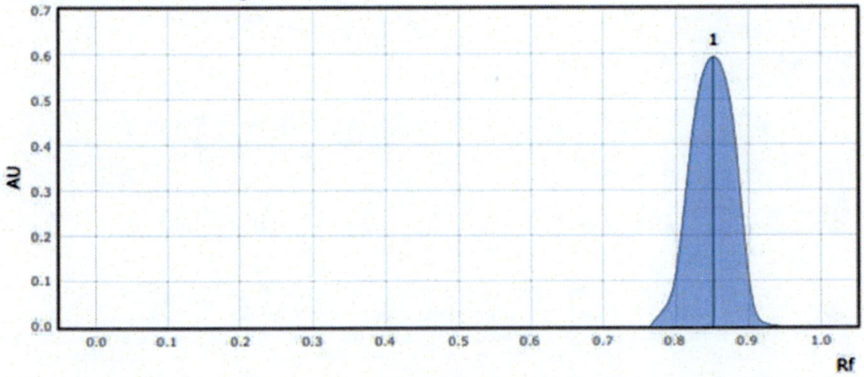

Peak #	Start Rf	Start H	Max Rf	Max H	Max %	End Rf	End H	Area A	Area %	Manual peak	Substance Name
1	0.765	0.0000	0.852	0.5933	100.00	0.932	0.0019	0.04485	100.00	No	vanillin

Calibration results:

Area calibration for substance vanillin @ 275 nm:

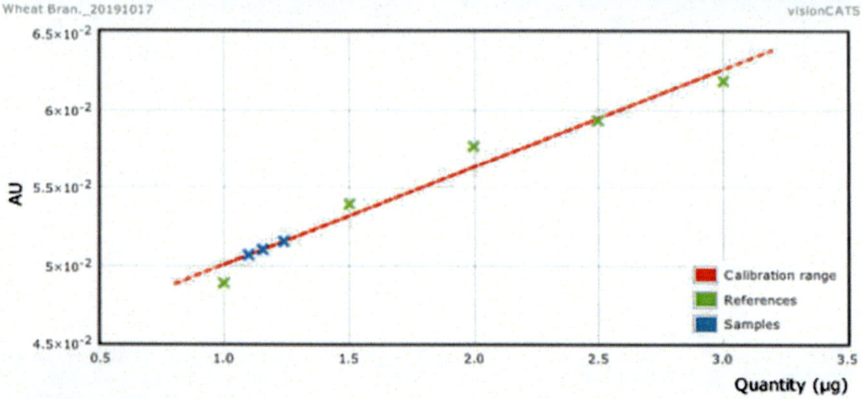

Regression mode	Linear-2
Range deviation	10.00 %
Related substances	Default
Number of references	5
Calibration function	$y=6.282\times10^{-9}x+4.378\times10^{-2}$
Coefficient of variation	CV 1.65 %
Correlation coefficient	R=97.876234 %

Results:

vanillin	(3 sample assignments) @ 275 nm		
Sample 'Rice'	298.7 µg/ml	CV=5.00 %	(2 applications)
			298.7 µg in 1.000 mg
Volume: 4.0 µl	298.7 µg/ml	CV=5.00 %	(2 replicas)
Track 10	309.3 µg/ml	1.237 µg	
Track 11	288.1 µg/ml	1.153 µg	
Sample 'Wheat Bran'	219.5 µg/ml	(CV unavailable)	(1 applications)
			219.5 µg in 1.000 mg
Volume: 5.0 µl	219.5 µg/ml	(CV unavailable)	(1 replicas)
Track 14	219.5 µg/ml	1.097 µg	

3.7 *System Suitability*

Framework reasonableness tests are the centre of a chromatographic examination and, furthermore, it confirms that the goal and reproducibility of the chromatographic frameworks are sufficient for investigation. Framework reasonableness tests were done on a newly arranged vanillin stock arrangement that builds up the viability of this technique.

3.8 Linearity

A calibration plot was constructed by taking peak area beside vanillin concentration was linear within the range of 150–500 μg/spot. The linear regression equation $Y = 6.282 \times 10^{-9}x_{x+}$ 4.378×10^{-2} represents the calibration lines, in this, X is concentration and y is the peak area. The correlation coefficient was observed to be R = 97.876234.

3.9 Detection and Limit of Quantification

The limit of detection(LOD) and limit of quantification (LOQ) were determined by utilization of the conditions LOD = 3 × N/B and LOQ = 10 × N/B where N is the standard deviation of the pinnacle territory of the medication, taken as a proportion of the commotion and B is the slant of the relating alignment bend. The constraint of evaluation and the restriction of discovery for vanillin were discovered to be 1.65 and 3.25 mg separately.

3.10 Accuracy and Precision

The exactness and accuracy of the strategy were concentrated by accomplishing tests by ordinary expansion techniques. The exactness of the strategy was dictated by recuperation tests. The recuperation of the strategy was resolved at one nearby adding a known amount of vanillin to the side-effects of pre-brokedown examples and the blends were investigated by the suggested technique. The normal recuperation got from each example was somewhere in the range of 98.16–99.35% and is appeared in the outcome. From the information acquired, supplementary recuperation of stock vanillin was discovered to be exact.

3.11 Roughness and Toughness

Toughness is the proportion of the reproducibility of a test result below ordinary, anticipated that working circumstances from instrument should instrument and from expert to the investigator. The heartiness of the strategy was dictated by rolling out slight improvements in the chromatographic circumstances. No stamped variations in the chromatograms showed that the HPTLC techniques created are tough and powerful.

4 Results and Discussion

The chose portable period of glacial acidic corrosive: methanol: water [02:20:05] settled vanillin proficiently and is appeared in the outcome. The RF estimation of vanillin was discovered to be 0.848. Vanillin arrangement provided an absorbance most extreme at 275 nm and was chosen for recognition. The strategy was utilized to decide vanillin content in the derivatives chose for investigation.

5 Conclusion

The HPTLC method proposed for the determination of vanillin in four different waste products were accurate, precise, rapid, selective, and sensitive and, therefore, can be conveniently adopted for the routine analysis of vanillin in waste products.

References

1. Frenkel C, Ranadive AS, Vázquez JT, Havkin-Frenkel D (2011) Curing of vanilla. In: Havkin-Frenkel D, Belanger FC (eds) Handbook of vanilla science and technology. Wiley-Blackwell, Oxford, pp 84–96
2. Priefert H, Babenhorst J, Steinbuchel A (2001) Biotechnological production of vanillin. Appl Microbiol Biotechnol 56(3–4):296–314
3. Santos AS, Pereira N Jr, Silva II, Antunes OAC (2004) Peroxidase catalyzed Piperonal. Progress Biochem 39:2269–2275
4. Klinke HB, Ahring BK, Schmidt AS, Thomsen AB (2002) Characterization of degradation products from alkaline wet oxidation of wheat straw. Bioresour Technol 82:15–26
5. Parikshit G, Gogoi, A, Rajbangshi C, Bhuyan P, Bhuyan BK (2017) Extraction, purification and spectroscopic characterization of ferulic acid by alkaline hydrolysis from brans of Assam, India. IJIR 3(1):358–361
6. Kaur B, Chakraborty D (2013) Statistical media and process optimization for biotransformation of rice bran to vanillin using *Pediococcus acidilactici*. Indian J Exp Biol 51:935–943
7. Krishna veni N et al (2013) Analysis of vanillin in food products by high performance thin layer chromatography. J Adv Sci Res 4(1):48–51

Phytochemical Studies and In Vitro Biological Activity of Citrus and Pomegranate Peel Extracts

K. Gnaneswari, M. Vani, and B. Lakshmi Thanuja

Abstract The study has demonstrated the occurrence of a significant amount of secondary metabolites such as tannin, steroids, reducing sugar, proteins and high content of carbohydrates from the acetone and chloroform extracts of *Citrus reticulata* and *Punica*. The fruit peel acetone and chloroform extract of *Citrus reticulata* and *Punica granatum* exhibited potent anti-oxidant activity. The present study has shown the usefulness of the extraction methodologies adopted for efficient extraction, processing and utilization of these citrus fruit peel wastes and also to characterize the phytochemicals, antioxidant property and anti-inflammation activities of fruit peel wastes of oranges and pomegranate.

Keywords *Citrus reticulata* · *Punica granatum* · Phytochemicals · Antioxidant · Anti-inflammation

1 Introduction

Citrus reticulate is too called as sweet orange, which is used as a fruit, and in some nations; it is taken as a medicinal herb. It is classified under *Rutaceae* family and grown in tropical and subtropical regions. Peels of the orange are usually thrown out as waste, leading to serious disposal problems that may damage the environment. Considering the huge quantity of "waste" that is produced in the food supply chain, orange peels offer a huge potential to be exploited as a value-added product, including for the recovery of natural antioxidants, pectin, enzymes or for the production of ethanol, organic acids, essential oils and prebiotics single-cell protein [1]. In addition, the *C. reticulata* peel is a rich source of vitamin C, fiber, and many nutrients, including phenolics and flavonoids. The *C. reticulata* peel has been used as a traditional medicine in certain parts of the world for relieving stomach discomfort,

K. Gnaneswari (✉) · M. Vani · B. Lakshmi Thanuja
Department of Applied Microbiology & Biochemistry, Sri Padmavati Mahila Visvavidyalayam, Tirupati, AP, India
e-mail: gnaneswarij@gmail.com

S. Jyothi et al. (eds.), *Proceedings of the 2nd International Conference on Computational and Bio Engineering*, Lecture Notes in Networks and Systems 215,
https://doi.org/10.1007/978-981-16-1941-0_21

skin inflammation, ringworm infections, aiding in neuroprotection and improving heart health [2, 3].

Punica granatum Linn. (Pomegranate) is a plant that belongs to Punicaceae family locally called as Anar, a fruit of great antiquity. This plant species has been cultivated in the Middle East for more than 5000 years ago and it is found in Iran, Egypt, India, Bangladesh, Srilanka, North Africa, California and Arizona [4]. Pomegranate peels are used as a popular remedy throughout the world and exploited in traditional medicine because of their strong mordancy properties [5]. Bark, leaves, immature fruits and fruit rind of different parts of pomegranate have valuable therapeutic potential.

2 Materials and Methods

2.1 Collection of Orange Peels and Pomegranate Peels

Punica granatum and *Citrus reticulate* fruits were taken from local market of Tirupati and Hyderabad and shifted to the laboratory for further analysis. The fruits were rinsed in running tap water and with double distilled water. The cleaned fruits were kept in RT for 5 min in open air. Fruit peel was carefully removed by blade and kept for dry. For further work, dried orange and pomegranate peel were ground and kept in freezer.

2.2 Preparation of Extracts

Acetone and chloroform extracts of pomegranate and orange peel extracts were prepared by soaking 30 g of dried pomegranate and orange peel separately for 24 h in acetone and chloroform, followed by filtration. The extracts are stored in clean dry Eppendorf tubes and label it.

2.3 Phytochemical Screening

Orange and pomegranate peel extracts were assessed for phytochemical screening by standard protocols described [6, 7] to analyze the presence of saponins, flavonoids, alkaloids, terpenoids, steroids, phenols and tannins.

2.4 *Fourier Transform Infrared Spectrometer (FT-IR)*

FTIR is an excellent technique for knowing the functional groups in peel extracts. The absorbance of the light is specific for identifying functional groups. Dried powder of different solvent extracts is used for FTIR analysis. Dried powder extract (10 mg) was ground with potassium bromide pellet for preparing translucent discs. The scan ranges from 400 to 4000 cm^{-1} with a resolution of cm^{-1}.

2.5 *Antioxidant Activity*

2.5.1 DPPH Radical Scavenging Activity

Different concentrations of peel extracts are taken and made up to 1ml by ethanol by taking ascorbic acid as standard. The reaction was started by the addition of DPPH solution dissolved in ethanol. The mixture was stored at room temperature and absorbance was read at 517 nm by using a spectrophotometer. By the following equation, radical scavenging activity was determined.

$$\text{Scavenging activity of DPPH (\%)} = \frac{1 - (\text{Absorbance sample})}{\text{Absorbance of control}} \times 100.$$

2.5.2 Total Antioxidant Activity

Total antioxidant activity was assayed by the phosphomolybdenum method where molybdenum is reduced to phosphate. To 0.1 ml of OAPE and PPAE extracts, add 1 ml reagent solution containing 28 mM sodium phosphate, 0.6 M sulphuric acid and 4 mM ammonium molybdate. The mixture was incubated at 95 °C for 90 min and read the absorbance spectrophotometrically at 765 nm by using ascorbic acid as standard [8].

2.6 *Anti-inflammatory Activity*

2.6.1 Protein Denaturation Inhibition

The peel extracts were taken at different concentrations and add reaction mixture containing bovine albumin and adjust the pH to 6.3 by adding 1 N HCl. The contents were mixed thoroughly and incubate for 15 min at room temperature followed by heating at 57 °C for 3 min. Then the mixture was cooled and add phosphate buffer. Turbidity was read spectrophotometrically at 660 nm by using milli-q water as control. The denatured protein inhibition percentage was calculated. Dispirin is used as standard.

$$\text{Percent inhibition} = \frac{(100 - (\text{O.D. of test} - \text{O.D. of control}))}{\text{O.D. of control}} \times 100.$$

2.6.2 HRBC Membrane Stability

The healthy human blood was collected and centrifuged with an equal quantity of Alsever solution. The packed cells were thoroughly washed with isosaline solution and finally 5% HRBC suspension was made. The assay mixture contains an equal volume of extracts at different concentrations (10–8050 μg/ml), HRBC suspension, hyposaline and phosphate buffer solution was added and kept at 30 °C for 30 minutes and centrifuged. Collect the supernatant and read the absorbance at 560 nm, by taking diclofenac sodium as standard. The percentage of hemolysis was calculated

$$\text{Haemolysis inhibition} = \frac{(100 - (\text{O.D. of test} - \text{O.D. of control}))}{\text{O.D. of control}} \times 100.$$

3 Results

3.1 Phytochemical Studies

The phytochemical analysis of citrus and pomegranate peel was carried out individually with acetone and chloroform organic extracts. This study showed the presence of secondary metabolites such as triterpenes, phenol, proteins, alkaloids and flavonoid. The acetone extract of both peels qualitatively proved the presence of a greater number of phytochemical constituents (Table 1). Hence, the further analysis

Table 1 Phytochemical studies of orange and pomegranate peel extracts

S. no.	Phytochemicals	OPAE	OPCE	PPAE	PPCE
1.	Alkaloids	–	+	+	+
2.	Flavonoids	+	–	+	–
3.	Saponins	+	–	–	+
4.	Tannins	+	+	+	–
5.	Phenols	+	+	+	+
6.	Terpenoids	–	+	+	+

of antioxidant and anti-inflammatory was continued with acetone organic extracts of OPAE and PPAE.

3.2 *FTIR Analysis*

The FTIR analysis of OPAE exhibited the peaks at 3832 and 1302 cm^{-1} and PPAE showed at 3689 and 2316.10 cm^{-1}. The OPAE peaks correspond to the functional groups such as alcohols (O–H stretch) and carboxylic acids (C=O stretch). The PPAE peaks correspond to the functional groups like alcohols (O–H stretch) and carbon dioxide (O=C=O stretch). The peaks are shown in Figs. 1 and 2.

3.2.1 Total Antioxidant Activity

Total antioxidant activity of PPAE, OPAE and ascorbic acid was measured by phosphomolybdenum method. Among the two samples, PPAE showed higher phosphomolybdenum reduction followed by OPAE. Total antioxidant activity is shown in Fig. 3.

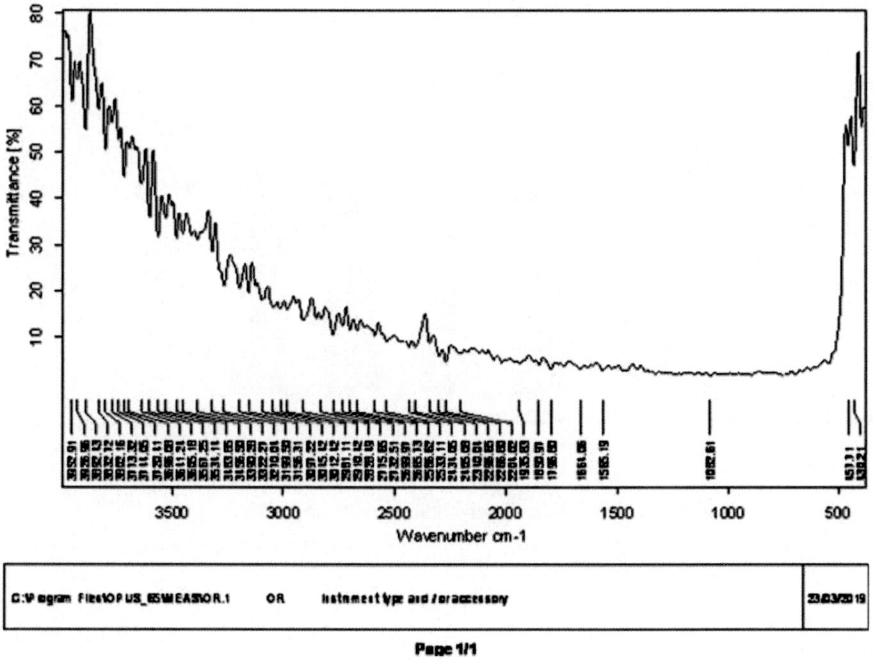

Fig. 1 FTIR of orange peel acetone extracts

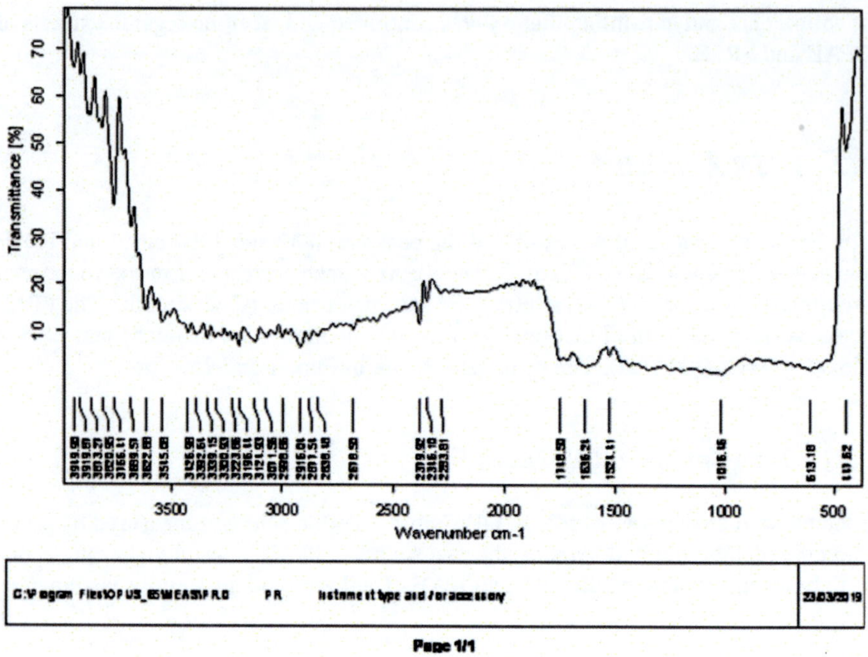

Fig. 2 FTIR of pomegranate peel acetone extracts

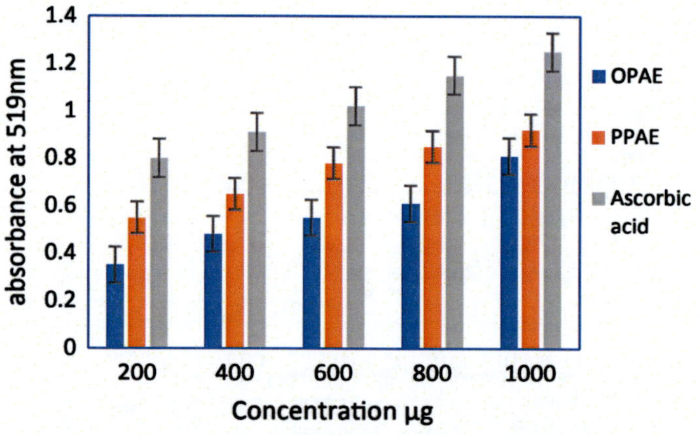

Fig. 3 Total antioxidant activity of orange and pomegranate peel extracts

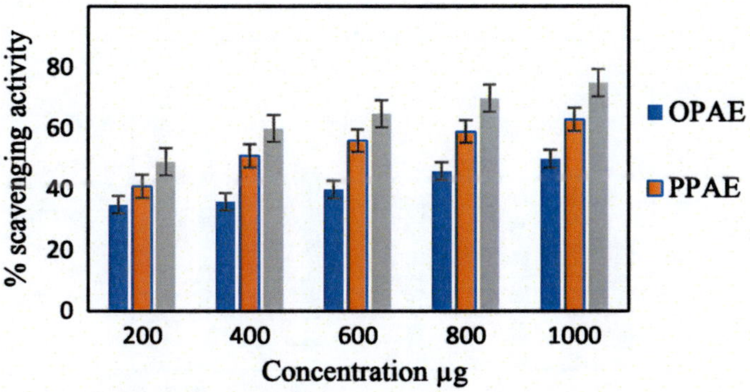

Fig. 4 DPPH free radical scavenging activity

3.2.2 DPPH Free Radical Scavenging Activity

DPPH free radicals are mostly used to access antioxidant activities. As represented in Fig. 4, the scavenging ability of PPAE, OPAE and ascorbic acid was dependent on concentrations termed as IC_{50} values. Lower IC_{50} values show higher antioxidant activity. DPPH free radical scavenging activity is more in PPAE when compared to OPAE.

3.3 Anti-inflammatory Activity

3.3.1 Anti-protein Denaturation Activity

In this activity, PPAE shows significant anti-protein denaturation activity. Dispirin is used as standard. The results are shown in Fig. 5.

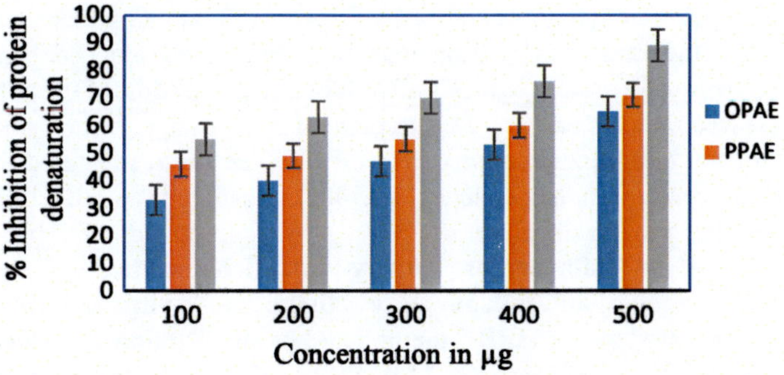

Fig. 5 Anti-protein denaturation of orange and pomegranate peels

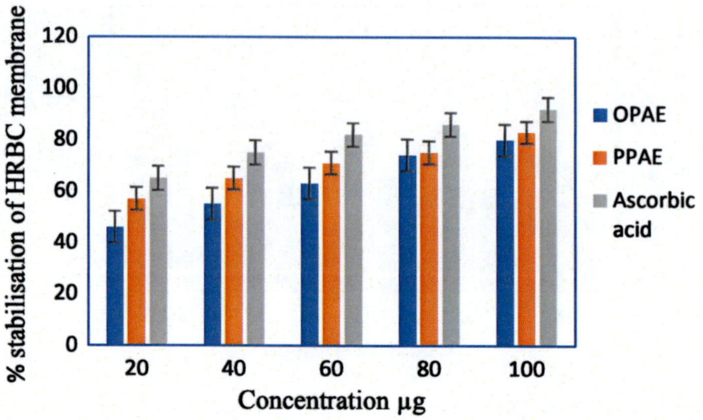

Fig. 6 HRBC membrane stabilization of orange and pomegranate peels

3.3.2 HRBC Membrane Stability

Among the two samples, the PPAE showed significant inhibition of hemolysis of the human erythrocyte membrane induced by heat (Fig. 6).

4 Discussion

The research work was carried out in PPAE and OPAE, which shows significant secondary metabolites that provide vital pharmaceutical properties, which play an important role in human health. Secondary metabolites such as terpenoids, flavonoids, phenols and alkaloids are treated as medicines or as food supplements, which will prevent or heal various diseases, and particularly few of these metabolites will inhibit numerous cancer types. By FTIR analysis, we can confirm each band represents characteristic absorption peaks of functional group, which shows the presence of carboxylic acid, alcohols and carbon dioxide. Total antioxidant activity can be stated as that the transfer of hydrogen/electrons from antioxidants depends on the antioxidant chemical structure, which will reduce stress [9]. Most of the antioxidants fight against free radicals and are used to prevent cell damage. In this study, PPAE shows significant antioxidant activity than OPAE. The result indicated that PPAE at different concentrations has excellent anti-inflammatory properties. Most of the flavonoids inhibit protein denaturation. Most of the compounds inhibiting denaturation will exhibit anti-inflammatory activity [10]. HRBC method was selected for the in vitro evaluation of anti-inflammatory activity. The in vitro anti-inflammatory activity was evaluated by the HRBC method and its stabilization suggested that the peel extract may stabilize lysosomes [11]. Due to rich in flavonoid, proteins lose their secondary and tertiary structure and exhibit anti-inflammatory activity.

5 Conclusion

Citrus and pomegranate are medicinal plants mainly its peels contain many therapeutic uses. Phytochemical screening, antioxidant and anti-inflammation efficacy of orange and pomegranate peel were performed. Due to the presence of phytoconstituents in OPAE and PPAE, it is used in treating different diseases, and also it serves as drug, which is essential for human use. The antioxidant activity of PPAE is higher than OPAE. Antioxidant-rich extracts serve as a source of nutraceuticals that alleviate the oxidative stress and therefore prevent or slow down the degenerative diseases. The anti-inflammation activity is more in PPAE than OPAE. Further investigations are required to fully understand the mode of action of the active constituents and to exploit orange and pomegranate peel as preventive and therapeutic potential.

Acknowledgements The authors are thankful to Department of Applied Microbiology and Biochemistry in Sri Padmavati Mahila Visvavidyalayam, Tirupati for supporting to do our research work.

References

1. Mamma D, Christakopoulos P (2014) Biotransformation of citrus by-products into value added products. Waste Biomass Valor 5:529–549. https://doi.org/10.1007/s12649-013-9250-y
2. Li S, Lo C-Y, Ho C-T (2006) Hydroxylated polymethoxyflavones and methylated flavonoids in sweet orange *Citrus sinensis* peel. J Agric Food Chem 54(12):4176–4185
3. Ghasemi K, Ghasemi Y, Ebrahimzadeh MA (2009) Antioxidant activity, phenol and flavonoid contents of 13 citrus species peels and tissues. Pak J Pharm Sci 22(3):277–281
4. Akter S, Sarker A, Hossain MD (2013) Antidiarrhoeal activity of rind of Punica granatum. Int Curr Pharm J 2(5):101–104
5. Nitave SA, Patil VA (2014) Study of antibacterial and antifungal activity of Punica granatum peel and its phytochemical screening. World J Pharm Res 3(10):505–512
6. Savithramma N, Linga RM, Bhumi G (2011) Phytochemical screening of Thespesia populnea [L.] Soland and Tridax procumbens L. J Chem Pharm Res 3(5):28–34
7. Selvaraj S, Chittibabu CV, Janarthanam B (2014) Studies on phytochemical screening, antioxidant activity and extraction of active compound (Swertiamarin) from leaf extract of Enicostemmalittorale. Asian J Pharm Clin Res 7(4):240–244
8. Prieto P, Pineda M, Aguilar M (1999) Spectrophotometric quantitation of antioxidant capacity through the formation of a phosphomolybdenum complex: specific application to the determination of vitamin E. Anal Biochem 269:337–341
9. Loo AY, Jain K, Darah I (2008) Antioxidant activity of compounds isolated from thepyroligneous acid, *Rhizophora apiculate*. Food Chem 107:1151–1160
10. Gandhudasan R, Thamaraichhelvan A, Babraj S (1991) Anti-inflammatory action of *Lannea-cormandelica* by HBRC membrane stabilization. Fitoterapia LXII:81–83
11. Williams LA, Latore et al (2008) Anti denaturation effects induced by natural product. Pubmed 57:327–349

Present and Future Prospectives of Microbial Fibrinolytic Enzyme Production and Its Applications

K. Gowthami and R. Jaya Madhuri

Abstract Cardiovascular diseases (CVDs) are the group of diseases that occur in blood vessels and heart, the most common cause of death worldwide. The treatment of this cardiovascular disease is using as an enzyme treatment such as streptokinase, urokinase, tPA, etc. This type of enzymatic drugs has some side effects. To overcome this problem, novel therapeutic enzymes are isolated from different sources. This review focused on novel microorganisms fibrinolytic enzymes produced from different sources and different types of thrombosis. This review is also focused on fibrinolytic enzymes or thrombolytic agents act on novel coronavirus.

Keywords Cardiovascular disease · Streptokinase · Fibrinolytic enzyme · Therapeutic agent, etc.

1 Introduction

Cardiovascular diseases (CVDs) are the group of diseases that occur in blood vessels and the heart. CVDs are the almost frequent cause of mortality worldwide, with 17.9 million deaths each year (https://www.who.int). Reduced blood flow and stagnation result in the formation of fibrin thrombus [1]. Thrombosis is manifested in the form of embolism, myocardial infarction, stroke and other cardiac disorders [2–6].

Medication and surgery are the major treatment modes for cardiovascular diseases. The treatment of acute thrombosis involves the use of thrombolytic agents intravenously. They hydrolyze the fibrin clot and restore blood supply to the ischemic regions of the heart. Even though there are a vast number of fibrinolytic agents such as the tPA, urokinase are available, bacterial plasminogen activator streptokinase are

K. Gowthami · R. Jaya Madhuri (✉)
Department of Applied Microbiology, Sri Padmavathi Mahila Visvavidhyalayam, Tirupati, India
e-mail: drjayaravuri@gmail.com

K. Gowthami
e-mail: cutejasmi818@gmail.com

S. Jyothi et al. (eds.), *Proceedings of the 2nd International Conference on Computational and Bio Engineering*, Lecture Notes in Networks and Systems 215,
https://doi.org/10.1007/978-981-16-1941-0_22

being extensively investigated, as the current fibrinolytic agents are associated with showing different side effects [7–10].

Fibrinolytic enzymes are gaining importance in recent years, because of their therapeutic applications as anticoagulants, oncolytics, thrombolytics, anti-inflammatories, fibrinolytics, etc., the fibrinolytic enzymes are being explored from microbes, animals, plants and fermented foods [11–14]. Literature survey reveals isolation and purification of fibrinolytic enzymes from a deluge of microorganisms.

Microorganisms are advantageous over the other source due to the feasibility in laboratory culture and manipulation. The production of microbial enzymes is relatively inexpensive and easy to scale up requiring minimum space and time.

2 Types of Cardiovascular Diseases

See Fig. 1.

2.1 Mechanism of Fibrinolytic Enzyme

See Fig. 2.

2.2 Applications of Fibrinolytic Enzyme

Microbial fibrinolytic enzymes separate from microorganisms were recovered to be much effective when compared with those isolated from non-microbial sources like earthworms and snakes. Several fibrinolytic enzymes have recovered application not only in blood clot removal those drugs are treated as blood pressure regulators, plant virus inhibitors, and detergent additives, antimicrobials, in the prevention of dental caries, as anti-bacterial, anti-inflammatory, analgesic, and mucolytic expect or antigens. However, the mechanics behind the irregular refer of fibrinolytic enzymes are presently still puzzling [15–18].

2.3 Fibrinolytic Agents

Based on their route of administration, fibrinolytic enzymes are categorized into two types. One is an oral anticoagulant and the second type of fibrinolytic agents is intravenously administrated. Different types of anticoagulants are tabled below (Table 1).

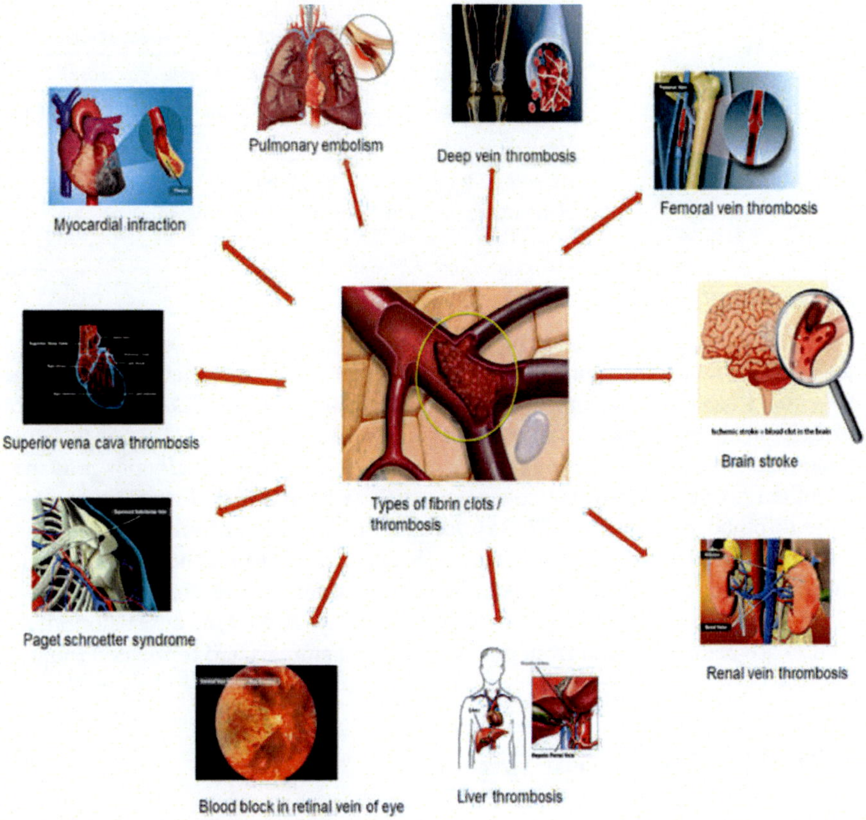

Fig. 1 Types of cardiovascular diseases

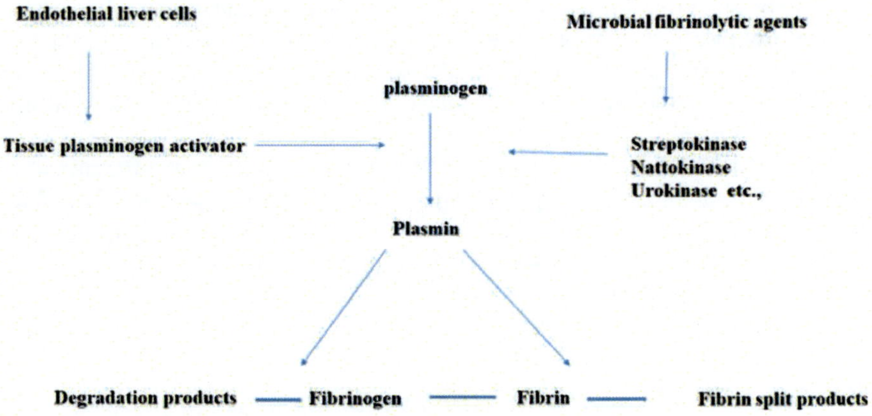

Fig. 2 Mechanism of fibrinolytic enzyme activity

Table 1 Types of anticoagulants

Oral anticoagulants	Intravenous anticoagulants
Warfarin (Coumadin)	Streptokinase (streptase)
Heparin (freedman)	Nattokinase (microbial origin)
Dabigatran (pradaxa)	Urokinase (Abbokinase)
Rivaroxaben (xarelto)	Altepase
Nattokinase (microbial origin)	Tenecteplase

2.4 Demerits of Anticoagulants

Oral anticoagulants are limited particularly in the elderly populations due to the intra-cranial bleeding associated with their administration [19–22]. Currently used oral anticoagulants such as warfarin and heparin have delayed onset of action and drug interactions along with diverse pharmacodynamics and pharmacokinetics property.

Synthetic compounds of anticoagulants such as coumarin and warfarin are used for the thrombosis treatment [23–25]. Intravenous injection of urokinase obtained from urine is used to treat the thrombosis. This urokinase enzyme has low specificity to fibrin and also expensive. This permits the search for a novel fibrinolytic enzyme with higher efficacy, safety, specificity and stability. Although, novel fibrinolytic enzymes are being explored from plants, microbes, animals and fermented foods.

3 Origin of Fibrinolytic Enzymes

Most essential occurrent for fibrinolytic agents, Streptokinase from *streptokinase hemolyticus* and staphylokinase from *staphylococcus aureus* were, in the beginning, proved to be effective in clot buster therapy. Over the past decades, more fibrinolytic enzymes from various microbes have been revealed in, progression, such as Nattokinase enzyme from *Bacillus natto*. On the whole, fibrinolytic enzymes are being produced by Bacteria, Actinomycetes, algae, Fungi, snake venom and earthworm [26–28].

A large number of fibrinolytic enzymes have been reportable from different bacterial species, including *streptococcus, staphylococcus, vibrio, paenibacillus, chryseobacterium, pseudomonas* [29–32]. A large amount of fibrinolytic enzymes have been reported from different microbial Sp., from the collection of sources, such as food, marine, soil mangrove sediment, etc. [33–35].

3.1 Recent Status in Fibrinolytic-Producing Microorganisms

See Table 2.

Table 2 List of microorganisms producing fibrinolytic enzymes

Source of fibrinolytic organisms	Organism	Isolates	References
Soil-slaughter houses	*Bacillus subtilis*	I-2	Bajaj et al. (2014)
Soybean grits	*Bacillus Firmus*	NA-1	Ji-Hyun (2004)
Dry douchi	*Bacillus subtilis*	DC-27	Hu et al. (2018)
Garden soil	*Bacillus infantis*	SKS1	Sandeep et al. (2017)
UV mutagen	*Bacillus subtilis*	WR350	Ruiwu (2019)
Fermented rice water	*Bacillus Haldoduran*	IND18	Ponnuswamy et al. (2016)
Rice water	*Bacillus cereus*	IND1	Ponnuswamy et al. (2014)
Dairy products	*Aspergillus brasiliensis*	AUMC9735	Kotb et al. (2015)
Fermented rice water	*Xanthomonas oryzae*	IND3	Ponnuswamy et al. (2018)
Germplasm collection center	*Citrobacter braakii*	BGCC#2123	Anjana Sharma et al.
Milk sample	*Pseudomonas aeruginosa*	CMSS	Subathra devi et al. (2015)
Chicken dump Waste	*Bacillus cereus*	SRM-001	Narshiman et al. [36]
Mangroove sediments	*Bacillus circulans*	Fibrinolytic enzyme	Sandeesh kumar et al.
Bovine milk	*Streptococcus equinus*	VIT-VB2	Vaishnavi and Subatra devi (2014)
Soil samples	*Aspergillus* sp.	Fibrinolytic enzyme	Shipa et al. (2014)
Mushroom	*Auricularia polytricha*	Fibrinolytic enzyme	Mohamed et al. (2014)
Soil from vegetable market	*Pencillium chrysogenum*	SGAD12	Gopinath and Sneha (2012)
Traditional fermented soybean food	*Lactobacillus* sp.	Fibrinolytic enzyme	Tho k chom and Joshi (2014)
Fermented food	*Paenibacillus* sp.	IND8	Vijayaragavan and Prakash Vincent (2014)
Marine halophilic	*Bacillus licheniformis*	KBDL4	Pathak and Deshmukh [37]
Recombinant mutants	Nattokinase *Bacillus subtilis*	WB800	Liu et al. [38]
Soil sample	*Bacillus amyloliquefaciens*	Mj5-41	Jo et al. [39]

(continued)

Table 2 (continued)

Source of fibrinolytic organisms	Organism	Isolates	References
Traditional Indonesian fermented food (gumbus)	*Bacillus pumilus*		Afifah et al. [40]
Food sample	Bongkrek	Fibrinolytic enzyme	Susmita et al. (2018)
Chinese traditional medicinal mushroom	*Cordyceps militaris*	Fibrinolytic enzyme	Cui et al. (2007)
Marine	*Bacillus subtilis*	ICTF-1	Mahajan et al. [3]
Nematode symbiotic organism	*Xenorhabdus indica*	KB-3	Pranaw et al. (2014)
Fermented shrimp paste	*Bacillus* sp.	Fibrinolytic enzyme	Anh et al. (2015)
Endophytes	Endophytic strain	EJS-3	Lu et al. (2007)
Marine green algae	*Codium latum*	Fibrinolytic enzyme	Matsubara et al. (1992)
Marine fish	*Urechis unicinctus*	Fibrinolytic enzyme	Wang et al. (2007)
Fermented pine needle extract	*Acetobacter* sp.	FP1	Jaeyoung et al. (2012)
Earthworm	*L. bimatus* *Eisenia anderi* *L. rubellus* *Neanthes japonica*	Fibrinolytic enzyme	Kotb [12]
Milk sample	*Lactobacillus plantarum*	KSK-II	Kotb [7]
Edible Mushroom	*Pleurotus ostreatus*	Fibrinolytic enzyme	Liu et al. [4]
Snake venom	*A. halys brevicaudus, A. contortrix, A. piscivorus piscivorus, Agkistrodon rhodostoma, A. piscivorus conami, A. acutus, A. caliginosus contortrix, A. halys pallas, A. halys ussuriensis, A. halys blomhoffii, A. saxatilis, Bothrops atrox, Cerastes cerastes, Crotalus atrox, Gloydius shedaoensis, G. ussuriensis, B. jararaca, B. alternates, B. asper, B. jararacussu, B. lanceolatus, B. neuwiedi, Trimeresurus elegans, T. flavoviridis, T. jerdonii, T. mucrosquamatus, and T. stejnegeri*	Fibrin(ogen)olytic venoms	Kotb [12] Kotb [20]

(continued)

Table 2 (continued)

Source of fibrinolytic organisms	Organism	Isolates	References
Plant	*Plant Allium tuberosum*	Fibrinolytic enzyme	Chung et al. [16]

3.2 *Disadvantages of Fibrinolytic Enzyme*

Disadvantages of using these thrombolytic agents include internal bleeding side effects, short half-life in the blood circulation and are expensive, thinning of blood vessels.

4 Future Prospectives

Fibrinolytic enzymes produce with minimal side effects, reduction in production and purification cost and augmentative therapy with other agents. Although the literal mechanism is well defined, COVID-19 appears to cause enhanced activation of cells that are essential in initiating the clotting process. This may lead to an increased risk of blood clots. In order to overcome this problem, future prospective is targeted to treat blood clots by using microbial fibrinolytic drugs that are more safe and economically viable https://www.healthline.com/health/coronavirus-and-blood-clots.

5 Conclusions

This review is especially focused on the broad aspects of the fibrinolytic enzyme giving particular intensity to the medical applications of the enzyme, challenges and future scope. Fibrinolytic enzyme plays a major role in the medical and pharmaceutical industries. With the advances in biotechnology, it has been aimed to overcome the unexpected obstacles by exploiting more potent fibrinolytic agents from marine microbiota and to provide improvement in quality of human life after medication.

References

1. Lu F, Lu Z, Bie X et al (2010) Purification and characterization of a novel anticoagulant and fibrinolytic enzyme produced by endophytic bacterium *Paenibacillus polymyxa* EJS-3. Thromb Res 126:349–355
2. Mahajan PM, Gokhale SV, Lele SS (2010) Production of nattokinase using B. natto NRRL 3666: media optimization, scale up and kinetic modeling. Food Sci Biotechnol 19:1593–1603

3. Mahajan PM, Nayak S, Lele SS (2012) Fibrinolytic enzyme from newly isolated marine bacterium *Bacillus subtilis* ICTF-1: media optimization, purification and characterization. J Biosci Bioeng 113(3):307–314
4. Liu X, Zheng X, Qian P et al (2014) Purification and characterization of a novel fibrinolytic enzyme from culture supernatant of *Pleurotus ostreatus*. J Microbiol Biotechnol 24(2):245–253
5. Mohammed AS, Ling TC, Muniandy S, Tan YS, Raman J, Sabaratanam V (2014) Recovery and partial purification of Fibrinolytic enzymes of *Auricular polytricha* (mont) sall by an aqueous two-phase system. Sep Purif Technol 122i:359–366
6. Mienda BS, Yahya A, Galadima IA, Shamsher MS (2014) An overview of microbial proteases for industrial application. Res J pharm Biol. Chem. Sci 5:388–396
7. Kotb E (2015) the biotechnological potential of subtilisin-like fibrinolytic enzyme from a newly isolated *Lactobacillus plantarum* KSK-II in blood distaining and antimicrobials. Biotechnol Progr 31(2):316–324
8. Kotb E (2012) Fibrinolytic bacterial enzymes with thrombolytic activity. Springer, New York
9. Pammi N, Chaitanaya KI, Mahmood SK (2015) *Servattia marcseens* OUSOT. SP Nov1. A cellulose and pharma. Producing bacterium isolated from polluted water. Int J Environ Biol 5:32–36
10. Tamura K, Stecher G, Peterson D, Filipski A, Kumar S (2013) MEGA 6: Molecular Evolutionary Genetics analysis version 6. Mol Biol Evol 36:2725–2729
11. Kotb E (2013) Activity assessment of microbial fibrinolytic enzymes. Appl Microbiol Biotechnol 97:6647–6665
12. Kotb E (2014) the biotechnological potential of fibrinolytic enzymes in the dissolution of endogenous blood thrombi. Biotechnol Progr 30(3):656–672
13. E silva FD, de Azevedo CA (2016) The Assistant software version 7-7 its isle in the analysis of experimental data. Afr J Agric Res 11:3733–3740
14. Wu R, Chen G, Pan S (2019) Cost-effective fibrinolytic enzyme production by *Bacillus subtilis* WR 350 using medium supplemented with corn steep powder and sucrose
15. Kotb E (2014) Purification and partial characterization of a chymotrypsin-like serine fibrinolytic enzyme from *Bacillus amyloliquefaciens* FCF-11 using corn husk as a novel substrate. World J Microbiol Biotechnol 30(7):2071–2080
16. Chung DM, Choi NS, Maeng PJ, Chun HK, Kim SH (2010) Purification and characterization of a novel fibrinolytic enzyme from Chive (*Allium tuberosum*). Food Sci Biotechnol 19:697–702
17. Lu X, Kopparapu NK, Yao L, Deng Y, Zheng X (2017) Bio-chemical characterization of a novel fibrinolytic enzyme from *Cordyceps militaris*. Int J Biol Macromol 94:793–801
18. Vijayaraghavan P et al (2017) Novel sequential sequential screening and enhanced production of fibrinolytic enzyme by *Bacillus* SP. IND12 using response surface methodology in solid state fermentation. Biomed Res
19. Kotb E (2015) Purification and partial characterization of serine fibrinolytic enzyme from Bacillus megaterium KSK-07 isolated from Kishk, a traditional Egyptian fermented food. Appl Biochem Microbiol 51(1):34–43
20. Kotb E (2017) Microbial fibrinolytic enzyme production and applications. In: Microbial Functional Foods and Nutraceuticals, pp 175–213
21. Yogeh D, Halami PM (2017) Fibrinolytic enzymes of *Bacillus SPP*. an overview. Int Food J 24(1):35–47
22. Almalki MA (2018) Solid state fermentation of agro-residues for the production of analyse from *Bacillus subtitlis* for industrial application. Int J Curr Microbiol Appl Sci 7(3):1341–1348
23. Green MR, Sambrook J (2012) Molecular cloning a laboratory manual. Cold Spring Harbor Laboratory Press, New York
24. World Health Organization (2019) The world health report. Cardiovascular diseases (CVD's)
25. Ningthoujam DS, Thokchom S (2016) Screening of fibrinolytic enzyme from microorganisms especially actinomycites from different biotypes in Manipur 7(3):21
26. Thokchom S,. Joshi SR (2014) Screening of fibrinolytic enzymes from lactic acid bacterial isolates associated with traditional fermented soybean foods. Food Sci Biotechnol 23:1601–1604

27. VijayaRaghavan P, Prakash Vincent SG (2014) Medium optimization for the production of fibrinolytic enzyme by *paenibacillus* sp., IND8 using response surface methodology scientific world 2014:276942
28. Krishnamoorthy A, Belur PD, Subramanaya SB (2018) Methods available to assess therapeutic potential of fibrinolytic enzymes of microbial origin a review 9:10
29. Cui H, Yang M, Wang L, Xian CJ (2015) Identification of new marine bacterial strain SD 8 and optimization of its culture conditions for producing alkaline protease. PLOS One 10:eo14606
30. Sakpal HC, Narayan G (2015) Thermostable alkaine protease from *Bacillus* Sp., and its potential application. J Pharm Biol Sci 10:58–67
31. Vuillemin M, Malbert Y, Laguerre S, Remaud-simeon M, Moulis C(2014) Optimizing the production of an α-(1-22) branching. Sucrose in Escherichia coli using statistical design appl. Microbiol Biotechnol 98:5173–5184
32. Chandrasekharan SD, Vaithilingam MS, Shankar R (2015) Exploring invitro thromblytic activity of natto kings from a new strain pseudomonas aeruginosa CMSS 8(10). https://doi.org/10.5812/jjm.23567
33. El Hidri D, Guesmi A, Najjari A, Cherif H, Ettomuni B, Hamdi C et al (2013) Cultivation-dependent assessment, diversity, and ecology of halo alkhalophilic bacteria in arid saline systems of southern Tunisia. Biomed Res Int 648141
34. Devi CS, Sinha D, Sharma V, Mohana Srinivasan V (2012) Screening for staphylokinase producing staphylococcus spp. from different environmental sample. Asian J Pharm Clin Res 5(4):125–128
35. https://www.healthline.com/health/coronavirus-and-blood-clots
36. Narshiman MK, Chandrasekharan M, Rajesh M (2015) Fibrinolytic enzyme, production by newly isolated blood clot lysis potential. Appl Microbiol 61:157–164
37. Pathak AP, Deshmukh KB (2012) Alkaline protease production, extraction and characterization from alkaliphilic *Bacillus licheniformis* KBDL4: a lonar soda lake isolate. Indian J Exp Biol 50:569–576
38. Liu Z, Zheng W, Ge C, Cui W (2019) High-level extracellular production of recombinant nattokinase in *Bacillus subtilis* WB 800 by multiple tandem promoters
39. Jo HD, Lee HA, Jeong SJ, Kim JH (2011) Purification and characterization of a major fibrinolytic enzyme from *Bacillus amyloliquefaciens* MJ 5-41 isolated from Meju. J Microbiol Biotechnol 21(11):1166–1167
40. Afifah DN, Scikhan M, Syah D, Yanti, Suhartono MT, Kim JH (2014) Purification and characterization of a fibrinolytic enzyme from bacillus pumilus 2-g. Isolated from Gembus an Indonesian fermented food. Prev Nutr Food Sci 19:213–219
41. Mohanasrinivasan V, Subatra Devi C, Biswas R, Paul F, Mitra M, Selvarajan E, Suganthi V (2013) Enhanced production of natto kinase from UV mutated Bacillus S.P. Bangladesh J Pharmacol 8:110–115
42. Chateterjee S (2015) Production and estimation alkaline protease by immobilized *Bacillus licheniformis* isolated from poultry farm SOTL of 24 parganas and its reusability. Adv Pharm Technol Res 6:2–6
43. Sivanandhini T, Subbaiya R, Gopinath M, Mahavined Angrash JKV, Kabilah T, Selvam M (2015) An investigation on morphological characterization of actionmycetes isolated from marine sediments Res J Pharm Biol Chem Sci 6i:1234–1243
44. Kim DW, Choi JH, Park SE, Kim S, K S et al (2015) Purification and characterization of a fibrin lytic enzyme from *Petasites japhonicas*. Int J Biol Macromol 72:1159–1167
45. Rhasmi B, Liny P (2013) Production and characterization of novel fibrinolytic enzyme from different soil fungal SP. Int J Pharma Bio Sci 43:454–464
46. Phanksri T, Luxanani P, Panyim S, Tirasophon W (2015) Synergism of regulatory elements in sigma (B) and sigma (A)—dependent promoters enhances recombinant protein expression in Bacillus subtilis. J Bio Eng 20(4):470–475
47. Xin X, Ambati RR, Cal ZW, Lei B (2019) Development of universal purification protocols for fibrinolytic enzyme – producing bacilli. CYTA J Food 17(1):112–120. ISSN: 1947-6337

Molecular Docking of Bacterial Laccases for the Decolorization of Textile Reactive Azo Dyes

Uma Maheswari Devi Palemaplli and Vani Mathakala

Abstract An integrated approach comprising computational analysis and in-vitro experimental analysis was carried out to the decolorization and degradation of two reactive azo dyes such as orange, RO-107 and Red, RR-5 by *Streptomyces bikeniensis* and *Streptomyces Sviceus.* Homology models for laccase enzyme were constructed based on the three-dimensional structures of *Streptomyces coelicolor* (PDB ID 3CG8) and *Streptomyces suicues* (PDB ID 4M3H). The docking analysis performed by Accelrys Discovery Studio 3.5 demonstrated the accommodation of the reactive azo dyes in the deep crevice of laccase active site. CDOCKER energies Reactive Orange, RO-107, showed two hydrogen bonds with ARG 140 of 3CG8 with bond energy of –28.9 and RR-5 exhibited three hydrogen bonds with ARG 167, GLY 224, and His 223. Reactive Orange RO-107 showed strong interaction with laccase, which correlated with the decolorization efficiency of *Streptomyces bikeniensis* and *Streptomyces Sviceus.* The results from the computational analysis were in good conformity with the decolorization assay. Thus, the results implicate that in-silico docking analysis will provide a prospective initial tool for the selection of potential actinomycetes that can be adopted in textile dye decolorization and degradation studies.

Keywords Laccase · In-silico analysis · Reactive azo dyes · Protein–ligand interaction · CDOCKER

1 Introduction

Currently, there is a global concern about the adverse effects of the textile industry as they discharge an enormous amount of colored effluent into water. Synthetic dyes have been abundantly used in textile dyeing due to cost economy in synthesis, color fastness, and availability of a wide spectrum of colors [1, 2]. More than 80% of

U. M. D. Palemaplli (✉) · V. Mathakala
Department of Applied Microbiology & Biochemistry, Sri Padmavati Mahila Visvavidyalayam, Tirupati, AP, India
e-mail: umadevi66@gmail.com

S. Jyothi et al. (eds.), *Proceedings of the 2nd International Conference on Computational and Bio Engineering*, Lecture Notes in Networks and Systems 215,
https://doi.org/10.1007/978-981-16-1941-0_23

textile effluents from dying industries are discharged into receiving water bodies without any pollution control [3]. The excessive release of synthetic dyes from textile industries adversely affects the natural resources and disturbs the integrity of the ecosystem [4]. During textile dyeing, 10–15% of synthetic dyes dumped into the effluent and greatly affect the water quality with the increase in biological and chemical oxygen demand [5]. On the basis of chromophore, dyes can be classified as azo, anthraquinone, nitro, nitroso, xanthenes, acridine, indigoid, thiazole, etc. Among these different classes, azo dyes demarcated the largest class of synthetic dyes with varied chromophore structures such as acidic, reactive, basic, disperse dyes. Azo dyes constitute more than 50% of the dyes produced annually and are the most important group of synthetic colorants that are extensively used in textile, food, printing, and pharmaceutical industries [6, 7]. As per the literature, 10–50% of the reactive dyes have not adhered with fabric and found as hydrolyzed products in the water bodies. Thus, the discharge of reactive dyes through the textile effluents is the major concern in the wastewater treatment process. Bioremediation-based technology, for the treatment of textile dye effluents, has got global attention due to its functional simplicity, low cost, minimal sludge quantity, and eco-friendly approach. Microorganisms like bacteria, actinomycetes, and fungi have been shown to be involved in the degradation and biotransformation of azo dyes. Several enzymes purified from bacteria and fungi have been reported to play a major role in the process of wastewater treatment. Among the decolorizing enzymes, several studies reported the catalytic function of laccases in the degradation of azo dyes [8, 9]. Laccases are copper phenol oxidases, catalyze the degradation of reactive dyes through the transfer of electrons from dyes to the copper ions particularly through a free radical mechanism with the release of nontoxic phenol compounds and water. Keeping in view the scope of laccase in the degradation of azo dyes, in-silico analysis was performed to analyze the impact of bacterial laccases in the decolorization of reactive azo dyes.

2 Methodology

2.1 Dyes and Their Properties

Azo dyes such as Reactive Orange 107 (RO-107) and Reactive Red (RR-1) grouped under vinyl sulphone dyes were chosen for experimental analysis. General properties of Azo reactive dyes used in this study were tabulated in Table 1 and chemical structures as shown in Fig. 1.

Table 1 General properties of reactive azo dye

Name of the commercial dye	Golden Yellow RNL	Remazol Red-1
Color index name	Reactive oxygen-107	Reactive Red-107
Number of azo dyes	Single azo	Single azo
Λ max (nm)	410 nm	540 nm
Molecular weight (g/mol)	566.49	717.38
Chemical formula	$C_{16} H_{16}N_4Na_2O_{10}S_3$	$C_{19} H_9Cl_2N_6Na_3O_{10}S_3$

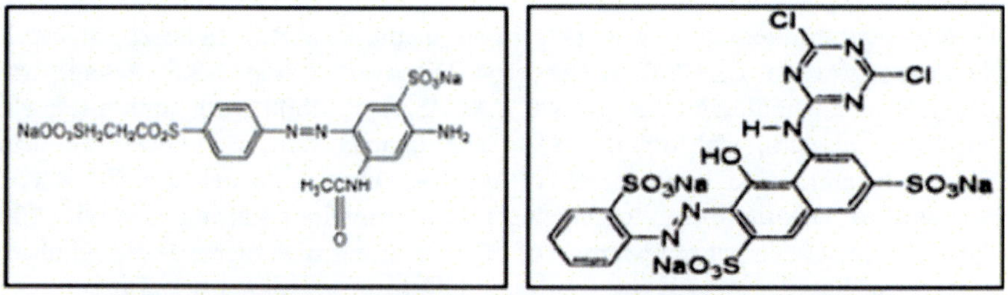

A: Reactive Orange 107 B: Reactive Red –1

Fig. 1 Chemical structures of Reactive Orange, RO-107 and React

2.2 *Actinomycetes for Decolorization of Reactive Azo Dyes*

Two cultures of actinomycetes, *Streptomyces bikeniensis* [10] and *Streptomyces sviceus* [11] were selected for the experimental work.

2.3 *In-Silico Studies for Degradation of Reactive Azo Dyes*

2.3.1 Software

Laccase receptor was identified with Accelrys Discovery Studio 3.5 and the structure of the ligand was generated using ChemSketch software. The molecular docking was performed with C DOCKER to observe the binding poses and calculate the docking energy of protein–ligand interaction.

2.3.2 Ligand Generation

The chemical structures of Reactive Orange RO-107 and Reactive Red RR-1 were sketched and optimized using chem sketch software. The ligands were stored in MOL2 format with the help of mercury software.

2.3.3 Structure of Laccase Protein

The crystal structures of laccase (PDB ID 3CG8 and 4M3H) were derived from PDB database. The validity of the 3D structure was also assessed by observing verify score value closest to verify expected high score and it was considered as the best model. CAST P tool was used to detect the active site of laccase enzyme.

2.3.4 Molecular Docking

Protein–ligand interaction was performed using CHARM m-based molecular docking algorithm, CDOCKER (Accelrys Discovery studio v 3.5 package from Biosystems Technologies, San Diego, CA, USA) on Z800 work station (Hewlett Packard, PaloAlto, California, USA). The ligands were minimized by using CHARMm and molecular force field of Accelrys Discovery Studio 3.5. CDOCKER was used to dock the ligand to generate 10 conformations starting each with 1000 dynamic steps at target temperature of 1000 K along with electrostatic repulsion. Docking of RO-107 and RR- was performed against three-dimensional structures of laccase (3CG8) and (4M3H) using C DOCKER discovery studio.

2.4 *Assay for Laccase Enzyme*

Qualitative screening approach was utilized to detect the extracellular production of laccase by using ABTS as substrate. The qualitative plate was prepared by mixing 100 ml nutrient glucose agar with 10 ml of 0.2 mM ABTS and 1 ml of CuSo4. The appearance of green color halos around the bacterial colony after 48 h incubation demonstrates the extracellular production of laccase.

2.5 *Decolorization Assay*

Docking analysis of RO-107 and RR reactive azo dyes with laccases was performed and the docking energies of the reactive dyes and laccase were further confirmed by decoloriztion assay using UV-spectral analysis. Streptomyces bikeniensis and Streptomyces sviceus were inoculated into 100 ml of mineral medium amended with one of the dyes such as RO-107 (400 mg/L), RR-1 (400 mg/L) and incubated at 37 °C. The collection of the decolorized samples was done at regular intervals. These samples were then subjected to centrifugation for a period of 10 min at 10,000 rpm. The supernatant was collected after centrifugation and was used to observe the change in the absorption at 410 nm for RO-107, 540 nm for RR-1 using UV–VIS Spectrophotometer (Schimadzu UV 1601). The percent decolorization of reactive azo dyes was determined as per the standard method [10].

3 Results and Discussion

The structural–functional relationship of laccases with azo dyes through homology modeling reported that the domains of laccases are highly conserved even though they are derived from fungi, plants, and bacteria [12]. The laccase protein sequences of Streptomyces bikeniensis and streptomyces sviceus were acquired from protein database of NCBI. The three-dimensional structures of Streptomyces coelicolor (PDB ID 3CG8) and Streptomyces suicues (PDB ID 4M3H) were obtained from protein data bank and were used as templates to build the homology model. Laccase from Streptomyces coelicolor (PDB ID 3CG8) and Streptomyces suicues (PDB ID 4M3H) were identified as the best homology modeled structures. The boundaries and substrate binding site of the active site were detected by CASTp (Computer atlas of surface topography of protein). This program measures and provides information on both surface accessible and internal accessible properties of the active site crevice to accommodate the substrate in deep binding mode. As per the literature, the fungal laccase comprises hydrophobic amino acids such as PRO184, PHE185, ILE474, PRO228, ALA261, PRO412, GLY413, PHE414, PRO415, PHE476, and acidic and basic amino acids such as HIS473, ASP227, ASP475, ARG182, and HIS477 in the solvent accessible region of the catalytic site of the laccase enzyme [13]. In our study, the laccase of Streptomyces coeli color demonstrated the presence of ARG119, ARG119, ASP141, ASP142, ARG146, PRO461. Research reports from earlier studies also demonstrated the role of computational tools in the prediction of azo dye detoxification potential of laccase [14]. Prasad and co-workers identified the weak interaction between laccase of Pycnoporus cinnabarinus and reactive acid blue and black dyes due to the planar ring confirmation of the dyes [14].

Molecular docking interactions were performed to understand the affinity between reactive azo dyes and laccase enzyme. Laccase of Streptomyces coelicolor (PDB ID 3CG8) and Streptomyces suicues (PDB ID 4M3H) was docked with Reactive Orange RO-107 and Ractive Red RR-5 by CDOCKER. Reactive Orange, RO-107 showed two hydrogen bonds with ARG 140 of 3CG8 with bond energy of –28.9 (Fig. 2). On the other hand, RR-1 exhibited three hydrogen bonds with ARG 167, GLY 224, and His 223 (Fig. 3). Hydrogen bonding patterns of reactive dyes with amino acid residues of the active site are depicted in Table 2. The molecular docking of RO-107 and RR-1 into the active site of 3CG8 and 4M3H laccase showed good binding energy –28.9 and –50.5 respectively. Our results are supported by the binding of Pycnoporus cinnabarinus laccase acid blue, reactive black, and reactive blue. The confirmations indicated the free access of the ligand into the protein active site and facilitate the oxidative degradation of reactive azo dyes into colorless intermediates. Thus, the laccases derived from streptomyces species are potential candidates in the decolorization and degradation of toxic reactive azo dye.

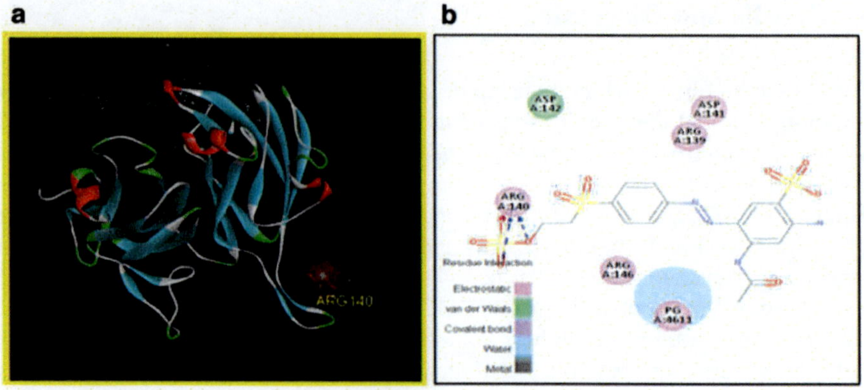

Fig. 2 Molecular docking of reactive orange, RO-107 into the active site of Laccase. **a** Hydrogen interaction of Reactive orange, RO-107 with laccase (3CG8)

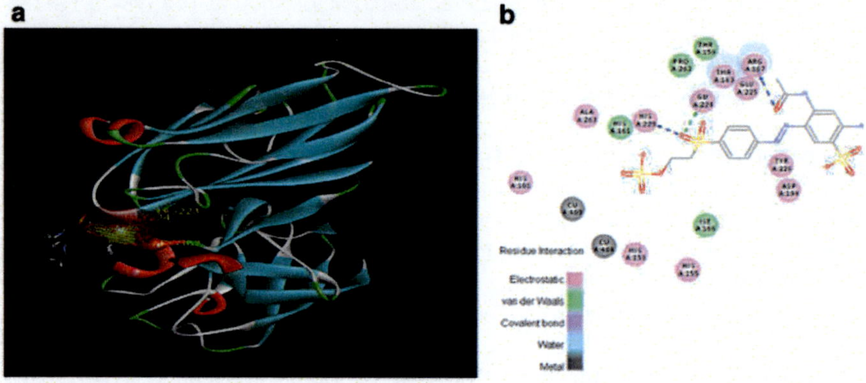

Fig. 3 **a** Docking interaction of Reactive Red, RR-1 azo dye ligand into the active site of Laccase (4M3H). **b** profile of hydrogen bonds between amino acids Laccase and Reactive Red azo dye

Table 2 Hydrogen bonding pattern and docking energy of reactive azo dyes with laccase protein

Laccase receptor	Ligand	Number of hydrogen bonds	Amino acids involved in hydrogen bonding	
3CG8	RO-107	Two	ARG 140	– 28.9
4M3H	RR-1	Three	ARG 167	– 50.5
			GLY 224	
			HIS 223	

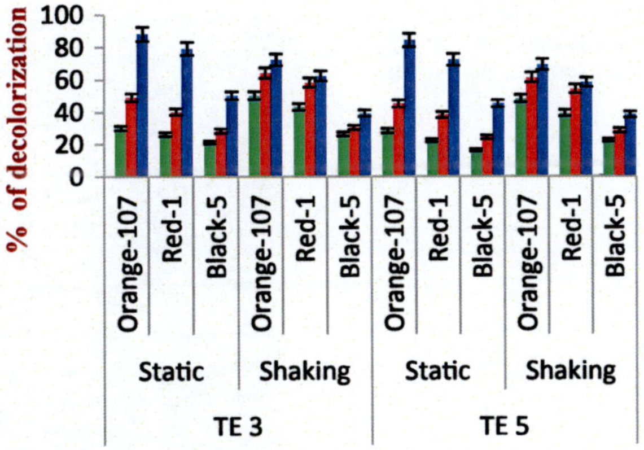

Fig. 4 Decolorization of azo reactive dyes by Streptomyces bikeniensis (TE3) and Streptomyces sviceus (TE5) at static and shaking conditions. Values are expressed in the mean ± SEM of three independent experiments statistical significance ** 1% level ($p \leq 0.01$)

3.1 *Decolorization of Azo Dyes by Actinomycetes*

The isolated actinobacteria Streptomyces bikeniensis (TE3) and Streptomyces sviceus (TE5) demonstrated the decolorization and detoxification of all the three reactive dyes and decolorization were noticed within 48 h of incubation under static conditions RO-107(94%), RR-1(91%) and RB-5 (58%) and RO-107(92%), RR-1(88%) and RB-5 (51%) under shaking conditions (Fig. 4). The actinobacteria were found to be more effective against reactive orange and red dyes on comparison with black dyes. Reactive black represented less than 50% of degradation with reactive black RB-5 and very weak interaction was noticed between laccase protein and the ligand reactive black RB-5.

Streptomyces bikiniensis and Streptomyces sviceus were exploited for the detection of extracellular enzymes. Plate assay is used to confirm the production of laccase into the extracellular fluid when grown in the malt extract medium. The qualitative screening exhibited the extracelluar production of laccase the concentrated Streptomyces bikiniensis as visualized by the appearance of halo around mycelia after 24 h of incubation (Fig. 5). The data confirm the production of laccase by Streptomyces bikiniensis. These results also indicated that the qualitative assay is suitable for visual identification of laccase enzyme [15] reported the production of laccase through bromophenol plate assay. The decolorization of azo dyes with laccase enzymes depends on the nature of enzyme sources and chemical structure of the dyes [16].

Our results are in tune with the degradation of azo dyes with Rigidoporus sp. Laccase [17].

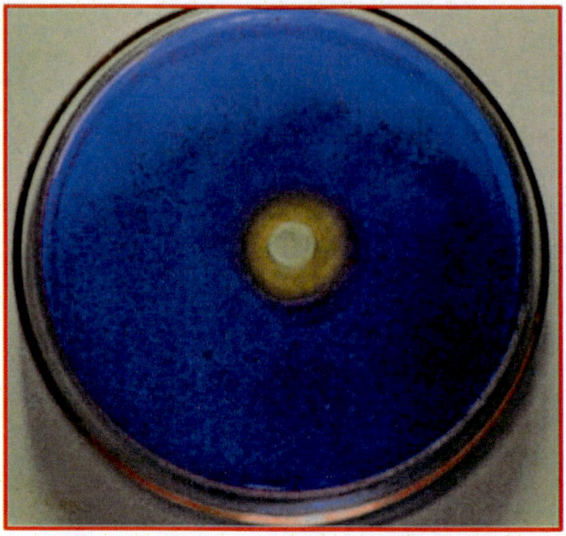

Fig. 5 Extracellular production of laccase by Streptomyces bikiniensis

4 Conclusion

In the current research paper, the prospect of computational tools and bioinformatics software was analyzed in the bioremediation of textile effluents enriched with reactive azo dyes. Molecular docking analysis was performed between laccase protein and reactive azo ligands. The strong interaction between the laccase enzyme and reactive azo dyes was demonstrated through hydrogen bonds, hydrophobic, electrostatic, and Van der Waals interactions. The docking score of the reactive azo dyes with the enzyme laccase resembled the in-vitro decolorization percentage pattern, thereby confirming the fact that the in-silico analysis of the azo dye—laccase interface may help us know the enzyme efficacy for bacteria-mediated decolorization and hence making the expansive preliminary screening of microbial strains obsolete.

References

1. Mohan SV, Rao NC, Srinivas S, Prasad KK, Karthikeyan J (2002) Treatment of simulated Reactive Yellow 22 (azo) dye effluents using Spirogyra species 22(6), pp. 575–582. Waste Manag, New York, NY
2. Sathiya Moorthi P, Periyar Selvam S, Sasikalaveni A, Murugesan K, Kalaichelvan PT (2007) Decolorization of textile dyes and their effluents using white rot fungi. Afr J Biotechnol 6:424–429
3. Mani P, Fidal VT, Bowmann K, Mark Breheny, Chandra TS, Keshavarz T, Kyazze G (2019) Degradation of Azo Dye (Acid Orange 7) in a microbial fuel cell: comparison between anodic microbial-mediated reduction and cathodic laccase-mediated oxidation. Front Energy Res 7:1–12

4. Sudha M, Saranya A, Selvakumar G, Sivakumar N (2014) Microbial degradation of azo dyes: a review. Int J Curr Microbiol App Sci 3:670–690
5. Puvaneswari N, Muthukrishnan J, Gunasekaran P (2006) Toxicity assessment and microbial degradation of azo dyes. Indian J Exp Biol 44:618–626
6. Chang MH, Huang CJ, Hwang SP, Lu IC, Lin CM, Kuo TF, Chou CM (2004) Zebrafish heparin-binding neurotrophic factor enhances neurite outgrowth during its development. Biochem Biophys Res 321:502–509
7. Amar T, Dayanand K, Jyoti J, Sanjay G (2008) Kinetics and mechanism of reactive Red 141 degradation by a bacterial isolate rhizobium radiobacter MTCC 8161. Acta Chim Slov 55:320–329
8. Kirby N, Marchan R, McMullan G (2000) Decolorization of synthetic textile dyes by Phlebia tremellosa. FEMS Microbiol Lett 188:93–96
9. Novotny C, Svobodova K, Kasinath A, Erbanova P (2004) Biodegradation of synthetic dyes by Irpex lacteus under various growth conditions. Inter Biodeter Biodegr. 54:215–223
10. Rizwana Parvez S, Uma Maheswari Devi P (2015) Decolorization and detoxification of Reactive azo dyes by *Saccharothrix aerocolonigenes* TE5. J Appl Environ Microbiol 3:58–62
11. Bhoodevi Chakravarthi, Vani Mathkala, Uma Maheswari Devi P (2017) Development of aerobic reactors for the remediation of textile effluents by marine *Streptomyces.* Asian J Appl Sc 6:1123–1129
12. Dwivedi LK, Singh DD (2011) Aflatoxin: an introduction and methods of biological control. Int J Pharm Bio Sci 2:400–406
13. Prasad NK, Vindal V, Narayana SL, Ramakrishna V, Swaraj PK, Srinivas M (2012) In-silico analysis of Pycnoporus cinnabarinus laccase active site with toxic industrial dyes. J Mol Model 18:2013–2019
14. Sadasivam SK, Srinivasan S, Thanavel M, Easwaran M, Rane N, Govindwar SP (2016) Bioinformatics aided microbial approach for bioremediation of wastewater containing textile dyes. Ecol Inform. 31:112–121
15. Shah H, Yusof F, Alam Z (2014) Production of laccase by Resinicium bicolor in submerged cultures: application of the Plackett-Burman experimental design to screen major factors. Int J Environ Eng 1:146–150
16. Simoes MF, Maiorano AE, dos Santos JG, Peixoto L, de Souza RFB, Neto AO (2019) Microbial fuel cell-induced production of fungal laccase to degrade the anthraquinone dye Remazol Brilliant Blue R. Environ Chem Lett 17:1413–1420
17. Sridhar R, Sivakumar V, Thirugnanasambandham K (2016) Response surface modeling and optimization of upflow anaerobic sludge blanket reactor process parameters for the treatment of bagasse based pulp and paper industry wastewater. Desalin Water Treat 57:4345–4356

Classification and Detection of Brain Tumor Using Segmentation Techniques on MRI Images

Gira Aparna, Gollapalli Sumana, and Gade Anitha Mary

Abstract In health care monitoring and observations, medical imaging is playing a vital role. Intelligent machine learning techniques for computer-aided diagnosis for automatic identification and detection of infected tissues, defected organs require accurate image segmentation algorithms. In medical image analysis, brain tumor is of high priority for the better diagnosis to improve the quality of the treatment, which increases the existing rate of the patient. MR (magnetic resonance) scans organs, bones and tissues by using immense magnetic forces and radio waves to generate images, which gives clear comprehension that provides a powerful medical diagnostic tool especially for soft tissues like brain. It is a challenging and time-consuming process to identify brain tumors from MRI scans manually in medical routing. Hence computational image segmentation involves systematic and more simplifying anatomical organs, which gives complete information of three-dimensional brain tissue and provides accurate volumetric measurements and other regions of interest (ROI). In this paper, the methodology of implementation of three techniques namely K-Means clustering, thresholding and region growing segmentation algorithms for a spontaneous segmentation of white matter (WM), cerebrospinal fluid (CSF) and gray matter (GM) are presented. The add-on cranial regions and the existence of tumors and also described for the performance evaluation of the three different segmentation algorithms for accurate and significant quantitative search for classification, identification and detection in biomedical images. The experiments are carried out on MRI brain images, the classification accuracy on both training and test data images demonstrates the comparative investigation with the other existing machine learning techniques.

G. Aparna (✉)
ECE Department, University College of Engineering, OU, Hyderabad, Telangana, India
e-mail: aparnaece27@gmail.com

G. Sumana
Command Control Center, Sri Padmavati Mahila Visvavidyalayam, Tirupati, AP, India
e-mail: sumanaspmvv@gmail.com

G. A. Mary
MCA Department, Loyola College, Hyderabad, Telangana, India
e-mail: anithareddi.gade@gmail.com

S. Jyothi et al. (eds.), *Proceedings of the 2nd International Conference on Computational and Bio Engineering*, Lecture Notes in Networks and Systems 215,
https://doi.org/10.1007/978-981-16-1941-0_24

Keywords MRI · ROI · Image segmentation · K-means clustering · Region growing · Threshold

1 Introduction

Segmentation is the basic step to be followed for any medical image to be visualized and analyzed for further processing. Image segmentation is basically applied for identifying things or objects and margins (lines, curves, etc.) of an image and assigning a label or tag with similar properties. The outcome of the image segmentation is a collection of segments that are jointly enclosed in the whole image present or a collection of contours taken out of that image. A set of rules present in general segmentation procedures is necessary. Homogeneous and uniform rules should be framed with respect to the characteristics of the image for segmentation. The characteristics may be of color, texture, region, etc. Segment margins or boundaries must be accurate. Segmentation has become a great challenge for medical images because of its complexity, especially for MRI scans. Physician generally suggests patients go for MRI scans to find the abnormalities in the brain tissue. Computerized support segmentation techniques will be helpful for medical practitioners to find the abnormalities of the brain perfectly.

2 Literature Review

2.1 Categories of Image Segmentation

Correct and specific segmentation applied for MR brain images is a compelling area in the research for different brain disorders. In the literature review, few trendy approaches applied for MR brain segmentations are given in three types based on

1. Classification
2. Region
3. Contour

2.1.1 Based on Classification Segmentation

Based on the classification segmentation approach [3], voxels are categorized and labeled as they exist in specific tissue belonging to a specific criterion. The most simple approach pertains to thresholding. In this algorithm, it tries to obtain a threshold value, based on the division of required classes. Repeated thresholding is applied to differentiate brain tissues in MR slices from others. Setting the values at the beginning, threshold for the brain and the head are repeatedly approximated based on the calculations on the geometry of obtained masks. Threshold approach is very simple and fast in computations but its features like components and noise in MR images make it delicate to INU (intensity nonuniformity) [5]. The evaluation of automatic threshold may be problematic in case of heavy overlapping among the intensities of various tissues based as intensity and noise are not homogeneities. Segmentation based on statistical classification is restored with a simple thresholding approach. Its robustness and stochastic theory for the mathematical base made it more advantageous.

In statistical classification approaches, the probability density functions of tissue intensity for various tissue classes are usually parametrically modeled. Next class of statistical [3, 10] based one is the voxel classification approach depending on the clustering method. Clustering [4, 10] is popular unsupervised classification approach that is applied in various applications like image segmentation and pattern classification. Clustering approach can be used to classify a voxel to tissue class based on similarity. The fuzzy c-means clustering (FCM) [3, 6] procedure is applied for the MRI segmentation. The various tissue classes partially can be evaluated using FCM procedure. In this procedure, it will not imitate the clustering approach. Therefore, FCM is applied to model PVA component [5]. In different cases, a voxel can have many tissue classes.

An approach of simultaneously calculating the INU component and achieving voxel classification depending on fuzzy clustering is emphasized, where in-between segmentation outcomes are used for INU calculation. The approach utilizes a transformed FCM cost function for modeling changes in intensity values and calculating bias fields generated as dynamic problems. In conventional FCM, clustering procedure will not consider spatial context among voxels as clustering is done alone in the feature scope. Recent adaptive fuzzy clustering procedure for MRI segmentation uses both INU components and spatial correlation among adjacent pixels.

2.1.2 Based on Region Segmentation

The object's shape can be defined by its edges or the area it occupies. Image area in an object usually has the same properties, e.g. same in texture, etc. Based on region segmentation [11, 13], approaches segment an image finding out the different same regions that relate to various objects in the image. Approaches based on regions take spatial interactions among adjacent voxels. In the simple form, area increasing approaches generally begin by pointing to a few seeds denoting a unique area present in the image. The points are increased until the entire area is covered. The area developed process is ruled by a constraint that develops a mechanism and a constraint that verifies the homogeneity of the area at every developed step. A semi-automatic, interactive MRI segmentation procedure was enhanced that employed an area-developing approach for lesion segmentation. In an automatic statistical area developing procedure rely on the robust evaluation of local area mean and variance for each voxel present in the image was given for MRI segmentation. The accurate area developing features are automated through cost function minimization. In addition, relaxation labeling, constrained region merging and region splitting were employed to enhance the quality segmentation in MRI. The finding of an accurate area homogeneity criterion is a base factor in the area developing segmentation approaches.

2.1.3 Based on Contour Segmentation

Based on contour segmentation [7] approach takes the various objects in an image and can be segmented by finding their boundaries. In edge detection segmentation or contour-based segmentation, edges are utilized to segment 3D MR images. A boundary finding approach is proposed, in which the evaluator clicks a pixel in an area to find edges and the approach searches the boundary beginning from that point. The approach is bound to segmentation in huge, well-defined structures, but not differentiating tissue varieties. Edge establish segmentation approaches generally have problems for extreme points segmentation, raised by not properly taking threshold value chosen. Adorned to the edges found are generally opened and complex edge linking approaches need to be improved.

2.2 *Block Diagram of the Segmentation Method*

The block diagram of the segmentation method is described in Fig. 1.

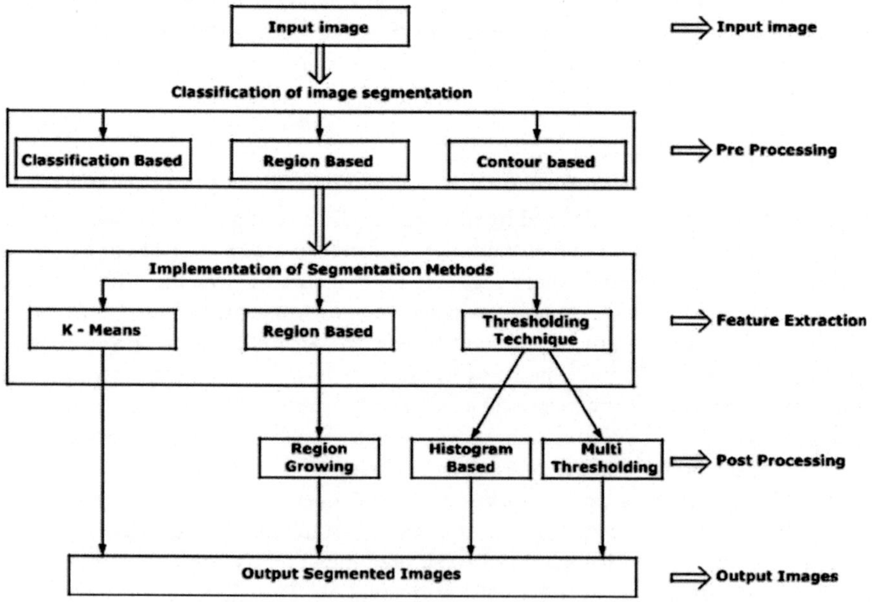

Fig. 1 Segmentation model

3 Implementation of Segmentation Methods

3.1 Implementation of K-means Clustering

K-means clustering depends on the count of other clusters and their location problems, which involves data grouping with similar feature vectors into individual clusters. K-means involves simple iterative technique to find the locally minimal solution. In K-mean techniques, input images are segmented based on pixel separation through cluster properties of interest. Gray value, contrast and local texture measures are the feature vectors of a pixel in the image. K-means is an unsupervised learning algorithm that is the best approach to make separation of d-dimensional data into k-clusters. For the basic cluster assignment and distance measure, the k-means clustering is best applicable, which follows the simple method that classifies the set of given data into certain sum of clusters (k clusters). It defines k centroids for k clusters respectively, which identifies different locations that give different outputs. The best way is to keep them away from each other and consider each point taken from each data set, which is nearest to the centroid and the process is continued until all the points are completed from all the clusters. The k number of the latest centroids has to be evaluated again, which acts as bray centers of each cluster. An update binding has to be made among the existing dataset and closer centroid. Continuous execution has to be made, as outcome k centroid keeps on changing the location until centroid stop to move further.

The objective function is as follows:

$$J = \sum_{j=1}^{k}\sum_{i=1}^{x}\left\|x_i^{(j)} - c_j\right\|^2,$$

where $\left\|x_i^{(j)} - c_j\right\|^2$ is the distance measure between a data point $x_i^{(j)}$ and the cluster center c_j, with n data points from their respective centroid.

Algorithm:

a. *Keep K points which are the initial group centroids in the space that are characterized by the objects that are clustered.*
b. Located every object in the cluster which has nearest centroid.
c. Locations of K centroids to be recalculated.
d. Steps b and c are repeated until the centroids stop to change further, which results in distinguishing the objects into clusters and henceforth minimized metric can be calculated.

Though it was substantiated that followed technique terminates, the k-means algorithm will not essentially locate the best optimal configuration, equivalent to the global objective function. The algorithm is applicable to new indiscriminate centroids also which can be looped any number of to minimize the outcome.

3.2 Implementation of Region Growing

Region growing technique steps are followed as below and explained with an example given in Fig. 2.

i. Similar sets of minimum locations are iteratively merged.
ii. Select random *seed pixel* and match with adjacent pixels.
iii. Region *grown* is observed in seed pixel combining with neighboring pixels, which are similar that increases the scope of the region. Observe until the growth of the region stops.
iv. Choose another seed pixel of the new region and start the procedure again.
v. Continue the process until all pixels fit any region.
vi. *This algorithm is bottom-up* method.

RG algorithm flow:

Volume_Grow ();

Let W be the set of dataset voxels and t be a threshold on. magnitude difference.

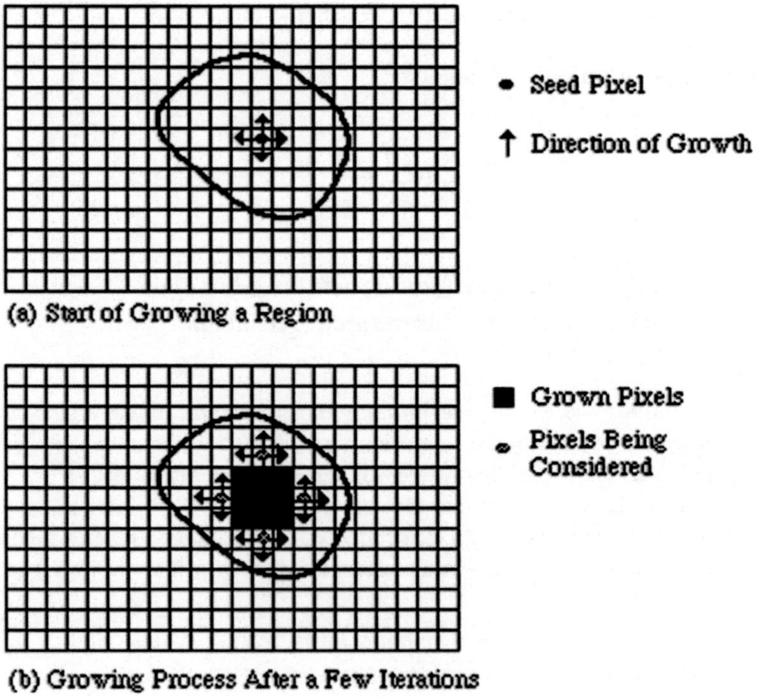

Fig. 2 Example of region growing

S = {};
Choose seed voxel w0 with intensity |w0|.
Determine median intensity I in the window of voxel about w0.
Recursive_Region_Grow (W, S, w0, I, t).
Remove Isolated Interior Voxel; Close.
Expand Region Boundary; Dilate One Voxel.
Recursive_Volume_Grow (W, S, w0, I, t):
S = S + W.
If wi belongs to W adjacent to w0, wi belongs to S, and.
If |I| - |wi| < t then.
Recursive_Region_Grow (W, S, wi, I, t).

3.3 Implementation of Thresholding Technique

3.3.1 Histogram Based:

Histogram-based [8] method is the best method when compared with other segmentation methods, which involves only one permit through the pixels. In this histogram technique, it computes all the pixels of the image. The pros and cons of the histogram are used to identify the clusters of the image. A refinement of using this histogram technique is to divide the image into smaller clusters with similarities. The process is continued until no more clustering is possible. The pixel-based direct classification method uses histogram statistics to define single or multiple thresholds to categorize an image pixel-by-pixel.

A pixel can be classified into one or more classes by comparing the gray value of each pixel to a particular threshold.

Where G (x, y) = 1 if f (x, y)>T &
G (x, y) = 0 if f (x, y) <T
where T = Threshold gray value.

3.3.2 Multi-Thresholding

$$
\begin{aligned}
g(i, j) &= 1 \quad \text{for } f(i, j) \in D_1 \\
&= 2 \quad \text{for } f(i, j) \in D_2 \\
&= 3 \quad \text{for } f(i, j) \in D_3 \\
&= 4 \quad \text{for } f(i, j) \in D_4 \\
&\ldots \\
&= n \quad \text{for } f(i, j) \in D_n \\
&= 0 \quad \text{otherise}
\end{aligned}
$$

In case of vision applications, it is suitable to discrete the regions of the image parallel to objects in which are interested, from the regions of the image that correspond to the background. Thresholding often provides an easy and convenient way to perform this segmentation on the basis of the dissimilar intensities or foreground or background colors of an image. It is also able to find the pixels with specified range values or band intensities of the image.

4 Experimental Investigations

The feature extraction techniques considered and their comparisons by considering input, output and outcome evaluated are tabulated in Table 1.

Table 1 Feature extraction technique comparisons

Method	Input image	Output image	Result
K-means clustering		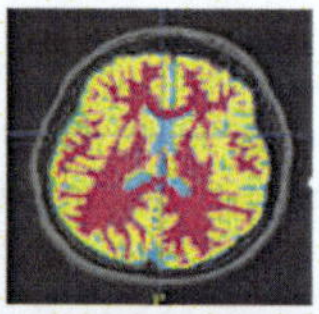	Segmented the input gray-scale image into three different regions using k-means clustering and assigned different colors to each tissue type
Region growing	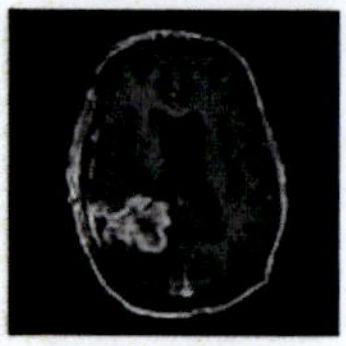	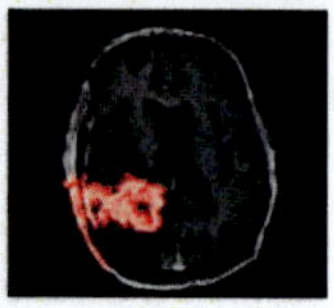	The results from K-means clustering are more useful in case of multi-tumor detection. The results from the region growing algorithm are more accurate compared to other methods
Histogram		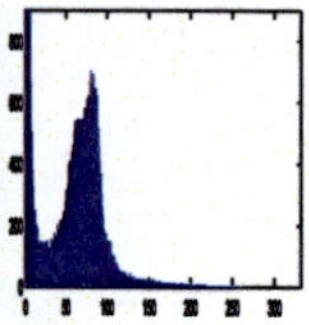	Histogram is a graph representing the pixels of different intensity values present in the collected image
Thresholding technique	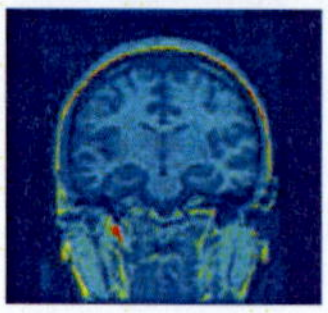	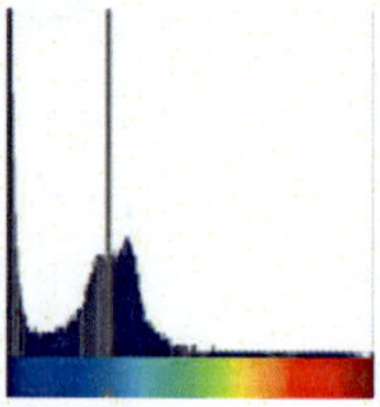	A thresholding procedure finds intensity value, which splits classes. Based on similar intensities higher than the threshold segmented into different classes. When more than one threshold is detected, it is called multi-thresholding

4.1 Results

From the results obtained and analyzed from Table 2 and Fig. 3, we can conclude Multi-Threshold technique is better in extracting the feature identification.

Table 2 Comparison of feature values

	K-means clustering	Region growing	Histogram	Multi threshold
Resolution	65,536	65,536	65,536	65,536
Number of Pixels	655,356	655,356	655,356	655,356
Gray matter	14,922	15,002	15,352	15,253
White matter	1523	1421	1450	1440
Back ground pixels	33,125	33,555	33,520	33,544
CSF (spinal Fluid)	14,000	15,000	14,500	14,600
Tumor pixels	728	929	824	780
Total measured pixels	67,976	66,976	66,978	66,750
Actual Number of pixels	65,536	65,536	65,536	65,545
% Error	0.03	0.02	0.03	0.02
% Error is ranging between	0.03–0.4	0.02–0.2	0.02–0.1	0.02–0.2

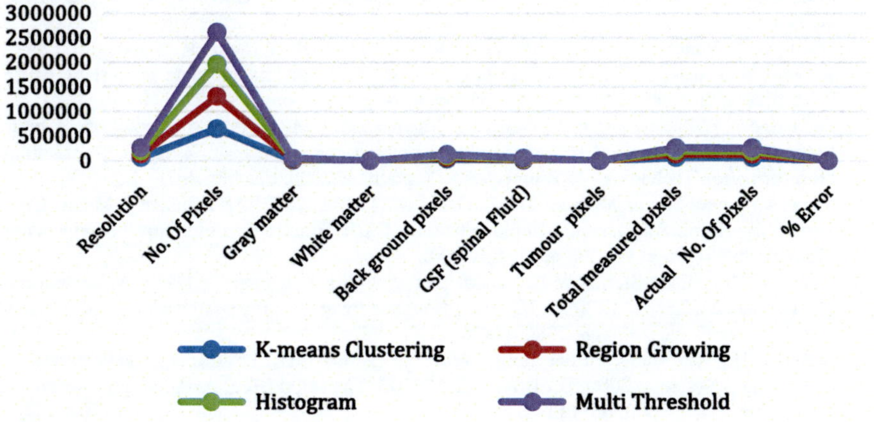

Fig. 3 Visual analysis

5 Conclusion and Future Scope

Automatic segmentation of tissues (GM, WM, CSF), the add-on cranial regions and the existence of tumor regions in MR brain image are effectively classified using the implementation of three segmentation techniques; K-means clustering, thresholding and region growing algorithms have provided the parameters that are chosen properly. In this study, locating tumors and proper investigations are carried out by using MRI data and 3D imaging analysis for the purpose of medical investigations.

The evaluation results from K-means clustering and thresholding are more useful in case of multi-tumor detection. The results from the region growing algorithm are more accurate compared to the other methods.

It has become more prominent to improve the accuracy, precision and computation speed of the segmentation process in identifying the abnormalities in the brain tissue to reduce human errors and increase the mortality rate. This is gaining importance as MRI image processing is evolving a routine diagnostic approach in clinical laboratory practices. As MRI data are 3D, it is compulsory that segmentation should follow the 3D volume segmentation as it gives the steady 3D boundaries of the segmented image.

References

1. Duška Kleut, Milorad Jovanović and prof. dr Branimir Reljin (2006) 3D visualisation of MRI images using MATLAB. In proc Journal of Automatic Control, University of Belgrade, vol. 16:1–3
2. Dillinger P, Vogelbruch JF, Leinen J, Suslov S, Patzak R, Winkler H, Schwan K (2006) FPGA-based real-time image segmentation for medical systems and data processing. IEEE Transactions on Nuclear Science, vol. 53, iss. 4, Aug. 2006, pp. 2097–2101
3. Shan Shen, William Sandham, Malcolm Granat (2005) MRI fuzzy segmentation of brain tissue using neighborhood attraction with neural-network optimization. IEEE Transactions on Information Technology in Biomedicine, September 2005, pp. 459–467
4. Tapas Kanungo, Senior Member, IEEE, David M. Mount (2002) An Efficient k-Means Clustering Algorithm: Analysis and Implementation. IEEE Transactions on Pattern Analysis and Machine Intelligence, vol. 24, no. 7, July 2002
5. David W. Shattuck,* Stephanie R. Sandor-Leahy, † Kirt A. Schaper, ‡ David A. Rottenberg, ‡,§, Richard M. Leahy* (2001) Magnetic resonance image tissue classification using a partial volume model", NeuroImage 13:856–876
6. Shi Juan He, Xia Weng, Yamei Yang, Weili Yan (2000) "MRI brain images segmentation", Circuits and Systems, 2000. IEEE APCCAS 2000. The 2000 IEEE Asia-Pacific Conference on Volume, Issue, 2000, pp. 113–116
7. Koen Van Leemput,* Frederik Maes, Dirk Vandermeulen, Paul Suetens (1999) Automated Model-Based Tissue Classification of MR Images of the Brain. IEEE Transactions on Medical Imaging, vol. 18, no. 10, October 1999
8. Stella Atkins* M, Mackiewich Blair T (1998) *Member, IEEE* "Fully automatic segmentation of the brain in MRI" IEEE Transactions on Medical Imaging, vol. 17, No. 1, February 1998
9. Kapur1∗, W. Eric L. Grimson1, William M. Wells III1, 2 and Ron Kikinis2 (1996) Segmentation of brain tissue from magnetic resonance images. In Proc Medical Image Analysis, vol. 1, number 2, pp. 109–127

10. Dana Cobzas, Neil Birkbeck, Mark Schmidt, Martin Jagersand "3D Variational Brain Tumor Segmentation using a High Dimensional Feature Set"
11. Dana Elena Ilea, Paul F. Whelan, Ovidiu Ghita Vision Systems Group, School of Electronic Engineering
12. Dublin City University, Dublin 9, Ireland Performance "Characterization of Clustering Algorithms for Colour Image Segmentation"
13. Chen1 HLJ, Samavati1 FF, Sousa1 MC, Mitchell1 JR (2006) Sketch-based volumetric seeded region growing. EUROGRAPHICS Workshop on Sketch-Based Interfaces and Modeling, Thomas Stahovich and Mario Costa Sousa (Editors)
14. Matei Mancas, Bernard Gosselin "Towards an automatic tumor segmentation using iterative watersheds", Signal Processing & Circuit Theory Lab, Faculté Polytechnique de Mons Bâtiment Multitel, Parc Initialis, avenue Copernic 1, 7000, Mons, Belgium Medical Image Analysis, Vol 7 (40, p. 513–527, Dec 2003)

Enhancing the Structure of Cyber Risks in Cloud Environment Using Cloud Forensics Technique

Vadetay Saraswathi Bai and T. Sudha

Abstract With a lively growth of cloud in all the sectors all over, cloud computing has become one of the reference structures in the field of war for cyber attackers, where one of the major problems of cloud computing is to protect data from multiple outbreaks. In security programs, cybersecurity encourages many compliance mechanisms to resolve many vendor management attackers. Multiple mechanisms are required in an enterprise to set up a supplier management system that involves determining the risks of confidentiality, credibility and accessibility associated with third parties in resource systems. The stance of cybersecurity depends on knowing the hazards that threaten them. A threat vector that requires access to confidential data and analytical systems is a third party. Together with an investigative team for cloud forensics, a substructure focused on trustworthy third-party proves to be a safer approach. To improve the service provider's productivity and thus allow cloud providers to collect strong evidence from security attackers could aid in further legal action.

Keywords Cloud territory · Hypervization · Trusted third force · Forensic obstacles

1 Introduction

A new IT environment for cloud computing where many people across the globe are widely adapted to this easily encourages the protection of stored data in the Cloud Territory. There are security attacks or regulation breaks that clamp down and carry out a forensic investigation. But for off-line investigation, current forensic standards,

V. S. Bai
Department Computer Science and Engineering, School of Engineering and Technology, SPMVV, Tirupati, India

T. Sudha (✉)
Department Computer Science, SPMVV, Tirupati, India
e-mail: saihumika9@gmail.com

S. Jyothi et al. (eds.), *Proceedings of the 2nd International Conference on Computational and Bio Engineering*, Lecture Notes in Networks and Systems 215,
https://doi.org/10.1007/978-981-16-1941-0_25

frameworks, procedures and tools are largely determined, these approaches assume, and in particular, that the storage media under investigation is entirely available.

Conducting studies, cloud environment poses the latest problems that could be identified. In order to make the cloud space stable and to maintaining the efficiency of cloud service providers, a substructure is proposed in this research.

2 Proposed One

Technology improvements-

(1) Greater memory, greater processing, power, greater speed
(2) Number of clients is increasing in great proportion where security matters.

- User Activity
- Valuable Data:

As we have more and more on our mobile access, personal computers in those systems, we store more personal information. Not only interactions with our associated individuals but also the social media sites they use. What is more, we store passwords for our apps, family images, business emails.

2.1 Profits from Substructure

Any outbreak can be successfully being collected using a teamwork with meticulous planning. For an outbreak, a model by a group of authenticated participants focused on collective conduct is clarified in this chapter to be defined and to make it ineffective. The primary segments with their locations are:

(a) **Customer of the Cloud**: The Ultimate Consumer who satisfies Cloud-based applications is the cloud client. The Customer cloud should specify a mark of identification to have the advantages of cloud services.
(b) **Trusted Third force**: With occasional support from the cyber forensics unit, TTF is involved in helping to ensure detection and acknowledgment to find out the holes in the defense.
(c) **Cloud Service Provider**: Cloud Service Provider a holding the tariff lending databases and servers, CSP might register for itself to TTF the purpose of providing Cloud-based applications.
(d) **Investigation Methodology for Cloud Forensics:** When an application from TTF is received, to deal with reluctant activities within the field of territory, the CFIT team may come into action. Investigation methodology for Cloud Forensics is still comfortable using the latest methods, as the new draft forensic software will still be required for TTF.

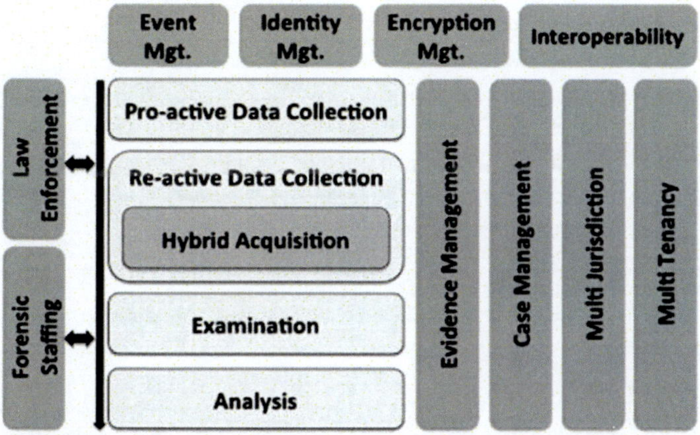

Fig. 1 Substructure management concepts

(e) **Digital Forensics Specification:** Digital forensics is the use of scientifically acquired methods to store, gather, verify, classify, examine, interpret, record and present digital evidence obtained from digital media in order to further promote the modification of events known as predicting criminal acts, which have been shown to interrupt planned operations (Fig. 1).

The major challenges of Cloud Forensics are grouped into the following seven major categories, according to the National Institute of Standards and Technology 4, which are summarized as

According to the National Institute of Standards and Technology, the major problems of Cloud Forensics are classified into the following nine major categories, which are summarized as

1. Architecture in the
2. Collecting data
3. Analization
4. Anti-forensics Programs
5. First Interviewee event
6. Management of positions
7. Legal to the Legal.

2.2 *Challenges in Cloud Forensics*

In several previous works, the benefits and fascinating directions in cloud forensic investigation have been incorporated together. Reilly et al. [19] concluded that one of the advantages of getting data in a central location is that the investigation can be carried out quicker. Cloud forensic challenges generally have to monitor the evidence, especially in the data collection, safeguarding and authentication process.

Table 1 Forensic challenging phases

Phases	The obstacle List
Recognition	Decentralized Data Replication Cross-Jurisdiction Dependency Chain Encryption Obscure Physical Region
Safeguarding	Chain of Custody Proof Isolation Volatility Knowledge Integrity of Details
Collections	Trust, Multi-Tenancy Time Synchronization Cross-Jurisdiction Erased Data Lack of instruments for investigation
Authentication	Absence of registration for logs Time Lining Proof Encrypted Information Coordination

The evidence in the cloud environment is difficult to define as the deployment of service models for further movement of the physical system that holds possible pieces of data evidence is also limited. According to the forensic cubes in a table format, the potential advantages of cloud forensics are primarily identified. The forensic cloud is specifically needed from the list of challenges (Table 1).

The researchers revisited the term suggested by several working definitions of cloud forensics based on the reference architecture of NIST cloud computing as the implementation of digital forensic science in cloud computing environments is cloud forensics. Technically, it consists of a hybrid forensic debate on digital evidence production and multi-tenant situations are also inferred.

3 Hypervization

As it inputs as production from TTF to deal with criminal activity in the territory, the Cloud forensic team carrying on IMCF will come into operation. As TT will still have the most updated versions of applications for forensics, the IMCF also has a strong advantage of using the updated software instruments. The model can be seen in the figure and the short description as follows.

- TTF corroborates the possibility of many CSPs and registers them as manual work as the trusted providers of service with mandatory fields. As TTF will categorize the Cloud-based application has the basis of the needs where provided and address where they are going. The customer will enter the TTF with an identification point if a CC wishes to use cloud services, along with a list of needs/services to be served by the cloud. The TTF then includes a concise overview of all the laws and regulations that CC has to follow and obtains all mandatory legal papers to get the affiliation signed by CC. Often the CC receives the written forms of the undertaking by TTF as per protocol. The TTF provides the client with login details to use the cloud services when all required proceedings are completed. The CC registration method is explained in Fig. 2.

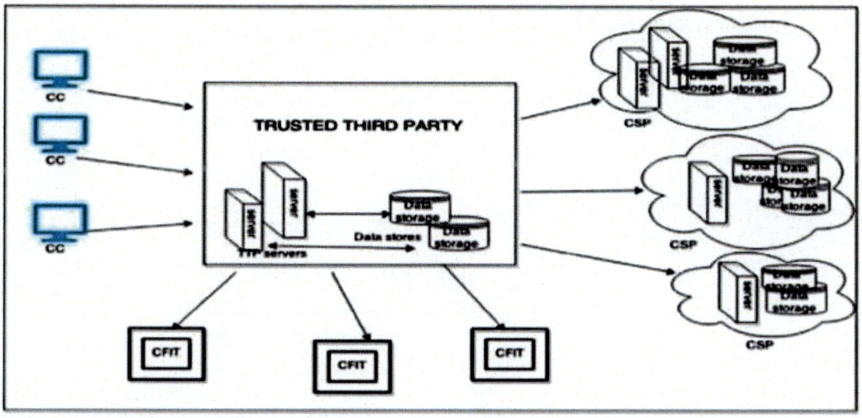

Fig. 2 Outline of the proposed model

- The TTF will notify a concern that the client should be allocated to registered CSP/CSPs referring to the specifications of CC and services rendered by several CSPs. On behalf of the CC, TTF establishes a new user account with the CSP and retains as shown in Tables 2 and 3 for further login procedures, the same information is in its Secure Client File along with Client Requirement Allocated.
- The TTF collides with a band of IMCF on the percentage of cases conducted in the investigation. Both the government and the private sectors should recognize CFITs. Now, since the CSPs are unaware of their CCs, and vice versa, the CSPs are unaware of their CCs. TTP becomes the middle authority of the cloud territory with full security and TTF transparency. The TTF will protect the CT's confidentiality. The first step if a CC wishes to engage in the territory authentication to the TTF and having the first checking time ticket produced and given by the TTF. The TTF provides the CSP with an output of the same. For further reference, it also records this duration information of the CT, generated data are for a period of time with TTF and expire within a specified amount of time, so leaks are not an easy task and thus authentication and protection are not easy tasks.
- To retain the identity information of CSP, 3 TTF is opened to an interface with the consumer on the side of a Cloud-based application so that a CT scan has a note of unknowns to which CSP is associated. The login information is forwarded to the CSP. Only via the TTF by validating STT will the CSP full login from the CT. From here, the TTF functions as a firewall to the CSP. This TTF offers a CSP time and addresses where the protection is preserved. IMCF will get the task of the related digital proof image information. All the mandatory details of the current version of software tools and recommendations for the IMCF are also given to the IMCF, which collides with several other IMCFs, making the authentication reachable. Once all the data needed to analyze the process is conducted in parallel by other IMCFs to speed up the investigation. Since IMCF understands the IP address and

Table 2 Document for protected client (DPC)

Login Costumer Id	Info TTF	I.P Ad-address	Location

Table 3 Prerequisite for client allocated

Custom-er Id	Required Id	allocated Login CSP info	CSP login infor-mation

location of the incident, the criminal act is highly likely to be recognized and presented in court.

4 Conclusion

While there are a large number of illegal cyber operations where cloud territory security is being extended, a lot of academic work is also going on with more difficult ideas coming up, while cloud territory has become a mandatory field for outbreaks where cloud forensic security areas have a significant impact. This paper introduced a cloud environment and digital forensics themes to explain their collision. Finally, the security level effect of cloud forensic preparedness was addressed in the paper and how cloud forensic preparedness would boost cloud security. Through this concept of TTF, a validated model for a safe solution is given. The TTF is defined by one of the key actors as it is a single point connecting the node to all other nodes. Nodes in cloud environments by backing up session-wise, this could be overcome.

References

1. Grispos George, Tim Storer, William Bradley Glisson (2013) Calm before the storm: the challenges of cloud. Emerg Dig Foren Appl Crime DetPrevent Security 4:28–48
2. Liu Fang et al. (2011) NIST cloud computing reference architecture. NIST special publication 500(2011):292
3. Cloud Computing Forensic Science Challenges (2014) National Institute of Standards and Technology Interagency or Internal Report 8006, June 2014
4. Ruan K, Carthy J, Kechadi T, Crosbie M (2011) Cloud forensics: an overview. Advances in Digital Forensics VII
5. Zawoad Shams, Ragib Hasan (2013) Cloud forensics: a meta-study of challenges, approaches, and open problems. Preprint arXiv:1302.6312

6. Jaya prakash Ramsaran (2014) Cloud Computing: Benefits and Challenges [ONLINE]. https://www.linkedin.com/pulse/20140921193928- 23699310-cloud-computing-benefits-and-challenges. Accessed 15 Dec 2016
7. NIST Cloud Computing Forensic Science Working Group. (Draft NISTIR 8006), 2014. NIST Cloud Computing Forensic Science Challenges
8. Ruan K, Carthy J, Kechadi T, Crosbie M (2011) Cloud forensics. Advances in digital forensics VII, IFIP advances in information and communication technology 361:35–46
9. Ruan K, Carthy J, Kechadi T (2011) Survey on cloud forensics and critical criteria for cloud forensic capability: a preliminary analysis. ADFSL Conference on Digital Forensics, Security and Law, pp. 55–70
10. Grispos G, Storer T, Glisson W (2012) Calm before the storm: the challenges of cloud computing in digital forensics. Int J Dig Crime Foren 4(2):28–48
11. Wolthusen SD (2009) Overcast: forensic discovery in cloud environments. In IMF 2009—5th International Conference on IT Security Incident Management and IT Forensics - Conference Proceedings, pp. 3–9
12. Yahya F. et al. (2014) Security challenges in cloud storage. 2014 IEEE 6th International Conference on Cloud Computing Technology and Science
13. Zawoad S, Hasan R (2013) Digital forensics in the cloud. CrossTalk (October), pp.17–20
14. Zawoad S, Hasan R (2013) Cloud forensics: a meta- study of challenges, approaches, and open problems. Preprint arXiv:1302.6312
15. Zhang Q, Cheng L, Boutaba R (2010) Cloud computing: State-of-the-art and research challenges. J Int Serv Appl 1(1):7–18
16. Zissis D, Lekkas D (2012) Addressing cloud computing security issues. Futur Gener Comput Syst 28(3):583–592
17. Heiser J, Nicolett M (2008) Assessing the security risks of cloud computing. Gartner Research (June), pp. 1–6
18. Haggerty J, Taylor M (2006) Managing corporate computer forensics. Comp Fraud Sec 2006(6):14–16
19. Hemdan EED, Manjaiah DH (2018) "Chapter 56 CFIM: Toward Building New Cloud Forensics Investigation Model", Springer Science and Business Media LLC, 2018

Facial Mask Detection Using Deep Learning and Machine Learning Approaches

P. Venkateswara Rao, K. Sai Susheel, and V. Sucharita

Abstract The COVID-19 pandemic is destructing the mankind. Until the discovery of the vaccine, precautionary measures have to be taken. The protection method that is effective is covering with a face mask in the areas where most of the people move around. The important utility is face masks in eliminating the COVID-19. In this paper, deep learning and machine learning are used for the detection of face masks. There are two components in the proposed work. The first one is the extraction of features and the second is the classification process of face masks by using Support Vector Machine (SVM). Real-world Masked Face Datasets (RMFD) have been selected for experimentation. The SVM classifier achieved an accuracy of 95.31% when tested.

Keywords Machine learning · Deep learning · Face mask · COVID-19 · Classifier

1 Introduction

In the year 2020, WHO, i.e. World Health Organization, has declared COVID-19 as pandemic [1, 2]. Before this pandemic situation, people protect themselves from pollution by wearing masks. The present situation of COVID-19 has made governments to declare lockdown and unlock. But certain sectors, all over the world, require manpower to perform the operations for basic necessities. Lockdown cannot be extended for so many days. To avoid the spread of coronavirus in such situations, wearing a face mask has become a basic requirement for all the people working in the organizations. There should be a system that provides the assurance regarding people wearing masks or not at the entry or exit points at various places. In this research paper, a model for face mask detection is based on deep learning and machine learning classifier. This model can be used with CC cameras for detecting the people not wearing the face masks to avoid the transmission of COVID-19. Deep learning is

P. Venkateswara Rao (✉) · K. Sai Susheel · V. Sucharita
Department of Computer Science and Engineering, Narayana Engineering College, Gudur, AP, India
e-mail: vrsairam23@gmail.com

S. Jyothi et al. (eds.), *Proceedings of the 2nd International Conference on Computational and Bio Engineering*, Lecture Notes in Networks and Systems 215,
https://doi.org/10.1007/978-981-16-1941-0_26

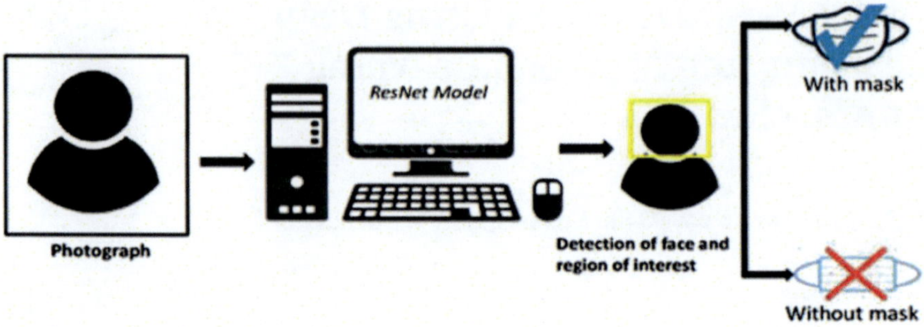

Fig. 1 Classification of the image

used for extracting the features. After the extraction of features, it is combined with the machine learning classifier for the detection of the face mask. Deep learning is a family of machine learning that causes the computation of different layer neural networks feasible. It instructs computers to do what comes naturally to human beings. The prediction accuracy and performance on a large amount of data can be increased more than the humans' IQ with the help of deep learning. It can be used for physical robots because they have neural networks and are also used for image recognition and compression, character recognition, stock market prediction, automatically detect objects, automation driving, aerospace and defense, medical research and industrial automation, etc. Artificial neural network algorithms are used [3, 4]. It can be worked with both structured and unstructured data and can perform complex operations easily than machine learning. Wearing facemask reduces the risk of coronavirus. Different types of masks are available like N95 masks, cloth masks and surgical masks. The objective of the proposed model aims at identifying the person whether he is wearing a mask or not. The results will be tested when an image is given as an input as shown in the figure. Face will be detected and the region of interest will be extracted. Then, the inputs will be classified with a mask or without a mask (Fig. 1).

2 Related Work

In this paper, the objective is to identify the people that are not wearing the masks for avoiding the COVID-19 transmission. The scientists and researchers proved that wearing the mask helps in reducing the spreading rate of the virus. Machine learning models are used to predict the COVID-19 outbreak and difference in the results from country to country by taking the comparative study on various datasets. The result of these models such as MLP and ANFS gives a more generalization capability for long-term prediction [1].

For seeing the health, geographical, travel COVID-19 patient's, demographic data to predict the seriousness of the case and possible results such as recovery or death by

Table 1 Comparison of the accuracy obtained of various methods in the literature

Method	Model	Accuracy
B. Quin et al.	SRCNet	98.7
Sabbir et al.	PCA	70%
Li et al.	FSA-Net, HGL method	74.97%
Khandelwal et al.	MobileNet V2	97.6%
Wang et al.	YOLOV 3	93.9%
Jiang et al.	ResNet and MobileNet	91%
Ristea et al.	ResNet	74.6%
Proposed method	ResNet	95.31%

using the random forest and AdaBoost models. After doing the analysis, a majority of death rates are male patients than female patients from various ages [2].

B. Quin et al. developed a method for identifying the way of wearing a face mask [5]. This was categorized in three ways. They are the correct way of face mask wearing, no face mask, incorrect wearing of face mask wearing. This method achieved 98.7% accuracy.

Sabbir and his coauthors [11] have suggested a technique based on principal component analysis to recognize whether the mask is there or not. The accuracy of recognition has been dropped to 70% when the face is masked. Li et al. used the CNN model for face detection quickly, which removes nonface sections and processes the regions at greater resolution for detection [6].

Ristea et al. proposed a novel data augmentation method for detecting masks from speech [10], which is used for communication between surgeons used in infectious diseases. The authors used multiple ResNet models and trained the network.

Wang et al. have used a YOLOV3 algorithm for the detection of face, which achieved 93.9% accuracy [7]. Nizam et al. [12] have suggested a new method that automatically removes masks by covering the area of the face. The output is the complete face image, which looks realistic and natural. Jiang et al. proposed a Retina face mask, which is of high accuracy by using ResNet and MobileNet [9].

Khandelwal et al. have given a model using deep learning that differentiates an image by identifying whether a mask is available or not [8]. Table 1 shows the comparison of the existing literature based on the various techniques that are used and the accuracies obtained.

3 Proposed Methodology

The detection of face masks has been done by using deep learning and machine learning methods. This work has been designed in two phases and testing of the face masks. The training and testing phases of face mask are shown in Fig. 2. The experiment was conducted on Real-world Masked Dataset (RMFD [7]). Figures 3

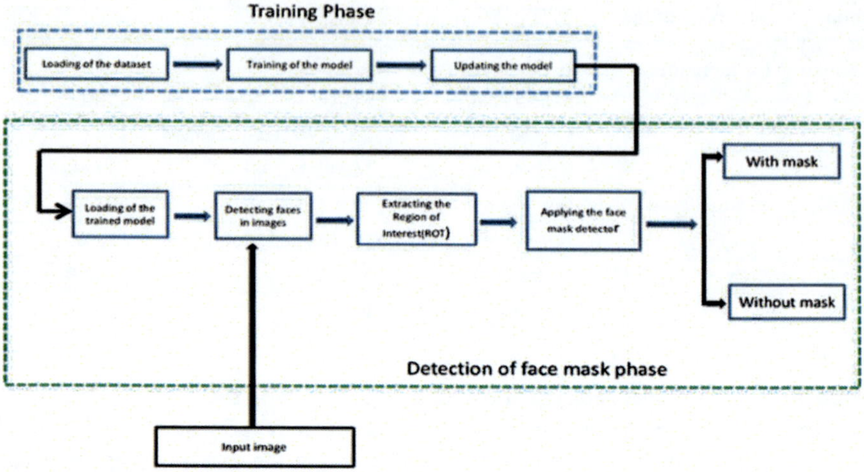

Fig. 2 Training and testing

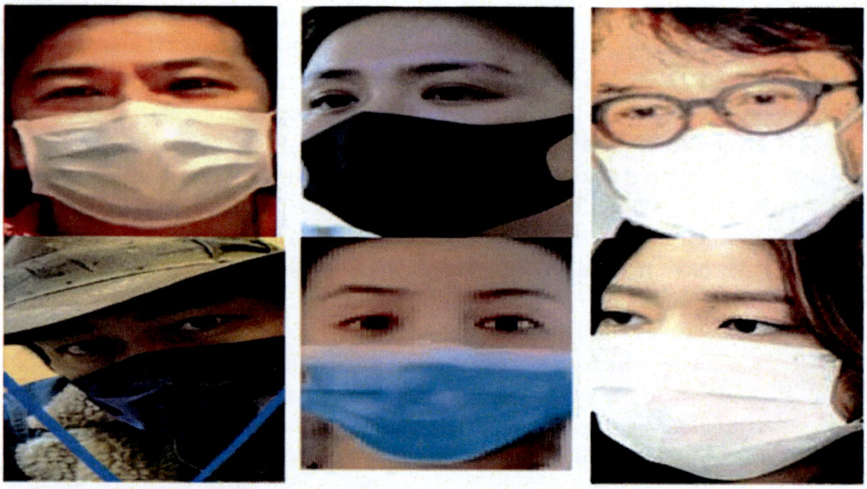

Fig. 3 RMFD dataset images with mask

and 4 show the samples of different faces with masks and without masks. For testing and training, RMFD dataset has been used. This method has two parts, first one is the extraction of features extraction using deep learning and the second one is machine learning using support vector machines. For feature extraction, residual neural networks are used, for training and testing, machine learning algorithms are used. A residual neural network is a type of deep transfer learning. Fifty layers were taken starting with convolution layer and ending with a fully connected layer. In

Fig. 4 RMFD dataset images without mask

the classification part, the last layer of the network has been removed and is being replaced with the support vector machines to improve the performance.

3.1 Support Vector Machines (SVM):

A supervised learning technique used for classification is support vector machines. A machine learning algorithm for classification using the Hinge function given in Eq. 1, P is the input vector, where a is label ranging 0 to 1, m.P-b is the output, the coefficients of classification are denoted by m and b.

$$h_j = \max(0,\ 1 - a_j(m.P_j - b)) \quad (1)$$

4 Results and Discussions

The experiments have been conducted using SVM classifier. It is implemented in python. A dataset of RMFD was taken with real face masks for training and testing phases. RMFD dataset was used by our model, which has resulted in an accuracy of 95.31% for identifying the images that are face masked or not. The results achieved with the model when it was tested on different images as shown in Figs. 5 and 6. SVM classifier achieved maximum accuracy of 95.31% and minimum accuracy of 88% respectively.

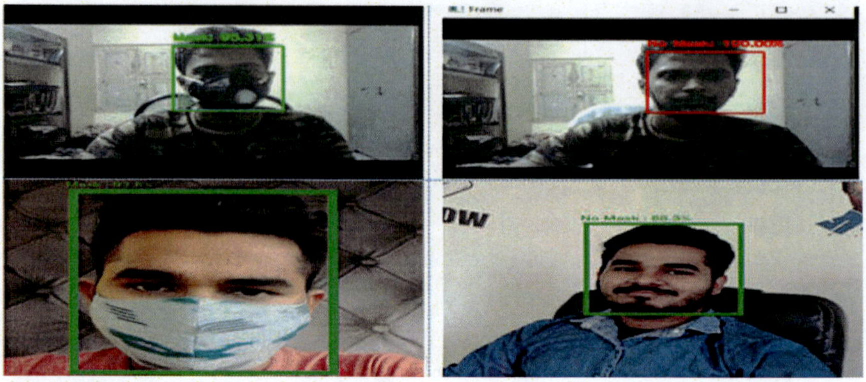

Fig. 5 Validation accuracy for different datasets

Fig. 6 SVM classifier testing accuracy

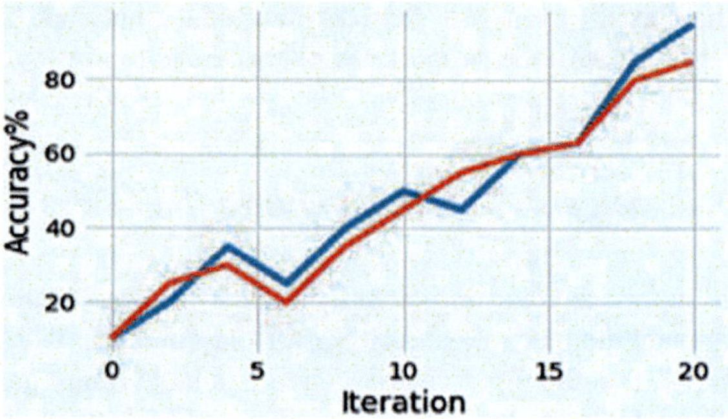

5 Conclusion and Future Works

In this paper, deep learning was used for detecting the face mask from images. The proposed methodology has two components, one is feature extraction and the second is the process of detecting the face mask using the machine learning algorithms. The machine learning algorithm used in this work is Support Vector Machines. The classifier was implemented to various images. Then our classifiers have been applied for achieving the results required. In this research, two datasets have been used for experimentation, one is for training and the second one is for testing. To prove the efficiency of the methodology proposed, the strategy includes training a particular dataset while testing over other datasets. With less training time consumed in the training process, this research work concluded that the SVM classifier has achieved the highest accuracy. The SVM classifier achieved 95.31% testing accuracy. The green and red rectangular frames have detected the face and the mask, respectively. Identification of the facemask in our work is least complicated in structure and is giving good results and, therefore, can be used in footages for identifying whether the individual wearing a face mask or not, so that he/she will not pose any danger to other people. The mass screen can be done in crowded places like bus stands, markets, malls, colleges, schools railway stations, etc. By detecting the facemask on the face, it can be made sure that individual is wearing it and helps to stop the spread of the coronavirus. The future work is to use deeper learning models for the extraction of features as it provides good results in classification and detection problems.

References

1. Feng S, Shen C, Xia N, Song W, Fan M, Cowling BJ (2020) Rational use of face masks in the COVID-19 pandemic, Lancet Respiration. Med 8(5):434–436. https://doi.org/10.1016/S2213-2600(20)30134-X
2. WHO Coronavirus Disease (COVID-19) Dashboard. https://covid19.who.int/. Accessed 21 May 2020. [4] Ting DSW, Carin L, Dzau V, Wong TY (2020) Digital technology and COVID-19. Nat Med 26(4):459–461. https://doi.org/10.1038/s41591-020-0824-5
3. Sucharita V, Jyothi S, Venkateswarrao P (2016) Comparison of machine learning algorithms for classification of Penaeid prawn species. 2016 3rd International Conference on Computing for Sustainable Global Development (INDIACom) IEEE 2016, pp. 1610–1613
4. Kotkar VA, Sucharita V (2017) A comparative analysis of machine learning based anomaly detection techniques in video surveillance. J Eng Appl Sci 12:9376–9381
5. Qin B, Li D (2020, May) Identifying facemask-wearing condition using image superresolution with classification network to prevent COVID-19. https://doi.org/10.21203/rs.3.rs-28668/v1
6. Li C, Wang R, Li J, Fei L (2020) Face detection based on YOLOv3, in recent trends inintelligent computing, communication and devices. Singapore, pp. 277–284. https://doi.org/10.1007/978-981-13-9406-5_34
7. Wang Z, et al. (2020) Masked face recognition dataset and application. Preprint arxiv: 2003.09093
8. Khandelwal P, Khandelwal A, Agarwal S (2020) Usingcomputer vision to enhance safety of workforce in manufacturing in a post covid world. Preprint arXiv: 2005.05287
9. Jiang M, Fan X (2020) Retinamask: A face mask detector. Preprint arXiv: 2005.03950
10. Ristea NC, Ionescu RT (2020) Are you wearing a mask? Improving mask detection from speech using aug-mentation by cycle-consistent gans. Preprint arXiv: 2006.10147
11. Ejaz MS, Islam MR, Sifatullah M, Sarker A (2019) Implementation of principal component analysis on masked and non-masked face recognition. 2019 1st International Conference on Advances in Science, Engineering and Robotics Technology (ICASERT), pp. 1–5, 10.1109/
12. Ud Din N, Javed K, Bae S, Yi J (2020) A novel GAN-based network for unmasking of masked face. IEEE Access 8:44276–44287. 10.1109/ ACCESS.2020.2977386

Application of Artificial Intelligence in Public Health Care in India

Sathvik Yerragolla, Sai Vennela Bandi, and Vijaya Lakshmi Peddiboyana

Abstract Artificial Intelligence (AI) is now rapidly used in conventional public health care systems to provide improved healthcare to all. The AI market for healthcare applications is predicted to grow from 663.8 million in the year 2014 to 6,662 million in the year 2021. One of the key features of healthcare transformation by Artificial Intelligence is improved diagnostics, helps in risk prediction, actively correlating the patient data collected on the field, and cross-verifying to initiate appropriate healthcare efforts to address the health issues. Artificial Intelligence is providing exceptional service in specialty care departments like radiology, generic pharmacy, and pathology. From drug discovery to diagnostics, chronic disease management to healthcare worker productivity, Artificial Intelligence is set to revolutionize the public healthcare system in India. AI can help public healthcare systems to have better visibility in the day-to-day operations of remote healthcare workers and subsequently improve the efficiency of healthcare.

This paper discusses the importance and challenges of public health and reviews the latest developments in the application of AI in public health care systems including disease diagnostics and prediction, field assistance, data documentation, biomedical information processing that will ultimately provide better health services to a larger section of people.

Keywords Artificial intelligence · Public health · Disease diagnostic · Better health care service

---

S. Yerragolla
CIDSE, Arizona State University, Phoenix, USA

S. V. Bandi
Department of Computer Science, SRM University, Chennai, India

V. L. Peddiboyana (✉)
Department of Communication and Journalism, Sri Padmavati Mahila Visvavidyalayam, Tirupati, India
e-mail: vijaya.cj@gmail.com

S. Jyothi et al. (eds.), *Proceedings of the 2nd International Conference on Computational and Bio Engineering*, Lecture Notes in Networks and Systems 215,
https://doi.org/10.1007/978-981-16-1941-0_27

1 Introduction

Even though artificial intelligence has been a concept that has been around for almost close to 70 years, but it is relatively a very modern concept for millennials. Alan Turing who is considered the father of artificial intelligence and computer science has shown the first records of using artificial intelligence in 1950, many scientists believe that it has not yet reached its full potential and still speculate that artificial intelligence will play a pivotal role in shaping the modern world. What might be the reason for scientists from all over the world pondering over the concept of artificial intelligence even when we as a human race have attained technological advancements in various fields like rocket science, computers, etc. It is simply because scientists believe that it will play a substantial part in the future and probably it will "be" the future. The software systems "make decisions that normally require [a] human level of expertise" and help people anticipate problems or deal with issues as they come up [1]. The most important and vital observation from their research is that artificial intelligence can make decisions that require human-level expertise without the involvement of humans directly. Another important aspect is that Artificial Intelligence can anticipate or deal with any problems that come up. This is the unique selling point for artificial intelligence and this is what makes it a very potential technological development that can shape the future.

Every few decades, humans come up with a technological development that will change the course of modern-day life. These paramount developments will influence the world to build itself around its principles and applications. Be computers, mobile phones, the internet, satellite communication, etc. are some of the discoveries that impact the world, and Artificial Intelligence (AI) will play such a role and especially in a country like India, where 1.3 billion people reside and the rate is increasing. AI will play a vital role in shaping the daily life of people. One of the most debated and an application of AI that the whole world is keen on is AI in public health. Scientists and governments believe that AI can change the public health structure in a much more efficient and effective manner. There have been some cases where countries have implemented AI in their public health sector. The most important question is how much Artificial Intelligence is utilized in the public health sector in the modern-day and what could Artificial Intelligence promise us for the future with the current development it is undergoing.

2 Why Public Health is Important

For us to understand why AI is important in public health, one should understand how important public health is for a particular society to function. If we ask anyone to define public health, it is difficult to explain and the most common definition it is associated with is building a safe and protective environment for the whole community and make sure that not only health care facilities but also the standard of living is safe and disease free. Public health transcends beyond hospitals or doctors, it applies to the whole community, city, state, or even the whole country. If the public health standards are substandard, then it will have a ripple impact on the whole community and it has a negative impact on the society's happiness index. India is ranked 145th out of 195 countries in the Healthcare access and quality index [2]. We can also draw parallels with the world happiness report 2020 where India was ranked 144th place [3]. Public health directly correlates with happiness, disease control, and life expectancy. For example, noncommunicable diseases (NCDs), which is a leading public health problem, encompass a vast group of diseases such as cardiovascular diseases, cancer, diabetes, and chronic respiratory disease contribute to around 5.87 million (60%) of all deaths in India [4]. India has a much-delayed response on NCD risk factor, surveillance and information of the same are sporadic and incomplete. It is reported that the chances of chronic diseases like diabetes and coronary-related diseases will increase if the community cannot uphold the public health standards [5]. But if we compare countries with better public health ranking, we can find some similar aspects like public awareness, but most importantly, they have certainly taken advantage of modern technology to enhance their public health care system.

3 Challenges of Public Health

India's public health standing in the world arena has been increasing steadily, but it does not mask the fact that India still has a long path to pave before itself. Researchers have tried to understand what exactly the aspects that India has to improve are and they have done extensive research regarding the respective topic. India is the fastest-growing major economy in the world and also a potential superpower in the world. Countries like The USA that is a superpower and China that is a potential superpower have done similar research over the years to find the probable improvements and many of the successful ones can be linked with better funding and extensive research on a solution for public health. According to the World Bank, only 3.53% of India's GDP is spent on health development. This is significantly less than countries like the USA where 17.06% of its GDP is on health and even China spends comparatively more with 5.15% [6]. The government healthcare system is free or, for some facilities, highly subsidized. Looking at the meager expenditure on healthcare by the Indian government, AI and medical technology help to maximize health outcomes.

India has the largest population in terms of the working force and hence proper health care can increase the yield of greater results. The major setbacks for public health that has to be dealt with are heart-related diseases and strokes, motor vehicle injuries, obesity, prescription overdose, diabetics, etc. [7]. These challenges result in a large number of deaths around the world and are also very recurrent in almost all nations. According to the US National Library of Medicine National Institute of health, India faces serious challenges with diseases like HIV/AIDS and malaria [8]. Their research also throws light on a very high mortality rate compared with other countries. Apart from these, India also lacks a rigid health care structure. The system is not rooted and the services provided to the public in a typical city in India like Mumbai or Hyderabad are not the same in the countryside. The availability of doctors and basic diagnosis is also lacking in rural India.

Another major challenge for public health in India is many citizens of India are not insured. Needless to say, the policies are different from other countries and some of the countries have a mandatory health insurance policy. Though the government provides subsidized health facilities, only 25% of the Indian population is insured [9]. This is comparatively less than many other countries like the UK where health insurance is compulsory for all permanent residents and the USA has over 92% of its population health insured as of 2019 according to the US Census [10]. The common notion about health insurance is not only about getting financial benefits when health issues arise but it also has a much advantage it in improving public health. Insurance ensures that health records are recorded and these records can be utilized for predicting any future complications. It is not just about insurance, it is moving more towards technology, and these health records are known as electronic health records, which have the records of all the tests and diagnostic sessions that a patient and doctor can employ this learning to provide a better solution or even predict problems in the near future. Health insurance can help in the transition of moving more towards technology in health care.

There is also the scarcity of health care workers and doctors though the ratio of doctors of 1000 people in India is 1.34 [11]. This indicates that there is approximately one certified doctor for 1000 people in India and this creates pressure on the health industry. This will lead to uncertified doctors diagnosing patients and people not being able to get proper consultancy from a certified doctor. Without proper diagnosis and consultation, it will lead to unprofessional prescription and detection.

Facts About Public Health In India
Socioeconomic Disparities • 275 million Indians are below the poverty line and an additional 300–400 million are vulnerable to economic stress because they work in the informal sector, usually on daily wages.
Health Infrastructure • Government sparing just 1.3% of the GDP for public healthcare, as opposed to the global average of 6%. • India's private healthcare system is estimated to be four times bigger in overall healthcare capacity, which operates in free market. • Only 10% of all registered doctors work in government hospitals and 90% are in the nongovernmental sector. • Number of urban health workers per 100,000 persons is less than 10% of the number of rural health workers per 100,000 persons.
Rural Scenario • Rural India is home to 750 million people living in 650,000 villages. • Most premature deaths in adults are due to chronic conditions such as heart disease. • Those at the highest risk are typically over 40—an age group that has doubled in size in India during the past 20 years. • More than 25 million rural people have diabetes and the number is rising rapidly

4 Artificial Intelligence in Public Health Care

The important question that needs to be asked is how to make public health more effective. The recent improvements and policies implemented by countries with successful health care system indicate the optimal use of technology. This is where Artificial Intelligence will play a pivotal role in developing public health care. What makes humans superior to machines is their capability to adapt and think according to the situation. The aspect where machines have the edge over humans is that humans tend to make errors. If developed properly, artificial intelligence has the ability to think and process and produce results without much of human errors that generally occur. Artificial Intelligence is anticipated to resolve some of the major issues regarding various divisions of the health care industry.

4.1 Artificial Intelligence in Medical Diagnostics

As professional medical expertise is very limited due to the large populations and a very small number of certified doctors, artificial intelligence has the capabilities to fill the gap in providing health care services. Research indicates that breast cancer detection technology powered by artificial intelligence has shown 95% accuracy, which can be compared with a normal cancer radiologist [12]. Apart from that AI-powered technology has shown the ability to detect and provide a diagnosis for day-to-day health issues. This is a key development as this will substantially decrease the workload pressure on medical practitioners. This will decrease the dependency on human expertise for the diagnostic of several diseases and it can be the way for the future as well. Human expertise and knowledge would be unparalleled in serious diagnosis, but it is possible for accurate diagnosis based on AI for a regular check-up as they have shown a lot of efficiency and promise.

Even though there has been a significant improvement in the diagnosis and treatment procedures in the cardiovascular field, it still is the cause of the highest mortality and morbidity across the world. In India, there are close to 3 million heart attack cases and over 30 million Indians are suffering from the cardiovascular diseases. For almost 85% of the time, cardiovascular diseases are due to the thrombus in the vessel called cerebral infarction. The challenging part is the identification of early stroke symptoms, which leads to several times where patients do not receive proper treatment before it becomes life threatening. As per a Harvard study, 5.2 million medical errors are happening in India annually [13]. Early detection and prognosis evaluations are the most important for improving cardiovascular outcomes. AI techniques can help doctors to detect early signs much more accurately and it reduces medical errors, which cost a lot of lives. AI can provide a precise diagnosis by utilizing the massive amount of clinical data available and enabling detection of subclinical organ dysfunction and this ensures that AI can give accurate results for the doctors to suggest a proper procedure of treatment [14].

Another major public health challenge in India is the rise of diabetes cases. In the last 4 years, almost 11.8% of the population of India is suffering from diabetes, according to the National Diabetes and Diabetic Retinopathy Survey report. Physicians have already predicted the complications such as nephropathy and retinopathy among diabetic patients. It is also alarming to know that diabetes and diabetic retinopathy are emerging as a significant noncommunicable disease, leading to ocular morbidity. The Union Health Ministry's first National Diabetes and Diabetic Retinopathy Survey (2015–19) revealed that the prevalence of Diabetic Retinopathy (DR) is 16.9% while the prevalence of sight-threatening DR is 3.6%. AI is already being used to predict the risk of complications of diabetes such as nephropathy and retinopathy. In April 2018, the US Food and Drug Administration (FDA) approved an AI algorithm, for DR identification [15]. There is a paucity of data in India on all these aspects and making use of these AI-driven technologies can substantially increase detection and early treatment of diabetic complications. The application of AI/ML has the ability to exceedingly enhance the potential of diabetes treatment, thereby making it more efficient [16].

Retinopathy of prematurity (ROP) is a leading cause of childhood blindness, which can be treated if diagnosed timely. This disease demands repeated follow-up and screening, which is very tedious and demanding and India is currently facing an ROP epidemic that has found its way into the rural part of the country. The incidence of ROP in different regions across India has been reported to range from 38 to 47% [17]. Making use of AI technology in ROP screening may expand the proficiency in ROP care. There have been encouraging results from various studies on the subject [18].

Artificial Intelligence has shown the capability to improve the decision-making process of doctors. Doctors make use of current knowledge about the disease and symptoms to analyze what ailment the patient is suffering from and suggest treat according to it. But the downside of it is the doctors have to always gain knowledge on various other sciences throughout their career and also the decision-making will be as good as the data and knowledge they have gone through. With artificial intelligence, the algorithm has the capability to consume a massive amount of information regarding symptoms, conditions, treatment procedures, and success ratio from various other doctors who have expertise in other fields and analyze and return us precise diagnostics. This will reduce research time and obstacles and hence the scientists and doctors can come up with better prescriptions and treatment procedures. When algorithms are designed by using pre-existing data regarding diseases, the software can predict and classify diseases using MRI scans, X-rays, and CT scans [12]. This will reduce the disease detection period and increase accuracy. Scientists are working to combine more present-day disease detection methods with AI to create a much more robust and accurate system.

5 Artificial Intelligence in Road Safety

Road accidents are one of the major challenges for public health and according to the World Health Organization over 1.35 million people die due to road accidents worldwide [19]. Not just that, road accidents constitute to 3% loss in the annual GDP of a country on average. This is a large amount of money comparatively where India is spending only 3% of its GDP on public health. According to the report on Road Accidents in India, there are close to half a million road accidents happen in India every year. The alarming information is that even though the USA and Japan have more accidents, India has the most number of deaths in the world, which is a public health concern [20]. The data on the cause of deaths in India during 2019 and 2018, as shown in Fig. 1, show that most number of deaths is due to speed driving, hit, and run and not following rules. In the present day, there has been the usage of cameras and technology to reduce and fine rule offenders, but AI offers a much more effective method to tackle the challenge and reduce the number of fatalities.

The data show that more than 100 thousand deaths were resulted due to speeding in India, but there is no clear technology for the police to identify speeding offenders. There have been instances of using cameras in expressways, but considering the size of roadways in India, it is insignificant. Dubai has launched an AI platform called Project Oyoon, which uses thousands of city cameras and also uses the past data regarding accident-prone zones to prevent collisions and a similar prototype technology has been launched in Bengaluru by Siemens Mobility [21]. India has

	2018	2019
Not wearing of helmet	43,614	44,666
Not wearing of seatbelt	24,435	20,885
Use of mobile phones	3,707	4,945
Speeding	97,588	101,723
Potholes	2,015	2,140
Driving under influence of alcohol/drugs	1,188	5,325
Hit & run	28,619	29,354
Driving on wrong lane/ breaking lane disciple	8,764	9,201
Jumping traffic light	1,545	1,797
Driving without licence	37,585	44,358

Source: MORTH

Fig. 1 Causes of death in road accidents. *Source* Ministry of Road Transport and Highways (https://morth.nic.in)

more deaths compared with any country even though the number of accidents is less. One of the major reasons for it is that the first responder not being able to reach the incident location. India being the second-most populous country, the traffic in India is harder to navigate. According to recent studies, Bengaluru has the worst traffic in the entire world [22]. AI-powered technology similar to what the Dubai government has implemented can employ thousands of cameras inside the city and the algorithm can predict the best possible route calculating the speed at which the traffic is moving. This will also ensure that medical support can reach the victims much quicker.

Employing technology in our traffic police department can reduce the number of hits and run's, jumping traffic light cases. Currently, cameras are widely utilized, but further improvements with AI can ensure that the offenders are caught much quicker and also the system can detect any collisions or accidents and alert the office in close proximity. Increasing the shift towards technology in the traffic sector can reduce human errors and elevate the competence of the system.

6 AI in Electronic Health Records

Electronic Health records have become extensively popular in recent time as it generates and makes use of the vast amount of information and records. Electronic Health records are not only vital for research purposes for discovering better solutions for diseases but also important for understanding the best treatment for a single individual as well. According to WHO, noncommunicable diseases like cardiovascular and diabetes and communicable diseases like lower respiratory infections account for the most number of deaths globally [23]. It was noted that the cardiovascular problems in India account for almost 25% of deaths annually [24].

Electronic health records are usually looked upon due to their complexity but AI-powered electronic health records with future developments can be used to predict any future complications. According to the US National Library of Medicine, in recent studies, they have utilized three AI techniques that are machine learning, deep learning, and natural language processing [25]. The primary goal is to combine the past information with the information received in the present to come up with predictive models. The algorithms developed can predict disease progression and devise an appropriate solution. With many people tending to heart-related diseases and diabetes due to family lineage, the pre-existing records can help the doctors to assess the situation much quicker. Electronic health records can also act as test data for the researcher to study and with the prediction from AI, they can produce much better treatment procedures in a shorter time.

7 AI in Preventing Pollution

Pollution constitutes a large number of diseases and is a major hazard for public health. Global warming is also a major challenge for not just public health but all of humanity. Technology can provide a solution for the problem or at least that what studies are suggesting. According to a survey by Intel and research firm Concentrix, they have found that 74% of the business decision-makers believe that artificial intelligence can provide solutions for environmental problems [26].

Air pollution is a major threat in India as 22 out of the 30 most polluted cities in the world are from India [27]. The method to tackle air pollution is by building air monitoring stations. To come up with solutions, researchers need predicted data by combining the carbon emission, traffic, and other sources to come up with accurate solutions. For scientists, it will take an ample amount of time to understand the root cause, trends and analyze why some locations have much more pollution compared with some other places. The root problem might not be the same for every place and it is difficult to categorize everything under one bracket. Artificial Intelligence combined with available data can give a specific solution for a specific problem. Artificial intelligence can also monitor what measures have yielded results and try to integrate them with other solutions to aggregate much better preventive measures. The algorithm keeps on evolving and improving and hence the results will become much more accurate with time.

According to National Geographic, researchers from Montreal Institute for Learning Algorithms (MILA), Microsoft, and ConscientAI Labs used GANs, a type of AI to create an accurate impression of what life would be like after global warming and increased storms [28]. Their idea is to make people understand how adverse life would be after climate change and they believe that this can bring awareness among people to work towards improving the environment. They are planning to develop an app that will show how our houses and streets would look after climate change and they are relying on AI-driven algorithms to produce them. Not just that, Google is working on AI that can monitor power plants that emit a lot of carbon by using satellites to send in alerts if emission is more than the recommended levels.

8 AI in Bridging Health Inequalities

India is the second most popular country and it has certain health inequalities that are specific to itself. Artificial Intelligence can play a role in helping to bridge those inequalities.

Two-thirds of India live in poverty and almost 68.8% of the Indian population lives on less than $2 a day [29]. Hence, not only public health as a community is important but also it is difficult for a large proportion to depend upon government medical facilities. On average, a government allopathic doctor attends to 10,926 people, which is 10 times more than what WHO recommends [30]. With the inclusion of AI, the number of doctors can ease, and, hence, people below the poverty line can receive basic health diagnoses. If the government can develop a robust system of an interface that can store the records of Indians, it will become much easier for the government to provide services to them. As the majority population of India lives in rural India, many of the people living in such remote areas do not have proper vaccination or even receive awareness regarding public health care. Artificial Intelligence can utilize electronic health records and Aadhaar card to consider the whole population and when AI has such big sample information, the algorithm can show the officials the best way to provide services to the population that is so diverse.

As the majority of the number of doctors who graduate in India is mostly part of the private hospitals rather than government hospitals. This is a major problem as the bulk of the Indian population cannot afford expenses at private hospitals. The government if embarks on Artificial intelligence would be able to provide better health services and eventually reduces health inequalities to an extent.

9 Conclusion

Artificial intelligence is relatively a very infant topic in the world right now. Only a handful of countries are utilizing AI in their public health system. But the amount of promise it shows is tremendous. Right now, AI is a technology that requires a meticulous amount of research and development. With time and resources, it can become more accessible and more effective. It is important for the governments of every country to invest in research and development of AI and its application in public health. A large country like India with visible problems in public health certainly requires technological advancement powered by AI. Artificial intelligence will certainly be integrated into the public health care system in the coming years and it will certainly have a significant role in shaping modern India.

References

1. Darell M. West (2018, October) What is artificial intelligence- a blueprint for the future of AI , a series from the brookings institution. https://www.brookings.edu
2. Lozano Rafael (2018, June 2) Measuring performance on the healthcare access and quality index for 195 countries and territories and selected subnational locations: a systematic analysis from the global burden of disease study 2016. Vol. 391:10136, pp. 2236–2271. https://www.thelancet.com
3. Helliwell John F, Richard Layard, Jeffrey Sachs, Jan-Emmanuel De Neve, eds. (2020) World happiness report 2020. New York: Sustainable Development Solutions Network
4. Nethan S, Sinha D, Mehrotra R (2017) Non Communicable Disease Risk Factors and their Trends in India. Asian Pac J Cancer Prev 18(7):2005–2010. https://doi.org/10.22034/APJCP.2017.18.7
5. WHO Global Report: Preventing Chronic Diseases:AVital Investment (2005). https://www.who.int/chp/chronic_disease_report/en/

6. The World Bank. https://data.worldbank.org/indicator
7. Center for Disease Control and Prevention. https://www.cdc.gov/psr/index.html
8. Narain JP (2016) Public health challenges in india: seizing the opportunities. Indian J Comm Med 41(2):85–88
9. Handbook on Indian Insurance Statistics-2016–17. https://www.irdai.gov.in
10. Katherine Keisler-Starkey, Lisa N. Bunch (2020, 15 September) Health insurance coverage in the United States: 2019. https://www.census.gov/library/publications/2020/demo/p60-271.html
11. Kumar R, Pal R (2018) India achieves WHO recommended doctor population ratio: a call for paradigm shift in public health discourse. J Family Med Prim Care.7(5):841–844. https://pubmed.ncbi.nlm.nih.gov
12. Rodriguez, Alejandro et al. (2019) Stand-alone artificial intelligence for breast cancer detection in mammography: comparison with 101 radiologists. JNCI J Natl Cancer Inst 111(9):djy222. https://academic.oup.com/jnci
13. Arvard Study In Aashita, et al. (2018, August) Basic concept of patient safety in healthcare. Int J Curr Res 10(08):72926–72932. http://www.journalcra.com
14. Jiang F, Jiang Y, Zhi H, et al. (2017) Artificial intelligence in healthcare: past, present and future. Stroke Vasc Neurol. 2(4):230–243. doi:https://doi.org/10.1136/svn-2017-000101. https://www.ncbi.nlm.nih.gov/pmc/articles/PMC5829945/
15. Food and Drug Administration (2018) FDA Permits Marketing Of Artificial Intelligence-Based Device To Detect Certain Diabetes-Related Eye Problems. https://www.fda.gov/NewsEvents/Newsroom/PressAnnouncements/ucm604357.htm
16. Padhy SK, Takkar B, Chawla R, Kumar A. Artificial intelligence in diabetic retinopathy: a natural step to the future. Indian J Ophthalmol 67(7):1004–1009. https://www.ijo.in
17. Bowe T, Nyamai L, Ademola-Popoola D, et al. (2019) The current state of retinopathy of prematurity in India, Kenya, Mexico, Nigeria, Philippines, Romania, Thailand, and Venezuela. Dig J Ophthalmol 25(4):49–58. https://www.ncbi.nlm.nih.gov/pmc/articles/PMC7001648/
18. Gilbert C, Foster A (2001) Childhood blindness in the context of VISION 2020-the right to sight. Bull World Health Organ 79:227
19. World Health Organization. https://www.who.int/health-topics/road-safety
20. Road Accidents in India Report. Ministry of Road Transport and Highways, Government of India (2019). https://morth.nic.in/sites/default/files/RA_Uploading.pdf
21. Baker Francesca (2018, 12 December). https://www.bbc.com/future/article/20181212-can-artificial-intelligence-end-traffic-jams
22. Traffic Index (2019). https://www.tomtom.com/en_gb/traffic-index/ranking
23. World Health Organization (2020, December). https://www.who.int/news-room/fact-sheets/detail/the-top-10-causes-of-death
24. What is the Leading Cause of Death in India?.Future Generali Total Insurance Solutions. https://life.futuregenerali.in/life-insurance-made-simple/cancer-heart-critical-illness-insurance/what-is-the-leading-cause-of-death-in-india
25. Lin WC, Chen JS, Chiang MF, Hribar MR (2020) Applications of artificial intelligence to electronic health record data in ophthalmology. Trans Vis Sci Technol 9(2):13. https://doi.org/10.1167/tvst.9.2.13Lin W-C, Chen JS, Chiang
26. Creative, Vox.How AI Can Help Us Clean Up Our Land, Air, and Water (2018, 26 October). https://www.recode.net/ad/18027288/ai-sustainability-environment
27. Mead, Nick Van. 22 of world's 30 most polluted cities are in India. https://www.theguardian.com/cities/2019/mar/05/india-home-to-22-of-worlds-30-most-polluted-cities-greenpeace-says
28. Snow Jacie (2019, 18 July) How artificial intelligence can tackle climate change. https://www.nationalgeographic.com/environment/2019/07/artificial-intelligence-climate-change/
29. Sos Children's Villages Canada. Poverty in India: Facts and Figures on the Daily Struggle for Survival. https://www.soschildrensvillages.ca/news/poverty-in-india-602
30. National Health Profile 2019 (2019) Central Bureau of Health Intelligence, Ministry of Health & Family Welfare, p. 220. www.cbhidghs.gov.in

Comparative Study of Enzyme Kinetics in Biotransformation of Loratadine by Selected Fungi

Keerthana Morusu and Vidyavathi Maravajhala

Abstract Loratadine is an antihistaminic nonsedating drug that acts on H1 receptors. Loratadine shows rapid absorption and subjected to pervasive first-pass metabolism and majorly converted to its metabolite descarboethoxyloratadine, which is also known as desloratadine. The Cytochrome P450-catalyzed reactions are sufficiently described by classical Michaelis–Menten kinetic parameters (e.g., K_m and V_{max}). In turn, these parameters may be used to predict pharmacokinetics. Microorganisms especially fungi can also have the capability to mimic the biotransformation reactions that occur in mammals. The present study was emphasized to estimate the enzyme kinetic parameters of loratadine biotransformation by fungi. In preliminary fungal screening studies, it was revealed that two *Cunnighmella* fungi and one *Aspergillus* fungus proved their capability for loratadine transformation. These three fungi were further used to evaluate the enzyme kinetics of loratadine biotransformation. The kinetic parameters K_m and V_{max} were obtained from Graph pad prism-5 software upon application of HPLC data of all samples. The results of these enzyme kinetic studies showed that the product (metabolite) formation by both *Cunnighamella elegans* and *Cunnighamella echinulata* was increased with increasing substrate concentration up to saturation level indicated that these follow classical Michaelis–Menten kinetics. The product formed by *Aspergillus niger* showed atypical kinetics due to substrate autoinhibition at high concentrations of substrates.

Keywords Loratadine · Enzyme kinetics · *Aspergillus niger* · *Cunnighamella elegans* · *Cunnighamella echinulata* · Michaelis–menten kinetics

K. Morusu · V. Maravajhala (✉)
Institute of Pharmaceutical Technology, Sri Padmavati Mahila Visvavidyalayam, Tirupati 517502, AP, India
e-mail: keethu.choudary16@gmail.com

S. Jyothi et al. (eds.), *Proceedings of the 2nd International Conference on Computational and Bio Engineering*, Lecture Notes in Networks and Systems 215,
https://doi.org/10.1007/978-981-16-1941-0_28

1 Introduction

Loratadine is a nonsedating antihistaminic drug that acts on H1 receptors. Autonomic anticholinergic effects were not observed with loratadine in humans [1–3]. Loratadine shows rapid absorption and subjected to pervasive first-pass metabolism and majorly converted to its active metabolite descarboethoxyloratadine, which is also known as desloratadine [4–6]. The hydroxylated metabolite of an active metabolite, i.e. 3-hydroxydesloratadine was also found in human plasma. It was established that CYP3A4 and CYP2D6 are the vital human liver enzymes involved in the metabolism [7, 8]. Most cytochrome P450-catalyzed reactions are sufficiently described by classical Michaelis–Menten kinetic parameters (e.g., K_m and V_{max}), which are usually determined by a saturation profile of velocity of product formation versus substrate concentration. In turn, these parameters may be used to predict pharmacokinetics. The variability of these enzymes may result in some unusual kinetic properties, some P450 enzymes exhibit atypical or non-Michaelis–Menten kinetics, mainly due to substrate inhibition at higher concentrations of substrate. CYP 450 enzymatic systems mediated the large number of xenobiotic-metabolic conversions, which contain one or multiple substrates and show wide substrate specificity [9, 10]. The values of Km and Vmax can be experimentally investigated to acquire adequate information about the enzyme under study.

The basic Michaelis–Menten equation is given as follows

$$[v_0 = V \max [S_0] \tag{1}$$

$$Km + [S_0]$$

where Km is the Michaelis–Menten constant, Vmax is maximum velocity of a reaction, v_0 is the initial velocity, and $[S_0]$ is the substrate concentration.

This relationship shows that the Michaelis–Menten constant is defined as the substrate concentration at which half of the maximal velocity is found. An enzyme with large value for Km will, therefore, need high concentrations of the substrate to achieve a given fraction of Vmax than an enzyme with a small value of Km [11]. A remarkable aspect of enzyme-catalyzed reactions is saturation, i.e., at increased concentrations of substrate, there is an increase in the rate and reaches a limit where there is a concentration-independent rate [12]. Kinetic modeling of metabolism holds great potential for metabolic engineering.

In preliminary microbial screening studies, two *Cunnighamella* fungi and one *Aspergillus* fungus proved their loratadine transformation capability [13]. The current study was emphasized to estimate the enzyme kinetic parameters of loratadine biotransformation by three selected fungi, *Cunnighamella elegans, Cunnighamella echinulata, and Aspergillus niger.*

2 Materials and Methods

a. **Materials:** *Cunninghamella elegans* (NCIM-689), *Cunninghamella echinulata* (NCIM-687), *Aspergillus niger* (NCIM-589) were acquired from the National Chemical Laboratory (NCL) Pune, India. Loratadine was a gift sample procured from Shasun pharmaceuticals, Guindy, Chennai, Tamil Nadu, India. The remaining chemicals were purchased from S.D. Fine Chemicals, Mumbai, India. HPLC-grade solvents were obtained from Indian Scientific Company Pvt Ltd, Tirupati, India.

b. **Methods:**

Subculturing: Three fungal cultures were maintained on agar slopes at 4 °C [14], and it is required to be subcultured every 6 months to preserve their viability. Potato dextrose broth was the suitable medium for the growth of fungal cultures [15]. The prepared medium was sterilized for 30 min in an autoclave and was used further in the study [16].

Enzyme Kinetics Study: These studies were performed in extension to the induction and inhibition studies [17] for the substrate drug Loratadine, to examine the affinity of the respective fungal enzyme to the given substrate involved in the fungal biotransformation. The three selected fungi, which were found to form metabolite in the preliminary screening studies, were inoculated in labelled sample flasks containing different concentrations of the Loratadine like 10, 20, 30, 40, 50, and 60 μg/ml and incubated under similar conditions like temperature was set at 28 °C; agitation speed was 120 rpm. But, 10 ml of sample was aseptically collected from each flask at different incubation time intervals of 24, 36, 48, 60, and 72 hrs to extract the metabolite and calculate the percentage of metabolite formed at each concentration of substrate and incubation time point [18].

Analysis Of Kinetic Data: The data of percentage metabolite formed during kinetic studies were fitted in MS EXCEL by plotting graph between time and percentage metabolite formed at different concentrations of substrate to get the velocity of biotransformation reaction from the slope of the graph. Then, the velocity obtained for each concentration by the above method was fitted into the Michaelis–Menten equation using nonlinear regression in GraphPad Prism 5.0 [19]. Best-fit values for Km and Vmax were reported to assess the possible extent of affinity of the fungal enzyme with the drug.

3 Results

Enzyme kinetic studies were conducted with three organisms *Cunninghamella elegans*, *Cunnighamella echinulata*, and *Aspergillus niger* for three formed metabolites M_1, M_2, and M_3, respectively. For each culture, six different substrate concentrations (10 to 60 μg/ml) were used. Samples were collected during incubation after regular intervals of 24, 36, 48, 60, and 72 hrs and analyzed by HPLC. The results of HPLC analysis obtained in the form of peak area were used to calculate the percentage metabolite formed at different time points at each concentration. For each concentration of substrate, a graph was plotted between incubation time v/s percentage metabolite formed. Six different plots were obtained for six concentrations of substrate for each metabolite. Slope values obtained from each plot were considered as velocity of the metabolic reaction by respective fungi.

Effect of Incubation Period on Percentage Metabolite Formation: During enzyme kinetic studies, samples were analyzed by HPLC after collecting at regular incubation time intervals of 24, 36, 48, 64, and 72. The graphs were plotted between incubation time and percentage metabolite formation to know the effect of incubation time on metabolite formation.

Metabolite-M$_1$: In the case of formation of metabolite (M_1) by *Cunnighamella elegans,* as the incubation time was increased, the percentage metabolite (M_1) formation was also increased up to 72 hrs of incubation, with all six concentrations of substrate and the effect of incubation time on percentage metabolite M_1 formation for six different concentrations of substrate is represented in Fig. 1.

Metabolite-M$_2$: In case of the formation of metabolite (M_2) by *Cunnighamella echinulata,* as the incubation time was increased, the percentage of metabolite formation was also increased with six concentrations of substrate. The effect of incubation time on percentage metabolite M_2 formation at six different concentrations of the substrate is shown in Fig. 2.

Metabolite-M$_3$: As the incubation time was increased, the percentage of metabolite M_3 formation was also increased with six concentrations of substrate. The effect of incubation time on percentage metabolite M_3 formation for six different concentrations of the substrate is shown in Fig. 3.

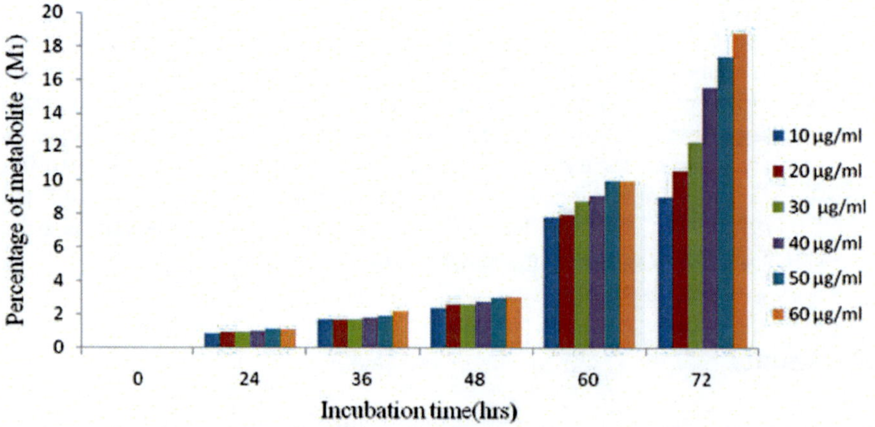

Fig. 1 Effect of incubation time on percentage metabolite M_1 formation for six different concentrations of substrate

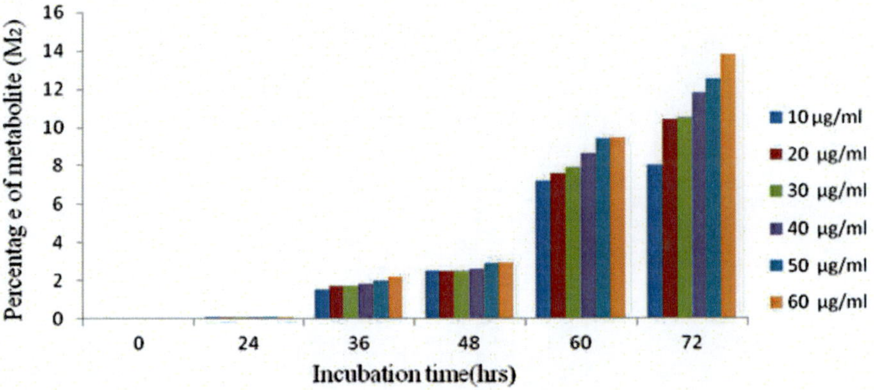

Fig. 2 Effect of incubation time on percentage metabolite M_2 formation for six different concentrations of substrate

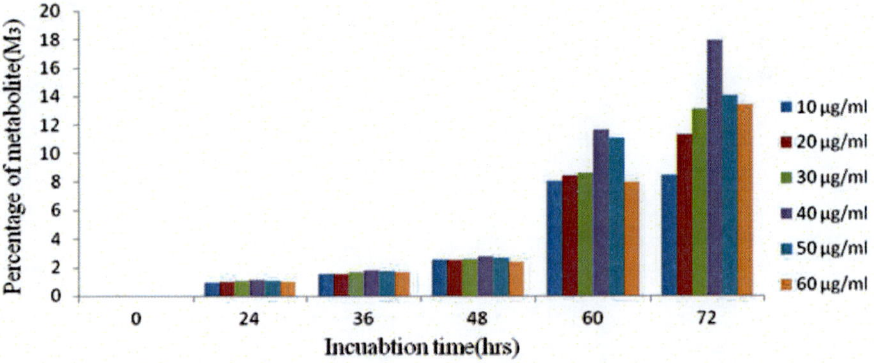

Fig. 3 Effect of incubation time on percentage metabolite M_3 formation for six different concentrations of substrate

The percentage of metabolites (M_1, M_2, and M_3) formed during kinetic studies of loratadine metabolism by three cultures at different incubation time points with six concentrations of substrate is represented in Table 1.

Slope values were obtained from the plots which were plotted between percentages of metabolites (M_1, M_2, and M_3 v/s incubation time are represented for three metabolites at six different concentrations, which are considered as velocity of metabolic reactions by respective organisms are represented in Table 2.

Effect of substrate concentration on percentage of metabolite formation: Six different substrate concentrations (10 µg/ml–60 µg/ml) were used to study the influence of substrate concentration on percentage metabolite formation. Effect of substrate concentration on percentage of three metabolites (M_1, M_2, and M_3) after incubation of 72 h is represented in Fig. 4.

Table 1 Percentage of metabolites (M_1, M_2, and M_3) formed during kinetic studies by three cultures

Substrate conc μg/ml	Incubation time (hrs.)														
	24			36			48			60			72		
	M_1%	M_2%	M_3%	M_1%	M_2%	M_3%	M_1%	M_2%	M_3%	M_1%	M_2%	M_3%	M_1%	M_2%	M_3%
10	0.84	0.08	0.91	1.69	1.56	1.49	2.4	2.5	2.5	7.8	7.2	7.97	9.0	8.02	8.4
20	0.95	0.08	1.0	1.70	1.72	1.50	2.6	2.5	2.5	8.0	7.6	8.4	10.6	10.4	11.3
30	0.96	0.09	1.08	1.7	1.75	1.66	2.6	2.5	2.6	8.8	7.9	8.6	12.3	10.5	13.1
40	1.04	0.10	1.15	1.8	1.83	1.76	2.8	2.6	2.77	9.13	8.6	11.6	15.6	11.8	17.9
50	1.11	0.10	1.08	1.94	2.0	1.72	3.0	2.9	2.67	9.96	9.4	11	17.4	12.5	14.0
60	1.16	0.11	0.99	2.22	2.2	1.62	3.04	2.9	2.4	10.0	9.46	7.9	18.8	13.8	13.4

Table 2 Slope values of three metabolites for six different concentrations

Substrate concentration (μg/ml)	Slope values(m)		
	M_1	M_2	M_3
10	0.131	0.123	0.127
20	0.148	0.147	0.157
30	0.169	0.150	0.175
40	0.202	0.167	0.238
50	0.224	0.179	0.198
60	0.237	0.191	0.173

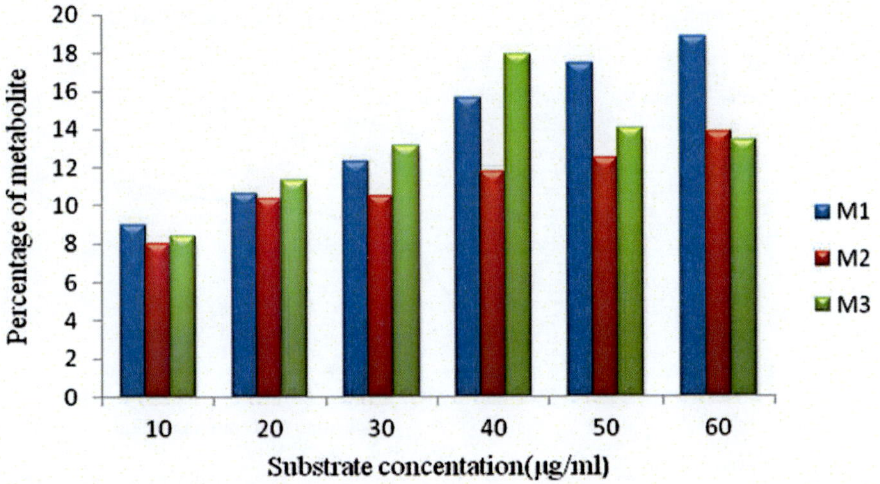

Fig. 4 Effect of substrate concentration on percentage of three metabolites (M_1, M_2, and M_3) after incubation of 72 h

Michaelis–Menten Kinetics: After 72 h of incubation, the formation of M_1 and M_2 was increased with the increase of substrate concentrations from 10 to 60 μg/ml. Perhaps, in case of $M_{3,}$ the percentage of formation was decreased with the rise of substrate concentration after 40 μg/ml. Kinetic parameters, Km and Vmax, can be obtained from graphs plotted between substrate concentration and the velocity of reaction (slope values) in Graph pad prism-5 software by applying Michaelis–Menten kinetics.

Metabolite-M_1: Velocity of reaction was increased with increase in concentration in case of M_1 formation by *Cunnighamella elegans.* Michaelis–Menten plot was plotted with velocity of M_1 formation v/s substrate concentration is shown in Fig. 5.

Metabolite-M_2: Velocity of reaction was increased with an increase in concentration in case of M_2 formation by *Cunnighamella echinulata.* Michaelis–Menten plot was plotted with velocity of M_2 formation and substrate concentration is shown in Fig. 6.

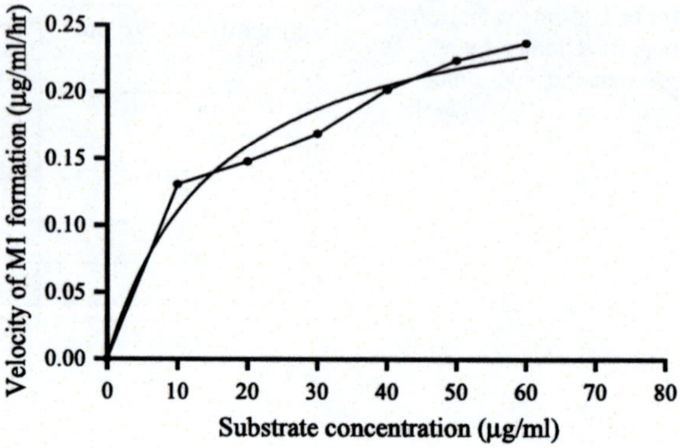

Fig. 5 Michaelis–Menten plot for M_1

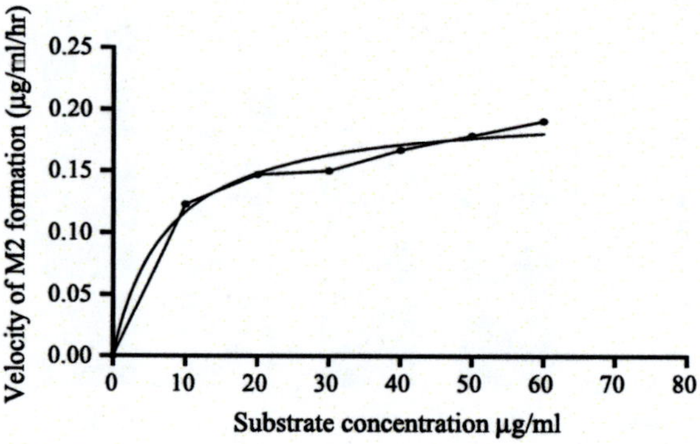

Fig. 6 Michaelis–Menten plot for M_2

***Metabolite-M*$_3$**: Velocity of reaction was decreased with an increase in concentration in case of M_3 formation by *Aspergillus niger.* So, the velocity of M_3 formation v/s substrate concentration data of M_3 was fitted to substrate inhibition kinetics to estimate inhibition constant (Ki), along with Km and Vmax. Substrate inhibition plot for M_3 formed by *Aspergillus niger* is shown in Fig. 7.

The kinetic parameters obtained for all three metabolites are mentioned in Table 3.

4 Discussion

The enzyme kinetic studies are crucial to trace out the affinity of the fungal enzyme towards the substrate by estimating kinetic parameters like Km and Vmax. Kinetics of microbial metabolism by three fungi was studied using substrate concentrations ranging from 10 to 60 μg/ml and estimating the percentage of metabolite formed at different incubation periods for each concentration.

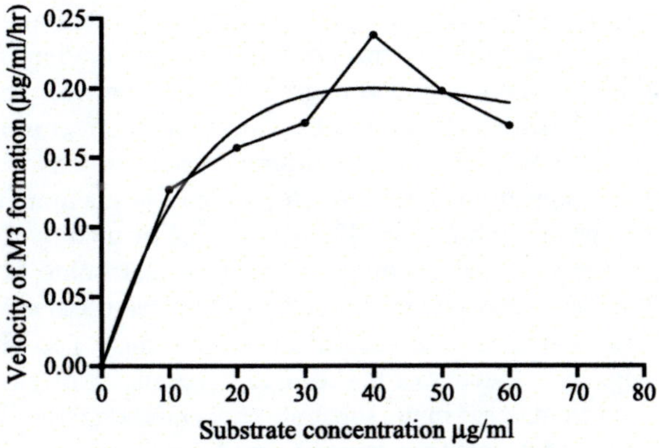

Fig. 7 Substrate inhibition plot of M_3

Table 3 Kinetic parameters of loratadine metabolism by three fungi

Kinetic parameters	M_1 (by *Cunnighamella elegans*)	M_2 (by *Cunnighamella echinulata*)	M_3 (by *Aspergillus niger*)
Km (µg/ml)	15.96	7.47	39.68
Vmax (µg/ml/hr)	0.28	0.20	0.59
Ki (µg/ml)	–	–	40.11

Metabolite-M_1: Based on the results, it was noticed that, as the incubation time was increased, the percentage metabolite (M_1) formation was also increased along with an increase in the substrate concentration in the metabolism of loratadine by *Cunnighamella elegans.* Velocity of the metabolism reaction was obtained from slope values (Table 2) of lines found in graphs plotted between incubation time and percentage of metabolite (M_1) for each concentration of substrate (Table 1). The study of effect of incubation time on percentage of metabolite (M_1) for six different concentrations of substrate was showed that the percentage of metabolite M_1 formation was maximum after 72 h of incubation when compared with 24, 36, 48 and 60 h of incubation (Fig. 1). The study of effect of substrate concentration on percentage metabolite M_1 formation revealed that the maximum percentage of metablite (M_1) was observed at 60 µg/ml substrate concentration after 72 h of incubation. (Fig. 4) Then, Michaelis–Meneten kinetics was applied to data of velocity of metabolism reaction v/s substrate concentration in order to estimate Km and Vmax (Fig. 5). Km and Vmax values of metabolism reactions mediated by enzymes in *Cunnighamella elegans* to form M_1 were 15.96 µg/ml and 0.28 µg/ml/hr, respectively (Table 3). These values stated that metabolism reaction that formed M_1 proceeded with the maximum velocity (Vmax) of 0.28 µg/ml/hr, and half maximum velocity of reaction was found at the concentration (Km) of 15.96 µg/ml.

Metabolite-M_2: The studies of enzyme kinetics in fungal metabolism mediated by enzymes in *Cunnighamella echinulata* to form M_2 were revealed that, as the incubation time was increased, the percentage metabolite formation was also increased along with an increase in the substrate concentration in the metabolism of loratadine. Velocity of metabolic reactions for six concentrations was obtained from slope values (Table 2) of lines found in graphs plotted between incubation time and percentage of metabolite (M_2) (Table 1). The study of effect of incubation time on percentage of metabolite (M_2) for six different concentrations of substrate was showed that the percentage of metabolite M_2 formation was maximum after 72 h of incubation when compared with 24, 36, 48, and 60 h of incubation (Fig. 2). The study of effect of substrate concentration on percentage metabolite M_2 formation revealed that the maximum percentage of metablite (M_2) was observed at 60 μg/ml substrate concentration after 72 h of incubation (Figure no-). The Michaelis–Menten kinetics was applied to kinetics of M_2 formation because the velocity of reaction was increased with an increase in the substrate concentrations from 10 to 60 μg/ml, Km and Vmax values of metabolism reactions mediated by enzymes in *Cunnighmella echinulata*to form M_2 were7.47 μg/ml and 0.20 μg/ml/hr, respectively (Fig. 6 and Table 3). These values stated that metabolism reaction that formed M_2 proceeded with the maximum velocity (Vmax) of 0.20 μg/ml/hr, and half maximum velocity of reaction was found at the concentration (Km) of 7.47 μg/ml.

Metabolite-M_3: Interestingly, in case micobial metabolism mediated by enzyme in *Aspergillus niger* to form M_3, as the incubation time was increased, the percentage metabolite formation was increased. Velocity of the metabolism reaction was obtained from slope values (Table 2) of lines found in graphs plotted between incubation time and percentage of metabolite (M_3) for each concentration of substrate (Table 1). The study of effect of incubation time on percentage of metabolite (M_3) for six different concentrations of substrate was shown that the percentage of metabolite M_3 formation was maximum after 72 h of incubation when compared with 24, 36, 48, and 60 h of incubation (Fig. 3). The study of effect of substrate concentration on percentage metabolite M_3 formation revealed that the maximum percentage of metablite (M_3) was observed at 40 μg/ml substrate concentration after 72 h of incubation (Fig. 4 and Table 1). As substrate concentartion was increased, decrease in the metabolite formation was observed after a concentration of 40 μg/ml. It indicated that the percentage of metabolite (M_3) was decreased with increased concentration of 50 and 60 μg/ml, which might be due to some inhibitory effects. Hence, Michaelis–Meneten kinetics could not be applied to these results and substrate inhibition kinetics was applied to estimate kinetic parameters like Ki, Km, and Vmax. These values of metabolism reactions mediated by enzymes in *Aspergillus niger* to form M_3 were 40.11 μg/ml, 39.68 μg/ml and 0.59 μg/ml/hr, respectively (Fig. 7 and Table 3). It stated that metabolism reaction that formed M_3 proceeded with the maximum velocity (Vmax) of 0.59 μg/ml/hr, and half maximum velocity of reaction was found at the concentration of (Km) 39.68 μg/ml and metabolite (M_3) formation was inhibited at 40.11 μg/ml concentration of loratadine.

Comparison of kinetics of formation of M_1, M_2, and M_3: Km is the substrate concentration at which half of the active sites are occupied.An enzyme with a high Km has a low affinity for its substrate, and a greater concentration of substrate is required to achieve Vmax and vice versa [20]. When Km values of formation of M_1, M_2, and M_3 were observed, Km value for formation of M_2 was lesser than Km values of both M_1 and M_3. Hence, the enzymes present in *Cunnighmella echinulata,* which formed M_2, has shown more affinity towards loratadine and low concentrations of loratadine are sufficient to saturate the enzyme and to reach maximum rate of reaction that produces metabolite M_2. While the Km value of M_1 formation was higher than Km value of M_2 formation, it indicated that enzyme present in *Cunnighmella elegans* has less affinity towards loratadine, so more amount of loratadine was required to saturate the enzyme and to reach the maximum rate of reaction that produce metabolite M_1, whereas in case of formation of M_3, atypical kinetics (substrate/product inhibition kinetics) was observed. The percentage of metabolite M_3 formed at higher concentrations like 50 and 60 μg/ml of loratadine was lesser than percentage of metabolite formed at 40 μg/ml. It indicated that higher concentration of loratadine (>40 μg/ml) may inhibit the metabolite M_3 formation.

The maximum rate, Vmax, is the measure of a number of substrate molecules converted into product by an enzyme molecule in a unit time when the enzyme is fully saturated with substrate. It was found that the rate of reaction was increased with an increase in the substrate concentration, but when the rate of reaction reached to Vmax, further increase in the substrate concentration did not cause any increase in percentage of metabolites (M_1, M_2, and M_3). It indicated that the enzymes present in these three fungi, which are involved in metabolic reactions formation of M_1, M_2, and M_3, were fully saturated with given substrate concentration at Vmax.

5 Conclusion

Enzyme kinetic studies revealed that metabolism mediated by enzymes present in *Cunnighamella echinulata* possessed low Km value than enzymes present in *Cunnighamella elegans* and *Aspergillus niger.* So the enzymes present in *Cunnighamella echinulata* has more affinity towards loratadine when compared with the other two enzymes in *Cunnighamella elegans* and *Aspergillus niger*. So, maximum metabolites can be produced by *Cunnighamella echinulata* with less quantity of loratadine.

References

1. Bedard PM, Del Carpio J, Gutkowski A (1985) Comparison of efficacy and safety of SCH 29851, terfenadine and placebo in treatment of seasonal rhinitis. Ann Allergy 55:233
2. Batenhorst RL, Batenhorst AS, Graves DA, Foster TS, Kung M, Gural RP, Amkraut HJ (1986) Pharmacologic evaluation of loratadine (SCH 29851), chlorpheniramine and placebo. Eur J Clin Pharmacol 31:247–250
3. Villani FJ, Magatti CV, Vashi DB, Wong J, Popper TL (1986) N-substituted 11-[4- piperidylene)-5,6-dihydro-11H-benzo-[5,6]-cyclohepta-[1,2-b] pyridines: Antihistamines with no sedating liability. Arzneim-Forsch 36:1311–1314

4. Katchen B, Cramer J, Chung M, Gural R, Hilbert J, Luc V, Mortizen V, D'Sousa R, Symchowicz S, Zampaglione N (1985) Disposition of 14C-SCH29851 in humans. Ann Allergy 55:393
5. Hilbert J, Radwanski E, Weglein R, Luc V, Perentesis G, Symchowicz S, Zampaglione N (1987) Pharmacokinetics and dose proportionality of loratadine. J Clin Pharmacol 27:694–698
6. Simons FER, Simons KJ (1999) Clinical pharmacology of new histamine H1 receptor antagonists. Clin Pharmacokinet 36:329–352
7. Yumibe N, Huie K, Chen K, Clement RP, Cayen MN (1995) Identification of human liver cytochrome P-450s involved in the microsomal metabolism of the antihistamine drug loratadine. Int Arch Allergy Immunol 107:420–429
8. Ghosal A, Gupta S, Ramanathan R (2009) Metabolism of loratadine and further characterization of its in vitro metabolites. Drug Metab Lett 3(3):162–170
9. Ault A (1974) An Introduction to Enzyme Kinetics. J Chem Edu 51:381
10. Magang S, Qin Mei, Michael W, Ettore JR, Renke D, Thomas A. Baillie, Thomas HR (1999) Sigmoidal kinetic model for two co-operative substrate-binding sites in a cytochrome P450 3A4 active site: an example of the metabolism of diazepam and its derivatives. Biochem J 340:845–853
11. Northrop DB (1998) On the meaning of Km and V=K in Enzyme Kinetics. J Chem Edu 75:1153
12. Segal HJ (1959) The development of Enzyme kinetics in The enzymes, 2 nd ed. Boyer PD, Lardy J, Myrback K, eds., vol. 1, Academic press, pp. 141–147
13. Keerthana M, Vidyavathi M (2018) Screening and evaluation of fungal resources for loratadine metabolites. J Biosci 43:823–833
14. Azerad R (1999) Microbial models for drug metabolism. Adv Biochem Biotechnol 63:169–218
15. Moody JD, James P. Freeman, Peter P. Fu, Cerniglia CE (2002) Biotransformation of Mirtazapine by Cunninghamella elegans. Drug Metb Dispos 30(11):1274–1279
16. Smith RV, Rosazza JP (1975) Microbial models of mammalian metabolism. J Pharm Sci 64:1737–1758
17. Keerthana M, Vidyavathi M (2019) Enzyme Induction and inhibition studies for fungal biotransformation of loratadine. Asian J Microbiol Biotechnol Environ Sci 21(1):96–101
18. Yuh L, Ping L, Cuyue T, Qin M, Grit Sandig A, David R, Thomas H. Rushmore, Magang S (2001) Substrate inhibition kinetics for cytochrome P450-catalyzed reactions. Drug Metab and Dispos 29(4):1, 368–374
19. Magang S, Yuh L, Ping L, Cuyue Tang, Qin Mei, Dan Cui, Wei Tang, Jason S. Ngui, C., Charles Lin, Rominder Singh, Bradley K. Wong, James A. Yergey, Jiunn H. Lin, Paul G. Pearson, Thomas A. Baillie, A. David Rodrigues, Thomas HR (2000) Enzyme kinetics of cytochrome P450-mediated reactions. Curr Drug Metab 2:17–36
20. Berg JM, Tymoczko JL, Stryer L (2002) The Michaelis-Menten Model Accounts for the Kinetic Properties of Many Enzymes, Biochemistry. 5th edn, New York: W H Freeman, 8.4. Available from https://www.ncbi.nlm.nih.gov/books/NBK22430

A Study Report on Introducing Sanskrit Language Based on the Technologies Related to Current Trends

G. Nagalakshmi and T. Sarath

Abstract In the ancient period, on words, Sanskrit devises some features that remain most useful in dissimilar areas. The Sanskrit stands a natural language presence, the finest selection to be transformed into the computing platform for robotic controller and artificial intelligence knowledge. The study focuses on Sanskrit between the pool of numerous human languages classifying it as one of the best suitable ones for computing methods. The computing has two aspects like logic and amp and language. The technical language employed by Indian Shastra disciplines is a derivation of Sanskrit leaving all disqualifications of natural language are taken qualities of an artificial language. Hence, it is a natural human being and as good as artificial for a machine. Logically, it has prevalent faith in natural languages that are incompatible for the diffusion of various concepts that artificial languages might extract with the finest accuracy and calculated consistency. In this, studying of Sanskrit is an honor for the Indo-Aryan language and it is fantastic to see a machine program with natural language and communication in Sanskrit.

Keywords Sanskrit · Computing · Natural language · Artificial intelligence · Indian shastra · Machine

1 Introduction

The Divine and Holy language of India is Sanskrit, which is scribbled in the script of Devanagari, which is famous for its beauty and clarity. This language is considered as a mother of all languages, which belongs to Indic set of language and family of Indo European and its progenies, which stand in Indo-Aryan and Indo Iranian.

G. Nagalakshmi (✉)
Department of CS, Rashtriya Sanskrit Vidyapeetha, Tirupati, India
e-mail: agnl.lakshmi@gmail.com

T. Sarath
Department CSE, Siddhartha Institute of Science and Technology, Puttur, India

S. Jyothi et al. (eds.), *Proceedings of the 2nd International Conference on Computational and Bio Engineering*, Lecture Notes in Networks and Systems 215,
https://doi.org/10.1007/978-981-16-1941-0_29

The importance of Sanskrit is cultured, adorned or formed in flawless procedure. It has a similarly with language of Jainism, Buddhism and Hinduism.

In present days, Sanskrit is drawing the modern people. The Hindus uses this language as in ceremonies and functions of Holy and it is treated as a sacred language in the religious conviction. Mantras in Sanskrit delivered in mishmash with the sound vibration have an unambiguous result on the mind and the essence of the individual.

To mentors of spiritual bodies, this language has been introduced by the creator god known as Brahma. So, this language is also named as Dev Vani that means the God's language. In the period of eighteenth century, the difference between Greek, Sanskrit and Latin is derived to give the reason to discover and study the relationship between all Indo-European languages.

Around 1500–200 BC, Vedic Sanskrit has come from the earliest form of Sanskrit language. This existed as the era when knowledge was communicated vocally through the groups.

Though not for all the languages, Sanskrit is surely the mother of many languages, especially languages spoken in Northern India. Even many words from Dravidian languages are derived from Sanskrit as shown in Fig. 1. Almost all of the ancient kinds of literature such as the Vedas, Upanishads, Epics, Shastras, Puranas and many works on medicine, astronomy, astrology and mathematics of those times were written in Sanskrit.

Sanskrit has a similarity between Latin and Greek languages. In William Jones Sir's view, "Sanskrit is flawless than Greek, more bountiful than Latin and more divinely cultured than any."

The three C's such as Culture, Civilization and Cultivation of East countries were all well denoted by this language, which was evolved even before many countries did not have a specific language. The script, grammar and phonetics of Sanskrit language are such that most of the neurons of the human body get activated. Sanskrit is being taught and learned still in many parts of the world.

क	ka [kʌ]	ख	kha [kʰʌ]	ग	ga [gʌ]	घ	gha [gʱʌ]	ङ	ṅa [ŋʌ]
च	ca [cʌ]	छ	cha [cʰʌ]	ज	ja [ɟʌ]	झ	jha [ɟʱʌ]	ञ	ña [ɲʌ]
ट	ṭa [ʈʌ]	ठ	ṭha [ʈʰʌ]	ड	ḍa [ɖʌ]	ढ	ḍha [ɖʱʌ]	ण	ṇa [ɳʌ]
त	ta [t̪ʌ]	थ	tha [t̪ʰʌ]	द	da [d̪ʌ]	ध	dha [d̪ʱʌ]	न	na [nʌ]
प	pa [pʌ]	फ	pha [pʰʌ]	ब	ba [ʋʌ]	भ	bha [bʱʌ]	म	ma [mʌ]
य	ya [jʌ]	र	ra [ɾʌ]	ल	la [lʌ]	व	va [ʋʌ]		
श	śa [ɕʌ]	ष	ṣa [ʂʌ]	स	sa [sʌ]				
ह	ha [ɦʌ]	ळ	ḷa [ɭʌ]						

Fig. 1 Shows the Sanskrit words

Sanskrit has two types in it, such as the Vedic Sanskrit in which Vedas and Puranas, etc. were written and the Classical Sanskrit from which the grammatical rules were derived by Panini. Vedic Sanskrit is the pre-classical form that is dated early to the second millennium BCE and classical Sanskrit was later derived, which is dated around the fourth century BCE.

Sanskrit is the oldest survived language on earth and is a sacred language of Hindu, Buddhist and Jain traditions. Many of the ancient literary works including Jain and Buddhist works and even Yoga Shastra were written in Sanskrit. It might be because of its great heritage and cultural hegemony Sanskrit is called as the mother of many languages.

2 Sanskrit in Ancient Indian Sciences

Scientific study is needed multi-disciplinary in appeal as it includes computer science, cognitive sciences, etc. Language ability needs human capabilities at numerous stages such as syntactic, lexical, phonetic, semantic, cognitive, socio-contextual, prosodic, etc. Earliest Indian scientific discourses compact through all issues inherently, ensuing in devising of an inclusive scheme of language depiction for accurate practice. The eldest occurrence of such a scheme relates to the Vedic (sacred) fiction courting back to numerous periods, which remained communicated merely verbally drawer a rare century past, but is incredibly distortion free.

This was proficient through the invent like:

(a) Lexicons of Standard (for illustration Yaska's Nirukta for Vedic relations (through exegetical and etymological particulars and Amarakoshna, etc. for worldly literature), if a collection of arguments.
(b) A grammar is well structured (for illustration Panini's Ashtadhyayi) to produce uncountable grammatically usable word methods after a limited set of origins over rules.
(c) Arrangement of logical compatibility and validity rules (for illustration Gautama's Nyaya sutras), to yield dialogue with importance.
(d) Appropriate phonetic arrangement of noises and definite euphonic grouping methods counting drawls (illustration Panini's Shiksha) aimed at the attractiveness set of the language.
(e) An inflection assembly of speech for the statement of sensations (illustration Pingala's Chandas sutras) concluded prosody and metrics.
(f) A malicious to realize the information produced in the complete sense (illustration sutras Vyasa's Brahma) at rational level and strategies for useful efficacy and knowledge (illustration Sutras Jaimini's Mimamsa for Vedic practice laukika nyayas for earthly custom) in the procedure of morals and axiom codes of morality.

3 Salient Features of Sanskrit

A few points are worth considering here regarding the nature of Sanskrit.

- All arguments are prepared either by characters, vowels or consonants. Vowels exist independently, while consonants depend on vowels. The process of Sandhi (euphonic combination) is defined.
- Arguments are collected in two portions, a static base portion and a mutable affix portion, both forming an integral unit. The mutable portion alters the denotation of the word base, dependent on a group of specified relations. Thus the process of derivation is properly defined.
- Arguments are either nominal form or verbal form, i.e., denoting either entities or actions, except particles that are not composite in their final form. Adjectives, adverbs, etc. are not independent of these.

4 Computational Database

Sanskrit is based on three major components like nominal stems, affixes and verbal stems familiarized when the first two produce further stems as well as completed arguments [4]. The Dhatupatha catalogs all accessible verbal stems distributed into 10 explicit sessions (the Gana patha history sets of small stems, which undertake definite grammatical procedures).

Panini's methodology is to generalize nominal and verbal affixes by defining a set of archetypal affixes for these which are introduced after all stems or roots and then specifying various allomorphs for different endings and gender for nominal and for different classes and tense/mood for roots. Here 21 archetypal affixes are present for nominal declensions (symbolized by "sup") and 18 for verbs (indicated by "tin").

In general, the knowledge represents in the computer? This is the crux of the whole exercise and we describe the ancient Indian method in this regard. This is necessarily empirical in nature. The development of a classification scheme for all substances in a structured form to explain meanings facilitates machine understanding [9].

Reasons for Choosing Sanskrit are

- Richness of scientific literature with in-depth and exhaustive analysis.
- Systematic approach meant for cognitive knowledge description.
- Expendability to other Indian languages (Pali and Prakrit).
- The phonetic basis of the language being useful for speech analysis as well.

Understanding a Natural Language Input

- Paraphrasing changing the voice, synonymous explanation, description or in a classical format.
- Answering queries involving Inference, Connection, Reference, etc.
- Summarizing by picking the key thoughts in the input sentences.

Computational

- Developing custom-made tools for natural language process creation of Dictionary, Thesaurus, Concordance, etc., creation of function and data libraries in standard formats.
- Developing a language-independent knowledge representation scheme based on ancient Indian scientific works.
- Design of a "Linguistic Work station" aimed at helping language research through computers.

5 How the Sanskrit is Worked as Another Language?

There is no denial that Sanskritam is perhaps one of the oldest of languages. It is also one of the first structured and organized languages with a strict set of syntax and grammar. Most of its credit goes to Panini, a Sanskrit grammarian, provided an inclusive and technical model of morphology, phonology and phonetics during a time when people in other parts of the world were drawing on caves. Panini's works distinguish [7] among the language of holy scripts and the normal language of announcement. To describe the grammar of Sanskrit, he has given proper creation guidelines and descriptions, beginning through 1700 simple features such as vowels, verbs, consonants, nouns that are put in classes. The sentences' creation is in simple nouns, etc. and is described as ordered guidelines working on fundamental arrangements in a method parallel to the current scheme.

In Panini's creations [6], it is parallel to the technique that a mathematical function is distinct currently. His phenomenal work has influenced modern linguists like Chomsky and De Saussure.

Grammatical classes help to definite awareness about the sphere. Panini's method of knowledge demonstration is created in the theory of karaka. The karaka is profound assembly association that facilitates mappings from relations of semantic (like goal, location, agent) to phonological signs (in relations to situation culminations that might direct voices) through surface assemblies (in relations to morphological groups like small cases, prepositions, and verbal speeches) [6]. On the level of morphological, the karakas are characterized by six troikas of case endings, collectively it resembles through single karaka [5]. The karaka guidelines are useful through the adhikara (governing) sutra Prote.2.3.1: an abhihite, (enhance a case ending when a verbal element to take a karaka solitary) if it remains not voiced (previously). Karakas have two, karman and kartr can be stated by verbal culminations, while some former karakas can be stated by main and minor suffixes. The theory of karaka is evident attention of the computer expert involved in natural language handling. The reader involved in the facts of this concept would perceive the themes by Kiparsky and Joshi7 and Staal [8].

An inclusive study of a perception computing science would contain algorithmic, structural and linguistic phases. Likewise, a reading needs to be founded on the long custom of scrutiny drives spinal around 2,500 ages. Difficulties of specific attention to the computer expert contain "A" procedure of the instructions and the minutest group of procedures would remain equal to A. Reordered procedures, by Bhattojı Diksita in his Kaumudi Siddhanta, would deliver a precious structure of evaluation. But earlier, an evaluation can remain completed after an algorithmic viewpoint one desires to [3] define A's instruction in a method suitable for exploration by computer. Thru this in mind, we converse in this preliminary paper convinced proper features of grammar of Panini's. In specific, we study the ensuing two features:

1. The style of sutra and the environment of procedures
2. The construction of the regulatory scheme

We resolve expression how the guidelines can remain certainly cast in acquainted transformational or algebraic methods. An obviously algebraic depiction is vital, earlier rules of Panini's are stated on a computer so that their cognitive and computational associations can accurately measure.

6 Challenges for Computer Language Conversion

The Sanskrit word for artificial is "Kritrima" "Kalpita" and "Kritaka." The words Prajna and Dhee stand for intelligence. Thus, "Kritaka Prajna" or "Kritaka Dhee" or "Kalpita Prajna" "Prajna Kalpa" means artificial intelligence.

In olden origin, the language of Sanskrit consumes some remarkable features that remain measured obliging in dissimilar areas. It remains used for sessions of therapy in **psychology** and for **remissions of spiritual**. Then in current connection through artificial intelligence is an honor verifying its control for existence an appreciated progress of literature.

Sanskrit takes a rich antiquity and remained used aimed at initial Indian science and mathematics. The Sanskrit grammar is logical, formula-bound and rule-bound, which creates it extremely fitting to write procedures.

The Sanskrit grammar also marks appropriate aimed at machine learning and even artificial intelligence. For historians besides irregular peoples, the prospect of consuming Sanskrit to improve machines in artificially intelligence is exciting since it feats the earlier inventively to distribute resolutions for the forthcoming.

The optimistic consequences have numerous experiments that would be retained into attention. Primary it has uncertainty of any natural language which means that the advices which procedure a verdict in a natural language might **observe numerous denotations**. It must remain capable to change the language in such a method that it decreases the abstruseness, creating it extra verbatim so that robots through artificial intelligence might recognize it remains an experiment.

The accurateness of verbal scrutiny is an actual significant influence that is tough for AI [1].

For illustration, **statements of sarcastic** are tough to understand precisely since accepting sarcasm needs expressive intelligence.

An appearance alike **"That's impartial what I desired currently!"** can remain inferred as both mordant or exultant scream reliant on tonal variations and sentiments stated beside with the declaration.

Accurate analysis of linguistic [2] reports with passions and minor tonal variations needs the discovery of the pointers. This is problematic for AI, mainly if the appearance is carried through a microphone or deliver out lurid by some other person.

However, this is the intention why Sanskrit looks more appropriate than former languages. The **rules of grammar are strict**, words and syllables are **reduced vagueness** making the accurate meaning of sentence and word. This absolutely lessens the ratio of abstract significances in the language.

The basis behind these dramatists is built on language that needs to be developed mainly to procedure **relations of logic with scientific correctness**. This means a logical connection in the situation of scientific exactness can simply be established by Sanskrit.

The perceptive of the research is inherent in a style of language that is developed specially to express logical relations through scientific precision. It is a concise, a compressed form of Sanskrit, which, illogically at periods, becomes so obscure that an explanation is needed to explain it.

The grade to which a semantic net (or any unequivocal, nonsyntactic demonstration) is awkward and odd-sounding in a language of natural is the grade at which that language is "natural" and diverges from the specific or "artificial." As we will realize, there was a spoken language between an ancient scientific communal that takes an aberration of zero. Since natural languages actuality the finest choice to be transformed into the computing program for robotic controller and Artificial Intelligence knowledge.

7 Conclusion

This article study states Sanskrit is the main language for the birth of the remaining language due to its grammar rules, various explanations are given to state Sanskrit is the language that represents knowledge. The research for feature will focus on Sanskrit between the group of various human languages, clarifying that **computing methods** are **one of the best suitable one**.

References

1. Rick Briggs (1984) Knowledge representation in Sanskrit and Artificial Intelligence, AI Magazine, sping 1985, pp. 32–39
2. Rick Briggs (1985) Shastric Sanskrit as Machine Translation Interlingua
3. Vineet Chaitanya, Ramakrishnamacharya KV (1987) SARALASAMSKRITAM as a knowledge representation language

4. Dr. Sinha RMK (1989) A Sanskrit based word Expert model for machine translation among Indian languages, proc.of Regional Workshop on Computer Processing of Asian Language (CPAL), pp. 82–89
5. Akshar Bharati, Vineet Chaitanya, Rajeev Sangal (1989) A Karaka based approach to parsing of Indian languages. TRCS-89–88, I.I.T.Kanpur
6. Subhash C. kak (1987) The paninian approach to Natural Language Processing. Intnl J Approxim Reas 1:117–130
7. Kanthan KL (1989) Formal language system of panini, pre-conference papers of national conference on 'Samskritam and Computers', New Delhi
8. Veda Varidh, Veda Vijnana Siromanu, Vaidika Bhusanam P. Ramanujam (1993) Issues in computer-based Sanskrit studies, presented at the conference on AI& Sanskrit teaching held at S.J.S.V, Puri
9. Shashank Saxena, Raghav Agarwal (2013) Sanskrit as a programming language and natural language processing. Global J Manag Busi Stud 3(10):1135–1142. ISSN 2248–9878

Celecoxib Topical Nanoemulgel: Formulation, *Ex-Vivo*, Pharmacodynamic, and Pharmacokinetic Studies

Y. Indira Muzib, Y. Sarah Sujitha, and Y. R. Ambedkar

Abstract Celecoxib is a COX-2 inhibitor used for anti-inflammatory activity and to reduce the oral side effects, it is formulated in topical gel form. The method used for the development of the emulsomes was emulsion solvent diffusion method. The formulated celecoxib emulsomes were optimized, and the optimized celecoxib emulsomes were converted into topical gel. The physicochemical properties like SEM, DSC, FT-IR, results indicate that there is no chemical interaction between the drug and polymer. The zeta potential, particle size, entrapment efficiency of the optimized emulsomes were found to be 119.3252 nm and –16.3 mV. Optimized X1 gel by physical examination appears to be homogenous, with pH 7 ± 0.2 and viscosity 8,640cps, and with 98.3% of drug content. The cumulative percentage of drug release was found both between the *in-vivo* and *ex-vivo* levels and *ex-vivo* skin permeation studies found that the release was 95% for 4 h. The gels that were prepared were subjected to pharmacokinetic and pharmacodynamic studies. The in-vivo pharmacokinetic and dynamic results showed that the emulsomal gel has shown effective inhibition of inflammation. The significant activity ($p < 0.05$) was exhibited by Celecoxib-loaded emulsomal gel when it is compared to the marketed formulation (Voveran Emulgel).

Keywords Emulsomal gel · Emulsomes · *In-vitro* · *In-vivo* · *Ex-vivo*

1 Introduction

The cyclooxygenase is of two isoforms as cox-1, cox-2 which catalyze the synthesis of prostaglandins from arachidonic acid [1]. In most of the cases, cycloxygenase-2 is not detected by more healthy patients but induced by inflammatory stimuli,

Y. I. Muzib (✉) · Y. S. Sujitha
Sri Padmavati Mahila ViswaVidyalayam, Tirupati, India
e-mail: yindira1415@gmail.com

Y. R. Ambedkar
Sri Venkateswar Veterinary University, Tirupati, India

S. Jyothi et al. (eds.), *Proceedings of the 2nd International Conference on Computational and Bio Engineering*, Lecture Notes in Networks and Systems 215,
https://doi.org/10.1007/978-981-16-1941-0_30

which finally results in the increased production of prostaglandins [2]. The drug celecoxib is an NSIAD, which is a specific inhibitor of cyclooxygenase(cox-2), which is used to treat patients suffering from rheumatoid arthritis [3]. Though celecoxib has side effects, it has less toxicity when compared to other cox1,2 inhibitors. Applying nanotechnology to medicine is known as nanomedicine, it concerns the manipulating of material to develop novel therapeutic drug delivery system. The first synthesized non-steroidal anti-inflammatory drug was found to be celecoxib. It is a cox-2 inhibitor drug with high permeation capacity and a poorly soluble drug with low molecular weight. Due to the side effects, all the above reasons restrict for oral use of celecoxib and making it a suitable candidate for the topical drug delivery system and making it more safe and effective [4]. Nanotechnology-based drug delivery system includes SLNS, nanocrystals, emulsomes, nanosponges, transferosomes. Formulation using celecoxib for incorporation into emulsomes shows increased permeation and efficacy of the drug. These emulsomes are vascular drug delivery system like ethosomes, transferosomes [5]. These emulsomes are colloidal carriers, which are composed of solid lipid core and surrounded by phospholipid bilayer with increased drug entrapment efficacy [6]. The key feature of the emulsomes is that they can incorporate water poorly soluble and water highly soluble drugs in their lipid core and between the phospo lipid by layer, the objective of the present study was to find out the efficacy of the emulsomes by loading the celecoxib in the emulsomes and formulating in the topical drug delivery system (Gels) [7].

2 Materials

A gift sample of celecoxib was received from Kekule Pharma LTD, Hyderabad, India, Polyvinyl alcohol, Ethyl Cellulose, Carbopol 934, Hydroxy Propyl Methyl Cellulose, which was purchased from Hi-Media Laboratories Mumbai.

3 Methods

3.1 Preparation of Celecoxib Emulsomes

For the formulation of celecoxib emulsomes, ethanol injection method was used. 10 ml of ethanol was taken into which celecoxib, soyalecethin, cholesterol, stearic acid were added and considered as organic solution (Table 1). Phosphate buffer of pH 7.4 was prepared to which organic ethanol solution was added under continuous stirring. The stirring was continued at 1200 rpm until the organic phase gets evaporated. The resulting solution was centrifuged and the sediment was collected. Then the sediment that was collected was subjected to lyophilization. The prepared emulsomes were found to be spherical in shape.

Table 1 Formulation composition of celecoxib-loaded emulsomes

S. no	Formulation code	Stearic acid (mg)	Cholesterol (mg)	Soyalecithin (mg)	Drug (mg)
1	F1	100	50	25	100
2	F2	100	50	50	100
3	F3	100	50	75	100
4	F4	100	50	100	100
5	F5	100	50	125	100
6	F6	100	100	100	100
7	F7	100	125	100	100
8	F8	100	200	100	100
9	F9	125	100	100	100
10	F10	200	100	100	100

3.2 *Preparation of Celecoxib-Loaded Emulsomal Gels*

The optimized emulsomal formulation F5 having a high percent of drug release was selected and was formulated into gel using gelling agents, carbopol 934 and HPMC (Table 2). The polymers carbopol 934 and HPMC were weighed and soaked for 24 hr prior and stirred at 1000 rpm for 30 min using mechanical stirrer. The gel was neutralized by adding 1–2 drops of triethanolamine. To the gel methyl paraben and propyl paraben and glycerine were added with continuous and slow stirring.

Table 2 The composition of topical gels using different concentrations of polymers (Carbopol 934 and HPMC)

S.No	Formulation code	Composition	Ratios of polymers
1	X1	Carbopol: HPMC	1:1
2	X2	Carbopol: HPMC	1:2
3	X3	Carbopol: HPMC	1:3
4	X4	Carbopol: HPMC	1:4
5	X5	Carbopol: HPMC	2:1
6	XP	Carbopol: HPMC(pure)	1:1

4 Evaluation of the Emulsomes

4.1 SEM Analysis

The size morphology and shape of the emulsomes were examined by using SEM [8]. Aluminum stubs were used and the celecoxib-loaded emulsomes were placed on the alumina stubs and examined.

4.2 Differential Scanning Calorimetry

Celecoxib-loaded emulsomes were loaded in a small aluminum pans and samples were weighed below 5 mg and the temperature was raised at a rate of 10 °C/min from 25 to 390 °C under constant heating [7].

4.3 Zeta Size and Zeta Potential

Dry celecoxib-loaded emulsomes were taken in test tube and made into the solution using double distilled water and the solution was sonicated for 30 s in an ice bath. Measurement of size was performed in triplicates [9]. Zeta potential was also measured in triplicates by using the same equipment at 25 °C.

4.4 In-Vitro *Drug Release Studies*

The amount of drug release at in-vitro level was done with dialysis membrane having molecular weight around 12,000 daltons, which was soaked 12 h before study. Accurately weighed 100 mg of the celecoxib emulsomes were weighed and placed in the dialysis membrane to which 2 ml of the phosphate buffer of pH 7.4 was added and placed in the beaker having 100 ml of phosphate buffer [10]. The beaker was kept on magnetic stirrer and maintained at 100 rpm and at 37 °C temperature. At regular time intervals, 0, 0.25, 0.5, 0.75, 1, 2, 3, 4, 5, 6, 7, 8, 12, 24 h samples were taken out from the beaker and the same quantity of buffer was poured into beaker to maintain sink conditions and samples were measured at 252 nm.

5 Evaluation of Gels

The gels that were prepared were evaluated for physical evaluation, homogeneity, drug content, pH measurement In vitro, In vivo studies.

5.1 Determination of Drug Content

Magnetic stirrer was used and 10 mg of gel in 10 ml of buffer was taken ad stirred all night [11]. The obtained solution was filtered and the absorbance was checked at 252 nm.

5.2 Homogeneity

By visual inspection, the homogeneity of the prepared emulsomes celecoxib gel was examined for the presence of any aggregates before and after transferring the gel into a container.

5.3 5.3. pH

The prepared gels were tested by pH meter at 1st, 15th and , 30th day to find any changes in pH.

5.4 5.4. Viscosity

Brook field viscometer was used to find out the viscosity of the gel. 5 gm of gel was transferred into a container and viscosity of the prepared gel was measured with the spindle number 64 at various points and the viscosity value was noted [12].

6 Pharmacokinetic Study

For this study, mobile phase acetonitrile and water were used in the ratio 25:65 using HPLC (Shimadzu) [13]. Samples were collected from the retiro-orbital route and immediately sodium citrate was added, which act as anticoagulant. Centrifugation was performed for all the samples for 30 min at 15,000 RPM. Internal standard of

Flutamide (0.25 μg) was added to the solution. Using Hamilton syringe, 25 μl of the sample was injected into HPLC column and measured at wave length 252 nm using RP-HPLC method. Rats were divided into three groups as (Group-I as control and Group-II pure celecoxib, Group-III emulsomes loaded with celecoxib Gel).

7 *In-Vivo Testing of* Anti-inflammatory Activity

Experimental animals: For the experimental study, albino rats were used. Animals were obtained from Sri Venkateswara Enterprises Bangalore. The environmental conditions for the study were maintained like temperature (25 ± 1 °C) and humidity (55 ± 5%) and light–dark cycles for housing the animals [14]. The anti-inflammatory activity procedure and protocol were carried out under the guidance of CPCSEA. This study was approved by animal ethical committee bearing Reg. Number 1677/PO/Re/S/2012/CPCSEA/IAEC/25 dt.3.05.2018. The efficacy of the prepared celecoxib emulsomes gel was studied by using the pharmacodynamic study in animals. The carrageenan as inducer was induced to the hind paw of the rat by using hind paw method. All the animals were divided into five groups, Group I—Control, Group II—Only carragennan induced, Group III—Marketed gel, Group IV—Treated with plain celecoxib gel, Group V—Celecoxib emulsomal gel. By using the plethysmometer, the paw volume was measured at regular time intervals.

Inhibition Percentage of edema =

$$\frac{(V_{Control} - V_{test})}{V_{Control}} \times 100$$

7.1 SEM Analysis

The obtained emulsome particles were found to be spherical and smooth texture. The morphology and shape of the emulsomes are shown in Fig. 1a, b.

7.2 DSC Thermogram

The pure celecoxib has shown the melting point at 163 °C and celecoxib emulsomes have shown the peak of melting point at 157 °C. The same melting point peaks denote no chemical interaction. The thermograms are shown in Fig. 2a, b.

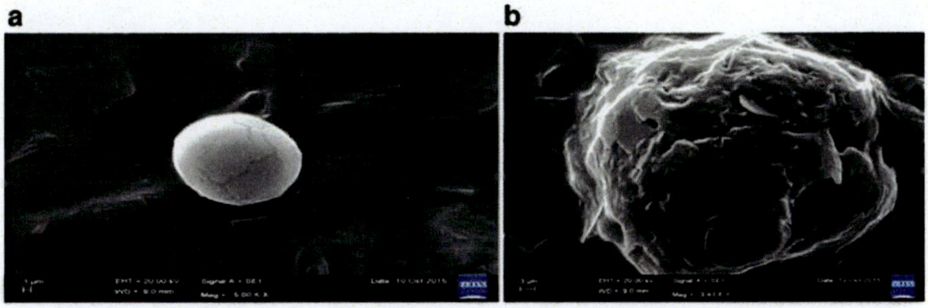

Fig. 1 (a, b) Scanning electron microscopy images of celecoxib-loaded emulsomes

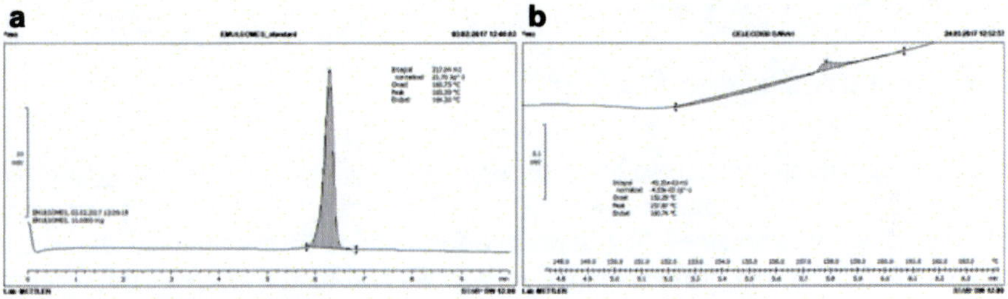

Fig. 2 (a, b) DSC thermogram of pure celecoxib and celecoxib emulsomes

7.3 Zeta Size and Zeta Potential

The optimized formulation of emulsomes was found to be F5 and the zeta size was found to be 119.32 nm. The zeta potential of the F5 formulation was –16.3 mV and is given in Fig. 3a, b.

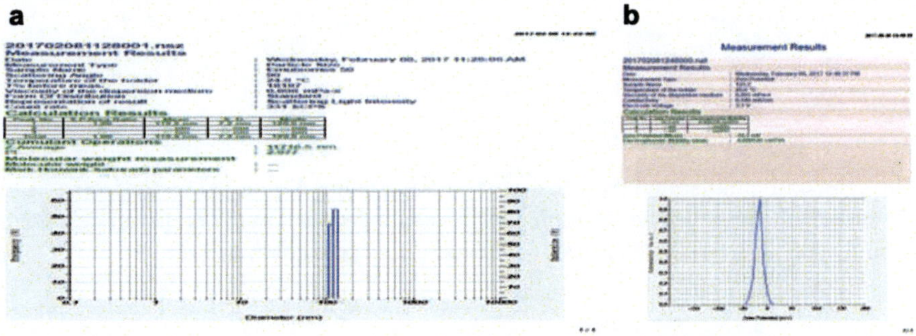

Fig. 3 (a, b) Optimized F5 formulation of celecoxib-loaded emulsomes zeta potential, zeta size

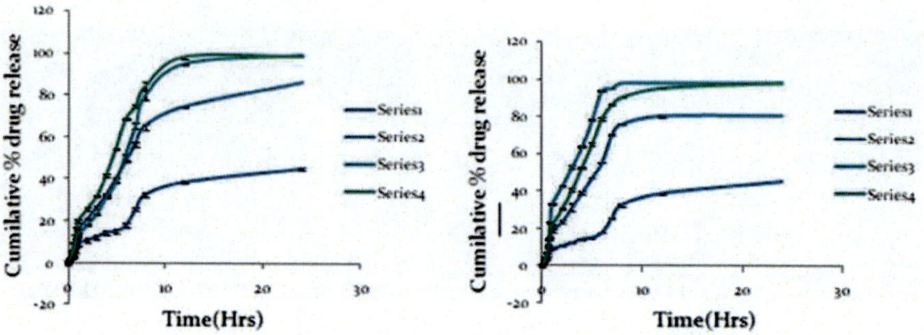

Fig. 4 Celecoxib-loaded emulsomal gel F0 (pure celecoxib)-F5 formulations. All values as mean ± SD, n = 3

7.4 In-Vitro *Release Studies:*

All the preparations from F1 to F6 were subjected to release studies. The graph was obtained by taking time on x-axis and CDR on the y-axis, the cumulative percentage of drug release was obtained. The optimized formulation F5 drug release was found to be 98% for 8 h. Pure celecoxib, the release was estimated as 45% at the end of 24 h, which is shown in Fig. 4.

8 Evaluation of Physicochemical Properties

8.1 Homogeneity

Prepared gels when observed visually show no aggregates, and the gel formed was uniform and smooth.

8.2 PH

The celecoxib emulsomal gel pH was calculated using digital pH meter from X1 to X6 and results were shown in Table 3. The results imply that the prepared gels were neutral with no irritation to the skin. The readings are shown in Table 3.

Table 3 Physicochemical evaluation of gels from X1-X5 and pure gel

S.no	Carbopol:HPMC	DRUG	Viscosity(cp)	Appearance	Drug Content	pH
1	X1	HPMC:Carbopol(1:1)	8,640	Homogenous	98.5 ± 0.69	7.0 ± 0.2
2	X2	HPMC:Carbopol(2:1)	8,820	Homogenous	95.8 ± 1.38	7.0 ± 0.1
3	X3	HPMC:Carbopol(3:1)	9,600	Homogenous	96.2 ± 0.87	7.0 ± 0.2
4	X4	HPMC:Carbopol(4:1)	12,456	Homogenous	95.0 ± 3.54	6.8 ± 0.3
5	X5	HPMC:Caroapol(5:1)	27,256	Homogenous	94.2 ± 0.32	7.0 ± 1.2
6	X6	Celecoxib Pure (1:1) Gel	8,569	Homogenous	97.5 ± 0.53	7.0 ± 0.53

8.3 Viscosity

Brook field viscometer was used to find out the viscosity of the prepared gels. The gels viscosity from X1 to X6 was in the range of 7000–39,000cps as HPMC concentration increases the viscosity increased. Table 3 shows the viscosity of the prepared gels by using various ratios of the carbopol 934 and HPMC.

8.4 Drug Content

The prepared emulsomal gels from X1 to X6 were in the range of 94.2–98.5%. The optimized celecoxib-loaded emulsomal gel X1 has shown the drug content 98.5%. The drug content of all the prepared emulsomal gels is shown in Table 3.

Values were given as mean ± SD, n = 3.

9 Pharmacodynamic Studies

For the pharmacodynamic study, albino rats were used for both pure and emulsomal gel. The inhibition of percentage was calculated by using percentage of inhibition formula and the results were found to be 76.2% for Group III, 57.6% for Group IV rats, and 98.3% for Group V and have significant difference ($p < 0.05$) between the pure and emulsomal gel. Celecoxib loaded emusomal gel is more efficient when compared to the pure celecoxib gel, marketed gel (celecoxib emulsomes gel > Marketed gel > pure celecoxib gel).

10 Pharmacokinetic Study

Plasma concentrations of the celecoxib-loaded emulsomes gel and pure gel were used for the kinetic study. Kinetic parameters of plain gel have the T_{max}, C_{max}, $t_{1/2}$, AUC as follows 15.3 mcg/ml, 4 h and 9.4 h, 90 µg.hr/ml. Celecoxib-loaded emulsomal gel has shown the C_{max}, T_{max}, $t_{1/2}$, AUC was found to be 22.4 mcg/ml, 2.45 h, 8.5 h and AUC as 140 µg/ml, which is shown in Fig. 5. The celecoxib emulsomal gel has shown greater AUC values 140 µg/ml and absorption is fast with a maximum concentration 22.2 µg/ml as it contains tiny size of emulsomes (119.325 nm), which makes drug high passage vide the skin when it is comparatively seen with to the pure gel. The pharmacokinetic parameters like (AUC), C_{max}, T_{max}, AUC, $T_{1/2}$ of celecoxib-loaded emulsomal gel has statistically significant difference ($p < 0.05$) when compared to the pure celecoxib gel, marketed gel.

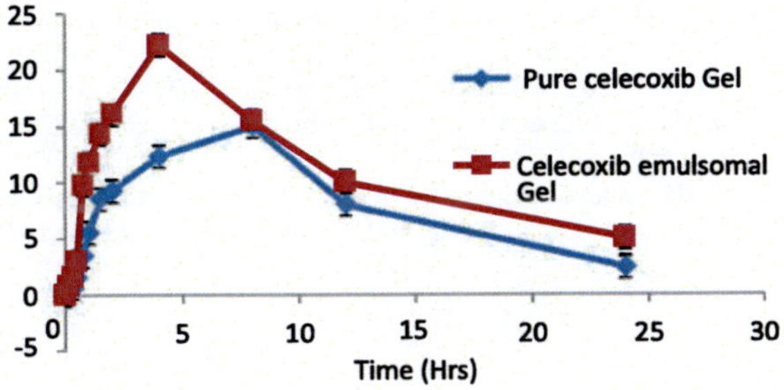

Fig. 5 Plasma profile of optimized emulsomal gel X1 and pure celecoxib gel

11 Conclusion

The better carrier for topical drug delivery system was shown by novel promising celecoxib nanoemulgel. The better stability throughout the study was exhibited by optimized formulation.

References

1. Flückiger FA, Hanbury D (1996) Pharmacographia: a history of the principal drugs of vegetable origin. Macmillan and Co., London, UK
2. Barthe GA, Jourdan PS, McIntosh CA, Mansell RL (1988) Radioimmunoassay for the quantitative determination of celecoxib and analysis of its distribution in Citrus sinensis. Phytochemistry 27:249–254
3. Kokkalou E, Kapetanidis I (1988) Flavonoids of the arial parts of Acinossuaveolens. Pharm Acta Helv 636:170–173
4. Kawaguchi K, Kikuchi S, Takayanagi K, Yoshikawa T, Kumazawa Y (1999) Colony stimulating factor inducing activity of celecoxib. Planta Med 65:365–366
5. Cao M, Ren L, Chen G (2017) Formulation optimization and ex vivo and in vivo evaluation of celecoxibmicroemulsion-based gel for transdermal delivery. AAPS PharmSciTech 18:60–71
6. Bettini R, Catellani PL, Santi P, Massimo G, Peppas NA, Colombo P (2001) Translocation of drug particles in HPMC matrix gel layer: effect of drug solubility and influence on release rate. J Contr Release Off J Contr Release Soc 70(3):383–391
7. Blanco-García E, Otero-Espinar FJ, Blanco-Méndez J, Leiro-Vidal JM, Luzardo-Álvarez A (2016) Development and characterization of anti-inflammatory activity of Curcumin-loaded biodegradable microspheres with potential use in intestinal inflammatory disorders. Int J Pharm 518(1):86–104
8. Tuncay M, Calis S, Kas HS (2000) M 2010: In vitro and in vivo evaluation of diclofenac sodium loaded albumin microspheres. J Microencapsul 17:145–155
9. Blanco-García E, Otero-EspinarFJ, Blanco-Méndez J, Leiro-Vidal JM, Luzardo-Álvarez A (2016) Development and characterization of anti-inflammatory activity of Curcumin-loaded biodegradable microspheres with potential use in intestinal inflammatory disorders. Int J Pharm 518(1):86–104
10. Dr PBS, Mohite SK (2016) Formulation design & development of artesunate nanosponge. Eur J Pharm Med Res 3(5):206–211
11. Khalid. A. Anasari, Pradeep .R. Valli (2011S) fransciscotrota, Roberta cavalla Cyclodextrin based emulsomes for delivery of resebertol: invitro characterization, stability, cytotoxicityand permeation study .AAPS Pharmasci Tech 12:279–286

12. Swarupa Arvapally, Harini M, Harshita G, Arun Kumar A (2017) Formulation and invitro evaluation of glipizide emulsomes. AJPTR 7(3):341–361
13. Koçkaya EA, Selmanoğlu G, Kısmet K, Akay MT (2010) Pathological and biochemical effects of therapeutic and supratherapeutic doses of celecoxib in Wistar albino male rats. Drug and ChemToxicol. 33:410–414
14. Sudharshini S, Stephen Rathina Raj B, Prasantha Kumar Choudary (2010) Quantitative analysis of flutamide in bulk drug and its pharamaceutical dosage formby by using HPLC

Cloud-Based SAAS Applications with Better Solution Using Fiori

M. Usha Rani and Narasimhappagari Chandana Priya

Abstract The aim of this paper is to provide the visualization of data in different cloud-based SAAS applications, by choosing ERP System in SAP focusing on cloud technologies. SAP S/4 HANA database is designed to process both structure and relational databases. Integration with these systems is highly challenging, as information is gathered from different modules of an organization data such as sales and distribution, product management and finance accounting. By analysis all these scenarios, proposed comes up with an approach to reach consistency, reliability and reusability of the results.

Keywords Cloud-based SAAS applications · On-premises system using SAP S/4 HANA · Fiori

1 Introduction

An ERP system stands for enterprise resource planning applications that share a database and communicate with each other, which means it can eliminate information data between departments and gives everyone a single source of information. The system can automate core business processes and help reduce risk, ensure regulatory compliance and fast reporting. SAP is focusing on the in-memory computing and cloud technologies of SAP S/4 (HANA). Real-time delivery of new products is aiming data representation using HANA Databases. High-performance analytic appliance is HANA, which is designed to "process structured data from the relational databases", in both SAP and non-SAP applications. HANA can run in parallel with SAP ERP applications. Analysts access the real-time transactional and operational data [1]. Here by choosing Fiori and SAP S/4 HANA, in current paper which cover

M. U. Rani · N. C. Priya (✉)
Department of Computer Science, SPMVV, Tirupati, India
e-mail: priyachandana.n2@gmail.com

M. U. Rani
e-mail: musha_rohan@yahoo.com

S. Jyothi et al. (eds.), *Proceedings of the 2nd International Conference on Computational and Bio Engineering*, Lecture Notes in Networks and Systems 215,
https://doi.org/10.1007/978-981-16-1941-0_31

the identification of different characteristic and value fields from Sales and Distribution and Material Management modules to come with the Profitability Analysis Document showing in UI as solution in the Fiori.

The current paper describes the history of SAP S/4 HANA, Fiori in the review, later chapters cover the choice of Fiori for the solution, next chapter visualizes data representation with Fiori using SAP S/4 HANA with the real-time figures to formulate the solution of Profitability Analysis Document and finally concluding with relevant Fiori elements to handle UI code and future scope and improvements.

2 Review of Literature

2.1 Sap

Brief history SAP company founded in 1972 by five former employees of IBM in Germany. The meaning of SAP is an abbreviation for Software, Anweldungen und Produkte. This translates to English as Software, Applications and Products. The founders wanted to develop a product that could cover all processes and manage whole economics of an organization. The first standard software for financial accounting was released in 1972. This product was the foundation of SAP R/1 system, still not fully featured ERP system. The letter R refers to the real-time processing. SAP R/2 was released in 1979 and was the first real ERP system. The breakthrough was SAP R/3 in 1992, which was based on client–server architecture. It used relational databases and could be run on any hardware and software. SAP R/3 leads SAP to the first position on the market of standard software for management of business economics [2].

SAP S/4 HANA "SAP is focusing on the cloud technologies and in-memory computing SAP S/4 (HANA). The newest products are aiming for the real-time delivery of data using HANA databases. HANA stands for high-performance analytic appliance and is designed to process structured data from the relational databases, both SAP and non-SAP applications. HANA can run in parallel with SAP ERP applications. Analysts can access real-time operational and transactional data. They do not have to wait for a run of daily or weekly reports. SAP HANA can be deployed on-site as an appliance, or purchased as a managed cloud service. SAP HANA Cloud Platform (HCP) is a cloud platform based on open standards. It provides access to a feature rich, easy-to-use development environment in the cloud. The platform includes a comprehensive set of the services for integration, enterprise mobility, collaboration and analytics". SAP HANA Cloud Platform enables the customers and partners to rapidly build, deploy and manage cloud-based enterprise applications that complement and extend SAP or non-SAP solutions, either on premise [3], or on-demand. As a Platform as a Service operated by SAP, HCP frees from any infrastructure and IT costs and offers advanced quality of service—availability, scalability, multitenancy. These are the programming models that could be used to build highly scalable applications:

- Java—SAP HANA Cloud Platform is Java EE 6 Web Profile certified. Development of Java applications for any application server is possible. The existing Java applications can be also run on the platform.
- HTML5—lightweight HTML5 applications can be easily developed and run in a cloud environment.
- SAPUI5—use the UI Development Toolkit for HTML5 (SAPUI5) for developing rich user interfaces for modern Web business applications.

The applications developed on SAP HANA Cloud Platform run in a modular and lightweight run time container. The platform provides a secure, scalable run time environment with reusable platform services. The preferred technologies to develop and deploy applications are: WEB IDE, SAPUI5 and HANA Cloud Platform. The main aim of the paper is to analyze Sales and Distribution and Material Management modules in SAP ERP system to give the profitability analysis document in Fiori.

2.2 SD and MM Modules

In order to better understand SD and MM modules as the process overlaps between one another, so it is briefly described.

Sales and Distribution Module This module handles the activities connected to processing orders, invoice generation and delivery. To understand it better here is a use case, initially customer sends sale order, based on the order received, purchase order is generated. This includes date of delivery of order, required materials and includes prices of order.

Material Management Module To accomplish business process, different materials are needed. MM module deals with material flow of access from organization until it is transformed into final product. So MM module is the backbone of SAP logistics and modules, which depend on it. Core is the procurement that deals with quantity, timing and appropriate value.

3 Choice of Fiori for the Solution

Fiori is the newest concept of delivering and processing SAP data. Fiori is suitable for desktop as well as for mobile devices. Fiori is built on JavaScript framework SAPUI5 using HTML5. The data for the applications are delivered by SAP services through OData protocol. There are several advantages: building the application from templates, mobile usage, OData protocol or nicer UI. Main disadvantage is no interaction with other SAP transactions and a new way of development for SAP consultants and developers. It is important to select the right form of delivering the functionality because it could improve the user experience and maintenance of the developed programs. Fiori could be seen as the first choice because of its modern UI, the ability to work on both desktop and mobile devices, and no need to sign into the SAP system [4].

As shown in Fig. 2, is the SAP Fiori UX with SAP Cloud Platform Portal that contains SAP Fiori Launch pad and SAP Business Client here we use SAP S/4 HANA. For the UI, we use SAPUI5 and OData services to map S/4 HANA database with Fiori by extracting data from the HAANA database with the help of OData Services. To show the report, we prefer Fiori UI by developing from WEB IDE to map database with UX and final report is shown in Fiori where data are represented. Fiori is preferred as it the first choice of modern UI to show the data either in desktop or mobile devices. Fiori launch pad includes personas in the screens (Fig. 1).

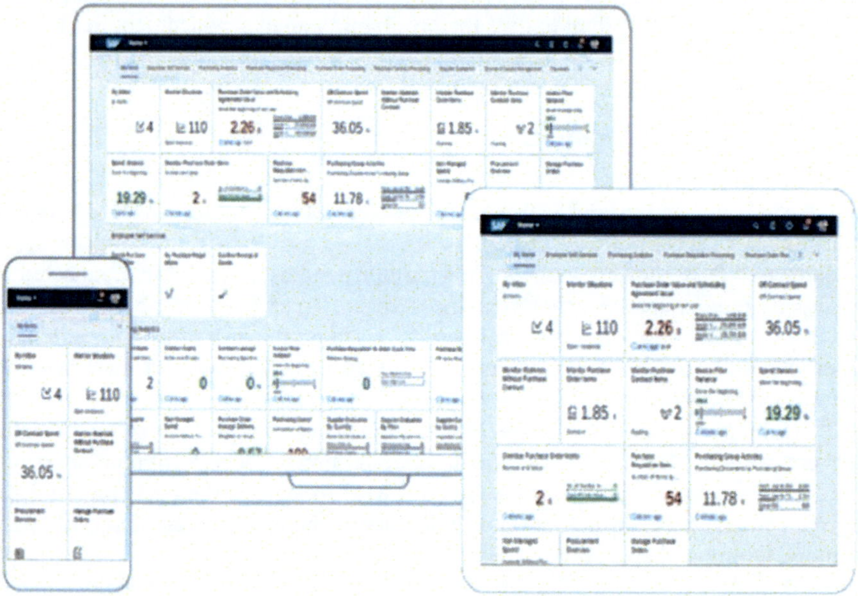

Fig. 1 SAP Fiori multi-device apps captured from the site online website link https://experience.sap.com/fiori-design-web/ [online]

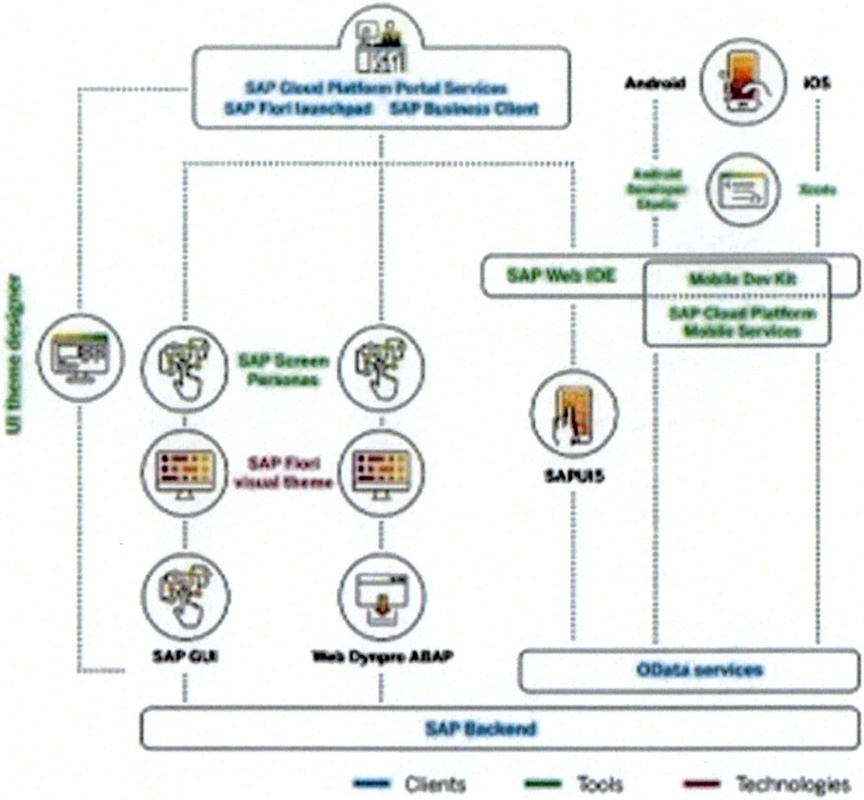

Fig. 2 SAP fiori UX

4 Data Representation with Fiori Using SAP S/4 HANA

4.1 Data Representation in SAP S/4 HANA

SAP S/4 HANA screens are shown in Fig. 3, which represent SAP Easy Access screen whereby executing ke24 tcode the line items display of initial screen is launched, where transactional code is used to fetch line items using ke24 tcode, after executing ke24, standard program for RKEB0601 is executed in the background to fetch the SAP actual data. Figure 4 is the initial screen after executing ke24tcode, includes selection conditions with different fields like currency type, record type, period/year and soon. To fetch the data choosing with year in the selection condition. After selecting period/year and result in the next screen is data extracted and represented as shown in Fig. 5 displays actual line items list is shown for the selected period with the currency type, record type, company code and other list of line items are displayed.

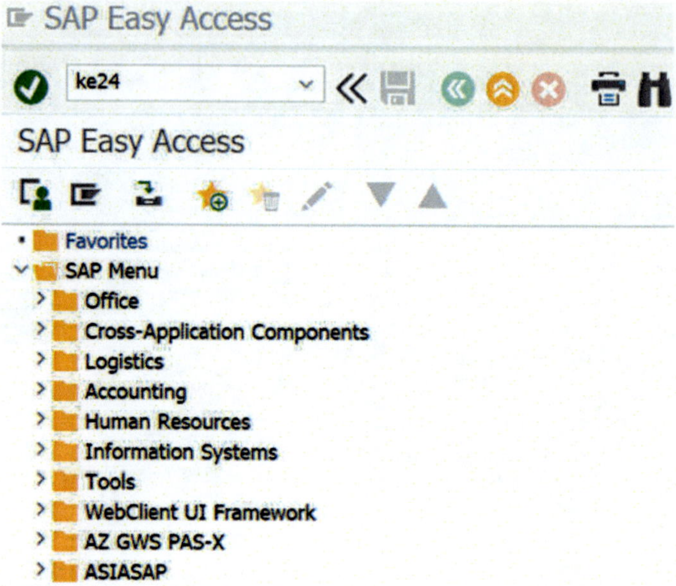

Fig. 3 SAP easy access

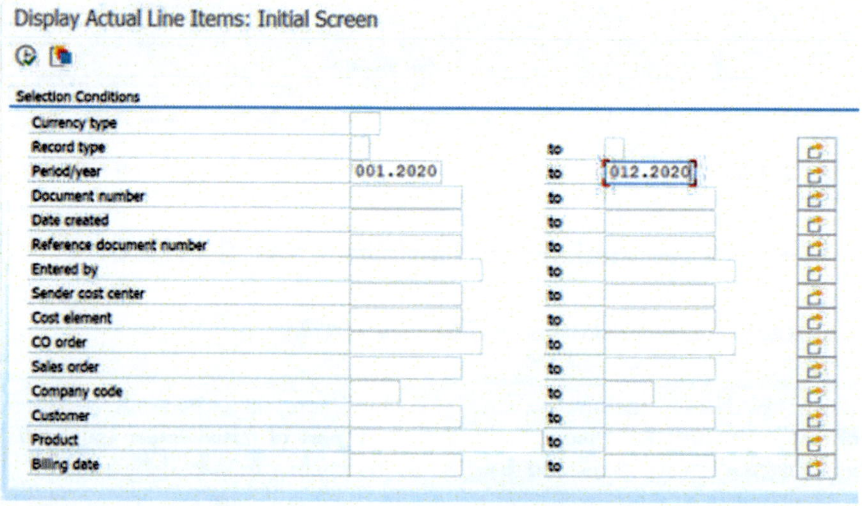

Fig. 4 Display actual line items

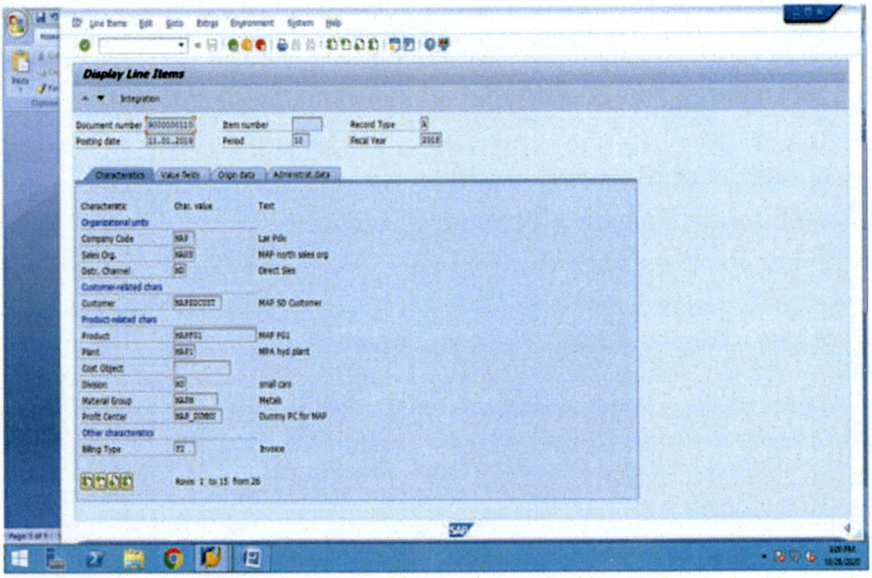

Fig. 5 Display actual line items list

4.2 *Profitability Analysis Document*

Profitability Analysis Document is prepared based on document number of line items in the HANA Database by selecting any single record in the list (Fig. 4). To prepare the profitability analysis, characteristic and value fields are considered. Characteristic fields include organization units, customer-related characteristics and product-related characteristics.

Value fields include sales quantity, cost of goods sold, gross profit or loss is considered in the analysis report as shown in Fig. 6.

4.3 *Use Cases of Preparing Profitability Analysis Document*

Use cases includes, let's consider document includes plant and company code as the selection conditions, following are the scenarios for the output for the fields revenue, COGS, GP are generated from the analysis.

Use case 1: When no selection, output report generated with the sum of revenue of all companies, sum of COGS, GPs are generated in the report.

Cost of Goods Sold (COGS), Gross Profit (GP) is calculated like this

$$\text{GP} = \text{Revenue} - \text{COGS}$$

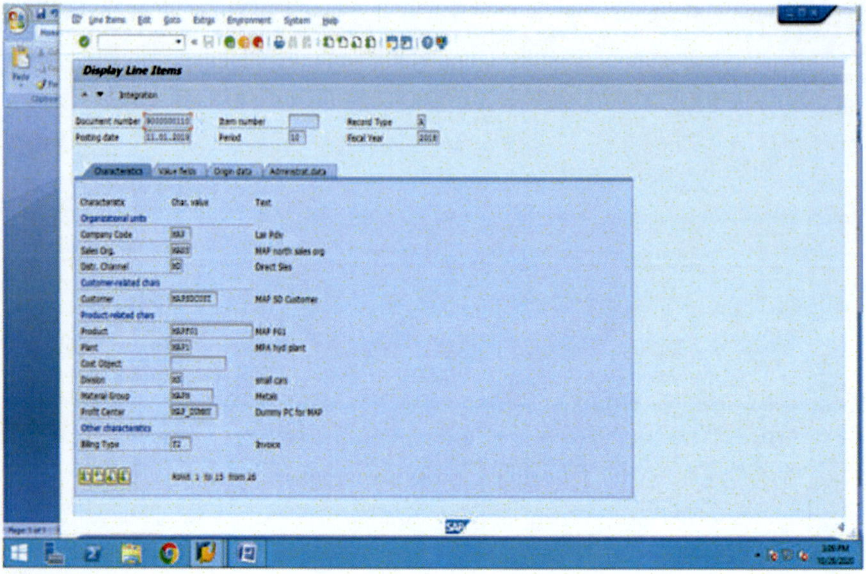

Fig. 6 Display line items

Use case 2: When company code is selected, output report result includes revenue is the sum of revenue of plants under that company, sum of COGS, GP is the same calculation.

Use case 3: When plant is selected, if output fields are considered as revenue, COGS, GP, Finance Expenses and Net Profit (NP) are generated.

$$\text{NP} = \text{GP} - \text{Finance Expenses}$$

In the use case 3, output results in the Revenue, COGS, GP, Finance Expenses, NP are the individual value of a particular plant and are returned in the report.

4.4 Data Representation in Fiori with S/4 HANA Database

Data are extracted from the S/4 HANA database using OData Services with annotations mapping in the service code, as shown in Fig. 7 is the Fiori Launch Pad, which is launched from the SAP Easy Access screen. Launch Pad includes the personas and inboxes, which are moved to the report page containing profitability report including all the selection conditions of Company Code, Sales Organization, Distribution Channel, Customer, Country, Product, Plant, Division, Profit Center, Sales Order and Sales Order Item. Selection results' report includes Revenue, Cost of Goods Sold (COGS), Gross Profit, Finance Expense, Net Profit (Fig. 8).

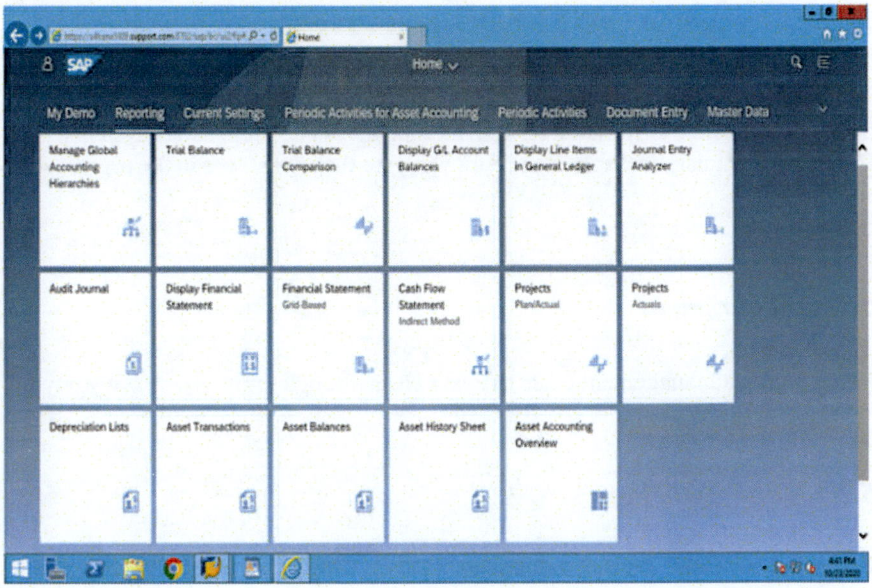

Fig. 7 Fiori LaunchPad

Fig. 8 Optimize business process
captured from online business app optimization processing

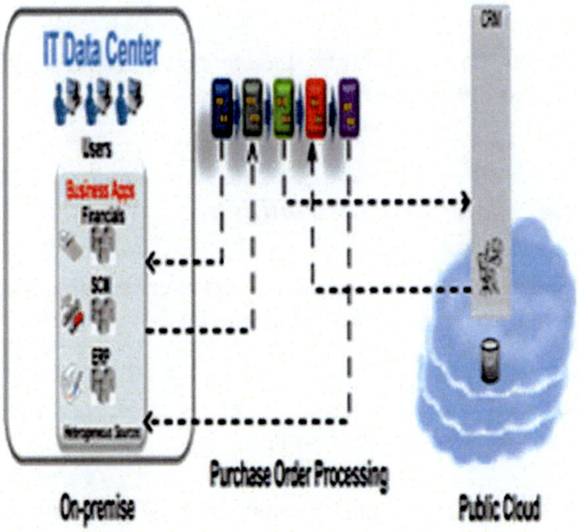

5 Advantages in Optimize Business Process Using SAP S/4 HANA

Sensitive data are transferred over the internet, across heterogeneous systems, SAP Cloud extends analytical solutions. Following are the advantages in the organization supply.

5.1 Procurement

Better contract management leads to lower costs through improved strategy buying.

5.2 Manufacturing

Product planning increases productivity.

5.3 Maintenance

Higher productivity is attained by better asset management and maintenance results are less during unexpected downtime.

5.4 Sales and Distribution

Updating resources based on customer requests, identifying earlier order requirements and appropriate actions are taken.

5.5 Customer Service

Return, replacement and repair after sales process are more efficient management activity, which is providing for the customer.

5.6 Product Management

Mobile device management processes lead to high-quality master data. With SAP S/4 HANA pays the new way for business, this enables to implement innovative technology. SAP Fiori gives a better mobile user experience, increases productivity and easy to use. As a future scope, this profitability analysis report can be designed in such a way, where the output report can be transformed rows into columns and columns into rows based on selection criteria.

6 Conclusion

To conclude, this paper provides the visualization of data in different cloud-based SAAS applications, by choosing ERP System in SAP focusing on cloud technologies. SAP S/4 HANA database is designed to process both structure and relational databases. Integration with these systems is highly challenging, as information is gathered from different modules of an organization data such as sales and distribution, product management and material management. Using Fiori elements front-end magic happens, these elements handle the UI code, so business logic and back-end services are handled. The Profitability Report generated by SAP S/4 HANA with Fiori UX can be used by any company that has an SAP system implemented.

References

1. SAP (2016) What is ERP. http://www.sap.com/solution/ what-is-erp.html
2. Venki Krishnamoorthy, Alexandra Carvalho (2014) Discover SAP: An Introduction to SAP, Beginner's Guide (3rd Edition). SAP Press. ISBN: 978–1592299874
3. SAP (2016) Integration Services. https://hcp.sap.com/capabilities/integration.html
4. Radovan Nieslanik (2016) Optimization of order and purchase process in SAP and integration with print system, Brno, Autumn
5. SAP Answers (2016) WEB IDE Cordova Print Plugin. https://answers.sap.com/questions/37377/web-ide-cordova-print-plugin. html
6. Matt Chudy, Luis Castedo (2014) Procurement with SAP MM - Practical Guide, SAP Materials Management. SAP Press. ISBN: 978-1592298402
7. Matt Chudy, Luis Castedo, Ricardo Lopez (2015) Sales and distribution in SAP ERP practical guide (2nd Edition), SAP SD. SAP Press. ISBN: 978-1493210541
8. Cloud9 (2016) Supported languages - Cloud9. https://docs.c9.io/docs/supported-languages
9. Panorama Consulting (2015) ERP REPORT. http: //go.panorama-consulting.com/rs/panoramaconsulting/images/2015%20ERP%20Report.pdf
10. SAP HANA Cloud Documentation (2016) Cockpit. https://help.hana.ondemand.com/help/frameset.htm?e47748b5bb571014afedc70595804f3e.html
11. SAP HANA Cloud Documentation (2016) November 2016. https://help.hana.ondemand.com/webide/frameset.htm?283a483cb4ef475f953e0066c5bfb791.html
12. SAP HANA Cloud Documentation. SAP HANA Cloud Connector. https://help.hana.ondemand.com/help/frameset.htm?e6c7616abb5710148cfcf3e75d96d596.html

13. SAP HANA Cloud Documentation (2016) Services in the Cockpit. https://help.hana.ondemand.com/help/frameset.htm?7613d9ce711e1014839a8273b0e91070.html
14. Frank Forndron, Thilo Liebermann, Marcus Thurner, Peter Widmayer (2006) mySAP ERP Roadmap: Business Processes, Capabilities, and Complete Upgrade Strategy. SAP PRESS. ISBN: 978- 1592290710

A Modified BAT Optimization Algorithm to Segment MRIs of Brain Subregions for Early Detection of Alzheimer's Disease

Sai Sindhuri Nasina and A. Rama Mohan Reddy

Abstract Alzheimer's disease (AZD) is considered as the most common neurodegenerative disease that generally affects people older than 65 years. It causes irreversible brain commotion that belatedly partitions the memory and thinking abilities on progressively collapsing the capability of accomplishing daily activities. Early detection of AZD symptoms slows down its progression to the next stage and eventually protects the patient from severity of the disease. It is evidenced from various research studies that, application of nature-inspired algorithms in the context of image processing, have depicted decent results in analyzing complex image datasets related to the medical domain. This paper presents a modified BAT optimization algorithm integrated with a novel histogram model to extract ROI (Region of Interest) from brain scan MRIs (Magnetic Resonance Images) to automatically detect and predict the progression of AZD. The experimental analysis presented in the paper indicates that the proposed modified BAT algorithm outperforms state-of-art optimization algorithms.

Keywords BAT optimization · Alzheimer's disease · Tensor-based feature extraction · MRI segmentation · Image processing · CNN

1 Introduction

In a progressive era, most people experience declines in memory, which trigger misplacement, breakdown of daily activities, language problems, etc. Examining the brain development and operation of the brain of elderly adults reveal that a significant brain alteration triggers changes in memory. The variations in neurological and neurobiological disorders are related to the era that triggers Alzheimer's disease (AZD) [1]. The ongoing deterioration of tissue contributes towards AZD and atrophy [2] in the brain subareas. In the first instance, the degradation of tissue occurs from the grey

S. S. Nasina (✉) · A. R. M. Reddy
Computer Science and Engineering, Sri Venkateswara University College of Engineering, Tirupati, India
e-mail: sindhuphd123@gmail.com

S. Jyothi et al. (eds.), *Proceedings of the 2nd International Conference on Computational and Bio Engineering*, Lecture Notes in Networks and Systems 215,
https://doi.org/10.1007/978-981-16-1941-0_32

matter (GM) and progresses to white matter (WM). The accelerated tissue disruption induces anatomical modifications in the brain, such as ventricles/extension, decreased Hippocampus (HC) and GM, which allows to identify AZD. HC distinguishes the right and left hemispheres of the cerebral cortex and allows contact between the hemispheres.

HC is a key region of the brain placed in the medium period lobes. HC generates fresh memories and restores existing memories and is called the memory portal. HC's key practice is to gather knowledge and transmit it to a long-term memory bank. HC is an important aspect of memory progress since it is linked to the memory directly. The reductions in HC are serious and are one of the most commonly diagnosed causes of AZD [3–6]. The study then has to be conducted by means of picture segmentation techniques to find the scale of HC. Different researchers tried many approaches, including multi-level thresholds and two-level thresholds for the HC sector [7–9]. Multi-level levels are a more sophisticated solution to addressing complicated challenges than the bi-level levels.

There are many articles on Alzheimer's disease detection using multiple imaging approaches. The patterns of patients and healthy individuals included in the volumetric MRI are calculated by feature-dependent morphometry (FBM) [9]. Another research suggested a picture analysis approach that indicates the existence of neurodegenerative disorders focused on the technique of fusion in the descriptive brain patterns [9]. The classification efficiency of a support vector machine (SVM) was evaluated on numerous data sets until the descriptive trends were determined. An additional research developed a system of classification dependent on multi-level brain partitions [10]. MRI data have been used for the classification of various brain stages using an SVM histogram-based feature. A new research has established a method for anatomical improvements in the hippocampus, amygdala, temporal planum and thalamus regions to be automatically detected [11]. Differences and correlations were recognized on MRI data dependent on grey histogram levels. In comparison to others, a recent research proposes to create cascading convolution neural networks (CNN's) to learn MRI and PET brain photos for Alzheimer's disease recognition on individual- and multi-level characteristics [12]. The selection of these methods influences the efficiency of machine-learning approaches. Most of the previous research implemented multiple methods on image samples. Few experiments used several data sets to train and test templates, which may decrease the model prediction efficiency due to a small number of samples. Fewer trials have sought immediately relative to more trials on the management of Alzheimer's disease.

The main intention of this study is to accurately predict Alzheimer's disease during its early stages on analyzing MRI data of the dementia patients using an optimization algorithm. The main contribution of this study is the application of tensor-based feature selection method along with a modified BAT optimization algorithm in the context of identifying the early stage of AZD. The rest of the paper is organized as follows: Section 2 provides a brief overview of various state-of-art machine-learning and optimization techniques utilized in the context of predicting AZD in the early stages. Section 3 details the proposed tensor-based feature extraction integrated modified optimization algorithm. Section 4 provides the details about the experimental analysis. Section 5 summarizes and concludes the contribution of this article and briefs about the scope of extending this research study.

2 Literature Study

AZD is a widespread dementia, contributing to increasing memory failure, reduced cognitive ability, forgetfulness etc. AZD influences the HC area and brain cortex in general. Because of brain tissue damage that begins from frontal and temporal lobes, anomalies are noticed and advance slowly through other areas of the mind. AZD is strongly related to amyloid b-protein aggregation [2, 13]. This relationship affects the composition of the brain and induces atrophy. The atrophy begins with GM and extends to WM and HC steadily. The decrease is large in the HC relative to the other areas. HC atrophy influences people in different forms as they may be closely related to memory [14, 15]. In the early phases of AZD, linear techniques are being attempted to avoid brain deterioration [16]. Therefore, the segmentation of the subzones to diagnose AZD at an early stage is important to analyze. Active appearance models, optimizing perceptions and Bayesian appearance methods, were aimed at the subcortical section framework, and 18 subcortical frameworks were applied [17]. Villain et al. clarify the interaction between AZD and brain subregions [18]. A variety of wrapping techniques include Free Surfer and HC-segmentation multi-atlas mark fusion, BET extraction of the mind and Quick segmentation of the brain tissue.

High efficiency was sought and seen [19]. A major source of study of AZD is the shrinkage of the HC area. Thus, the shifts of HC in terms of segmentation must be quantified. The strategies for segmentation may be classified into multi-level thresholds, for example. Multi-level thresholding approaches, in particular, have been commonly used in recent decades and are most popular [20–24]. Wind-Driven Optimization (WDO), PSO, cuckoo scan (CS), GA, etc. are relevant thresholding technologies [25, 26]. The explanation for preferring this approach of optimization is rather easy and local, scalable and feasible, optimal solution is avoided. The above-mentioned procedures for optimization are influenced by hunting but will not be leading. The natural lifestyle of LOA is focused on its character of cooperation [27]. Standard photos and AZD photos are listed with the CNN plus the CNN network [28, 29]. The HC area is divided into many optimization strategies, including GA, ABC, BAT, PSO and LOA, in this report. The findings of sectional areas are confirmed by overlapping and resemblance tests and are checked by the GT pictures. Finally, the AZD and routine tests are listed with the DL Classifier and are used for statistical measurements (SM) as a validation method.

Few papers are eligible for AZD as an ongoing study subject using MRI static photos identification and image recovery CBIR, for sure. CBIR applications of medical science capacities. The area is indeed doubtful and a big obstacle for science. The explanation is linked to the precise existence and subtlety of medical photos Modifications involving identification and analysis of this [19]. The key goal is to find a strong image of the quality of the picture by using extraction techniques that properly reflect scientifically essential knowledge. It afterwards allows the use of CBIR medical systems more relevant, more effective and relevant scientifically. A neural network was a multi-class network Ref. [20] marker to identify numerous AZD actions. At the first time, this numerous methods exposing the shape and contour have been used and extracted from MRI scans of the hippocampus area that used brain back regions to verify AZD [21] length. A multi-species technology [22] enabling experimental versatility accuracy of 82.45% has different characteristics, including grouping, reviews, distribution, and rating. The functionality is an input description not only contained the phases number disease but also led to a rating change.

In an earlier work, several classifiers to be included in the identification and were tested AZD progression estimation, utilizing RIM photography, with the development of a more stable CBIR [23, 24] method. The SVM was used for the early diagnosis of dementia and Alzheimer's stages by the application of biomarker brain function [23, 24]. Three were included sectional pictures cross-section, hippocampus, co-axial to the aspects of the cortex to classify From AZD. The outcome precision of this methodology amounted to 90.65% [21]. Deep learning algorithms were able to learn those depictions data from CNN, in particular [28]. Another study was recently carried out [22]. The usage of deep CNNs for problems was investigated and used CAD in the world of pharmacy.

3 Proposed Work

The proposed work includes five phases: image collection, pre-processing, segmentation, ground truth validation and classification.

Phase 1: Acquire brain MR pictures. In general, the standard of the hospital photos is low and pre-processing is necessary.

Phase 2: Next is the threshold and histogram equalization methods in which tensor-based feature selection method is used to remove the skull and improve picture consistency.

Phase 3: The third is to employ the segmentation mechanism in the HC field segment with a modified BAT optimization algorithm.

Phase 4: The divided HC has then been validated with the GT using overlapping steps, comparisons and the coefficient of association.

Phase 5: Ultimately, the categorization of typical patients with AZD who help to identify the disorder is achieved using the CNN classifier.

3.1 Tensor-Based Feature Selection

From the observation of the traditional feature extraction/selection method, SVD de- composition X = UVW, the data(columns) of U are the Eigen vectors of the matrix XX^t, whereas the data (rows) of W are the Eigen vectors of the matrix X^tX. Eigen vectors are orthogonal and, therefore, $U^tU = I$ and $WW^t = I$. by applying the dimensionality reduction, the approximation X = UnVnWn is minimizing the matrix representation error . $\in = \|X - X^\wedge\|^2 = \text{trace}\ \{(X - X^\wedge)(X - X^\wedge)^t\}$. Generalized version of SVD modifies the orthogonal constraints such that UtCuU = I and WCwWt = I, where Cu and Cw are the two constraint matrices. By performing the reduction process, the approximation is the one minimizing {trace {Cu(X-X^) Cw(X-X^)^t}.

Thus, generalized SVD is a very versatile tool by the consideration of constrain matrices that can be particularized to correspondence analysis. For the analysis of categorical variables (a generalized version of factor analysis), C_u is the relative frequency of the rows of the data matrix, X, and C_u is the relative frequencies of its rows, as in the discriminant analysis and canonical correlation analysis—is a procedure for analyzing two groups of continuous variables and performing simultaneously two-dimensionality reductions, so that the two new sets of features have maximum cross-correlation.

3.2 *Multi-Linear Singular Value Decomposition (MLSVD)*

Multi-Linear Singular Value Decomposition (MLSVD) [3] is one of the tensor-based decomposition method for the generalization of pairwise symmetric tensors through a symmetric Eigenvalue decomposition strategy. The concept of MLSVD is an appropriate generalization of the link the row (column) vectors and the right (left) singular values of the matrix. A simple representation of MLSVD is depicted as in Fig. 4.2. Let T be the tensor of size *I1X I2X I3 ……X IN* as a multilinear tensor matrix product of core tensor S with size *R1X R2XR3…… XRN*, *U(n)*factors of N, size *InXRn*

$$T \approx S.1^1\ U.\ 2^2\ U.\ 3^3\ U\ \ldots N^N\ U.$$

where N is a tensor matrix product of mode n (Fig. 1).

One of the fruitful developments of linear algebra in image processing is the prime concept of SVD representations for the analysis of image patterns in terms of features. Multi-Linear SVD [3] is one of the concepts of SVD with generalization of Tucker decomposition method (tensor decomposition). The MLSVD is a basis of an approximate generalization of the relation between two vectors column/row and left/right singular vectors of the matrices. In the concept of tensors, the matrix representations of the tensor are in the form of column/row vectors are stacked one after another, i.e. matrix unfolding.

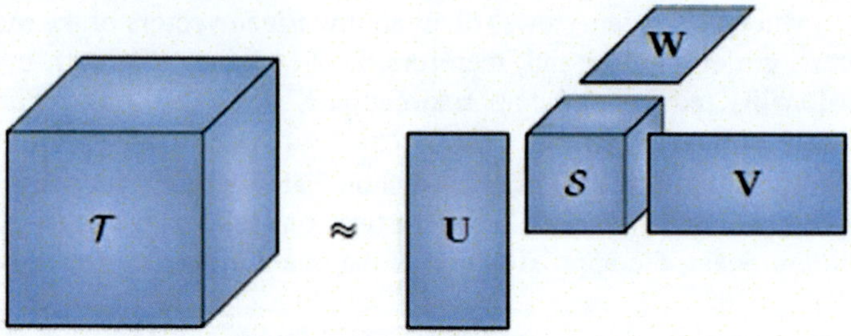

Fig. 1 Multi-linear singular value decomposition

3.3 Modified BAT Optimization Algorithm

In this article, we suggested an updated FBA Fuzzy Bat Algorithm to increase consistency and prevent the local solution from dropping. By removing all fats, we did not adjust its fitness value four times in a series by a new solution, provided by calculating the average of the five best solutions. Any bat has in FBA:

- The bat's location is the Xi (c* matrix N).
- The velocity of the bat is Vi (c*N matrix).

Other than fi, Ai, ri are the frequency, loudness and the respective emission amounts. repi parameter to count as much the same parameter repeated health benefit.

ALGORITHM: FUZZY BAT FOR SEGMENTATION

```
Initialize the bat population randomly with Xi I = 1 , 2,.,, n) as in fuzzy position
matrix;
Standardize Xi for each bat;
Defining Xi pulse frequency fi;
Initiate ri and Ai loudness pulse rates;
t < Full number of iterations or stop conditions
        Every bat does Bi
                Build a new approach by frequency change and speed upgrade.
                position;
                connected and normalized;
                If then rand > ri
                        Choose one of the best solutions;
                        Generate a local solution for the best solution
                        selected;
                        Bound and standardize;
                End
                Offer random flight a new solution;
                Achieving a defuzzing Xi linking matrix;
                If < Ai & Util(Xi) > Util(X) then Util(X)
                        Accept the modern route;
                        Rip up and decrease Ai;
                End
                Rate the bats and finding the strongest X of today;
        End
```

4 Experimental Analysis

Our analysis incorporates evidence from three public datasets: the neuroimaging program for Alzheimer's disease. The AIBL research and the Open Access Collection of the Australian Neuroscience, Biomarker and Lifestyle Tests. (ADNI) Studies in Photography (OASIS). Ses data sets are listed in eMethod 4 in addition. The T1w MRI was used. In both of these experiments, they are available. We will see comprehensive MRI protocols (Samper-González et al. 2018). Our studies have 1500 ADNI data sets for which a T1w MR picture is used. At least one visit had been possible. Conversion of the data sets obtained into a standard data structure is considered as a major task during the implementation.

In theory, the capacity of CNNs to remove automatically requires just minimum preprocessing functions on medium to a high standard. In AD, however, where data sets are minimal and, therefore, profound. It is not obvious that networks may be

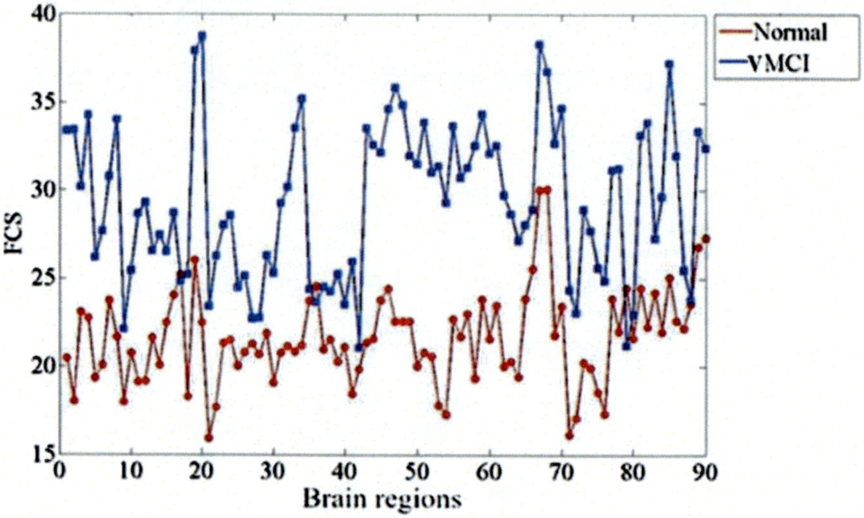

Fig. 2 Functional cognitive analysis of normal and VMCI patients

hard to train that they will gain from a larger network preprocessing. In comparison, previous experiments used different preprocessing techniques but without them. Assessment of their effects routinely. So we contrasted two separate pictures in the current analysis procedures for preprocessing: the "Minimal" protocol and the more detailed one. Both approaches contained racism correction of field and rescaling (optional) pressure. The "Minimal" procedure was also a linear method The "Extensive" entailed non-linear registration and skull-strip registration. The functional cognitive strength between the normal and VMCI subjects is denoted in Fig. 2.

BrainNet Viewer (https:/www.nitrc) in Matlab R2012a, Normal and VMCI (Jiao et al. 2018) are now shown in the.org/projects/bnv/) toolkit. There are a lot of individuals Full CFN relations and results are not obviously they 're all visualized after that. That is why just the first 200 maximum association practical relations As seen in Fig 7. It is difficult to locate no apparent distinction between Normal was noticed VMCI and The important functional partnerships are largely stable and localized which demonstrates strong internal DMN coupling network. Ground truth validation is presented in Fig. 3.

Scatter plot is used to verify the connection between The GT zone and the divided. The connection in general. The diagram meanings are −1 and +1. When the values are near 1 the segmented pictures are strongly associated with GT images; otherwise, they do not correlate. The best match values between GT and segmented HC area variables are represented by the FBA method in Fig. 3.

From Fig. 3, almost 96% of values are very similar to 1, which are seen in red line. 3, it is observed. From the photo, the ACC is strongly linked with GT values, which is 96% of the FBA segmented region. It is clear from this exam that the FBA technology is talented to precisely segment the exact HC area. This technique proves

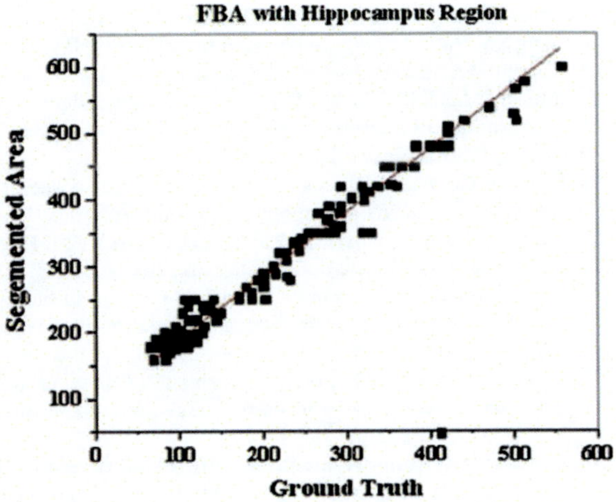

Fig. 3 Ground truth validation for the proposed algorithm

that the predicted approach is able to demonstrate the best accuracy relative to all other approaches in the HC segmentation field.

5 Conclusion

This paper presents a modified BAT optimization algorithm integrated with a novel histogram model to extract ROI (Region of Interest) from brain scan MRI's (Magnetic Resonance Images) to automatically detect and predict the progression of AZD. The experimental analysis presented in the paper indicates that the proposed modified BAT algorithm outperforms state-of-art optimization algorithms. The presented work could be extended the device envisaged would also do for MCI that enables the physician to At an early point, analyze AD. The scientific literature states that DA is not curable while the lifespan of patients may be improved by delaying the disease's progress. The clinical validation method is used to evaluate the AD prediction and finding. The emergence of intellectual disability will then be detected and controlled.

References

1. Frisoni GB, Fox NC, Jack CR, Scheltens P, Thompson PM (2010) The clinical use of structural MRI in Alzheimer disease. Nat Rev Neurol. https://doi.org/10.1038/nrneurol.2009.215
2. Younes L, Albert M, Miller MI (2014) Inferring change point times of medial temporal lobe morphometric change in preclinical Alzheimer's dis- ease. Neuro Image Clin. https://doi.org/10.1016/j.nicl.2014.04.009
3. Kurban T, Civicioglu P, Kurban R, Besdok E (2014) Comparison of evo- lutionary and swarm based computational techniques for multilevel color image thresholding. Appl Soft Comput J.

https://doi.org/10.1016/j.asoc.2014.05.037
4. Yang XS, Deb S (2009) Cuckoo search via Levy flights, in: 2009 World Congr. Nat Biol Inspired Comput NABIC 2009 – Proc. http://dx.doi.org/10. 1109/NABIC.2009.5393690
5. Suresh S, Lal S (2016) An efficient cuckoo search algorithm based multi- level thresholding for segmentation of satellite images using different objective functions. Expert Syst Appl. https://doi.org/10.1016/j.eswa.2016.03.032
6. Oliva D, Cuevas E, Pajares G, Zaldivar D, Perez-Cisneros M (2013) Multilevel thresholding segmentation based on harmony search opti- mization. J Appl Math 2013(1–24):230
7. Oliva D, Cuevas E, Pajares G, Zaldivar D, Osuna V (2014) A multi- level thresholding algorithm using electromagnetism optimization. Neurocomputing 139:357–381
8. Mousavirad SJ, Ebrahimpour-Komleh H (2017) Multilevel image thresholding using entropy of histogram and recently developed pop- ulation-based metaheuristic algorithms. Evolut Intell 10(1–2):45–75
9. Sarkar S, Das S, Paul S, Polley S, Burman R, Chaudhuri SS (2013) Multi-level image segmentation based on fuzzy-Tsallis entropy and diferential evolution. In: 2013 IEEE Int Conf Fuzzy Syst (FUZZ). IEEE, pp. 1–8
10. Bhandari AK, Kumar A, Singh GK (2015) Modifed artifcial bee col- ony based computationally efcient multilevel thresholding for satellite image segmentation using Kapur's, Otsu and Tsallis functions. Expert Syst Appl 42(3):1573–1601
11. Zhang J, Li H, Tang Z, Lu Q, Zheng X, Zhou J (2014) An improved quantum-inspired genetic algorithm for image multilevel thresholding segmentation. Math Probl Eng 2014:1–12
12. Mala C, Sridevi M (2016) Multilevel threshold selection for image segmentation using soft computing techniques. Soft Comput 20(5):1793–1810
13. Singh VK, Rashwan HA, Akram F, Pandey N, Sarker MM, Saleh A, Abdulwahab S, Maaroof N, Torrents-Barrena J, Rom- ani S, Puig D (2018) Retinal optic disc segmentation using conditional generative adversarial network. CCIA

Morphological and Molecular Identification of l-Asparaginase-Producing Marine Fungus Sarocladium kiliense

M. Bhargavi and R. Jaya Madhuri

Abstract The identification of L-asparaginase-producing fungal organisms has gained considerable awareness because of its less allergic properties. The current investigation deals with morphological and molecular identification of L-asparaginase-producing aquatic fungi confined from marine water. Samples were collected from Kerala coastal areas (Shangumugham Beach and Veli Back Water). It was confirmed that the colony appeared large, with filamentous shape, filamentous margin, convex elevation and whole colony filamentous. It is found to be filamentous fungi having conidia on staining with lactophenol. It was confirmed that the fungal isolate obtained from marine water has the potential to produce L-asparaginase, and the fungal species were identified as *Sarocladium kiliense*.

Keywords L-asparaginase · *Sarocladium kiliense* · Marine water · Kerala and filamentous fungi

1 Introduction

Pharmaceutical industries are in need of a new class of anti-cancer drugs from cheaper and safer sources due to the alarming occurrence of multiple drug-resistant pathogens. The terrestrial organisms acquired 50% of the existing drugs, which are used to care for human diseases, are derivatives of natural products [1]. However, due to nonstop and exhaustive study, water-based natural compounds have become a more promising source compared to land-based natural compounds, not only for industrial and commercial applications but also a pharmacological observation [2].

The bacterial L-asparaginase causes allergic reactions. Because of their continuous administration, several side effects arise, like anaphylaxis, pancreatitis, neurological disorders, leucopenia, hemorrhage or thrombotic events, coagulation abnormalities such as stroke [3]. Therefore, the discovery of a new L-asparaginase immunologically different from that of bacteria has been greatly desired. The eukaryotic

M. Bhargavi · R. J. Madhuri (✉)
Department of Applied Microbiology, Sri Padmavati Mahila Visvavidyalayam, Tirupati, India
e-mail: drjayaravuri@gmail.com

S. Jyothi et al. (eds.), *Proceedings of the 2nd International Conference on Computational and Bio Engineering*, Lecture Notes in Networks and Systems 215,
https://doi.org/10.1007/978-981-16-1941-0_33

microorganisms like yeast and filamentous fungi genera such as *Aspergillus, Penicillum* and *Fusarium* are generally described in scientific literature to produce L-asparaginase with less undesirable effects compared to bacterial L-asparaginase [4].

The L-asparaginase-producing fungal source is more suitable than other microbial sources because of its less allergic properties when it is employed in leukemia treatment. Acute lymphoblastic leukemia is the situation where the blood cells turn into cancerous and in this case, L-asparaginase is a wonderful curative measure for the cure of leukemia. The detection of L-asparaginase as an anti-cancer drug and its further investigation began in 1953 [5].

The identification of L-asparaginase-producing fungi at the molecular level is very essential in anti-cancer research and it will be helpful for the enormous manufacture of L-asparaginase. The present investigation deals with the morphological and molecular identification of fungus isolated from the marine water sample.

2 Materials and Methods

2.1 Sample Collection

Two marine deep water samples from a depth of 10 cm were collected at Shangamugam and Backwaters of Veli Beach in Kerala state, with a view to isolate potent L-asparaginase-producing microorganisms, as these areas have not been explored previously in this direction.

2.2 Isolation and Screening for L-Asparaginase

For the L-asparaginase-producing marine bacteria and fungi isolation, samples were subjected to tenfold serial dilution and plated on three different media, namely starch casein agar medium, glycerol asparagine agar medium and glucose asparagine agar (ISP-5 medium) [6]. The modified glycerol asparagine agar is added with 0.3 ml of 2.5% phenol red at pH 6.5 as an indicator which is used in the screening for L-asparaginase. The formation of pink color around the colonies indicated the production of L-asparaginase [7].

2.3 Identification of Fungal Species

Proper classification and characterization of microorganisms are very essential because it expands the scope for exploitation of industrially significant products.

A combination approach includes classical and molecular methods used for the identification of promising isolates.

2.4 *Classical Method*

To establish the identity of the current isolate with those reported in the literature, the classical approach is very important. The classical method includes the cultural, morphological and biochemical studies of a strain for initial classification. The strains were grown in ISP-5 broth for the following cultural [8] and morphological studies [9].

The following taxonomic properties were investigated

Colony characters: Size, shape, elevation, margin, appearance, color, consistency, etc. Morphological characters: Lactophenol cotton blue staining for fungi.

2.5 *Molecular Characterization*

As per [10], the entire DNA was isolated from new mycelia by means of a procedure. Primer pair off ITS4 (5-TCCTCCGCTTATTGATATGC-3) and ITS5 (5-GGAAGTAAAAGTCGTAACAAGG-3) were utilized to expand the ITS1, ITS2 regions and 5.8S gene. The process of enhancement was carried out in a 50 μl reaction volume. The thermal cycle consists of 5 min primary distortion at 95 °C, proceeded by 30 cycles of 1 min denaturation at 95 °C, 1 min primer annealing at 55 °C, 1 min extension at 72 °C and ending 10 min extension at 72 °C. The purification of PCR products was done by means of Gel Band Purification Kit and GFXTM PCR DNA following the protocol. The Macrogen Sequencing System (Korea) is used for the sequencing of purified PCR products with ITS4 and ITS5 as sequencing primers. Nucleotide sequences were aligned. By using CLUSTAL W, the neighbor-joining phylogenetic tree was constructed.

3 Results and Discussion

3.1 *Isolation and Screening for L-asparaginase*

Nine different isolates were obtained from the two marine water samples collected from coastal areas of Kerala (Shangumugham Beach and Veli Back Water). Out of these one strain was potent to the production of L-asparaginase and was screened for L-asparaginase production using modified glycerol asparagine agar. In accordance

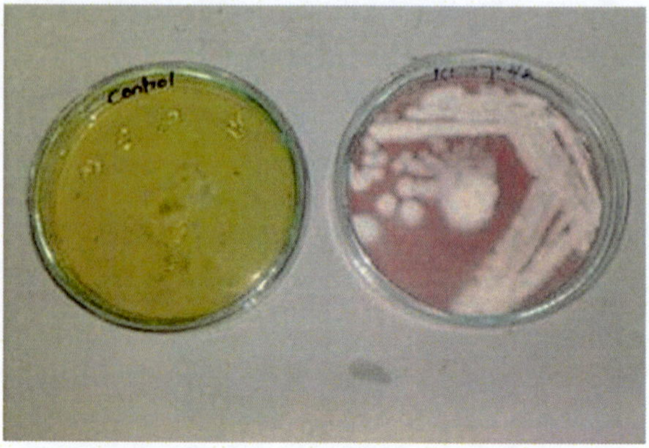

Fig. 1 Pink colour zone in the entire plate for L-Asparaginase production by the marine fungal isolate

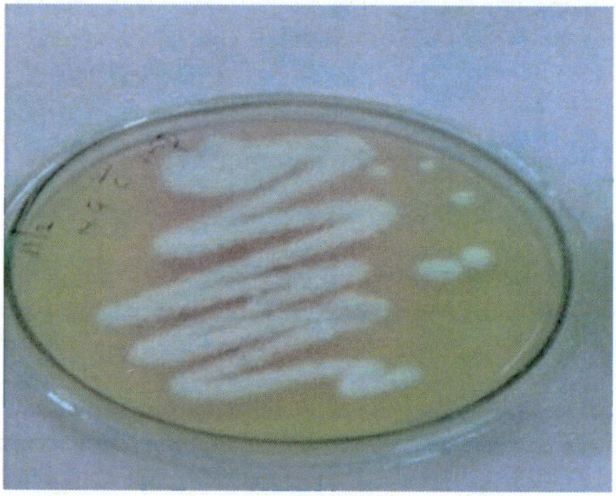

Fig. 2 Pink colour zone zone around the colony for L-Asparaginase production by the marine fungal isolate

with the progress of pink color around the colonies assessed, the isolated fungal organism from marine water has the possibility to produce the anti-cancer enzyme L-asparaginase [11]. The production of L-asparaginase was investigated in filamentous fungi such as *Aspergillus tamarii* and *Aspergillus tereus* (Figs. 1 and 2).

3.2 *Identification of Fungal Species*

Colony morphology:

At the early stage, colonies of marine fungus showed a large, filamentous, off-white color and on the reverse side of the plate appears yellowish-orange with convex elevation, filamentous margin and whole colony filamentous (Table 1 and Fig. 3).

Table 1 Colony morphology of potent marine fungi

S. no	$BKJM_2$	Colony morphology
1	Size	Large
2	Shape	Filamentous
3	Color	Off-white, reverse plate appeared yellowish orange
4	Margin	Filamentous
5	Elevation	Convex
6	Appearance	Whole colony filamentous

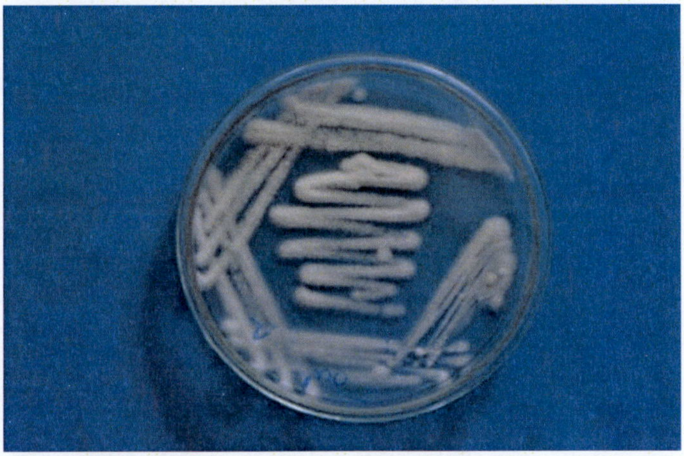

Fig. 3 Colony morphology of the positive marine fungal isolate

The authors of [12] studied a total of 364 fungal cultures that are morphologically distinct and identified them belonging to the genera *Aspergillus, Fusarium, Pencillium, Cladosporium* and *Streptomyces.*

3.3 Microscopic Characteristics of Marine Fungi

Microscopic observation of L-asparaginase-producing marine fungus found to be filamentous fungi having conidia (Figs. 4 and 5) on staining with lactophenol. According to [13] the L-asparaginase-producing fungal spp. (SMGR-F1) was isolated from the medicinal plants. The colony morphology of the lactophenol stained SMGR-F1 shows oval-shaped spores and belongs to the genus *Fusarium.*

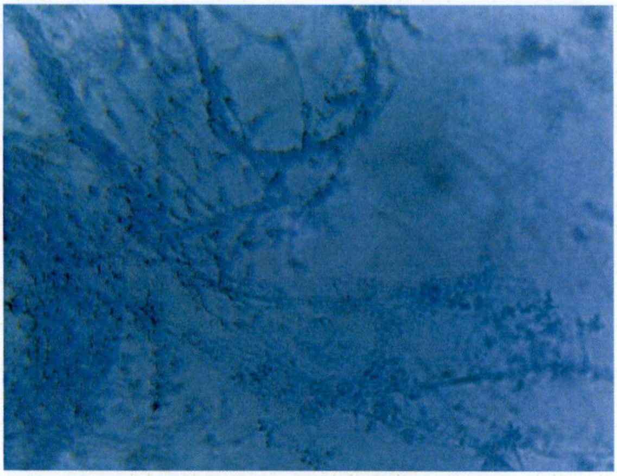

Fig. 4 Micrograph showing the mycelial appearance of fungi

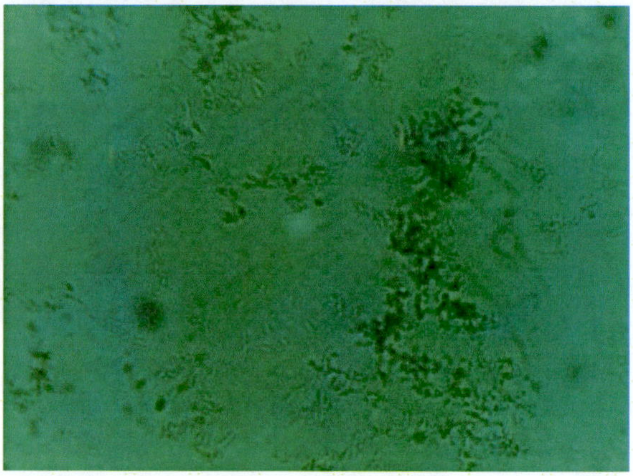

Fig. 5 Micrograph showing individual conidia

3.4 Molecular Characterization of the Potent Isolate

The comparison of the marine fungi search has shown no similarity with the sequence of the reference species of bacteria and actinomycetes. The molecular characterization of marine fungi isolate was done by using the molecular approach, i.e., internal transcribed spacer (ITS) sequencing. The ITS region of marine fungi was sequenced with primers ITS1 and ITS4 that resulted in around 596 bp of a fragment. The outcome sequences were compared with those in the National Center for Biotechnology Information (NCBI) Nucleotide Sequence Database by using the Basic Local Alignment Search Tool (BLAST) algorithm. The sequence data of marine fungi were 99% homologous with the sequence data of *Sarocladium kiliense*. The bioinformatics tool CLUSTAL W was used for the phylogenetic tree construction by the neighbor-joining method. The sequence data of marine fungi were submitted to NCBI with accession number KY970463 (Fig. 6). In the present study, 589 bp amplified PCR product was obtained. A slightly high molecular weight PCR product (661 bp) was achieved by Abbas [14]. Unlike the previous experiment internal transcribed spacer (ITS)-4 and 5 primers were selected by [15].

GenBank flate file of *Sarocladium kiliense*

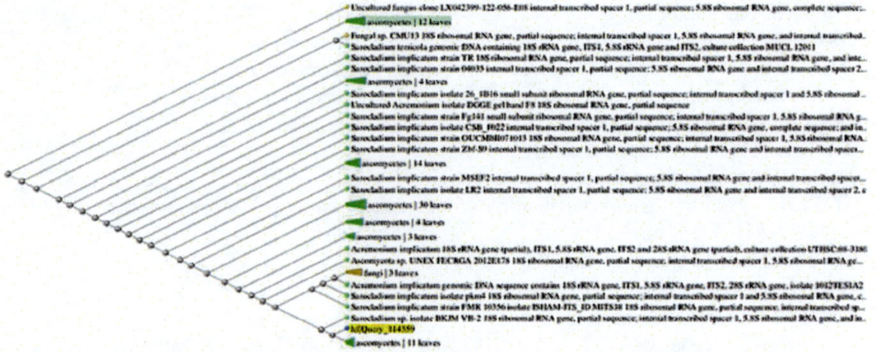

Fig. 6 *Sarocladium kiliense* phylogenetic tree constructed by the neighbor-joining method

LOCUS KY971463 574 bp DNA linear PLN 07-JUN-2017
DEFINITION Sarocladium sp. isolate BKJM VB-2 18S ribosomal RNA gene, partial sequence; internal transcribed spacer 1, 5.8S ribosomal RNA gene, and internal transcribed spacer 2, complete sequence; and 28S ribosomal RNA gene, partial sequence.
ACCESSION KY971463
VERSION KY971463.1
KEYWORDS.
SOURCE Sarocladium sp.
ORGANISM Sarocladium sp.
Eukaryota; Fungi; Dikarya; Ascomycota; Pezizomycotina; Sordariomycetes; Hypocreomycetidae; Hypocreales; Hypocreales incertae sedis; Sarocladium.
REFERENCE 1 (bases 1 to 574)
AUTHORS Madhuri,R.J.
TITLE Production and purification of anti-cancer enzyme L-Asparaginase
JOURNAL Unpublished
REFERENCE 2 (bases 1 to 574)
AUTHORS Bhargavi,M.S.
TITLE Direct Submission
JOURNAL Submitted (20-APR-2017) Applied Microbiology, Sri Padmavati Mahila Visvavidyalayam, Tirupati, Andhra Pradesh 517501, India
COMMENT ##Assembly-Data-START##
Sequencing Technology :: Sanger dideoxy sequencing
##Assembly-Data-END##
FEATURES Location/Qualifiers
source 1..574
/organism="Sarocladium sp."
/mol_type="genomic DNA"
/isolate="BKJM VB-2"
/isolation_source="marine water"
/db_xref="taxon:1707703"
misc_RNA<1..>574
/note="contains 18S ribosomal RNA, internal transcribed spacer 1, 5.8S ribosomal RNA, internal transcribed spacer 2, and 28S ribosomal RNA"
ORIGIN
1 tgtctccctt ctatggaagt aaaagtcgta acaaggtctc cgttggtgaa ccagcggagg
61 gatcattacc agagtgcctt ttggctctcc aacccactgt gaacatacct acgtttccct
121 cggcgggctc agcgcgttgc ggttccgccg cctcgcgtcc gccgggggca cccaaactcg
181 aatttacatc gtgtatctct gaggggcgaa agcccgtaaa acaaatgaat caaaactttc
241 aacaacggat ctcttggctc tggcatcgat gaagaacgca gcgaaatgcg ataagtaatg
301 tgaattgcag aattcagtga atcatcgaat ctttgaacgc acattgcgcc cgccggcact
361 cggcgggcat gcctgtccga gcgtcatttc aaccctcggg cccacccctc gcggggaacg
421 ggcccggcgt tggggaccgg aggccgcccc ggcggcaccc gccccctaaa ttcagtggcg
481 gtcgcgccgc agcctcccct gcgtagtagc acacctcgca ccggagagcg gcacggccac
541 gcctcgaaac cccccaattt tcaggttgac ctcg

4 Conclusions

The present study concluded that the coastal ecosystem water samples can provide a potential source of L-asparaginase-producing fungi and identified microscopically. It is a marine filamentous fungus, white in color, convex, entire plate filamentous, and on staining with lactophenol, aerial mycelia with individual conidia are observed. The marine fungi were molecularly identified by internal transcribed sequencing (ITS). It shows 99% homology with the genus Sarocladium species kiliense. KY97463 is the accession No. allotted by Genbank. It is the very first report of L-asparaginase production by the marine fungus *Sarocladium.*

References

1. Singh Y, Gundampati KR, Jagannadham MV, Srivastava SK (2013) Extracellular L- asparaginase from protease deficient *Bacillus aryabhattai* ITBHU02: purification characterization, and evaluation of antineoplastic activity in vitro. Biotechnol Appl Biochem 171:1759–1774
2. Venkata Siva Lakshmi T, Siva Mallika D, Jeevan aoms S, Kasturi K (2015) Marine L-asparaginase: a novel microbial therapeutic approach for cancer. Int J Pharm Sci Rev Res 32(2): 228–234
3. Shah AJ, Karadi RV, Parekh PP (2010) Isolation, optimization and production of L-asparaginase from Coliform Bacteria. Asian J Biotechnol 2(3):169–177
4. Boopathy SN, Kathiresan K (2010) Anticancer drugs from marine flora: an overview. J Oncol 21:41–86
5. Benchamin D, Sreejai R, Jensy Roshan F, Sujitha S, Kurup BS, Albert C (2019) Morphological and molecular identification of L-Asparaginase producing fungus from *Annona muricata.* J Pharm Innov 8(4): 18–20
6. Hyde KD, Jones EBG, Leaño E, Pointing SB, Poonyth AD, Vrijmoed LLP (1998) Role of fungi in marine ecosystems. Biodivers Conserv 7(9):1147–1161
7. Gulati R, Saxena RK, Gupta R (1997) A rapid plate assay for screening L-asparaginase producing microorganisms. Lett Appl Microbiol 24:23–26
8. Abdel-Fatah MK (1996) Studies on the asparaginolytic enzymes of *Streptomycetes.* Culture conditions for the production of L-asparaginase enzyme from *Streptomyces longsporus fl*avus (F-15) strain. Egypt J Microbiol 30: 247–260
9. Samson NP, Campa C, Legal L, Noirot M, Thomas G, Lokeswari TS, Kochko A (2006) Effect of primary culture medium composition on high frequency embryogenesis in different coffee species. Plant Cell Tissue Organ Cult 86(1):37–45
10. Jeewon R, Liew ECY (2004) Phylogenetic evaluation of species nomenclature of pestalotiopsis in relation to host association. Fungal Divers 17: 39–55
11. Sarquis MIM, Oliveira EMM, Santos AS, Costa GL (2004) Production of asparaginase by filamentous fungi. Mem Inst Oswaldo Cruz 99(5):489–492
12. Anil Kumar M, Savitha J (2015) (Purification and characterization of L-asparaginase therapeutic enzyme from a tropical soil Fungal isolate *Fusarium culmorum ASP-87*. MOJ J Proteomics Bioinform 2(6):1–6

13. Kumar R, Sedolkar VK, Triveni AG, Kumar MS, Shivannavar CT, Gaddad SM (2015) Isolation, screening and characterization of L-asparaginase producing fungi from medicinal plants. Int J Pharm Pharm Sci 8(1): 281–283
14. Abbas Ahmed MM, Dahab NAF, Taher TM, Fareed Hassan SM (2015) Production, purification and characterization of L-asparaginase from marine endophytic *Aspergillus* sp. ALAA-2000 under submerged and solid state fermentation. J Microb Biochem Technol 7(3): 165–172
15. Lapmak K, Saisamorn L, Sutheera T, Pairote W, Uraporn S (2010) L-asparaginase production by *Bipolaris* sp. BR438 Isolated from brown rice in Thailand. Chiang Mai J Sci 37(1): 160–164

Elicitation of Induced Systemic Resistance in Groundnut (*Arachis Hypogaea* L.) Plants by *Trichoderma* Spp. Against *Sclerotium Rolfsii*

Saraswathi Maddu and Jaya Madhuri Ravuri

Abstract *Sclerotium rolfsii* causes stem rot infection to groundnut with severe economic losses. *Trichoderma* spp. are free-living, non-virulent plant fungi living nearly in all agricultural soils and other environments. During this examination, the ability of *Trichoderma* spp. on induction of resistance against stem rot infection and their abilities to elicit induced resistance in groundnut plants were investigated. If the seeds are treated with the *Trichoderma* formulations, it enhanced the vigor index compared with the untreated seeds. In limited field trials, the seeds soaked with *Trichoderma* formulations show enhanced shoot and root length and also the yield of the plant compared with the untreated plants. The study of defense enzymes by RT-PCR revealed that *Trichoderma* administration considerably heightened the actions of chitinase, glucanase, phenylalanine ammonia-lyase, and peroxidase.

Keywords *Sclerotium rolfsii* · *Trichoderma* · Groundnut · Induced systemic resistance · Defense enzymes · Stem rot

1 Introduction

Groundnut (*Arachis hypogaea* L.) is a self-pollinating legume, intensively being cultivated in Andhra Pradesh which orders second, both in terms of land and yield. Though our country is the prime producer of this crop, harvests are low due to different types of diseases, and major losses are caused by fungal diseases—both aerial and soil-borne, which occur during various growth phases. *Sclerotium rolfsii*, the causative agent of stem rot infection in groundnut, is a facultative organism that lives in soil usually as *sclerotia* and represents the prime inoculum. It persists sustainably for numerous years and causes severe damage during any phase of plant development. It attacks the entire plant, but shoot infection is dangerous with losses

S. Maddu (✉) · J. M. Ravuri
Department of Applied Microbiology, Sri Padmavati Mahila Visvavidyalayam, Tirupati, A.P., India
e-mail: saraswathiphd@gmail.com

S. Jyothi et al. (eds.), *Proceedings of the 2nd International Conference on Computational and Bio Engineering*, Lecture Notes in Networks and Systems 215,
https://doi.org/10.1007/978-981-16-1941-0_34

over 10–25%, and yield damages up to 80% have been reported under severe conditions [5, 19]. This loss leads to less quantity foods, low-worth foodstuffs, financial lack for farmers and processors, and finally higher costs. *Trichoderma* spp. are free-living, plant-associated non-virulent fungi residing nearly in all agricultural soils and other environments [3, 8, 21]. In addition to disease-controlling capabilities, these fungi also have extra advantageous effects on plants; they inhabit and enter into the host roots, start a sequence of morpho-biochemical modifications that are considered to be part of the plant defense reaction, and lead to induced systemic resistance (ISR). In turn, the state of enhanced plant defense, appearing as a consequence of the activity of a biotic or abiotic factor, is termed induced resistance. Usually enhanced resistance is not limited to the site at which a particular agent acts. A signal induced at one site spreads systemically to other organs, as a result of which the whole plant is prepared for an attack by a pathogen. Two kinds of induced resistance with systemic range can be distinguished: systemic acquired resistance (SAR) and induced systemic resistance (ISR). SAR is induced by pathogens stimulating necrotic changes in the plant tissues. Their presence within tissues results in increased synthesis of salicylic acid (SA), which in turn activates the manifestation of pathogenesis-related (PR) genes [12].

In addition to the disease management, *Trichoderma* spp. also triggers the resistance (ISR) in the entire plant for various phytopathogens and reduces the infection rate [4]. *Trichoderma*-mediated ISR looks like pathogen prompted SAR, reduces uninfected plant portions, and is more resistant to wide-ranging microbes. Numerous *Trichoderma* spp. activate the salicylic acid (SA)-dependent SAR pathway by generating SA at the root exteriors, whereas other rhizobacteria activate various signaling pathways. Self-regulating salicylic acid (SA), jasmonic acid (JA) and methyl jasmonate (MeJA) play the main role in plant growth and in the induction of resistance to several stress influences [32]. SA and MeJA participate in several signal transduction mechanisms associated with plant defense and stimulate many enzymes and establishment of different types of defense compounds, including polyphenols, alkaloids, and pathogenesis-related (PR) proteins [6], conferring plant protection from pathogen-attack [7, 15]. A combination of ISR and SAR can raise a defense against pathogens that are resisted through both pathways, besides prolonged defense to a wider variety of pathogens than ISR/SAR alone. ISR-inducing *Trichoderma* spp. displays host specificity in regard to stimulating resistance [16, 29], which indicates that definite recognition between defensive bioagent and the plant is a requirement for the initiation of the signaling cascade leading to ISR.

The key objective of the current work is to evaluate *Trichoderma* spp. that is effective to stimulate systemic resistance in groundnut plants against stem rot disease by stimulation of defense enzymes.

2 Materials and Methods

2.1 *Isolation of* S. Rolfsii *and* Trichoderma

A total of 76 *S. rolfsii* isolates were isolated from the area of Sadhanavaripalem, Chittoor District from the fields of groundnut using potato dextrose agar (PDA) media, and pure cultures were preserved. All isolates were subjected to virulence tests by following Koch's postulates using TMV 2 groundnut plants as a host and highly virulent Sr 44 isolate selected for future use. A total of 20 *Trichoderma* isolates were isolated from disease-suppressive rhizospheric soil of groundnut fields of Chittoor district by using a modified *Trichoderma* specific medium. All the *Trichoderma* isolates (T_1-T_{20}) were screened against the pathogen by dual-culture technique and the *Trichoderma* isolate (T_{10}) was identified as a highly potent *Trichoderma* isolate, which inhibited the development of pathogen (100%) when related with the control and stored for further use to carry out all experiments.

2.2 Trichoderma *Growth for Inoculation*

Seven days-aged *Trichoderma* spore suspension was prepared by adding the sterile distilled water, and the suspension was filtered through muslin cloth and diluted to maintain 1×10^5 spores/ml.

2.3 *Pathogen Culture for Inoculation*

Three-day-old culture of a highly infectious isolate of *S. rolfsii* (S_{44}) which was maintained on PDA, was inoculated to the sterilized oatmeal sand medium and incubated at 29 ± 2 °C for 10 days, by which time the fungus shows well-developed visible growth.

2.4 *Groundnut Seed and Development*

High-quality groundnut TMV-2 seeds were collected from Sri Venkateswara Agricultural University, Tirupati and superficial disinfected with 0.1% mercuric chloride followed sterile distilled water and propagated in seed pans. After 10 days of germination, seedlings were used for further experimental studies.

2.5 Study for Induction of ISR by a Specific Strain of Trichoderma

To study the *Trichoderma* talc formulations effect on induction of resistance in groundnut plants, three treatments were maintained. The treatments are as follows:

Tt_1: Groundnut plants neither pathogen nor antagonist,
Tt_2: Groundnut plants treated with *S. rolfsii* only and.
Tt_3: Plants treated with *S. rolfsii* + *Trichoderma* isolate.

The tests were conducted twice and entirely randomized block design. In each treatment, 10-day-aged seedlings were planted per pot and triplicates were maintained for each test. Pathogen and antagonists were inoculated to the pots as mentioned in the above treatments, and samples from treatments (Tt_1, Tt_2, and Tt_3) were taken at different time intervals (0, 2, 5, 9, and 11 days) to detect the induced defense enzymes by RT-PCR.

2.6 Gene Expression of Defense-Related Enzymes in Groundnut Plants

Extraction of total RNA from plant tissues: To assess the up and downregulation of defense genes in five stages of disease development, three treatment samples of groundnut stems were collected and stored at −20 °C. RNA was purified separately from the triplicate samples of all treatments. Total RNA was purified with RNeasy Qiagen kit according to the directions of manufacturers and exposed to DNase (Ambion) to test genomic adulteration by PCR.

cDNA production using purified RNA: Reverse transcription reactions were done in a reaction mixture of 25 μl. The reaction mixture contains 2.5 μl of 5X buffer with $MgCl_2$, 2.5 μl of 2.5 mM dNTPs, 4 μl from oligo (dT) primer (20 pml/μl), 2 μg RNA, and 200 U reverse transcriptase enzyme (MLV, Fermentas, USA). RT-PCR augmentation was executed in a thermal cycler (Eppendorf, Germany), automated at 42 °C for 1 h and 72 °C for 10 min, and the cDNA was then stored at −20 °C for further use.

RT-PCR: The following primers were used: CHIT, forward 5’-TGGTCACTGCAACCCTGACA-3’ and reverse 5’-AGTGGCCTGGAATCCGACT-3’; forward 5’-TGGGAAGATTCTTATGAAGTGC-3’ and reverse 5’-CTCGTCAAATTTA CACATTGGT-3’; β-1,3-Glucanase, forward 5’-TCAATTATCAAAACTTGTTCGATGC-3’ and reverse 5’-AACCGGTCTCGGATACAACAAC-3’; PALase, forward 5’-ATGGAGGCAACTTCCAAGGA-3’ and reverse 5’-CCATGGCAATCTCAGCACCT-3’; PR1, forward 5'-AGTCGAGGTTTGGCCACCAT-3' and reverse 5'-CCACCTGATGCGCTCGATG-3'. PCR produces primers between 100 and 200 bp with TM = 60 °C. PCR reactions were run in 96-well plates. The reaction mix contained 2 ll of diluted cDNA, 10 ll SYBR_ Green mix (AB Applied Biosystems, Foster, CA), and 8 ll primers mix. Quantifiable real-time PCR was done by GeneAmp 7900HT Sequence Detection System (AB Applied Biosystems) at 95 °C for 10 min followed by 40 cycles at 95 °C for 15 s and at 60 °C for 1 min. Housekeeping gene tubulin (tub) was used as a reference gene in the experiments.

2.7 Preparation of Talc-Based Formulations of *Trichoderma* (T_{10}) Isolate for Seed Treatment

The talc-based preparation of *Trichoderma* (T10) isolate was prepared as per the method of Vidhyasekharan and Muthamilan [30]. Three-day-old culture of *Trichoderma* (T_{10}) isolate was inoculated to potato dextrose broth in a conical flask, kept 150 g at 28 °C for weekdays under shaking conditions. The total contents were stirred and propagules were set to be 2 × 108 cfu/ml, and the suspension was used for making formulations. pH of 1 kg talc powder was maintained by mixing the $CaCO_3$ @15 g/kg. The mixture of carboxymethyl cellulose (CMC) (10 g) and talc powder (1 kg) was sterilized for 30 min on alternative days. 400 ml of the antagonist culture comprising 1.5×10^8 cfu/g were added to the sterile talc powder under sterilized environments. Then the mixture was shade dried for 24 days at 28 °C and dried formulations were standardized; 50 g was distributed into each polypropylene bag as described [13].

2.8 Testing *Trichoderma* (T_{10}) Isolate Against the Disease Under Controlled and in the Limited Field Trials

Surface-sterilized groundnut seeds were soaked for 30 min in the slurry of the *Trichoderma* (T_{10}) and dried at 37 °C. Both under field conditions and in vitro conditions tested for the capability, *Trichoderma* isolate decreases stem rot infection, promotes the rise of seedlings, enhances plant height, fresh and dry weight, and crop yield.

3 Results and Discussion

In the current investigation among 76 isolates of *S. rolfsii*, S_{44} was identified as more virulent by pathogenicity test compared to other isolates of *S. rolfsii* isolated from the infected groundnut plants. When the plants were inoculated with S_{44} isolate, symptoms were more prominent and death of the plants occurred within 11 days of inoculation. All the 20 *Trichoderma* (T_1–T_{20}) isolates inhibited the growth of the *S. rolfsii* by dual culture method. T_{10} isolate greatly inhibited (100%) the growth of pathogen compared to all other isolates, possibly may be due to the secretion of huge amounts of antifungal metabolites by the fungal agent. Many workers reported that the biological control activity of the *Trichoderma* against *S. rolfsii* varied in their effectiveness to control *S. rolfsii* [11, 31, 32] by the mechanisms of mycoparasitism and excretion of bioactive products [9].

3.1 Testing of *Trichoderma* (T_{10}) Isolate Against Stem Rot Disease in Limited Field Trials

Three treatments were followed to study the in vitro and in vivo effect of *Trichoderma* (T_{10}) isolate talc formulations on stem rot disease incidence. *Trichoderma* (T_{10}) isolate inoculation improved the seed development compared to treatment 2 (Tt_2) with the maximum rate of seed germination (Fig. 1). The incidence of stem rot was maximum in the treatment 2 (Tt_2) and minimum in seeds soaked with the *Trichoderma* (T_{10}) isolate. In the Tt_3 treatment where the seeds were soaked in the slurry of talc formulations of *Trichoderma* (T_{10}) isolate, all the seeds were germinated. In the Tt_2 treatment, the seeds were rotten and all the seeds failed to germinate. *Trichoderma* has been shown to improve nutrient uptake by plants. Crops just grow bigger and healthier. Host establishment by *Trichoderma* spp. is recognized to augment plant development and guard host plants against phytopathogens by stimulation of systemic acquired resistance [10].

Shoot and root measurements were augmented related to control plants and also had advantageous effects on the fresh and dry shoot and root weights in treated seeds (Fig. 2 and Table 1). The vigor index was highest in Tt_3 and lowest in Tt_2 treatments. Vitti et al. [31] reported that *Trichoderma* application increased the plant progression, due to its strong antipathogenic action, and directly through the (i) stimulation of the biosynthesis of phytohormones, (ii) enhanced solubilization and intake of nutrients, (iii) root toughening and expansion, (iv) improvement in the

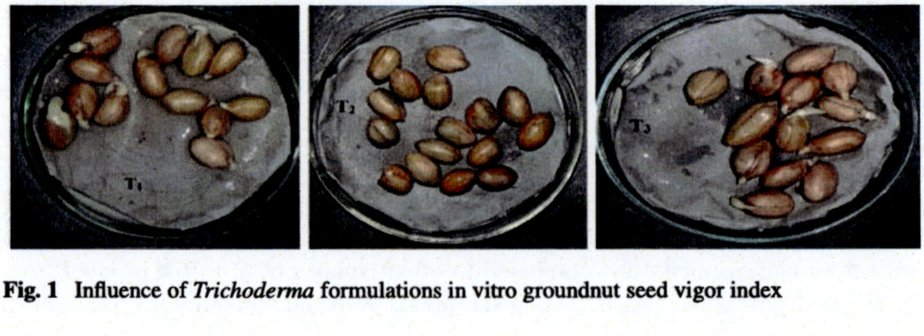

Fig. 1 Influence of *Trichoderma* formulations in vitro groundnut seed vigor index

Fig. 2 Influence of *Trichoderma* formulations in vitro groundnut seedling growth

amount of carbohydrate metabolism, and (v) photosynthesis increase [25, 27]. In our study, *Trichoderma* isolate induced growth in Tt_3 treatment due to an increase of PR-proteins and defense enzymes [4, 23, 26].

Under greenhouse conditions, *Trichoderma* isolate was shown to be efficient in decreasing the disease effectiveness in groundnut when applied along with the *S. rolfsii* compared to soil infested only with the pathogen (Fig. 3). *Trichoderma* also significantly showed the effect of the treated plants on yield, and the number of pods per plant compared to untreated plants. Several researchers reported that the application of *Trichoderma* talc formulation inhibits various pathogens [1, 11]. Marco and Osti [17] stated that mortality of vine plants improved because application of *Trichoderma* induces morpho-physiological features of grapevine and improved its tolerance to stress-related diseases. Singh et al. [24] conveyed that the efficacy of a *Trichoderma harzianum* formulation in reduction of disease incidence enhanced crop yield against wilt disease of cumin.

3.2 Related Gene Expression of Defense Enzymes in Groundnut Inoculated with S. Rolfsii *and* Trichoderma *Sp.*

Differential display and real-time PCR analysis with four different primers carried out for host samples procured at prescribed time intervals to evaluate the appearance of augmented enzyme plant samples. As shown in Fig. 4a–d, expressional variation was observed among the treated and untreated host samples and stated that various upregulated genes were detected in treated and untreated plants.

Table 1 Influence of *Trichoderma* in vitro on growth parameters of groundnut

Treatments	Shoot length (cm)	Shoot fresh weight (mg)	Shoot dry weight (mg)	Root length (cm)	Root fresh weight (mg)	Root dry weight (mg)	Vigor index (%)	Yield (No. of pods/ plant)
Untreated	9.44 ± 0.38c	290.67 ± 12.58c	19.47 ± 1.00c	2.05 ± 0.19c	11.93 ± 1.80c	2.33 ± 0.32c	100	64
Treated with *Trichoderma* (T_{10}) isolate	12.83 ± 0.47a	583.27 ± 7.13a	44.40 ± 3.14a	3.18 ± 0.39a	33.83 ± 2.08a	4.02 ± 0.25a	60	18

All the standards characterize the mean of triplicates ± standard deviation

Fig. 3 Influence of *Trichoderma* formulations on groundnut growth and yield in limited field trials

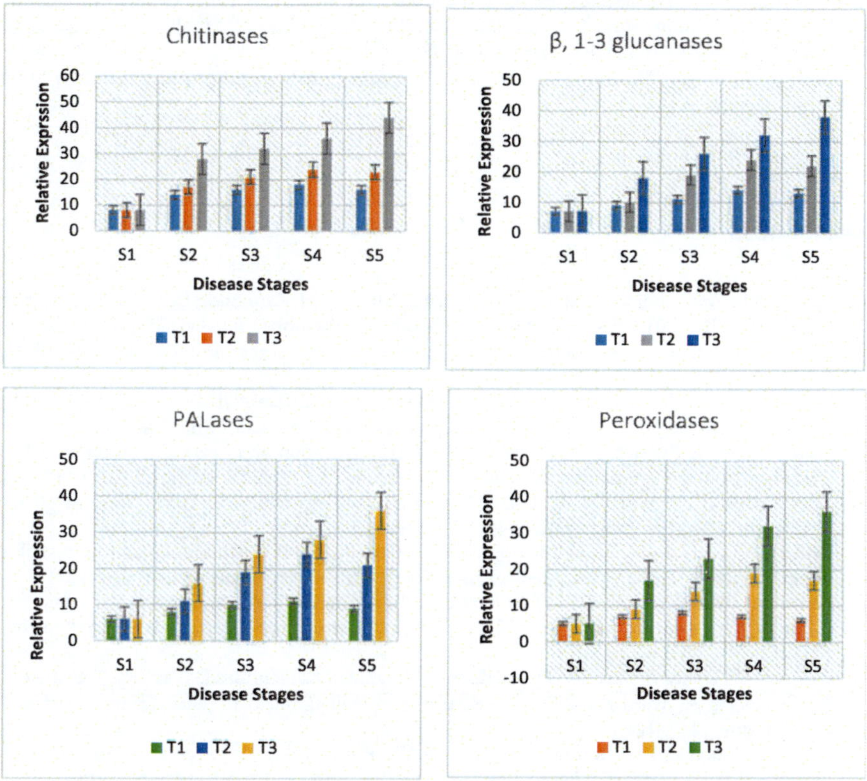

Fig. 4 Relative expression of defense enzymes in groundnut stems treated with *Trichoderma* sp

For analysis of enhancement of defense genes triggered by *Trichoderma* sp., post inoculated host samples were harvested from three treatments (Tt_1, Tt_2, and Tt_3) at five disease stages and quantified by real-time PCR. In this investigation chitinases, glucanases, PALases, and peroxidases encoding gene expression levels were studied.

Several workers reported that infection of *Trichoderma* spp. host encourages various plant defensive mechanisms and triggers the resistance against pathogen infection [18, 20, 22, 28]. Benito et al. [2], who considered that *Botrytis cinerea* triggers the differential display expression in tomato plants, also stated that various pathogens stimulate the enzymes which fight randomly against plant pathogens. Kiba et al. [14] applied the differential display technique to separate fragments of genes that are controlled in tobacco, by pathogenic strains of *R. solanacearum* OE1-1. Alizadeh et al. [1] described that the mixture of *Trichoderma harzianum* Tr6, and *Pseudomonas* sp. Ps14, Tr6 and Ps14 considerably induced the highest resistance in the host upon challenge with *Fusarium oxysporum f.* sp. *radices cucumerinum.*

Acknowledgements The authors are very much thankful to the University Grants Commission (UGC) of India for fellowship. They are also thankful to Sri Padmavati Mahila Visvavidyalayam for giving the amenities to complete research work.

Conflict of Interest On behalf of all authors, the corresponding author states that there is no conflict of interest.

References

1. Alizadeh H, Behboudi K, Ahmadzadeh M, Javan-Nikkhah M, Zamioudis C, Corné MJ, Pieterse P, Bakker AHM (2013) Induced systemic resistance in cucumber and Arabidopsis thaliana by the combination of *Trichoderma harzianum* Tr6 and *Pseudomonas* sp. Ps14. Biol Control 65(1): 4–23
2. Benito EP, Prins T, van Kan JA (1996) Application of differential display RT-PCR to the analysis of gene expression in a plant-fungus interaction. Plant Mol Biol 32:947–957
3. Budiarti SW, Widyastuti SM, Margino S (2009) Purification and characterization of β-1, 3 glucanase from the antagonistic fungus *Trichoderma reesei.* Hayati J. of Biosci 16(3):115–119
4. Choudhary DK, Prakash A, Johri BN (2007) Induced systemic resistance (ISR) in plants: mechanism of action. Indian J Microbiol 47: 289–297
5. Cilliers AJ, Herselman L, Pretorius ZA (2000) Genetic viability within and among mycelial compatibility groups of *Sclerotium rolfsii* in South African. Phytopath 90(9):1026–1031
6. Creelman RA, Mullet JE (1995) Jasmonic acid distribution and action in plants: regulation during development and response to biotic and abiotic stress. Proc Natl Acad Sci USA 92:4114–4119
7. Delaney TP, Uknes S, Vernooij B, Friedrich L, Weymann K, Negrotto D, Gaffney T, Gut-Rella M, Kessmann H, Ward E (1994) A central role of salicylic acid in plant disease resistance. Science 266: 1247–1249
8. Ellen CG, Robert FHD, Barbosa AM, Maria LCS, Roberto S (2011) Production of β-(1, 3)—glucanases by *Trichoderma harzianum* Rafai: Optimization and application to produce gluco-oligosaccharides from paraqmylon and pustulan. J Ferment Technol 1(1):1–5
9. Hafez EE, Hashem M, Mahmoud M, Balbaa El-Saadani MA, Ahmed A (2013) Induction of new defensin genes in tomato plants via pathogens-biocontrol agent interaction. J Plant Pathol Microbiol 4(3):1–9
10. Harman GE, Petzoldt R, Comis A, Chen J (2002) Interactions between *Trichoderma harzianum* strain T22 and maize inbred line Mo17 and effects of these interactions on diseases caused by *Pythium ultimum* and *Colletotrichum graminicola.* Phytopathol 94(2):147–153
11. Hoitink HAJ, Madden LV, Dorrance AE (2006) Systemic resistance induced by *Trichoderma* spp: interactions between the host, the pathogen, the biocontrol agent, and soil organic matter quality. Phytopathol 96(2):186–189

12. Jankiewicz U, Kotonowicz M (2012) The involvement of *Pseudomonas* bacteria in induced systemic resistance in plants. Appl Biochem Microbiol 48(3):244–249
13. Jayaraj J, Radhakrishnan NV, Velazhahan R (2006) Development of formulations of *Trichoderma harzianum* strain M1 for control of damping-off of tomato caused by *Pythium aphanidermatum*. Archieves Microbiol 39(1):1–8
14. Kiba T, Henriques R, Sakakibara H, Chua NH (2007) Targeted degradation of Pseudo-response regulator5 by an SCFZTL complex regulates clock function and photomorphogenesis in *Arabidopsis* thaliana. Plant Cell 19:2516–2530
15. Kozlowski G, Buchala A, Metraux JP (1999) Methyl jasmonate protects Norway spruce [*Picea abies* (L.) Karst.] seedlings against *Pythium ultimum* Trow. Physiol Mol Plant Pathol 55: 53–58
16. Leeman M, Ouden FM, Pelt JA, Cornelissen C, Matamala-Garros A, Bakker PAHM, Schippers B (1996) Suppression of *Fusarium* wilt of radish by co inoculation of fluorescent *Pseudomonas* spp. and root-colonizing fungi. Eur J Plant Pathol 102: 21–31
17. Marco SDI, Osti F (2007) Applications of *Trichoderma* to prevent *Phaeomoniella chlamydospora* infections in organic nurserie. Phytopathol Mediterr 46:73–83
18. Meyer GD, Joseph B, Elad Y, Monica H (1998) Induced systemic resistance in *Trichoderma harzianum* T39 biocontrol of *Botrytis cinerea*. Eur J Plant Pathol 104:279–286
19. Okereke VC, Wokocha RC (2007) *In vitro* growth of four isolates of *Sclerotium rolfsii* Sacc in the humid tropics. Afr J Biotechnol 6(16):1879–1881
20. Pieterse CMJ, van Wees SCM, van Pelt JA, Knoester M, Laan R, Gerrits H, Weisbeek PJ, van Loon LC (1998) A novel signaling pathway controlling induced systemic resistance in arabidopsis. The Plant Cell 10: 1571–1580
21. Sahin S, Ismail O, Haci HB (2013) Purification and characterization of endo-β-1, 4 glucanase from local isolate *Trichoderma ouroviride*. Inter J Biosci Biochem Bioinfor 3(2):129–132
22. Shoresh M, Harman GE (2008) The molecular basis of shoot responses of maize seedlings to *Trichoderma harzianum* T22 inoculation of the root: a proteomic approach. Plant Physiol 147:2147–2163
23. Siddaiah CN, Satyanarayana NR, Mudili V, Gupta VK, Gurunathan S, Rangappa S, Huntrike SS, Srivastava RK (2017) Elicitation of resistance and associated defense responses in *Trichoderma hamatum* induced protection against pearl millet downy mildew pathogen. Sci Rep 7: 1–18
24. Singh HB, Shishirsrivastava Anand S, Katiya RS (2007) Field efficacy of *Trichoderma harzianum* application on wilt disease of cumin caused by *Fusarium oxysporum* f. sp. *Cllnlini.* J Biol Control 21(2): 317–319
25. Sofo A, Tataranni G, Scopa A, Dichio B, Xiloyannis C (2012) Direct effects of *Trichoderma harzianum* strain T-22 on micro propagated GiSeLa6(R) (*Prunus* spp.) rootstocks. Environ Exp Bot 76: 33–38
26. Souza TP, Dias RO, Silva-Filh MC (2017) Defense-related proteins involved in sugarcane responses to biotic stress. Genet Mol Biol 40(1): 360–372
27. Stewart A, Hill R (2014) Applications of *Trichoderma* in plant growth promotion. In: Gupta VK, Schmoll M, Herrera Estrella A, Upadhyay RS, Druzhinina L, Tuohy MG (eds) Biotechnology and biology of trichoderma. Amsterdam: Elsevier, pp 415–425
28. Vallad G, Goodman RM (2004) Systemic acquired resistance and induced systemic resistance in conventional agriculture. Crop Sci 44:1920–1934
29. Van Wees SCM, Van der Ent S, Pieterse CMJ (2008) Plant immune responses triggered by beneficial microbes. Curr Opin Plant Biol 11:443–448
30. Vidhyasekaran P, Muthamilan M (1995) Development of formulations of *Pseudomonas fluorescens* for control of chickpea wilt. Plant Dis 79:782–786
31. Vitti A, Pellegrini E, Nali C, Lovelli S, Sofo A, Valerio M, Scopa A, Nuzzaci M (2016) *Trichoderma harzianum* T-22 induces systemic resistance in tomato infected by *Cucumber mosaic virus*. Front Plant Sci 7:1520
32. Zehra A, Meena M, Dubey MK, Aamir M, Upadhyay RS (2017) Synergistic effects of plant defense elicitors and *Trichoderma harzianum* on enhanced induction of antioxidant defense system in tomato against *Fusarium* wilt disease. Bot Stud 58:44

Applications of Anomaly Detection in Manufacturing

P. Balasubramanyam and K. Sreenivasa Murthy

Abstract Anomaly detection is one of the very predominant activities in manufacturing and it has many practical applications, like scrap reduction, quality testing and compliance testing. Many machine learning applications are available for anomaly detection in the literature and they are performing a decently good job. In the current study, we are describing various applications of anomaly detection in manufacturing domains, like automotive, oil and gas, pharma and consumer goods, and also some of the latest methods available in the literature to perform anomaly detection. We also highlighted some of the gaps and future scope of work in anomaly detection topics.

Keywords Anomaly detection · Machine learning models · Applications in manufacturing · Random forest and K-means clustering

1 Introduction

As per the Gartner predictions, by 2020, the number of connected IIoT devices will reach $20.8 billion. McKinsey confirms this with a report that predicts the potential economic impact of IIoT being $4 trillion to $11 trillion a year by 2025. Therefore, it is very evident that billions of connected things produce trillions of GBs of data at our disposal for analysis. This huge explosion of data presents different types of challenges as well as very good opportunities for us to implement innovative processes around it. Here comes the key analytical technique called "anomaly detection" which will, in turn, change the way manufacturing is operating for longer periods [1].

Anomaly detection is one of the very popular techniques that has many applications in the manufacturing domain. The main output of applying these techniques

P. Balasubramanyam (✉)
Goa Institute of Management, Goa, India
e-mail: balasubramanyam@gim.ac.in

K. Sreenivasa Murthy
Department of Management, Central University of Tamil Nadu, Thiruvarur, India
e-mail: sreenivasamurthy@cutn.ac.in

S. Jyothi et al. (eds.), *Proceedings of the 2nd International Conference on Computational and Bio Engineering*, Lecture Notes in Networks and Systems 215,
https://doi.org/10.1007/978-981-16-1941-0_35

is mostly in the form of cost reduction or improvement in productivity. This topic found its root for many years in the literature and a lot of research happened about the machine learning algorithms for anomaly detection.

1.1 Definition of Anomaly

An anomaly is defined as an event, process, or item that behaves entirely in a different manner from the rest of the group [1]. It shows some special characteristics that are not present in the remaining group. The process of identifying such special, surprise, and unexpected patterns in the data is known as anomaly detection.

1.2 Difference Between Anomaly and Outlier

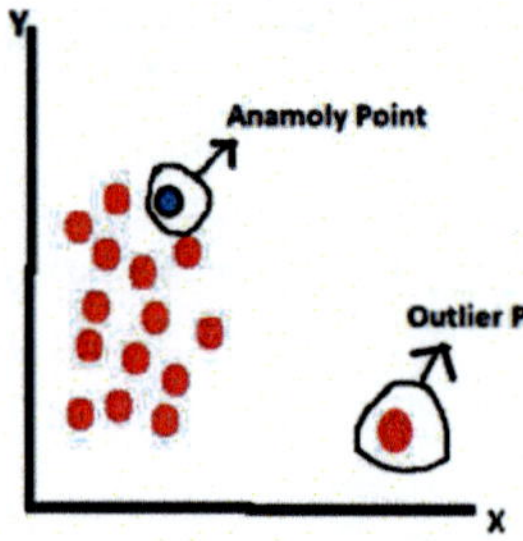

An anomaly exhibits some special property or characteristics that distinguish it from the rest of the group, and the outlier is some value that is numerically away from its central value. In most of the literature, these two words are used as synonyms to each other and did not highlight much difference between them. Most of all outliers are treated as anomalies but the converse may not be true always. A plot showing anomaly and outliers might look as follows.

2 Types of Anomalies

2.1 Point Anomaly

If a single point from the entire data is lying very far from the rest of the data points, then it is called point anomaly. This is very easily observed in scatter plots or time series plots. It is just a single observation whose distance from the center or mean is very huge.

2.2 *Contextual Anomaly*

If a single point in the data is a little far from the mean, only at certain points of context, like certain timeframe, certain length of a process, etc., then it is called contextual anomaly. In other words, this is an anomaly only to that particular context and in other times, it is not an anomaly in the data.

2.3 *Collective Anomalies*

A sequence/group of data points within a data range is called collective anomalies if these values as a group slightly move away from the normal center point, but the values of each point in this group of points may not be an anomaly when they are considered alone. Usually, this might occur due to a sudden shift in the process or temporary change in the machine working or due to faulty supply of power to the respective machine.

These three types of anomalies can be viewed in the below time-series event over a period of time (Fig. 1).

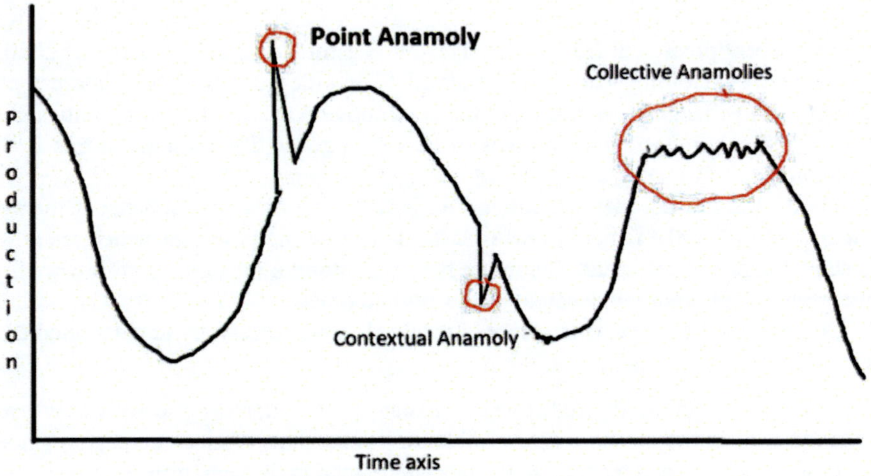

Fig. 1 Point, contextual and collective anomalies

3 Applications of Anomaly Detection in Manufacturing

Manufacturing and product development industries, such as iron and steel, automotive, chemicals and fertilizers, involve many sequential and parallel processes, and those processes generate a huge amount of data. In this environment, they always target at productivity improvement and reduction in scrap through several advanced analytical tools. One more key area is predictive maintenance (or just-in-time maintenance) where the maintenance schedule is decided based on data it generates through its continuous functioning. In most of the above practical scenarios, anomaly detection is one of the very popular analytical techniques being used successfully and it results in higher productivity at a low cost of investment.

3.1 *Applications of Anomaly Detection in Automotive*

From a recent prediction from the analysts, it is estimated that by the end of 2020, more than 0.25 billion cars are connected to internet and industrial internet of things (IIoT) changing everything in the automotive domain. The opportunities are grown bigger through huge volumes of new data generation throughout this IIoT technologies.

For the last few decades, the automotive industry has created excellence in manufacturing through the fantastic new industrial engineering processes which are now assumed to be not sufficient. The future here is to monitor the part's performance by themselves and order replacement at appropriate times.

Some of the key areas where anomaly detection is successfully implemented are as follows:

- Anomaly detection for predictive maintenance is the topmost priority area where anomalies in the machine data can identify the signals of machine performance through which maintenance schedules are defined at the right time.
- Anomalies in the quality data are one more application where the defects in products can be easily separated out through anomaly detection and the highest quality can be ensured in the output.

- Sales and purchase data also contain some anomalies in terms of price anomaly, vendor anomaly or quantity anomaly, and identification of these anomalies results in bigger cost-saving for the automotive domain.

  Approximate value-driven benefits in the automotive industry are:

- Optimizing the spare part availability by 13%
- Predictive diagnostics of machine by 75% faster
- Machine uptime improvement by 35%.

3.2 Applications of Anomaly Detection in Oil and Gas

Oil and gas industry is one of the highly dynamic industries, where a very small improvement in productivity results in billions of dollars of revenue for the organization. The latest revolution in the drilling processes comes with huge sensor-equipped machines that will send continuously big volumes of data to the database.

The several thousands of sensors that were deployed in the drilling machines send a huge volume of data where anomaly detection can help in

- Identifying the faulty signals from the machine operation data in order to prevent huge machine downtime.
- Anomalies in environmental parameters like pressure and temperature indicate the probable hotspots for accidents and can be fixed in advance to prevent them.
- Alarming the faulty decisions made by the machines/humans in the process will improve productivity.

3.3 Applications of Anomaly Detection in Pharma Manufacturing

There are several applications of anomaly detection in the pharmaceutical life sciences domain, which includes process monitoring in the production line and quality control using statistical process control through deploying the quality control charts. Some of the key examples are:

- In the retail data of pharma industry, anomaly detection can help to detect the fraud or misuse of prescription over-the-counter transactions
- Perfect and quick identification of abnormal events in the results data can help to adhere to critical safety standards.
- Finally, the real-time or near real-time detection of anomalies in multinomial clinical trial data helps to ensure the success of clinical trials.

3.4 Applications of Anomaly Detection in Consumer Goods Manufacturing

Consumer goods companies have been at the forefront of analytics implementation in specific domains such as marketing and sales. Supply chain and operations are also gaining the advantage of huge data availability and realizing the maximum benefits.

Some of the key areas where anomaly detection becomes very critical are:

- Quality testing department widely applies anomaly detection techniques to identify the faulty goods out of bulk manufacturing.
- Packaging abnormalities are another key area where if not identified on the floor may lead to damage of the consumer goods and thereby several legal issues arise from consumers on the field. These abnormalities are identified through anomaly detection techniques more efficiently.

4 Machine Learning Methods for Anomaly Detection

With the huge availability of data and advancement in the applications of machine learning algorithms, several techniques were identified to detect anomalies in the data [2]. These techniques are broadly divided into supervised and unsupervised ML techniques [3] and a few of these techniques were briefly discussed in the following section.

4.1 Supervised Learning Methods for Anomaly Detection

Supervised methods for anomaly detection require the presence of a labeled data set that contains both normal as well as anomalous data points.

Supervised methods always have an upper hand in terms of anomaly detection accuracy percentage. Few examples of supervised methods include anomaly detection using logistic regression, the K-nearest neighbors (or k-NN) method and random forest methods.

4.1.1 Logistic Regression

Logistic regression is generally applied to a binary-dependent variable [4]. The two possible dependent variable values are often labeled as "0" and "1", which represent outcomes such as pass/fail, win/lose, or in our case "legitimate" or "fraudulent". The binary logistic model is used to estimate the probability of a binary response based on one or more predictor (or independent) features.

The functional form of logistic regression is usually represented as follows.

$$\mathrm{Ln}\left(\frac{P}{1-P}\right) = \beta_0 + \beta_1 X_1 + \beta_2 X_2 + \ldots + \beta_k X_k$$

where P denotes the probability of dependent variable assuming one of the outcomes in the binary experiment. In many labeled data where anomalies are clearly identified in the train data, this model will produce at most good accuracy up to 98% in all kinds of experiments.

4.1.2 k-NN

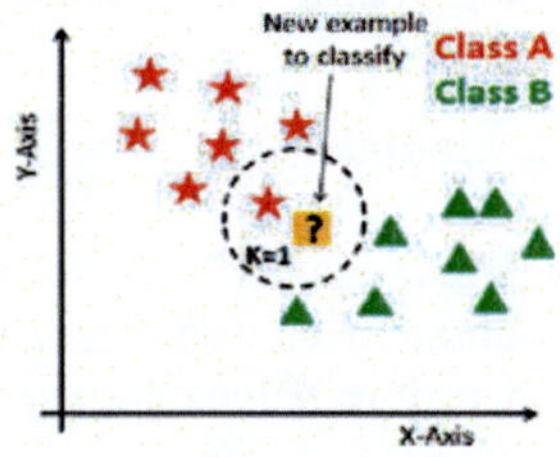

The k-nearest neighbors (k-NN) algorithm is a simple, easy-to-implement supervised machine learning algorithm that can be used in many anomaly detection problems [5]. The k-NN method captures the similarity between the new point and its k-nearest neighbors and classified them into one of the classes.

It arrives at the approximate distances using any distance measures between various points on the input vectors and then assigns the unlabeled new data point to the class of its k-nearest neighbors. In the process of creating k-NN classifier, k is an important parameter and different k values can produce different accuracies. If k is a very big number, the neighbors, when used for prediction, might consume a larger amount of time and it also impacts the accuracy of the calculation.

4.2 Unsupervised Machine Learning Methods for Anomaly Detection

Unsupervised methods of anomaly detection do not depend on any training data that contains labeling information. These methods are based on the statistical assumption that most of the points in the flowing data are normal and only a minor percentage would be anomalous data. These methods also estimate that any such anomaly data surely would be different statistically from normal data. Some of the unsupervised methods include the k-means method and principal component analysis.

4.2.1 K-Means Clusters

The main idea behind using clustering for anomaly detection is to identify the normal groups of identical characteristics and then to check whether the new point falls into any one of the formed groups [6]. This is a very good basic method and we cannot assume that all anomalies are identified in this method as it is just based on distance measure only. The basic steps in k-means clustering are:

- Divide the data into random k groups and find the centroid of each group
- Now for each data point, find its distance from k centroids and find which is the closest centroid. Allocate the point to that centroid cluster
- Repeat the procedure until re-grouping does not occur after all points allocation.

4.2.2 Principal Component Analysis

Principal component analysis (PCA) usually looks for correlations among the variables and determines the combination of values that best captures differences in outcomes [7]. These combined feature values are used to create a more compact feature space called the *principal components.*

When a new observation comes for an anomaly test, it is analyzed, and the anomaly detection algorithm computes its projection on the eigenvectors, together with a normalized reconstruction error. The normalized error is used as the anomaly score. If the value is too high, we can say that it is a more typical anomaly in the given data.

5 Summary and Conclusions

Anomaly detection in manufacturing is one of the key activities that result in the highest quality products and huge productivity gains in a longer period. There are several applications in the manufacturing (or) production processes, like faulty item detection, process abnormality findings, defects in product quality, predictive maintenance and preventive diagnostics are very good examples where anomaly detection methods are successfully applied and gained huge benefits for the respective organizations.

The huge data creation in manufacturing and a lot of research in the area of machine learning algorithms laid a good path for anomaly detection topics and many domains like automotive, consumer goods, pharma and iron and steel are getting benefits through this application. A couple of areas like detection accuracy and sophisticated ML algorithms require a lot more research to register more concrete successes in the future times.

References

1. Chandola V, Banerjee A, Kumar V (2009) Anomaly Detection: a survey. ACM Comput Surv 41.https://doi.org/10.1145/1541880.1541882
2. Liu J et al (2018) Anomaly detection in manufacturing systems using structured neural networks. In: 2018 13th world congress on intelligent control and automation (WCICA). Changsha, China, pp 175–180.https://doi.org/10.1109/WCICA.2018.8630692
3. Pittino F, Puggl M, Moldaschl T, Hirschl C (2020) Anomaly detection on in-production manufacturing machines using statistical learning methods. Special issue data analysis for smart sensor systems
4. Rosario DS (2004) Highly effective logistic regression model for signal (anomaly) detection. In: 2004 IEEE international conference on acoustics, speech, and signal processing. Montreal, Que, pp V-817. https://doi.org/10.1109/ICASSP.2004.1327236
5. Omar S, Ngadi A, Jebur HH (2013) Machine learning techniques for anomaly detection: an overview. Int J Comput Appl (0975 – 8887) 79(2)
6. Mehrotra KG, Mohan C, Huang H, A text book "Anomaly Detection Principles and Algorithms". ISBN 978-3-319-67526-8
7. Carletti M, Masiero C, Beghi A, Susto GA (2019) A deep learning approach for anomaly detection with industrial time series data: a refrigerators manufacturing case study. In: 29th international conference on flexible auotmation and intelligent manufacturing (FAIM2019). Limerick, Ireland

Pollution Monitoring System Using Raspberry Pi and Php Web Server

T. Monica and A. Sai Suneel

Abstract In India and around the world, day-to-day pollution is increasing, which severely affects the entire living beings and damages our earth. Because of the rise in vehicles, there is a rise in air and sound pollutions. On the other hand, factories also release waste gases into the air. Because of pollution there is a rise in harmful ultraviolet rays. Light intensity during night times causes accidents because light reflections from either side of vehicles may cause a blurred sensation in the eyes. In this study, we propose an embedded device, which is a combination of embedded system and PHP web server, and measure the parameters, like CO gases, sound intensity, UV, and light intensities. If any of these parameters cross their limit, then it will send data to the PHP server and the server message will be sent to the authorities to take action in that area to control pollution.

Keywords UV · Light · Air and sound intensities · PHP web server · Embedded application

1 Introduction

Air, sound, and UV rays are some of the major issues growing these days. The main objective of this project is to monitor and alert the authorities to control the pollution because of air, sound, UV, and light. There is an increase of rapid population and factories, and vehicle pollution increases day by day which needs to be monitored. It is necessary to keep an eye to track the parameters mentioned above to control the pollution. Otherwise, there will be many ears, nose, and eye diseases. Because of light intensity accidents may occur. Technology has advanced a lot today, so we can find solutions or preventing ways to solve these pollution problems. It creates a link

T. Monica (✉) · A. S. Suneel
Department of Electronics and Communication Engineering, School of Engineering and Technology, Sri Padmavati Mahila Visvavidyalayam, Tirupati, India
e-mail: thotimonica@gmail.com

A. S. Suneel
e-mail: saisuneel.adem@gmail.com

S. Jyothi et al. (eds.), *Proceedings of the 2nd International Conference on Computational and Bio Engineering*, Lecture Notes in Networks and Systems 215,
https://doi.org/10.1007/978-981-16-1941-0_36

between people and the environment. In this study, we are implementing embedded and PHP web servers.

2 Existing System

Air and sound pollutions are the major issues that have bad effects on the environment as well as on human beings. Monitoring pollution [1] is a difficult and risky task. Generally, these parameters are calculated by the authorities related to pollution monitoring. Each and every time they need to visit the places, check the pollution due to air and sound, enter into the books, and consult higher authorities to take action. This is a lengthy, time-taking, and costly task.

3 Proposed System

In this implemented system we are using gas, sound, DHT11, UV, and LDR sensors to monitor gases, sound and humidity temperature, UV rays, and light intensity in the environment or atmosphere [2, 3]. In this study, we are using gas sensors that detect gases in the environment so that we can estimate the air quality[4–6] in the environment. DHT11 is used to check temperature and humidity content in the atmosphere and sound sensors to check sound intensity. Light intensity is checked by using LDR and UV radiations by a UV sensor. The sensors continuously monitor the parameters and send them to Raspberry Pi. These values are displayed on LCD. Any of the values exceeded its threshold value then the raspberry pi will send the information to PHP web server and from there message will be sent to the concerned authorities for necessary precautions. The values of sensors are displayed on the LCD continuously.

3.1 *Pollution Monitoring System Using Raspberry Pi and PHP Web Server*

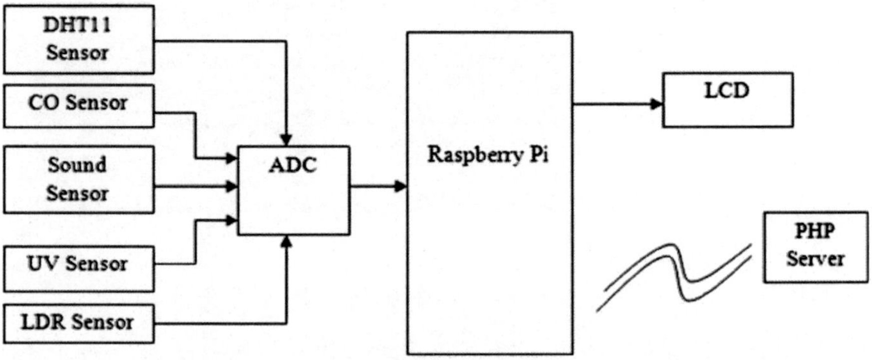

The DHT11, CO, sound, UV, and LDR sensors are connected to ADC which in turn is connected to Raspberry Pi. The sensors detect the amount of humidity, temperature, sound, UV rays, and LDR values and give them to ADC, which converts the values into digital form and gives to the Raspberry Pi, and the values were displayed on the LCD screen. With the help of Raspberry Pi and in-built Wi-Fi, the measured values will be sent to the PHP web server. If any of the measured values cross the threshold value or limit, then a message will be sent to the concerned authorities to take necessary action.

3.2 *Hardware Requirements*

Raspberry Pi

Raspberry Pi is a credit card-sized microprocessor and microcontroller that have 40 pins. It works as a microcomputer if we connect that with essential HDMI and audio cables. It works and responds fast when compared to other microcontrollers (Fig. 1).

LCD

The acronym of LCD is liquid crystal display which is an electronic device used to display letters and digits. The most common type of LCD used here is 16*2, where we have two lines and 16 characters per line, so a total of 32 characters for two lines (Fig. 2).

DHT11 Sensor

The DHT11 sensor is for temperature and humidity. It can be connected to various microcontrollers like Arduino, Raspberry pi, Nodemcu, etc.; it has three pins connected to a microcontroller and power supply.

Fig. 1 Raspberry Pi

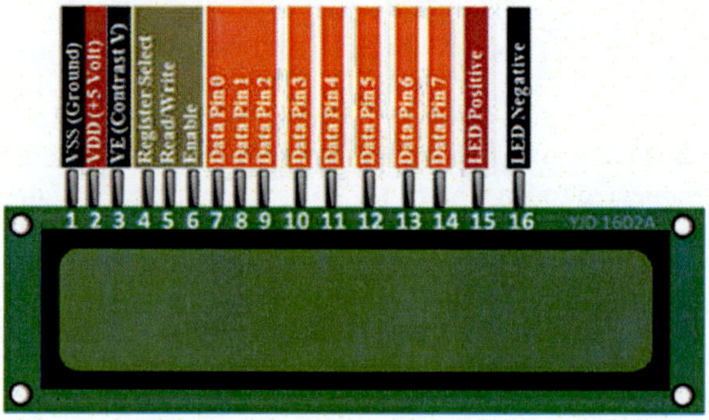

Fig. 2 LCD

Pin Description
See Fig. 3.

Gas (CO) Sensor
Gas sensor is an electronic sensor that detects gases, like propane, LPG, alcohol, and other gases. These are mainly used in the detection of harmful gases. Based on the concentration of the gases in the places, the potentiometer in the sensor increases internally. The sensor can be interfaced with microcontrollers, like 8051, Arduino, and Raspberry Pi.

Pin Description
See Fig. 4.

Sound Sensor
It is an electronic sensor that detects sound. This sensor is mainly used to detect sound intensity. If this sensor senses the sound, it activates and processes the signal to the microcontroller.

Pin Description
See Fig. 5.

Fig. 3 DHT11 sensor

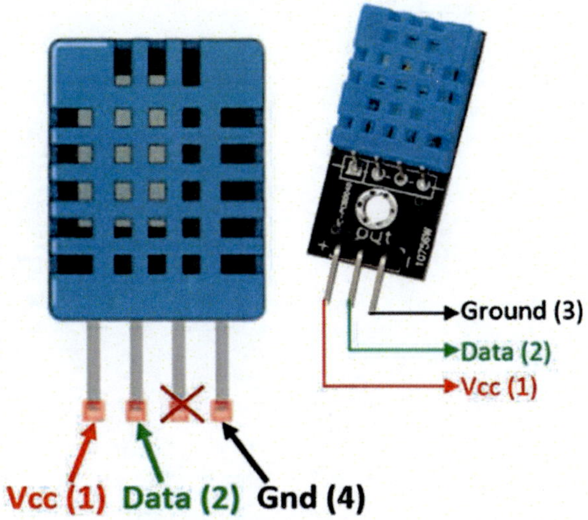

Fig. 4 CO sensor

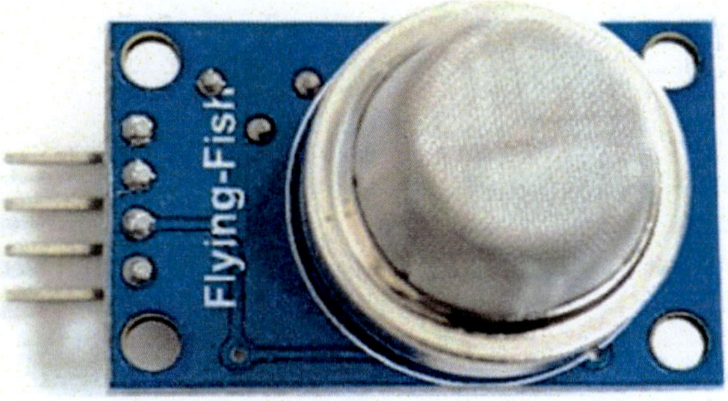

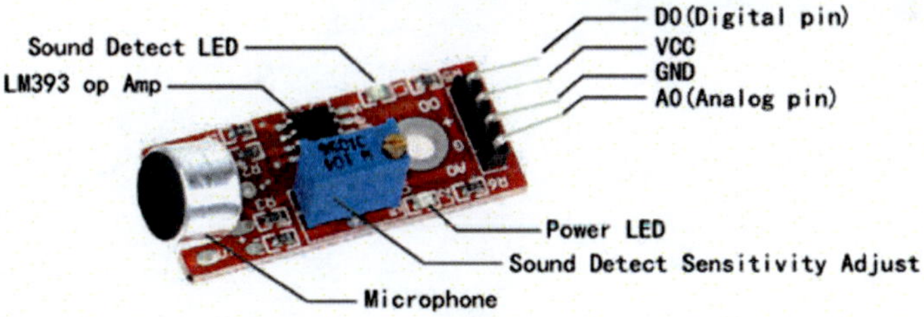

Fig. 5 Sound sensor

UV Sensor

It is an electronic device used for detecting ultraviolet radiation. It can detect radiation up to 240–370 nm range of light (Fig. 6).

Fig. 6 UV sensor

LDR Sensor

The acronym of LDR is light-dependent resistor. As the name indicates, it is a variable sensor that varies its resistance based on the light intensity that falls on it. The resistance of the sensor decreases with an increase in light intensity and vice versa (Fig. 7).

ADC

ADC is a module that converts an analog signal to a digital signal. It has 16 pins IC on the module and ADC is an 8-channel, 10 bit ADC. It works under the operating voltage of 2.7–5 V (Fig. 8).

Fig. 7 LDR sensor

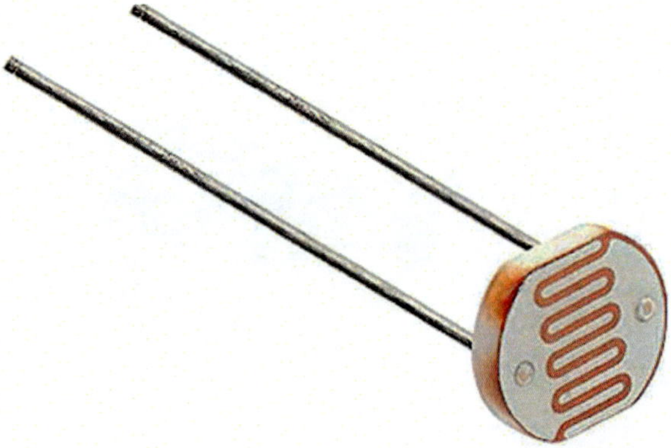

Fig. 8 ADC IC

3.3 *Software Requirements*

Python

Python is a programming language by which we can work in a Raspberry Pi environment. In 1985–1990 Guido van Rossum has developed this programming language. The programming language name was given after a TV show called "Monty Python's Flying Circus" but not by the name of a snake (Fig. 9).

3.4 *Applications*

- To detect and secure from harmful gases at any place
- Indoor air quality monitoring
- Used in weather monitoring centers.

3.5 *Advantages*

- Easy to install
- Updates on a mobile phone directly
- Accurate pollution monitoring and remote location monitoring.

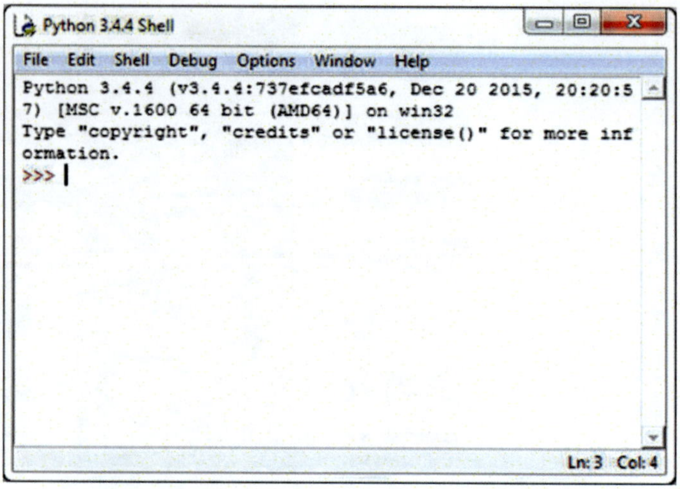

Fig. 9 Python shell

4 Results

4.1 Outputs from Hardware

Displaying of Project Title on LCD Screen

At first, the title will be displayed on the LCD screen as "Air and Sound Monitoring system". Next, the measured values will be displayed on the LCD screen. Then the values or parameters measured and displayed on the LCD screen will be uploaded to the PHP web server and this will also be displayed on the LCD screen as "Data Updated to server". If any of the measured parameters is high or crosses its threshold value, then the message will be displayed on the LCD screen (Fig. 10).

Displaying of UV Index and Light Intensity on LCD Screen

The ozone layer which absorbs the cancer-causing UVB rays can be depleted by air pollutants. UV rays lead to eye problems, immune system problems, and skin cancer. Here we are using a UV sensor and also the LDR sensor to measure the UV rays and also light intensity. The measured values will be displayed on the LCD screen and the data will be updated to the PHP web server (Figs. 11 and 12).

Displaying of sound, gas, humidity, and temperature values on the LCD screen

Sound pollution is caused due to elevated levels of noise caused by human activities such as construction works, traffic, household noise. Other reasons for noise pollution are industrialization, social events, animals, and air traffic noise, which have a very harmful impact on humans, wildlife, and the environment. This causes noise-induced hearing loss (NIHL), memory loss, disturbance of sleep, heart disease, high blood

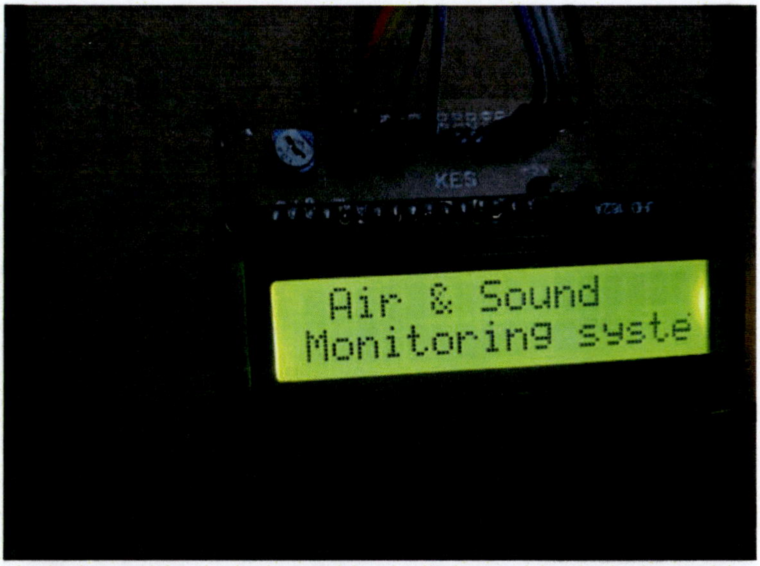

Fig. 10 Displaying project title on the LCD screen

Fig. 11 Displaying values of the LDR sensor

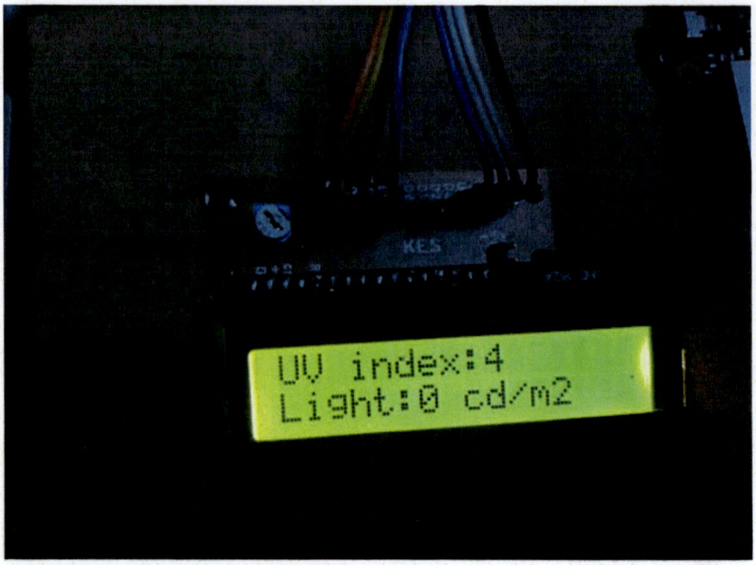

Fig. 12 Displaying values of the UV sensor

pressure, and stress. Here we are using a sound sensor to measure the sound pollution present and the measured value will be displayed on the LCD screen in decibels. The data measured will be updated to the PHP web server (Fig. 13).

Air pollution is caused by human activities and also by natural sources. It may be caused due to traffic such as vehicle emissions, ozone, noxious gases like nitrogen oxides and sulfur oxides, particulate matter, and volatile organic compounds. This causes chronic obstructive pulmonary disease (COPD), chronic bronchitis, cardiovascular diseases such as hemorrhagic stroke, increased risk of breast cancer, lung cancer, etc. The co-sensor is interfaced with Raspberry Pi to measure the air pollution present and to monitor the value, and the measured value is updated to the PHP webserver.

Humidity affects air pollution that causes very adverse effects on the human respiratory system. High humidity rises the level of toxic chemicals in the air. Low humidity results in airborne germs.

Temperature affects air pollution. Surface inversion produces smog which traps the pollutants that contain hydrocarbons and nitrogen oxides which were converted into ozone by sunlight, and this harmful ozone reduces air quality.

The DHT11 sensor is used to measure the humidity along with the temperature. The measured humidity and temperature values will be displayed on the LCD screen and the data will be updated to the PHP server.

Hence the sound, gas, humidity, and temperature values will be displayed all at once on the LCD screen. The sound intensity will be displayed in terms of decibels. The CO gas present will be displayed in terms of parts per million (ppm); humidity

Fig. 13 Displaying values of the sound, CO, humidity, and temperature values

will be displayed in terms of percentage, and temperature will be displayed in terms of centigrade on the LCD screen.

Hence the outputs of all the ultra-violent sensors, light-dependent resistor sensors, CO sensor, sound sensor, and temperature and humidity sensors will be displayed on the LCD screen after running the program code using the run module option with the help of Raspberry Pi and VNC viewer. The program code was written in Python language and after running the code, the output will be displayed on the Raspberry Pi desktop and also on the liquid crystal display screen. Also, if any of the measured parameters is high or exceeds its limit or threshold value, then the LCD will display which measured parameter is high.

Uploading of Data to PHP Web Server

The monitored pollution parameters along with temperature, humidity, light intensity are updated through Raspberry Pi to the PHP webserver. Since Raspberry Pi has an in-built Wi-fi, it will be more applicable for IoT applications, and with the help of Wi-Fi, this Raspberry Pi transmits the data to the web server and the data updated on the server can be seen with the link http://cloudmonitoring.dbandroid.online/ with the help of user name and password. After forwarding the data to the PHP web server, a message will be displayed on the LCD screen as "Data Updated to server" (Fig. 14).

Displaying of Measured Parameter if it Exceeds Threshold value

If any of the measured parameters exceeds its threshold value, then it will be shown on the LCD screen and the value will be sent to the phone through message and the data will be updated to the PHP web server and from there this data will be sent to

Fig. 14 Data is uploaded to the PHP server through Raspberry Pi

the concerned authorities to take necessary actions in that particular area to control pollution (Fig. 15).

Fig. 15 Displaying when humidity is high

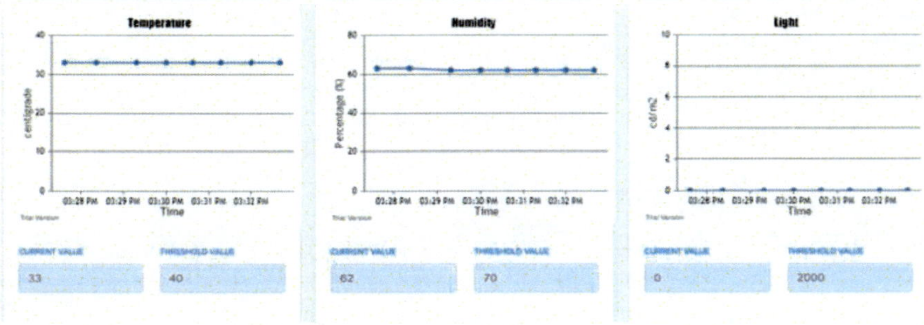

Fig. 16 Graphical view of temperature, humidity, and light

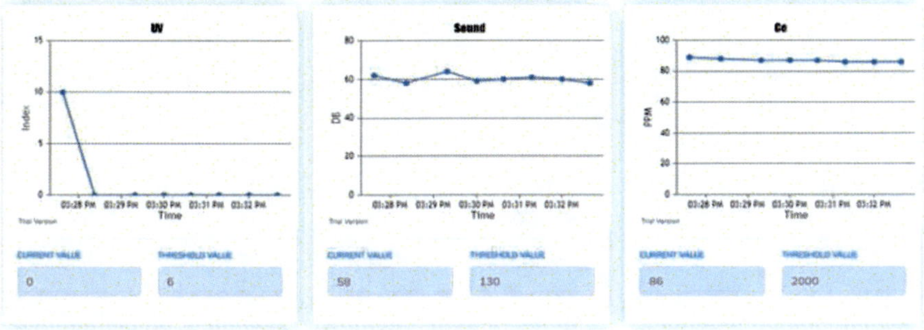

Fig. 17 Graphical view of UV, sound, and CO

4.2 Outputs from PHP Server

Since Raspberry Pi has an in-built Wi-fi, it will send the monitored data to the PHP server. The outputs from the PHP server will be shown in the form of graphs. The threshold values for temperature, humidity, light, UV, sound, and CO can be adjusted through threshold value settings, and also the mobile number can be given to which the message will be sent (Figs. 16, 17 and 18).

5 Conclusion and Future Scope

The air, sound, radiation, and light intensity are some of the major threats to the environment that cause various health issues to human beings. In this study with the combination of embedded and PHP web servers, we can control the pollution within time by intimating to the authorities of using technology. This project creates an interaction between the environment and human beings.

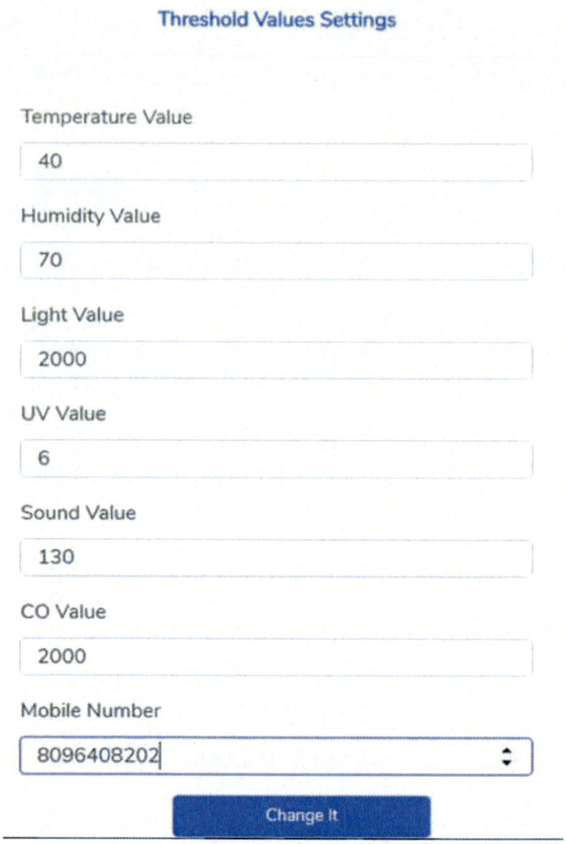

Fig. 18 Setting up of threshold values and the phone number to which message has to be sent

In the future, we can add GPS, camera, and other more powerful gas sensors to get the information of different harmful gases so that if any values are exceeded then the corresponding location along with the exceeded parameter values and the images of the place can be sent through mail or as an MMS to the mobile number.

References

1. Santosh GB, Somani SB (2017) A aurvey on WiFi based air pollution monitoring system. Int J Innov Res Comput Commun Eng 5(3)
2. Rewatkar NR, Khatri DM (2017) A review: cost effective IOT based air pollution monitoring and air quality analysis. Int J Recent Innov Trends Comput Commun 5(1). ISSN: 2321–8169
3. Tapashetti A, Vegiraju D (2016) IoT-enabled air quality monitoring device—a low cost smart health solution. IEEE global humanitarian technology conference
4. Phala KSE, Kumar A, Hancke GP (2016) Air quality monitoring system based on ISO/IEC/IEEE 21451 standards. IEEE Sens J 16(12)

5. Shaban KB, Kadri A, Rezk E (2016) Urban air pollution monitoring system, with forecasting models. IEEE Sens J 16(8)
6. Wang D, Jiang C, Dian Y (2016) Design of air quality monitoring system based on internet of things. In: 10th international conference on software, knowledge, information management & applications (SKIMA)

Genetic Diversity of Rice Cultivars (*Oryza Sativa L.*) Assessed Through Simple Sequence Repeat Marker

R. Usha and Ch. Indhravathi

Abstract Among the many cereal crops of the world, rice (*Oryza sativa* L.) is one of the commonest, with extensive genetic variance. The great heterogeneity of rice genome can be utilized to produce cultivars with many different traits through crop improvement programmes. The traits and genetic variance need to be analysed, for which simple-sequence repeats (SSRs) are preferred. In our current study, we evaluated the genetic variance of 10 selected rice varieties utilising 13 SSRs as markers. Furthermore, dendrogram was constructed through cluster analysis using unweighted pair-group method with arithmetic average (UPGMA) approach. The analysis revealed variation among the varieties. The varieties were grouped under two clusters based on similarity levels. Based on this study, Katan Dora Sanalu, Pushyami and Nellore varieties were found genetically different from other varieties studied. Markers of genetic variance are essential for the selection of superior varieties for breeding programmes and their commercial success. Marker-assisted selection (MAS) in rice improvement is increasingly emphasising gene mapping across the world.

Keywords Genetic diversity · SSR markers · Unweighted pair-group

1 Introduction

Rice is a staple food crop for about 60% of the world population [1]. Rice belongs to Poaceae and Orazoidea family and sub-family, respectively. One-fifth or 1.5 billion hectares of the land under cereal cultivation is allotted for rice, yielding about 645 million tons annually. Asia produces and consumes most of the world's rice, constituting more than half of the world population. Rice ranks next only to wheat in arable land proportion annually, with approximately 11% of the total land cultivated [2]. Furthermore, the availability of high-precision sequence data and markers gives

R. Usha (✉) · Ch. Indhravathi
Department of Biotechnology, Sri Padmavati MahilaVisvavidyalayam, Tirupati 517502, Andhra Pradesh, India
e-mail: ushatirupathi@gmail.com

S. Jyothi et al. (eds.), *Proceedings of the 2nd International Conference on Computational and Bio Engineering*, Lecture Notes in Networks and Systems 215,
https://doi.org/10.1007/978-981-16-1941-0_37

an added advantage [3]. Besides, almost all the rice varieties can be categorized into Indica, Japonica, Aromatic and Glutinous. Morphological traits are not always consistent in heritability, thus may not serve the purpose of crop variety selection for further improvement. The variability among different genotypes of a species is known as genetic diversity. Genetic variance imparts traits that provide defence against genetic vulnerabilities, a long known knowledge of traditional farmers cultivating landraces, which needs to be introgressed into modern cultivars to make them sustainable. A genetic marker, a fragment of DNA naturally accruing, might be used as a marker to trace the desired gene in a test genotype, and these markers are the basis for approaches to explore plant breeding. These markers are present in the proximity of a target gene, thus serves as a selection tool too for genetic improvement. DNA variation among rice genotypes and genetic markers presents tremendous potential for rice cultivars/variety [4]. Assessment of genetic basis for phenotypic attributes and establishment of degree of genetic variance are effective in the genetic development of a crop. This calls for the genetic diversity evaluation and the detection of crossover or hybridization of parental lines. Information on genetic diversity within and among closely related crop varieties is important for the improvement of crops and the diver's objectives, such as the development of higher yield cultivars, wider adaptability, desirable quality and resistance to pests and diseases. Di-, tri- and tetra-nucleotide tandem repeats containing loci from eukaryotic genomes are basic sequence repeats (SSR) markers. In fact, these are non-coding regions that have remained preserved throughout evolution. SSR markers have been successfully used in rice for many purposes, including (i) genome mapping [5]; (ii) genetic diversity assessment and interaction between different cultivars, including both aromatic and non-aromatic rice [6]; (iii) variety recognition and purity checking [7]; and (iv) variety determination [7, 8]. The genetic relationship of popular rice cultivars was determined by using SSR markers [9]. To evaluate the genetic diversity among common rice cultivars in Iran, 50 SSR markers were used.

2 Objectives

With the following objectives, the present study is formulated

- To determine the extent of genetic diversity within some local rice varieties by the SSR technique.
- Genetic diversity data analysis, which includes:
- Scoring of data and similarity matrix
- Cluster analysis.

3 Materials and Methods

The experiment was carried out at the Biotechnology Department, Sri Padmavati Mahila Visvavidyalayam, to study the genetic diversity of 10 local rice varieties. The details of different materials used and methodologies followed for the study have been described below:

Climate and soil
The study site comes under the sub-tropical climate. It is characterized by a cloudless sky, better sunshine fall and lower winter temperature during the earlier crop growth stage. The soil in the experimental pots was textured clay loam with a pH of 6.2.

Collection of seeds and rising of seedlings
From the Agri station, Thadepalligudam, Andhra Pradesh, the seeds of 10 rice varieties were collected. To break the seed dormancy, the seeds were held at 50 °C for 3 days in a convention oven. For pre-germination, the oven-treated seeds were soaked with tap water for 24 h. On the surface of the soil, the pre-germinated seeds of the test rice varieties were sown in pots filled with water. The seedlings grew 1–2 cm above the surface of the soil after 14 days of sowing, and the plants were protected from pests and diseases. Healthy leaf fractions were harvested using sterile scissors and surface-sterilized with sterile water followed by ethanol to remove any source of foreign DNA and blot-dried.

4 Genomic DNA Extraction and Principle

With some modifications, DNA was extracted based on a previously recorded method of cetyl trimethyl ammonium bromide (CTAB). The nano drop spectrophotometer from Nano Drop Technologies is intended to measure the concentration of nucleic acid in 1 μl sample volumes. To measure the respective nucleic acid concentrations automatically, the DNA experiments used the absorbance at 260 nm. In order to isolate, classify and purify DNA fragments, 1% agarose gel mix was prepared.

4.1 PCR Analysis for SSR Markers

1. **Primers used for amplification**

A total of 13 SSR markers distributed on chromosomes 1, 2, 3, 4, 5 and 6 were used to analyse the genetic diversity. All of the SSR markers were obtained from IRRI-India. Polymerase chain reaction amplification of SSR markers was carried out using 13 primers listed in Table 2. PCR was carried out in a volume of 12 μl containing

Table 1 Components of PCR cocktail

S. no	Component	Quantity (for single reaction)
1	ddH_2O	2 μl
2	Emerald master mix	7 μl
3	Primer forward	0.5 μl
4	Primer reverse	0.5 μl
5	DNA	2 μl
	Total	**12 μl**

2 μl of DNA templates from each variety, 7 μl of emerald master mix, 0.5 μl of each forward and reverse primer with the required amount of sterile deionized water (Table 1).

Amplification was carried out in oil-free Biorad thermal cycler with the following program:

- Profile 2 was programmed to run for 35 cycles
- Electrophoresis of Amplified Products

Following amplification, 12 μl of aliquot-amplified SSR samples was combined with 3 μl loading buffer (0.4% (w/v) bromophenol blue, 0.4% (w/v) xylene cyanole and 5 ml glycerol) and electrophoresed at 4 V/cm for 3.5 h on 4% (w/v) agarose gel in 0.5xTBE buffer (10 mM Tris–borate, 1 mM EDTA) containing 0.5 μg ethidium bromide per ml. As a standard size marker, a 100 bp DNA Ladder (Qiagen GmbH, Germany) was used to compare the molecular weights of amplified products. The gels were documented using a UVI Doc Gel Documentation System after electrophoresis (UVITEC, Cambridge, UK).

4.2 *Genetic Diversity Data Analysis*

1. **Scoring of data and similarity matrix**

The size of the amplification products was calculated following electrophoresis by comparing the migration of the amplified fragment with that of the identified fragment size of 100 bp DNA Ladder molecular weight marker. Bands were then graded for the SSR bands to be present. The information was inserted as a discrete variable into a binary matrix, where 1 represents the present band and 0 represents the absence of a band, and the data was further analysed. The 0/1 matrix was used to determine the coefficients of Jaccard's similarity matrix. After Jaccard (1908), the coefficients were determined using the following formula:

$$\text{Similarity coefficient} = a/n.$$

Table 2 SSR primers used in the present study for diversity analysis

Name of the markers	Chromosome	Position start	Position end	Motif	No. of repeats	Forward primer (5'-3')	Reverse primer (3'-5')
RM562	1	14,610,402	14,610,440	AAG	13	CACAACCCACAAACAGCAAG	CTTCCCCCAAAGTTTTAGCC
RM543	1	32,776,801	32,776,830	CCG	10	CTGCTGCAGACTCTACTGCG	AAATATTACCCATCCCCCCC
RM12434	2	2,072,425	2,072,502	AAT	26	TTGGTCATCTTTGGTTGGTTCAGG	GGACAATTCACACAGCAGAATTGG
RM300	2	13,190,528	13,190,590	AAC	21	GCTTAAGGACTTCTGCGAACC	CAACAGCGATCCACATCATC
RM497	2	29,052,446	29,052,478	ACC	11	TCCTCTTCACCTATGGGTGG	GCCAGTGCTAGGAGAGTTGG
RM7332	3	390,594	390,641	ACAT	12	ACACTGTACACCACACTTCAGC	CAGGGAAATGACACTGTCCC
RM470	4	28,248,941	28,248,982	AAG	14	TCCTCATCGGCTTCTTCTTC	AGAACCCGTTCTACGTCACG
RM440	5	19,830,882	19,830,947	AAG	22	CATGCAACAACGTCACCTTC	ATGGTTGGTAGGCACCAAAG
RM587	6	2,291,878	2,291,931	AAG	18	ACGCGAACAAATTAACAGCC	CTTTGCTACCAGTAGATCCAGC
RM583	1	8,328,958	8,329,017	AAG	20	AGATCCATCCCTGTGGAGAG	GCGAACTCGCGTTGTAATC
RM592	5	2,736,620	2,736,697	AAT	26	TCTTTGGTATGAGGAACACC	AGAGATCCGGTTTGTTGTAA
RM411	3	21,231,694	21,231,717	AAC	8	ACACCAACTCTTGCCTGCAT	TGAAGCAAAAACATGGCTAGG
RM16369	4	2,531,191	2,531,229	AGG	13	CTGATCGAGAGCGTTAAGGG	GGGATCAAACCACGTTTCTG

where *a* is the number of matching band for each pair of comparisons; *n* is the total number of bands observed in two samples.

2. **Cluster Analysis**

The resulting similarity matrix was used to build a dendrogram UPGMA (unweighted pair-group method with arithmetic means) showing the relationship between the varieties computed from polymorphic marker frequencies to estimate the genetic relationship between the 20 rice germplasm studied using the arithmetic means unweighted pair-group method (UPGMA). Using SPSS tools, the dendrogram was designed (version 20).

5 Results and Discussion

5.1 *Confirmation and Quantification of Genomic DNA*

Total genomic DNA was isolated by CTAB method and the purity of the extracted DNA was quantified by taking readings at wavelength 260/280 nm. The DNA samples showed the ratio of absorbance between ~1.8 and 2.0 (Table 1). After checking the purity of DNA, the samples were further used for SSR analysis. The isolated DNA samples were further rechecked by running samples on 1% agarose gel electrophoresis (Figs. 1 and 2).

5.2 *Genetic Diversity Analysis Using SSR Markers*

The assessment of genetic diversity is a very important factor in the improvement of rice that can be obtained through approaches based on DNA markers, which can reveal a large number of loci for extensive variability. Using a highly repeatable PCR-based assay known as the simple sequence repeat (SSR) or microsatellite marker, DNA was collected from selected rice varieties. With respect to rice varieties, these SSR markers generate greater levels of DNA polymorphism.

Electrophoresis of Amplified Products

PCR-based molecular screening was conducted using the standard procedures. In this study, 10 varieties of rice were analysed using 13 primers (Table 2). The bands were found in all 10 rice varieties (Figs. 3, 4, 5, 6, 7, 8 and 9). Amplified SSR loci were analysed and the results revealed that among the rice varieties analysed, nine primer pairs detected polymorphism. The SSR loci were also polymorphic, indicating their dominance in the detection of DNA polymorphism, and the alleles were co-dominant (Table 5).

Fig. 1 Ten rice varieties grown in pots collection of leaf sample

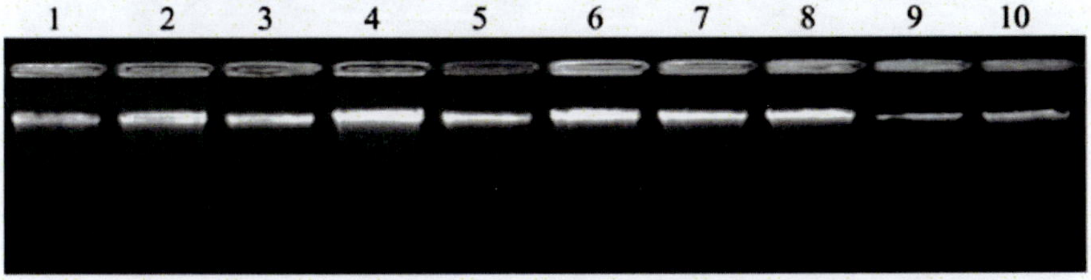

Fig. 2 Genomic DNA quantification of 10 rice varieties using agarose gel

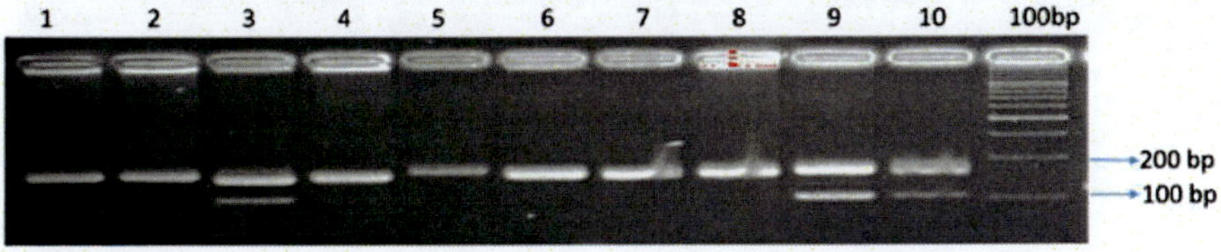

Fig. 3 DNA profiles of the 10 rice varieties with SSR marker RM562 with ladder 100 bp

Fig. 4 DNA profiles of the 10 rice varieties with SSR marker RM543 with ladder 100 bp

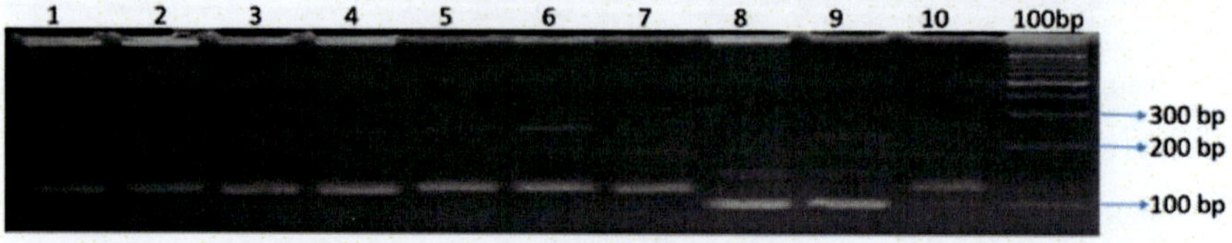

Fig. 5 DNA profiles of the 10 rice varieties with SSR marker RM12434 with ladder 100 bp

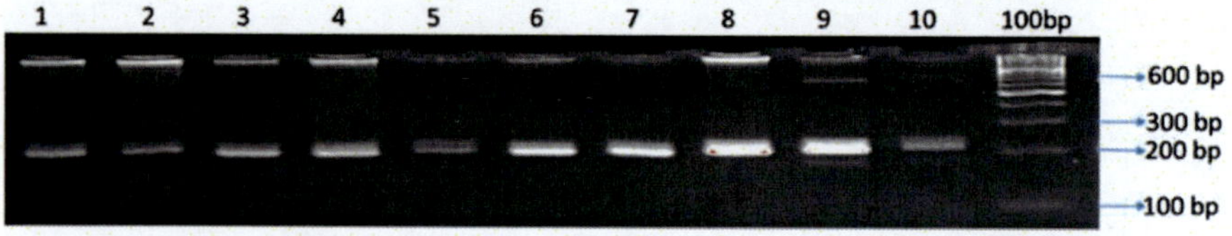

Fig. 6 DNA profiles of the 10 rice varieties with SSR marker RM497 with ladder 100 bp

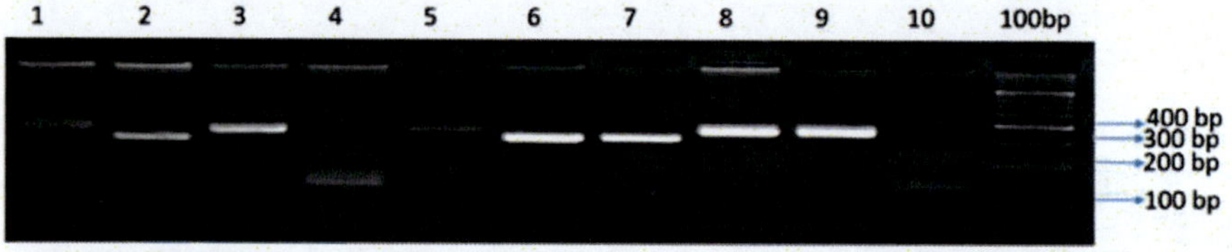

Fig. 7 DNA profiles of the 10 rice varieties with SSR marker RM440 with ladder 100 bp

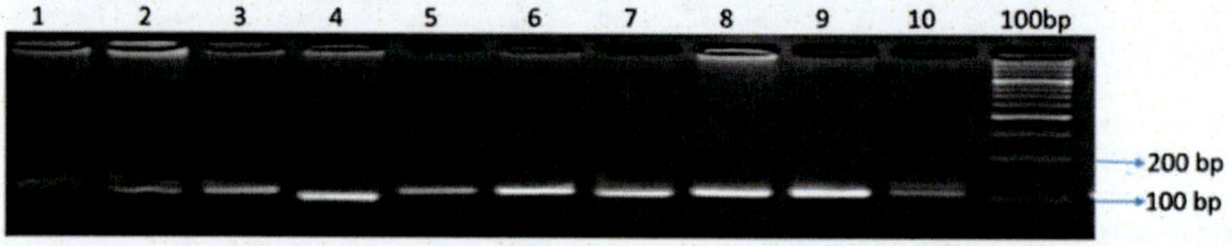

Fig. 8 DNA profiles of the 10 rice varieties with SSR marker RM470 with ladder 100 bp

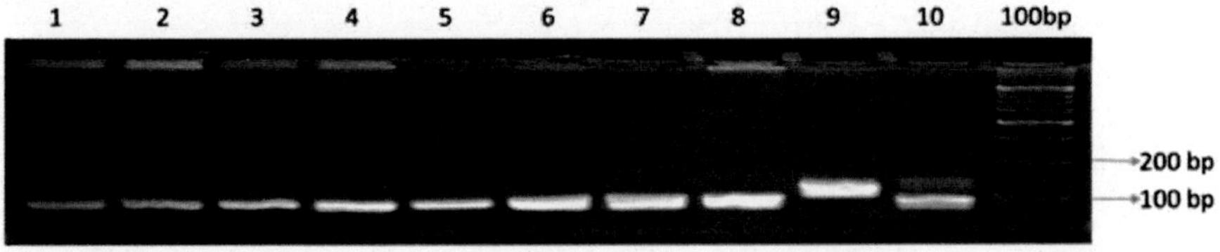

Fig. 9 DNA profiles of the 10 rice varieties with SSR marker RM587 with ladder 100 bp

The amplicon size of all 10 varieties for each marker allele varied from 100 bp produced by all primers to 400 bp produced. The highest range of band size was found for SSR marker RM12434 (100–400 bp) followed by RM470 (100–300 bp) and RM7332 (100–250 bp), respectively. The number of alleles per locus ranged from 2 to 5 alleles, with an average of 3.5, which is strikingly similar to that reported by Siwach et al. (2004), even with higher allele per SSR locus (1–8 alleles per locus) and (2–7 alleles per locus) [2]. A comparable result was reported. The same results were also reported when investigating the genetic variability of 12 rice landraces of India using five primers. However, this value is quite low compared with those reported in other places of the collections. The number of alleles per locus varied from 3 to 8, with an average number of alleles per locus at 4.86 and reported a range of 6–22 numbers of alleles per primer with an average of 14.7 alleles per locus. This indicates a high magnitude of diversity concerning the fewer markers among the plant materials in the present study.

5.3 *Genetic Diversity Data Analysis*

1. Scoring of data and similarity matrix

A total of 27 useful bands were scored from the amplification products with the 10 random SSR primers from 10 rice varieties in binary form: 0 and 1. The number of amplification products generated by each primer ranged in size from 100 to 400 bp followed by a genetic variance assessment from the generated scores. Binary SSR data were subjected to the genetic similarity matrix calculation (Table 3) based on which genetic variation was assessed. The pair-wise similarity estimates ranged from 0.14 to 1.00 and the average similarity was 0.57 for all 10 varieties. Saini and colleagues (2004) have also reported almost identical coefficient similarity values between 18 varieties of basmati and non-basmati using molecular markers. Similarly, in eight basmati accessions originating in Pakistan and one single Indica accession for the SSR analysis by similarity coefficients ranging from 0.24 to 0.92 were observed (2007). For basmati and non-basmati long-grain Indica rice varieties, a greater similarity was reported by Siwach et al. using SSR markers (similarity coefficient 0.67–091) (2004). The intra-specific variation in the germplasm used may be

Table 3 The temperature profile for PCR

Profile	Step	Temperature (°C)	Duration
1	Initial denaturation	94	5 min
2	Denaturation	94	40 s
	Annealing	55	40 s
	Primer extension	72	45 s
3	Final extension	72	7 min
4	Dump	4	

one of the reasons for this high level of similarity reported in the present and previous studies (Table 6).

This is supported by the results of the study done by Bakari (2010) on the characterization of rice landraces collected from various places in Tanzania using SSR markers, concluding that some rice landraces exhibited low genetic variance (0.76–0.98 distant coefficients). A narrow genetic background of Korean rice germplasm was documented, as revealed by DNA fingerprinting with SSR markers and their pedigree details. In Japan, a report showed that sake-brewing rice had low genetic diversity as compared to the cooking rice cultivars. The results from our study show that significant genetic differences among 10 varieties of rice as the coefficient of similarities was concentrated at a range of 0.14–1.0 genetic distances for the most of varieties studied.

6 Cluster Analysis

Cluster analysis was performed using an unweighted pair-group approach with arithmetic average method to establish the genetic dendrogram relationship between the studied rice varieties (UPGMA). The software used was SPSS. The dendrogram-based UPGMA analysis revealed the existence of two clusters with sub-clusters based on genetic similarity level (Fig. 10).

All two clusters have seven varieties as summarized in Table 7.

Cluster 1 is divided into two sub-clusters having three varieties, including Swarna, Amhra and Vijetha (Table 4). The genetic similarity distance is 0.85 between Swarna and Amhra and 0.64 between Vijetha and Amhra, indicating 85 and 64% similarity. Cluster 2 is divided into three sub-clusters having four varieties, including Prabath, Indra, Samba Masuri and Badava Masuri (Table 4). Genetic similarity of 83% is present between Prabath and Indra, 64% between Samba Masuri and Prabhat and 82% between Badava Masuri and Indra. Even though Badava Masuri and Swarna are placed in different clusters, somehow they are related based on their similarity level, whereas Katan Dora Sanalu, Pushyami and Nellore varieties are distantly grouped from the clusters. Based on their coefficient of similarity, the SSR analysis showed that most of the rice varieties studied were genetically related.

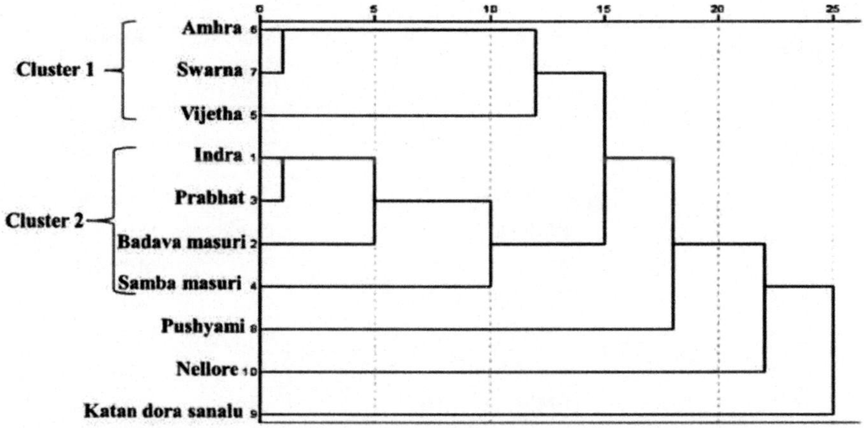

Fig. 10 Dendrogram showing cluster analysis of 10 rice varieties using SSR markers

Table 4 Results of DNA quantification of 10 varieties

S. no	Rice variety name	Quantity of DNA (ng/μl)	Absorbance (260/280 nm) ratio
1	Indra	3115.3	1.88
2	Badava Masuri	1855.4	1.95
3	Prabhat	2394	2.01
4	Samba Masuri	1175.5	1.95
5	Vijetha	3632.2	1.82
6	Amhra	3780.3	1.91
7	Swarna	3736.8	1.85
8	Pushyami	3256	2.09
9	Katan Dora Sanalu	3299.2	1.84
10	Nellore	2430.9	1.98

Results from the dendrogram produced from a genetic distance similarity matrix based on UPGMA showed an overall variation pattern as well as the degree of relation between rice varieties. The dendrogram produced by cluster analysis in this study resulted in two clusters with sub-clusters of seven rice varieties that had a genetic distance ranging from 1.0 to 0.14 of coefficient of similarity. These varieties showed a higher genetic similarity coefficient at level 1.0, indicating that they belong to a similar genetic background. The explanation for this similarity may be due to farmers choosing from a single landrace. A high coefficient of similarity suggests

Table 5 List of nine markers showing polymorphism

Name of the markers	Chromosome	Annealing temperature (°C)	Expected PCR product (bp)
RM562	1	55	100–200
RM543	1	55	100–400
RM12434	2	55	100–400
RM300	2	55	100–150
RM497	2	55	100–200
RM7332	3	55	100–250
RM470	4	55	100–300
RM440	5	55	100–500
RM587	6	55	100–200

that cultivars belonging to a similar genetic history can be caused by occasional outcrossing events that may occur between cultivars randomly or selectively [2]. Also by introducing different varieties to the local rice gene pool, the cultural practices of exchanging seeds between farm families, villages, or regions often contribute to the improvement of genetic diversity within the regions. The independent grouping of varieties Katan Dora Sanalu, Pushyami and Nellore indicated that they are genetically different from other varieties. These divergent varieties may have many good agronomically important traits including pests and disease resistance, tolerance to biotic and abiotic stress. Further examination of these varieties may therefore be important for exploring their distinct characteristics from other materials that are included in this analysis, which can be used as parents of crosses with other accessions from different clusters and for direct selection.

7 Conclusion

In conclusion, our current study showed significant variance amongst the assessed varieties. The varieties were grouped under two clusters based on similarity levels. Based on this study, Katan Dora Sanalu, Pushyami and Nellore varieties were genetically different from other varieties studied. Furthermore, the molecular diversity of the 10 rice cultivars was accessed through SSR markers. In order to create a dendrogram, diversity analysis of 10 different cultivars evaluated based on cluster analysis was performed. Unweighted pair-group method with arithmetic average (UPGMA) method analysis established significant genetic similarity among the assessed cultivars. In order to provide a more definitive report on the selective power of microsatellite markers, genotypes from the same geographical region fall into different clusters and vice versa, indicating that parent choices for hybridization should be about

Table 6 Genetic similarity pair values among studied 10 rice varieties

Similarity matrix										
Varieties	Indra	BadavaMasuri	Prabhat	Samba Masuri	Vijetha	Amhra	Swarna	Pushyami	Katan Dora Sanalu	Nellore
Indra	1.000									
BadavaMasuri	0.818	1.000								
Prabhat	0.833	0.692	1.000							
Samba Masuri	0.615	0.615	0.643	1.000						
Vijetha	0.615	0.500	0.643	0.571	1.000					
Amhra	0.571	0.692	0.600	0.643	0.643	1.000				
Swarna	0.467	0.571	0.500	0.533	0.533	0.846	1.000			
Pushyami	0.538	0.429	0.467	0.400	0.500	0.467	0.375	1.000		
Katan Dora Sanalu	0.200	0.143	0.300	0.190	0.250	0.300	0.368	0.412	1.000	
Nellore	0.333	0.333	0.444	0.389	0.250	0.368	0.300	0.143	0.273	1.000

Table 7 Distribution of varieties to clusters based on the UPGMA method

Cluster	No. of variety	Name of varieties
1a	7,6	Swarna, Amhra
1b	5,6	Vijetha Amhra
2a	3,1	Prabath, Indra
2b	4,3	Samba Masuri, prabhat
2c	2,1	Badava Mashuri, Indra

genetic variation rather than geographical regions. Further research should be carried out using a greater number of samples and markers.

References

1. Rahman MS, Molla MR, Alam MS, Lutfur R (2009) DNA fingerprinting of rice (Oryza sativa L.) cultivars using microsatellite markers. Austr J Crop Sci 3(3): 122–128
2. Chakravarthy BK, Naravaneni R (2006) SSR marker based DNA fingerprinting and diversity study in rice (Oryza sativa L.). Afr J Biotech 5:684–688
3. Causse MA et al (1994) Saturated molecular map of the rice genome based on an inter-speci®cbackcross population. Genetics 138:1251–1274
4. Jones N, Oughum H, Thomas H (1997) Genome mapping molecular markers and marker assisted selection in crop plants. New Phytol 137:165–177
5. Temnykh S, Park WD, Ayers N, Cartinhour S, Hauck N, Lipovich L, Cho YG, Ishii T, McCouch SR (2000) Mapping and genome organization of microsatellite sequences in rice (Oryza sativa L.). Theor Appl Genet 100: 697–712
6. Ravi M, Geethanjali S, Sameeyafarheenand F, Maheswaran M (2003) Molecular marker based genetic diversity analysis in rice (Oryza sativa L.) using RAPD and SSR markers. Euphytica 133:243–252
7. Joshi RK, Behrea L (2006) Identification and differentiation of indigenous non-Basmati aromatic rice genotypes of India using microsatellite markers. Afr J Biotech 6:348–354
8. Ni J, Colowit PM, Mackill DJ (2002) Evaluation of genetic diversity in rice subspecies using microsatellite markers. Crop Sci 42:601–607
9. Ali BY, Samadi J, Wu L, leung H, Wu JL, Leung H (2003) Genetic diversity and relatedness of selected Iranian rice cultivars and disease donors assayed by simple sequence repeats and candidate defence gene markers. Euphytica 113(3): 275–284

Classification of Diabetic Retinopathy Using PSO Clustering and Raspberry Pi

Bhimavarapu Usharani, Raju Anitha, and Ravi Kumar Tata

Abstract Diabetic retinopathy is a disorder that causes diabetes, which includes abnormalities in the retina that trigger vision problems. This paper presents a technique by recognizing the lesions present in the retina and afterwards surveys the influenced retina as indicated by the diabetic retinopathy of various sorts. The current approach identifies the symptoms associated with retinopathy for diabetics. The novelty of the proposed technique is the execution of the particle swarm optimization (PSO) clustering algorithm for detecting diabetic retinopathy in raspberry pi. The outcomes guaranteed the identification and the grade-severity categorization of the diabetic retinopathy execution of the presented PSO clustering model on the publicly available dataset with extreme accuracy, recall and specificity of 99.99%, 99.8% and 99.1%, respectively.

Keywords Diabetic retinopathy · Diabetes classification · Microaneurysms · Haemorrhages · Exudates

1 Introduction

Diabetic retinopathy is a gradual increasing disorder. The main issue of diabetic retinopathy is that there are almost no traces of eyesight fluctuations at the early stages of diabetes, and the seriousness of the diabetic retinopathy is determined by the type and the number of scratches and gashes existing on the retina. The human retina comprises different elements such as optic disc (OD), macula, fovea and blood vessels. Diabetic retinopathy is classified into two types:

(1) Non-proliferative diabetic retinopathy (NPDR);
(2) Proliferative diabetic retinopathy (PDR).

B. Usharani (✉) · R. Anitha · R. K. Tata
Department of Computer Science and Engineering, Koneru Lakshmaiaha Education Foundation, Vaddeswaram, Andhra Pradesh, India
e-mail: ushareddy@kluniversity.in

S. Jyothi et al. (eds.), *Proceedings of the 2nd International Conference on Computational and Bio Engineering*, Lecture Notes in Networks and Systems 215,
https://doi.org/10.1007/978-981-16-1941-0_39

In NPDR the different signs are microaneurysms, haemorrhages, exudates (hard and soft), cotton wool spots and inter retinal microvascular abnormalities. Microaneurysms are the early sign of diabetic retinopathy. Microaneurysms are small red lesions typically measuring 10–125 μ in diameter located at the posterior pole [1, 2]. Exudates are bright, reflective, white or cream-coloured lesions that appear on the retina. Generally, the pixel values of the exudates are brighter than the other region pixels.

Proliferative is a complicated stage of diabetic retinopathy. During this stage new abnormal blood vessels are growing in several regions of the retina. This stage may cause the entire blindness.

The novelty of this research is a classification model for grading the level of the disease by using the raspberry pi. The final classification model is in a position to foresee the seriousness level depending on the pixel of the info fundus picture. Assurance of the seriousness level upheld the pixel of the information picture to frame an official choice, and hence the results are accounted for inside the raspberry pi. The commitments of this examination are summed up as follows:

- An innovative raspberry pi-based model is proposed for the identification and the grade severity categorization of diabetic retinopathy.
- A model for prognosis and classification of the retina fundus images is proposed.
- Implementation of preprocessing to reinforce the quality of the image using raspberry pi.
- Presentation of the PSO clustering to segment the preprocessed images using raspberry pi.
- Our model used the PSO classification model to categorize the severity levels of diabetic retinopathy using raspberry pi.
- Experiments have been carried out using DiaretDB0 and Messidor dataset using the raspberry pi.

The rest of the paper is coordinated as follows: The proposed methodology is examined in Sect. 2. The outcomes are discussed in Sect. 3. Finally, this paper is concluded in the conclusion section.

1.1 PSO Clustering

Clustering is a technology that groups similar data points into the same group from the given dataset [3]. Some applications of the clustering algorithms are image segmentation [4], pattern analysis [5] and image compression [6]. Clustering is based on swarm intelligence, i.e. PSO can be used for clustering analysis [7]. PSO-based algorithms consider datapoints as particles. The initial clusters of particles are collected from other clustering algorithms. The cluster of particles is updated continuously based on the centre of the clusters, velocity, and position of the cluster until the cluster centre converges.

PSO image clustering algorithm is as follows [8]:

1. Initialize every particle of N_c to randomly selected cluster means.
2. For every iteration

   i. For every particle i
   ii. For every pixel Z_p
   iii. Calculate $d(Z_p,m_{ij})$ for all clusters C_{ij}
   iv. Assign Z_p to C_{ij}, $d(Z_p,m_{ij}) = \min\{d(Z_p,m_{ij})\}$
   v. calculate the fitness, $f(x_j(t),Z)$.

3. Find the global best solution (G-Best) $y(t)$
4. Update the cluster centroids and use the following two equations:

   i. $v_i(t+1) = wv_i(t) + c_1r1(t)(y_i(t) - x_i(t)) + c2r2(t)\,(y(t) - x_i(t))$
   ii. $X_i(t+1) = x_i(t) + v_i(t+1)$.

The vantage of PSO clustering is it reduces the impact of preliminary conditions and parallel look for an optimal clustering.

2 State-of-The-Art

The automatic DR grading system automatically concentrates on the detection of the microaneurysms, haemorrhages, exudates, cotton wool spots, abnormal growth of the blood vessels. Some automatic grading system classifies the patient as "disease/no disease" or "referral/no referral". Some systems provide detailed levels of diabetic retinopathy.

Shanthi and Sabeenian [9] performed the preprocessing by using the green channel, used the classifier alexnet and identified the features microaneurysms, microvascular abnormality, neovascularization with an accuracy of 96.6%, 96.2, 95.6, 96.5%, sensitivity of 95.3, 88.0, 90.1, 96.0%, and specificity of 97.3, 98.3, 97.6, 96.6% on MESSIDOR dataset.

de La Torre et al. [10] used the classifier deep learning interpretability and identified the features microaneurysms, haemorrhages, microvascular abnormalities, neovascularisation with an accuracy of 85.7, 91.0%, sensitivity of 90.6, 91.1% and specificity of 84.7, 90.8% on EyePACS, MESSIDOR2 dataset.

Hemanth et al. [11] performed the preprocessing using the resize, r channel, b channel, g channel, HE, CLAHE, concatenation and used the classifier softmax and identified the features microaneurysms, haemorrhages, microvascular abnormalities, neovascularization with an accuracy of 97%, recall of 94% and specificity of 98% on MESSIDOR publicly available dataset.

Wan et al. [12] performed the preprocessing using the normalization, data augmentation, cropped, non-local means denoising, used the classifier softmax and identified the features microaneurysms, haemorrhages, mascular abnormality, neovascularization with an accuracy of 89.75, 95.68, 93.17, 93.73, 93.36, 90.40%, sensitivity of 81.27, 86.47, 90.78, 89.31, 77.66, 88.78% and specificity of 94.07, 97.43, 94.32, 96.49, 93.45, 95.56% on KAGGLE dataset.

3 Methodology

The present study research intends to classify the diabetic retinopathy fundus images into various severity levels with high accuracy. We discussed a computerized model for classifying the severity of diabetic retinopathy. The proposed methodology flow is given in Fig. 1.

3.1 Preprocessing

The preprocessing phase helps to eliminate the imperfections in the retina images and improves the quality of the retina image. In this study, the preprocessing step uses histogram processing, as this technique works on pixels. The stages in preprocessing used are resizing, conversion of RGB image to the green channel, applying the CLAHE algorithm [13], background exclusion, and filtration techniques. The preprocessing phase of the retina image is presented in Fig. 2.

Fig. 1 Proposed methodology

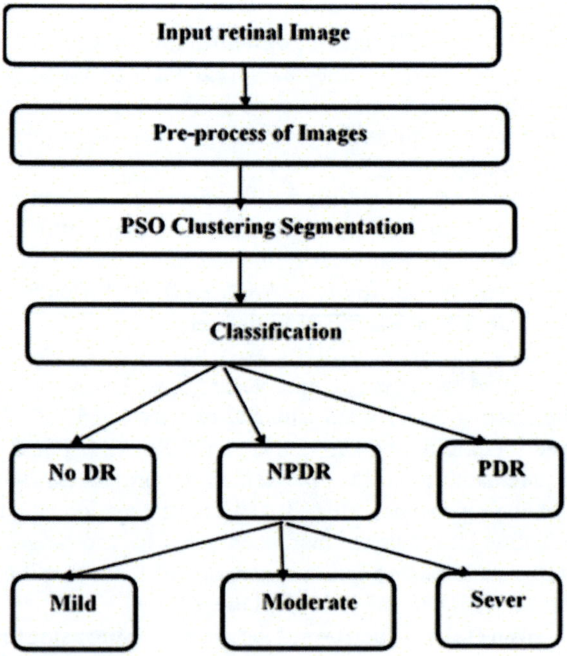

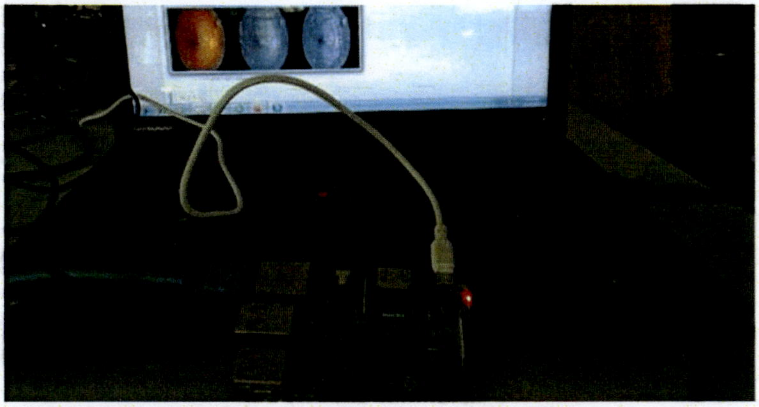

Fig. 2 **a** Resized image, **b** grey image, **c** after applying CLAHE

3.2 *Segmentation*

The main aim of the segmentation is to rearrange or change the retina image's appearance to make it more significant, simple to differentiate and extract the features of the retina image. For better segmentation and to increase the accuracy of the segmented image, optimization algorithms are used. In this work, we are using PSO clustering [3–5] to segment the preprocessed image. The segmentation phase helps in the detection of the microaneurysms from the retina image. We are using two different PSO continuous clustering variants: new continuous and improved continuous PSO clustering algorithms to segment the preprocessed image.

3.3 *Classification*

In this work, we used the convolution neural network (CNN) to diagnose diabetic retinopathy. CNN is a multilayer structure, which employs layers and all layers are interconnected. The ReLU activation function is used and applied with a value of 0.001. The loss function used is the categorical cross-entropy function. During feature extraction two steps have been followed in this paper. The first one is the true detection of microaneurysms, exudates, haemorrhages and abnormal vessel growth candidate extraction using the above discussed various features. The second one is the identification of the mild NPDR, moderate NPDR, severe NPDR and PDR via the usage of the PSO classifier. In PSO, the particles discover the search space by using the Pbest and Gbest experience. A number of tests have been performed by using the feature set to identify the best classification approach.

4 Results and Analysis

Performance evaluation of the diabetic retinopathy detection technique was performed on publicly available datasets—DIARETDB1 and MESSIDOR. DIARETDB1 database based on 129 images consists of mild non-proliferative diabetic retinopathy. MESSIDOR database consists of 1200 fundus images which are collected from the three ophthalmologic departments. For experimentation, raspberry pi with a processor of 1 GHz is used and the programs are written in python on an Intel core i5 3.4 GHz processor. LAN ethernet and Wi-Fi are used for internet connection for the raspberry module (Table 1 and Fig. 3).

Table 1 Comparison of different metrics on different severity levels of DR

S. no	Project	Accuracy	Sensitivity	Specificity
1	Chen et al. [15]	89.9%	N/A	N/A
2	Liu et al. [16]	90.3%	N/A	N/A
3	Ting et al. [17]	N/A	90.5%	91.6%
4	Gulshan et al. [18]	N/A	90.3%	90%
5	Pratt et al. [19]	75%	95%	N/A
6	Wang et al. [20]	91.1%	N/A	N/A
7	Current technique	99.99%	99.8%	99.91%
8	Current technique	99.28%	99.27%	95.86%

N/A: Not applicable

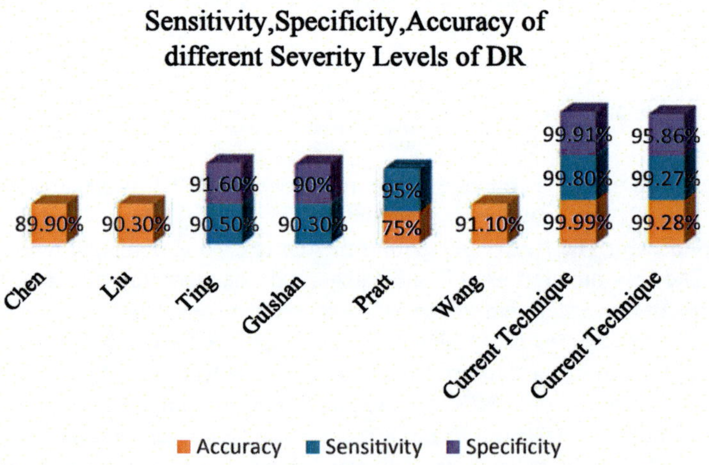

Fig. 3 Sensitivity, specificity and accuracy of different severity levels of DR

5 Conclusion

Unsupervised five-class classification of diabetic retinopathy technique is presented in this paper. This paper has introduced and presented the detection standard for the fundus images using the PSO clustering algorithm. According to the performance analysis, the PSO classifier surpasses all the other classifiers for the MESSIDOR, DIARETDB0 publicly available datasets. Later on, the current framework can be used to consequently recognize the other retinal problems at the beginning phase of diabetic retinopathy and naturally distinguish the other retinal issues at the beginning phase of diabetic retinopathy.

References

1. Mazlan N, Yazid H, Arof H, Isa HM (2020) Automated microaneurysms detection and classification using multilevel thresholding and multilayer perceptron. J Med Biol Eng 1–15
2. Eftekhari N, Pourreza H-R, Masoudi M, Ghiasi-Shirazi K, Saeedi E (2019) Microaneurysm detection in fundus images using a two-step convolutional neural network. Biomed Eng Online 18(1):67
3. Anderberg MR (1973) The broad view of cluster analysis. Cluster Anal Appl 1–9
4. Ray S, Turi RH (1999) Determination of number of clusters in k-means clustering and application in colour image segmentation. In: Proceedings of the 4th international conference on advances in pattern recognition and digital techniques. Calcutta, India, pp 137–143
5. Li X, Fang Z (1989) Parallel clustering algorithms. Parallel Comput 11(3):275–290
6. Lee RC (1981) Clustering analysis and its applications. In: Advances in information systems science. Springer, pp 169–292
7. Kennedy J (2000) Stereotyping: improving particle swarm performance with cluster analysis. In: Proceedings of the 2000 congress on evolutionary computation. CEC00 (Cat. No. 00TH8512), vol 2. IEEE, pp 1507–1512
8. Omran MG, Engelbrecht AP, Salman A (2004) Image classification using particle swarm optimization. In: Recent advances in simulated evolution and learning. World Scientific, pp 347–365
9. Shanthi T, Sabeenian R (2019) Modified alexnet architecture for classification of diabetic retinopathy images. Comput Electr Eng 76:56–64
10. de La Torre J, Valls A, Puig D (2019) A deep learning interpretable classifier for diabetic retinopathy disease grading. Neurocomputing
11. Hemanth DJ, Deperlioglu O, Kose U (2020) An enhanced diabetic retinopathy detection and classification approach using deep convolutional neural network. Neural Comput Appl 32(3):707–721
12. Wan S, Liang Y, Zhang Y (2018) Deep convolutional neural networks for diabetic retinopathy detection by image classification. Comput Electr Eng 72:274–282
13. Zuiderveld K (1994) Contrast limited adaptive histogram equalization. In: Graphics gems IV. Academic Press Professional, Inc., pp 474–485
14. Kennedy J, Eberhart R (1995) Particle swarm optimization. In: IEEE international of first conference on neural networks
15. Chen Q, Peng Y, Keenan T, Dharssi S, Agro E et al (2019) A multi-task deep learning model for the classification of age-related macular degeneration. AMIA Summits Transl Sci Proc 2019:505
16. Liu L, Dou Q, Chen H, Olatunji IE, Qin J, Heng P-A (2018) Mtmr-net: multi-task deep learning with margin ranking loss for lung nodule analysis. In: Deep learning in medical image analysis and multimodal learning for clinical decision support. Springer, pp 74–82

17. Ting DSW, Cheung CY-L, Lim G, Tan GSW, Quang ND, Gan A, Hamzah H, Garcia-Franco R, San Yeo IY, Lee SY et al (2017) Development and validation of a deep learning system for diabetic retinopathy and related eye diseases using retinal images from multiethnic populations with diabetes. JAMA 318(22): 2211–2223
18. Gulshan V, Peng L, Coram M, Stumpe MC, Wu D, Narayanaswamy A, Venugopalan S, Widner K, Madams T, Cuadros J et al (2016) Development and validation of a deep learning algorithm for detection of diabetic retinopathy in retinal fundus photographs. JAMA 316(22):2402–2410
19. Pratt H, Coenen F, Broadbent DM, Harding SP, Zheng Y (2016) Convolutional neural networks for diabetic retinopathy. Procedia Comput Sci 90:200–205
20. Wang Z, Yin Y, Shi J, Fang W, Li H, Wang X (2017) Zoom-in-net: deep mining lesions for diabetic retinopathy detection. In: International conference on medical image computing and computer-assisted intervention. Springer, pp 267–275

A Study on the Sericulture Farm Automation System Using the Internet of Things (IoT)

S. Jyothi, D. M. Mamatha, P. J. Raju, Sufia Sultana, and J. Seetharamulu

Abstract Sericulture is production of silk from silkworm. Central Silk Board (CSB) reports that India stands second in silk production next to China with 15–85%, respectively. This huge difference in silk production between these countries is the lack of automation in the Indian sericulture process. This is one such sector where automation has not been adopted in full-fledged use. The production of high-quality silk depends on the maintenance of environmental factors, such as temperature, humidity, light intensity, and gases in the threshold range in every rearing stage. "internet of things" (IoT) helps in maintaining the rearing environment. It provides the interconnection and tracking among the real-time objects through a network system. The sensors play a vital role in logically sensing the input and send data within the connected machines. This paper consolidates data from various research works on the implementation of smart automation techniques and methodology using IoT platforms in sericulture.

Keywords Sericulture · Internet of Things (IoT) · Sensors · Microcontroller · Automation

S. Jyothi
Department of Computer Science, Sri Padmavati Women's University, Tirupati, India
e-mail: jyothi.spmvv@gmail.com

D. M. Mamatha (✉)
Department of Biosciences and Sericulture, Sri Padmavati Women's University, Tirupati, India
e-mail: drdmmamatha@gmail.com

S. Sultana
JRF-DBT, Department of Biosciences and Sericulture, Sri Padmavati Women's University, Tirupati, India
e-mail: sufiabiotech@gmail.com

P. J. Raju · J. Seetharamulu
A. P. State Sericulture Research and Development Institute, Hindupur, India
e-mail: apssrdidirector@gmail.com

J. Seetharamulu
e-mail: sithara_jolapuram@rediffmail.com

S. Jyothi et al. (eds.), *Proceedings of the 2nd International Conference on Computational and Bio Engineering*, Lecture Notes in Networks and Systems 215,
https://doi.org/10.1007/978-981-16-1941-0_40

1 Introduction

India is the second-biggest producer of silk, the first being China providing 15% of total silk produced. Sericulture is an economical and labor-based business, especially for women living in rural environments. Silk is known for centuries for its luster, and luxury look also had its roots as a suture material in traditional medicine. The usage of silk as biomaterial attracted the world's focus on this biodegradable and biocompatible material with its usage as implants in the medical field. The silkworm larval development depends mainly on environmental parameters, such as temperature, humidity and sterile environment. The production of high-quality silk depends upon these factors. Hence maintaining the optimal conditions in an accurate time is very crucial. The seasonal fluctuation causes an environmental change in the rearing house, which affects the cocoon weight and quality, shell ratio, and finally silk quality. The growth of each molt necessitates a standard set of environmental factors for optimum and high-quality silk. At present, seri-farming practices manually maintain these parameters which lack the automation-sensor node system like IoT. The automated smart sericulture system fills the drawbacks of manual sericulture practices. The IoT plays an important role in maintaining these parameters in an error-prone way with sensors and actuators. An ideal temperature of 23–28 °C and humidity of 65–85% are to be maintained in the silkworm rearing system. The "automated smart sericulture system" is a sensor-based IoT technology, where the network used is smart sensor nodes interfaced with parameters, like temperature, light intensity, gases and humidity that collect readings of the larval lifecycle.

2 Smart Automation Using IoT in Sericulture

There are a wide variety of smart automation model studies performed in seri-farming using various microcontroller boards like Arduino, raspberry pi and NodeMCU depending on the application. A model by Nivaashini et al. [1] that included the internet of things (IoT) empowered wireless personal area network (WPAN) system to maintain the environmental parameters in the rearing house and image processing to identify the condition of rearing house, thus distinguishing various phases in the silkworm lifecycle based on the color images. This entire work model based on an Arduino Board stack integrated with both moisture and temperature sensors along with a camera to capture pictures and songs examines the various phases of sericulture life. The raw data from the serial camera is processed for pre-processing, enhancement and display, and information extraction. For mathematical calculation, Matlab is used by implementing various implicit algorithms.

Arduino is an open-source platform with both physical programmable circuit board or *microcontroller* and a software component such as an integrated development environment (IDE) that enables us to write and upload computer code to the physical board simply using the USB cable. Arduino Uno is a microcontroller board

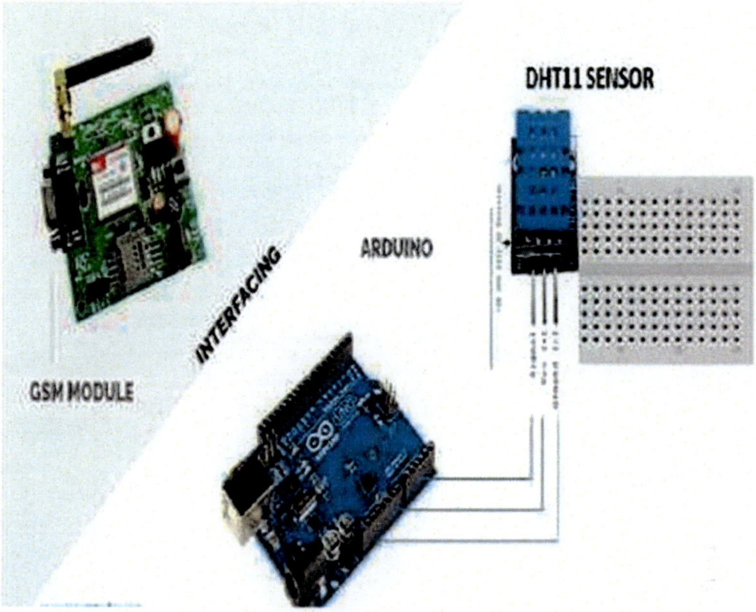

Fig. 1 Components are Ardunio, sensor, GSM module interface [1]

with components such as six analog inputs, USB cable, 14 digital input/output pins, a power jack and reset. The DHT11 sensor reads both temperature and moistness with a row of three-pin package, 3.5–5.5 V power supply. The temperature with a range between 0 and 50 °C, correctness of ±2 °C, moistness ranging from 20 to 95% and a precision of ±5% are sensed by the DHT11 sensor. Figure 1 presents the components of the proposed work model.

GSM Sim900A within frequency 900/1800 MHz with RS232 interface enables connection between the microcontroller and PC with RS232 Chip (MAX232). The GSM/GPRS Modem used has an internal TCP/IP stack that connects the internet via GPRS and facilitates sending SMS, receiving incoming audio calls and web through simple AT commands. With this trio setup with Ardunio, temperature and moistness sensor, GSM module is interfaced which senses the reading of the temperature (<22 °C or >28 °C) and the moistness (<65% or >85%) at a particular threshold range. An automated message via the GSM Sim900A goes to the concerned person when the sensors sense the reading that goes beyond the threshold range. The set of these Arduino with sensors and GSM SIM900 is represented in Fig. 2.

Divya Darshini et al. [2] developed a smart sericulture automation model using IPv6 over low-power wireless personal area network (6LoWPAN) and image processing technique. In the same lines, Arun et al. [3] proposed IoT-based smart sericulture monitoring and automated actuation system using LM35 temperature sensor and DHT11 dual parameter temperature sensor and humidity sensors (20–80% with 5% accuracy). Both the sensors provide an analog output. The three relay circuits

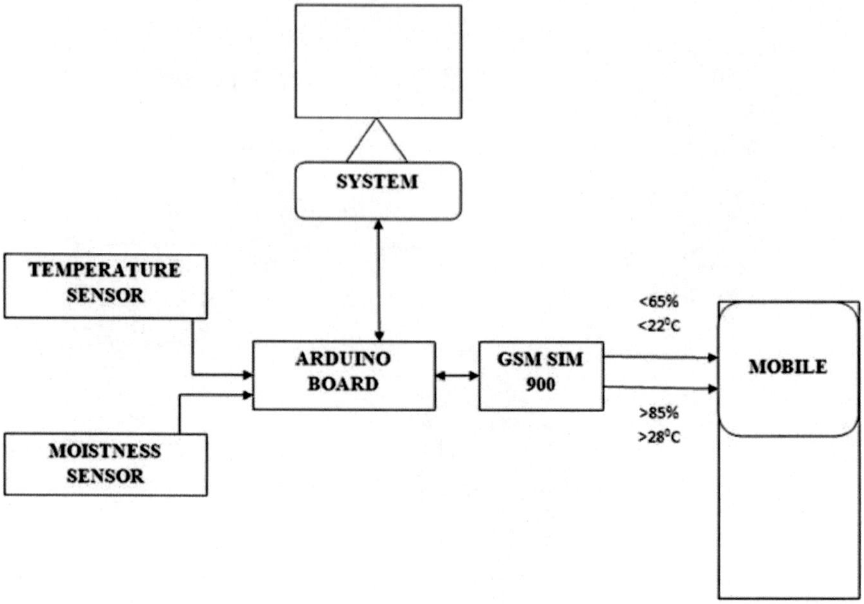

Fig. 2 Systematic design of system [1]

for the three actuation systems, such as air cooler, disinfection system and heater, perform the switching operation, thus enabling the actuator to run for the desired time. The proposed work uses a border router such as TelosB mote connected to raspberry pi, which in turn is connected to the internet. The end-user accesses the data via CoAP Copper (Cu) Mozilla Firefox add-on and stores it in a database for further analysis. The ADC converted data of sensors look for the comparison with the prefixed threshold values of environmental parameters. If the reading of the temperature crosses above the threshold range the cooler will be switched on and if it is below the threshold range the heater actuation system will be switched on. The reading of humidity is also checked against the threshold values and will fit within the threshold range (Fig. 3).

The two image processing techniques are analog and the other digital image processing where analog techniques are used for printouts and photographs processing while digital processing techniques carry out modification to the digital images. The raw data from the camera go for further processing. Matlab is used for various operations such as mathematical computation, development of an application for signal processing with the usage of multiple built-in algorithms. This system uses the change in the body color of a worm as an indicator to identify the different phases where light yellow color denotes the cocoon stage (Fig. 4).

The IoT-based automated sericulture system proposed by Srinivasa et al. [4] uses NodeMCU board, sensors and actuators. The sensor senses the fluctuation in the environmental parameters, like temperature, humidity and light intensity, and thereby

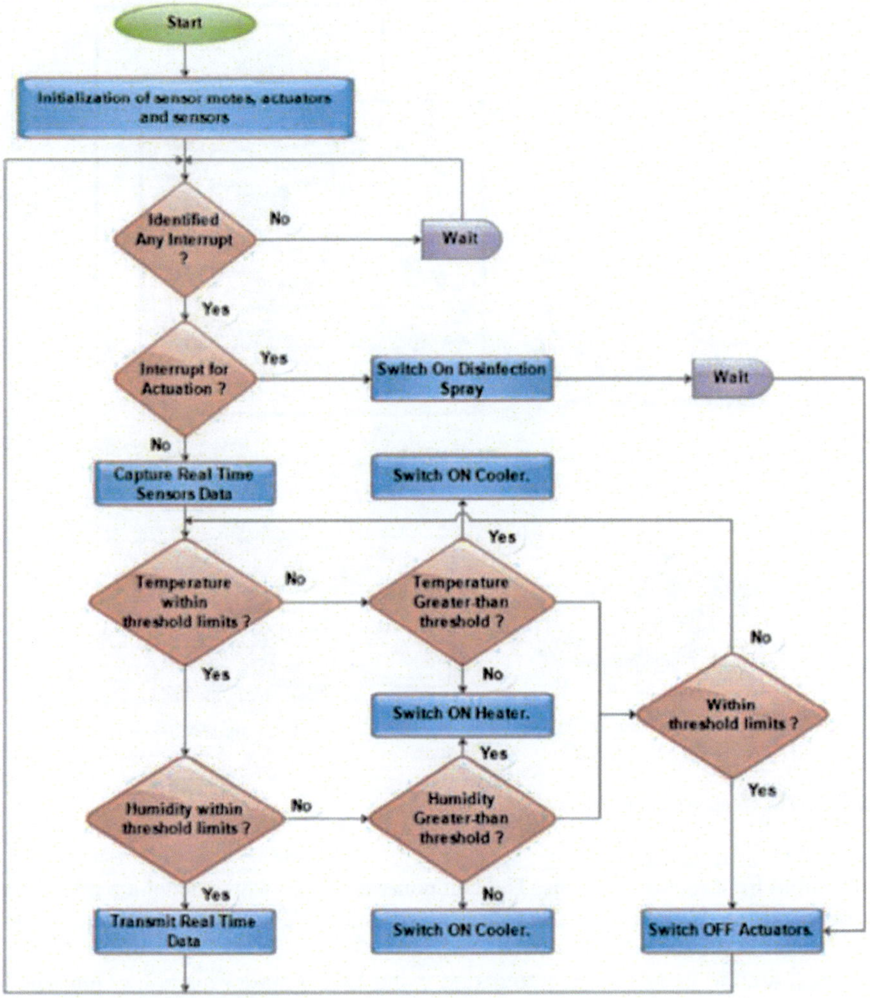

Fig. 3 Flowchart of automation system [3]

sends a notification to the end-user. The code is loaded in the circuit board using the selected ports. The microcontroller receives data from the sensors such as temperature, humidity and LDR and compares it with the threshold values saved in the cloud. The simple C++ program is written in Arduino integrated development environment (IDE) for maintaining the predefined environmental parameters in the silkworm rearing house. The DHT11 is a simple and cost-effective dual-mode digital humidity and temperature sensor that reads the humidity and temperature present in the silkworm rearing house. LDR is a light sensor system that measures the light intensity present in the silkworm rearing house. The LED light turns off and on based on the

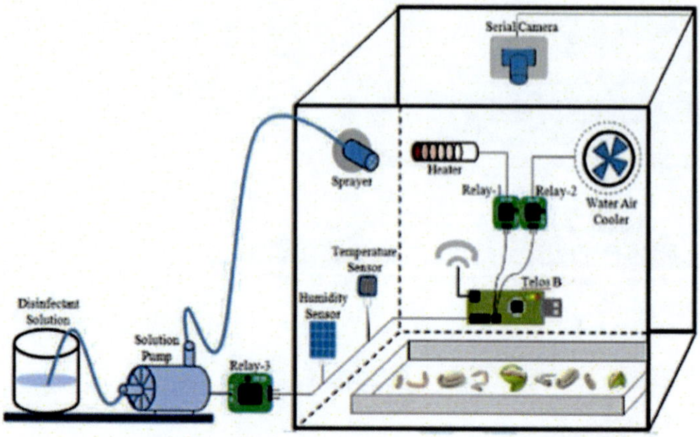

Fig. 4 Block representation of the sericulture monitoring and actuation system [3]

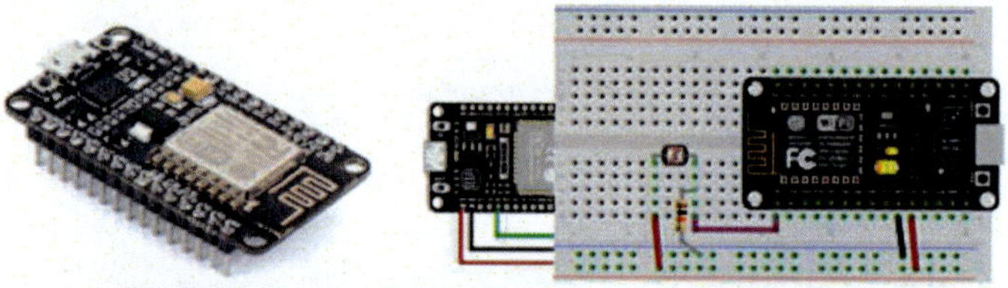

Microcontroller
LDR Sensor

DHT11 sensor

Fig. 5 The components of system [4]

threshold light intensity values. The components of this work model are microcontroller, DHT11 sensor and LDR sensor represented in Fig. 5, and the front view of the model is shown in Fig. 6.

In a similar study on the automation using IoT, a model made by Gunasheela et al. [5] in the seri-farming has both sensing part and actuation part. The sensing components are microcontroller, ADC, all sensors, GSM mobile and the actuation components are microcontroller, motor driving circuit (MDC), temperature control unit (TUC), sprinkler, light source, medicine sprayer unit (MSU) and finally food feeding tray. The proposed framework screens and controls the fluctuating environmental factors on customary premise. The framework has all the essential components for the sensing and actuation of the system. It has sensors, microcontroller, ADC and actuators. This study uses a fire sensor that provides assurance against the flame, along with other basic sensors like temperature, humidity and CO_2. The LM35, a temperature sensor, and DHT11, a dual temperature and humidity sensor, are used in this study to sense the temperature and moistness in the rearing house. The MQ-9 is gas sensor made of Sno2. The overall screening of the edge values

Fig. 6 The front view of prototype [4]

and their control is done by the microcontroller via sensor and actuators. The real-time condition of the rearing house will be sent to the end-user through GSM. The GSM based on AT commands and uses universal asynchronous receiver/transmitter (UART) to send the message. This UART module gets the information from GSM module and transmits it via the SIM card. The LCD in the framework cautions the client about the conditions of the rearing house. This framework used W78E052D microcontroller with low power usage and with in-system programmable (ISP) 2KB LDROM for loader program, using a voltage of 3.3–5.5 V. The Keil micro vision 3 compiler used in this system is the most prevalent C compiler to load the code in the microcontroller (Fig. 7).

There are many such studies conducted in similar lines on smart automation using the internet of things in agriculture and sericulture. The data collected from the sensors and automatic control performed by the microcontroller act as the brain of the whole setup. The interface between the hardware and software occurs giving the end-user about the condition of the setup or rearing house. A wireless sensor network (WSN) with sensor nodes is equipped with a sensor to read the environmental parameters of the environment. The system consists of three components wireless sensor modes (LAU-WSN), wireless sensor management node (LAU-WMN) and connected to the server. A WSN facilitates the end-user to monitor and control the environmental parameters, like temperature, humidity, gases like CO_2 using the fire alarm sensor [6–8]. The applications of IoT are used in the development of a multi-sensor system and along with image processing it is used in the classification of the gender in the silkworm cocoon [11]. The consolidated data collected from the

Fig. 7 Messages to the end-user [5]

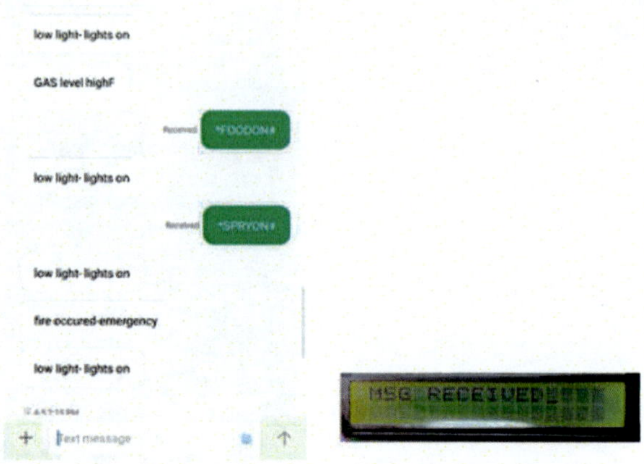

various research studies on the implementation of IoT in the seri-farming along with the technology used are listed in Table 1.

3 Methodology

The proposed methodology envisages benefiting the rural poor with emphasis on women. The on and off farm activities and enormous employment generation potential in this sericulture industry has attracted the attention of both the planners and policymakers to recognize the industry. This industry stands as one among the most appropriate avenues for socio-economic development of a largely agrarian economy like India. The project involves extension-oriented activities to be taken up in the sericulture potential areas of Anantapur and Chitoor districts of Andhra Pradesh. It is an IoT-enabled training and extension intensive project to be jointly implemented by the APSSRDI, Hindupur in Anantapur district and SPMVV, Tirupati in Chitoor district under the following cluster development approach linking end-to-end activity of all the phases of sericulture industry. There are certain areas where technological interventions and related skill up-gradation covering mulberry cultivation and young age silkworm (Chawki) rearing and late age silkworm rearing with supportive prophylactic activities play vital role toward production and productivity improvement. Further, a few new subsidiary areas covering value addition products and seri-business through seri-bazaars have now been added for promoting entrepreneurship, especially suited for the educated women farmers who can participate and promote self-reliance and sustainability in rural society, and with overall socio-economic upliftment with associated financial growth.

Andhra Pradesh State Sericulture Research and Development Institute and Sri Padmavati Mahila University sericulture and Computer Science Departments has come out with several user-friendly ICT, and IoT technologies particularly with

Table 1 Consolidated data of the implementation of IoT in the seri-farming [9–27]

S. no	Smart automation in sericulture	Technology used	Reference
1	Monitoring and controlling the temperature, humidity, light power	NodeMCU microcontroller sends the notification of variations of parameters to the user's mobile	Bhargava Ram et al. [9]
2	Monitoring/controlling environmental factors such as temperature, relative humidity, CO_2 and light power	Arduino microcontroller-based automation system, using ng wireless sensors, GSM and microcontroller for monitoring and controlling environmental factors	Manjunatha and Neelagar [10]
3	Development of multi-sensor system for the gender classification of silkworm cocoon	Based on the image processing and used the support vector machine	Joseph Raj et al. [11]
4	Wireless sensor network sensing/monitoring the conditions of the rearing process	Sensor-based utilizing the BME 280 sensor for temperature and relative humidity and BH17 50 sensor for light intensity with ATmega328/P interface	Doloi et al. [12]
5	Sensing the abiotic parameters in a silkworm incubator	Raspberry pi microcontroller-based system fixed in in the silkworm incubator	Duque-Torres et al. [13]
6	Sericulture plant automation system	Zone-based cascade to monitor and control of environmental parameters	Dixit et al. [14]
7	Development of automatic mulberry leaf feeder for silkworm larvae	Used rover for the automatic mulberry feeder	Ranjith kumar et al. [15]
8	Analysis and implementation of the cloud computing for seri-monitoring application	Cloud computing-based module for the monitoring the parameters used in the field of sericulture	Thirumeni and Vairamani [16]
9	The auto-controlled actuation of exhaust fan, sprinkler and heater in the silkworm rearing house	NodeMCU microcontroller with Wi-Fi connected directly sends notification to the user mobile application	Ashwitha et al. [17]
10	The monitoring and counting of the silkworm egg	Image processing and internet of things (IoT) in egg counting	Jayaraghavan et al. [18]
11	The automation of temperature moistness and feed supplement in the rearing house	The temperature and moisture sensors connected with PIC16F877A microcontroller for automation	Vijayanand et al. [19]
12	Silkworm male and female gender detection using microcontroller	The components used are light emitting diodes as indicators interfaced with microcontroller that are connected with internet	Murlidhar et al. [20]

(continued)

Table 1 (continued)

S. no	Smart automation in sericulture	Technology used	Reference
14	Forecasting the potential/suitable areas for cultivation of mulberry	Remote sensing techniques along with image processing applied. Soil pH and thermal conductivity estimated to find potential areas for mulberry cultivation	Bhandari and Bhandari [21]
15	Silkworm gender separation system	Used optical penetration method and image processing	Kamtongdee et al. [22]
16	Non-destructive testing system for estimation of the quality of silkworm chrysalis	Estimating the silkworm chrysalis quality by analyzing the vibration of acceleration sensor	Li and Shen [23]
17	In the automatic artificial diet feeding of the silkworm larvae	A three-axis robot used to deliver the artificial feed at the predesignated point in the rearing house	Ohura and Li [24]
18	ARM7-based real-time tracking of silkworm in the rearing system	ARM7 controller and image processing-based techniques to plot the real-time track of the silkworm and infections in the rearing house, respectively	Yashaswini et al. [25]
19	Automated monitoring and controlling of the environmental parameters in sericulture system	Sensor-based tracking of the temperature and relative humidity in the rearing for sustainable silkworm cocoon production	Madihalli and Ittannavar [26]
20	Artificial intelligence in the silkworm rearing house control system	The silkworm control system intelligently implements specific changes in the rearing house using artificial intelligence techniques. The data from the respective zones collected by the data acquisition unit (DAU)	Raste et al. [27]

breeds suited to high temperature, high humidity and also to rear in adverse conditions, which need to be transferred to the field much more rapidly through technological interventions and training associated with organized support for market growth and money flow so as to support empowerment of the sericulturists economically and socially with overall development of the industry and to uplift the living standards among the rural masses. The activities involved in seri-farm require proper time-to-time attention fitting into the instinctive qualities of rural folk. Upgradation of knowledge on certain important activities of sericulture, especially covering the new and scientific technologies such as ICT and IoT for cocoon production and related activities of silk production, is the need of the hour. IoT in agriculture has become one of the fastest growing fields (pun intended) in the machine to machine (M2M) space. As it creates current solutions and addresses several business challenges, today, more than ever before, farmers, researchers, and conservationists need a method to more effectively utilize and conserve resources. The most effective way to do is through actionable data, and using M2M communication makes the ongoing collection of that data simple and affordable. One of the critical stages in silkworm rearing is disinfection as it determines the success rate of the crop. IoT-based approach can be applied here by designing real-time sericulture monitoring and disinfection actuating system with image processing technology to identify the stages of silkworm lifecycle. The complete system is designed to control the atmospheric condition inside the sericulture system as per the requirement in each stage of the silkworm larvae. The whole built is interfaced with temperature and humidity sensors with a disinfection actuation system and a serial camera to auto capture the pictures and to analyze it using image processing analysis to check the status on the progress of sericulture. Hence, a project/program encompasses skill upgradation components along with exclusive training programs in sericulture starting from mulberry cultivation, pest management to cocoon production with due emphasis for on-farm training and entrepreneurship development through ICT, IoT and mobile app.

The proposed methodology seeks to bridge the gap between research knowledge and its adoption by promoting chawki resource centers. Despite extensive research and development work, there is a considerable gap in the technologies generated by the research institutes and their adoption by the farmers in the field. The gap has thus affected the production and profitability of the sericulture sector, making it a sector with high untapped potential. Therefore, the new proposal seeks to take the CRCs to the next level of meeting the farmers' needs through ICT-enabled platform of not only providing chawki worms but also other critical inputs such as bio-agents for pest and disease management, disinfectants, implements, and more importantly, provide a platform for sharing new knowledge and experience. Thus the formed new CRCs will grow from being production centers to knowledge centers with a "Cluster Development Approach" by linking end-to-end program of rearing, reeling, spinning, weaving and dyeing with new and innovative designs, which could attract the market to promote indigenous and expert knowledge through well-developed outlets. Internet connectivity handheld devices are developed. Seri-chawki mobile apps can be developed and given as a handy tool to all the sericulturists and government employees. The selected farmers will be imparted with technical knowledge and necessary skill involved in practicing different technologies at the institute by using different methods such as teaching, demonstration of state of art technologies, hands-on training, experience sharing and enterprise development skills in the local language.

Thus the newly formed CRCs will grow from being production centers to knowledge centers with a "Cluster Development Approach" by linking end-to-end program of rearing, reeling, spinning, weaving and dyeing with new and innovative designs, which could attract the market to promote indigenous and expert knowledge through well-developed outlets. Internet connectivity handheld devices are developed.

4 Conclusion

All these automation models in the field of sericulture screened and controlled the environmental factors in the rearing house using various board such as NodeMCU, Arduino and raspberry depending on the application and usage of the sensors for sensing the temperature, humidity, gases and fire. The internet of things (IoT)-based smart automation maximizes crop yield as it reduces human interference, facilitating the end user with timely monitoring of the conditions of the rearing house and with fast actuation process to control the conditions. Controlling the fluctuation of environmental factors in the rearing house in real time under the supervision of human may not be that accurate in comparison with automation and monitoring under the control of IoT. The full benefits to IoT to the seri-farmers is obtained only when the awareness of such automation technology is spread in the seri-farm by conducting workshops and training programs. As these IoT system works on internet connection, the end-user should be in a condition to access to the broad brand.

References

1. Nivaashini M, Soundariya RS, Dinesh kumar A (2018) Silkworm growth monitoring smart sericulture system based on Internet of Things (IoT) and image processing. Int J Comput Appl (0975–8887) 180(18)
2. Divya Darshini B, Adarsh B, Shivayogappa HJ, Navya KN (2016) Automated smart sericulture system based on 6LoWPAN and image processing technique. In: International conference on computer communication and informatics (ICCCI-2016). Coimbatore
3. Arun R, Sandhya P, Kiran DL, Monika SG, Chiranjeevi S (2019) Automated smart sericulture system. J Emerg Technol Innov Res 6. www.jetir.org. ISSN-2349-5162
4. Srinivas B, Kumari K, Goverdhan Reddy H, Niranjan N, Hariprasad SA, Sunil MP (2019) IoT based automated sericulture system. Int J Recent Technol Eng (IJRTE) 8(2). ISSN: 2277-3878
5. Gunasheela TJ, Tali RV, Prathiba SN, Shilpa AP (2018) Implementation of sericulture farm automation using sensor network and GSM technology. Int J Pure Appl Math 119(14): 13–20. ISSN 1314-3395
6. Dixit M, Kulkarni A, Raste N, Bhandari G (2015) Intelligent control system for sericulture. In: International conference on pervasive computing (ICPC). ISBN:978-1-4799-6272-3
7. Le TD, Tan DH (2015) Design and deploy a wireless sensor network for precision agriculture. In: 2nd national foundation for science and technology development conference on information and computer science. ISBN: 978-1-4673-6640-3
8. Manikantan P, Puneet R (2016) An economical wireless sensor network (WSN) based solution for precision agriculture. Capstone Research Project Proposal, ISSN 2319-7242
9. Bhargava Ram P, Sreedhar G, Thulasi Priya N, Makam LB, Reddy M (2020) Intelligent control system for sericulture using IoT. J Xi'an Univ of Arch Technol XII(IV). ISSN 1006-7930
10. Manjunatha, Neelagar MB (2018) Arduino based automated sericulture system. Int J Comput Sci Mob Comput 7(7): 88–95
11. Joseph Raj AN, Sundaram R, Mahesh VGV, Zhuang Z, Simeone A (2019) A multi-sensor system for silkworm cocoon gender classification via image processing and support vector machine. Sensors 19(12): 2656. https://doi.org/10.3390/s19122656

12. Doloi A, Barkataki N, Saikia M, Saikia D (2019) Development of a wireless sensor network based smart multiple ambient conditions sensing system for the rearing process of ERI silkworm. Int J Adv Technol Eng Explor 6(52). https://doi.org/10.19101/IJATEE.2019.650010. ISSN 2394-5443; 2394-7454
13. Duque-Torres A, Rodriguez-Pabon C, Ruiz-Rosero J, Zambrano-Gonzalez G, Almanza-Pinzon M, Rendon OMC, Ramirez-Gonzalez G (2018) A new environmental monitoring system for silkworm incubators. F1000Research, 7: 248
14. Dixit MA, Kulkarni A, Raste N, Bhandari G (2015) Intelligent control system for sericulture. In: 2015 international conference on pervasive computing (ICPC). ISBN 978-1-4799-6272-3
15. Ranjith kumar R, Aravindh P, Kavinesh T, Sathyanarayanan Y, Sedhumaadhavan P (2019) Automatic feeder for sericulture. Int J Innovative Technol Explor Eng (IJITEE) 9(2). ISSN 2278-3075
16. Thirumeni E, Vairamani K (2017) Application of cloud computing in sericulture monitoring. Int J Comput Sci (IJCS J) 5(1), No 26. ISSN 2348-6600 Reference ID: IJCS-283 PAGE NO: 1839-1848
17. Ashwitha R, Vikraman V, Shashank S, Angadi VM, Sindh J (2019) WSN based intelligent control system for sericulture. Int J Res Eng, Sci Manage 2(12). www.ijresm.com. ISSN 2581-5792
18. Jayaraghavan V, Manjunath CR, Soumya KN (2018) A survey: silkworm egg counting and monitoring based on image processing and Internet of Things (IoT). Int J Trend Sci Res Dev (IJTSRD) 2(4). ISSN 2456-6470
19. Vijayanand S, Immanuvel Bright E, Vijay Anand L (2016) Automation in sericulture farm. Int Res J Adv Eng Sci (IRJAES) 1(2)
20. Bairagi MN, Bhosle AK, Valkunje NB, Gawade PL (2016) Detection of male & female silkworm using image processing. Int Res J Multi Stud (IRJMS). ISSN 2454-8499
21. Bhandari G, Bhandari S (2014) Soil based remote sensing techniques for forecasting potential areas for mulberry cultivation. Int J Sci, Eng Technol 2(8). www.ijset.in. ISSN 2348-4098
22. Kamtongdee C, Sumriddetchkajorn S, Chanhorm S (2015) Highly accurate and fast optical penetration-based silkworm gender separation system. In: Proceedings of the SPIE 9659, international conference on photonics solutions, 96590P, 29 July 2015. https://doi.org/10.1117/12.2196082
23. Li X, Shen T (2017) Nondestructive testing system design for biological product based on vibration signal analysis of acceleration sensor. J Vibroengineering 19(3): 2164–2173. https://doi.org/10.21595/jve.2017.17173
24. Ohura M, Li MZ (2001) Automatic artificial diet feeding system for rearing silkworm, Bombyx mori. J Insect Biotechnol Sericol 70: 59–62
25. Yashaswini B, Madhusudhan N, Suresh D (2020) Automated smart sericulture based on IoT and image processing technique 10(6). ISSN 2321 3361 © 2020 IJESC
26. Madihalli DB, Ittannavar SS (2017) Arduino based automated sericulture system. J Adv Sci Technol 14(1). ISSN 2230-9659
27. Raste N, Kulkarni A, Bhandari G (2014) Optimizing yield of silk by zone based intelligent control system 2(7). ISSN 2348-4098

Secured Data Transfer Between Fog Nodes Using Blockchain

R. Priyadarshini and N. Malarvizhi

Abstract Fog computing, being the underlying network between edge devices and cloud data centre, has high security vulnerability. In the existing literature, security concerns in fog computing are been specified in terms of user authentication, secured data exchange, data privacy, and DoS attack. In this paper to address the security issues between fog nodes, a novel approach to introducing blockchain technology has been proposed. Usage of blockchain in fog nodes brings novel architecture as well as storing and transmitting data to trusted fog nodes and keep their references as blocks in all the connected fog nodes in the network. Results from analytical evaluation demonstrate that our proposed work provides security features that include tamper-proof and attacks resiliency in the fog computing environment.

Keywords Fog computing · Fog nodes · Security · Blockchain

1 Introduction

Fog node being the building block of fog computing is the infrastructure implementing the cloud in relation to the edge devices over specific use case of an application. Fog computing emerged as an extension of cloud computing. This mainly concerns a large number of IoT devices and big data volumes of real-time low-latency applications. The computing power of both cloud and fog remains the same, but the fog has a wider distribution. Being close to the end devices, fog computing provides applications, storage and data services to the end-user acting as an intermediate between the end-users and the cloud. It computes and responds to the edge devices earlier, and the computed data are sent to the cloud for future analysis.

R. Priyadarshini (✉) · N. Malarvizhi
Veltech Rangarajan Dr. Sagunthala R&D Institute of Science and Technology, Chennai, India
e-mail: darshini.sr@gmail.com

N. Malarvizhi
e-mail: drnmalarvizhi@gmail.com

S. Jyothi et al. (eds.), *Proceedings of the 2nd International Conference on Computational and Bio Engineering*, Lecture Notes in Networks and Systems 215,
https://doi.org/10.1007/978-981-16-1941-0_41

Security seems to be an important factor to achieve the implementation of the communication between the fog nodes. It needs data confidentiality and authentication in the majority of data transfer and network formation among the nodes. This is required to improve the integrity, confidentiality and privacy of both nodes. A fog node, be a service provider, has to trust the other fog node which is a service acceptor, and vice versa, as a dynamic network is established for the exchange of the data and its computing [1]. To maintain the secured connection, a secret key is maintained by the fog node for mutual authentication, but for a massive number of fog nodes hash invocations and symmetric key encryption and decryption are required to overcome the malicious or malfunctioning of the fog nodes.

In recent times, blockchain technology comes into action in many areas of implementation. In blockchain technology, every transaction is a block with a unique hash value, and adding those blocks in chronological order makes it tamper-proof. Blockchain comes with the characteristics of decentralization, trust and transparent [2], making a way for utilizing the technique in fog computing for achieving trust among the fog nodes. It also supports data confidentiality, data integrity, data validity and its authentication [3, 4].

In this paper, we propose the utilization of blockchain technology in fog computing to enhance and establish a secured connection between the fog nodes for data transfer. Using blockchain will have the trust information tamper-proof and reliable data integrity, enhancing the information availability in the completely computing process. To achieve this, each fog node acts as a miner and mines the block to the blockchain and maintains the chain of blocks. This solution will be effective only when all the fog nodes are consistent.

The rest of the paper is organized as a problem statement in detail in Sect. 2, where the literature review is specified briefly. Then in Sect. 3 the proposed approach is explained in detail. The paper is concluded with the future work for the proposed solution.

2 Problem Statement

Security standards for fog computing have not been characterized and need more research [5]. The general standard principles for any computing environment are integrity, availability, confidentiality and authenticity. On considering the same for the fog computing environment, the following possible malicious behaviors are defined for fog nodes:

- **Packet dropping**: Packets transmitted between fog nodes can be dropped while transmitted from one node to other.
- **Delay in packet**: A fog node can send the packets in delay after completing the computation.
- **No authentication nodes**: A rouge node can enter the fog computing environment with no proper autenthication.

- **Message tampering**: Before sending to the cloud the data can be tampered at the fog node.
- **Stealing of information**: Storing the unauthorized data in the fog node.
- **Consumption of bandwidth**: Wasting of the bandwidth with unnecessary data.
- **Draining of battery**: One node can drain the battery of other fog node, even by knowingly keeping the other node in high energy-consuming state.

Since fog computing is designed upon the traditional networking components, it is also highly vulnerable to security attacks [6]. Also, fog computing will surely inherit security concerns from the virtualized environments as cloud computing, with the understood fact that the execution is done at the network edge for fog computing with a combination of highly heterogeneous devices. Undoubtedly, cloud computing security solutions will not be suitable for fog computing. Fog computing is dynamic in nature, where authentication is still an unresolved issue at different levels [7]. Consider an example, where a fake one can compromise a gateway, which is serving as fog node. The situation gives rise to a man-in-the-middle attack problem [8]. In addition, the edge devices do some of the computations and they have to be shared with sensitive data, where privacy and trust become the major considerations. Thus, a secured communication has to be granted to guarantee data privacy at the edge of the network for the fog nodes.

3 Proposed Approach

The objective of the proposed work is to make a secured connection and data transfer based on blockchain technology between the fog nodes. This framework aims at the implementation of the blockchain technique by considering each node as the miner themselves. Figure 1 gives the overall outline of the proposed system.

System model: In Fig. 1, the considered system is composed of a number of fog nodes. The entire node network is made of a set of physical resources depending upon the applications like sensors, IoT devices, data servers etc. Each node acts as an authenticator to authenticate every other node and also as a miner to mine the new data transaction as a block and add that to the existing blockchain. In detail, the blockchain being a decentralized ledger for all the transactions across the network can perform a transaction without a central authority authorization. When a request for data transfer from one fog node to another fog node is generated, it will be broadcasted to all the participant nodes in the blockchain. The fog node accepts the transaction signed with the device's private key for the authentication of the device. The participant nodes are also the miners, which verify the transaction and will create a new block of transaction data that will be added as a new block to the existing blockchain ledger. The blocks in the blockchain are permanent and tamper-proof.

Miner: The main task of this entity is to check and verify the authentication of the requested node and integrating these values to the new transaction block created.

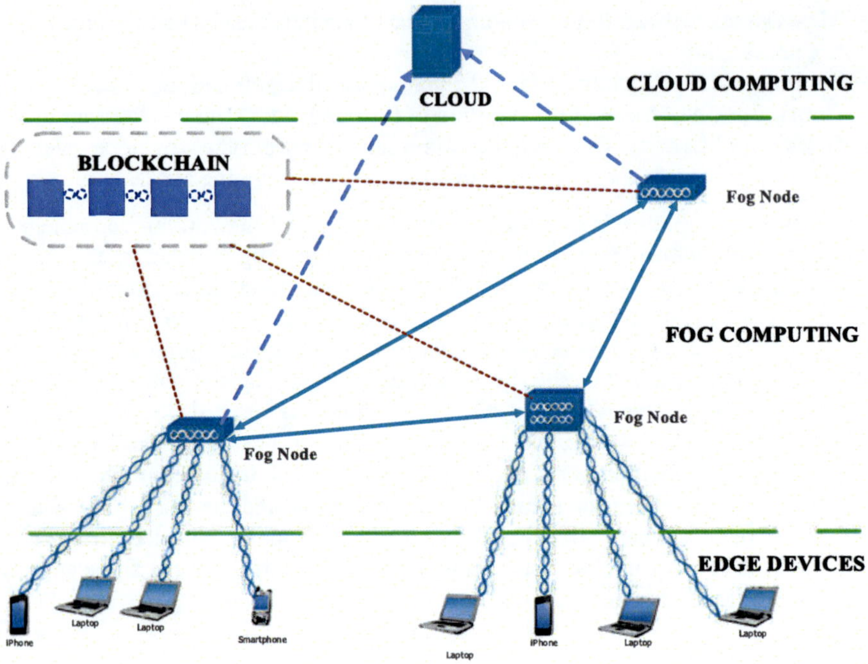

Fig. 1 Fog nodes as miner mining the blocks

The created block is been broadcasted to all the miners in the network, where they all will check the newly created block's validity, and on successful validation, the block is added to the existing blockchain ledger. Understandably, all the miners in the network will have the same complete copy of the blockchain ledger.

In the proposed framework, the exchange of messages and interactions among the fog nodes are considered in three stages (Fig. 2). In the first stage, the fog node authentication of credentials for the device authorization is performed. In the second stage, the data requested is considered as a transaction block and sent to all miners for validation. Thirdly, the validated block will be added to the entire fog nodes blockchain ledger. On unsuccessful authentication of the device, it leads to fog node request rejection.

4 Analytical Evaluation

The proposed framework Metamask using Ethereum is assumed for analytical analyses. Metamask is believed to be used for accessing the Ethereum-enabled distributed applications. Ethereum is one type of blockchain built on distributed web applications to store data to run programs. The data are stored in the form of software called a smart contract, which is shared among computer owners who are part of the Ethereum network. In the Metamask extension, n number of accounts are created, which was considered as a fog node individually. To create a transaction block for the blockchain, it undergoes many key steps. On account creation, each account holds a public key and a private key. When a transaction is initiated for a data request, then they are used for secured encryption and decryption. A smart contract is written on mutual agreement about transactions among the fog nodes. A transaction detail will hold the requested data, requesting fog nodes private key and responding fog nodes public key. Considering these details for a transaction, authentication is done to the requested fog node, and on successful authentication, a block will be generated using a hash algorithm. SHA-256 is been implemented in this procedure. The cryptographically converted hash strings are transmitted in the network. This generated hash string will be representing a block. It is tamper-proof on the basis of utilizing the previous block hash value as a part of input to the new block, while creating a chain of blocks. Each account being the fog node acts as a miner performing this task. The receiver fog node on submission of the private key can view the data block. The data encrypted is immutable, and also can be securely shared among the fog nodes performing the secured data transaction.

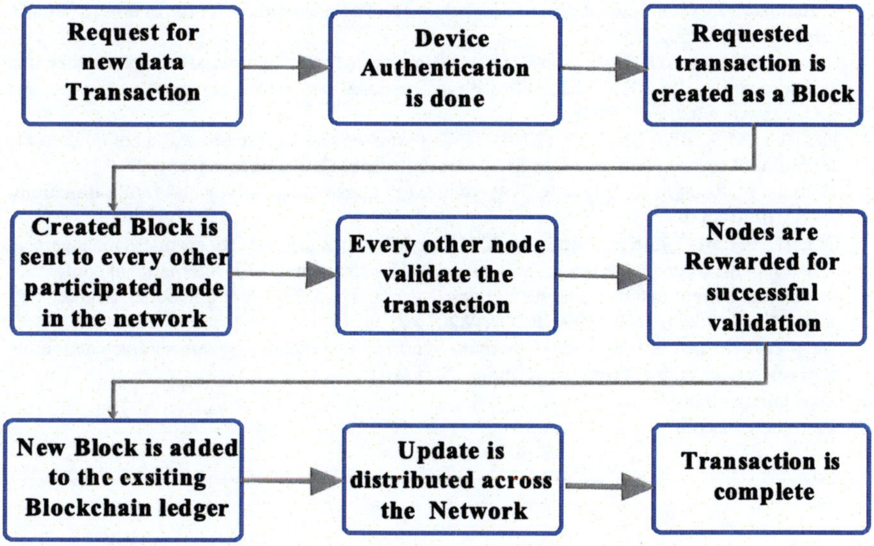

Fig. 2 Blockchain technique

5 Conclusion and Future Work

In this paper, we have proposed the implementation of blockchain technology in the fog computing environment between the fog nodes to ensure secured connection and data transfer. This enables reliable data confidentiality and integrity verification for data storage and sharing among the fog nodes. Also, the blocks generated are tamper-proof and not subjected to modification in the distributed fog network. The analytical evaluation shows the proposed model is deployable and feasible in the fog computing environment. In the future, the implementation of the same scenario can be analyzed in real time for the accuracy of analytics done.

References

1. Mukherjee M, Matam R, Shu L, Maglaras L, Ferrag MA, Choudhury N, Kumar V (2017) Security and privacy in fog computing: challenges. IEEE Access. https://doi.org/10.1109/ACCESS.2017.2749422
2. Yan Z, Zhang P, Vasilakos AV (2014) A survey on trust management for Internet of Things. J Netw Comput Appl 42:120–134
3. Christidis K, Devetsikiotis M (2016) Blockchains and smart contracts for the internet of things. IEEE Access 4:2292–2303
4. Conoscenti M, Vetro A, De Martin JC (2016) Blockchain for the Internet of Things: a systematic literature review. In: 2016 IEEE/ACS 13th international conference of computer systems and applications (AICCSA). IEEE
5. Vaquero LM, Rodero-Merino L (2014) Finding your way in the fog: towards a comprehensive definition of fog computing. ACM SIGCOMM Comput Commun Rev 44(5): 27–32
6. Mahmud R, Kotagiri R, Buyya R, Fog computing: a taxonomy, survey and future directions. arXiv:1611.05539
7. Cirani S, Ferrari G, Iotti N, Picone M (2015)The IoT hub: a fog node for seamless management of heterogeneous connected smart objects. In: 2015 12th annual IEEE international conference on sensing, communication, and networking—workshops (SECON Workshops). Seattle, WA, pp 1–6. https://doi.org/10.1109/SECONW.2015.7328145
8. Wang Z, Jin H, Dai W et al (2021) Ethereum smart contract security research: survey and future research opportunities. Front Comput Sci 15:152802
9. www.ethereum.org
10. www.metamask.io

Computational Screening of Potential Phenolic Phytochemicals Against Targets of SARS-CoV-2 Main Protease (Mpro), S-Protein and RdRp by Molecular Docking

Sarath Nalla, Ramesh Eluri, and Vijetha Pendyala

Abstract Coronavirus rapidly evolved into a pandemic and presents a possible threat to human lives worldwide. As no approved drug is currently available, an urgent need has been developed for COVID-19. For the molecular docking study, 20 polyphenolic phytochemical molecules have been used. All these phytochemicals were selected based on their antiviral potential. The crystallized form of novel therapeutic drug targets SARS-CoV-2 main protease (Mpro), Spike protein (S-protein) and RNA-dependent RNA polymerase (RdRp). This study was conducted to evaluate the efficacy of medicinal plant-based bioactive compounds against SARS CoV-2Mpro, S-protein and RdRp by molecular docking study. Molecular docking study was performed using 1-click docking of mcule software. The target proteins were selected from protein data bank RCS-PDB. Out of 20 selected phytochemicals, Quercetin showed good binding efficacy with a docking score of −9.80 kcal/mol with MPro. Myricetin showed good binding affinity with S-protein with a docking score of −9.5 kcal/mol, whereas, Apigenin showed good binding affinity with a docking score of −9.6 kcal/mol with RdRp. Therefore, Quercetin, Apigenin and Myricetin may be considered to be potential inhibitors of Covid-19 but need to be explored for further drug development process.

Keywords Coronavirus · SARS-CoV-2 · Docking · Myricetin · Apigenin · Quercetin · Polyphenols

---

S. Nalla (✉)
Department of Pharmaceutical Chemistry, University College of Pharmaceutical Sciences, Andhra University, Visakhapatnam 530003, AP, India
e-mail: nallasarat@gmail.com

R. Eluri
Department of Computer Science and Engineering, RVR & JC College of Engineering, Guntur 522019, AP, India
e-mail: eluri.r@gmail.com

V. Pendyala
Department of Pharmacognosy& Phytochemistry, Chebrolu Hanumaiah Institute of Pharmaceutical Sciences, Guntur 522019, AP, India
e-mail: drvijethapendyala@gmail.com

S. Jyothi et al. (eds.), *Proceedings of the 2nd International Conference on Computational and Bio Engineering*, Lecture Notes in Networks and Systems 215,
https://doi.org/10.1007/978-981-16-1941-0_42

1 Introduction

Around five million patients are affected each year by viral infections. Whilst antivirals commonly used to show limited effectiveness and severe adverse effects, herbal extracts have long been recognized for their antiviral properties and less side effects for medicinal purposes. Polyphenolic compounds were reported as potent antiviral agents and many of them were reported to show antiviral effect on influenza virus, coronavirus and other viruses. SARS-CoV-2 (severe acute respiratory syndrome coronavirus-2) is an acute respiratory disorder caused by novel coronavirus (nCoV) and has emerged as a public health issue at the end of 2019 [1–3]. SARS-CoV-2 marked the history in the twenty-first century, after SARS-CoV in 2002 and Middle East respiratory syndrome coronavirus (MERS-CoV) in 2012 [4]. To date, this potentially fatal virus killed thousands of people while millions of people were infected in more than 250 countries around the globe and SARS-CoV-2 cases are still steadily growing due to its rapid human to human transmission (WHO, 2020) [3]. Although few antiviral drugs are being used to treat patients, lack of vaccine or specific drug candidate against SARS-CoV-2 is further worsening the situation. Thus, there is an urgency to identify and develop specific and effective antivirals against SARS-CoV-2 to fight with this deadly virus.

SARS-CoV-2 is a β-coronavirus, a non-segmented enveloped positive-sense RNA virus, with 29.9 kb genome [5]. SARS-CoV-2 utilizes angiotensin-converting enzyme 2 (ACE2) receptors and causes severe respiratory tract infection in humans [6]. Scientists isolated SARS-CoV-2 and sequenced the genome of SARS-CoV-2 on January 7, 2020 [7]. Liu et al. (2020) have demonstrated that principal protease (Mpro) is a target protein with potential for inhibiting SARS-CoV-2 replication. Mpro is the main protein required to mature the virus proteolytically [8]. Mpro inhibitors may have the ability, through inhibition of the cleavage viral polypeptide, to provide effective treatment against SARS-CoV-2.

During the introduction of the virus into host cells, the S-Protein, type 1 glycoprotein in the coronate surface plays an important role. S protein can assist with the viral binding of the host accepter, which is especially important because of its receptor binding role [9]. In SARS-CoV-2, however, a distinct loop with flexible glycyl residues can replace rigid prolyl residues [10]. A single loop of phenylalanine (F486) can enter the hydrophobic pocket of ACE2 and can play a key part in the identification of the recipient [11]. Nsp12 is a viral RNA-dependent RNA polymerase (RdRp) with significant polymerase activity. Three proteins, Mpro (the main protease), RdRp and S-protein, were studied as potential drug targets and selected for the study.

There was no approved drug, currently available for COVID-19, therefore, an urgent need has been developed. The interaction of different phytocompounds with the selected target proteins can be examined using molecular docking analysis. Then, the structure–activity relationship can lead to the development of new products with higher inhibitor activity against SARS-CoV-2. Polyphenolic compounds generally possess the antioxidant potential and all the selected phytochemicals were reported with antiviral activity [12], hence these constituents were selected as ligands for in silico docking studies. In silico screening was a useful tool in meeting the antiviral drug discovery challenges. Screening of virtual compound libraries of natural or synthetic compounds by computational screening methods such as docking saves resources in terms of money as well as time [13]. Plant-based compounds served humans as safer and cheaper candidates against several diseases [14, 15]. Thus, we have screened a small library of natural compounds against selected targets by in silico-based screening and in this study; we report the identification of natural compounds (Quercetin, Apigenin and Myricetin) as potent inhibitors of Mpro, S-protein and RdRp by using molecular docking approach.

2 Materials and Methods

Selection of Ligand:

A total number of 20 phytochemicals that contain polyphenols in their structure were selected as ligands based on previous literature [12]. The list of compounds selected is given in Table 1.

Preparation of target proteins:

All 20 compounds were docked to the selected target proteins. The structures of the target proteins were obtained from RCS PDB and were prepared for docking using the tool prepare protein from the software PlayMolecule, an online tool for the preparation of proteins for docking. The target protein Mpro is designated with PDB ID 6LU7, S-protein is designated with PDB Id 6VXX and the PDB ID of RdRp is 6XQB. All the proteins were validated as therapeutic targets for Covid-19.

Preparation of ligand:

The structures of the ligands were prepared to isolate the proteins, water and diverse substances from each target macromolecule. The structures of the 20 ligands were refined by mcule software.

Table 1 List of Phytochemicals selected

Name of the selected Phytochemicals		
Apigenin	Curcumin	Gallic acid
Arjunolic acid	Ellagic acid	Genistein
Ascorbic acid	Epigallocatechin gallate	Hesperidin
Baicalein	Eugeniflorin	Honokiol
Catechin	Eugenol	Myricetin
Chebulagic acid	Eugenol-o-methyl ether	Quercetin
Chrysin	Forsythoside	

Docking method validation:

The docking methods and parameters used were tested by redocking experiments so as to ensure that the docking studies were accurate and reflecting a fair possible binding model. In order to decide whether the software can replicate guidance, the copy of each native ligand is docked into the native protein

ADME properties:

The pharmacokinetic profile of all the molecules was analyzed in silico by Swiss ADME tool. It also provides information of drug likeliness, lead likeliness and bio-availability. The Lipinski rule notes that the therapeutical activity is measured and calculated by the features like molecular weight, log p, number of donors and acceptors of hydrogen bonding. The selected phytocompounds were analyzed, using the same tool to check the Violations of Lipinski's rules [16]. This rule predicts the molecular properties, which are important for drug pharmacokinetics (ADME) within the human body. As per Lipinski's rule, any component to become an orally active drug should have not more than 5 H bond donors, not more than 10 H bond acceptors, molecular weight less than 500 daltons, partition coefficient A Log P less than 5. Topological polar surface area is another descriptor used to assess the drug transporter properties. TPSA is the sum of surfaces of polar atoms such as oxygen, nitrogen and attached hydrogen. Furthermore, the number of rotatable bonds is important for a conformational flexibility of the molecule in turn decides the binding of receptors. Another descriptor used to measure the properties of the drug transporter is the polar surface region. TPSA is the number of polar atom surfaces such as oxygen, nitrogen and hydrogen connection. In addition, for a conformational flexibility of the molecule, the number of rotatable bonds is essential to decide on binding the receptors.

3 Results and Discussion

The purpose of the ligand and protein docking is to predict a ligand with a protein with known three-dimensional structures in the prevailing binding model [17]. Intermolecular versatile docking simulations were performed and energy values were determined on the basis of the docked conformation of the receptor-ligand complexes for the analysis of binding modes of bioactive compounds at the binding site of target proteins [18]. Lipinski's rule five is always considered in molecular docking for drug development. All the components chosen met the properties of Lipinski.

The phytocompounds had the potential to dock with the target protein and their interaction details are listed in Table 2 and Fig. 1. The phytochemicals exhibited the docking score between −6.4 and −9.8 k.cal/mol against the target proteins. Reports unveil that flavanoids have anti-viral properties and increased intake of flavonoids increased immunity [12]. Furthermore, these are indigenous compounds that have good antioxidant property and help in boosting immunity. Our results are consistent

Table 2 Docking scores of selected phytochemicals against Mpro, S-protein and RdRp

S.No.	Name of the Phytochemical	Docking score on Mpro (6LU7) (k.cal/mol)	Docking score on S-protein (6vxx) (k.cal/mol)	Docking score on RdRp (6xqb) (k.cal/mol)
	Remedesivir	**−8.2**	**−7.8**	**−8.0**
1.	Quercetin	**−9.8**	−9.2	−9.1
2.	Chrysin	−9.3	−9.2	−9.5
3.	Genistein	−9.3	−9.2	−8.6
4.	Myricetin	−9.2	**−9.5**	−8.5
5.	Apigenin	−9.2	−9.3	**−9.6**
6.	Catechin	−9.2	−9.1	−9.3
7.	Curcumin	−9.1	−8.7	−8.5
8.	Honokiol	−8.7	−8.6	−8.4
9.	Hesperidin	−8.6	−8.5	−8.4
10.	Ascorbic acid	−8.5	−8.7	−8.0
11.	Baicalein	−8.4	−8.4	−7.8
12.	Ellagic acid	−8.4	−8.0	−6.9
13.	Epigallocatechin gallate	−8.0	−7.8	−6.8
14.	Forsythoside	−7.8	−6.9	−6.8
15.	Chebulagic acid	−6.9	−7.8	−6.5
16.	Eugeniflorin	−6.8	−7.8	−6.5
17.	Gallic acid	−6.8	−7.5	−6.5
18.	Arjunolic acid	−6.5	−6.5	−6.5
19.	Eugenol	−6.5	−6.5	−8.5
20.	Eugenol-o-methyl ether	−6.5	−6.5	−8.5

with the previous reports stating that a diverse range of flavonoids occur in traditional medicines that exert potent anti-viral activity; the lesser the docking score, the greater the binding energy and hence more effective in blocking activity of a protein. Ligand binding interaction with the target proteins and their analysis can be useful for invention of new drugs from natural resources for COVID-19.

The target protein 6LU7, when counteracted with the phytoconstituent, Quercetin exhibited a least score of −9.8 k.cal/mol. The phytoconstituents, Chrysin, Apigenin and Myricetin, exhibited the docking scores of −9.3 k.cal/mol, −9.2 k.cal/mol respectively. Remedesivir, an FDA-approved anti-viral drug exhibited a dock score of −8.2 kcal/mol with Mpro, which is greater than Quercetin, indicating that Quercetin is having more binding affinity compared to Remedesivir. Myricetin showed good binding affinity with the target 6vxx i.e. S-protein, with a docking score of −9.5 kcal/mol. Apigenin showed the least docking score of −9.6 kcal/mol with

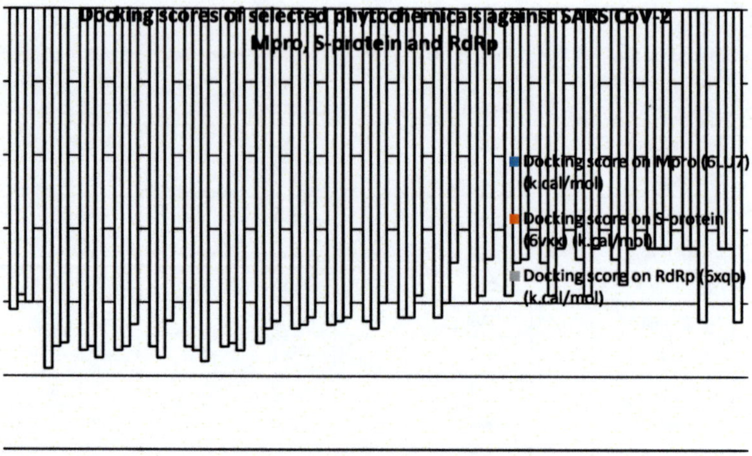

Fig. 1 Docking scores of selected phytochemicals against Mpro, S-protein and RdRp

RdRp (6xqb), which is also lesser than that of Remedesivir (8.0 kcal/mol). The 2D interactions of these protein ligands are depicted in Figs. 2 and 3.

The results of ADME studies are depicted in Table 3. From the results, it was observed that all the phytochemicals, except Chebulagic acid, Hesperidin and Forsythoside were predicted to have a high passive human gastrointestinal absorption (GI) and less BBB permeability. All these phytochemicals were also predicted to inhibit the major CYP isoforms (CYP1A2, CYP2D6 and CYP3A4), which are important in drug elimination through metabolic transformation. The study findings will be useful both in understanding the inhibitory mode and in predicting new inhibitors' operations on the basis of docking results quickly and accurately.

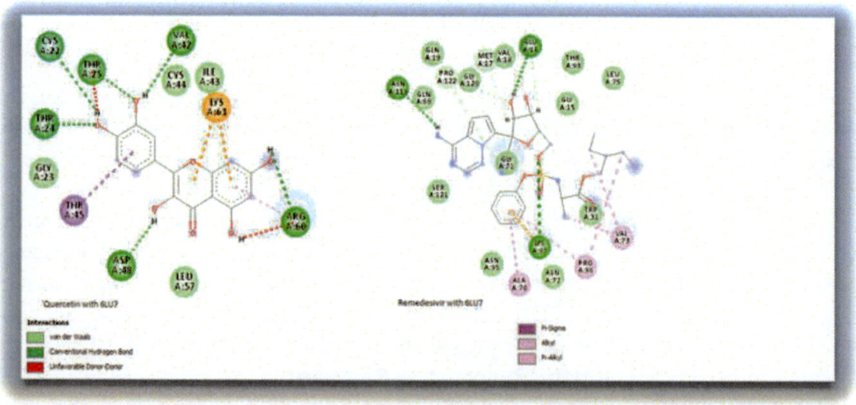

Fig. 2 2D interactions of Quercetin and Remedesivir with the target protein 6LU7

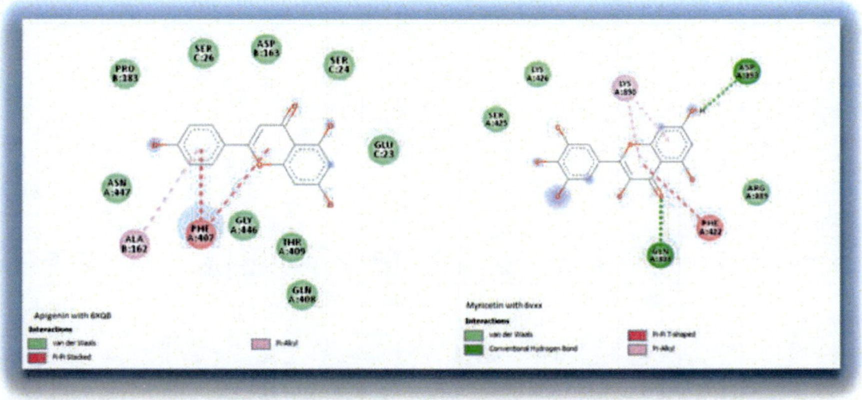

Fig. 3 2D interactions of Apigenin with 6QXB and Myricetin with the target protein 6VXX

New SARS-CoV-2 today poses a serious threat to human health. The treatment of SARS-CoV-2-mediated infections is currently available with no specific clinical therapies. Identifying and characterizing the newer drugs to overcome the health losses caused by SARS-CoV-2 are the needs of the hour. Considering the immediate need for drugs against COVID-19 and the importance of natural products in drug discovery, we screened polyphenolics against selected novel drug targets, Mpro, S-protein and RdRp of nCovid-19 to provide therapeutics of natural origin. Previous studies have documented antiviral effects of flavonoids, catechins, isoflavones on various viral strains [19]. Two previous coronaviral SARS-CoV and MERS-CoV outbreaks demonstrated nature's ability to provide treatment for the continuing COVID-19 epidemic, as they proved to be successfully treated with natural medicines [20–22].

4 Conclusions

The present study envisages the use of plant phenols as new potent anti-viral compounds. The results hoist the compound Quercetin, as potential lead molecules for developing novel anti-viral drugs, which can confer better acceptability.

Table 3 ADME properties, toxicity and drug likeliness of selected phytoconstituents

Molecule	#H-bond acceptors	#H-bond donors	TPSA	LOGP	ESOL Class	GI absorption	BBB permeant	Lipinski #violations	CYP2C19 inhibitor	Bioavailability Score
Apigenin	5	3	90.9	3.02	Soluble	High	No	0	No	0.55
Arjunolic acid	5	4	97.99	5.84	Poorly soluble	High	No	0	No	0.56
Ascorbic acid	6	4	107.22	−1.64	Highly soluble	High	No	0	No	0.56
Baicalein	6	5	110.38	0.36	Soluble	High	No	0	No	0.55
Catechin	5	3	90.9	3.16	Moderately soluble	High	No	0	No	0.55
Chebulagic acid	27	13	447.09	0.36	Moderately soluble	Low	No	3	No	0.11
Chrysin	4	2	70.67	3.52	Moderately soluble	High	Yes	0	No	0.55
Curcumin	6	2	93.06	3.2	Soluble	High	No	0	No	0.55
Ellagic acid	8	4	141.34	1.1	Soluble	High	No	0	No	0.55
Epigallocatechin gallate	9	5	131.34	1.1	Soluble	High	No	0	No	0.55
Eugenol	2	1	29.46	2.27	Soluble	High	Yes	0	No	0.55
Eugeniflorin	27	13	447.09	0.36	Moderately soluble	Low	No	3	No	0.11
Eugenol-o-methyl ether	3	1	38.69	2.22	Soluble	High	Yes	0	No	0.55
Forsythoside	15	9	245.29	−0.5	Soluble	Low	No	3	No	0.17
Gallic acid	7	3	117.2	1.46	Soluble	High	No	0	No	0.55

(continued)

Table 3 (continued)

Molecule	#H-bond acceptors	#H-bond donors	TPSA	LOGP	ESOL Class	GI absorption	BBB permeant	Lipinski #violations	CYP2C19 inhibitor	Bioavailability Score
Genistein	5	3	90.9	3.02	Soluble	High	No	0	No	0.55
Hesperidin	15	8	234.29	−0.14	Soluble	Low	No	3	No	0.17
Honokiol	2	2	40.46	4.98	Moderately soluble	High	Yes	0	Yes	0.55
Myricetin	8	6	151.59	1.18	Soluble	Low	No	1	No	0.55
Quercetin	7	5	131.36	1.54	Soluble	High	No	0	No	0.55

References

1. Coronaviridae Study Group of the International Committee on Taxonomy of V (2020) The species Severe acute respiratory syndrome-related coronavirus: classifying 2019-nCoV and naming it SARS-CoV-2. Nat Microbiol
2. Lee PI, Hsueh PR (2020) Emerging threats from zoonotic coronaviruses-from SARS and MERS to 2019-nCoV. J Microbiol Immunol Infect
3. Zhu N, Dingyu Z, Wenling W, Xingwang L, Bo Y et al (2020) A novel coronavirus from patients with pneumonia in China, 2019. N Engl J Med
4. Wu F, Su Z, Bin Y, Yan-Mei C, Wen W, Yi Hu et al (2020) A new coronavirus associated with human respiratory disease in China. Nature 579(7798):265–269
5. Li Q, Xuhua G, Peng W, Xiaoye W, Lei Z, Yeqing T, Ruiqi R et al (2020) Early transmission dynamics in Wuhan, China, of novel coronavirus–infected pneumonia. N Engl J Med
6. Qiu Y, Yuan BZ, Qiong W, Li JY, Zhou ZJ et al (2020) Predicting the angiotensin converting enzyme 2 (ACE2) utilizing capability as the receptor of SARS-CoV-2. Microbes Infect
7. Lu R, Xiang Z, Li J, Peihua N, Wang W et al (2020) Genomic characterisation and epidemiology of 2019 novel coronavirus: implications for virus origins and receptor binding. The Lancet 395(10224):565–574
8. Structure of Mpro from COVID-19 virus and discovery of its inhibitors.http://www.rcsb.org/structure/6LU7.Ibrahim, IM Abdelmalek DH
9. Elshahat ME, Elfiky AA (2020) COVID-19 spike-host cell receptor GRP78 binding site prediction. J Infect
10. Sun J, He WT, Wang L, Lai A, Ji X, Zhai X (2020) COVID-19: Epidemiology, evolution, and cross-disciplinary perspectives. Trends Mol Med 26(5):483–495
11. Chen Y, Guo Y, Pan Y, Zhao ZJ (2020) Structure analysis of the receptor binding of 2019-nCoV. Biochem Biophys Res Commun
12. Ben-Shabat S, Yarmolinsky L, Porat D, Dahan A (2020) Antiviral effect of phytochemicals from medicinal plants: applications and drug delivery strategies. Drug Delivery Transl Res:1–14
13. Murgueitio MS, Bermudez M, Mortier J, Wolber G (2012) In silico virtual screening approaches for anti-viral drug discovery. Drug Discovery Today Technol 9:219–225
14. Shen B (2015) A new golden age of natural products drug discovery. Cell 163:1297–1300
15. Thomford NE, Senthebane DA, Rowe A, Munro D, Seele P, Maroyi A, Dzobo K (2018) Natural products for drug discovery in the 21st century: innovations for novel drug discovery. Int J Mol Sci 19
16. Lipinski CA (2000) Drug like properties and the causes of poor solubility and poor permeability. J Pharmacol Toxicol Methods 44:235–249
17. Mittal RR, McKinnon RA, Sorich MJ (2009) Comparison data sets for benchmarking QSAR methodologies in lead optimization. J Chem Inf Model 49(7):1810–1820
18. Srivastava V, Gupta SP, Siddiqi MI, Mishra BN (2010) Molecular docking studies on quinazoline antifolate derivatives as human thymidylate synthase inhibitors. Bioinformation 4(8):357
19. Song JM, Lee KH, Seong BL (2005) Antiviral effect of catechins in green tea oninfluenza virus. Antiviral Res 68:66–74
20. Li SY, Chen C, Zhang HQ, Guo HY, Wang H, Wang L, Zhang X, Hua SN, Yu J, Xiao PG, Li RS (2005) Identification of natural compounds with antiviral activities against SARS-associated coronavirus. Antiviral Res 67(1):18–23
21. Lin CW, Tsai FJ, Tsai CH, Lai CC, Wan L, Ho TY, Hsieh CC, Chao PD (2005) Anti-SARS coronavirus 3C-like protease effects of Isatisin digotica root and plant-derived phenolic compounds. Antiviral Res 68:36–42
22. Zhou Y, Hou Y, Shen J, Huang Y, Martin W (2020) Cheng FNetwork-based drugrepurposing for novel coronavirus 2019-nCoV/SARS-CoV-2. Cell Discovery 6

The Research Importance and Possible Problem Domains for NoSQL Databases in Big Data Analysis

G. Somasekhar, Raj Kumar Patra, and K. Srujan Raju

Abstract The days are gone where we depend on traditional database techniques to solve the contemporary database problems. With the evolution of unstructured and semi-structured data forms and due to the frequent gathering of voluminous data from time to time, new techniques to store, process and manage data are becoming most crucial nowadays. NoSQL forms are emerging as the basic source for researchers to invent new database techniques, which can be applied in addition to the conventional techniques in order to solve the existing real-time database problems today. Occasionally, NoSQL can also be termed as Not Only SQL. It is a modern approach to handle data without organizing the source data in the form of tables and sidesteps the use of SQL. In this article, the advantages and disadvantages of NoSQL forms are discussed. The possible research directions in NoSQL domain are outlined to facilitate NoSQL research from the big data perspective.

Keywords Consistency · Database · Datastore · NoSQL · Scalability · Schema

1 Introduction

The state and features of the data are varying nowadays. Most data are unstructured entailing photographs, video, text data, speech recognition or voice commands, sensor data, streaming data, medical or health records, social media blogs and so on. The traditional (RDBMS) techniques failed to fulfill the needs of the above data forms where NoSQL forms came into existence. The invention of new database

G. Somasekhar (✉) · R. K. Patra · K. Srujan Raju
Department of Computer Science and Engineering, CMR Technical Campus, Hyderabad 501401, Telangana, India
e-mail: giddalurisomasekhar@gmail.com

R. K. Patra
e-mail: patra.rajkumar@gmail.com

K. Srujan Raju
e-mail: ksrujanraju@gmail.com

S. Jyothi et al. (eds.), *Proceedings of the 2nd International Conference on Computational and Bio Engineering*, Lecture Notes in Networks and Systems 215,
https://doi.org/10.1007/978-981-16-1941-0_43

techniques has emerged as a primary need in recent times. There is a lot of scope for research in this area. It is a basic requirement for any data scientist or data analyst to be aware of the fundamental characteristics and functionalities of NoSQL forms. The article focuses on the discussion about important aspects and research scope in the NoSQL domain. It discloses various research gaps and problem domains so that any researcher can pick up a particular NoSQL research problem and concentrate on finding the solution.

2 Literature

Many research works were carried out on NoSQL forms. An efficient schema design advisor model for RDBMS to NoSQL transfer was recommended by Namdeo and Suman [1]. It uses the past SQL queries as input and results in all probable mixtures of NoSQL schemas. The cost of each resultant schema is estimated. The performance of each schema is weighed on the basis of select and update queries. A complete study on NoSQL design aspects was done by Davoudian et al. [2]. A systematic and thoughtful analysis of NoSQL stores was made. The quality features of NoSQL databases were assessed for the selection of apt NoSQL form by Lourenco et al. [3]. Certain unknown or unanticipated connections flanked by quality traits of the NoSQL stores were revealed. The exhaustive analysis on RDBMS and NoSQL was made by Sahatqija et al. [4]. The authors highlighted the need for hydrid system, which gains an advantage from both NoSQL and RDBMS compensating for their downsides by each other. A comparison of functional and non-functional attributes among various NoSQL forms was done by Gupta et al. [5]. The enactment of NoSQL data stores in the distributed setting was addressed. Different restrictions for usage of each specific NoSQL form were well explained. The pros and cons of NoSQL espousal for big data applications were stated by Pivert [6]. A number of implementation issues in handling a huge amount of data in cloud were effectively stated. Current trends and challenges of NoSQL data models were discussed by Bhogal and Choksi [7]. With the help of the improved prototype, the significance of NoSQL over SQL was accentuated. The security issues of two NoSQL databases were effectively stated by Okman et al. [8]. Various security aspects of Cassandra and MongoDB were tabulated and described with feasible solutions. A comparative analysis of Cassandra, MongoDB and Couchbase was performed by Bazar and Iosif [9]. The essential things to dwell while working with big data were explained. Different security mechanisms for NoSQL databases were suggested by Kadebu and Mapanga [10]. Several ways to improve the data security of NoSQL environment were suggested. Various opportunities and limitations of NoSQL stores were stated by Liu and Gawlick [11]. A new term "Flexible Schema Data" (FSD) was coined with a valid explanation of better engineering practices to support FSD. A thorough survey on relational versus NoSQL was conducted by Mohamed et al. [12]. The security susceptibilities of NoSQL forms were quantified. The document database and column-oriented database were compared by Manoj [13]. More in-depth analysis of

Cassandra, MongoDB and Hbase was made. The limitations of NoSQL databases in the area of big data were discussed by Zvarevashe and Gotora [14]. Many problems lead by the implementation of NoSQL databases were reiterated.

3 Categorization of NoSQL Forms

NoSQL forms are basically categorized into four types.

- Key-Value stores (e.g., DynamoDB, Riak, Berkeley DB and Redis): In these databases, the data are stored in two portions namely "Key" and "Value".
- Document databases (e.g., MongoDB and CouchDB): These are the extended version of Key-Value stores where values exist in the form of documents.
- Column family stores (e.g., Cassandra, HBase and Kudu): Here, data storage is in the form of segments of data columns.
- Graph databases (e.g., Neo4J and Giraph): The origin of these databases is graph theory (Fig. 1).

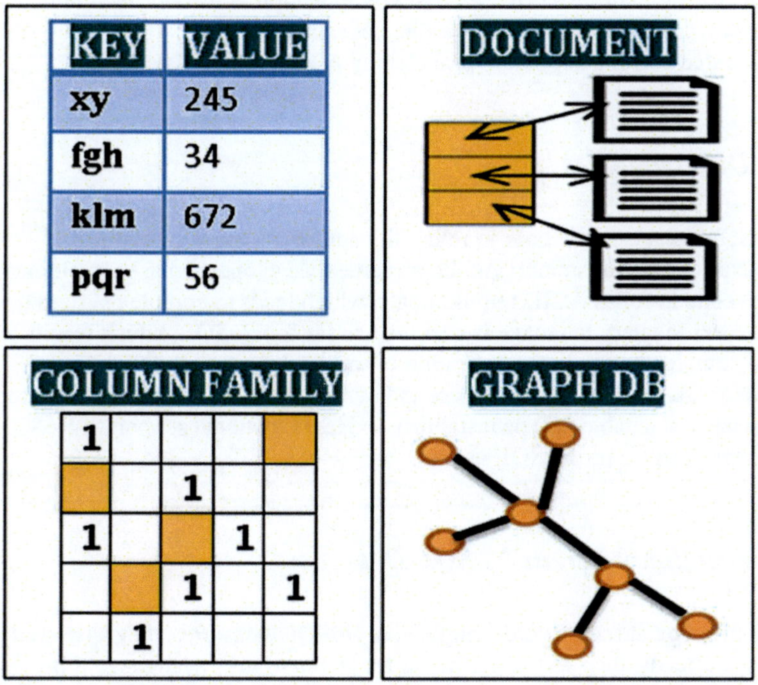

Fig. 1 Types of NoSQL forms

All the NoSQL forms are suitable for big data applications. If we focus on these NoSQL forms, many of the recent real-time problems associated with data could be identified and possible solutions could be investigated.

4 Importance of NoSQL Forms

4.1 Advantages

- Cheap commodity computers can be used for storing and processing NoSQL data stores by which we can minimize cost and improve scalability.
- NoSQL databases are very much suitable for big data applications.
- NoSQL databases need no schemas to start working with data.
- NoSQL databases support auto-sharding. When the data size to be processed increases, the data are spread automatically to store on a random number of natively available servers. The user can not notice auto-sharding effect directly.
- The data recovery, failure handling and data availability are automatically provided by many NoSQL databases. Replication is the key feature of NoSQL databases, which gives the above three advantages.
- Integrated caching support is provided by many NoSQL databases.

4.2 Disadvantages

- NoSQL databases are poor in reliability and consistency aspects.
- As NoSQL do not support ACID properties, developers have to implement their own code to fulfill ACID requirements, which leads to complex system design.
- No specific query language is designed so far for NoSQL, which results in slow and complex nature in the execution of transactions.
- As NoSQL is newly evolved, more and more development is yet needed compared to relational databases. The instability and less functional support of NoSQL have become hurdles for NoSQL users.

4.3 Relational Versus NoSQL Databases

As relational databases are old compared to NoSQL datastores, they have readymade solutions already such as the provision for standard query language, high-security concerns, strong consistency levels and so on. But the NoSQL forms exhibit their own advantages over relational databases. Many recent problems were solved by the NoSQL deployment in the industry.

Table 1 Relational versus NoSQL databases

Characteristic	Relational DBs	NoSQL DBs
Performance	Poor in handling large volumes of data	By using auto-sharding facility, huge data can be spread across servers and efficiently processed
Flexibility Data model	Less flexible due to static schema Can handle structured data only	More flexible due to dynamic schema Can handle all kinds of data including structured, semi-structured and unstructured
Scalability	Can be expensive	Can be achieved with low-cost
Data availability and retrieval Consistency Need for license Security	Replication leads to performance issues Follow ACID model Closed source and license is required High	Replication Improves data availability Follow BASE model Generally open source and may not require license Low

Today, data scientists are searching for better solutions in order to manage the vulnerabilities caused by big data. NoSQL concept should be viewed from the perspective of distributed environment.

The major differences and importance of NoSQL databases over relational databases are shown in Table 1. Various characteristics of relational and NoSQL databases are compared with valid explanations.

5 Research Directions for NoSQL Forms

NoSQL forms are unstable and immature. Still, further development is needed in various facets.

- At the beginning, NoSQL is arranged to satisfy the needs of Web 2.0 applications. As a result, the initial development is in favor of the same. The regular commercial problem scenarios need enhancements further than the basic NoSQL operations. Corresponding research in Business Intelligence tools (BI tools) is more desirable.
- Generation of standard programming language or interface for NoSQL family.
- Investigation on NoSQL security issues.
- Regulation mechanisms to determine the order of execution for concurrent write requests on same database object.
- The efficient data recovery schemes to recover the data whenever the topologies of NoSQL-based clusters are subjected to frequent modifications.
- Efficient backup store procedures to keep track the failure data nodes throughout the large NoSQL-based cluster environment for flexible data management.

- Investigation towards effective deduplication strategies to overcome the dynamic nature of updates across various nodes within the cluster to make NoSQL-deduplication simplified.
- Scalability and performance issues to meet service level agreements (SLAs).
- Data integrity issues pertaining to NoSQL-distributed environment.
- Study of trade-offs among ACID properties for each NoSQL form.
- Study on data consistency issues.
- Each NoSQL form is different from one another. The end-user has more flexibility to choose a suitable NoSQL form according to the situation. A proper NoSQL form can be selected and effectively utilized to create a worthy and innovative product with less engineering cost and more returns. Data scientists may investigate the methods of apt NoSQL form selection depending on the situation.
- The syntax and notation for non-relation schema can be developed.
- Efficient schema development for semi-structured data.
- Efficient schema development for key-value data.
- Support of multiple schemas simultaneously.
- Fast and scalable key-value or wide-column stores that serve as the foundation for multi-modal databases (e.g., Cassandra with Titan, FoundationDB, MarkLogic's document + graph, Sqrrl, etc.).
- Working on resilient distributed datasets in Apache Spark.
- Automatic conversion of unstructured data into structured data.
- Investigation of new transaction processing techniques in distributed environment.
- Efficient data partitioning methods to maintain big data.
- One may work on the flaws of any NoSQL form of his/her own choice depending on the necessities.
- Study for the generation of hybrid datastores consisting of the benefits of both relational and NoSQL databases.

6 Conclusions

An attempt is made to draw the attention of the present-day research community to focus on up-to-date real-time data handling issues. Yet, more financial support and motivation are needed to accomplish the NoSQL research objectives. We hope better database solutions can be found in near future.

7 Future Scope

In future, we will focus on specific problems in each NoSQL form such as Cassandra, Hbase, MongoDB, etc.

References

1. Namdeo B, Suman U (2020) Schema design advisor model for RDBMS to NoSQL database migration. Int J Inf Technol
2. Davoudian A, Chen L, Liu M (2018) A survey on NoSQL stores. ACM Comput Surv 51:1–43
3. Lourenco JR, Cabral B, Carreiro P, Vieira M, Bernardino J (2015) Choosing the right NoSQL database for the job: a quality attribute evaluation. J Big Data 2:1–26
4. Sahatqija K, Ajdari J, Zenuni X, Raufi B, Ismaili F (2018) Comparison between relational and NOSQL databases. In: 41st International convention on information and communication technology, electronics and microelectronics (MIPRO), pp 0216–0221. Croatian Society for Information and Communication Technology, Electronics and Microelectronics, MIPRO, Opatija, Croatia
5. Gupta A, Tyagi S, Panwar N, Sachdeva S, Saxena U (2017) NoSQL databases: Critical analysis and comparison. In: 2017 International conference on computing and communication technologies for smart nation (IC3TSN), pp 293–299. IEEE, Gurgaon, India
6. Pivert O (2018)NoSQL data models: trends and challenges. Wiley publications
7. Bhogal J, Choksi I (2015) Handling big data using NoSQL. In: IEEE 29th International conference on advanced information networking and applications workshops, pp 393–398. IEEE, Gwangiu, South Korea
8. Okman L, Gal-Oz N, Gonen Y, Gudes E, Abramov J (2011) Security issues in NoSQL databases. In: International joint conference of ieee trustcom-2011, pp 541–547. Changsha, China
9. Bazar C, Iosif CS (2014) The transition from RDBMS to NoSQL-a comparative analysis of three popular non-relational solutions: Cassandra, MongoDB and Couchbase. Database Syst J:49–59
10. Kadebu P, Mapanga I (2014) A security requirements perspective towards a secured NoSQL database environment. In: 2014 International conference of advance research and innovation (ICARI-2014), pp 472–480. New Delhi, India
11. Liu ZH, Gawlick D (2015) Management of flexible schema data in RDBMSs–opportunities and limitations for NoSQL. In: 7th Biennial conference on innovative data systems research (CIDR'15). Asilomar, California, USA
12. Mohamed MA, Altrafi OG, Ismail MO (2014) Relational vs NoSQL databases a survey. IJCIT 3:598–601
13. Manoj V (2014) Comparative study of NoSQL document, column store databases and evaluation of Cassandra. IJDMS 6:11–26
14. Zvarevashe K, Gotora TT (2014) A random walk through the dark side of NoSQL databases in big data analytics. IJSR 3:506–509

Classification of Normal/Abnormal Heart Sound Recording Through Convolution Neural Network Through the Integration of Baseline and AdaBoost Classifier

P. Jyothi and G. Pradeepini

Abstract Automatic detection of heart sound detects heart pathology accurately, which can also be examined in non-clinical environments. The goal of normal classification and abnormal classification for heart sounds extracted from phonocardiogram (PCG) and given as input to baseline model and AdaBoost classifier through convolution neural network (CNN) trained using phonocardiogram signals. Efficient improvements in robustness achieve through the integration of AdaBoost classifier and baseline model. Sensitivity can also be improved through integrating the above two models during classification. PCG represents the graphical view of heart sounds produced including the murmurs in which the human ear or trained physician may miss those sounds due to inaudibility.

Keywords Myocardial · AdaBoost classifier and baseline model · Segmentation · Classification

1 Introduction

Heart plays a vital role among other organs in the human body. Cardiac activities and their functionality are examined physically through cardiac auscultation. Cardiac valves' flow of blood makes audibility of sounds. Two normal sounds can be heard from healthy adults, which are S1 and S2. Duration of S1–S2 defines systole and duration of S2–S1 defines diastole. During the start of systole phase of ventricular, S1 will occur and produce sound of low pitch with longer duration, and during the

The original version of this chapter was revised: The authors "P. Jyothi and G. Pradeepini" affiliations have been updated with new affiliation. The correction to this chapter is available at https://doi.org/10.1007/978-981-16-1941-0_79

P. Jyothi (✉) · G. Pradeepini
Department of CSE, Koneru Lakshmaiah Education Foundation, Vaddeswaram, India
e-mail: jyophani.reddy@gmail.com

G. Pradeepini
e-mail: pradeepini_cse@kluniversity.in

S. Jyothi et al. (eds.), *Proceedings of the 2nd International Conference on Computational and Bio Engineering*, Lecture Notes in Networks and Systems 215,
https://doi.org/10.1007/978-981-16-1941-0_44

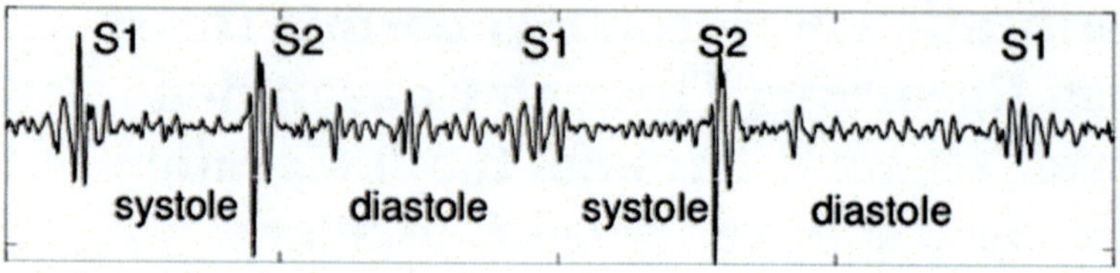

Fig. 1 Heart Sounds of S1 and S2

start of diastole phase of ventricular sound, S2 occurs and produces sound of high pitch with shorter duration. The other heart sounds like S3 and S4 are heard during the interval of systolic and interval of diastolic segments, normally, these murmurs can be heard from children and healthy aged people. Murmurs and acoustic sounds of heart values produce through mechanical events of blood and muscular vibrations can be recorded through phonocardiography (PCG). The PCG signal information can be utilized by physician for diagnosing various abnormalities related to heart of patient, which helps in detecting diseases related to heart [1]. Cardiac auscultation is inexpensive and reliable diagnosis at early stage of detection. Auscultation diagnosis using stethoscope leads to low accuracy and sensitivity due to the dependence of experienced physician [2]. PCG information can be viewed graphically related to murmurs and sounds of heartbeats with digital stethoscope over chest wall of various parts. Sounds of heart S1 and S2 are shown in Fig. 1.

2 Phonocardiography (PCG)

PCG is a device that records the pumping action of the heart [11] and indicates rate of the heart and its rhythmicity along with pumping of the blood and its effectiveness. Stethoscopes that are electronic can determine the heart sounds whose frequency is too high and intensity is too low. Heart cycle can also consist of sounds that are produced from mechanical events [12] as blood flow leakage. The first heart sound represented in the wave is longer in duration and with low frequency and whose intensity is greater when compared to the second sound. Usually, the frequency of sounds ranges from 30 to 100 Hz with 50 to 100 ms of duration. S2 is having higher pitch than the S1 whose frequencies 100 Hz above and whose duration between 25 ms and 50 ms. The other sounds of the heart are S3 and S4 having low intensity, which are inaudible. Blood flow to the ventricles produced through S3 and atria contraction produces sound through S4. These sounds commonly heard among children and from normal adults, it is inaudible [3].

3 Electrocardiogram (ECG)

Heart functioning [12] is recorded through an electrical activity known as electrocardiogram. Electrocardiogram (ECG) displays heart functioning recording of electrical activity synchronizes rhythmically the repeated signal [4]. Heart consists of four chambers, upper contains two layers and lower contains two layers whose chambers in which upper layer called atria and lower layer called ventricles. Related signals from the heart can be reflected through electronic signals called ECGs and heart sound signals through PCGs.

ECG approach is economical and convenient, which is used for detecting various cardiac diseases as diagnosing myocardial ischemia [16] and arrhythmia through clinical medical cleaning, which makes clinicians take reference, which reflects the structure of the heart. ECG automation will monitor the situation of the heart for 24 h along with diagnostic information, which in turn will benefit remote diagnosis and mobile medical applications.

4 Method

Flow chart as shown below

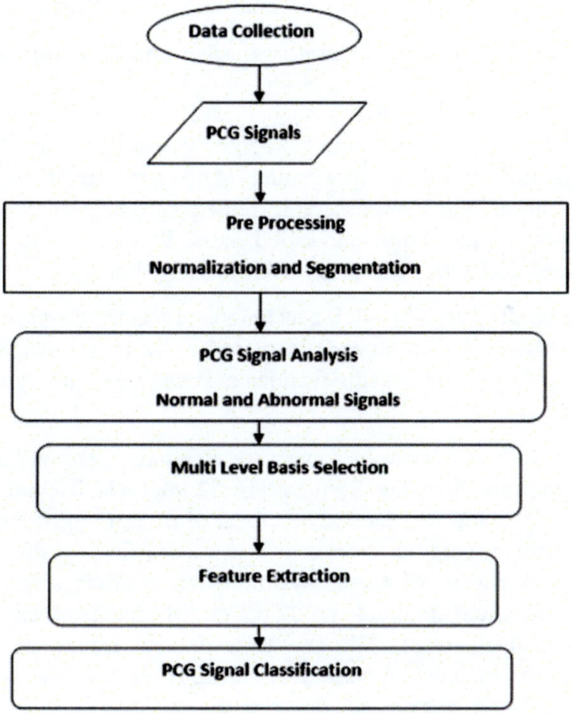

(A) **Pre-Processing**:

To remove spikes in PCG, preprocessing is used to resample [14] each PCG to 1000 Hz and bandpass filter to 25 Hz and 400 Hz. Segmentation method is applied to segment PCGs, which are preprocessed into states of four heart sounds proposed by Springer et al. [5].

(B) **Feature Extraction**

To raw signal, input features are proportional to time-frequency features. In order to capture the pitch of sound and intensity, an efficient method called spectrograms was chosen due to its interpretability and robustness.

(i) **Spectrograms**—Fundamental heart sound (FHS)-based expected duration extraction features of Power Spectral Density (PSD) having window size of 150 ms and stride size of 50 ms. Expected S1 sound is 122 ms and S2 sound expected is 92 ms whose shortest expected FHS is 50 ms

(ii) **Mel Frequency Cepstrum Coefficients (MFCCs)**—MFCC is designed to the perception of a human sound to mimic and used widely for recognition of speech. Extracted coefficients of 13 using window analysis of 25 ms with 10 ms stride.

(C) **Classifiers—Integration of Baseline models and AdaBoost abstain classifier**

(i) **Baseline Models**—Classifiers include Support Vector Machine (SVM), Random Forests and Logistic Regression. Using features of power spectral density and features of energy distribution, logistic regression is trained [6]. Three different kernel functions used are polynomial, linear and radial basis. Random forest is trained with forest size 100 and 200 with package of scikit-learn Python.

(ii) **AdaBoost abstain classifier**—It is a technique of machine learning for the effective building of classifier [13] from weak learners. To classify normal/abnormal PCGs AdaBoost Classifier [7] trained using features of domain Frequency and features of Time domain [8].

- Time-domain features—It involves mean and standard deviation (SD). PCG intervals such as systolic, diastolic, S1 intervals, S2 intervals, RR intervals and ratio of systolic interval and RR interval heartbeat of each, ratio of diastolic and RR interval heartbeat [15] of each and ratio of interval of systolic to diastolic heartbeat. PCG amplitudes consist of during systole mean absolute amplitude ratio and diastole mean absolute amplitude on S2 period for each heartbeat
- Frequency domain features—Discrete-Time Fourier Transform and Hamming Window are used to estimate heart sound state of each power spectrum. Heart sound state can be S1, systole, S2, diastole [9].

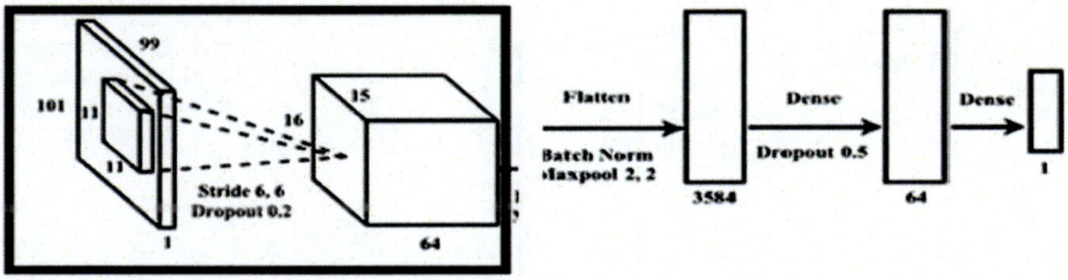

Fig. 2 Convolution neural network architecture

(D) Convolution Neural Network (CNN)-based approach

Convolution layer is used to capture changes made locally. The objective function used is binary cross-entropy and output layer activation [11] function is sigmoid with single neuron. Models are trained through gradient descent stochastic using Adam [10] optimizer with python library Theano and Keras. Adam is used to minimize the loss function and stochastic optimization. The architecture of CNN is shown in Fig. 2.

5 Discussion

The performance classification signal having poor quality is slightly worse than good signal. By integrating baseline models with AdaBoost abstain classifier, the performance of prediction is improved as shown in Table 1 when compared to classifying through AdaBoost classifier alone as described [7]. Normal and good diagnosis the accuracy value is good when compared to other predictions.

Table 1 Normal and abnormal prediction comparison is represented graphically in Fig. 3, Record-wise majority voting is predicted through CNN model, integration of Baseline model and AdaBoost classifier model.

Table 1 Prediction of normal and abnormal diagnosis

Diagnosis SQI	Prediction		
	Abnormal	Normal	%Correct
Abnormal with Good	95	41	0.85
Abnormal with Poor	6	10	0.76
Normal with Good	20	97	0.98
Normal with Poor	10	14	0.92

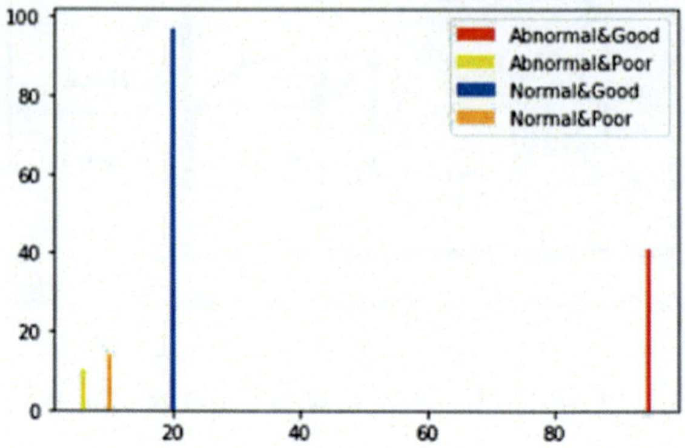

Fig. 3 Record wise prediction through majority voting

6 Conclusion

Classification of normal and abnormal recordings through convolution neural network by integrating AdaBoost abstain classifier and baseline models providing more efficient at normal and good prediction when compared to AdaBoost abstain classifier alone or Baseline model classifier alone performed for classifying the signals for normal and abnormal. During PCG signal analysis, signals are categorized into normal and abnormal signals. Further classification can be further extended to improve abnormal with good prediction using CNN.

References

1. Wen YN et al (2014) Beyond auscultation: acoustic cardiography in clinical practice. Int J Cardiol 172:548–560
2. Sapire DW (1992) Understanding and diagnosing pediatric heart disease: Heart sounds and murmurs. Connecticut Norwalk Applcton & Langc: 27–43
3. Carr JJ, Brown JM (2001) Introduction to biomedical equipment technology
4. Khandpur RS (2003) Hand book of biomedical instrumentation. Tata Mc Graw Hill publication, New Delhi
5. Springer DB, Tarassenko L, Clifford GD (2015) Logistic regression-HSMM-based heart sound segmentation. Springer, IEEE Trans Biomed Eng 63(4):822–832
6. Bangare SL, Pradeepini G, Patil ST (2017) Brain tumor classification using mixed approach. In: International conference on information communication and embedded systems (ICICES 2017)
7. Bryan C, Larry E, Cristhian P, Minnan XW (2015) A dynamic ensemble approach to robust classification in the presence of missing data. Mach Learn
8. Jayanthi N, Babu BV, Rao NS (2017) Survey on clinical prediction models for diabetes prediction. J Big Data 4(1):26
9. Hasane Ahammad SK, Rajesh C (2018) Image processing based segmentation techniques for spinal cord in MRI. Ind J Publ Health Res Develop 9(6):317–323
10. Inthiyaz S, Madhav BTP, Kishore PVV (2018) Flower image segmentation with PCA fused colored covariance and gabor texture features based level sets. Ain Shams Eng J 9(4):3277–3291

11. Kuchibhotla S, Niranjan MSR (2018) Emotional classification of acoustic information with optimal feature subset selection methods. Int J Eng Technol (UAE) 7(2):39–43
12. Narasinga Rao MR, Venkatesh Prasad V, Sai Teja P, Zindavali M, Phanindra Reddy O (2018) A survey on prevention of overfitting in convolution neural networks using machine learning techniques. Int J Eng Technol (UAE) 7(2,32 Special issue 32):171–180
13. Patro P, Kumar K, Kumar GS (2018) Applications of three layer CNN in image processing. J Adv Res Dyn Control Syst 1:510–512
14. Bulla S, Nageswara Rao J, Mothukuri R (2019) Analysis of unsupervised dicretization methods impact on c4.5 classification. Int J Adv Sci Technol 29(5):802–815
15. Syamala M, Nalini NJ (2019) LDA and deep learning: a combined approach for feature extraction and sentiment analysis. In: 10th International conference on computing, communication and networking technologies (ICCCNT) (pp 1–5). IEEE
16. Bodiga VL, Eda SR, Bodiga S (2014) Advanced glycation end products: role in pathology of diabetic cardiomyopathy. Heart Fail Rev

Integration of Artificial Intelligence and the Internet of Things with Blockchain Technology

K. Saritha, Muralidhar Kurni, K. Madhavi, and D. Nagadevi

Abstract Artificial intelligence (AI), blockchain, and the Internet of Things (IoT) give a unique opportunity to the public sector and enterprises. All institutions that leverage these innovations develop advanced goods and services for new customer generation by associating streamline and improving modern processes and building entirely new business models. The internet, mobile devices, and personal computing put a supercomputer in our hands. We have been exposed to distracting technology earlier and recognized that it changes the world as we know it. However, several of these examples from past times appeared in isolation steadily and objectively. Before the origination of the internet, we used personal computing, and the game changed again. We were internet-savvy lay before smartphones put the internet in our pocket. Several new developments are evolving at once: 5G broadband, 3D printing, virtual and augmented reality, plus more. As these technologies provide the business with exciting opportunities and major challenges, we consider three to be truly transformative: IoT, AI, and blockchain. All three alone would have the ability to alter industry, leisure, and culture as a whole. However, together they will have an unparalleled transformational effect. First time in civilized history, the very generation has created three revolutionary innovations; furthermore, it is a sight of what is coming. The digital revolution means technological transformation has shifted from a phenomenon once in a lifetime to a continuous phase. Now any company that falls behind will soon be able to close an insurmountable distance. In

K. Saritha
Sri Venkateswara Degree and PG College, Anantapuramu 515001, AP, India
e-mail: ksaritha.phd@gmail.com

M. Kurni (✉)
JNTUA, Anantapuramu 515002, AP, India
e-mail: muralidhar.kurni@ieee.org

K. Madhavi
Department of CSE, JNTUACEA, Anantapuramu 515002, AP, India
e-mail: kasamadhavi.cse@jntua.ac.in

D. Nagadevi
Department of ECE, CBIT, Hyderabad 500075, Telangana, India
e-mail: dnagadevi_ece@cbit.ac.in

S. Jyothi et al. (eds.), *Proceedings of the 2nd International Conference on Computational and Bio Engineering*, Lecture Notes in Networks and Systems 215,
https://doi.org/10.1007/978-981-16-1941-0_45

this paper, we will look into how you can develop your company for AI, IoT, and blockchain and enable your business to use them both forthwith and in the future. We will explore how the cloud functions as both an enabler and an accelerant—paving the way to adopt any new technology.

Keywords Internet of things (IoT) · Artificial intelligence (AI) · Blockchain · Digital revolution · Transformational technologies

1 Internet of Things: Sense the World

To understand AI, IoT, and blockchain, we accommodate to consider them as interconnected organic processes. IoT is similar to the human nervous system. It senses that a universe of new data recorded with billions of connected devices worldwide [1]. AI is like the brain's imagining part. It thinks by a review of previously reserved data and decisions for humans [2]. Blockchain, furthermore, is like our memory—it provides a secure and persisting record of data transfers and transactions [3]. Figure 1 shows the synthesis of AI, blockchain, and IoT.

IoT transforms a world of things into a world of data. Everything practically from an intelligent look that tracks heart rate, blood pressure, and blood sugar levels to

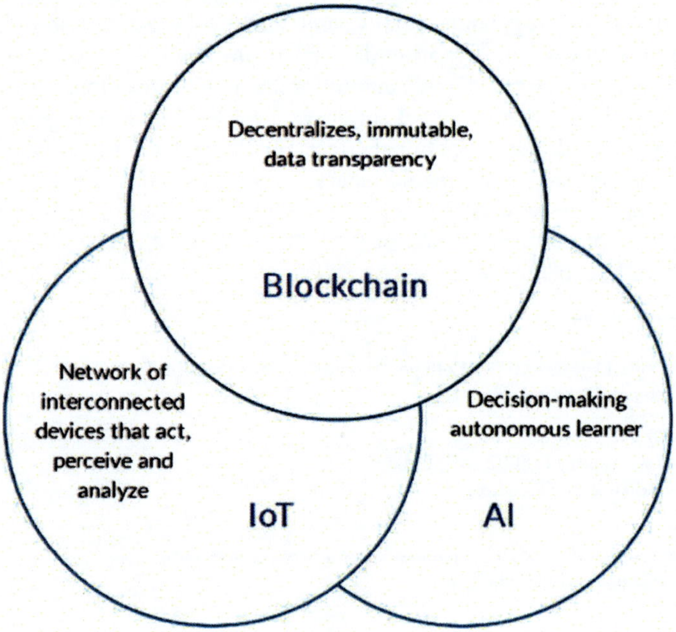

Fig. 1 Synthesis of AI, Blockchain, and IoT [4]

a connected plant that monitors all manufacturing stages can be equated with and made smart.

1.1 IoT in Action

IoT has a more comprehensive impact on some sectors than others [5]. When IoT considers public markets, transportation, manufacturing, consumer goods, automobiles, and healthcare are not identical. It enables collecting new information and enhancing current business processes, swift promotion of inventive goods and services, and collecting information on customer preferences and patterns.

It is easy to imagine IoT in and around the home for consumer applications. A host of personal electronic devices and smart home appliances will lead to consumer products' transformation to revolutionize customer experience and provide retailers with a spate of valuable knowledge. However, IoT's influence is felt far beyond the home.

In automotive manufacturing, IoT allows manufacturers to make connected, autonomous, shared, and electric (CASE) vehicles [6] a usable reality and secure. Data created by these connected vehicles, including numerous other items, to track their driving patterns, create custom coverage options, may be used by insurers, and claim process accurately. Moreover, manufacturers can produce connected, intelligent factories that track health equipment, reduce downtime, production cost, and optimize productivity.

1.2 Going There

In IoT, there is great potential. An estimated 20.4 billion IoT devices will be in circulation, offering higher than 14 zettabytes of data annually. However, both IoT initiatives have been hard to execute and underused; only a small part of the data deluge is processed and implemented.

To achieve the market advantages of IoT, the organization must [7]:

- Develop and adopt an IoT-enabled device ecosystem
- Source, handle, and store large quantities of data
- Introduce advanced analytics and learning capacities for machines
- Build innovative IoT applications using data perspicacity
- Incorporate IoT with current workflows and applications used
- End-to-end security implementation
- Track and control the whole chain of values

In IoT, there is a big chance. Nevertheless, the company's hurdle is to carry on existing infrastructure, technologies, and skills to increase time to assess and lessen

IoT implementation costs and complexity. The rewards for good people would also be important.

2 Artificial Intelligence: Thinking About Data

If IoT is the sensory portion of transformation technology, AI is the thinking component. In the wide field of artificial intelligence, machine learning is ready to have the greatest effect [8]. It has the potential to allow fast, smart decision-making, either to help or replace human intelligence. Businesses can assign mundane or complex tasks beyond workers' capacity to achieve a degree of precision and productivity.

However, it has significant legal, ethical, and even theological ramifications to bring decisions into the hands of smart machines. Whilst AI and machine learning are now conducting intelligent interventions on behalf of people.

2.1 Machine Learning in Action

The acumens of machine learning help the organization to recognize consumer choices better and market dynamics and enable automated, personalized commitments. These help to develop new products and services designed to meet modern customers' needs and efficiently reduce market gaps. Moreover, it will improve business operations with informative strategic feedback and guidance.

Machine learning can transform human capital through increased recruiting, retention of workers, and maximization of productivity. It is the driving force in the automotive industry behind autonomous vehicles. It recognizes and fixes network errors in the telecommunications industry and enables financial service institutions to make customers more accurate. Customer Service chatbots [9] include machine learning capabilities, marketing expertise, personalized products and services, and much more like recognizing cybersecurity vulnerabilities. It is not easy to imagine a technology with comprehensive transformational potential now or across the past.

2.2 Going There

There is little distrust that machine learning would have a profound effect on business. However, why are we not seeing more innovative machine learning products, services, and business models today on the market? Like IoT, the response is to optimize the company.

Machine learning advantages can be more complex than they seem. Implementation is not going to be a case of tossing and watching a turn come to life. It is a progressive operation.

To make use of the true value of machine learning in the real world, the enterprise must [7]:

- Identify opportunities for improved company or supply chain intelligence
- Accumulate and store huge quantities of data—internal and external, structured and unstructured
- Recruiting and maintaining the best talent to make machine learning work
- Fully incorporate machine learning into old and modern applications.
- Set up it over a range of functions alternative of narrow cases
- Use it for present-day infrastructure and capabilities rather than starting from scratch
- Consider legal and ethical concerns surrounding the introduction of AI

The initial challenge was to introduce and leverage machine learning in the business. However, the advantages of performing so are accessible. Enterprises have no alternative but to manage to incorporate machine learning into their business model processes or risk falling behind agile competitors.

3 Blockchain: Committing Transactions to Memory

Blockchain technology is regarded as the underlying Bitcoin and other cryptocurrencies, yet it is much, far more. Imagine blockchain as the framework for trustworthy computing, it provides all forms of data exchanges with reliability, transparency, security, financial transactions, legal and contractual arrangements, and ownership changes.

A blockchain adopts a categorized peer-to-peer network to register all the exchanges, eliminating the need for granted arbiter go-betweens in digital transactions. The result is faster operations, visibility of real-time transactions, and decreased costs in all sectors.

There are few places in which the transforming power of blockchain is not felt. Gartner predicts that by 2025, blockchain could generate $176 billion in value-added revenue, revolutionize the supply chain, encourage new business models, and upset established models [10].

3.1 Blockchain in Action

Blockchain in many sectors and industries, including insurance, financial services, health, human resources, e-commerce, can prove itself a game-changer [11]. Typically, digital information is transmitted everywhere. Blockchain will provide transparency across the entire supply chain in the consumer goods industry by streamlining product recalls, asset monitoring, improving accountability, and improving consumer confidence. It helps to ensure that intellectual property rights are safeguarded in science and education. Moreover, in banking, the rocket fuel that powers the revolution in fintech is blockchain.

3.2 Getting There

As with AI, blockchain and IoT would have embraced and applied more compressively by now, especially given new media speculation on cryptocurrencies and blockchain. However, it is still a comparative rareness outside these agile fintech startups. The issue is well known; perceived danger and uncertainly prevent extensive adoption.

Limits to adoption by blockchain comprise [7]:

- Cost and device resource availability.
- Failure to control blockchain miners.
- Anyone who enters must consent rules of the blockchain network.
- Currently, blockchain contracts are unchecked in court.
- Blockchain shall be compatible with current recording systems.
- Few cases of prevailing use offer a convincing or direct investment return.
- Conventional stakeholders continue to be risk-averse.

There is no question that blockchain is revolutionary. Much of its effect, even on a theoretical level, has yet to be investigated. However, these stumbling blocks must be solved before the organization can explore the blockchain potential's scope.

4 Making Transformation Real

In each of us, feeling, thinking, and remembering combine these three skills to allow all the human effort's wonderful possibilities. IoT, machine learning, and blockchain are both theoretically self-transforming but exponentially more efficient when combined. These three disruptive technologies will change the world together [7].

- ***At home***: Even the new technology today is typically more reactive than constructive. Consider a virtual assistant like Alexa, Siri, or Cortana. Send them a command, and they will answer, order a product, say, or call on your behalf. However, these virtual assistants, powered by transformative technologies, become even more proactive. Shortly, your virtual helper might discover that you are using a specific product and recommend that it tell you how to find the best value by changing purchasing habits and places an order for you. Your refrigerator's IoT data will decide that your milk is inadequate. Machine learning figures out where you get the favorite brand, cheapest, and distributor list. Moreover, blockchain ensures to get precisely what you paid for, and the transaction is safely processed.

- ***Healthcare***: Consider a healthcare study in its current practice that depends on the patient today. You know when you have to see your doctor, usually only if there are noticeable signs, an accident has happened. Moreover, have to make an appointment. Moreover, they also have to recall the important specifics of medical history. Moreover, of course, human beings are frail and distracted. Alternately, assume a nation of patients with sensor equipment. Machine learning tracks IoT sensor information and can assess when something is wrong at an early level. The virtual assistant will reference your calendar with your physician and automatically schedule an appointment. Moreover, blockchain ensures that the doctor has a safe, precise digital medical history once the patient arrives.
- ***Cyber protection***: Modern, more stringent laws, such as the GDPR (General Data Protection Regulation), the mutating threat of cybercrime, and the rising importance and proliferation of consumer data, made cybersecurity universally urgent. However, even here, AI, IoT, and blockchain can transform. These systems largely take the human factor out of cybersecurity processes and, thus, human error. In a centralized repository, log and audit data can be obtained when virtually everything is sensor-fitted. Machine learning can interpret these data much quicker and more reliably than any human being can, take rational decisions, and take autonomous action. Moreover, every critical proof is safely registered through blockchain. It is a system that effectively circumvents the most common reasons for carelessness, data violations, human error, and malicious interference.
- ***In production***: Let us understand how the manufacturing industry reinvents these transformational innovations. Production managers obtain a remote digital view of any machine in each factory via a production management cloud solution. Using machine-learning proactive maintenance, they can see output data for each piece of equipment and predict malfunction in advance. The right technician may be allocated the right job at the appropriate time, a work order can be put, and replacement parts safely purchased via blockchain.

5 Realizing Tomorrow, Today

Disruptive technologies affect the change in our professional lives, personal lives, businesses, and society as a whole. However, their influence does not belong to an ambiguous and faraway future. Most of these features are currently available. So how you utilize the venerable perspective and enable companies to benefit from AI, IoT, and blockchain disruptive impacts? The solution is in the cloud. The truth is that it would be incredibly difficult to introduce any new technology without a cloud. It equalizes the field of play. Moreover, small businesses can reach the same computing power and resources as comprehensive companies via the cloud, while conventional companies can embrace the same versatility as agile technology startups. Cloud facilitates creativity, minimizes IT investment, and speeds up introducing emerging innovations such as those three game-changers.

5.1 Going There

Things to remember in planning your company for technology transformation [7]:

- Focus on IoT, AI, and blockchain.
- Note that the playfield level is the cloud. Without it, you cannot get there.
- Recognize that massive volumes of data underpin these technologies.
- Appreciate that the use and realization of transformative technology can involve a change in business processes or models.

6 Conclusion

With IoT's intuitive ability, AI's and machine learning cognitive strength, and the unfailing blockchain memory, everyone can develop and restore business processes and standards. However, collectively, they are revolutionary, adapting all markets, sectors, and companies into their heads. There has been a technologically powered future. It is time to sculpt your niche. Start today with your transformation.

References

1. Kurni M (2021) Internet of Things (IoT) in Agriculture. In: Multidisciplinary functions of blockchain technology in AI and IoT applications. IGI Global, pp 88–117. https://doi.org/10.4018/978-1-7998-5876-8.ch005
2. Liu J, Kong X, Xia F, Bai X, Wang L, Qing Q, Lee I (2018) Artificial intelligence in the 21st century. IEEE Access 6:34403–34421. https://doi.org/10.1109/ACCESS.2018.2819688
3. Zheng Z, Xie S, Dai H, Chen X, Wang H (2017) An overview of blockchain technology: architecture, consensus, and future trends. In: Proceedings of the 2017 IEEE 6th international congress on big data, bigdata congress 2017, pp 557–564. https://doi.org/10.1109/BigDataCongress.2017.85
4. Parker B, Bach C (2020) The synthesis of blockchain, artificial intelligence and internet of things. Eur J Eng Res Sci 5(5):588–593. https://doi.org/10.24018/ejers.2020.5.5.1912
5. Forbes (2018) How IoT Is impacting 7 key industries today. Forbes. https://www.forbes.com/sites/insights-inteliot/2018/08/24/how-iot-is-impacting-7-key-industries-today/#6d06ac11a845
6. Afshar V (2020) The car of the future is connected, autonomous, shared, and electric. ZDNet. https://www.zdnet.com/article/the-car-of-the-future-is-connected-autonomous-shared-and-electric/
7. Oracle (2018) Transformational technologies : today How IoT, AI, and blockchain will revolutionize business. http://www.oracle.com/us/solutions/cloud/tt-technologies-white-paper-4498079.pdf

8. Balasubramanyam S (2012) New technologies and environments. In: The IFPUG guide to IT and software measurement, pp 385–386. https://doi.org/10.1201/b11884-28
9. Appmomos (2020) AI tools to improve your mobile application. Appmomos. https://appmomos.com/ai-tools-to-improve-your-mobile-application/
10. Gartner (2019) Newsroom Contacts. Gartner Newsroom (August 2019), 2019–2021. https://www.gartner.com/en/newsroom/press-releases/2019-07-03-gartner-predicts-90–of-current-enterprise-blockchain#:~:text=By2025%2Cthe business value, a recent forecast by Gartner.&text=Gartner clients can learn more,by Adrian Lee and others
11. CB Insights (2020) The blockchain report 2020. In: CB Insights. https://www.cbinsights.com/reports/CB-Insights_Blockchain-Report-2020.pdf?utm_campaign=marketing_blockchain_2020-03&utm_medium=email&_hsmi=84940793&_hsenc=p.2ANqtz-8r6RWZHBTipnpnf9o5z8YAo2Mq-9tyJp_jE0YS7MRNFqtjaCWOopxvlgTfZ_98oqHo1er2A08wyuVJnSP4M3HlPsGXAg

Prominent Features in Sleep Disorder Disease in Bioinformatics Using Relevant Data Sets

P. Naga Deepthi, Naga Madhavi Latha Kakarla, and K. Swathi

Abstract In bioinformatics, prominent features in sleep disorders are yet to receive in-depth analysis. However, most of the current literature suggests that the disorders accrue from contributors existing in the form of molecular abnormalities and interconnected aberrant pathways. By establishing some of the pathways and genes with which sleeping disorders are associated, this paper is poised to increase the understanding of some of the risk factors with which the conditions are associated. This paper has integrated data from different investigations in different settings; including genome-wide association studies, the protein-to-protein interaction (PPI) data, and genome-wide expression data. To discover some of the networks with which the sleep disorders are associated (and thus predict prominent features using informatics), some investigations have focused on the seeding of searches via commercial software while others have gained data from PPI databases and gene expression profiles that are publicly available. From the findings, it is evident that several proteins from visceral and subcutaneous fats accounting for the occurrence of sleep disorders. Apart from the visceral and subcutaneous fats, some parameters have also been found to cause sleeping disorder-related stresses. These parameters include insulin signaling, the STAT protein family, and PI3K. As such, the paper concludes that bioinformatics leads to the discovery of prominent features in sleeping disorders due to its capacity to reveal the disorders' predisposing pathways.

Keywords Bioinformatics · Protein-to-protein interaction (PPI) data · Genome-wide expression data · Sleeping disorder

P. Naga Deepthi (✉) · K. Swathi
Department Of Computer Science and Engineering, Koneru Lakshmaiah Education Foundation, Vaddeswaram 522502, AP, India
e-mail: deepthi.viggeswarapu@gmail.com

N. M. L. Kakarla
Department Of Computer Science and Engineering, Sir C R Reddy College of Engineering, Vatluru 534007, AP, India

S. Jyothi et al. (eds.), *Proceedings of the 2nd International Conference on Computational and Bio Engineering*, Lecture Notes in Networks and Systems 215,
https://doi.org/10.1007/978-981-16-1941-0_46

1 Introduction

Globally, sleep disorders have been documented to approach epidemic proportions and affect the elderly and the youth. To advance the field of sleep research, some of the practices that have been embraced include bioinformatics and associated epidemiological, digital, molecular, and behavioral big data [1]. For research efforts aimed at unearthing prominent features behind the occurrence of sleep disorders, major conditions that have received growing interest in bioinformatics include pain, anxiety disorders, sleeplessness, sleep apnea syndromes, and depressive disorder [2]. For sleep disorders, major pathways that have been researched include menopause, secretion, pathogenesis, aging, and circadian rhythm. From a biological perspective, most of the recent studies document that the sleep phenomenon forms one of the least understood areas [3]. However, a discovery that the regulation of sleep occurs genetically has led to growing scholarly interest; especially in bioinformatics. Two broad areas through which prominent features of sleep disorders have been unearthed in bioinformatics include genetic underpinnings linked to human sleep disorders and the establishment of model organisms that aid in discerning molecular mechanisms of sleep [2–4]. The purpose of this paper is to examine prominent features in sleep disorder disease in bioinformatics using relevant data sets. To provide an in-depth analysis of the prominent features, the paper focuses on genetic model systems and molecular model systems that will aid in uncovering some of the regulatory mechanisms that underlie sleep.

2 Understanding Prominent Features in Sleep Disorder Disease

Until about 10 years ago, most of the studies are focusing on sleep emphasized avian and mammalian species. The main reason accounting for this emphasis was that the definition of sleep was based on EEG-recorded altered brain electrical activity [4, 5]. Through EEG signals, the data surrounding the concept of sleep reveals three behavioral states. These states include non-REM (NREM) sleep, rapid eye movement (REM) sleep, and wake [6]. Through the night, humans have been documented to experience 90-min cycles of NREM and REM sleep [5]. As stage 3 of NREM sets in, animal studies suggest that humans are likely to experience sleep's deepest stage in terms of a slow-wave sleep [7]. During REM sleep, the studies avow that the majority of brain areas tends to be active; which implies that NREM sleep accounts for sleep's associated quiescence in neural activity [4].

To address molecular-level biological problems, bioinformatics uses techniques such as artificial intelligence (AI), computer science, statistics, and informatics. In recent decades, sleep medicine has expanded significantly. With an increase in the prevalence of sleep disorders, most of the hospitals' sleep centers have expanded to ensure that they meet the demands of the healthcare industry [5–7]. Thus, bioinformatics had aided in establishing prominent features of sleep disorders; especially with the aim of predicting treatment and diagnosis processes that could be applied to the disorders. Through the wireless recording of sleep and signals with which it is associated, bioinformatics has paved the way for the remote and wide provision of therapy follow-up and diagnostic tools [8].

Despite significant effort relative to the research about sleep disorders, molecular mechanisms underlying these conditions are yet to receive an in-depth analysis [8]. According to most of the recent scholarly investigations, the molecular mechanisms with which sleep disorders are associated remain unclear because the disorders do not emerge as simple conditions linked to a few proteins or genes [5–8]. Rather, the disorders have been documented to exist as manifestations of numerous molecular abnormalities [9], as well as multiple aberrant pathways that remain interconnected [10, 11]. Furthermore, the failure to clarify molecular mechanisms behind the existence of sleep disorders has led to the conditions' emergence as risk factors for other disorders; including oxidative stress and inflammatory states [8, 9, 12]. Hence, bioinformatics has evolved as an important field through which the sleep disorders' associated genetic variants could be identified. Through approaches such as case–control, genome-wide and family studies, bioinformatics has attempted to offer prognostic and diagnostic data associated with sleep disorders [13].

Through network modeling, sleep disorder disease-related genes accruing from protein–protein interactions have been investigated [14]. Particularly, these investigations have aided in uncovering sleep disorder-associated protein-to-protein sub-networks. Two major techniques through which these goals have been achieved include adipose microarray data-integrated genome-wide procedure and adipose tissue microarray data-integrated candidate gene procedure [15]. In the latter approach, p-values are mapped to adipose tissue-specific interactive nodes, which arise from gene expression data [15]. On the other hand, the first approach is achieved via the use of sleep disorders' (and their associated co-morbidities') associated proteins, which lead to the seeding searches of human PPIs to ensure that the disease phenotype's underlying candidate sub-networks are identified [16]. To ensure that the sleep disorders' significantly associated sub-networks are identified, most of the investigations in bioinformatics have also relied on Cytoscape-based tools [17].

During the network construction processes, the PPI networks are established using two methods. Initially, knowledge regarding previous genetic characteristics and the sleep disorders' underlying biology is used to inform the seeding of proteins or genes [18]. For pathways that represent the sleep disorders' intermediate phenotypes, most of the current literature on bioinformatics suggests that the majority of the target genes tends to fall in one or more of these pathways [16, 17]. These pathways include ventilator control pathways, inflammation, obesity, and craniofacial morphology [18]. To assure successful construction of networks, the software that has received increasing application entails Ingenuity Pathway Analysis (IPA).

In bioinformatics seeking to determine the sleep disorders' prominent features, the IPA software procedure involves:

- Combining seed proteins into the rest of the networks through which specific connectivities of those proteins are maximized.
- Adding more proteins obtained from the Ingenuity Knowledge Base to ensure that combinations or connections of at least two smaller networks lead to the realization of larger networks [19, 20].

In the data processing of gene expression, the cDNA microarray has been employed toward measuring the visceral and subcutaneous fat tissues' mRNA expression data [21]. Also, the Bioconductor's aroma package has been used to generate expression values while background normalization and correction have been achieved using quantile normalization and robust multichip average (RMA) methods [20, 21].

Given a protein set $S = \{g1, g2,...,gk\}$, with $gi \in S$ denoting the expression level of mRNA and mRNA depicted by $e(s) = \sum_{i=1}^{k} z_i/\sqrt{k}ei$, the resultant sub-network activity (*S*) has been established as:

The procedure above is followed by p-value mapping to the adipose tissue-related ineractome map's nodes or proteins. For the respective networks' protein (*gi*), a p-value that is deemed the most significant tends to be designated as that is associated with sleeping disorders [22, 23]. With each network represented by $S = \{g1, g2, \ldots, gk\}$, the score of such networks, through which sleep disorders are predicted, becomes:

$$A(s) = \sum_{i=1}^{k} z_i/\sqrt{k}.$$

2.1 Scholarly Findings

To ensure that sleep disorder-related seed proteins are used toward the generation of networks, IPA is employed. With each network exhibiting proteins, the broad categories into which networks are grouped include respiratory disease, neurological disease, genetic disorder, organismal abnormalities and injury, and neurological disease [24, 25]. As mentioned earlier, visceral and subcutaneous fat tissues aid in providing mRNA expression data; with these data aiding further in transcriptional dysregulation. For the case of the subcutaneous fat tissue's expression data from which sub-networks and their associated nodes have been obtained, the IPA-generated network becomes:

On the other hand, the visceral tissue's sub-network and its associated nodes become:

In the IPA-generated networks above, seed proteins are represented by the blue protein names. During the analysis of such sub-networks, the frequency with which proteins occur is calculated and aids in understanding and predicting parameters associated with sleep disorders. As documented by most of the previous scholarly investigations, a central assumption governing this calculation is that the most frequently appearing proteins tend to prove significant toward the definition of differences between the controls and phenotypes of sleep disorders [22–25].

Indeed, the results demonstrated in most studies are in concurrence that there is a direct relationship between certain proteins and their associated sleep phenotypes and sleep disorders [26]. For instance, obstructive apnea apex as a sleep disorder exhibits a correlation with the follicle-stimulating hormone (FSH); especially when the second network above is analyzed. Regarding animal studies that have examined the rats' response to hypoxia, findings illustrate that disordered breathing accrues from PDGF BB, which reflects the first network's subcutaneous fat protein (see Fig. 1 above). In flies, sleep patterns have also been documented to be altered by the first network's visceral fat protein that occurs in the form of the extracellular signal-regulated kinase [27]. It is also worth noting that human studies established in contexts such as those involving European Americans, sleep disorder has been established to arise from SNP missense coding in Endothelin 1 EDN1. The latter (EDN1) also holds that when studies focus on the mice, it accounts for increased cases of respiratory distress; a predictor of obstructive sleep apnea (OSA) [27, 28]. It is also worth indicating that hypertension is associated with one of the potent vasoconstrictors in the form of EDN1 and that this protein causes OSA because hypertension with which it correlates is a predictor of the sleep disorder [29, 30].

Notably, the proteins indicate above exhibit a direct relationship with sleep disorders. In sub-networks such as those established above, some proteins are associated with processes that cause sleep disorders. For example, inflammatory responses tend

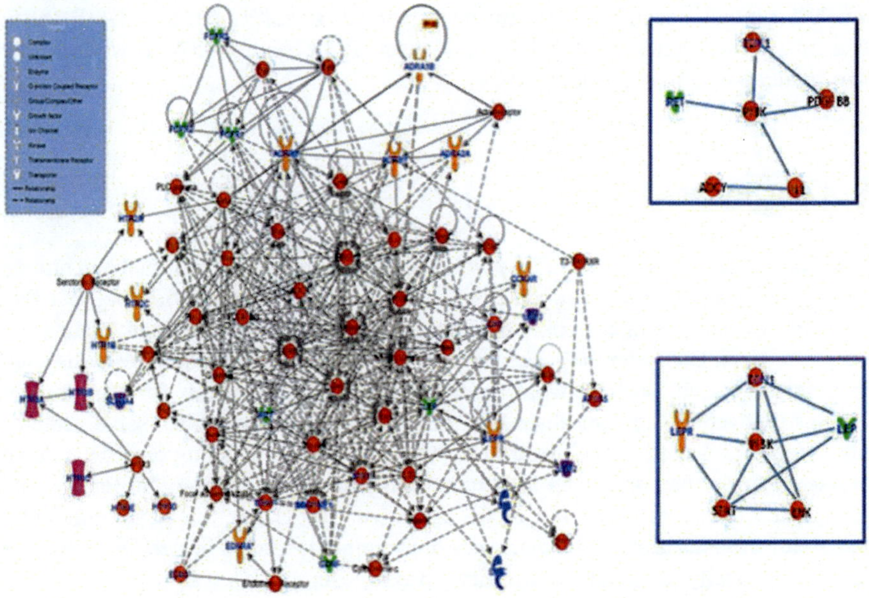

Fig. 1 Subcutaneous tissue's sub-networks and nodes (*Source* Liu, Patel and Nibbe et al. (2011))

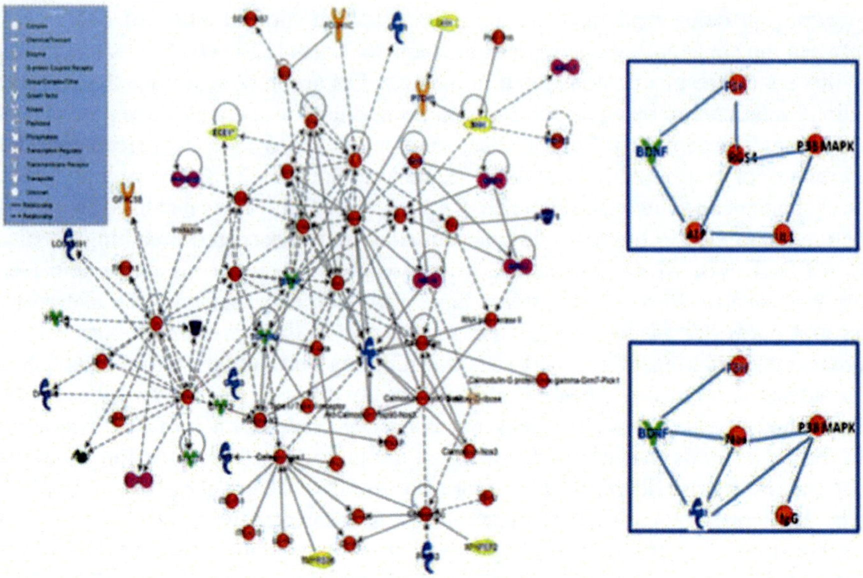

Fig. 2 Visceral tissue's sub-networks and nodes (*Source* Liu, Patel and Nibbe et al. (2011))

to be signaled by P38 MAPK (a nitrogen-activated protein kinase) and this protein, which belongs to the second sub-network (see Fig. 2 above), has its pathways (with which it interacts) accounting for the occurrence of oxidative stress [28–30]. Also, data from bioinformatics suggest that 3-kinases (PI3K) cause OSA, as well as fatty acid-induced insulin resistance [26].

3 Conclusion

Based on the data from SNO, PPI networks and adipose tissues in animal and human studies, analyses of mRNA expressions, a product of bioinformatics, suggest that data from disparate sources could be combined to determine prominent features in sleep disorders. Based on the scholarly findings emanating from bioinformatics and documented above, proteins from subcutaneous and visceral fats form prominent features that reflect sleeping disorders' disposition pathways. Also, some proteins have been established to be linked to pathways reflecting sleeping disorder-related stresses. With parameters such as insulin signaling, the STAT protein family, and PI3K found to account for sleeping disorders (such as OSA), this paper concludes that bioinformatics leads to the discovery of prominent features in sleeping disorders due to its capacity to reveal the disorders' predisposing pathways.

References

1. Liu Y, Patel S, Nibbe R et al (2011) Systems biology analyses of gene expression and genome wide association study data in obstructive sleep apnea. In: Pacific symposium biocomputing, pp 14–25
2. Sehgal A, Mignot E (2011) Genetics of sleep and sleep disorders. Cell 146(2):194–207
3. Aldabal L, Bahammam AS (2011) Metabolic, endocrine, and immune consequences of sleep deprivation. Open Respir Med J 5:31–43
4. Alsubie HS, BaHammam AS (2017) Obstructive sleep apnoea: children are not little adults. Paediatr Respir Rev 21:72–79
5. Garbarino S, Lanteri P, Durando P, Magnavita N, Sannita WG (2016) Co-Morbidity, mortality, quality of life and the healthcare/welfare/social costs of disordered sleep: a rapid review. Int J Environ Res Public Health 13:831
6. Garbarino S, Sannita WG (2017) Poor sleeping has underrepresented medical, healthcare, and social costs? Eur J Intern Med 38:15–16
7. Gadie A, Shafto M, Leng Y, Kievit RA, Cam-CAN (2017) How are age-related differences in sleep quality associated with health outcomes? an epidemiological investigation in a UK cohort of 2406 adults. BMJ Open:7
8. Crowley SJ, Van Reen E, LeBourgeois MK et al (2017) A longitudinal assessment of sleep timing, circadian phase, and phase angle of entrainment across human adolescence. PLoS ONE:9
9. Holst SC, Valomon A, Landolt HP (2016) Sleep Sleep pharmacogenetics: personalized sleep-wake therapy. Annu Rev Pharmacol Toxicol 56:577–603
10. Shafto MA, Tyler LK, Dixon M, Duncan J, Dalgleish T et al (2014) The Cambridge Centre for Ageing and Neuroscience (Cam-CAN) study protocol: a cross-sectional, lifespan, multidisciplinary examination of healthy cognitive ageing. BMC Neurol 14:204
11. Goel N (2017) Neurobehavioral effects and biomarkers of sleep loss in healthy adults. Curr Neurol Neurosci Rep 17:89
12. Sehgal A, Mignot E (2011) Genetics of sleep and sleep disorders. Cell 146:194–207
13. Ragia G, Archontogeorgis K, Simmaco M et al (2019) Genetics of obstructive sleep apnea: vitamin D receptor gene variation affects both vitamin D serum concentration and disease susceptibility. OMICS 23:45–53
14. Doherty A, Smith-Byrne K, Ferreira T et al (2018) GWAS identifies 14 loci for device-measured physical activity and sleep duration. Nat Commun 9:5257
15. Nilsson EK, Boström AE, Mwinyi J, Schiöth HB (2016) Epigenomics of total acute sleep derivation in relation to genome-wide DNA methylation profiles and RNA expression. OMICS 20:334–342
16. Anafi RC, Lee Y, Sato TK et al (2014) Machine learning helps identify CHRONO as a circadian clock component. PLoS Biol:12
17. Dean DA, Goldberger AL, Mueller R et al (2016) Scaling up scientific discovery in sleep medicine: the national sleep research resource. Sleep 39:1151–1164
18. Yetton BD, McDevitt EA, Cellini N et al (2018) Quantifying sleep architecture dynamics and individual differences using big data and Bayesian networks. PLoS ONE:13
19. Alizadeh Savareh B, Bashiri A, Behmanesh A et al (2018) Performance comparison of machine learning in sleep scoring based on wavelet features and neighboring component analysis. Peer J:6
20. Khandoker AH, Palaniswami M, Karmakar CK (2009) Support vector machines for automated recognition of obstructive sleep apnea syndrome from ECG recordings. IEEE Trans Inf Technol Biomed 13:37–48
21. Zalesky A, Breakspear M (2015) The connectomics of brain disorders. Nat Rev Neurosci 16:159–172
22. Tagliazucchi E, von Wegner F, Morzelewski A et al (2012) Automatic sleep staging using fMRI functional connectivity data. Neuroimage 63:63–72

23. Wu Y, Liu M, Zeng S et al (2018) Abnormal topology of the structural connectome in the limbic cortico-basal-ganglia circuit and default-mode network among primary insomnia patients. Front Neurosci 12:860
24. Kaufmann T, Elvsåshagen T, Alnæs D et al (2016) The brain functional connectome is robustly altered by lack of sleep. Neuroimage 127:324–332
25. Cheng W, Rolls ET, Ruan H, Feng J (2018) Functional connectivities in the brain that mediate the association between depressive problems and sleep quality. JAMA Psychiat 75:1052–1061
26. Pavel M, Jimison HB, Korhonen I, Gordon CM, Saranummi N (2015) Behavioral informatics and computational modeling in support of proactive health management and care. IEEE Trans Biomed Eng 62:2763–2775
27. Althouse BM, Scarpino SV, Meyers LA et al (2015) Enhancing disease surveillance with novel data streams: Challenges and opportunities. EPJ Data Sci 4:17
28. Bragazzi NL, Dini G, Toletone A, Brigo F, Durando P (2016) Leveraging big data for exploring occupational diseases-related interest at the level of scientific community, media coverage and novel data streams: the example of silicosis as a pilot study. PLoS ONE:11
29. Harari GM, Lane ND, Wang R, Crosier BS, Campbell AT, Gosling SD (2016) Using smartphones to collect behavioral data in psychological science: opportunities, practical considerations, and challenges. Perspect Psychol Sci 11:838–854
30. Garbarino S, Scoditti E, Lanteri P et al (2018) Obstructive sleep apnea with or without excessive daytime sleepiness: clinical and experimental data-driven phenotyping. Front Neurol 9:505

High-Performance Encoded-Driven LFSR for Improved Fault Coverage

Avvaru Sai Mounika, Chilakapati Sripriya, and V. Sarada

Abstract Due to the emergence of new digital transformation and recent advancements in nanotechnology, testing becomes a difficult task to accomplish in many circuits. There are several numerous investigations done by many researchers well prove that Linear Feedback Shift Register (LFSR) is the predominant test pattern generator for BIST. However, it requires some unique modifications for reduced power consumption and improved fault coverage and each causes some significant performance tradeoffs. The proposed TPG uses encoder-driven Linear Feedback Shift Register (BS-LFSR) architecture, which replaced dynamic polynomial computation and periodic reseeding mechanism with some decomposition and concatenation techniques to overcome the limitations while retaining the inherent characteristics. This paper considers the complexity and energy-related issues in LFSR-based BIST implementation without compromising the fault coverage and tends to narrow down the complexity and energy overhead for improved system performance. The objective of this paper is twofold. One, to prove the effectiveness of proposed encoded-driven LFSR over existing multi-polynomial and reseeding LFSR. It is shown that the proposed LFSR-TPG consumes lesser hardware and outperforms other LFSR models in terms of operating speed. Second, to carry out experimental evaluation over ISCAS85 and 89 benchmark datasets and to validate the fault coverage and energy efficiency measures.

Keywords Test pattern generator (TPG) · Built-in self-test (BIST) · Linear feedback shift register (LFSR) · Fault coverage

---

A. Sai Mounika (✉) · C. Sripriya
Department of ECE, Sri Padmavathi Mahila University, Tirupati 517502, Andhra Pradesh, India
e-mail: avvarusai456@gmail.com

C. Sripriya
e-mail: Sriya34@gmail.com

V. Sarada
Department of ECE, SRMIST, Chennai 603 203, Tamil Nadu, India
e-mail: saradav@srmist.edu.in

S. Jyothi et al. (eds.), *Proceedings of the 2nd International Conference on Computational and Bio Engineering*, Lecture Notes in Networks and Systems 215,
https://doi.org/10.1007/978-981-16-1941-0_47

1 Introduction

In recent years, due to the emergence of complex digital systems, the demands for built-in self-test (BIST) have been increased to ensure reliability [1, 2]. This is due to the fact that for proper and reliable operation, all digital designs need to be tested using some random test inputs, which can be generated by test pattern generator. The test vectors generated by random generator should check all internal elements to verify their functionality and should ensure fault-free operations. In general, BIST architecture includes two basic functional blocks, namely test pattern generator (TPG) and output response analyzer (ORA). During BIST incorporation for testability, it is essential to note down some measures as follows: (i) the complexity overhead should be minimum, (ii) additional energy consumption during testing should meet the design power constrains and (iii) the overall testing time should be minimum. These factors motivate several models of TPG's for BIST architecture to improve the effectiveness through randomness, test pattern length and fault coverage etc. among various methods investigated for TPG, LFSR is considered as pre-dominant methods due to its simplicity. In contrast, the test patterns from LFSR are always lack in all these measures.

Digital transition due to switching activity during BIST implementation leads to major concern in many real-time applications [3]. Test mode always increases the energy consumption considerably, which is several times higher than that of normal mode. Moreover, enormous energy consumption during fault detection affects the lifetime of the design under test and that may damage with excessive high temperature. Though keeping the power consumption as stable one as like general design operation in functional mode, by reducing the switching activity during testing one can narrow down this penalty gap. This can be done by using an appropriate test pattern generator (TPG) with some post-processing design methodologies like weighted driven through modulo operation [4] and test compression [5] etc.

This paper combines pre-computed substitution-based encoding modeling with LFSR to exploit energy and area efficiency in test vector computation with the following advantages.

(1) Pre-computation-driven encoding model is investigated using appropriate values that improve the performance of LFSR generated sequence.
(2) LFSR is configured with fixed polynomial to avoid complex computation and consists of minimal number of seeds for reseeding.
(3) Performance metrics of encoding-based LFSR model is evaluated over various ISCAS benchmark circuits.

In addition, to prove the area and energy efficiency of TPG, the proposed LFSR modules are synthesized and its resource utilization rate and energy levels are compared to other counterpart LFSR models that include conventional LFSR reseeding and computationally complex multipolynomial LFSR configuration process. The remaining of the article is organized as follows. Section 2 discussed various methods of test pattern generator (TPG) modeling for low complexity BIST and includes detailed analyses of all reference models. Section 3 explains the basic methodologies involved in the proposed LFSR, transition reduction techniques and its BIST system integration and also presents the validated experiment results and discussion. Finally, the conclusion has been drawn.

2 Related Works

For any digital design, the inclusion of built-in self-test (BIST) implementation will increase the computational complexity and energy consumption rate. It is essential to bring down this performance trade-off and the difficulties without compromising the testing features. In most cases, test pattern generators are widely investigated to accomplish this task. In [6], Bit-swapping LFSR is developed using multiplexer-based test pattern integration to narrow down the maximal possible digital that occurs during testing operation. The results proved that 50% switching activities are reduced as compared to conventional LFSR. In addition to these LFSR test pattern modifications, a new scan-chain-ordering is introduced for each scan cells to optimize both the average and peak energy consumption. In [7], combined the reconfigurable Johnson counter with the scalable single input change counter to generate test vectors with the least possible digital transition over successive sequences. This method also includes test-per-clock and the test-per-scan in BIST scheme for improved fault coverage for given test vector length. LFSR-reseeding model proposed in [8] is used to compress the test vector size and optimize the number of feedback elements. With appropriate path search methodology, the number of stored seeds required for 100% fault coverage is optimized, which increases memory efficiency considerably.

In [9], the overall test time of memory core design is reduced by implementing the BIST in parallel mode. Sine memory tests required some unique models to validate the functionality, which includes address generation, address decoder, data coupling, data fetching, etc. The generated test pattern sequences need to examine various functions from input to output data propagation path in addition to the random struck at faults. The LFSR reseeding-based test compression model developed in [10] improves the BIST system performance by optimizing the redundant switching activities during scanning using some random changes in seed inputs. An optimized encoding model is integrated with LFSR TPG to mitigate the thorny problem, which is the major concern for low power BIST implementations. Vashist et al. [11] proposed a unified BIST system for detecting the various types of attack detections to ensure valid protection over on-chip wireless interconnects. The adopted testing model allows online monitoring of the wireless data transmission to trigger security threats. This system also helps to detect and explore the default random faults in addition to the security measures in digital wireless communication systems.

In [12], the performance trade-off in BIST due to the complex Berlekamp–Massey (BM) algorithm-based polynomial computation is analyzed and its penalty gaps are noted down. The efficiency of BM to cover hard-to-detect faults with their respective optimization level requirements has experimented. The system begins with test vector generation using fixed polynomial and improvements are done using BM hierarchical manner. The results proved that LFSRs with BM improve the randomness but require unique optimization and independent polynomial expressions for hard-to-detect faults. The low-cost TPG introduced in [13] reduces the complexity overhead of the well-known MSIC-TPG by incorporating broadcaster between the MSIC-pattern generator and the design under test. The introduced broadcaster helps to distributing test vectors randomly for each MSIC-generated test pattern sequences and accommodates large sets of scan chains for given test patterns. In [14] to optimize the energy consumption, modified linear feedback shift register (LFSR) is developed that can reduce the switching activity dynamically. Some address sequences are generated from two dedicated address blocks, each operates by two different clock signals running at different speeds. The combination of these two sequences helps to control the transition between two successive test patterns. The performance metrics are validated using memory built-in self-test (MBIST) of 64 k $\times$ 32 static random access memory (SRAM).

3 Pseudo-Random Number Generator

Pseudo-random number generator (PRNG) is used to generate all required test input stimulus for testing the CUT. Randomness is the predominant and potentially useful attribute used to measure the performance metrics of TPG. In this work, a linear feedback shift register (LFSR) with some pre- and post-processing methodology is used as PRNG for BIST implementation. In BIST, the test patterns applied need to be pseudorandom patterns that can be generated typically by a linear feedback shift register (LFSR).

Requirement for TPG: High randomness, memory and area-efficient implementation.

Requirement for BIST: High fault coverage, detect hard-to-detect faults and low testing power.

3.1 Proposed Encoded-Driven LFSR TPG

This paper overcomes the aforementioned limitations of existing LFSR-based TPG with the considerable improvements on the BIST methodology using appropriate design optimization. A novel pre-computed encoding technique is proposed, which acts as substitution block for the existing LFSR reseeding scheme as shown in Fig. 1.

Here by utilizing the randomness provided by encoder, the dynamic polynomial configuration is excluded that makes system suitable for low complexity BIST implementation and reduces the seed size that optimizes the memory space requirements for seed storage. This method also provides finite control over switching activity over generated test patterns. This is based on the principle of given priority to MSB bit

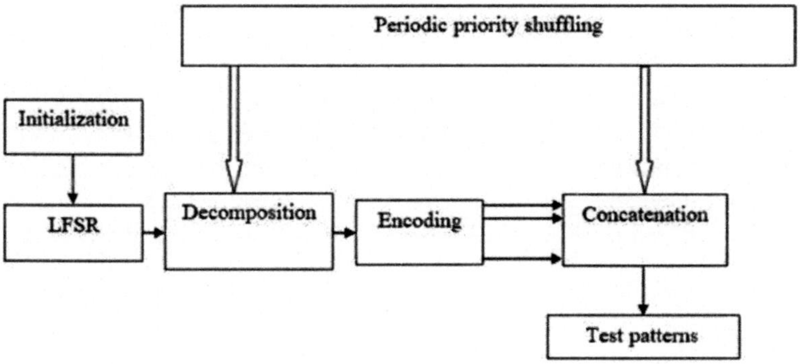

Fig. 1 Proposed LFSR test pattern generator

changes during pattern generation for all successive uninterrupted data progression and also allows to construct the test pattern of any given length using randomized concatenation. This work also presents the post-processing model for minimum digital transition between the test sequences. The proposed TPG generates random test sequences and downconverts them to various functional blocks. Each converted sub-blocks are applied to encoder and concatenation of these sub-blocks provides test pattern of any length. The transition of these generated test patterns is decreased with appropriate data concatenation, since each encoder has different priority levels towards binary changes. The proposed encoded-driven LFSR TPG has numerous metrics as follows:

Narrowed digital transitions: Here each LFSR generated test sequence is decomposed into multiple sections and applied to encoder, which can minimize the actual transition that happened in LFSR generated patterns.

Randomness: The generated encoded test sequence does not allow any repetitions. This is achieved with random changes of encoded bits and random concatenation of encoded sequences. This will increases the fault coverage with least possible test patterns and testing time.

Non-Uniform distribution: The conventional LFSR reseeding and periodic polynomial configurations allow some uniform changes over the test patterns. Each requires some pre-computed seed values and polynomial expressions that not only cause additional hardware and some manual data interpretations. However, the attainable randomness in these models is always predictable, which results in poor fault coverage and the required maximum number of test patterns for some of the hard-to-detect faults. It is mitigated in this proposed LFSR by some non-linear transformation and distribution of test patterns.

4 Performance Analysis

To validate the performance metrics of the proposed LFSR-TPG, testing is carried out over ISCAS'85 and standard full-scan ISCAS'89 benchmarks with random faults. The functional verification is carried out with the models in simulator using exhaustive test bench input stimulus as shown in Fig. 2, and hardware synthesis is carried out using QUARTUS II FPGA EDA tool based on 45-nm device model as shown in Fig. 3.

For ISCAS'85 and 89 benchmarks, the test pattern reductions of 6–14% and 9–39% respectively as shown in Table 1 as compared with LFSR reseeding-based TPG. In terms of energy efficiency, the proposed LFSR-TPG achieves a maximum of 12.86% against the state-of-the-art LFSR reseeding as shown in Table 2. Experimental results show that the proposed LFSR outperforms the existing benchmark LFSR reseeding and multi polynomial configuration both in terms of randomness and testing power contributes as shown in Figs. 4 and 5.

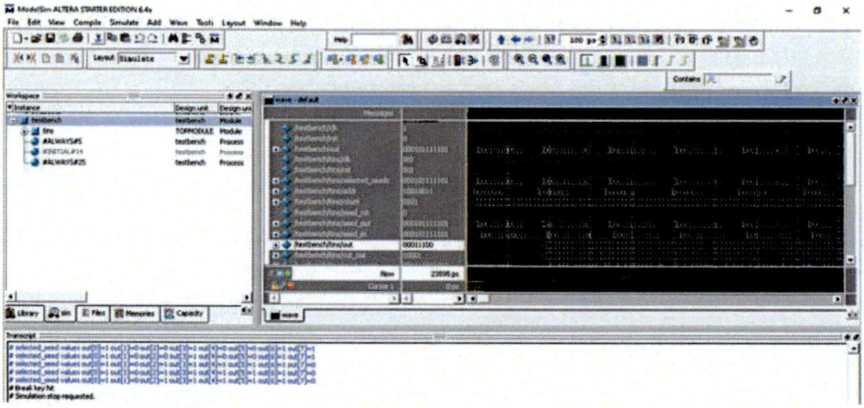

Fig. 2 Simulated output

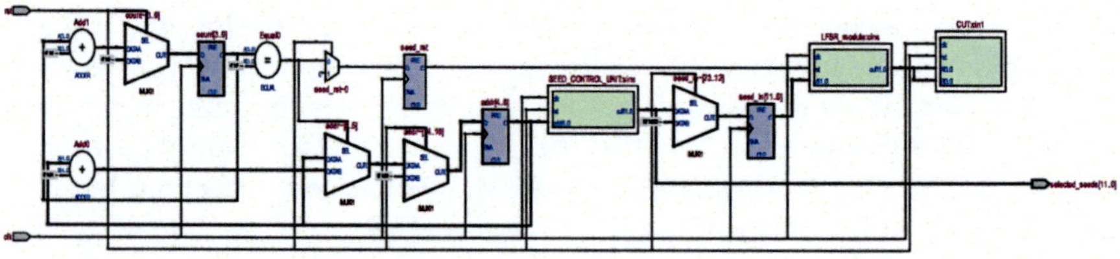

Fig. 3 Hardware synthesize report

Table 1 Performance metric comparison of proposed TPG in test volume reduction

Circuit	Number of test patterns generated			
	LFSR reseeding	LFSR multi polynomial	Proposed LFSR	Test pattern reduction (%)
c499	432	456	403	6.71
c1355	2038	1967	1854	9.02
c2670	3045	2984	2458	19.27
c6288	5786	5897	4978	13.96
s386	198	203	180	9.09
s641	670	602	587	12.38
s713	1572	1287	956	39.18
s953	2001	1944	1643	17.89

Table 2 Performance efficiency comparison of proposed TPG over energy measure

Circuit	Related power consumption (mW)			Energy efficiency (mW) (%)
	LFSR reseeding	LFSR multi polynomial	Proposed LFSR	
c499	64.56	64.03	63.19	2.14
c1355	69.45	71.23	67.98	2.13
c2670	78.09	74.33	71.37	8.61
c6288	82.89	80.07	72.34	12.86
s386	61.89	60.54	58.32	5.85
s641	72.43	72.40	68.56	5.37
s713	77.29	73.54	72.08	6.76
s953	80.23	78.09	74.60	7.03

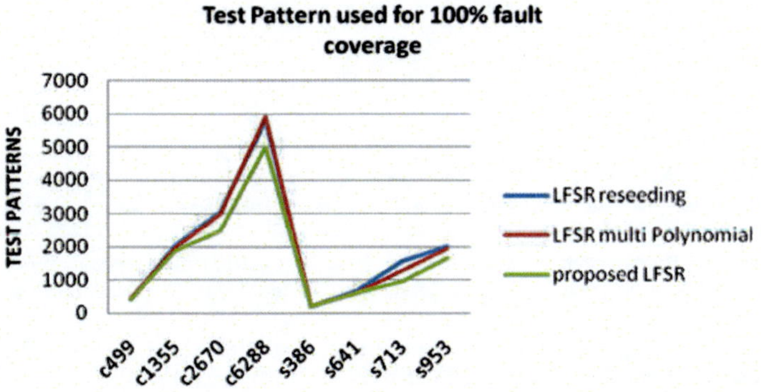

Fig. 4 Test pattern length comparison of proposed LFSR scheme for 100% fault coverage

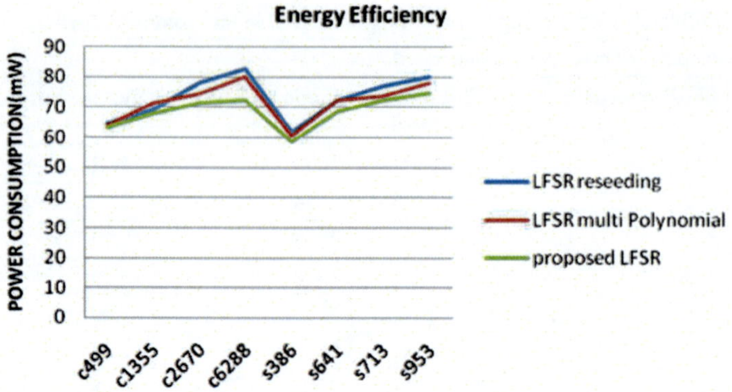

Fig. 5 Energy efficiency comparison of proposed LFSR scheme during testing

5 Conclusion

In this work, we developed a unified encoder-driven LFSR TPG for cost and energy-efficient BIST implementation. While the existing LFSR reseeding and dynamic polynomial configuration can improve the TPG system performance, they need appropriate pre-computation for random functioning. Moreover, the complexity overhead causes make the system unsuitable for many embedded and wireless applications. The proposed TPG can also be used to generate non-test cube vectors with appropriate encoded values in any BIST circuits. Experimental results prove that the proposed LFSR TPG schemes can achieve 100% fault coverage with least number of test patterns for all of the benchmark circuits and testing power is also reduced considerably. This reduction in test patterns and energy level increased linearly to the complexity level benchmark circuits and associated test scan inputs.

References

1. Xiang D, Wen X, Wang LT (2016) Low-power scan-based built-in self-test based on weighted pseudorandom test pattern generation and reseeding. IEEE Trans Very Large Scale Integr (VLSI) Syst 25(3):942–953
2. Kanimozhi P, Ramalingam S, Murugesan S, Ramya N (2017) Input vector observing simultaneous BIST architecture utilizing SRAM cells. Int J Sci Res Sci Technol (IJSRST) 3(8):506–514
3. Wang S, Gupta SK (2006) LT-RTPG: a new test-per-scan BIST TPG for low switching activity. IEEE Trans Comput Aided Des Integr Circuits Syst 25(8):1565–1574
4. Bagalkoti A, Shirol SB, Ramakrishna S, Kumar P, Rajashekar BS (2019) Design and implementation of 8-bit LFSR, bit-swapping LFSR and weighted random test pattern generator: a performance improvement. In: 2019 International conference on intelligent sustainable systems (ICISS). IEEE, pp 82–86
5. Tyszer J, Filipek M, Mrugalski G, Mukherjee N, Rajski J (2013) New test compression scheme based on low power BIST. In: 2013 18th IEEE european test symposium (ETS). IEEE, pp 1–6

6. Abu-Issa AS, Quigley SF (2009) Bit-swapping LFSR and scan-chain ordering: a novel technique for peak-and average-power reduction in scan-based BIST. IEEE Trans Comput Aided Des Integr Circuits Syst 28(5):755–759
7. Liang F, Zhang L, Lei S, Zhang G, Gao K, Liang B (2012) Test patterns of multiple SIC vectors: Theory and application in BIST schemes. IEEE Trans Very Large Scale Integr (VLSI) Syst 21(4):614–623
8. Liu T, Kuang J, Cai S, You Z (2014) An effective logic BIST scheme based on LFSR-reseeding and TVAC. Int J Electron 101(9):1217–1229.
9. John PK (2017) BIST architecture for multiple RAMs in SoC. Proc Comput Sci 115:159–165
10. Yuan H, Zhou C, Sun X, Zhang K, Zheng T, Liu C, Wang X (2018) LFSR reseeding-oriented low-power test-compression architecture for scan designs. J Electron Test 34(6):685–695.
11. Vashist A, Keats A, Dinakarrao SMP, Ganguly A (2019) Unified testing and security framework for wireless network-on-chip enabled multi-Core chips. ACM Trans Embed Comput Syst (TECS) 18(5s):1–20.
12. Martínez LH, Khursheed S, Reddy SM (2019) LFSR generation for high test coverage and low hardware overhead. IET Comput Digital Techn 14(1):27–36
13. Zhang G, Yuan Ye, Liang F, Wei S, Yang C-F (2019) Low cost test pattern generation in scan-based BIST schemes. Electronics 8(3):314
14. Zhi-guo YU, Qing-qing LI, Yang FENG, Xiao-feng GU (2020) An LFSR-based address generator using optimized address partition for low power memory BIST. J Measure Sci Instrum 3

Novel Hadamard Transform-Based Companding Technique for PAPR Mitigation in OFDM IDMA

Korada Lavanya and Recharla Divya

Abstract In present situations, the applications based on Wireless Communication Techniques have huge popularity. Because of this, the development of various Wireless Communication Applications using different techniques is needed. One of the most adorable techniques to develop Wireless Communication Applications is OFDM. Although this Orthogonal Frequency Division Multiplexing has a wide range of advantages, there is one problem with this technique. OFDM produces a significant distortion when signal passes through non-linear device due to its high PAPR (Peak to Average Power Ratio). There are a number of techniques realized to suppress the OFDM PAPR. Among all, the best optimal method to narrow down the Peak to Average Power Ratio in multicarrier system is Companding Technique. Our research is to narrow down the PAPR in CP-OFDM system through the novel Hadamard Transform-based Companding Technique. This system also includes Interleave Division Multiple Access (IDMA) technique to overcome the shortcomings that usually come from TDMA and CDMA Techniques. In this paper, reduction of PAPR, error rate performance and Power Spectral Density accomplished with Hadamard Companding Technique in OFDM IDMA will be discussed and simulated.

Keywords Wireless communication systems · OFDM · PAPR companding technique · Hadamard transform · BER · PSD

K. Lavanya (✉) · R. Divya
Department of ECE, Sri Padmavathi Mahila University, Tirupati 517502, AP, India
e-mail: koradalavanya30@gmail.com

R. Divya
e-mail: dharani.divya14@gmail.com

S. Jyothi et al. (eds.), *Proceedings of the 2nd International Conference on Computational and Bio Engineering*, Lecture Notes in Networks and Systems 215,
https://doi.org/10.1007/978-981-16-1941-0_48

1 Introduction

First of all, communication is the process of transmitting any type of information from one place to other places. There are several types of communication systems available now. Basically, the common communication systems are wired and wireless. But wireless communication is emerging steadily due to the invention of forms based on digital transmission techniques and applications such as digital audio broadcasting (DAB) and digital terrestrial mobile broadcasting (DTMB), etc. Moreover, it is also suitable for long-distance transmission. One of the huge popular wireless communication system techniques is OFDM. OFDM stands for Orthogonal Frequency Division Multiplexing. Robert W. Chang who belongs to Bell Labs has introduced this OFDM Technique in 1966. This OFDM is one of the digital transmission techniques. In this technique, digital data are encoded into multiple carrier frequencies. This wireless OFDM Technique is broadly used in wideband Digital communications, which are audio broadcasting, digital television, internet access, 4G/5G Techniques, etc. In this OFDM, the digital data are transmitted in parallel through Orthogonal subcarrier signals that are closely spaced along with overlapping spectra. Orthogonal Frequency Division Multiplexing has a high amount of spectral efficiency compared with other schemes [1, 2]. The best thing with this wireless communication technique is it can easily adapt to any critical channel circumstances. Also, this technique shows resistance to narrowband cochannel interference. Fading that is caused by multipath propagation does not occur through this technique. Also, OFDM is less affected by errors of time synchronization. There is no requirement of Tuned Sub Channel Filters like other conventional techniques [3]. Although it has huge advantages more than other techniques, it also has few disadvantages. OFDM is very sensitive to Doppler shift. OFDM produces maximum Peak Power when signal is submitted to liner transmitter. There are several researches going on the path in order to minimize this PAPR rate in OFDM. Among them, our Novel Hadamard Transform based on Companding Technique for OFDM PAPR Reduction can give the sure result.

In our design, we also included the IDMA with OFDM. IDMA stands for Interleaving Division Multiple Access Technique. IDMA is one of the advanced multiple access techniques. In this technique, user separation is done by random interleaves. OFDM IDMA is an advanced technique of OFDM CDMA. Compared with Code Division Multiple Access Technique, IDMA has more advantages, which are IDMA Technique has a high rate multiple access communication through MUD (Multi User Detector) Algorithm. Therefore, OFDM IDMA Technique has the advantages, which are combined from these two individual techniques. OFDM IDMA Technique gives reduction of ISI, high rate of multiple access, high efficiency, 5G mobile communication access, etc. The main problem with this technique is PAPR. Our design aim is to optimize the PAPR in OFDM IDMA Technique through the Hadamard Transform using Companding Technique.

This paper is organized as follows: Sect. 2 discusses various PAPR reduction implementations, which include Piecewise Liner Companding Technique and

Hadamard Transform-based approaches and its potential merits. The proposed framework of PAPR computation and characteristics has been analyzed in Sect. 3. And then, Sect. 4 explains the evaluation of the proposed models used for PAPR reduction process and its associated BER performance, followed by the conclusion.

2 Literature Survey

2.1 Related Works

Many existing methods investigated for PAPR reduction solve the problems related to power measures using scrambling or phase modification, which is manipulated statistically based on observation made from different modulations and its associated dynamic ranges. In [4] developed combined the Companding and Hadamard transform for PAPR in multicarrier OFDM system and the experimental results well proved that this novel integration offers significant PAPR optimization without causing any notable losses in error rate performance in terms of BER at the receiver side. Ref [5] proposed inverse Fourier transform (IFFT) to formulate the unitary matrix of the OFDM system to reduce the peak power. Here, the tradeoff between error rate performance and the PSD is reduced as compared with other state-of-art competitive PAPR reduction techniques.

Based on the correlative measure observation between Golay complementary sequences (GCS) and second-order Reed–Muller codes (RMC), new alleviated PAPR reduction is carried out in [6]. Here, the configurable code rate and its associated error correction capability enabled improved flexibility over higher range of PAPR. The performance metrics of the overall system is moderated using appropriate decoding algorithm, which includes several forms of fast Hadamard transforms.

In [7], discrete multitone modulation (DMT) is investigated to mitigate the PAPR problems over multicarrier baseband models. The penalty gap that exists in analog component design and power amplifier characteristics due to PAPR is analyzed. The influence of oversampling in overall attainable PAPR reduction is also studied. Finally, the essential system requirements for notable PAPR tolerance and its sensitiveness in any multicarrier systems are validated.

2.2 Piecewise Liner Companding Technique

Our OFDM IDMA PAPR Reduction System is designed based on Piecewise Linear Companding technique and Hadamard Transform as shown in algorithm 1. This is one of the best techniques designed to reduce the OFDM IDMA PAPR rate. Linear Piecewise Companding Technique is the immediate operation after the OFDM IDMA modulation. The specialty of this Companding Piecewise Technique is the signal

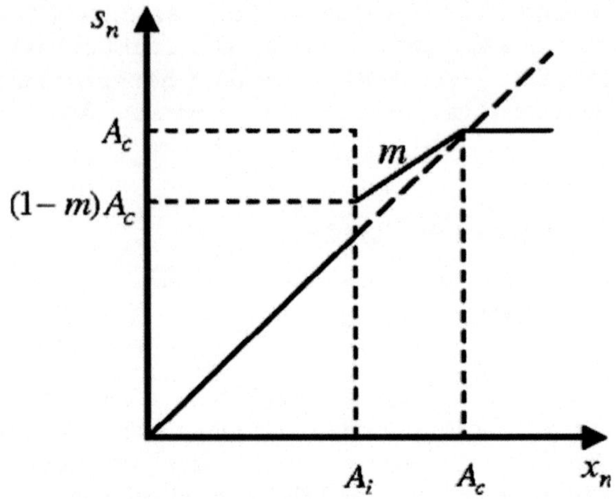

Fig. 1 linear characteristics of Piecewise Companding scheme

average power is still same even companding is done. Piecewise Companding Technique optimizes the Peak Power Ratio without compromising the error rate performance. In Companding Technique, low amplitude signals are expanded and high amplitude signals are contracted in order to obtain BER optimization. The references of BER optimization rating is taken from CCDF (Complementary Cumulative Distributive Function) PAPR Readings. In this section, Liner Piecewise Companding Technique is theoretically explained.

For example, assume the original signal as Xn, which is going to compand with peak values of Ac. Then the linear function characteristics of Piecewise Companding are analyzed statistically as shown in Fig. 1.

The Companding function of given signal is represented as follows

$$s(n) = c\{x(n)\} = \begin{cases} x(n) & x(n) \leq A_i \\ mx(n) + (1-m)A_c & A_i < |x(n)| \leq A_c, \\ \operatorname{sgn}(x(n))A_c & |x(n)| > A_c \end{cases} \tag{1}$$

Here, the sign(x) represents the sign either it may be positive or negative of signal. At the receiver side, reverse Companding will be performed. The reverse piecewise linear Companding function is represented as follows

$$y(n) = C^{-1}(r(n)) = \begin{cases} r(n) & |r(n)| \leq A_i \\ (r(n) - (1-m)A_c)/m & (1-m)A_c < |r(n)| \leq A_c, \\ \operatorname{sgn}(r(n))A_c & |r(n)| > A_c \end{cases} \tag{2}$$

Here, Ai, Ac, and m represent the specified parameters of OFDM PAPR. The value of PAPR and BER of OFDM IDMA is determined by using these parameters. Ac value can be found through the following formula

$$A_c = \sigma_x 10^{PAPR_{preset}/20}. \tag{3}$$

Later, m and Ai values are determined through the following formula

$$\int_{A_i}^{A_c} (mx + (1-m)A_c)^2 f_{|x_n|}(x)dx + \int_{A_c}^{\infty} A_c^2 f(x)dx = \int_{A_i}^{\infty} x^2 f_{|x_n|}(x)dx. \tag{4}$$

2.3 Algorithm

Step 1 Initializing parameters (size of OFDM of symbol(N), Number of OFDM symbols to be simulated m, Enter modulation order (e.g. 8,16,32,64), sampling factor.

Step 2 At transmitter section to the generated data apply interleaving and given to the mapper and after that serial to parallel then apply the Hadamard Transform. These data are modulated by IDFT section and parallel to serial then apply the piecewise Companding technique after that apply the cyclic prefix.

Step 3 These data are transmitted through the channel in the channel noise is added.

Step 4 At receiver, remove the cyclic prefix and remove piecewise Companding after serial to parallel conversion. Then demodulate the data and remove Hadamard transform then parallel to serial conversion of data then demapping and deinterleaving is done.

Step 5 Calculate PAPR, BER, SNR, PSD, CCDF.

Step 6 plot the PSD, SNR versus BER and CCDF

2.4 Hadamard Transform

One of the best transforms to narrow down the PAPR in any form of multicarrier system is Hadamard Transform. Previously there are transforms used in OFDM PAPR reduction. They are not much satisfied PAPR mitigation in OFDM Technique. In our research, we have used the Hadamard-Based Companding Technique to overcome the drawbacks in previous suggestions of OFDM PAPR reduction. Hadamard Transform performs the autocorrelation to input sequence in order to reduce PAPR in OFDM IDMA. Also, this Hadamard Transform does not require any side band signal to transmit to receiver. In this section, Hadamard Transform is explained theoretically.

Let us assume Hadamard Matrix as H for N orders. The Hadamard Matrix is a orthogonal matrix such that every element would be either 1 or -1. Therefore, the two-order H is defined as

$$H_2 - \frac{1}{\sqrt{2}}\begin{bmatrix} 1 & 1 \\ 1 & -1 \end{bmatrix} \tag{5}$$

Above equation will be reconstructed as follows

$$H_{2N} = \frac{1}{\sqrt{2N}}\begin{bmatrix} H_N & H_N \\ H_N & -H_N \end{bmatrix} \tag{6}$$

Since HN is complement to HN as well it has to satisfy the following equation

$$H_{2N}H_{2N}^T = H_{2N}^T H_{2N} = I_{2N} \tag{7}$$

3 PAPR Mitigation in OFDM IDMA Using Hadamard Transform-Based Companding Technique

Our proposed system, The Novel Hadamard Transform based on Companding Technique for PAPR Mitigation in OFDM IDMA block diagram, is shown in figure III.1. The block diagram is displaying the step by step operation of proposed Hadamard Transform-based Companding technique for optimizing the PAPR in OFDM IDMA system. Our proposed system is an advanced technique for DCT Companding Technique. The changes in conventional DCT Companding Technique for PAPR Reduction system are included with the existing structure. The explanation of each step is as follows.

Initially, the binary form of input bits is submitted to channel coding. The data are encoded and mapped at Mapper section. Later, the signal will be processed and given to Hadamard transform. At this section, the PAPR will be reduced. Then, the out coming signal will be separated into k interleaves. Later, the signal will digitally modulate and then IFFT will be applied. Once the cyclic prefix is done, the signal will be transmitted to receiver through the certain channel. At the receiver section, the opposite actions will be done to obtain the original signal.

3.1 Peak-Average Power Ratio (Papr)

In CP-OFDM system, the power-related problems are analyzed by computing peak power associated to each antenna. Since OFDM is the multicarrier system, the accumulation of modulated subcarriers leads maximum power in the time domain. In general, the PAPR of the conventional OFDM system is computed as a ratio between the peak versus average power as shown in Eq. (8)

$$PAPR = 10\log\frac{P_{PEAK}}{P_{AVG}}dB \tag{8}$$

During evaluation, the complex baseband representation of Eq. (8) is formulated using time interval as follows:

$$PAPR = 10\log\frac{\max|X(t)|^2}{E|X(t)|^2}dB \tag{9}$$

where $\max|X(t)|^2$ is maximum signal power and $E|X(t)|^2$ is the average signal power.

In general, PAPR performance is evaluated in Complementary Cumulative Distribution Function (CCDF). CCDF explores the probability of exceeded PAPR in a certain PAPR threshold bound as shown in Eq. (10)

$$CCDF(\gamma) = \Pr(PAPR > \gamma) = 1 - (1 - e^{-\gamma})^N \tag{10}$$

The complementary cumulative distribution function (CCDF) is used to validate the performance metrics of the proposed method in terms of peak power suppression, which is the major cause in multicarrier modulation techniques like CP-OFDM system. Due to the inclusion of the time-varying phase components among subcarriers, the overall power accumulation during subcarrier mapping is regulated and its peak to average power is reduced considerably (Fig. 2).

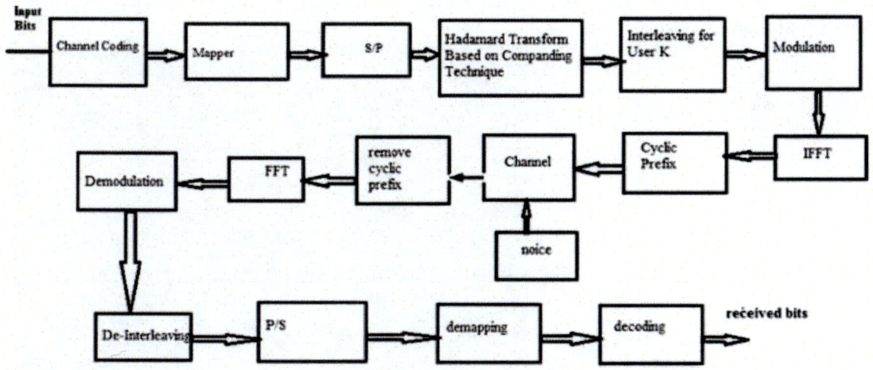

Fig. 2 Block Diagram for OFDM IDMA PAPR Reduction using Hadamard Transform based on Companding Technique

4 Result and Simulation

The success of any system will be decided based on their out and simulation. Like that, our proposed OFDM IDMA PAPR reduction Technique also had simulated and checked the following result. We can observe the success of our research through the following result graphs. Initially, we have applied the digital data to the system. The original sequence of the given input is as shown in Fig. 3.

Later, we have simulated the Power Spectral Density measures of CP OFDM system with the inclusion of the DCT and piecewise Companding and results show that the proposed IDMA scheme in OFDM system suppresses the energy ripples as shown in Fig. 4. Based on the PSD plot of modulated OFDM signal processed using the Hadamard transform-based Companding transform shows that the proposed method leads to little spectral regrowth compared with the conventional CP-OFDM signal and, hence, the PAPR reduction using the proposed scheme is effective for wireless systems.

As shown in Fig. 5, the proposed IDMA scheme shows superior error rate performance as compared with conventional DCT Hadamard OFDM system over a wide range of SNR, and its penalty gap is widened when SNR is beyond 8 dB. From the experimental results, it is well proved that the proposed method is outperformed in terms of BER performance as compared with the existing methods as shown in Table 1.

Finally, the complementary cumulative distribution function PAPR is simulated using the QAM modulation technique. The resultant graph is shown in Fig. 6.

From the above simulation result, clear PAPR Reduction in OFDM IDMA can be observed.

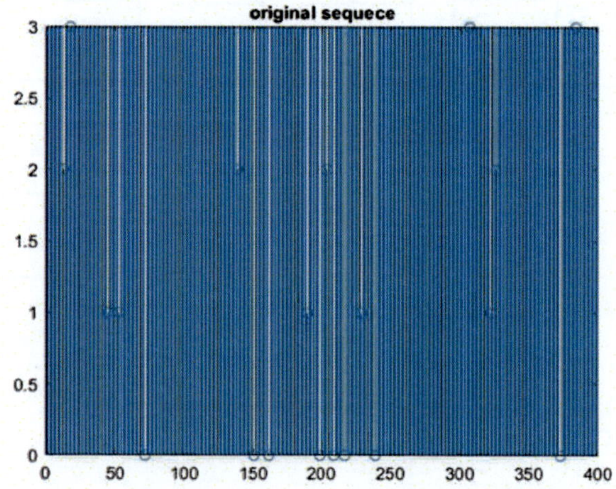

Fig. 3 Original sequence

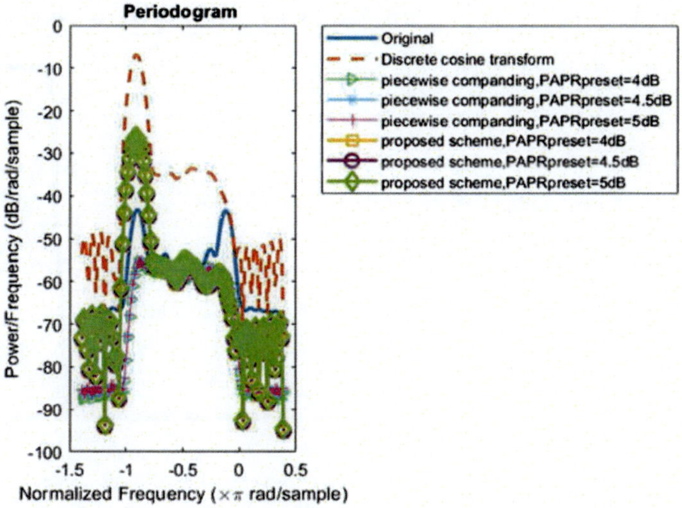

Fig. 4 Out of band suppression measures DCT Hadamard OFDM versus IDMA scheme

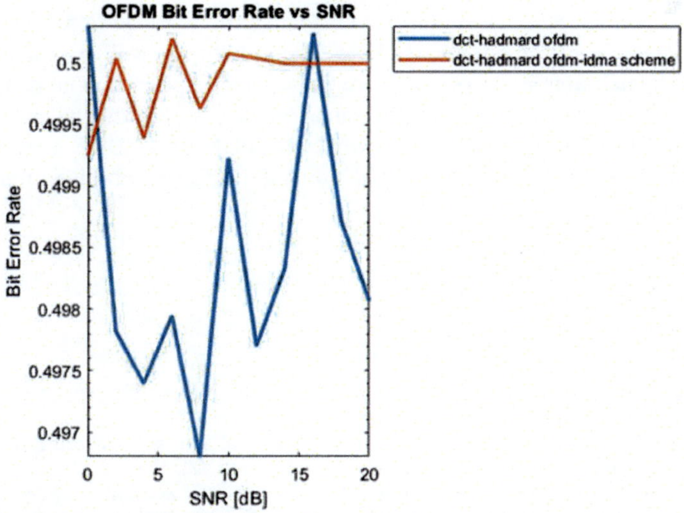

Fig. 5 Error rate performance comparison between DCT Hadamard OFDM versus IDMA scheme

Table 1 Performance comparison with various modulation orders

METHOD	OFDM Symbol	BER	SNR	PAPR	PSD (power frequency)
Old Method	32 QAM	0.599	4	13	50
	64QAM	0.6	8	12.7	55
Proposed method	32QAM	0.5	9	11.17	25
	64QAM	0.5985	12	11.10	24

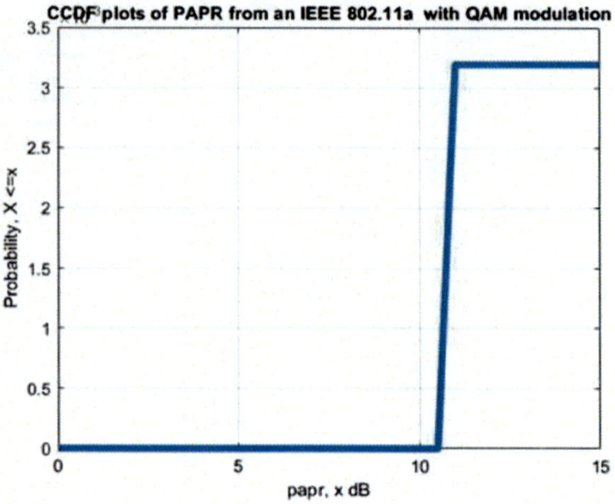

Fig. 6 CCDF plots for PAPR

5 Conclusion

In this paper, our research Novel Hadamard Transform based on Companding Technique for PAPR Mitigation of OFDM IDMA is explained as well as designed. Also, the output signal simulation results were drawn and explained. The result of our proposed system has compared with conventional systems and advantages have been noted. The PSD and BER versus SNR of OFDM and CCDF were observed. Hence, PAPR of OFDM IDMA System was reduced based on given input signal characteristics.

References

1. Speer R, Chin J, Havasi C (2016) Concept net 5.5: An open multilingual graph of general knowledge. arXiv:1612.03975
2. Naidu GBSR, Rao VM (2017) Comparative analysis of OFDM with reduced PAPR based on companding techniques. Int J Pure Appl Math 114(10):363–371
3. Wylie-Green MP, Perrins E, Svensson T (2011) Introduction to CPM-SC-FDMA: A novel multiple-access power-efficient transmission scheme. IEEE Trans Commun 59(7):1904–1915
4. Wang Z, Zhang S, Qiu B (2010) PAPR reduction of OFDM signal by using Hadamard transform in companding techniques. In: 2010 IEEE 12th international conference on communication technology. IEEE, pp 320–323
5. Zhu X, Zhu G, Jiang T (2009) Reducing the peak-to-average power ratio using unitary matrix transformation. IET Commun 3(2):161–171
6. Davis JA, Jedwab J (1999) Peak-to-mean power control in OFDM, Golay complementary sequences, and Reed-Muller codes. IEEE Trans Inf Theory 45(7):2397–2417
7. Krongold BS, Jones DL (2004) An active-set approach for OFDM PAR reduction via tone servation. IEEE Trans Signal Process 52(2):495–509

Automatic Vehicle Number Plate Recognition System (AVNPR) Using OpenCV Python

K. Suneetha and K. Mounika Raj

Abstract Nowadays with the rapid increase in vehicles, identification of the vehicle and control the traffic become challenging issues in every country. Also, it is very difficult for a traffic police to identify the person who drives too fast, who violates the traffic rules and to identify whether the person is owner or thief from the moving vehicles. So, unable to catch and punish those kinds of people. Hence, there is a need for Automatic Vehicle License Plate Recognition (AVNPR) system. Though many LPR systems are available, still it is a challenging task due to unstructured license plate formats, language on plate, speed of vehicle and lightning effects. The main objective of AVNPR is to recognize the license plate using image processing techniques or optical character recognition by applying *pytesseract* OpenCV Python package. The main focus is on the detection of vehicle license plate, character segmentation and character recognition.

Keywords Automatic vehicle license plate recognition(AVNPR) · Optical character recognition (OCR) · License plate (LP) · Image processing

1 Introduction

1.1 Overview

Automatic vehicle number plate recognition was introduced in 1976. The main purpose of using this image processing technology is to identify and prevent a wide range of criminal activities and for security purposes. Over a decade, it finds very significant in the field of research in many applications [1] with the introduction of digital camera and with the increase in computational capacity. This system automatically captures the image of the vehicle plates without human intervention. AVNPR is

K. Suneetha (✉)
School of CS & IT, Jain (Deemed-to-be) Univesity, Bangalore, Karnataka, India

K. Mounika Raj
Sree Vidyanikethan Engineering College, Rangampet 517102, AP, India

S. Jyothi et al. (eds.), *Proceedings of the 2nd International Conference on Computational and Bio Engineering*, Lecture Notes in Networks and Systems 215,
https://doi.org/10.1007/978-981-16-1941-0_49

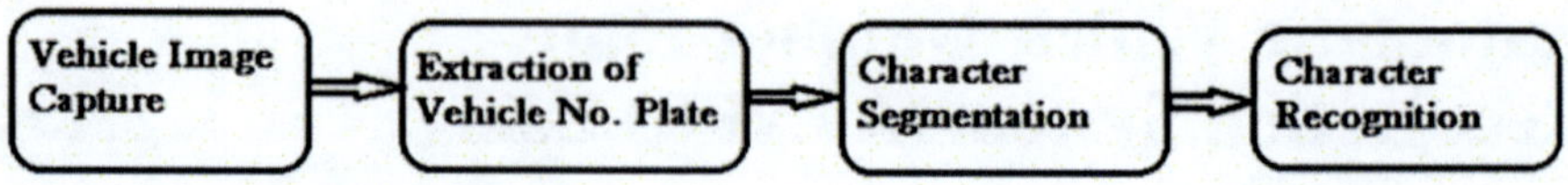

Fig. 1 AVNPR workflow

mainly used to support the citizens' safety, security to keep track of criminal behavior and to detect the average speed calculation especially for police forces. The recognition of license plates becomes easier for a standard plate, but in India, it becomes a challenge due to unstructured plate formats and environments.

The AVNPR work is categorized into several steps: Vehicle image capture, identification of vehicle license plate, character segmentation and character recognition.

As shown in Fig. 1, the first step is to snare the image of a given vehicle. It seems that it is easy to capture, but the real challenge is how to capture the moving vehicle image. Many existing systems follow different approaches to identify the license plate like Artificial Neural Network, Optical Character Recognition, Support Vector Machine, Lease Square Method, trichromatic imaging, etc. In India, most of the car number plates composed of six numbers and four letters, assuming that the first two letters mention the state of vehicle registration.

To recognize the number plate, many researchers applied various algorithms and techniques to detect the license plate by considering several factors like size of plate, location of plate, background color, etc. An ımage binarization method can be used to detect the vehicle image, which converts the colored picture to grayscale using Otsu's method.

Generally, the number plate extraction is very typical due to its smaller occupancy of the entire image, un-uniform structure of number plates and environmental conditions. With improper placement of camera, climatic illumination conditions [3] and tracking of vehicles make the proposed system very challenging.

2 Literature Survey

Nazmus Saif et al. the authors presented a methodology for Bangla automatic vehicle license plate recognition system [1]. They used approximately 200 multi-class image vehicle data sets and applied Convolutional Neural Network (CNN) to find out the license plates. The car image is given as an input for input layer and later applied CNN for segmentation and recognition of number plates. They obtained the results with 99.5% accuracy. The authors found that CNN is more dependent on data sets, they need further data sets to train their CNN model effectively for better results. Also, they believe that with more data sets, they could able to classify vehicles without major change in their model.

The researches, Prashengit Dhar et al. [2], outlined an LPR framework for Bangladeshi license plate with white plate black fonts and with the white background. To localize the plate, they used edge properties, verification of shape by DtB method, horizontal tilt correction by extrema points; character segmentation is implemented using region and morphological operations and finally characters in the license plate are identified using CNN deep learning algorithm. They have shown the results with 99.6% accuracy in character classification.

The authors, Md. Zainal Abedin et al. designed a system [3] based on deep supervised machine learning and computer vision models. They focused on three factors, namely, (i) detection of license plate with image preprocessing, binary image conversion, detection and filter of contours, (ii) segmentation of cropped LP to extract the characters on the license plate and (iii) identification of characters by applying deep convolutional neural network. The authors implemented the system using Python OpenCV and showed the significant results with respect to vehicle detection, segmentation, detection and execution rates.

In this paper, the authors, Thanh-Nga Nguyen & Duc-Dung Nguyen, combined the three methods, namely, detection, segmentation and recognition to identify the full sequence of license plates for Vietnamese [4]. They implemented the system using SHL-CNN model with shared hidden layers for recognizing Chinese characters and two unique softmax layers for alphanumeric characters. During the training process, they used early stopping rule to avoid over fitting issues.

Their experimental results have shown better results when compared with CNN model due to their maximum utilization of data sets and shared hidden layer structure.

3 Proposed System

The proposed system identifies the characters/numbers on the license plate using image processing concepts. It helps to extract useful and meaning information from digital images. The proposed system typically includes the following steps:

(i) Generation of synthetic license plate images with the help of high-speed cameras.
(ii) Image preprocessing—to analyze the images to enhance the quality for effective recognition.
(iii) Vehicle license plate detection—to detect the license plate, contour option in OpenCV is used for rectangular objects. Various processes that we are going to apply are:

- Converting the given input vehicle image to the required size and then to the grayscale format.
- Remove the blurring and noisy image data, bilateral filter is applied.
- Edge detection using Canny edge method from OpenCV.
- Identification of rectangle shape contour on images.

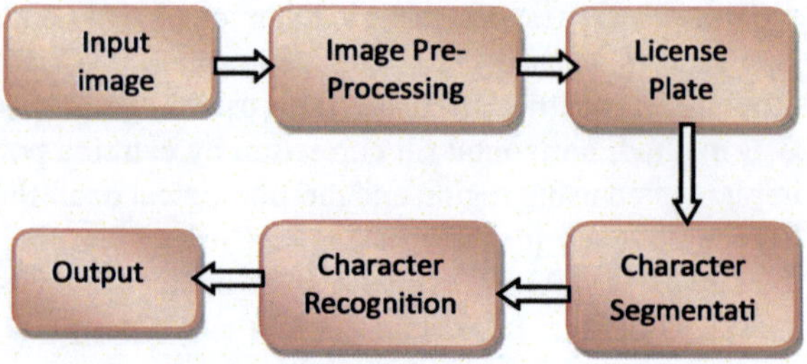

Fig. 2 AVNPR methodology

Except the number plate, mask the total vehicle image.

(iv) Character segmentation—The detected License Plate has to crop it and save it as a new image using OpenCV.

(v) Character recognition—An Optical Character Recognition (OCR) is used to identify the numbers or alphabets on the new image detected in the previous step (Fig. 2).

1. **License Plate Detection**: The main aim is to identify the exact location of the license plate in the given input image by using edge statistics, texture techniques, etc. To recognize the license plate initially, the input image is converted into grayscale image with the required size using the following Python code (Figs. 3 and 4).

```
img1 = cv.resize(img1,(620,480))
gray1=cv.cvtColor(img1, cv.COLOR_BGR2GRAY)
```

Resizing helps to resolve the issues related to high-resolution images, so that the license plate will fit into the frame. To speed up the image processing, a synthetic image data set is taken as input image and then converted into grayscale image as the detection of license plate is not focusing on details of color. In step 2, every input

Fig. 3 Original image of car

Fig. 4 Gray scale image of car

Fig. 5 Bilateral filter image cf Car

image consists of some irrelevant information called noise that needs to be blurred using a bilateral filter (Fig. 5).

```
gray1 = cv. bilateralFilter(gray1, 13, 15, 15)
```

In step 3, edge detection can be done using Canny edge method from OpenCV as shown below:

```
edged1= cv. Canny(gray1, 30, 200)
```

This command will identify the edges with highest gradient, which is more than minimum threshold and less than maximum threshold value (Fig. 6).

Step 4 checks the rectangle contour-shaped image with closed four-sided shape. A rectangle box is drawn around the license plate to detect the vehicle license plate location correctly (Fig. 6).

Fig. 6 Canny edges of car

```
contours1=cv.findContours(edged1.copy(),cv.RETR_TREE,
cv1.CHAIN_APPROX_SIMPLE)
contours1 = imutils.grab_contours1(contours1)
contours1 = sorted(contours1,key=cv.contourArea, reverse = True)[:10]
screenCnt = None

    for i in cnts:
        peri = cv.arcLength(i, True)
        approx = cv.approxPolyDP(i, 0.018 * peri, True)
        if len(approx) == 4:
        screenCnt = approx
        break
```

In Step 5, after finding the location of the license plate, the rest of image is not required for analysis, so masking is done except the license plate location.

```
mask1 = np1.zeros(gray.shape,np.uint8)
new_image1 = cv.drawContours(mask1,[screenCnt],0,255,-1,)
new_image1 = cv.bitwise_and(img,img,mask1=mask1)
```

4 Character Segmentation

Here license plate is segmented by cropping the input image and save as new image to detect the characters or numbers in the license plate. The main image that is cropped to region of interest (ROI) code is as follows:

```
(x1, y1) = np1.where(mask == 255)
(topx1, topy1) = (np1.min(x1), np.min(y1))
(bottomx1, bottomy1) = (np1.max(x1), np1.max(y1))
cropped = gray[topx1:bottomx1+1, topy1:bottomy1+1]
```

5 Character Recognition

Here, ***pytesseract*** package is used to read the characters/numbers from the segmented number plate.

```
Text1 = pytesseract.image_to_string(Cropped, config='--psm 11')
print("Detected license plate Number is:",text1)
```

Fig : License Plate of car 1

See Table 1.

6 Conclusion

An Automatic Vehicle License Plate Recognition (AVNPR) system has been proposed to recognize the vehicle license plate using image processing techniques or optical character recognition by applying OpenCV Python package. The detection of car license plate on generated synthetic image data set works on image preprocessing, detection, segmentation & recognition of characters/numbers on the license plate using ***pytesseract*** package from the segmented number plate.

Table 1 Sampled results detected and non-detected plates using AVNPR system

Car image	Plate detected?	Result	Latency(sec)	Characters detected?	Result	Latency (sec)
	Yes	DL8CAF5030	0.146	Yes	DL8CAF5030	0.062
	Yes	HR26BP3543	0.153	Yes	HR26BP3543	0.053
	Yes	BG 224-NZ	0.220	Yes	BG#224-NZ	0.067
	Yes	PG MN112	0.193	Yes	PGeMN112	0.076

(continued)

Table 1 (continued)

Car image	Plate detected?	Result	Latency(sec)	Characters detected?	Result	Latency (sec)
	Yes	3JOH22A	0.140	No	–	–
	Yes	AP 02 B 7767	0.251	Yes	AP 02> B-7767	0.082
	Yes	VS 001-BU	0.201	No	–	–
	Yes	YJ65Y0M	0.226	No	–	–

References

1. Saif N et al (2019) Automatic license plate recognition system for bangla license plates using convolutional neural network. In: IEEE, pp 926–930
2. Dhar P et al (2018) A system design for license plate recognition(LPR) by using edge detection and convolution neural network. In: IEEE, pp 221– 277
3. Abedin MZ et al (2017) License plate recognition system based on contour properties and deep learning model. In: 2017 Region 10 humanitarian technology conference (R10-HTC), 21-23 Dec 2017, Dhaka, Bangladesh. IEEE, pp 590–593
4. Nguyen TN, Nguyen DD (2018) A new convolutional architecture for vietnamese car plate recognition. In: IEEE, 10th International conference on knowledge and systems engineering (KSE), pp 7–12
5. Suneetha K, Sreekanth M Smart Home Monitoring and Automation Energy Efficient System using IoT devices", Emerging Research in Data Engineering Systems and Computer Communications. Advances in Intelligent Systems and Computing, Vol 1054. Springer, Singapore, pp. 627 – 637,2020
6. Suneetha K, Sreekanth M, Sankara K (2018) An approach for ıntelligent traffic signal control system for ambulance using IoT. In: SSRN Elsevier, scopus indexing, IADS ınternational conference on computing, communications & data engineering (CCODE), vol 01(03)
7. Abolghasemi V, Ahmadyfard A (2009) An edge-based color-aided method for license plate detection. Image Vis Comput 27(8):1134–1142
8. Ozbay S, Ercelebi E (2005) Automatic vehicle identification by plate recognition. Proc PWASET 9(4):222–225
9. Wang R, Sang N, Huang R, Wang Y (2013) License plate detection using gradient information and cascade detectors. Optik Int J Light Electron Opt:1–10 (in press)
10. Yu S, Li B, Zhang Q, Liu C, Meng MQ-H (2015) A novel license plate location method based on wavelet transform and EMD analysis. Pattern Recogn 48(1):114–125

A Hybrid Framework for Prediction of Heart Disease Using Rough Set and Fuzzy Set Approach

Srikanth Meda and Raveendra Babu Bhogapathi

Abstract Data mining is the procedure utilized for segregating the cardinal data from humongous amounts of data or data warehouses. The data comprises outliers which are deviations from the regular data. The outlier identification is a significant preprocessing strategy to discard the data that strays away from ordinary information. The anomaly recognition which is present in the huge database, is broadly utilized in different applications like medical diagnosis, human reliability analysis in health care, flaw identification. The primary target of the outlier identification is to extricate the strange information that has abnormal characteristics. The data gathering from the various systems is ambiguous in nature. The data ambiguity leads to irregularity and vagueness. This paper proposes a novel Hybrid Fuzzy Rough Set Classifier (HFRSC) for an effectual classification with less, analytical and computing effort.

Keywords Heart disease · Fuzzy logic · Classification · Prediction · Rough set · Fuzzy set

1 Introduction

The practice of fuzzy logic [1] and the refinement of atypical data with less complexity settle the outliers [2] in the huge database. With the advent of the concept of fuzzy sets, the ambiguity issues in the data have been resolved successfully. Each item in the

S. Meda (✉)
Acharya Nagarjuna University, Guntur, India
e-mail: medasrikanth@gmail.com

R. B. Bhogapathi
Chalapathi Institute of Engineering and Technology, Lam, Guntur, India

S. Jyothi et al. (eds.), *Proceedings of the 2nd International Conference on Computational and Bio Engineering*, Lecture Notes in Networks and Systems 215,
https://doi.org/10.1007/978-981-16-1941-0_50

input data is mapped to a value between 0 and 1. The fuzziness with the appropriate boundary confirmation is used to assess the intensity of the event rather than its level of occurrence. On the other hand, the matrix-based outlier classification approach using Triangular Boundary is affected with complications in computations.

Fuzziness is a technique that implies the embodiment of human-like thinking into the decision-making systems and helps in categorizing the belongingness of an object in a set. The conglomeration of the medical data obtained by multiple means is unsure and unresolved. Lack of certainty in data leads to various shortcomings like indeterminateness and obscurity.

Rough set theory [3] is utilized to categorize the data. In precise, this approach is employed to discover structural relationships within imprecise and noisy data. To improve the computational efficiency in classification, a fuzzy_rough_set classifier is proposed. The fuzzy logic usage along with the restoration of dubious records largely helps in bringing out the deviants from the huge databases. As the clinical data sets consist of massive volumes of data, it is prone to noises that are labeled as feature noise or class noise.

2 Methodology

The presence of outliers within the real-world dataset is highly probable and may mislead the training phase employed by various machine learning algorithms. The exceptions which are available inside the data may, of course, lead to inaccurate predictions. So in order to overcome this problem, a fuzzy-roughset classifier is proposed. A rough set [4] class is used to segregate the data into a lower, upper and boundary approximation. By using this approach, superfluous attributes are reduced and the most significant attributes are found by looking at the records at the borderlines. By using this approach, out of the 14 attributes, the most impact creating attributes are taken, and they are used in building very good fuzzy rules, which is the most significant factor for building a good model. Based on the outcome obtained from sensitive attributes, rules have been designed and deployed to test the input data.

The fuzzy rough set [5] amalgamation approach improves the computational capabilities without compromising the time accuracy and maximizes the ability to detect the outliers efficiently. AS the input space is mapped to any intermediate value between 0 and 1, the ambiguity problem that occurs with the distorted data items is completely avoided. The following mechanism reviews the functioning of fuzzy logic data classification [6].

3 Fuzzification

The fuzzy technique is primarily employed as a part of preprocessing to categorize the data into low, moderate and high classes. With the aid of membership functions [7], the records are translated into linguistic variables. The UCI heart data set [8] is taken into consideration. Among various features available in the dataset 14 features are considered input for testing and building the model, and the target classes are taken as output. Out of the considered attributes, four are assigned to the fuzzy variable [9] using the membership functions listed below. The range of values chosen for the features is also listed. By considering the traits of membership, the collected records are changed into fuzzy linguistic [10] variables.

3.1 *Membership Function for the Attribute Age:*

Attribute	Range
Age	<35 33–45 40–59 55>

$$\mu_{\text{young}}(a) = \begin{cases} 1 & a < 30 \\ \frac{(35-a)}{5} & 30 \le a \le 35 \end{cases}$$

$$\mu_{\text{middle}}(a) = \begin{cases} \frac{(a-33)}{7} & 33 \le a < 40 \\ 1 & a = 36 \\ \frac{(45-a)}{295} & 36 \le a \le 45 \end{cases}$$

$$\mu_{\text{old}}(a) = \begin{cases} \frac{(a-40)}{5} & 40 \le a < 5 \\ 1 & a = 45 \\ \frac{(59-a)}{6} & 45 \le a \le 59 \end{cases}$$

3.2 Membership Function for the Attribute Cholesterol

Cholesterol is characterized by fuzzy terms, such as low, medium, excessive and quite high.

Attribute	Range
chol	<190 180–245 210–300 270>

$$\mu_{\text{low}}(c) = \begin{cases} 1 & c < 150 \\ \frac{(190-c)}{40} & 150 \le c \le 190 \end{cases}$$

$$\mu_{\text{medium}}(c) = \begin{cases} \frac{(c-33)}{7} & 180 \le c < 220 \\ 1 & c = 220 \\ \frac{(245-c)}{25} & 220 \le c \le 245 \end{cases}$$

$$\mu_{\text{high}}(c) = \begin{cases} \frac{(c-210)}{40} & 210 \le c < 250 \\ 1 & c = 250 \\ \frac{(300-c)}{50} & 250 \le c \le 300 \end{cases}$$

3.3 Membership Function for the Attribute Thalach

Attribute	Range
Thalach	<141 111–194 >152

$$\mu_{\text{low}}(t) = \begin{cases} 1 & t < 100 \\ \frac{(141-t)}{41} & 100 \le t \le 141 \end{cases}$$

$$\mu_{\text{medium}}(t) = \begin{cases} \frac{(t-111)}{41} & 111 \le t < 152 \\ 1 & t = 152 \\ \frac{(194-t)}{42} & 153 \le c \le 194 \end{cases}$$

$$\mu_{\text{high}}(t) = \begin{cases} \frac{(t-152)}{64} & 152 \le t < 216 \\ 1 & t \ge 216 \end{cases}$$

3.4 Membership Function for the Attribute Blood Pressure

Attribute	Range
BP	<130 120–150 140–170 >151

$$\mu_{\text{low}}(b) = \begin{cases} 1 & b < 100 \\ \frac{(130-b)}{30} & 100 \le b \le 130 \end{cases}$$

$$\mu_{\text{medium}}(b) = \begin{cases} \frac{(b-120)}{30} & 120 \le b < 135 \\ 1 & b = 135 \\ \frac{(150-b)}{40} & 135 \le b \le 150 \end{cases}$$

$$\mu_{\text{high}}(b) = \begin{cases} \frac{(b-140)}{15} & 140 \le b < 155 \\ 1 & b = 155 \\ \frac{(170-b)}{15} & 155 \le b \le 170 \end{cases}$$

$$\mu_{\text{veryhigh}}(b) = \begin{cases} \frac{(b-160)}{18} & 151 \le \theta < 169 \\ 1 & b \ge 169 \end{cases}$$

3.5 Training Algorithm

INPUT: Dt – Trained dataset
OUTPUT: Rl – Rules
Step I : Load the training dataset.

Step II: Isolate the tuples as records (x) and features (a) in the dataset
Dt(x) = {Dt(x1), Dt(x2),……Dt(xn)}
Dt(a) = {p id, p age,p sex, p heartrate(phr)}

Step III: Assign the threshold value (thhr) for different ages

Age (variable A)	Threshold limit of Heart Beat Rate	Age (variable A)	Threshold limit of Heart Beat Rate
$a \leq 20$	200	$40 < a \leq 45$	175
$20 < a \leq 25$	195	$45 < a \leq 50$	170
$25 < a \leq 30$	190	$50 < a \leq 55$	165
$30 < a \leq 35$	185	$55 < a \leq 60$	160
$35 < a \leq 40$	180	$60 < a \leq 65$	155

Step IV: Break the steps depending on the membership functions of age, blood pressure, rate of heart beat and
lipid factor

Step V: Calculate the probability of a given record (P (xi))

i. Find the Ha for a given Ai which is the maximum value.
ii. Find the current value of Ai.
iii. Calculate the probability of ai as mentioned below

$Pro_{indiv}(Ai) = \frac{Ha(A_i)}{Va(A_i)}.$

$Pro(xi) = \{Pro_{indiv}(A1), Pro_{indiv}(A2) \ldots\ldots\ldots Pro_{indiv}(An)\}$

Step VI: Calculate mean of all the records (Po (xi)).

$Po(xi) = 1/n \sum_{i=1}^{n} Pro(x_i)$

Step VII: Attain Upper Limit and Lower Limit for mean (Po(xi)) at every level

Lo1 → Lower Boundary of position 1
Lo2 → Lower Boundary of position 2
Up1 → Upper Boundary of status 1
Up2 → Upper Boundary of status 2

Step VIII: Lower Estimated Region Rl(x) → Lo1≤ R(x) < Lo2
Upper Estimated Region Ru(x) → Up1 < R(x) ≤ Up2
Border Region Rb(x) → Rl (x) ∩ Ru(x) → L02 ≤ R(x) ≤ Up1

Step IX: Obtaining Delicate Features in Ru,Rl records
aL_l → Freq item Count [Rl(a)]
aL_u → Freq item Count [Ru (a)]
Sat → $aL_l \cap aL_u$

Step X: Set rubrics constructed on attributes which are sensitive Sat
Sat (a) = St1, St2……… StN {for each value of attribute {1,2,3}
Assigned frame of rules are r= {r1, r2….rz}

Step XI: Compute an estimate index for the above result

Step XII: Final Output from this process is the set of rules .
RulesR = {Ru_1, Ru_2, ……Ru_i}

Once the rules have been generated and the final fuzzy model is built and the data for testing is given to the model with the reduced number of attributes and the model is compared with other existing techniques.

4 Experimental Results

4.1 Average Under Curve

Average under Curve = $\frac{S0-n_0\frac{(n_0+1)}{2}}{n_0.n_1}$

Where, S0—The addition of grade of irregular occurrences forecast.
n_0—The no. of test instance fits to normal
n_1—The no. of occurrences which fits to the outliers (Table 1 and Fig. 1)

Table 1 Average under curve

% of noise	NSVDD	SSLM	R-SVDD	N-SVDD	FRSC
0	0.809	0.741	0.704	0.692	0.842
10	0.804	0.729	0.692	0.69	0.827
20	0.785	0.720	0.691	0.687	0.814
30	0.773	0.708	0.69	0.682	0.778
40	0.762	0.701	0.68	0.675	0.761

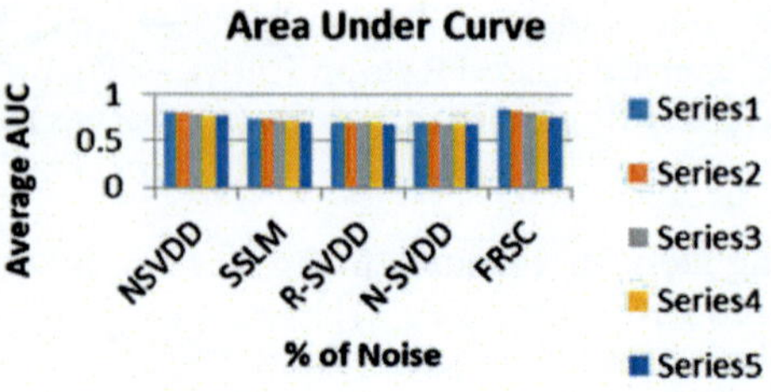

Fig. 1 Average of the area under the curve

4.2 ROC Curve

The ROC curve with its complete working range can then be involved to estimate the efficiency of the classifier. This curve will then be involved to select the approach to correlate the behavior of the classifiers with the prevailing classifiers (Fig. 2).

Figure 3 depicts the ROC with the mix-up matrix. The graph is then described with the content of proper tremendous, true poor, false effective and false negative values.

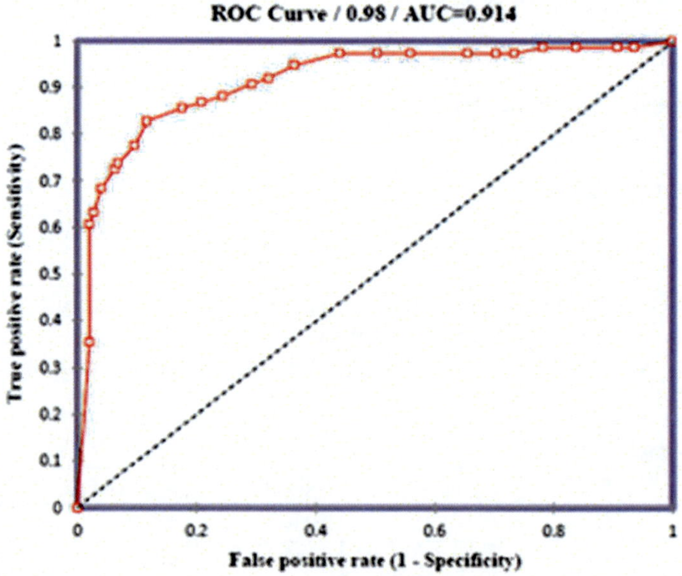

Fig. 2 ROC curve with confusion matrix

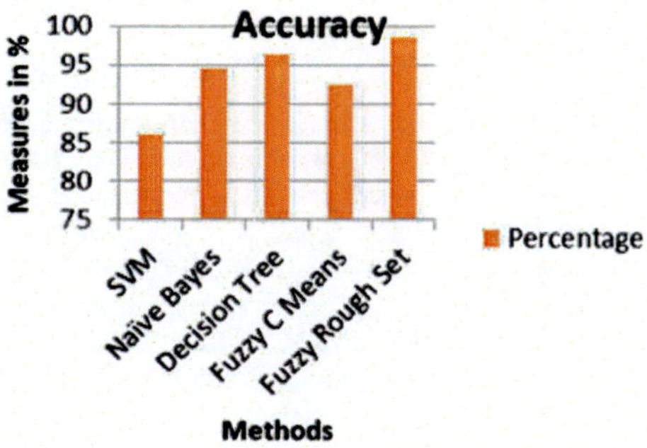

Fig. 3 Comparison of various methods accuracy

Table 2 Accuracy

Strategy	Percentage
Fuzzy C means	92.5
Naïve Bayes	94.5
Decision tree	96.4
Fuzzy C means	92.5
Fuzzy rough set	98.5

4.3 Accuracy of Classifier

The accuracy of the classifier relies totally at the variety of examples which is correctly labeled. The exactness of the approach is shown in the Figure 3 and values are systematized in Table 2.

$$\text{Accuracy} = \frac{\text{Number of Predicated Outliers}}{\text{Total number of Outliers}}$$

5 Conclusion

The proposed Hybrid Fuzzy Rough Set Classifier (HFRSC) is trained and tested successfully. It is observed that the HFRSC improves the computational capabilities of the model without compromising on the time accuracy and also maximizes the ability to detect the outliers in an efficient manner. The HFRSC showed considerable improvement in performance when compared with various existing models

References

1. Kim J, Lee J, Lee Y (2015) Data-mining-based coronary heart disease risk prediction model using fuzzy logic and decision tree. Health Inform Res
2. Rajeswari N, Nachammai S, Jemima PE, Rajeswari AM (2019) Unexpected health ıssues prediction in medical data using apriori rare based outlier detection method. In: Proceedings of the 2019 ınternational conference on vision towards emerging trends in communication and networking (ViTECoN). Vellore, India, pp 1–6
3. Azar AT, Bouaynaya N, Polikar R (2015) Inductive learning based on rough set theory for medical decision making. Fuzzy Systems (FUZZ-IEEE)
4. Bal M (2013) Rough sets theory as symbolic data mining method: an application on complete decision table. Inform Sci Lett Int J 2(1):35–47
5. Armento E, Vluymans S, Verbiest N, Caballero Y, Bello R, Cornelis C, Herrera F (2014) IFROWANN: imbalanced fuzzy- rough ordered weighted average nearest neighbor classification. IEEE Trans Fuzzy Syst 23(5):1622–1637
6. Reddy GT, Khare N (2017) An efficient system for heart disease prediction using hybrid OFBAT with rule-based fuzzy logic model. World Scientific Publishing Company, vol 26, no 4
7. Zarandi MHF, Zolnoori MM, Heidarnejad H (2010) A fuzzy-rule based expert system for diagnosing asthma. Trans E: Indus Eng 17(2):129–142
8. https://archive.ics.uci.edu/ml/datasets/heartdisease
9. Jodoin E, Pena Reyes CA, Sanchez E (2006) A method for the fuzzification of categorical variables. In: 2006 IEEE ınternational conference on fuzzy systems. Vancouver, BC, pp 831–838. https://doi.org/10.1109/FUZZY.2006.1681807
10. Adeli A, Neshat M (2010) A fuzzy expert system for heart disease diagnosis. In: Proceedings of the ınternational multiconference of engineers and computer scientists, vol I

Coastal Indian Prawn Species Recognition System Through Molecular Identification and Neural Networks

D. M. Mamatha, S. Jyothi, K. Hari Priya, and S. Sharmila

Abstract Species delineation is one of the strenuous tasks to be done. In Crustaceans species, which are lookalike in appearance, is very intricate and need much effort for the right species identification. Prawns & Shrimps are being consumed widely and have more health benefits. According to Holthius (1980), prawns include 33 genera with about 2,500 species globally in which only a few hundred have their commercial importance. In such case it is very important to differentiate the trade from important prawns to noncommercial ones. For parting such species, a special technique is being applied named DNA barcoding, which is a fresh, fast and accurate technique for the species delineation. DNA barcoding concept uses nucleotide sequences of precise mitochondrial marker (Cytochrome Oxidase I gene) to generate the DNA barcode. The nucleotide sequences are compared in parallel to the sequences in the BOLD (Barcode of Life Data system) reference library for appropriate recognition of the species. This is further followed by phylogenetic analysis to know their genetic distance. This study addresses the difficulties in species discrimination using morphology-based taxonomy through molecular identification. These barcodes are preserved in a large-scale system. In order to identify the species, it is essential to measure and recognize the features in detail. Feature extraction of barcode is an important problem to identify the prawn species through barcodes. On the computational part, the focus is on the feature extraction of prawn species using image processing. Image pre-processing is the main step where filtering, interpolation and Sobel edge detection are followed. From the pre-processed image, the segmentation process is employed. The results of the analyses are the time, accuracy and evaluation measures. The provisional results are evaluated by the segmentation adaptability for pre-processing and width magnitude for feature extraction under each family thereby, facilitating automatic identification of prawn species.

D. M. Mamatha · K. Hari Priya · S. Sharmila
Department of Biosciences and Sericulture, Sri Padmavati Women's University, Tirupati, AP, India

S. Jyothi (✉)
Department of Computer Science, Sri Padmavati Women's University, Tirupati, AP, India
e-mail: jyothi.spmvv@gmail.com

S. Jyothi et al. (eds.), *Proceedings of the 2nd International Conference on Computational and Bio Engineering*, Lecture Notes in Networks and Systems 215,
https://doi.org/10.1007/978-981-16-1941-0_51

Keywords Species delineation · COI (Cytochrome oxidase I) · BOLD (Barcode of life data system) · Pattern recognition · Edge detection · Segmentation · Feature extraction, classification

1 Introduction

Prawns, the aquatic shelled organism bundled with all rich nutrients belonging to the order Decapoda, have more than 2000 species under them. Only a few hundred species are known for their economical importance. Studies state that the prawns rich in selenium, calcium, Vitamin B12, proteins, etc., have incredible health benefits; selenium prevents the growth of cancerous cells, thereby preventing from cancer, high content of vitamin B12 gives cardiovascular support, other nutrients such as calcium and vitamin E are known for their bone and skin benefits. The taste and benefits of it made it into a major consuming product worldwide. India is one of the largest seafood suppliers globally. South India being covered all sides of it with water has the abundant occurrence of prawn species and occupies a major part in supplying the seafood. Homologous phenotypic characteristics of prawns make it a difficult task to identify and misleads the identification and separation of the economically important ones from uneconomical. At this point, there is a chance to defraud the high commercial prawns with low commercial ones. To overcome such straits, DNA barcoding is a great tool for species delineation. Accurate and rapid gene-based identification is performed using a short standardized universal mitochondrial marker gene-cytochrome C oxidase I of size 658 base pairs [1–4]; the diversity of prawns and shrimps in the south Indian region were studied using DNA barcoding tool [5, 6]. DNA barcode is coloured stripes assigned based on the nucleotide arrangement; all four nucleotides have different colour stripes, which represent their uniqueness and identity. The barcodes are finally issued by the BOLD database based on nucleotide setup. The BOLD database is termed as reference library as they hold all the information of the species such as geographical, voucher, pictorial details. This study enables us to overcome the difficulties in species identification using morphology-based classification.

2 Materials and Methods

2.1 Specimen Collection

Prawn Species have been collected from different major fish landing parts of South India such as Kakinada, Vizag covering the east coastal line of Andhra Pradesh, and few from Chennai of Tamilnadu. The caught specimens were morphological identified, labelled and the vouchers were raised for each specimen. Finally, they were preserved as a whole in 70% ethanol and placed at DNA barcoding lab, Sri Padmavati women's University, Tirupati, A.P, India for future use. Taxonomical identification was done using phenotypic characterization such as carapace length and

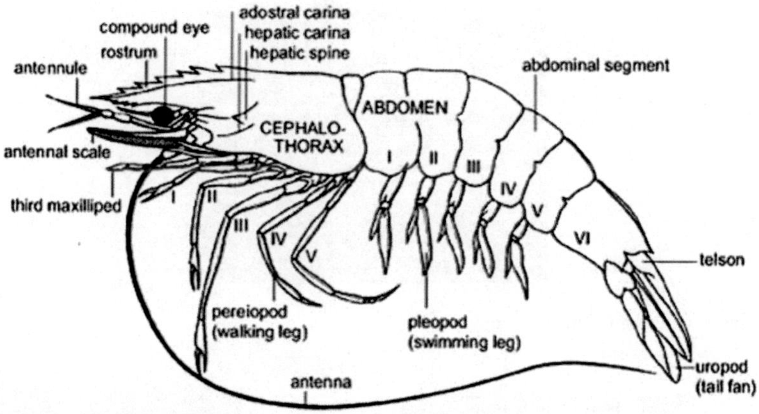

Fig. 1 External morphology of the Prawn (Google Courtesy)

mass, cheliped length, etc., Fig. 1, but this goes erroneously if external mutilation or malformation occurs. Therefore, this present study addresses this matter technically through DNA barcoding and image processing through neural networks.

Morphological classification:

Morphological classification is done studying the general characters of the prawns such as external morphology, including size, rostrum, shape, carapace length and mass, life cycle, etc.

Taxonomic classification:

See Fig. 1.

2.2 *DNA Extraction and Electrophoresis*

Total genomic DNA is pulled out from the ethanol preserved samples using a typical and standard method of DNA extraction named CTAB (Cetyl Trimethyl Ammonium Bromide) technique. The occurrence of extracted Genomic DNA is checked in 1% Agarose Gel electrophoresis (Fig. 2).

2.3 *Quality Assessment*

The purity of the DNA concentrate is checked by using a nano drop. Nanodrop reading clearly displays the purity range, protein contamination and concentration of the DNA in the sample; DNA quality check also plays a critical role in developing DNA barcodes as further steps in DNA barcoding depend on the quality of the DNA raised.

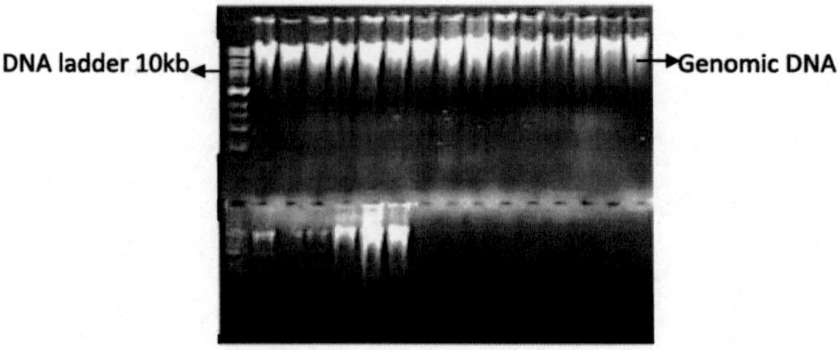

Fig. 2 Genomic DNA of Prawn samples

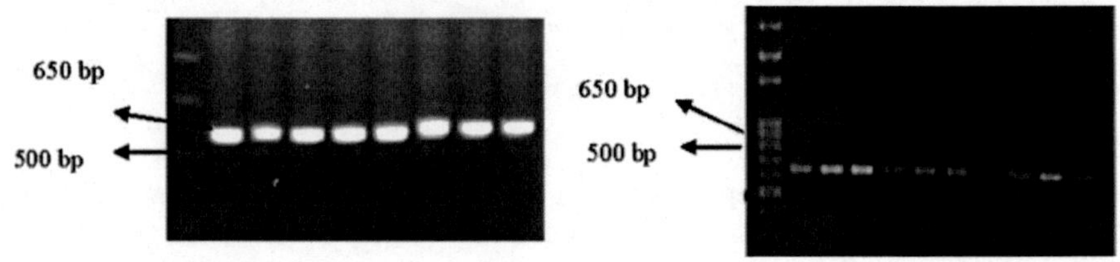

Fig. 3 Amplified CO-I gene DNA from Prawn samples

2.4 Amplicon Extraction and Visualization

The amplicon extracted for developing DNA barcodes is the mitochondrial COI gene. This gene marker is amplified by polymerase chain reaction using universal primer LCO 1490 & HCO 2198 [7]. PCR reaction is performed by using Emerald kit, which is a readily available mix. DNA template of 1 μl, forward and reverse primers of 0.5 μl each and molecular grade water are the add-on ingredients to be added to the PCR mix. PCR Cycling profile for COI amplification-denaturing temp of 94 °C for 3 min followed by 35 cycles at 94 °C for 30 s. Temperature ranging from 45 to 51 °C for 40 s is set as annealing temperature, followed by extension 72 °C for 1 min; finally, final extension done at 72 °C for 10 min, visualization of amplified DNA is done by using 1.5% Agarose gel (Fig. 3).

2.5 PCR Cleanup, Sequencing and Species Identification

PCR products were cleaned up using ExoSAP to remove unbound primers, etc., they were further taken for cycle sequencing and finally sent for sequencing at PHCDBS lab, Aurangabad, India. If the sequences' quality were good, they were taken for sequence editing and alignment using codon code aligner to remove noises and weak peaks. Further, the nucleotide sequence developed were blasted in NCBI, displaying the identification percentage of the sequence uploaded. The NCBI lists all the closely related sequences to the uploaded sequence; the exact identification and naming can be done by checking the identification percentage displayed (Fig. 4). Further, the sequences are submitted to the Barcoding online Database called the BOLD, which is designed to support the generation and application of DNA barcode data. BOLD is cloud-based data storage and analysis platform developed at the Centre of Biodiversity genomics in Canada. It consists of four main modules, a data portal, and educational portal, a registry of BINs, and a data collection and analysis workbench. The specimen voucher details, specimen picture and CO-I gene sequence, and trace files are uploaded to the BOLD database, where the DNA barcodes and specimen accession number are generated.

Specimen identification using Co-I gene DNA barcoding

See Fig. 4.

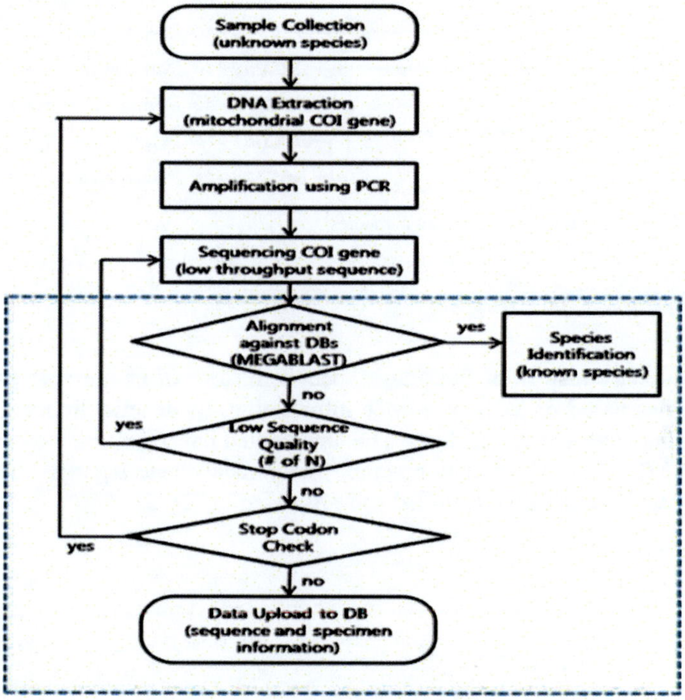

Fig. 4 Identification of Prawn species through DNA barcoding using Co-I gene

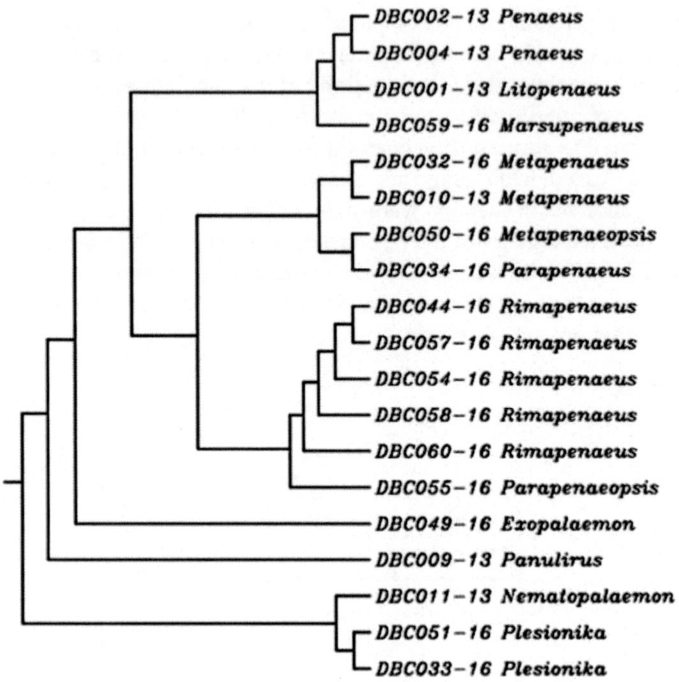

Fig. 5 Phylogenetic analysis of the Prawn species

2.6 *Phylogenetic Analysis*

The evolutionary history of the prawn species, history of extant species or taxa, morphologies based on DNA sequence information, its genetic distance has been studied using phylogenetic analysis. The unweighted pair group method with arithmetic mean (UPGMA) is an agglomerative hierarchical clustering method where the genetic distances can be well studied and presented in Fig. 5.

Tree upgma

See Fig. 5.

2.7 *Image Analysis and Neural Networks*

The images of different prawns are isolated from one another and from the background. The physical differences between the types of prawns are position of the mouth, number and shape of fins, length, width, body mass, etc. A single prawn information is sent to a feature extractor, then features are passed to a classifier and make a final decision. The automatic recognition of prawn species by their patterns is a very interesting field. Based on specific properties of prawn species, new image processing techniques are developed.

By using DNA-based molecular markers, biotypes, organism's species, subspecies and population differences can be identified. Also, the nature and extent of intraspecific or interspecific genetic variation can be examined for impact on taxonomic problems of organisms and their natural enemies. Surveys of natural populations of organisms for DNA variability can define population structure and measure gene flow. Molecular genetic data may enhance the understanding of insect biosystematics and evolutionary trends (Figs. 6 and 7).

Neural networks can detect patterns in their input and partition those patterns into categories without requiring that the number or types of categories be predefined. The inputs for the neural networks are physical features of prawn species extracted from feature extraction techniques and general characteristics of prawn species. Thus the neural networks have been proved to be very powerful tools for pattern recognition

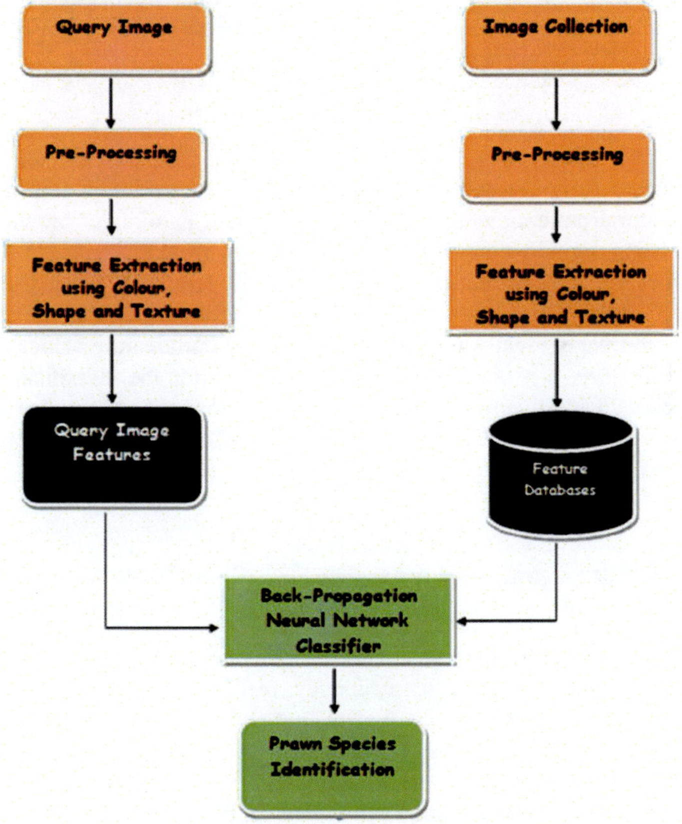

Fig. 6 Specimen identification through image processing and neural networks

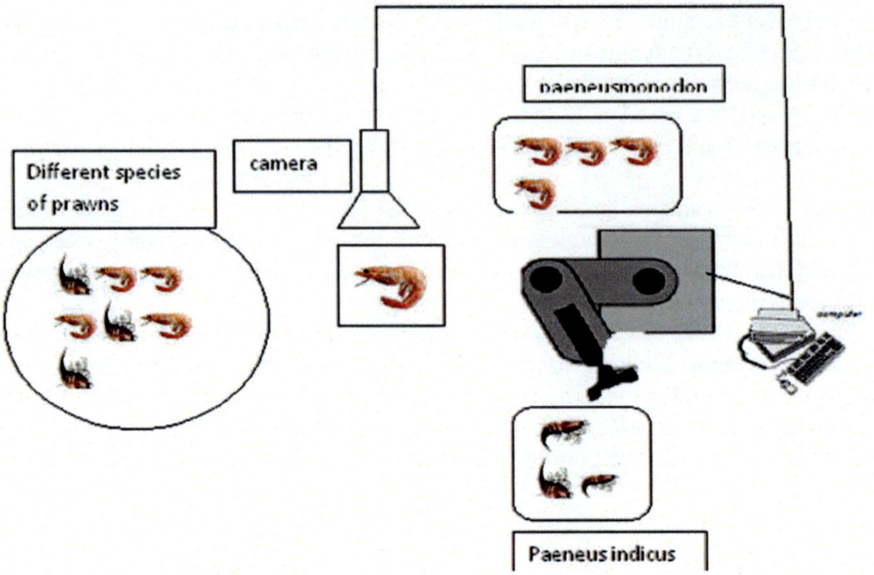

Fig. 7 Automated Identification and Segregation of prawn species (Ricardo Gutierrez-Osuna, Texas University)

and have successfully been used in the automated classification of acoustic signals including natural patterns which enables the different patterns of prawn structure efficiently using digital laboratory and wet laboratory (Fig. 8).

To extract the Texture features, an algorithm is developed using Gabor Filter Technique. Texture features are found by calculating the mean and variance of the Gabor filtered image. Finally, by considering the resultant feature values of Colour, Shape and Texture, a neural network is constructed for the classification of the prawn species. In addition to the classification, the proposed system also stores and retrieves the resultant values. The user-friendly classification system is designed and implemented in MATLAB 7.11 with GUI features.

3 Result and Discussion

The DNA barcodes have been generated for the prawn sequences using the COI marker gene by the BOLD. This clearly displays that COI mitochondrial marker gene carries a unique sequence for each species that shows the species level molecular divergence (Figs. 9, 10, 11 and 12). COI gene proved to be a robust evolutionary marker for the analysis of intraspecific and interspecific in many marine organisms [8–11]. The phylogenetic analysis clearly depicts the prawn species with evolutionary distance (Figs. 4 and 5).

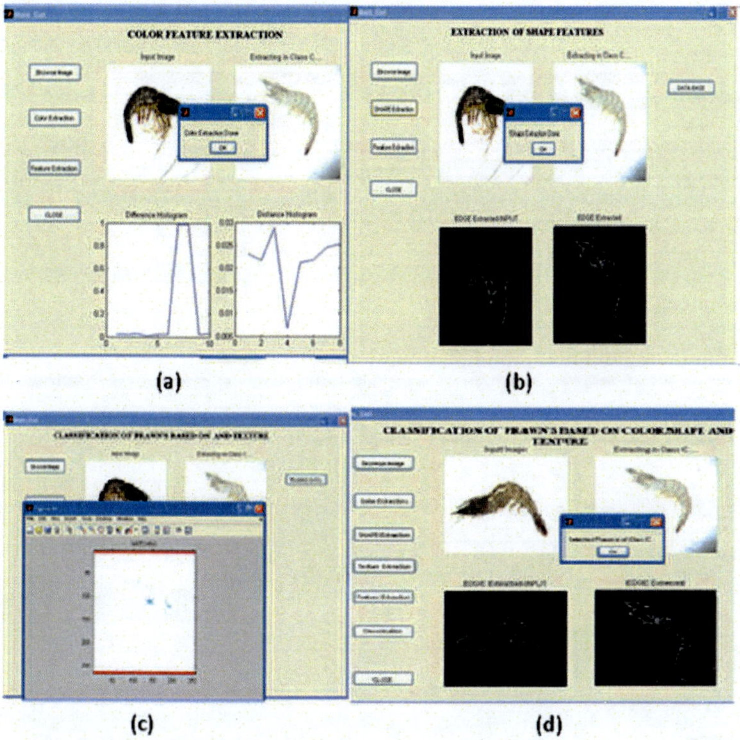

Fig. 8 **a** Colour feature extraction. **b** Shape feature extraction. **c** Texture feature extraction. **d** Prawn classification based on colour, shape and texture using neural networks

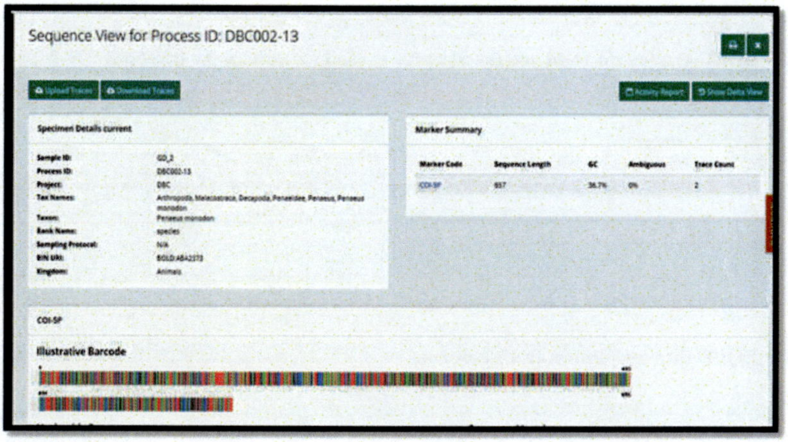

Fig. 9 BOLD display—the specimen details such as process ID, sequence length, species name and Barcode

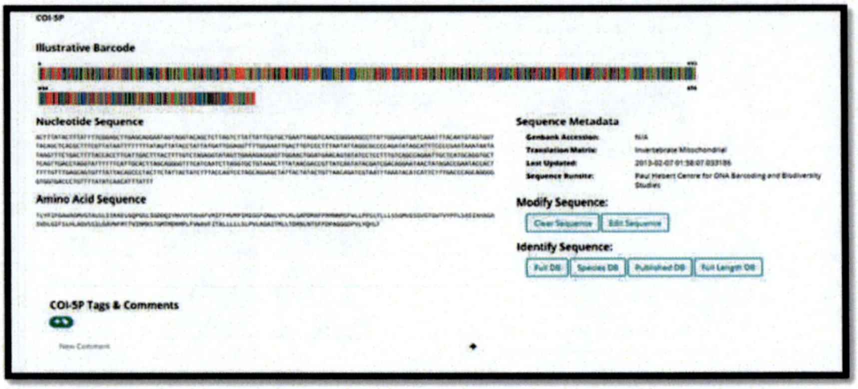

Fig. 10 BOLD display—nucleotide sequence and aminoacid sequences of the barcode

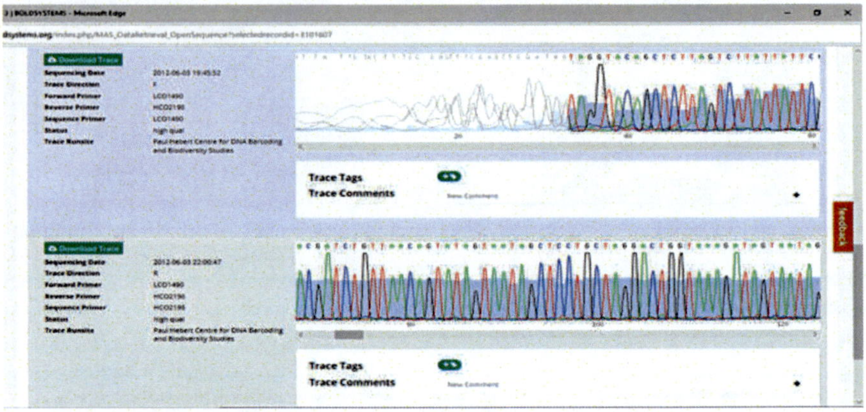

Fig. 11 BOLD display: forward and reverse primer used for developing barcodes

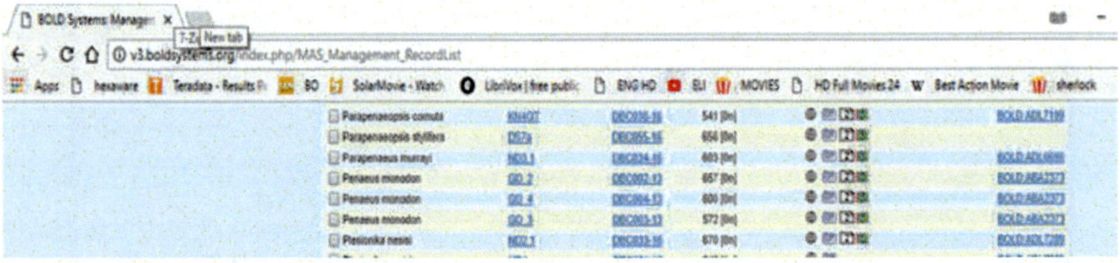

Fig. 12 BOLD display—Prawn DNA barcodes

The image analysis and the neural networks applied through the developed algorithm of Gabor filter methodology aided the automated feature extraction of the prawn species effectively and efficiently.

4 Documentation of Prawn Species with Their Respective DNA Barcodes

S. No	Specimen image	Specimen Details	DNA Barcode
1.		❖ **Name:** *Exopalaemon carinicauda* ❖ **Process id:** **DBC031-16** ❖ **Primers:** LCO1490 ,HCO2198 ❖ **Labelling: MJ1.1**	
2.		❖ **Name:** *Metapenaeus ensis* ❖ **Process id:** **DBC032-16** ❖ **Primers:** LCO1490 ,HCO2198 ❖ **Labelling: MJ5.1**	
3.		❖ **Name:** *Metapenaeus sp* ❖ **Process id:** **DBC032-16** ❖ **Primers:** LCO1490 ,HCO2198 ❖ **Labelling: MJ5.1**	
4.	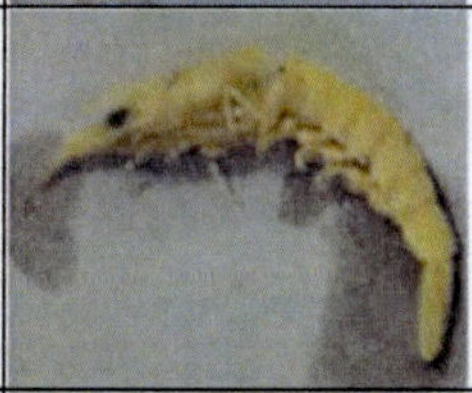	❖ **Name:** *Parapenaeopsis cornuta* ❖ **Process id:** **DBC041-16** ❖ **Primers:** LCO1490 ,HCO2198 ❖ **Labelling: KN4BT**	
5.		❖ **Name:** *Oratosquillina interrupta* ❖ **Process id:** **DBC029-16** ❖ **Primers:** LCO1490 ,HCO2198 ❖ **Labelling: KN3.1**	
6.		❖ **Name:** *Macrobrachium rosenbergii* ❖ **Process id:** **DBC030-16** ❖ **Primers:** LCO1490 ,HCO2198 ❖ **Labelling: KN2.1**	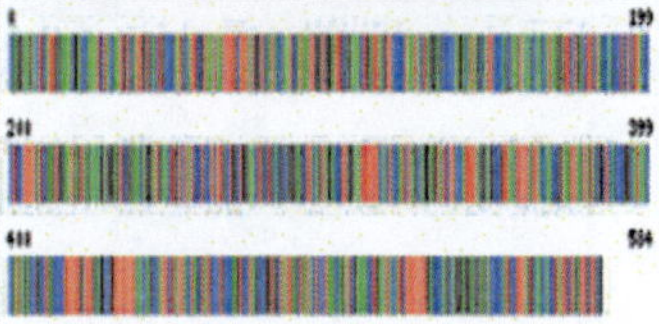
7.	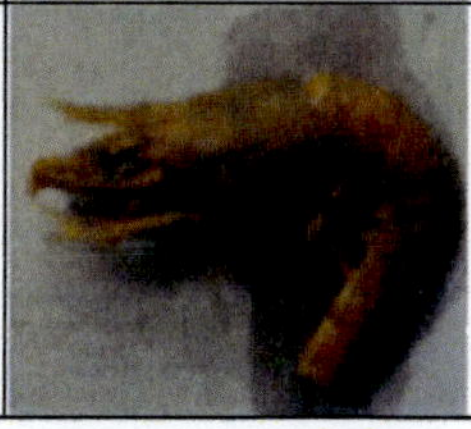	❖ **Name:** *Metapenaeopsis palmensis* ❖ **Process id:** **DBC047-16** ❖ **Primers:** LCO1490 ,HCO2198 ❖ **Labelling: DS4UL**	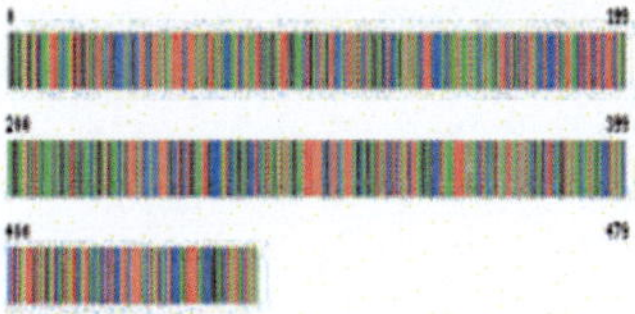

8.		❖ **Name:** *Rimapenaeus sps* ❖ **Process id:** **DBC026-16** ❖ **Primers:** LCO1490 ,HCO2198 ❖ **Labelling: DS2**	
9.		❖ **Name:** *Fenneropenaeus sps* ❖ **Process id:** **DBC036-16** ❖ **Primers:** LCO1490 ,HCO2198 ❖ **Labelling: DS6.1**	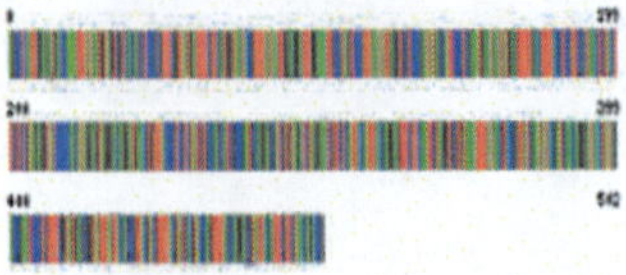
9.		❖ **Name:** ***Marsupenaeus japonicus*** ❖ **Process id:** **DBC059-16** ❖ **Primers:** LCO1490 ,HCO2198 ❖ **Labelling : MJ1GT.1**	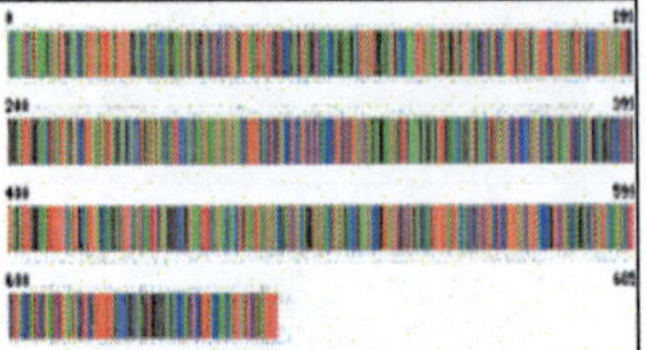
10.		❖ **Name:** ***Solenocera membranacea*** ❖ **Process id:** **DBC058-16** ❖ **Primers:** LCO1490 ,HCO2198 ❖ **Labelling: MJ4a**	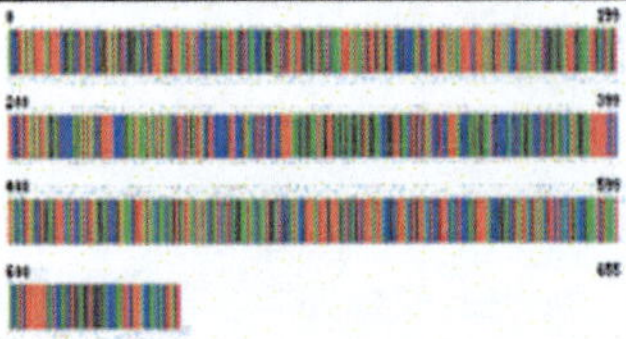
11.		❖ **Name:** ***Parapenaeopsis stylifera*** ❖ **Process id:** **DBC055-16** ❖ **Primers:** LCO1490 ,HCO2198 ❖ **Labelling: DS7a**	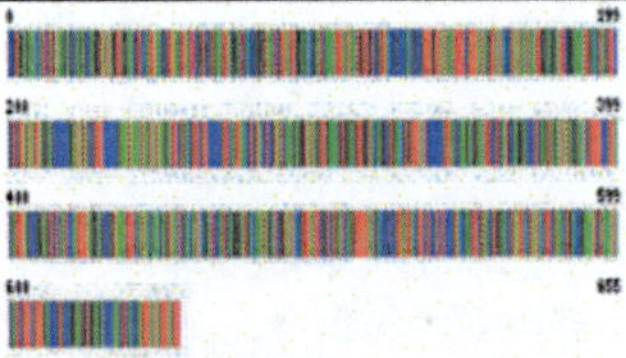
12.		❖ **Name:** ***Litopenaeus vannamei*** ❖ **Process id:** **DBC001-13** ❖ **Primers:** LCO1490 ,HCO2198 ❖ **Labelling: GD1**	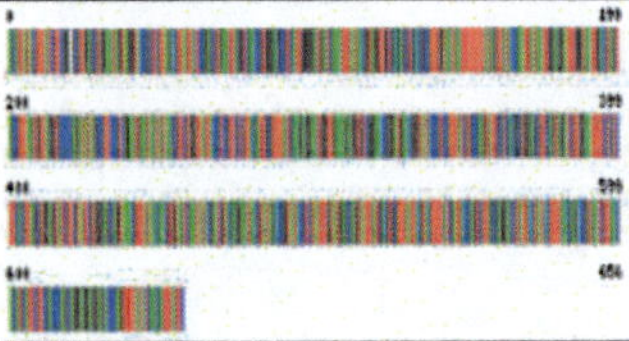

13.	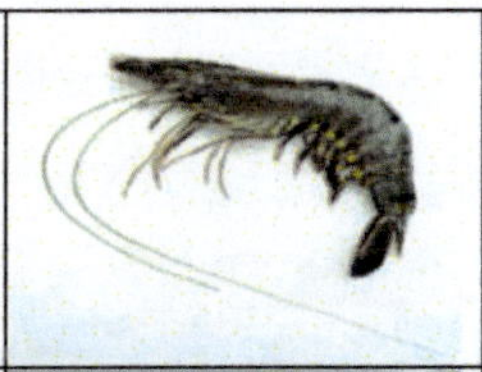	❖ **Name:** *Penaeus monodon* ❖ **Process id:** **DBC002-13** ❖ **Primers:** LCO1490 ,HCO2198 ❖ **Labelling:** **GD2**	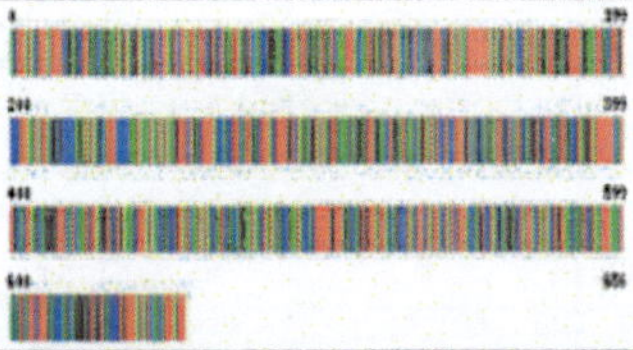
14.		❖ **Name:** *Metapenaeus brevicornis* ❖ **Process id:** ❖ **Primers:** LCO1490 ,HCO2198 ❖ **Labelling:**	
15.		❖ **Name:** ***Nematopalaemon tenuipes*** ❖ **Process id:** **DBC011-13** ❖ **Primers:** LCO1490 ,HCO2198 ❖ **Labelling:** **KAK5**	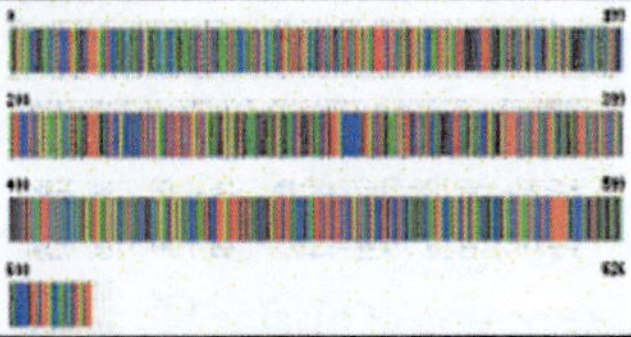
16.		❖ **Name:** ***Munida gordae*** ❖ **Process id:** **DBC006-13** ❖ **Primers:** LCO1490 ,HCO2198 ❖ **Labelling:** **VCR 6**	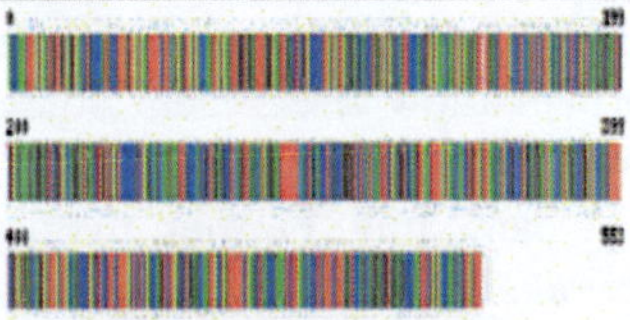
17.		❖ **Name:** *Exhippolysmata ensirostris* ❖ **Process id:** **DBC022-14** ❖ **Primers:** LCO1490 ,HCO2198 ❖ **Labelling:** KAK1	
18.		❖ **Name:** *Metapenaeus sp.* ❖ **Process id:** **DBC024-14** ❖ **Primers:** LCO1490 ,HCO2198 ❖ **Labelling KAK8**	
19.		❖ **Name:** *Macrobrachium asperulum* ❖ **Process id:** **DBC023-14** ❖ **Primers:** LCO1490 ,HCO2198 ❖ **Labelling: KAK5A**	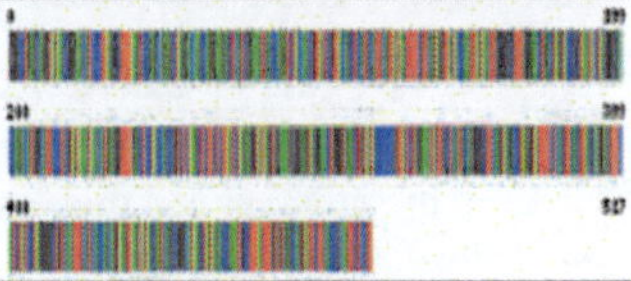

20.		❖ **Name:** *Fenneropenaeus indicus* ❖ **Process id:** DBC005-13 ❖ **Primers:** LCO1490 ,HCO2198 ❖ **Labelling: VCR_5**	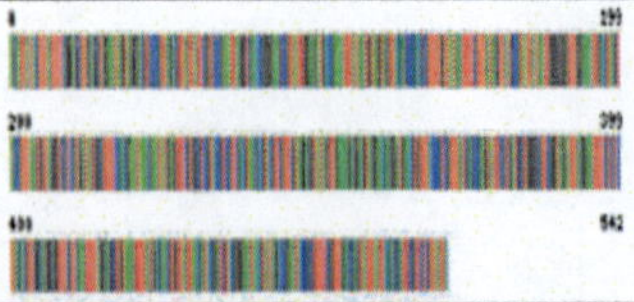
21.		❖ **Name:** *Metapenaeus monoceros* ❖ **Process id:** DBC007-13 ❖ **Primers:** LCO1490 ,HCO2198 ❖ **Labelling: VCR 7**	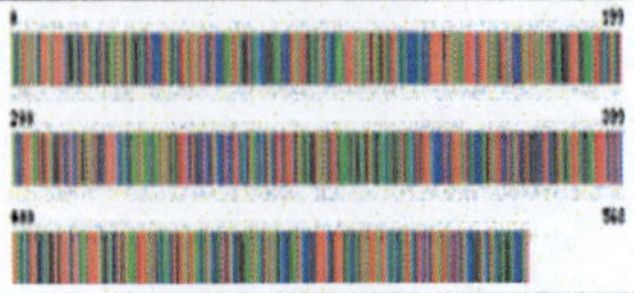

5 Conclusion

Conservation of marine biodiversity and its management is the major challenge for the Indian, particularly regarding species identification. The present study gives a gizmo that will allow researchers, students, aqua-techs and the public to better understand how to recognize the species, be it at any stage or form of their life cycle, through DNA barcoding. Further, the algorithms developed for image analysis and neural networks facilitate the automated identification of prawn species which comes as a machination. That's not all, and this study further helps to maintain quality control and ease of specimen identification at export and imports of the aqua foods.

References

1. Hebert PDN, Cywinska A (2003) Biological identification through DNA barcodes. Proc R Soc B Biol Sci 270(1512):313–321
2. Radulovici AE, Archambault P, Dufresne F (2010) DNA barcodes for marine biodiversity: moving fast forward. Diversity 2(4):450–472
3. Hebert PDN, Ryan GT (2005) The promise of DNA barcoding for taxonomy. Syst Biol 54(5):852–859
4. Casiraghi M, Labra M, Ferri E, Galimberti A, De Mattia F (2010) DNA barcoding: theoretical aspects and practical applications. In: Nimis PL, Vignes Lebbe R (eds) Tools for identifying biodiversity: progress and problems, pp 269–273
5. Mamatha Y, Murthy CK, Prakash BS (2016) Study of morphological variations of fissures and lobes of lung. Int J Anat Res 4(1):1874–1877
6. Ramasamy S, Andavar R, Shankar KR, Selvam M (2017) Biomimetic synthesis of silver nanoparticles from Streptomyces atrovirens and their potential anticancer activity against human breast cancer cells. IET Nanobiotechnol 11(8):965–972
7. Folmer O, Black MB, Wr H, Lutz R (1994) DNA primers for amplification of mitochondrial Cytochrome C oxidase subunit I from diverse metazoan invertebrates. Mol Mar Biol Biotechnol 3(5):294–9
8. Ward RD, Zemlak T, Innes BH, Last PR (2005) DNA Barcoding Australia's fish species. Philosop Trans R Soc B Biolog Sci 360(1462):1847–1857

9. Chae HS, et al (2016) Anti-inflammatory effects of 6,8-Diprenyl-7,4′-dihydroxyflavanone from *Sophora tonkinensis* on lipopolysaccharide-stimulated RAW 264.7 Cells. Molecules 21:1049
10. Baldwin B, Sanderson MJ, Mark PJ, Wojciechowski MF (1995) The its region of nuclear ribosomal DNA: a valuable source of evidence on angiosperm phylogeny. Ann Mo Bot Gard 82(2):247
11. Spies IB, Stevenson D, Orr J, Hoff GR (2011) Molecular systematics of the skate subgenus Arctoraja (Bathyraja: Rajidae) and support for an undescribed species, the leopard skate, with comments on the phylogenetics of Bathyraja. Ichthyol Res 58(1):77–83

A Substantial Approach to Predict Air Quality Using LVQ Neural Network

Sumaya Sanober and K. Usha Rani

Abstract Nowadays, air pollution is one of the major threats to the ecosystem and human existence. Addressing air pollution is the biggest environmental challenge in smart city environments. Therefore, monitoring of air quality has become an important task in urban and industrial areas. The primary focus to monitor air quality is a human-centred approach, as data collected from different monitoring stations are evaluated based on human experience and unreliable. Therefore, the emergence of machine learning technologies such as decision trees, neural networks (NN), and support vector machine (SVM) justifies the application of statistical methods to environmental modelling, especially in the forecasting of air quality. In addition, the existing techniques are insufficient to find the impact of every air pollutants for minimising air pollutions. An Artificial Neural Network (ANN) with Learning Vector Quantisation (LVQ) is implemented for prediction in this research study. The results showed that the ANN-LVQ achieved 97.58% of accuracy with a minimal error rate.

Keywords Air pollution · Air quality index · Artificial neural network · Learning vector quantisation · AQI prediction

1 Introduction

In recent year Carrying out the air in cities is of considerable importance. Air efficiency quality s, a serious air pollution problem is happening every day. In particular, smog on waste products such as nitric oxide (NO), carbon monoxide (CO), hydrocarbons and synthetic nectar from cellular sources have affected the world's prominent surroundings [1]. Not only are the above barriers harmful to human health, but PM 2.5 [2] and the main components of photochemical smog are also important for

S. Sanober (✉)
SPMVV, Tirupati, India
e-mail: sumayacsdept@gmail.com

K. Usha Rani
Department of Computer Science, SPMVV, Tirupati, India
e-mail: usharanikuruba@yahoo.co.in

S. Jyothi et al. (eds.), *Proceedings of the 2nd International Conference on Computational and Bio Engineering*, Lecture Notes in Networks and Systems 215,
https://doi.org/10.1007/978-981-16-1941-0_52

the formation of inflammation. The air quality is more contaminated with PM and many pollutants called CO, NO, NH. Particles in the air pose a serious threat to production, life, human health and other aspects. Exposure to high concentrations of particles is associated with increased mortality or morbidity, which is proven by the World Health Organization (WHO) [3, 4]. The emission standards are exceeded, and different types of pollutants are discharged, high concentration and rapid growth of mobile emissions [5]. For example, the vehicles provide the excess releases of CO $ NO2 are 4.5% and 3500 (ppm), correspondingly.

India is the second most country for high air pollution in the overall world due to the growth of urbanisation and industrialisation that results in high PM concentrations [6–8]. In addition, ambient concentrations of PM in many metropolitan cities, namely Kolkata, Chennai, Mumbai, Bangalore, Hyderabad, etc., frequently violates the guidelines of WHO [9, 10]. The problem of predicting air pollution using linear statistical methods has been considered in the past. However, due to the complexity and variability of the data in time series, this method provides poor estimation in air pollution prediction [11]. Various methods of machine learning have been developed over the past 60 years to solve the problem of complexity [6]. But, the main problem is handling the information and it is important to ensure that data analysis is efficient and reliable. Misinterpreting the data can lead to wrong decisions, which can be extremely dangerous [7]. The use of rigorous data and use of artificial intelligence are important to ensure the robustness and speed of communication sources. In addition, the existing machine learning techniques are insufficient to recommend highly pollutant particles to control air pollutions [8, 9]. Therefore, in this research study, an effective ANN with LVQ algorithm is used for predicting the Air Quality Index (AQI) on standard and collected datasets. In the feed-forward neural network classifier, typical features of a dataset are represented by the prototypes with the same feature space in the LVQ algorithm. The advantage of LVQ is that the algorithm is easy to implement for multiclass (ANN) and multi-cluster (SOM) classification problems, where the complexity of the algorithm is adjusted during training. The experiments are conducted to test the efficiency of two objectives in terms of false discovery rate (FDR), accuracy, sensitivity, specificity, error rate and False Omission Rate (FOR).

The organization of the research paper is as follows: Sect. 2 presents the study of existing techniques. Section 3 explains the two objectives of the proposed methodology and the validation of the proposed method with existing techniques on standard and collected datasets are presented in Sect. 4. In conclusion, the research study is discussed in Sect. 5.

2 Literature Review

In this segment, various existing methods that are developed to predict the air quality index using artificial neural network techniques were studied. In addition, the key benefits of existing techniques with its limitations are also presented from the year 2018 to 2020.

Zhou et al. [10] implemented a neural network, and it was boosted with a non-linear autoregressive neural network, where the method was specially intended for time series prediction. Two models were presented in this study; Examine the method grounded on preceding air quality index and atmospheric statistics to first estimate the pollution concentration and then calculate the AQI. The outcomes exhibit to verify the efficiency of the developed approach in terms of MAPE, RMSE and band accuracy. This approach achieved registration accuracy (over 81%) for all records. However, by further analysing the estimated values of individual pollutants, the experiments found that the method based on a neural network for estimating the particle matters 10 (PM 10) was not optimal.

Ameer et al. [11] solved the air pollution prediction issues by developing multi-stage classifiers such as Decision Tree regression, and Gradient Boosting regression. The optimal models were anticipated in terms of both processing time and least error rate in this research. The parameters such as RMSE and MAE were used to validate the presence of multi-stage classifiers. Gradient regression led to the worst of all algorithms because it has the longest processing time in almost all data sets, and in most cases, it has a very high error rate.

Jiang et al. [12] presented a novel hybrid wavelet packet decomposition (WPD), modified extreme learning machine (MELM), multi-dimensional scaling and K-means (MSK) as WPD-MELM-MSK-MELM learning attitude for predicting air quality index. The improved pigeon-inspired optimisation (IPIO) attitude was used to enhance the preliminary weights and thresholds of ELM. MAE, normalised RMSE (NRMSE), RMSE and directional symmetry (Ds) were considered as parameters to validate the effectiveness of the developed method. The method focused on short, middle and long-term prediction but is considered only limited factors for the prediction of air pollution concentrations.

Masmoudi et al. [13] proposed a new feature-ranking scheme, labelled as Ensemble of Regressor Chains-guided Feature Ranking (ERCFR) to predict manifold impurities concurrently for two monitoring stations of towns. ERCFRs have been incorporated into multipurpose regression methods to progress air quality forecasting. The outcomes proved that this design offers better performance and improved prediction values than other feature-ranking techniques. Consistency index and robustness performance trade-off were used to validate the performance of the ERCFR method. However, ERCFR was more time-consuming compared to Random Forest Predictive Clustering Tree (RFPCT) attitude due to the time difficulty of ERC. In addition, the method was insufficient to handle the large data sets.

Fazai et al. [14] proposed novel fault detection and isolation approaches that are applicable to the process monitoring using the Online Reduced Kernel principal component analysis (OR-KPCA) technique. The fractional OR-KPCA error was a beneficial technique used in isolation. From the OR-KPCA in a subset of values, conventional structural residues were obtained. The simulation was taken as a case study of live data. The error of the specified methods and fault isolation power sensor was checked on real-time air quality data by examining the process. Upon analysis, the OR-KPCA method was efficient to appropriately identify only simple faults.

Zhao et al. [15] designed a hybrid ensemble networks called CERL, which was a hybrid synchronisation model designed to manage temporary serial data to predict air feature over several hours. CERL was able to copy several basic models into data, surpassing them in comparison with other basic models. MAPE was used for 1, 3, 5, and 8-step forecasts, which were considered as parameters for validation. However, this work was based on hourly forecasts and did not have high accuracy for long-term forecasts.

Li et al. [16] predicted the air pollutant concentration by implementing the self-adaptive neuro-fuzzy weighted learning machine (ANFIS-WELM). The membership degree of every input variables were selected by Gaussian membership function parameters. The pollutants, namely CO, nitric oxide, Particulate Matter (2.5, 10), were obtained from a particular monitoring station and used for authentication. The experimental results showed that the ANFIS-WELM improved performance upon considering MAE, RMSE, time, MAPE, and correlation coefficient (R). The method works well only in single monitoring stations in the Datong, Taiwan region and provides poor performance while using the data of multiple monitoring stations in that region.

Soh et al. [17] developed a combined neural network such as adaptive ANN, CNN and LSTM for extracting the relations of spatial-temporal data and predicted the air quality of PM2.5 up to 48 h. When comparing with LSTM, CNN was useful for long-time predictions because the temporal delay factors were extracted from the surrounding features by learning the information. The experiments were conducted on real-time data of Taiwan and Beijing in terms of MAE, where the results proved that the combined neural networks achieved better performance. However, noises were occurred in the model due to the inclusion of all locations for verifying the AQI of PM2.5.

Feng et al. [18] enriched the accuracy of ANN for PM2.5 concentrations by developing a hybrid model that combined the analysis of air mass trajectory and wavelet transformation. The distinct corridors between dirty air and unpolluted air for the designated regions were identified by air mass trajectory, where direction and wind speed were used as the parameter for calculating these trajectory values. The wavelet transformation was used to decompose the novel time series of PM2.5 into sub-series with less variability. Finally, the multi-layer perceptron (MLP) was used to predict the AQI by using PM2.5 pollutant and regular climatological prediction values as input. The simulations were conducted on the 13 monitoring stations in Hebei, Beijing and Tianjin by means of RMSE and detection rate. The overall prediction accuracy of the hybrid model was very less due to the severe air pollution in the selected monitoring stations.

3 Proposed Methodology

Section below explained prediction as follows.

3.1 Predicting Air Quality Index (AQI)

The proposed method consists of dataset collection, data cleaning and prediction process, which is explained in Figs. 1 and 2.

The collected dataset has 12 pollutants, namely PM2.5, PM10, NO, NO2, NOx, NH3, CO, SO2, O3, Benzene, Toluene and Xylene remain used in this research study. The raw data are given to the data cleaning process for the enhancement proposed prediction technique.

3.2 Data Cleaning

Raw data takes a series of initial processing steps so that it can be used in the descriptive phase of the final analysis. The process of data cleaning is used to identify and eliminate the errors in the datasets that may negatively impact on the predictive model. The errors includes noisy data, imbalance data and missing data are eliminated in the data cleaning process and gives the cleaned data as input to ANN with LVQ technique for prediction of pollutants.

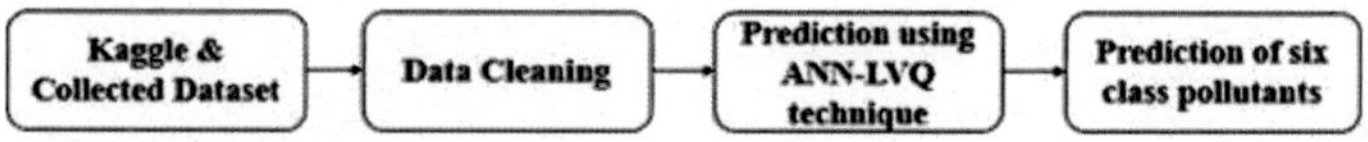

Fig. 1 Workflow of proposed method for AQI prediction

Fig. 2 Geographical location of 25 cities in India region [19]

Algorithm 1: Working Procedure of Proposed ANN with LVQ

Input: ProblemSize, InputPatterns, $iterations_{max}$, $learn_{rate}$; Set t=0;
Output: Network
Network ← ConstructNetworkLayer ();
$Network_{weight}$ ← InitializeWeights (Network, ProblemSize);
for $i = 1$ to $iterations_{max}$ do
    $Pattern_i$ ← SelectInputPattern (InputPatterns);
    Update the learning rate using LVQ as $\alpha(t) = \alpha_0(1 - {}^{t}/_{T})$
    Determine and record the winning unit of training vectors as x_k;
    Update the synaptic weight vectors of the winning unit according to the following eq;

$$v_i(t+1) = v_i(t) + \alpha(t)(x_k - v_i(t))o_i(x_k)$$

    $t = t + 1$; if t>T stop.
    UpdateWeight (Pattern, Output, Network, $learn_{rate}$);
end
return network;

4 Results and Discussion

In this section, two kinds of experiments are considered to validate the performance of the prediction and recommendation model on two datasets. Here, two algorithms are implemented in the computer with a Pentium IV processor of 1.8 GHz using MATLAB Version 18.a. The parameters such as accuracy, sensitivity, specificity, FOR, FDR and error rate are used for the validation process. Accuracy is the suitable evaluation metric to find the efficiency of AQI prediction and recommendation, and the error rate is defined as the inaccuracy of predicted output variables, which are illustrated in Eqs. (1 and 2).

$$Accuracy = \frac{TP + TN}{TP + TN + FP + FN} \times 100 \tag{1}$$

$$ErrorRate = 100 - Accuracy \tag{2}$$

Sensitivity and specificity is used to understand the relationships between the variables of input and output in a system and also used to test the robustness of model in the presence of uncertainty. The mathematical Eqs. (3) and (4) is expressed as follows:

$$Sensivitiy = \frac{TP}{TP + FN} \times 100 \tag{3}$$

$$Specificity = \frac{TN}{TN + FP} \times 100 \quad (4)$$

FDR defines the ratio of predicted false +ve standards to the total number of T and F +ve values, which shows in Eq. (5)

$$FDR = \frac{FP}{TP + FP} \times 100 \quad (5)$$

The ratio of predicted false negative values to the total no of T and F –ve standards called FOR that is illustrated in Eq. (6)

$$FOR = \frac{FN}{TN + FN} \times 100 \quad (6)$$

Total Positive and Total Negative are represented as T +ve and T −ve, FP and FN are depicted as False Positive and False Negative.

4.1 Quantitative Analysis of Proposed Prediction Model

The validation of ANN-LVQ is tested on overall collected datasets, which is explained in this section. The performance of ANN-LVQ is compared with traditional ANN and adaptive ANN on the collected dataset, which is presented in Table 1 and Fig. 3 shows the graphical representation for the proposed ANN-LVQ technique on the collected dataset.

The results showed that the adaptive ANN method provides poor performance than the other two techniques; for example, the accuracy of adaptive ANN is only 57.81% and the error rate is 42.19%. In addition, the traditional Feed-forward-ANN acquires high performance than the adaptive ANN and minimised the error rate, FOR and FDR. The sensitivity of traditional feed-forward ANN achieved 81.9% sensitivity and 92.41% specificity, where the adaptive ANN achieved only 60.48% of sensitivity and 70.55% specificity. However, the traditional ANN has error rate of 7.07% and consumed high computation time. In this research study, LVQ is used for the weighting function of ANN technique and achieved 97.58% accuracy, 0.07% FOR,

Table 1 Performance of ANN-LVQ

Methodology	Parameters (%)					
	Accuracy	Sensitivity	Specificity	Error rate	FDR	FOR
Adaptive ANN	57. 81	60.48	70.55	42.19	75.74	5.11
Traditional ANN	90.07	81.91	92.41	9.93	12.67	7.07
ANN-LVQ	**97.58**	**89.99**	**99.45**	**2.42**	**5.12**	**0.07**

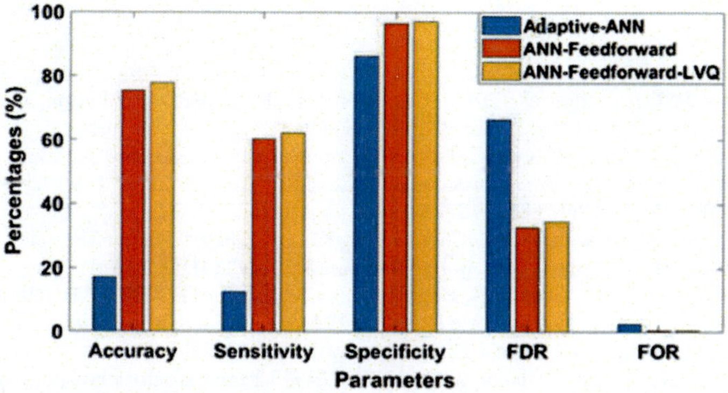

Fig. 3 Graphical illustration of performance of ANN-LVQ

2.42% of error rate, 99.45% of specificity, and 89.99% of sensitivity on collected data. Table 1 and Fig. 3 show the performance of ANN-LVQ for all 25 cities of India.

5 Conclusion

In conclusion, predicting air quality is a multifaceted task because of its dynamic nature, volatility, and high variability in space and time of pollutants and particulates in the air. At the same time, it is changing progressively, forecast and track air quality, especially in urban areas, due to the crucial effect of air pollution on populations and the environment. The results of the proposed models show that fusion models of Artificial Neural Network remain effective to detect air quality and predict the level of pollution in urban, ruler or industrial areas. ANN is implemented and the weighting function of ANN is optimised by LVQ on the collected datasets. Six classes are predicted from the cleaned data for the prediction of AQI. The validation of ANN with LVQ is checked on dataset of different cities in India for the period of 2015–2020. The results proved that ANN-LVQ achieved 97.58% accuracy, 89.99% of sensitivity and 99.45% of specificity, and ANN-LVQ is only 2.42% on the collected dataset. The accuracy achieved by the proposed model is quite enhanced when compared to the other existing research works. This piece of work would contribute greatly to society as the same model can be applied to the other state smart cities data.

References

1. Mannucci PM, Franchini M (2017) Health effects of ambient air pollution in developing countries. Int J Environ Res Public Health 14(9):1048
2. Klimont Z, Kupiainen K, Heyes C, Purohit P, Cofala J, Rafaj P, Borken-Kleefeld J, Schöpp W (2017) Global anthropogenic emissions of particulate matter including black carbon. Atmos Chem Phys Discuss 17(14):8681–8723
3. Fang D, Chen B, Hubacek K, Ni R, Chen L, Feng K, Lin J (2019) Clean air for some: unintended spillover effects of regional air pollution policies. Sci Adv 5(4):p.eaav4707
4. Glencross DA, Ho TR, Camiña N, Hawrylowicz CM, Pfeffer PE (2020) Air pollution and its effects on the immune system. Free Radic Biol Med 151:56–68. https://doi.org/10.1016/j.freeradbiomed.2020.01.179. Epub 2020 Jan 30. PMID: 32007522
5. Gately CK, Hutyra LR, Peterson S, Wing IS (2017) Urban emissions hotspots: quantifying vehicle congestion and air pollution using mobile phone GPS data. Environ Pollut 229:496–504
6. Gu K, Zhou Y, Sun H, Zhao L, Liu S (2019) Prediction of air quality in Shenzhen based on neural network algorithm. Neural Comput Appl: 1–14
7. Ghaemi Z, Alimohammadi A, Farnaghi M (2018) LaSVM-based big data learning system for dynamic prediction of air pollution in Tehran. Environ Monit Assess 190(5):300
8. Chen S, Mihara K, Wen J (2018) Time series prediction of CO2, TVOC and HCHO based on machine learning at different sampling points. Build Environ 146:238–246
9. Sahoo D, Hoi SC, Li B (2019) Large scale online multiple kernel regression with application to time-series prediction. ACM Trans Knowl Discov Data (TKDD) 13(1):1–33
10. Zhou Y, De S, Ewa G, Perera C, Moessner K (2018) Data-driven air quality characterization for urban environments: a case study. IEEE Access 6:77996–78006
11. Ameer S, Shah MA, Khan A, Song H, Maple C, Islam SU, Asghar MN (2019) Comparative analysis of machine learning techniques for predicting air quality in smart cities. IEEE Access 7:128325–128338
12. Jiang F, He J, Tian T (2019) A clustering-based ensemble approach with improved pigeon-inspired optimization and extreme learning machine for air quality prediction. Appl Soft Comput 85:105827
13. Masmoudi S, Elghazel H, Taieb D, Yazar O, Kallel A (2020) A machine-learning framework for predicting multiple air pollutants' concentrations via multi-target regression and feature selection. Sci Total Environ 715:136991
14. Fazai R, Abdellafou KB, Said M, Taouali O (2018) Online fault detection and isolation of an AIR quality monitoring network based on machine learning and metaheuristic methods. Int J Adv Manuf Technol 99(9–12):2789–2802
15. Zhao Z, Qin J, He Z, Li H, Yang Y, Zhang R (2020) Combining forward with recurrent neural networks for hourly air quality prediction in Northwest of China. Environ Sci Pollut Res: 1–18
16. Li Y, Jiang P, She Q, Lin G (2018) Research on air pollutant concentration prediction method based on self-adaptive neuro-fuzzy weighted extreme learning machine. Environ Pollut 241:1115–1127
17. Soh PW, Chang JW, Huang JW (2018) Adaptive deep learning-based air quality prediction model using the most relevant spatial-temporal relations. IEEE Access 6:38186–38199
18. Feng X, Li Q, Zhu Y, Hou J, Jin L, Wang J (2015) Artificial neural networks forecasting of PM2. 5 pollution using air mass trajectory based geographic model and wavelet transformation. Atmos Environ 107:118–128
19. Pant P et al (2019) Monitoring particulate matter in India: recent trends and future outlook. https://doi.org/10.1007/s11869-018-0629-6

An Automatic Identification of Diabetic Macular Edema Using Transfer Learning

Y. Nagendra Prasad, C. Shoba Bindu, E. Sudheer Kumar, and P. Dileep Kumar Reddy

Abstract The growth of fluid in the macular region is an area in the retina center, which causes Diabetic Macular Edema (DME). The early stages it is classified by extracting features from retinal images to grade DME. This paper acquaints a transfer learning approach for DME treatment from computerized fundus images and correctly grouping its reality. This paper introducing a Convolutional Neural Network (CNN) to deal with diagnosing DME from digital fundus images and precisely classifying its seriousness, and building a system with DenseNet architecture, which can convolute features engaged in the task of classification. This model is trained by utilizing an openly accessible Messidor and IDRiD dataset and shows the amazing outcome, especially for a significant level order task. The proposed model achieves 96.78% accuracy on Messidor and 96.51% on IDRiD datasets.

Keywords Retinal images · Diabetic macular edema · Convolutional neural network · Transfer learning · DenseNet

1 Introduction

Diabetic Macular Edema (DME) is a hassle of diabetic retinopathy (DR), and it generally happens while vessels inside the primary region of the macula are tormented by accretion of fluid [1]. Due to the less production of insulin in the pancreas and/or by the inheritance leads to diabetics, is long-term aliment which leads to the disease called DME. DME is a serious phase of DR; it leads to the loss of vision for a lifetime

Y. Nagendra Prasad (✉) · C. Shoba Bindu · E. Sudheer Kumar
Department of CSE, JNTUA, Ananthapuramu, Andhra Pradesh 515002, India
e-mail: nagendrayeddula580@gmail.com

C. Shoba Bindu
e-mail: shobabindhu.cse@jntua.ac.in

P. Dileep Kumar Reddy
Department of CSE, SV College of Engineering, Tirupati, Andhra Pradesh 517502, India

S. Jyothi et al. (eds.), *Proceedings of the 2nd International Conference on Computational and Bio Engineering*, Lecture Notes in Networks and Systems 215,
https://doi.org/10.1007/978-981-16-1941-0_53

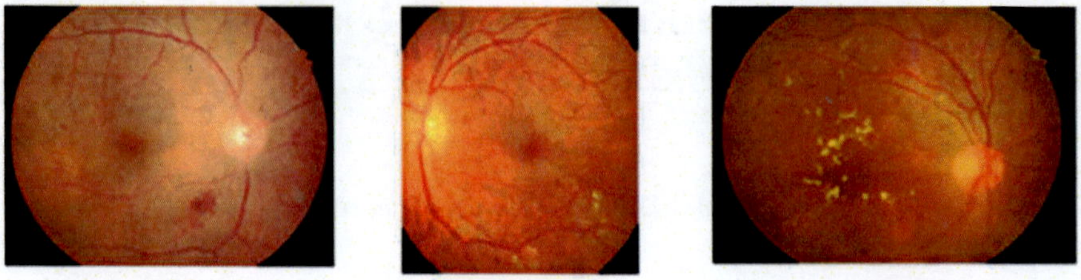

Fig. 1 Grading of diabetic macular edema

[2–4]. Grading of DME is in three levels—grade0, grade1, grade2, which is depicted in Fig. 1.

Diabetes presently influences people around 425 million throughout the world, and it is predicted, by the year 2025, the count will reach 520 million. It means that 10\% of folks are affected by any of the Diabetic forms, which probably leads to the DME. Presently, around 26.7 million people worldwide are affected by the DME, and the quantity may reach around 50 million by the end of 2025 [3, 5]. Detection of exudates is a popular approach for the classification of DME by using fundus snapshots [6, 7]. Based on the distance between macular and exudates, the severity of DME is determined, and the chance of getting DME is more when exudates present inside the macula; otherwise low chance of getting DME [8]. Quick therapy is essential for the person who is affected by the DME to save the vision.

Drive-by using these good opportunities, suggests developing a successful answer utilizing Deep learning strategies to routinely classify the fundus pics. In clinical investigations, the machine learning strategy is widely used. Deep learning is a powerful approach in the machine learning techniques in the field of the Medical Department. Deep learning strategies and particularly convolutional neural networks have hastily obtained fame for evaluating medical images [9]. The rest of the paper is organized as follows: Sect. 2 contains the related work on DME classification, Sect. 3 describes the dataset preparation, Sect. 4 describes the proposed transfer learning approach in Deep learning, Sect. 5 describes experimental evaluation, results, and draw some conclusions in Sect. 6.

2 Related Work

For the screening of DME from the last decades, many are tired of different approaches. Some of them were described below:

Lim [10] localized fovea and regions of the macula were marked on the reference of Early Treatment Diabetic Retinopathy Studies (ETDRS). For the extraction uses a watershed transformation. Based on the region of extracted exudates computed to classify the stages of DME on the dataset of Messidor. Magotra [11] utilizing

Table 1 Existing techniques for DME detection

S. no.	Author name	Methods
1	Lim [10]	Localization of fovea and macula using Early Treatment Diabetic Retinopathy Studies (ETDRS)
2	Magotra [11]	Hierarchical Classification and Grading of Diabetic Macular Edema using Texture Features
3	Sengar [12]	Region-Based Method
4	Thulka [13]	CLAHE & ASF, Optic disc mask, and exudates detection
5	Singh [14]	DMENet: DME diagnosis using Hierarchical Ensemble of CNNs

the texture features extracted from both spatial domain and wavelet domain from sub-regions of the mentioned radius near the macula. At first, they are finding the optic disc and the macular in a particular region from the optic disc. After finding the macular, a circular region of 1DD was extracted around the macular from the preprocessed images. Based on the ETDRS grading scale classifies the severity of the DME as stage 0, stage 1, stage 2. Sengar [12] proposed a Region-Based Method for the detection of Diabetic Macular Edema. The staging of DME is achieved by breaking the retina of the image by the International Standard. The severity of DME can be described by using the scaling of bright lesions in macular regions. The detection of macular is independent of the optic disc; this is efficient because optic disc location can be wrong due to the noise and reflection. Thulka [13] proposed a method to extract the vessels in retinal images by applying the CLAHE and Alternate Sequential Filtering. After removing the Residual vessels from the retinal images, the morphological operation dilation and erosion are applied to the preprocessed images. On the resultant images, the exudates are detected. Singh [14] utilizing a hierarchial ensemble of CNNs for the diagnosis of the Diabetic Macular Edema. Trained two groups of the cluster of CNNs from the retinal images and then apply gamma ensemble to group the predictions from the clusters to classify the DME positive or negative. If the output is DME is positive, then it is forward to the second stage where two clusters of CNNs are trained and applied a Delta ensemble technique to ensemble the predictions and classifies the DME grade of 1 or 2. The list of the existing approaches is listed in Table 1.

3 Dataset Preparation

3.1 Datasets

In this paper, the publicly available datasets Messidor and IDRiD were used. Both the dataset has the same grading labels that are grade0, grade1, grade2. The detailed information about the dataset is specified in Table 2.

Table 2 Number of obtained images from publicly available resources

Sno	Dataset	Grade 0	Grade 1	Grade 2	Resize
1	Messidor	974	75	151	224*224
2	IDRiD	222	41	243	224*224

3.2 Pre-processing of Dataset

The R, G, B color channels are extracted from the retinal images. To the G channel image, CLAHE is applied to enhance the image regions in the retinal images and it is cropped to avoid the background. The complete pre-processing is depicted in Fig. 2.

4 Deep Learning Framework

In the present day, Deep learning technology is working well for image classification [15]. It combines both feature extraction and classification. Deep learning techniques can provide better outcomes utilizing neural networks developed on large data. This developed model introduces a transfer learning approach for automatically grading the DME with precise performance. Transfer Learning is utilized for the extraction of features from the retinal images and the extracted features are forwarded as the input to the densely connected classifier. The proposed architecture is completely illustrated in Fig. 3.

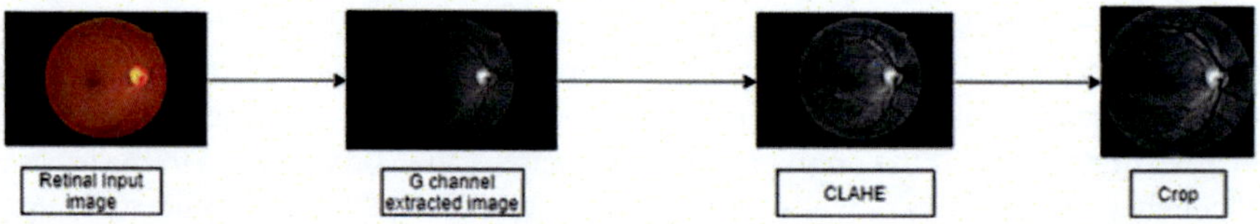

Fig. 2 Pre-processing of input retinal images

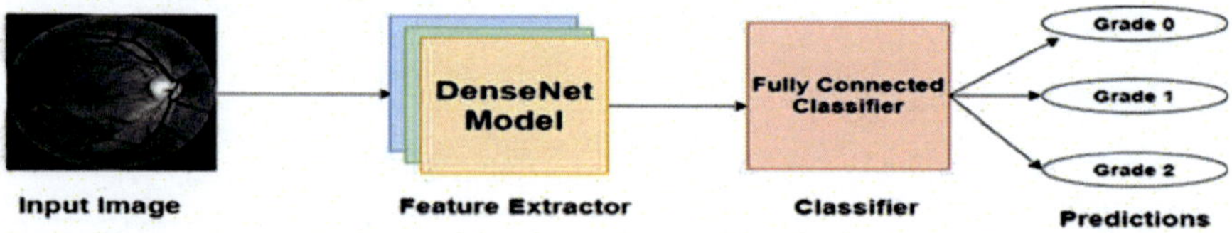

Fig. 3 Proposed architecture

4.1 Transfer Learning

Transfer learning is a technique to extract the features from the images by using the pre-trained model, which is already trained on the Imagenet dataset. The extracted features are given to the classifier to classify the images. In the transfer learning, the pre-trained weights are not modified, it will be freeze and use these weights to extract the features from the new dataset [16]. Fine-tuning is the concept of using the pre-trained model for the classification of a new dataset of similar tasks by adjusting the pre-trained model weights by backpropagation. In the fine-tuning model, the pre-trained model extracts features from the new dataset and it is flattened the features. The flattened features are connected to the fully connected dense layers, and based on the loss, the pre-trained model weights are adjusted based on the loss.

DenseNet Model. The issues emerge when the architecture of CNNs goes deeper, which is because the path to flow knowledge between the input layer & output layer becomes so large that they can get disappeared before reaching the other side. For this issue, DenseNet architecture rearranges the network design by interfacing each layer legitimately to one another. Rather than drawing representational power from very profound or wide structures, DenseNets utilize the network's potential through feature reuse. By the reuse of features, the DenseNet model utilizes the network's potential. By organizing layers in this way, DenseNet requires fewer parameters than conventional CNN's, and it is not required to learn redundant features [17]. The workflow of DenseNet and one level of DenseNet is depicted in Fig. 4.

Methodology. In this paper, the proposed architecture is a transfer learning architecture. The architecture mainly consists of two parts: (i) in the first part, the features are extracted from the retinal images by using the pre-trained model DenseNet model. (ii) in the second part, the extracted features are flattened and provided as input to the densely connected layer. The final dense layer with three output nodes with softmax activation used a dropout layer in between densely connected layers to overcome overfitting [18].

5 Experimental Evaluation

5.1 Network Training Setup

Hardware and Software. All experiments are performed on the Google Colab with Tesla K80 GPU machine with 12 GB of RAM, installed on Ubuntu 18.04 LTS. This approach is implemented in the Tensorflow 2.3.0 framework along with the Keras which is integrated with Tensorflow and the language used is Python 3.6.

Data Preparation. For training the model, the publicly available datasets IDRiD and Messidor are utilized and divided into training, validation, and test sets. For training 70% of images, 15% of images for validation, and 15% for testing.

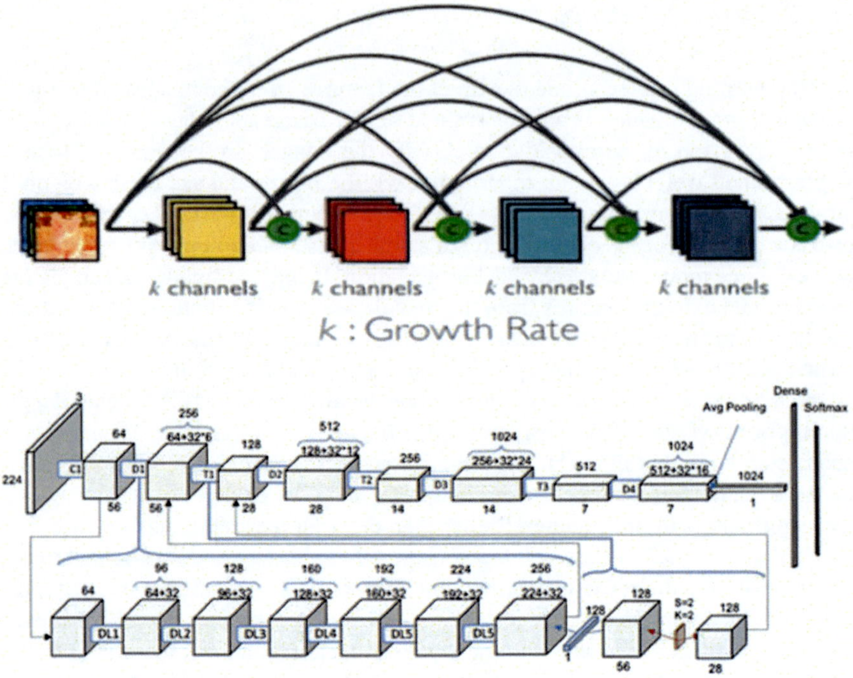

Fig. 4 The deeper look of one level in DenseNet. Dense block and Transition Block [19]

Learning Rate. If the learning rate is low, it will take a long time to train the network or model. But if it is high, the training may give a suboptimal result. Here the model is trained under the learning rate of 0.0001.

Optimizers. Optimizers are techniques or algorithms used to change the characteristics of your neural system, for example, learning rate and weights to diminish the losses. In this approach ADAM, Categorical Cross Entropy is used as optimizer & loss function.

5.2 Comparison & Results:

The accuracy and loss of the model on training and validation dataset are depicted in Fig. 5:

The comparison of the proposed approach with the existing approaches is listed in Table 3.

The evaluation metrics on the public datasets Messidor and IDRiD are calculated by the following formulas and listed in Table 4.

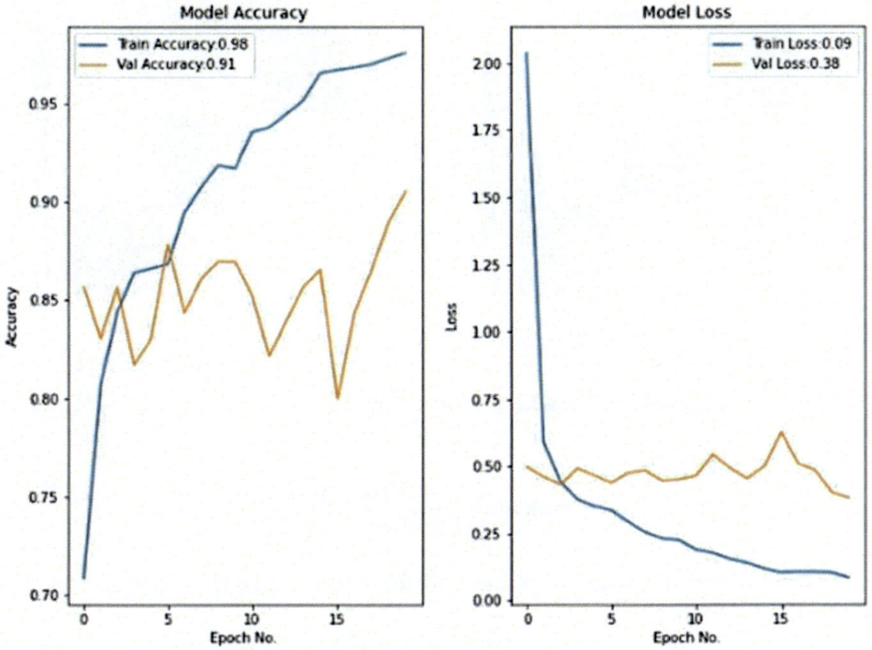

Fig. 5 Model accuracy and loss of training and validation data

Table 3 Comparison with the existing approaches

Approach	Accuracy (%)
Extraction method using marker-controlled watershed transformation [15]	85.2
Region-based method [16]	80–90
DMNET using hierarchical ensemble of CNNs [18]	96.15
Proposed approach	96.78

Table 4 Evaluation metrics on the public dataset

S. no.	Dataset	Accuracy (%)	F1 Score (%)
1	Messidor	96.78	96.72
2	IDRiD	96.51	96.54

$$\text{Accuracy} = \frac{\text{TP} + \text{TN}}{\text{TP} + \text{TN} + \text{FP} + \text{FN}} \tag{1}$$

$$\text{Precision} = \frac{\text{TP}}{\text{TP} + \text{FP}} \tag{2}$$

$$\text{Recall} = \frac{\text{TP}}{\text{TP} + \text{FN}} \tag{3}$$

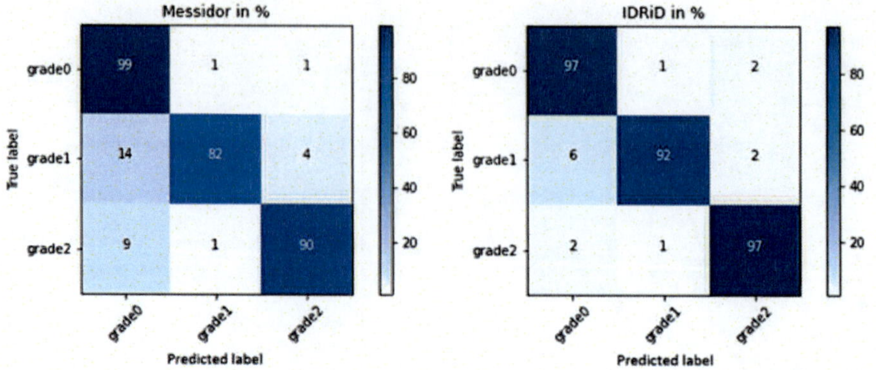

Fig. 6 Confusion matrices for Messidor and IDRiD Datasets

$$F_1\text{Score} = \frac{2 * \text{precision} * \text{recall}}{\text{precision} + \text{recall}} \tag{4}$$

The confusion matrices are plotted on the Messidor Dataset and IDRiD and illustrated in Fig. 6:

6 Conclusion and Future Work

Diabetic Macular Edema (DME) mainly causes by the growth of the fluid in the macular region, which is the center of the retina. This paper address the grading of DME by using transfer learning by using the DenseNet pre-trained model. The proposed model achieves an accuracy of an average of 96% accuracy and an F1 score on Messidor and IDRiD datasets. Future work, by collecting more data and applying different pre-trained models, may achieve higher accuracy.

References

1. Ciulla TA, Amador AG, Zinman B (2003) Diabetic retinopathy, and diabetic macular edema: pathophysiology, screening, and novel therapies. Diabetes Care 26(9):2653–2664. https://doi.org/10.2337/diacare.26.9.2653, PMID: 12941734
2. King H (1999) WHO and the international diabetes federation: regional partners. Bull World Health Organ 77(12):954, PMID: 10680241
3. Zhang X, Thibault G, Decencière E, Marcotegui B, Laÿ B, Danno R, et al (2014) Exudate detection in color retinal images for mass screening of diabetic retinopathy. Med Image Anal 18(7):1026–1043.https://doi.org/10.1016/j.media.2014.05.004, PMID: 24972380
4. Zheng Y, He M, Congdon N (2012) The worldwide epidemic of diabetic retinopathy. Indian J Ophthalmol 60(5):428. https://doi.org/10.4103/0301-4738.100542, PMID: 22944754

5. Sivaprasad S, Oyetunde S (2016) Impact of injection therapy on retinal patients with diabetic macular edema or retinal vein occlusion. Clin Ophthalmol (Auckland, NZ). 10:939. https://doi.org/10.2147/OPTH.S100168
6. Davidson JA, Ciulla TA, McGill JB, Kles KA, Anderson PW (2007) How the diabetic eye loses vision. Endocrine 32(1):107–116. https://doi.org/10.1007/s12020-007-0040-9, PMID: 179926084
7. Wilkinson C, Ferris FL III, Klein RE, Lee PP, Agardh CD, Davis M, et al (2003) Proposed international clinical diabetic retinopathy and diabetic macular edema disease severity scales. Ophthalmology 110(9):1677–1682. https://doi.org/10.1016/S0161-6420(03)00475-5, PMID: 13129861
8. Mookiah MRK, Acharya UR, Chua CK, Lim CM, Ng E, Laude A (2013) Computer-aided diagnosis of diabetic retinopathy: a review. Comput Biol Med 43(12):2136–2155. https://doi.org/10.1016/j.compbiomed.2013.10.007, PMID: 24290931
9. Sudheer Kumar E, Shoba Bindu C. Medical image analysis using deep learning: a systematic literature review. In: Emerging Technologies in computer engineering: microservices in big data analytics (ICETCE 2019). Communications in Computer and Information Science, vol 985, pp 81–97, 18th May 2019, Print ISBN: 978-981-13-8299-4, Online ISBN: 978-981-13-8300-7, https://doi.org/10.1007/978-981-13-8300-7_8. Springer, Singapore
10. Lim ST, Zaki WMDW, Hussain A, Lim SL, Kusalavan S. Automatic classification of diabetic macular edema in digital fundus images. https://doi.org/10.1109/CHUSER.2011.6163730
11. Magotra S, Kunwar A, Sengar N, Partha Sarathi M, Kishore Dutta M. Hierarchical classification and grading of diabetic macular edema using texture features. https://doi.org/10.1109/ICIIP.2015.7414763
12. Sengar N, Kishore Dutta M, Burget R, Povoda L. Detection of diabetic macular edema in retinal images using a region-based method. https://doi.org/10.1109/TSP.2015.7296294
13. Thulkar D, Daruwala R. Diabetic macular edema detection and severity grading. https://doi.org/10.1109/INDICON45594.2018.8987019
14. Singh RK, Gorantla R. DMENet: diabetic macular edema diagnosis using hierarchical ensemble of CNN's" https://doi.org/10.1371/journal.pone.0220677
15. Lawrence S, Giles CL, Chung Tsoi Ah, Back AD (1997) Face recognition: a convolutional neural-network approach. IEEE Trans Neural Netw 8(1):98–113. https://doi.org/10.1109/72.554195
16. Masood S, Luthra T, Sundriyal H, Ahmed M Identification of diabetic retinopathy in eye images using transfer learning. https://doi.org/10.1109/CCAA.2017.8229977
17. Review: DenseNet—Dense Convolutional Network (Image Classification)
18. https://towardsdatascience.com/review-densenet-image-classification-b6631a8ef803
19. Srivastava N, Hinton GE, Krizhevsky A, Sutskever I, Salakhutdinov R (2014) Dropout: a simple way to prevent neural networks from overfitting. J Mach Learn Res (JMLR) 15: 1929–1958
20. Understanding and visualizing DenseNets https://towardsdatascience.com/understanding-and-visualizing-densenets-7f688092391a

Pharmacological Evaluation and Molecular Docking Studies of *Sorghum bicolor* for Diuretic Activity

S. Shobha, A. Sreedevi, and K. Sai Sruthi

Abstract The current study was endeavoured to evaluate the diuretic activity of seeds of *Sorghum bicolor* and molecular docking studies of identified phytoconstituents. Ethanol extract of seeds was prepared using the hot extraction technique. The prepared extract was subjected to preliminary phytochemical screening and HPLC studies. The extract was evaluated for diuretic activity in male albino rats by Lipschitz method at doses of 200 and 400 mg/kg b.w. Diuretic activity is evaluated by assessing urine volume, urinary electrolytes, Lipschitz value, natriuretic index, saluretic index and ion quotient. Further molecular docking studies were performed for identified phytoconstituents in HPLC. The phytochemical studies revealed the presence of bioactive phytoconstituents. Pharmacological studies displayed that extract possess potent diuretic effect and molecular docking studies also evidenced that the identified phytoconstituents exhibited good binding affinity towards Human carbonic anhydrase enzyme II. Hence our study scientifically validates the folklore use of seeds of *Sorghum bicolor* as a diuretic.

Keywords *Sorghum bicolor* · Molecular docking · Carbonic anhydrase · Lipschitz value

1 Introduction

Diuretics are drugs that escalate renal excretion of water and electrolytes. The vital purpose of diuretic therapy is to dwindle the fluid volume in the body and to mend the water and electrolyte balance. They are beneficial in many life menacing ailments like oedema, congestive heart failure, nephritis, renal disorders, hypertension and pregnancy toxaemia [1]. Although diuretics play a crucial role in treating various disorders, many synthetic diuretics have adverse effects such as impotence, weakness, fatigue, arrhythmia, paralysis, coma, and hypotension [2]. Due to these varied side

S. Shobha · A. Sreedevi (✉) · K. Sai Sruthi
Institute of Pharmaceutical Technology, Sri Padmavati MahilaVisvavidyalayam, Tirupati, Andhra Pradesh 517502, India
e-mail: sreedeviadikay8@gmail.com

S. Jyothi et al. (eds.), *Proceedings of the 2nd International Conference on Computational and Bio Engineering*, Lecture Notes in Networks and Systems 215,
https://doi.org/10.1007/978-981-16-1941-0_54

effects of allopathic medicines, medicinal plants provide better therapy with lesser side effects.

Many medicinal plants were claimed to possess diuretic activity [3]. ***Sorghum bicolor*** is one such plant whose seeds are used by the folklore of African Negroes to treat kidney and urinary bladder disorders and as a cooling agent [4]. A perusal of literature revealed that there is no scientific data is available to verify the claim of *Sorghum bicolor* as a diuretic agent. To date, there were no reports on the diuretic activity of *Sorghum bicolor* plant. Hence the present study was planned for the evaluation of ethanol extract of seeds of *Sorghum bicolor* for diuretic activity.

2 Methodology

2.1 Collection and Authentication of Seeds

Sorghum bicolor seeds were purchased from Acharya Nagarjuna Agricultural Research centre, Nandyal, Kurnool, A.P. and authenticated by botanist Dr. K. Madhava Chetty (Voucher No. 2301).

2.2 Preparation of Extract

The seeds of *Sorghum bicolor* were powdered and defatted using petroleum ether (60°–80°) and the marc was macerated by utilising ethanol for 24 h and later refluxed for 3 h. Then the process was rerun three times, and the filtrate was concentrated under reduced pressure.

2.3 Phytochemical Studies

The prepared ethanol extract was initially subjected to preliminary phytochemical studies using standard methods. (Harborne) HPLC analysis was performed and Acetonitrile: Phosphoric acid in the ratio of 36:64 was used as mobile phase (Fig. 1).

2.4 Pharmacological Screening

Wistar Strain Albino rats of weight 150–200 g were used for study, and the experiment was carried out with prior permission from IAEC and as per CPCSEA guidelines.

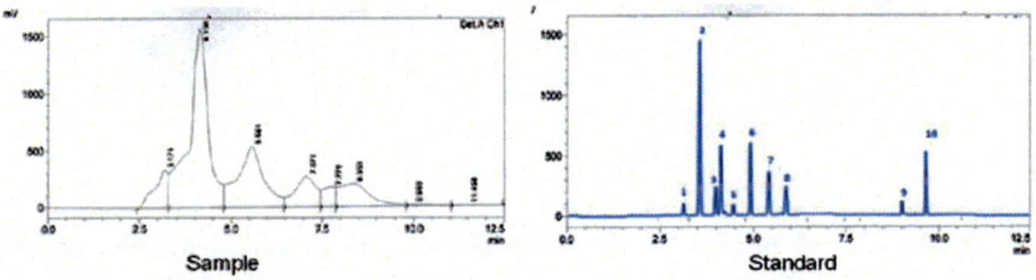

Fig. 1 HPLC chromatogram of EESB and standard

2.4.1 Acute Toxicity Studies

Acute Toxicity Screening Was Executed in Accordant with OECD Guidelines [5]

2.4.2 Evaluation of diuretic activity

Diuretic activity is assessed as per the methods of Lipchitz et al. [6]. Animals were separated into four groups, with Six each. Fifteen hours preceding to investigation, food and water are taken out. Group-I was served as normal control and administered with 5 mL of 0.9% NaCl solution per 100 g b.w. Group-II and Group-III were administered lower (200 mg/kg b.w) and higher (400 mg/kg b.w) doses of extract along with 5 mL of 0.9% NaCl solution/100 g b.w. Group-IV was served as standard and administered with Furosemide (10 mg/kg b.w) apart from 0.9%NaCl. Diuretic activity was assessed by measuring the parameters, namely urine volume, urinary electrolytes, diuretic index, Lipschitz vale, saluretic and natriuretic activity and ion quotient value as per standard methods [7]. Statistical analysis was executed as per one-way ANOVA succeeded by Turkey-Kramer comparison. A probability value ($p < 0.05$) was regarded as significant.

2.5 *In Silico Studies*

It was based upon the HPLC results of ethanol extract of seeds of *Sorghum bicolor*. The reported compounds were selected for docking. Drug likeness properties for the identified compounds and standard drug were determined using Molinspiration an online software [8]. The bioactivity scores of the compounds were calculated for their GPCR ligand, a kinase inhibitor, protease inhibitor, ion channel inhibitor, nuclear receptor ligand, enzyme inhibitor activities. Pre ADME properties were also evaluated [9]. Further molecular docking was performed to assess the binding scores of the compounds to human angiotensin-converting enzyme using Patch dock online server.

3 Results

3.1 Phytochemical Studies:

Preliminary phytochemical examination of ethanol extract of seeds of *Sorghum bicolor* (EESB) showed the presence of phenols, flavonoids, carbohydrates and steroids. HPLC analysis showed eight peaks and out of eight peaks 4 peaks were identified, comparing their retention times with the standard sample. The identified compounds were Vanillin acid (3.17), Myrecetin (4.10), Chrysin (5.56) and Isosakaruretin (9.98).

3.2 Pharmacological Studies

3.2.1 Acute Toxicity Studies

The animals had not shown any morbidity and animals did not show any changes in behaviour. Ethanol extract was found to be safe at 2000 mg/kg body weight dose.

3.2.2 Evaluation of Diuretic Activity

Effect on Urinary Volume

The animals which received extracts at 200 and 400 mg/kg b.w. exhibited dose-dependent increase in urine volume when compared with normal control group animals.

Effect on Electrolytes Excretion

The amount of urinary electrolytes excreted from all groups collected at 6 and 12 h are mentioned in Table 1. Oral administration of EESB at both doses resulted in a significant increase in sodium, potassium and chloride ions excretion when match up with normal control.

Effect on Natriuretic, Saluretic and Carbonic Anhydrase Inhibition Activity

Effect of seed extract on diuretic parameters upto 6 and 12 h are represented in Table 2. Diuretic index was increased in dose-dependent passion, which in turn indicates that extract increases the urine output. Lipsctz value in high dose treated animals is close to 1, showing that the extract exhibits a positive diuretic effect. Animals that

Table 1 Effect of ethanol extract of seeds of *Sorghum bicolor* on urinary parameters

Group	Urine volume (mL)		Na+ Ions (mMol/L)		K+ Ions (mMol/L)		Cl-IONS (mMol/L)	
	6 h	12 h	6 h	12 h	6 h	12 h	6 h	12 h
I	7.15 ± 0.15	10.50 ± 0.37	115.62 ± 4.75	129.19 ± 2.51	3.68 ± 0.19	4.75 ± 0.15	87.10 ± 3.94	78.42 ± 1.68
II	7.73 ± 0.08 ns	11.32 ± 0.37 ns	128.24 ± 2.04 [a]	156.22 ± 2.00 [a]	4.83 ± 0.30 [ns]	4.47 ± 0.12 [ns]	109.98 ± 1.95[a]	106.37 ± 1.94[a]
III	9.22 ± 0.30[a]	13.90 ± 0.33 [a]	169.55 ± 2.14[a]	186.00 ± 2.64 [a]	5.06 ± 0.20[a]	6.10 ± 0.56[a]	136.85 ± 0.96[a]	110.31 ± 3.55[a]
IV	10.34 ± 0.17[a]	15.66 ± 0.49 [a]	179.59 ± 2.29[a]	203.67 ± 2.18 [a]	5.75 ± 0.14[a]	6.25 ± 0.19[a]	137.40 ± 2.20[a]	130.34 ± 3.56[a]

Results are indicated in Mean ± SEM from six animals in each group
a = $p < 0.05$ when compared with control group
ns = non-significant when compared with control group

Table 2 Effect of ethanol extract of seeds of *Sorghum bicolor* on diuretic parameters

Group	Diuretic index		Lipschtz value		Na+/K^+ index (natriuretic)		$Na^+ + Cl^+$ index (saluretic)		Cl-IONS/$Na^+ + K^+$ (ion quotient)	
	6 h	12 h	6 h	12 h	6 h	12 h	6 h	12 h	6 h	12 h
I	–	–	–	–	–	–	–	–	1.151	0.59
II	1.08	1.07	0.73	0.72	0.85	1.14	2.36	2.57	0.66	0.66
III	1.28	1.32	0.86	0.85	1.07	1.05	3.03	2.85	0.63	0.57
IV	1.44	1.49	–	–	0.99	1.13	3.12	3.23	0.74	0.62

received low dose and high dose of extract exhibited natriuretic, aluretic and carbonic anhydrase inhibition activity.

3.3 *In Silico Studies*

The physicochemical (drug likenesses) parameters of the compounds were anticipated and set down in Table 3. The predicted values revealed that the compounds obeyed the Lipinski rule of five. The bioactivity score of the compounds was predicted and tabulated in Table 4. The results of Pre ADME were mentioned in Table 5. The results found that except D2 all the compounds possess good oral bioavailability. Except myricetin all the compounds obey Lipinski rule of five.

The result of the docking of the identified compounds with receptor 2VVB (Human carbonic anhydrase II) is mentioned in Table 4. The atom contact energy (ACE) score is the atomic contact energy score of the ligand with the receptor. Lower the ACE score values higher the activity will be. According to patch dock analysis, chrysine showed highest affinity to 2VVB followed by Methyl 4′-Naringenin and Isosakaruretin. Further Myricetin and Vanillic Acid showed the nearest score to that of Furosemide (Table 6).

4 Discussion

In the current research, evaluation of the diuretic action of ethanol extract of seeds of *Sorghum bicolor* was investigated by comparison with the effects of furosemide, the reference drug. The extract was directed orally at a dose of 200 mg/kg body weight and 400 mg/kg body weight and Urinary volume, the concentration of electrolytes excreted in the urine, Lipschtz value, natriuretic index, saluretic index, ion quotient were determined.

In previous studies, urinary volume plays a vital role in evaluating the diuretic efficacy of a drug [10]. The main effect of any diuretic drug is to improve urinary volumes and electrolytes excretion. Excretion of electrolytes is another important criterion in maintaining the body electrolyte balance in treating various disorders [11]. Urine volume and excretion of electrolytes were increased in a dose-dependent manner by both lower and higher doses of ethanol extract compared to control. In accordance with the earlier findings on various medicinal plants, the present study also lipscitz value, natriuretic, saluretic and ion quotient values evidenced that the extract has strong diuretic effect with long duration of action favourable potassium sparing effect and strong carbonic anhydrase inhibition [12].

HPLC analysis provides the most comprehensive resolution and achieves analysis in a single step. Upon HPLC studies of ethanol extract of *Sorghum bicolor* revealed that the presence of vanillic acid, Chrysin, Myricetin, Isosakaruretin. Earlier studies evince that various secondary metabolites like flavonoids, saponins, terpenoids or

Table 3 Physicochemical properties of identified phytoconstituents

Compound	Compound code	miLogP	TPSA	Molecular weight	nON	nOHNH	nvio	nrotb	Volume
Chrysine	D1	3.75	48.68	282.30	4	0	0	3	251.09
Myricetin	D2	1.39	151.58	318.24	8	6	1	1	248.10
Vanillic acid	D3	1.19	66.76	168.15	4	2	0	2	144.61
Methyl 4′-methyl ether	D4	2.65	76.00	286.28	5	2	0	2	247.79
Furosemide	S	1.77	122.63	330.75	7	4	0	5	249.48

miLogP (partition coefficient); nrotb (No of rotatable bonds); TPSA (Topological polar surface area); nVio (No of violations); nON (No of hydrogen bond acceptors); nOHNH (No of hydrogen bond donors)

Table 4 Bioactivity score of identified phytoconstituents

Compound	GPCR ligand	Ion channel inhibitor	Kinase inhibitor	Nuclear receptor ligand	Protease inhibitor	Enzyme inhibitor
Chrysine (D1)	−0.12	−0.22	0.11	0.15	−0.28	0.09
Myricetin (D2)	−0.06	−0.18	0.28	0.32	−0.20	0.30
Vanillic acid (D3)	−0.85	−0.42	−0.99	−0.61	−1.12	−0.35
Isosakaruretin (D4)	0.02	−0.26	−0.25	0.40	−0.12	0.15
Furosemide (S)	−0.39	−0.56	−0.37	−0.53	−0.40	−0.08

Table 5 Pre ADME properties of identified phytoconstituents

Compound	BBB	Caco-2	HIA	MDCK	PPB	Skin permeability
Chrysine (D1)	0.06	3.00	87.82	34.24	87.07	−3.87
Myricetin (D2)	0.11	0.99	40.96	4.88	96.78	−4.52
Vanillic acid (D3)	0.62	19.93	85.36	29.06	52.10	−2.15
Isosakaruretin (D4)	0.16	12.8	92.75	21.29	95.21	−3.40
Furosemide (S)	0.06	0.38	89.79	1.21	76.80	−3.02

Table 6 Molecular docking scores of identified phytoconstituents

Compound	Patchdock score	ACE score
Chrysine (D1)	3898	−143.46
Myricetin (D2)	3912	−120.85
Vanillic acid (D3)	2392	−117.89
Isosakaruretin (D4)	4006	−136.67

organic acids in plants will contribute to the diuretic effects [13]. Moreover, in our study, flavonoids and other phytoconstituents may be responsible for the efficient diuretic activity. To support the pharmacological findings in silico studies were performed. Molecular docking studies were performed for four compounds (identified by HPLC) with Human carbonic anhydrase II. Out of four compounds chrysin showed higher inhibitory action to HCA II. Hence, flavonoid Chrysin may result in the positive diuretic activity of extract, which comes in accordance with the previous study of Cherkaoui et al. [15], which stated that chrysin acts as a prominent diuretic [14]. Thus the Molecular docking studies evidenced that identified compounds exhibited good inhibitory action against HCA II.

5 Conclusion

The present study strongly suggested that the seeds of *Sorghum bicolor* can be used as a potential therapeutic alternative as diuretic. Further, our findings offer scientific validation to the tribal use of *Sorghum bicolor* as diuretic.

References

1. Snigdha M, Kumar SS, Deepa C, Lalith S, Thanuja S (2013) Review on recent advances in modern day treatment: diuretic therapy. Int Res J Pharm 6:25–30
2. Ryan MP (1984) Magnesium- and potassium-sparing effects of amiloride Review and recent findings. Magnesium 3:274–288
3. Ahmad W, Zeenat F, Ahmad M, Ansari N (2017) Medicinal plants as potent diuretic: a review. Int J Adv Pharm Med Bioallied Sci, 1–8
4. Kirtikar KR, Basu BD (1918) Indian medicinal plants, 2nd edn, vol 4, p 2723
5. Harbone JP (1973) phytochemical methods, a guide to modern techniques of plant analysis. Chapmann and Hall, London
6. Organization for Economic Cooperation and Development (OECD) (2001) Guideline 423 for testing chemicals. Paris, pp 1–14
7. Lipschitz WL, Hadidian Z, Kerpcsar A (1943) Bioassay of diuretics. J Pharmacol Exp Ther 79:97–110
8. Vogel hans G, wolfgang H (Eds) (1997) Drug discovery and evaluation, p 175
9. Lipinski CA, Lombardo F, Dominy BW, Feeney PJ (1997) Experimental and computational approaches to estimate solubility and permeability in drug discovery and development settings. Adv Drug Delivery Rev 23:4–25
10. Verdonk ML, Cole JC, Hartshorn MJ, Murray CW, Taylor RD (2003) Improved protein–ligand docking using GOLD. Proteins 52:609–623
11. Jayanthi MK, Siddamma A (2018) To evaluate the diuretic activity in ethanolic extract of leaves of delonix regia in wistar albino rats. Biomed Pharmacol J 11:959–963
12. Roumelioti ME, Glew RH, Khitan ZJ, Rondon-Berrios H, Argyropoulos CP, Malhotra D, Raj DS, Agaba EI, Rohrscheib M, Murata GH, Shapiro JI, Tzamaloukas AH (2018) Fluid balance concepts in medicine: Principles and practice. World J Nephrol. 7:1–28
13. Naveen A, Shankar J, Prakash M, Venkatanarayana N (2015) Evaluation of diuretic activity of alcoholic extract of cinnamomum zeylanicum in swiss albino rats. IJPRS 4:417–428
14. Jabeen Q, Aslam N (2013) Hypotensive, angiotensin converting enzyme (ACE) Inhibitory and diuretic activities of the aqueous-methanol extract of Ipomoea reniformis. Iran J Pharm Res. 12:769–776
15. Cherkaoui-Tangi K, Lachkar M ,Wibo M (2008) Pharmacological studies on hypotensive, diuretic and vasodilator activities of chrysin glucoside from Calycotomevillosa in rats. Phytother Res, 356–361

A Review on Bio-computational Measures for the Control of Invasive Pest: *Tuta absoluta* (Tomato Leaf Miner)

Kalpana Sriramadasu and D. M. Mamatha

Abstract *Tuta absoluta* (Lepidoptera: Gelechiidae), also called a tomato leaf miner, has been recorded as a worldwide serious insect pest on tomato and other crops of the Solanaceae family. It originated from South America and invaded to the other parts of the world, mostly in Africa, Asia, and Europe continents. In India, at first, *Tuta absoluta* was found in infested tomato crops of Maharashtra. The pest has a wide range of host plants but feeds exclusively on tomato (*Solanum lycopersicum*), one of the most economically and nutritionally important crop. The crop production is challenged because of the severe infestation by the pest. If proper control measures are not taken, the damage caused by the pest reaches up to 100% as it happened in recent history. Destruction of economically important crops may show a negative impact on the country's economy. In the present paper, we elucidated an overview of pest's control strategies viz., chemical, biological, molecular and computational that can be applied in different invaded areas. An integrated way of management method is necessary to control the pest in an efficient way.

Keywords *Solanum lycopersicum* · *Tuta absoluta* · Invasive pest · Lepidoptera · Gelechiidae

1 Introduction

Tuta absoluta [22], which is also called as South American pinworm or tomato leaf miner, is an oligophagus, nocturnal moth that belongs to the order Lepidoptera and family Gelechiidae. This insect pest is a native of South America since the 1950s [47]. Later, it was invaded to European countries and reported in Spain in the year 2006 [11]. At first this pest was identified from infested tomato crop in Peru and

K. Sriramadasu · D. M. Mamatha (✉)
Department of Biosciences & Sericulture, Sri Padmavati Women's University, Tirupati, AP, India
e-mail: prof.mamatha@gmail.com

K. Sriramadasu
e-mail: sriramadasu.kalpana@gmail.com

S. Jyothi et al. (eds.), *Proceedings of the 2nd International Conference on Computational and Bio Engineering*, Lecture Notes in Networks and Systems 215,
https://doi.org/10.1007/978-981-16-1941-0_55

named *Phthorimaea absoluta* [22, 53]. Later it was kept under various genera such as *Scrobipalpula absoluta* [67] *or Scrobipalpuloides absoluta* [17] *and Gnorimoschema absoluta* [37]. Finally, in 1994, Povolny [18] named the pest *Tuta absoluta* [58]. This pest has a wide range of host plants but feeds exclusively on tomato (*Solanum esculentum*) [68]. It also attacks other Solanacious plants such as eggplant (*Solanum melongena* L.), pepper (*Capsicum annum* L.) [71], alfalfa (*Medicogo sativa*), [51] etc. The tomato is a warm seasonal crop sensitive to freezing conditions, cultivated both in the open field and greenhouse condition, facing several constraints and the main one is infestation by pests such as South American tomato leaf miner (*Tuta absoluta*), become a major threat to agriculture because the damage caused by the pest reaches up to 100% if proper management was not taken [23]. Various control strategies are followed by the farmers against insect pests. The management strategies of the pest are mainly based on using chemical insecticides. However, excessive dependence and irrational use of synthetic insecticides for longer periods resulted in so many problems such as insecticide-resistant development, loss of biodiversity, the eruption of beneficial insects [34, 36]. Moreover, excess use of synthetic insecticides becomes a serious threat to the environment due to high toxicity levels [69]. Hence it is necessary to reduce the use of chemical insecticides. Integrated Pest Management (IPM) and biopesticide usage should be prioritized as alternative methods for sustainable agriculture, which in turn can at least protect the valuable life-systems associated therein. Several ecofriendly options were assessed against numerous pests in different crop ecosystems mainly in utilizing the biopesticides, which include novel molecular methods such as RNA interference (RNAi) [8, 9] and Juvenile hormone degradation pathway [20, 48]. Identification of pest attack in early stages can play a major role in describing the problem of Insect pest management and helpful to farmers to follow proper control methods. Early detection of *Tuta absoluta* attack on tomato plants is one of the most important factors for its control and prevention. However, the data regarding early quantification of *Tuta absoluta,* which gives a brief idea about plant disease and pest control methods to farmers was limited. In the present review paper, complete details on currently available management strategies of the toamto leaf miner (*Tuta absoluta*) were explained.

2 Effect of Environmental Parameters on the Life Cycle of *Tuta absoluta*

To eradicate or control the pest completely, it is necessary to acquire awareness on the effect of ecological parameters on lifecycle of the concerned pest. Temperature is one of the important abiotic factors for insects [29, 39, 63]. In view of the importance of the environmental factor in the life cycle of the insect pest, many studies were conducted under various temperatures such as 10, 15, 20, 25 and 30 °C and verified that *Tuta absoluta* completed its life cycle at all the temperatures provided but eggs laid at 10 and 30°C were infertile. A similar type of results was obtained in the study conducted by Cuthbertson et al. [4], which verified that only 3% larva underwent pupating and no adult emerged at 10 °C and was adapted to a wide range of temperatures. But its reproduction and survival rate was negatively affected at the extreme constant temperatures. The pest may take 76 days at 14 °C and 24 days at 20 °C and 24 days at 27 °C to complete its life cycle [58].

3 Control Measures

Tuta absoluta can easily spread few kilometers by flying or floating in the wind currents, causing damage of 50%-100% if proper preventive methods are not followed. If management measures are taken, the damage might decrease to 1%-5%. As per the severe damage of *Tuta absoluta* on tomato production, the affected countries are exploring different management strategies to control the insect pest, classified as Chemical, Biological, Molecular, and Computational methods.

3.1 Chemical Methods

Since the insect pest was reported, Insecticides have been used as the main control strategy because of its rapid action and strong toxicity against the target pest. Approximately 12 classes of insecticides, including synthetic insecticides, are being used for the chemical management of the pest. Commonly used chemicals are pyrethroids [56], organophosphates, spinosad, emamectin benzoate and abamectin, chloride channel activators, benzoylureas and diamide [24]. Indoxacarb, spinosad, imidacloprid, deltamethrin and *Bacillus thuringiensis var. kurstaki* have been used to control of larval infestations of *T.absoluta* in Spain [57]. Chlorpyrifos and pyrethrins are frequently used in the chemical control of tomato leaf miner in Italy [33]. In Malta, Abamectin, indoxacarb, spinosad, imidacloprid, thiacloprid, lufenuron and *B. thuringiensis* (*Bt*) have been recommended [16] to control *Tuta absoluta.* It is difficult to control the *Tuta absoluta* by chemical methods, as the pest lives inside the fruit, leaves and stems. Once the larvae entered into mesophyll, the effect of insecticides is relatively less. The insect pest has been developing resistance to insecticides led to the emergence of new insecticides such as insect growth regulators, indoxacarb, chlorfenapyr, spinosyns, and diamides. The research studies of Samir A.M. Abdelgaleil et al. [5] concluded that Spinosad worked as a most effective insecticide in controlling *T.absoluta*, followed by Abamectin + Thiamethoxam and Chlorpyrifos under the field conditions in the IPM program for tomato crop. Accordingly, chlorantraniliprole has shown the highest toxic effect on *T. absoluta* eggs followed by spinosad, abamectin, indoxacarb, acetamiprid, and diazinon [73]. Research studies of Saad et al. [59] suggested that emamectin benzoate, chlorantraniliprole and metaflumizone were the most effective insecticides in the management of *Tuta absoluta* and the efficacy of the above insecticide is increased by the addition of mineral oil (Kzoil@). Hence, these insecticides could be used in IPM program for better improvement in effective control of *Tuta absoluta.* Some research studies revealed that lambda-cyhalothrin was the most effective in the control of *T. absoluta* at its high concentration, which causes complete mortality of the larval stage of pest, especially inside and outside the treated fields [50]. Indiscriminate use of insecticides causes unpleasant environmental consequences, including water pollution, abolition of beneficial organisms and harming the health of humans. Therefore, integration with other alternative control methods (biological, biotechnological and biocomputational methods) against the pest becomes imperative [6, 7, 49]. Researchers turned toward implementing several auxiliaries of ecofriendly control measures, such as biological control, including the use of natural enemies to manage *T. absoluta.*

3.2 Biological Methods

The biological method is one of the promising control methods where natural enemies are released in invaded areas to control the pest [12, 44, 69]. Numerous natural enemies or antagonists such as parasitoids (80%), predators (20%) have been found on *Tuta absoluta.* The most effective and widespread natural enemies of parasitoids on eggs and larvae of *Tuta absoluta* are *Trichogramma* strains [42], *Pseudoapanteles dingus* [31], *Dineulophus phythorimaeae* De Santis [54], *Apanteles gelechiidivoris, Necremnus* sp., [44] *Neochrysocharis formosa, Stenomesius* sp., and *Bracon nigricans. Trichogramma evanescens* westwood (Hymenoptera: Trichogrammatidae) is currently used as a biological control agent in tomato crops cultivated in greenhouses [13]. A research study was conducted in India, to study the evaluation of the efficiency of *Trichogrammatids* strains (egg parasitoids) to control *Tuta absoluta* revealed that *Trichogramma pretiosum* Riley [27] was the most potential egg parasitoid. In the contrary *Trichogramma achaeae* was most efficient in the management of *Tuta absoluta* in tomato crops cultivated in Greenhouse [13]. Furthermore, among the several predator species preying on the *Tuta absoluta*, the most promising predators are *Nesidiocoris Tenuis* (Reuter) [12, 26], *Macrolophus pygmaeus* [26], *Engytatus varians* and *Campyloneuropsis infumatus.* These predators generally colonize on the crops and consume large number of *Tuta aboluta* eggs [65, 66]. Use of these predatory bugs is crucial in their efficacy because they are slowly spread in the early growth stages of the crop. The right time to release predatory bugs is when the larvae come out of the galleries [14]. Chailleux et al. [1] reported that when the combination of both parasitoid *Trichogramma achaeae* and predator *Macrolophus pygmaeus* are used in the biological control of *Tuta absoluta.* Some species of bacteria, fungi, nematodes are also used as biological control agents against *Tuta absoluta.* The combined use of *Bacillus thuringiensis* treatment and predatory bug *Nesidiocoris tenuis* against *Tuta absoluta* resulted in 97% reduction of the leaf damage. The other natural enemy that effectively controls *Tuta absoluta* is Pyemotes species, also known as "straw-itch mites." The entomopathogenic fungi *Beauveria bassiana* is another safe approach to biologically control the *Tuta absoluta.* It causes larvae mortality up to 95% compared to chemical treatment which causes 88% mortality. It is a cost-effective method and locally mass-produce [28], grows naturally throughout the world with no adverse effects on human health or the environment [38]. The biological control strategy is more focused than chemical control methods because of adverse effects of chemical methods such as resistance developed by the pest due to over usage of insecticides, high cost of insecticides and its negative effect on other beneficial insects.

3.3 *Use of Pheromone Traps*

One of the most promising approaches for the control of *T.absoluta* is the use of pheromone traps which helps to monitor the presence of the pest and population levels. This method has been referred as environmentally acknowledged pest management strategy and plays a major role in monitoring the population of *T. absoluta* [55, 70]. These traps have been reported to be effective in controlling tomato borer [2, 3, 46, 62]. For better results in the field, the sex pheromone traps are to be suitably hanged at correct positions depending on the height of plants and wind direction [10, 70]. The color of the trap is one of the important factors that influence the movement of the pest toward it, for trapping [15, 46]. Shining colors, particularly red color, have been reported to be the most attractive to *T. absoluta* [52]. By the addition of insect-killing ingredients to the above factors improves trap efficiency in the field [21, 64, 70]. The pheromone trap of the *T. absoluta*, attractive to males consists of (3E, 8Z, 11Z)-3, 8, 11-tetradecatrien-1-yl (90%) and (3E, 8Z)-3, 8-tetradecadien-1-yl (10%) acetate [60]. Accordingly, studies on chemo-attractants revealed that more male moths were caught by PAA (phenylacetaldehyde) than females and the mixture of AA (acetic acid) & MB(3-methyl-1-butanol) showed better results in luring and trapping of *T.absoluta*. A study was conducted to explain pheromone based control strategies such as mass trapping and mating disruption in controlling *T. absoluta* and concluded that the success of mass trapping strategies is usually low and the mating disruption technique was effective in high-containment greenhouses, when compared to low-containment greenhouses and open-field crops [25]. The pheromones are reported to perform well both in open fields and greenhouses [3, 41]. This control management reduced the dependence on insecticide treatment. In Mass trapping, traps are used to reduce the population of the pest. But these methods are not assured to reduce pest infestation. Pheromone traps may be expensive and not readily available.

3.4 *Molecular Methods*

Novel management methods for the control of *T. absoluta* at the molecular level include RNA interference (RNAi) and juvenile hormone strategy against the pest.

RNA of INTERFERENCE (RNAi)

The mechanism of RNA of interference (RNAi), an evolutionary conserved post-transcriptional gene silencing, became a potent tool for evaluating gene function in various organisms [30]. The control of Agricultural insect pests by RNAi-mediated mechanism has rapidly become apparent [43]. But production of dsRNA and its delivery has become a major challenge. In RNAi studies, microinjection method was used for delivering dsRNA into insects. Soaking used as a deliver method to individual larvae [19]. According to the literature survey, two delivery approaches are used for ingestion of dsRNA to *T.absoluta* larvae. First approach follows the pest larvae feed on tomato leaflets have dsRNA and second approach depends on the transient transcription of dsRNA by the host plant after agro-infiltration termed as "in planta-induced transient gene silencing" (PITGS). Thus the study recommends the RNAi technology as an alternative control method against *Tuta absoluta* [9].

Juvenile hormone degradation strategy

Identification of Juvenile hormone (JH) for the pest control strategies has been explored and initiated in mid 1980s as it modulates a crucial role in a variety of developmental and physiological processes in insects. Accordingly, the research report of ovicidal activity of pyriproxyfen, a Juvenile Hormone mimic against the eggs of *T. absoluta* revealed that it showed negligible effect on eggs of *T. absoluta* but causes a significant delay in hatching process and also observed 50% mortality in first instar larvae [20] (Fig. 1).

3.5 Computational Methods

Studies on environmental suitable factors for *Tuta absoluta* by using pest forecasting through mathematical and computer-based models:

The impact of climatic conditions on insect pest infestation is one of the serious factors to select the suitable season and right time for the pest management strategies. In many countries such as Egypt, variations in thermal temperature bring about deviation in *Tuta absoluta* life cycle to finish yearly generations. The demographic data that is obtained from the effect of climatic changes on Insect pest development made a way for the researchers to depict potent solution and the right time for insect pest control programs on tomato crops. In Egypt, a research study on *Tuta absoluta* infestation against tomato crops used the association between gathered thermal points articulated as degree days (DDs) and changes in Insect pest population and calculated yearly generations and estimated *Tuta absoluta* moth frequency in the experimental area of different tomato strain fields based on existing and probable upcoming climatic variations. According to Salama et al. [35] the annual generations of *Tuta absoluta* were varied with different climatic conditions in different experimental areas. They generated data in a mathematical model by using Jasic [74]

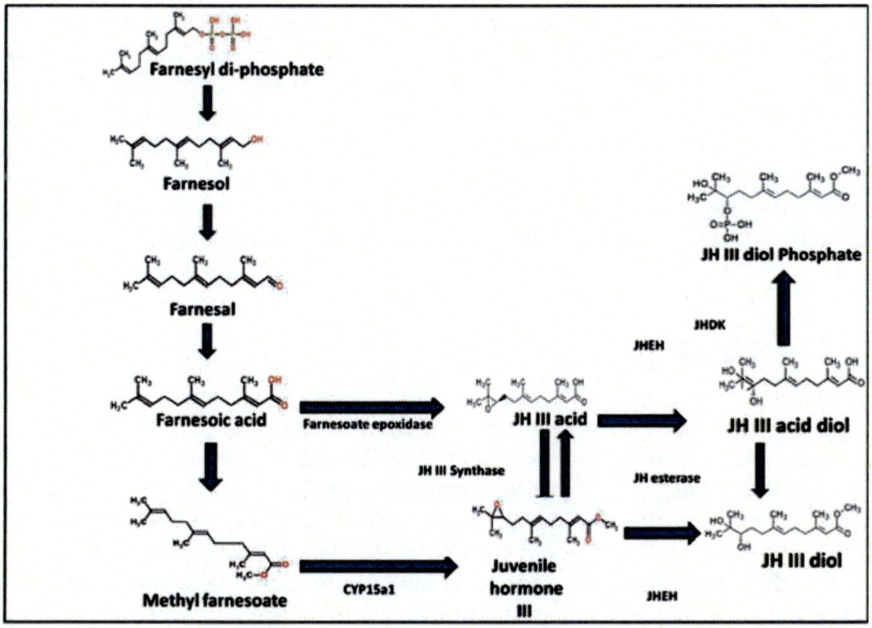

Fig. 1 Biosynthesis and Degradation pathway of the Juvenile hormone (*source* Arya et al. [61] Scientific reports (2018) 8:7573)

formula, which indicated that *Tuta absoluta* emergence occurs throughout the year. Insect development and its complete life cycle were prolonged when the temperatures were low, that is, in the range of 15oc, 16oc, etc. The prolongation of life cycle was decreased when the temperatures were high, that is, in the range of 26, 27, etc. The data produced by using thermal units will be an important tool for insect pest management in forecasting pest incidence observation and selecting proper timing for insecticide applications [72]. Farag et al. [45] revealed that the status of various insect development and emergence in different seasons were calculated based on the accumulated thermal units. In many countries, the noted data regarding climatic conditions for the occurrence *of Tuta absoluta* in the open fields have been used to update its allocation forecasting. Based on the environmental suitability version, the climatic conditions that are necessary for *Tuta absoluta* were observed in South and Central America, southern Europe, some places of Australia and East Africa. The entire collected data concludes that *Tuta absoluta* is spread to almost all tomato growing places. It can survive under high biotic factors and to acclimatize to different environmental factors and made various plants as their hosts. Not only tomato plants from the solanaceous family, but alternative host plants also may face the risk of attack and spread of *Tuta absoluta* in the presence of less suitable climatic conditions places [32, 56]. Because of advanced food techniques, *Tuta absoluta* was dispersed in many ways, which becomes a major problem to the model phenomenon for facing its fast invasion. Hence, machine learning techniques were used to analyse spatio-temporal distribution of *Tuta absoluta* as a multi-scale producing process based on environment, biology, seasonal incidence, line and population statistics over a period of time by implement framework. McNitt et al. [40] applied framework and revealed that even in modest self-mediated expansion abilities, *Tuta absoluta* could quickly extend its range through the local city to city trade by using previous invasive data and also explained expansion rate was less in selected interventions at main producing places and early detection of pest invasion was identified in high-consumption places.

4 Conclusion

Currently, different control strategies are followed against *Tuta absoluta*, such as chemical strategy, Biological strategy, pheromone traps-based strategy, molecular-based control strategy. Unfortunately, synthetic insecticides are used for the control of lepidopterans worldwide. The use of pesticides can effectively control pests, but it is always associated with negative concerns. On the other hand, biological control methods have been tried for the control of insect pests. However, when the threshold limits of pests exceed, these methods are not potential enough. Natural enemies show low parasitism on the pest and take time to adapt to the new host. Novel molecular methods against the pest are RNA interference (RNAi) and Juvenile hormone degradation strategy that specifically control the target pest without causing any harm to other beneficial insects. Applying machine learning techniques to identify early detection of *Tuta absoluta* leads a way to know its rapid invasion in an over-period of time to design a framework phenomenon for its control. The above studies throw light on the novel and innovative methods for the integrated Bio-computational control of the invasive pest *T. absoluta* at the field level.

References

1. Chailleux A, Bearez P, Pizzol J, Amiens-Desneux E, Ramírez Romero R, Desneux N (2013) Potential for combined use of parasitoids and generalist predators for biological control of the key invasive tomato pest *Tuta absoluta*. J Pest Sci 86:533–541
2. Cocco S, Deliperi G (2012) Delrio, Potential of mass trapping for Tuta absoluta management in greenhouse tomato crops using light and pheromone traps, IOBC/WPRS Bulletin. Integr Control Prot Crop, Mediterr Clim 80:319–324
3. Cocco A, Deliperi S, Delrio G (2013) Control of *Tuta absoluta* (Meyrick) (Lepidoptera: Gelechiidae) in greenhouse tomato crops using the mating disruption technique. J Appl Entomol 137:16Ð28
4. Cuthbertson AGC, Mathers JJ, Blackburn LFM, Korycinska WL, Luo W, Jacobson RJ, Northing P (2013) Population development of *Tuta absoluta* (Meyrick) (Lepidoptera: Gelechiidae) under simulated UK glasshouse conditions. Insects 4:185–197
5. Abdelgaleil SA, El-Bakary AS, Shawir MS, Ramadan GR (2015) Efficacy of various insecticides against tomato leaf miner, *Tuta absoluta*, in Egypt. Appl Biol Res 17(3):297–301

6. Birgücü AK, Bayındır A, Çelikpençe Y, Karaca I (2014) Growth inhibitory effects of bio- and synthetic insecticides on *Tuta absoluta* (Meyrick, 1917) (Lepidoptera: Gelechiidae). Turk J Entomol 38(4):389–400
7. Birgücü AK, Turanlı F, Gümüş E, Güzel B, Karsavuran Y (2015) The effect of grape cultivars on oviposition preference and larval survival of *Lobesia botrana* Den. & Schiff. (Lepidoptera: Tortricidae). Fresenius Environ Bull 24(1):33–38
8. Camargo AR, Herai HR, Santos NL, Bento MF, Lima EJ, MarquesSouza H, Figueira A (2015) De novo transcriptome assembly and analysis to identify potential gene targets for RNAi-mediated control of the tomato leafminer (*Tuta absoluta*). BMC Genom 16:635–642
9. Camargo AR, Guilherme O, Possignolo PI, Peres PL, Lam E, Lima EJ, Figueira A, Marques-Souza H (2016) RNA interference as a gene silencing tool to control *Tuta absoluta* in tomato (Solanum lycopersicum). Peer J 4:1–24
10. Soliman M, Hegazi M, Attaia I (2013) Imam, Evaluating the role of sex pheromone in monitoring and controlling tomato leaf miner, *Tuta absoluta* (Meyrick), under the Egyptian north western coast conditions. Egypt J Biol Pest Cont 23:169
11. Urbaneja A, Vercher R, Navarro Llopis V, Porcuna Coto JL, García Marí F (2007) La polilla del tomate, Tuta absoluta. Phytoma España 194:16–23
12. Urbaneja A, Montón H, Mollá O (2009) Suitability of the tomato borer Tuta absoluta as prey for Macrolophus pygmaeus and Nesidiocoris tenuis. J Appl Entomol 133:292–296
13. Ballal CR, Gupta A, Mohan M, Lalitha Y, Verghes A (2016) The new invasive pest *Tuta absoluta* (Meyrick) (Lepidoptera: Gelechiidae) in India and its natural enemies along with evaluation of Trichogrammatids for its biological control. India Curr Sci, Curr Sci 110(11):2155–2159
14. Van Lenteren JC, Hemerik L, Lins JC, Bueno VH (2016) Functional responses of three neotropical mirid predators to eggs of *Tuta absoluta* on Tomato. Insects 7(3):34
15. Megido RC, Haubruge E, Verheggen FJ (2013) Pheromone-based management strategies to control the tomato leafminer, *Tuta absoluta* (Lepidoptera: Gelechiidae). A review Synthèse bibliographique: les stratégies de lutte phéromonale utilisées pour contrôler la mineuse de la tomate, Tuta absoluta (Lepidoptera: Gelechiidae). Biotechnol Agron Soc Environ 17:475
16. Mallia D (2009) Guidelines for the control and eradication of *Tuta absoluta*. Ministry for Resources and Rural Affairs, Plant Health Department, Malta
17. Povolny (1987) Gnorimoschemini of southern South America III: the scrobipaluloid genera (Insecta, Lepidoptera, Gelechiidae). Steepnstrupia 13:1–91
18. Povolny D (1994) Gnorimoschemini of South America VI: Identification keys checklist of Neotropical taxa and general considerations (Insecta, Lepidoptera, Gelechiidae). Steenstrupia 20:1–42
19. Price D, Gatehouse JA (2008) RNAi-mediated crop protection against insects. Trends Biotechnol 26:393–400
20. Boina DR, Rogers ME, Wang N, Stelinski LL (2010) Effect of pyriproxyfen, a juvenile hormone mimic, on egg hatch, nymph development, adult emergence and reproduction of the Asian citrus psyllid, Diaphorina citri Kuwayama. Pest Manag Sci 66:349–357
21. Lobos E, Occhionero M, Werenitzky D, Fernandez J, Gonzalez LM, Rodriguez C, Calvo C, Lopez G, Oehlschlager AC (2013) Optimization of a trap for *Tuta absoluta* Meyrick (Lepidoptera: Gelechiidae) and trials to determine the effectiveness of mass trapping. Neotrop Entomol 42:448–457
22. Meyrick E (1917) Descriptions of South American Micro-Lepidoptera. Trans Entomol Soc Lond, 1–52
23. EPPO, Data sheets on quarantine pests: *Tuta absoluta*. OEPP/EPPO Bull 35:434–435
24. Roditakis E, Vasakis E, Grispou M, Stavrakaki M, Nauen R, Gravouil M, Bassi A (2015) First report of Tuta absoluta resistance to diamide insecticides. J Pest Sci 88:9–16
25. Refki E, Sadok BM, Ali BB (2016) Study of the biotic potential of indigenous predator Nesidiocoris tenuis on *Tuta absoluta* pest of geothermal culture in the south of Tunisia. J Entomol Zool Stud 4(6):692–695
26. Calvo FJ, Bolckmans K, Belda JE (2012) Release rate for a pre-plant application of Nesidiocoris tenuis for Bemisia tabaci control in tomato. Bio Control 57:809–817

27. Haji FNP, Prezotti L, Carneiro JS, Alencar JA (2002) Trichogramma pretiosum para o controle de pragasno tomateiro industrial. In: Controle biológico no Brasil: parasitóides e predadores—Manole, São Paulo, Brasil, pp 477–494
28. Qazzaz FO, Al-Masri MI, Barakat RM (2015) Effectiveness of Beauveria bassiana native isolates in the biological control of the Med-iterranean fruit fly (Ceratitis capitata). Adv Entomol 3:44–55
29. Hallman GJ, Denlinger DL (1998) Introduction: temperature sensitivity and integrated pest management. Westview Press, Boulder, pp 1–6
30. Hannon GJ (2002) RNA interference. Nature 418:244–251
31. Luna M, Wada VI, Sanchez NE (2010) Biology of Dineulophus phtorimaeae (Hymenoptera: Eulophidae) and field interactions with Pseudapanteles dingus (Hymenoptera: Braconidae), larval parasitoid of Tuta absoluta (Lepidoptera: Gelechiidae) in tomato. Ann Entomol Soc Am 103:936–942
32. Guimapi RY, Mohamed SA, Okeyo GO, Ndjomatchoua FT, Ekesi S, Tonnang HE (2016) Modeling the risk of invasion and spread of *Tuta absoluta* in Africa. Ecol Complex 28:1–17
33. Garzia T, Siscaro G, Colombo A, Campo G (2009) Reappearance of Tuta absoluta in Sicily, Rinvenuta in Sicilia Tuta absoluta. L'Informatore Agrario 65:71
34. Siqueira A, Guedes RN, Picanço MC (2000) Insecticide resistance in populations of Tuta absoluta (Lepidoptera: Gelechiidae). Agric For Entomol 2:147–153
35. Salama HS, Ismail AI, Fouda M, Ebada I, Shehata IB (2015) Thermal requirements and developmental zero of the tomato leaf miner *Tuta absoluta* (Meyrick), (Lepidoptera: Gelechiidae). Bull NRC, Egypt 40(1):1–16
36. Torres B, Evangelista WS, Barras R, Guedes RNC (2002) Dispersal of Podisus nigrispinus nymphs preying on tomato leaf miner: effect of predator release time, density and satiation level. J Appl Entomol 126:326–332
37. Clarke F (1962) New species of microlepidoptera from Japan. Entomol News 73:102
38. Haas-Costa J, Alves LFA, Daros AA (2011) Safety of Beauveria bassiana (Bals.) Vuill to Gallus domesticus L. Braz Arch Biol Technol 53(2):465–471
39. Ivanovic J, Nenadovic V (1999) The effect of diet and temperature on the life cycle of phytophagous insects. Pesticides 14:309–327
40. McNitt J, Chungbaek YY, Mortveit H, Marathe M, Campos MR, Desneu N, Brévault T, Muniappan R, Adiga A (2019). Assessing the multi-pathway threat from an invasive agricultural pest: *Tuta absoluta* in Asia. Proc R Soc B 286:20191159(2019)
41. Abbes K, Chermiti B (2011) Comparison of two marks of sex pheromone dispensers commercialized in Tunisia for their efficiency to monitor and to control by mass-trapping *Tuta absoluta* under greenhouses. Plant Prot 6:133–148
42. Ghoneim K (2014) Parasitic insects and mites as potential biocontrol agents for a devastative pest of tomato, *Tuta absoluta* Meyrick (Lepidoptera: Gelechiidae) in the world: a review. Int J 2:81–115
43. Gordon K, Waterhouse PM (2007) RNAi for insect-proof plants. Nat Biotechnol 25:1231–1232
44. Zappalà L, Biondi A, Tropea Garzia G, Siscaro G (2011) Studi sui parassitoidi indigeni di *Tuta absoluta* in Sicilia. In: Proc. XXIII Congresso Nazionale Italiano di Entomologia, pp 335
45. Farag MA, Shehata NF, Mahmoud YA (2009) Predicting the annual generation peaks of peach fruit fly, Bactrocera zonata (Saunders) (Diptera: Tephritidae) using heat units accumulation at Giza governorate, Egypt, The 4th Conference on Recent Technologies in Agriculture, Cairo University, Egypt
46. Braham M (2014) Sex pheromone traps for monitoring the tomato leafminer, *Tuta absoluta*: effect of colored traps and field weathering of lure on male captures. Res J Agric Environ Manage 3(6):290–298
47. García MF, Espul JC (1982) Bioecology of the tomato moth (Scrobipalpula absoluta) in Mendoza, Argentine Republic. Revista de Investigaciones Agropecuarias 17(2):135–146
48. Richardson ML, Lagos DM (2007) Effects of a juvenile hormone analogue, pyriproxyfen, on the apterous form of soybean aphid (Aphis glycines). J Appl Entomol 131:297–302

49. Lietti MMM, Botto E, Alzogaray RA (2005) Insecticide resistance in Argentine populations of *Tuta absoluta* (Meyrick) (Lepidoptera: Gelechiidae). Neotrop Entomol 34(1):113–119
50. El-Ghany NM, Abdel-Razek AS, Ebadah IM, Mahmoud YA (2016) Evaluation of some microbial agents, natural and chemical compounds for controlling tomato leaf miner, *Tuta absoluta* (Meyrick) (Lepidoptera: Gelechiidae). J Plant Prot Res 56(4)
51. Abdul-Rassoul MS (2014) A new host record for Tomato Leaf Miner *Tuta absoluta* (Meyrick, 1917) in Baghdad province, Iraq. Bull Iraq Nat Hist MusM 13(1):15–18
52. Taha M, Homam H, Afsah E, EL-Sharkawy M (2012) Effect of trap color on captures of *Tuta absoluta* moths (Lepidoptera: Gelechiidae). Int J Environ Sci Eng 3:43–48
53. Desneux N, Wajnberg E, Wyckhuys KAG, Burgio G, Arpaia S, Narváez-Vasquez CA, et al (2010) Biological invasion of European tomato crops by Tuta absoluta: ecology, geographic expansion and prospects for biological control. J Pest Sci 2(83):197–215
54. Sanchez NE, Pereyra PC, Luna MG (2009) Spatial patterns of parasitism of the solitary parasitoid Pseudapanteles dignus (Hymenoptera: Braconidae) on *Tuta absoluta* (Lepidoptera: Gelechiidae). Environ Entomol 38:365–374
55. Witzgall P, Kirsch P, Cork A (2010) Sex pheromones and their impact on pest management. J Chem Ecol 36:80–100
56. Guedes RNC, Picanc MC (2012) The tomato borer *Tuta absoluta* in South America: Pest status, management and insecticide resistance. EPPO Bull. 42:211–216
57. Russell IPM (2009) *Tuta absoluta*—insect profile, Russell IPM Ltd
58. Barrientos RZ, Apablaza HJ, Norero SA, Estay PP (1998) Temperatura base y constante térmica de desarrollo de la polilla del tomate, *Tuta absoluta* (Lepidoptera: Gelechiidae). Cienc Investig Agrar 25:133–137
59. Saad SA, Radwan MA, Mesbah HA, Ibrahim HS, Khalil MS (2017) Evaluation of some non-fumigant nematicides and the biocide avermactin for managing Meloidogyne incognita in tomatoes. Pak J Nematol 35(1):85–92. ISSN 0255-7576 (Print), ISSN 2313-1942
60. Svato SA, Attygalle AB, Jham GN, Frighetto RTS, Vilela EF, Saman D, Meinwald J (1996) Sex pheromone of tomato pest Scrobipalpuloides absoluta (Lepidoptera: Gelechiidae). J Chem Ecol 22:787–800
61. Arya SK, Dhar YV, Upadhyay SK et al (2018) De novo characterization of Phenacoccus solenopsis transcriptome and analysis of gene expression profiling during development and hormone biosynthesis. Sci Rep 8:7573
62. Vacas C, Alfaro J, Primo V (2011) Navarro-Llopis, Studies on the development of a mating disruption system to control the tomato leafminer, *Tuta absoluta* Povolny (Lepidoptera: Gelechiidae). Pest Manag Sci 67:1473–1480
63. Liu S, Zhang GM, Zhu J (1995) Influence of temperature variations on rate of development in insects: analysis of case studies from entomological literature. Ann Entomol Soc Am 88:109–117
64. Speranza S, Sannino L (2012) The current status of *Tuta absoluta* in Italy. EPPO Bull 42:328–332
65. Bueno VHP, van Lenteren JC, Lins Jr JC, Calixto AM, Santiago LD, Pérez LM (2013) New records of *Tuta absoluta* (Meyrick) (Lepidoptera: Gelechiidae) predation by Brazilian Hemipteran predatory bugs. J Appl Entomol 137:29–34
66. Bueno VHP, Montes FC, Calixto AM, Lins Jr JC, van Lenteren JC (2012) Can recently found Brazilian hemipteran predatory bugs control *Tuta absoluta*. IOBC-WPRS Bull 80:63–67
67. Becker VO (1984) Gelechiidae. In: Atlas of Neotropical Lepidoptera, Checklist: Part 1 Micropterigoidea-Immoidea. W. Junk, 112 pp
68. Giustolin A, Vendramim TJD, Alves SB, Vieira SA, Pereira RM (2001) Susceptibility of *Tuta absoluta* (Meyrick) (Lep., Gelechiidae) reared on two species of Lycopersicon to Bacillus thuringiensis var. kurstaki. J Appl Entomol 125(9/10):551–556
69. Bawin T, De Backer L, Dujeu D et al (2014) Infestation level influences oviposition site selection in the tomato leafminer *Tuta absoluta* (Lepidoptera: Gelechiidae). Insects 5(4):877–884
70. Kılıç T (2010) First record of *Tuta absoluta* in Turkey. Phytoparasitica 38(3):243–244

71. Bayram Y, Büyük M, Özaslan C, Bektas O, Bayram N, Mutlu C, Ates E, Bükün B (2015) New host plants of Tuta absoluta (Meyrick) (Lepidoptera: Gelechiidae) in Turkey. J Tekirdag Agric Fac 12(2):43–46
72. Zalom FG, Goodell PB, Wilson LT, Barnett WW, Bentley WJ (1983) Degree-days, the calculation and use of heat units in pest management: university of california division of agriculture and natural resources leaflet 21373. Univ. of California, Berkeley
73. Nozad-Bonab Z, Hejazi MJ, Iranipour Sh, Arzanlou M (2017) Lethal and Sublethal effects of some chemical and biological insecticides on *Tuta absoluta* (Lepidoptera: Gelechiidae) Eggs and Neonates. J Econ Entomol 110(3):1138–1144
74. Jasic I (1975) On the life cycle of Perillus bioculatus (Heteroptera: Pentatomidae) in Slovakia. Acta Ent. Bihemoslov 72:383–390

In Silico Studies on Juvenile Hormone Epoxide Hydrolase in *Spilarctia Obliqua* to Elucidate JH Analogs as Green Biopesticides

Swetha kumari Koduru and D. M. Mamatha

Abstract The battle between human and insect pests always faces unexpected problems. Identification of novel strategies from the Insect Endocrine system seems to be more essential for their biocontrol. One such approach is targeting Juvenile Hormone pathway. The Juvenile hormone epoxide hydrolase protein has an immense potential to apply as an insecticide due to its irreversible degradation of insect's JH, disrupting all the major physiological processes of the insect pests leading to the death of the pests. The present paper includes the in silico analysis of JHEH protein in *Spilarctia obliqua*, which is a serious agricultural pest on many food crops. The three-dimensional structure of JHEH protein is modeled based on Homology using MOE considering *Bombyx mori* JHEH as a template. The obtained model is minimized and evaluated. The inferred SoJHEH protein structure showed the conserved catalytic triad and oxyanion hole domains, indicating the well-preserved catalytic mechanism of epoxide hydration. Rigid receptor docking was performed with selected JH analogs to analyze the catalytic activity of JHEH. The docked complex showed stability with the Juvenile hormone structure with the lowest energy during 50 ns molecular dynamic simulations run. Thus, the proposed study represents a preface step to develop a biocontrol strategy where JH analogs can be used as third-generation pesticides.

Keywords Juvenile hormone epoxide hydrolase · Homology modeling · Molecular docking · *Spilarctia obliqua* · Third-generation pesticides

1 Introduction

The severe dependence and regular arbitrary utilization of pesticides resulted in the accumulation of their residues in soil, alarming a prominent negative impact on the environment. In particular, with respect to agriculture, crop pollination and beneficial

S. Koduru (✉) · D. M. Mamatha
Molecular Cloning Lab, Department of Biosciences and Sericulture, Sri Padmavati Mahila Visvavidyalayam, Tirupati 517502, India
e-mail: swethakumari.k@gmail.com

S. Jyothi et al. (eds.), *Proceedings of the 2nd International Conference on Computational and Bio Engineering*, Lecture Notes in Networks and Systems 215,
https://doi.org/10.1007/978-981-16-1941-0_56

insects are being destructed, thereby decreasing the maintenance of secondary pests for control. Therefore, there is a major emphasis on developing alternative strategies to use insecticides that are being concerned by many local, national, and international organizations involved in pest control. Moreover, the continuous dependence on insecticides developed insect resistance in many serious pests leading to serious threats to the environment. Relevant examples include *Spilarctia obliqua* (Bihar Hairy caterpillar) as one among them, where resistance has developed. Its moth is brown colored with a red abdomen and 40–50 mm wingspan. The adult lays eggs in 50–100 clusters on the lower side of leaves. The larvae are red and black colored with long yellowish black hairs. They are gregarious leaf feeders causing huge loss by defoliation of the total crop [1].

The metamorphosis in the life cycle of insects is kept in harmony by juvenile hormones (JHs) and ecdysteroids. Final instars of larvae are characterized by very low JH titers and high ecdysone titers, leading to molting. Juvenile hormone (JH), a highly pleiotropic sesquiterpenoid insect hormone, is secreted by the corpora allata (CA). JH esterase (JHE), JH epoxide hydrolase (JHEH) and JH diol kinase (JHDK) are three key enzymes involved in JH degradation [2]. JH is converted to JH acid in presence of JHE, a member of carboxylesterases and is a reversible reaction while, JHEH, belonging to microsomal epoxide hydrolases (EHs) family, converts JH to JH diol. Finally, the water-soluble metabolite is formed by the phosphorylation of JH diol in the presence of JHDK. Since JH is a specific hormone in insects and plays a significant role in the regulation of physiological processes in development and reproductive maturation. Hence, because of JHEH role in JH regulation and irreversible degradation, its characterization in pests is very important.

EHs convert epoxides to 1,2-trans-diols and belong to the ab-hydrolase fold superfamily. EHs play important roles in the toxic compound and allelochemicals metabolism and in JH regulation [3]. JHEH, a non-secreted enzyme, has two significant roles: decomposing JH balancing its concentration during each developmental stage. Second is JH diol formation, an irreversible hydrolysis reaction. For these two reasons, JHEH can be further explored for its role as a novel strategy in insect Biocontrol [4]. The current study is focused on the role of JH Epoxide Hydrolase (JHEH) ineffective degradation of JH and its analogs. The structural analyses together with molecular simulations and docking studies to reveal insights into the specific binding of JHEH in its active sight pocket are tried in the current study. Homology modeling and molecular docking studies of SoJHEH increase our understanding of the substrate binding and catalysis of EHs and might help to design JH-derived pesticides. This study explores the dual enzyme regulating system of JHEH, which indirectly controls varied developmental activities of the insect pest by regulating JH can throw insights for mounting further innovative biocontrol ways.

2 Methodology

The full-length gene sequence of JHEH from Bihar Hairy Caterpillar, *Spilarctia obliqua* (SoJHEH) was cloned and submitted to NCBI with GenBank Accession no. KC148541.1 [4]. It consists of 5' UTR with 27 bp followed by single ORF corresponding to 1323 bp and 3' UTR with 85 bp and Poly A tail (A_{25}) was considered.

2.1 *Primary Structure Analysis*

The primary structural analysis of *Spilarctia obliqua* juvenile hormone epoxide hydrolase protein using expasy tools, namely Compute pI/Mw [5, 6] was depicted. The amino acid composition, isoelectric point (*pI*) and molecular weight (*Mw*) were depicted

2.2 *Secondary Structural Analysis and Transmembrane Prediction*

The secondary structural analysis using Chou-fasman prediction method and PSIpred [7, 8] were depicted. The number of helices, sheets, coils, extended strands was predicted. The transmembrane topology and signal peptide prediction of SoJHEH were also determined using TMpred [9] and Phobius expasy tools.

2.3 *Homology Modeling*

Homology modeling of SoJHEH protein was carried out using MODELLER 9.13 software [10]. Protein BLAST (Blastp) was performed against PDB using SoJHEH sequence as a query, which showed the highest homology (Identity = 43%, Query coverage = 95%) to the crystal structure of JHEH from the silkworm, *Bombyx mori* (*Bm*JHEH) (PDB ID: 4QLA) out of all the other EH proteins reported in Protein Data Bank. Thus, the protein structure with PDB Id 4QLA was considered a template for modeling the SoJHEH structure using MODELLER 9.13. The respective script files, alignment files and segment files were written following the manual and were run on the command prompt of MODELLER. The final SoJHEH 3d structure was modeled with many intermediary files run in the program. The obtained model was subjected to energy minimization with free rotating Hydrogens and keeping the highest root mean square deviation (RMSD) value of 0.30 Å [11]. In a total of ten models generated,

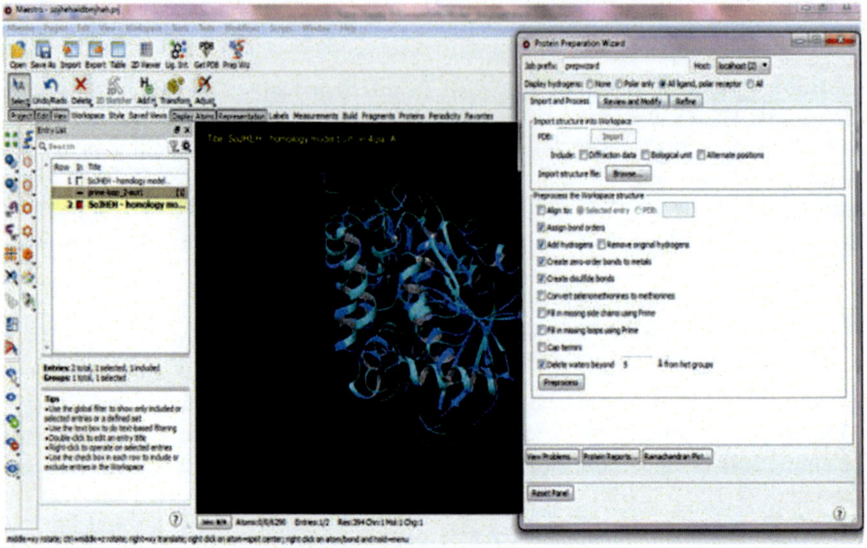

Fig. 1 Protein preparation of modeled SoJHEH protein

the model number SoJHEH.B99990010 was considered based on DOPE (Discrete Optimized Protein Energy) score.

2.4 Evaluation of Modeled SoJHEH Protein Structure

Three-dimensional structure of SoJHEH was evaluated for consistency, structural geometry for epitope prediction and lead discovery. The stereochemical quality of the obtained final protein model named SoJHEH.B99990010.pdb (Fig. 1) was checked using PROCHECK of ADIT Validation server [12], producing a number of PostScript plots analyzing its overall and residue-by-residue geometry [13, 14] and the Ramachandran plot was generated. The Z-score and energy plot were calculated using ProSa server [14] and the results were discussed.

2.5 Ligand Preparation

Juvenile Hormone III (JH-III), Cis stilbene oxide (c-SO) and trans-diphenyl propene oxide (t-DPPO) (Fig. 2) were the substrate or the ligand molecules used to dock with modeled SoJHEH structure. LigPrep v3.0 [LigPrep, Schrödinger, LLC, New York, NY, 2019.] module of Schrödinger was used for ligand preparation to attain chemical correctness by considering stereochemistry, protonation and tautomerism

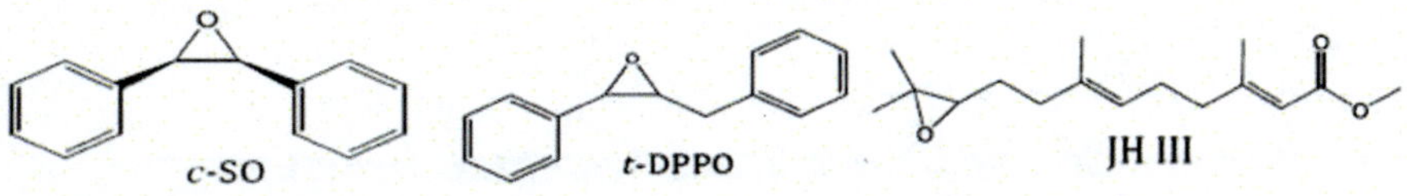

Fig. 2 Structure of selected substrates

characteristics. To utilize for further docking studies, the generated ligand molecules were subjected to energy minimization, maintaining pH (7.0 ± 2.0).

2.6 *Molecular Docking Studies*

The docking studies were performed using Glide interface of Schrödinger software [1, 12, 15]. A grid box of 10 × 10 × 10 Å surrounding the predicted catalytic triad of modeled SoJHEH protein was generated by importing it to Glide v.6.3. The two docking strategies namely Induced fit docking (IFD) and Quantum polarized ligand docking (QPLD) were applied to compute the docking score and binding affinity between catalytic triad and JH analogs. The prepared substrate molecules (JH-III and JH analogs) and the 3d protein structure were subjected to docking to screen the best ligand molecules and to calculate the binding affinities using XP-docking method. Binding affinity of the docked complex was termed as binding free energy (ΔG). The method used to compute the binding free energy (ΔG) of the docking complex was Prime/Molecular Mechanics (MM)—Generalized born surface area (GBSA) (Prime version 2 (2009) Schrödinger, LLC: New York, NY). With the application of OPLS-AA 2005 force field and GBSA continuum solvent model, the docking poses were minimized. The best lead resulting from QPLD with the lowest binding free energy (ΔG) was subjected to induced fit docking considering the flexibility of both lead and receptor. At this stage, to calculate the receptor lead docking scores, the parameters were set with 0.5 A° limit of Van der Waals radii scale [16].

2.7 *Molecular Dynamics Simulations*

Molecular dynamics (MD) simulations were executed to SoJHEH catalytic triad – Substrate complex for 50 ns (nanoseconds) using Desmond v3.8. The solvated model system included with Simple point charge (SPC) water molecules embedded with OPLS_AA-2005 force field was depicted and was neutralized and minimized before simulating. A 300 K and 1.0132 bar of temperature and pressure, respectively, was kept using Nose-Hoover temperature coupling and isotropic scaling [17]. The minimized system was passed through NVT ensemble for 12 picoseconds followed by NPT simulations for 50 ns, and the trajectories were recorded.

3 Results

3.1 *Primary, Secondary and Transmembrane Structural Analyses of SoJHEH*

A full-length JHEH encoded from cDNA of *Spilarctia obliqua* (KC148541) showed 1460 nts long and contained 1323 nts long open reading frame flanked by 5′ UTR sequence of 27 nts and 3′ UTR sequence of 85 nts with a poly A_{25} tail. 'Phobius' and 'TMpred' results showed that the first 1–20 amino acids constituted the signal peptide and 1–21 amino acids overlapping transmembrane region at N-terminus, respectively (Fig. 3). The deduced protein of SoJHEH showed 440 amino acid residues and had a predicted mass of 49475.9 Daltons and PI of 6.32. The secondary structural analysis of SoJHEH was performed and shown (Fig. 4, Table 1).

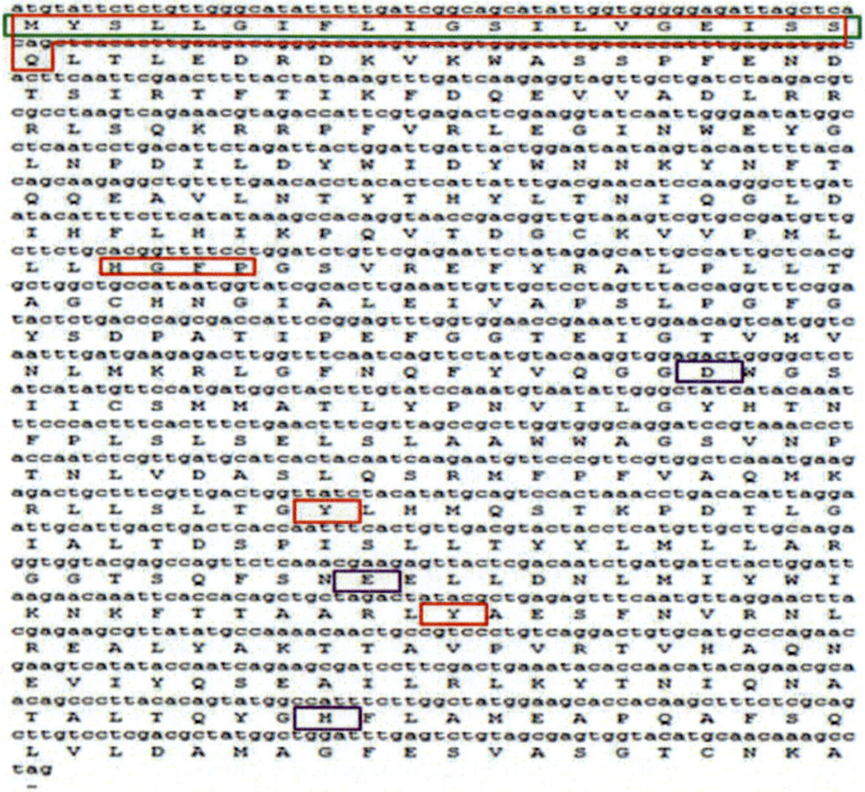

Fig. 3 Nucleotide and deduced amino acid sequence of *Spilarctia obliqua* JHEH. (Highlighted text: Transmembrane region (Red, 1-2aa), Signal peptide (Green, 1-20aa), Oxyanion Hole: Y_{288}–Y_{351}, red &HGFP motif (143–146), Catalytic triad (D_{217}–H_{408}–E_{329}, violet)

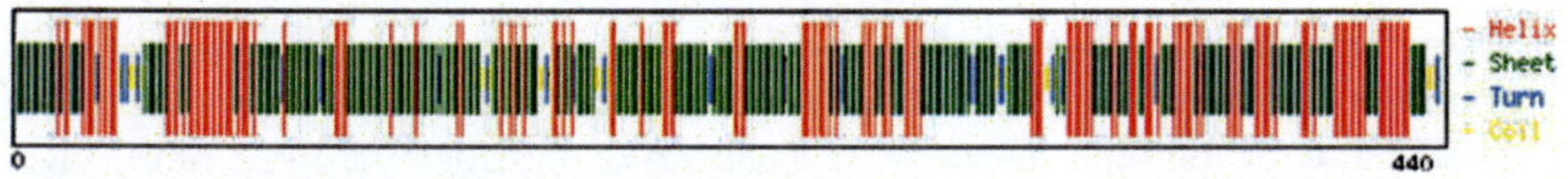

Fig. 4 Secondary structural features of SoJHEH

Table 1 Primary and secondary structural features of SoJHEH

S. no.	Parameter	Value	S. no.	Parameter	Value
1	Helices	168 (No. of aa)	5	Instability index	41.27
2	Extended strands	71 (-do-)	6	Grand average of hydropathicity	0.035
3	B-turns	17 (-do-)	7	Extinction coefficient	75540 M^{-1} cm^{-1}
4	Random coils	184 (-do-)			

3.2 *Prediction of Catalytic Site in SoJHEH*

Upon clustering and conserved Domain analysis of deduced SoJHEH protein, it has been clearly found that all the amino acids involved in the catalytic mechanism were conserved in r-SoJHEH. Among the protein structures reported in PDB, *Bombyx mori* JHEH (PDB ID: 4QLA.A) showed the highest sequence homology (Identity-43%, E-value: 9e-123) r-SoJHEH. The proposed catalytic triad of SoJHEH was located at a putative active site as shown in Fig. 5.

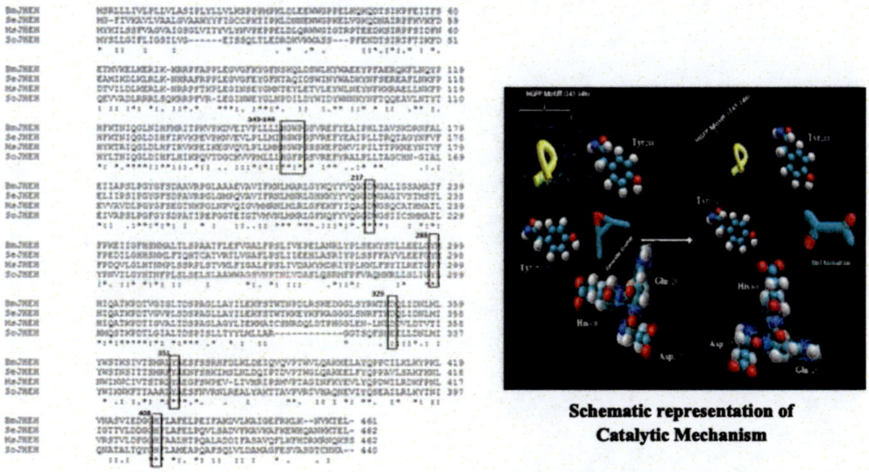

Fig. 5 Multiple Sequence alignment showing conserved catalytic amino acid residues

They were predicted to be Asp_{217}–His_{408}–Glu_{329}, and the other amino acids forming the Oxyanion hole, Y_{288}–Y_{351} and HGFP motif (143-146aa), were also found in SoJHEH, indicating the conserved mechanism of epoxide hydrolysis of the juvenile hormone with an intermediate formation of hydroxyl alkyl-enzyme intermediate. Like all epoxide hydrolases, SoJHEH is also characterized by a nucleophile-histidine-acid catalytic triad and had a two-step mechanism [18], as shown in Schematic representation.

Homology Modeling:

Using MODELLER 9.13, the tertiary structure of Juvenile hormone epoxide hydrolase in *Spilarctia obliqua* was predicted (Fig. 6). The predicted structure was evaluated using PDB-Ramachandran plot and Prosa server. Ramachandran plot (Fig. 6b) showed 99.2% of amino acid residues in all allowed regions, with only 0.8% in disallowed regions. It showed 89.7% of residues in most favored region, 7.7% in the additionally allowed region and 1.8% in the generously allowed region.

The Prosa server evaluated the protein structure that was within X-ray crystallography range. The Z-score of −6.79 was calculated for the modeled SoJHEH protein (Fig. 7a). The energy plot revealed that most of the residues were negatively charged (Fig. 7b).

Molecular docking:

The obtained protein structure was minimized using the energy minimization tool in Maestro interface of Schrödinger software. The ligands were prepared and minimized in Maestro using 'Ligprep'. The energy minimized protein was subjected to protein preparation as per the Glide software. Grid was generated by selecting the specified residues in the SoJHEH protein. The protein-ligand docking was performed using Glide docking by using Xtra Precision (XP) method. The docking poses were as

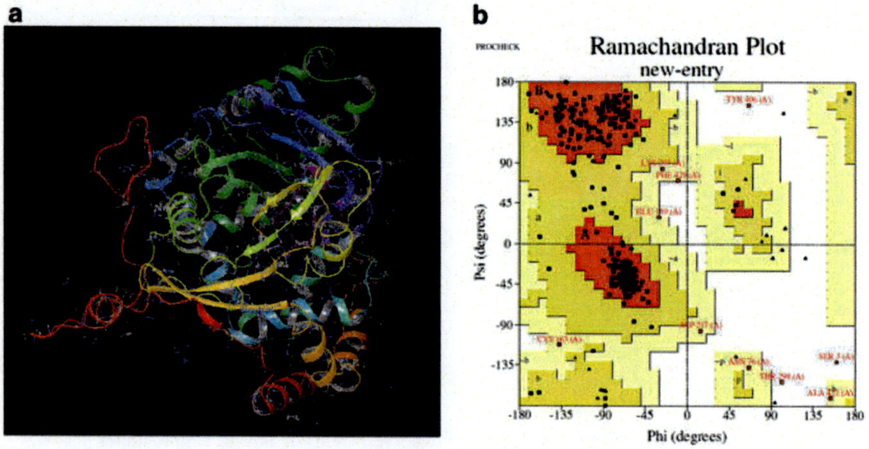

Fig. 6 **a** Modeled SoJHEH protein (Maestro Interface). **b** Ramachandran plot

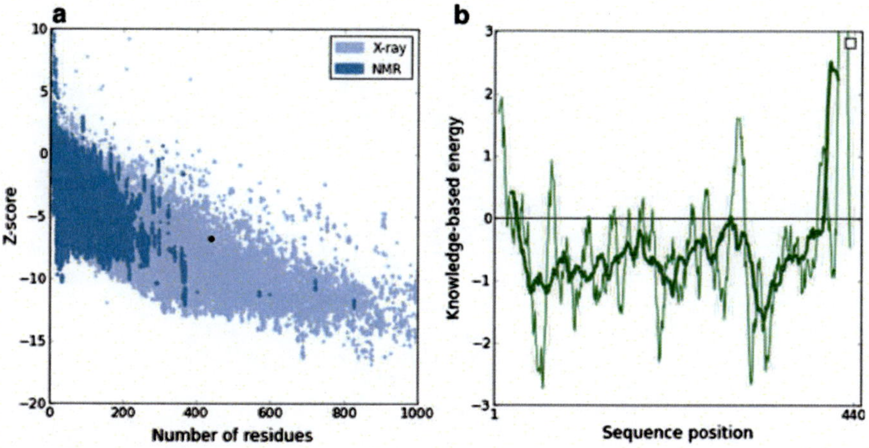

Fig. 7 **a** Prosa output for SoJHEH. **b** Energy plot

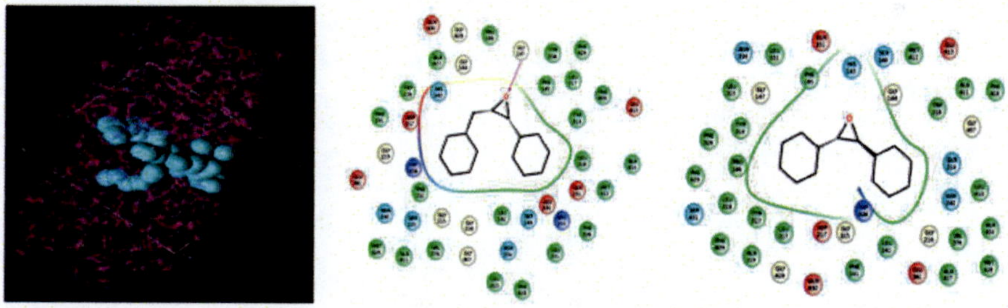

Fig. 8 Grid box and docking poses of t-DPPO and c-So within SoJHEH catalytic site

shown (Fig. 8). JH-III showed more binding affinity toward the SoJHEH protein. The docking scores for c-So, t-DPPO and JH-III were found to be −3.90465, −5.04655 and −6.36045, respectively (Table 2).

4 Discussion

In *Spilarctia obliqua* (SoJHEH), an economically significant major lepidopteron pest on mulberry and many other food crops in India, the juvenile hormone epoxide hydrolase gene was cloned and expressed. The in silico analysis of SoJHEH fulfilled many novel insights. On the basis of its high affinity for JH-III, its structurally conserved domains and trends in enzyme hydrolysis, it is understood that SoJHEH plays a crucial role in maintaining JH concentrations by converting it into JH-diols in an irreversible manner and is the biologically active JHEH of *Spilarctia obliqua*.

Table 2 Table showing docking parameters and scores

Title	RMS derivative-OPLS	Stretch energy-OPLS	Bend energy-OPLS	Vander Waals energy-OPLS	Electrostatic energy-OPLS	XP electro	Docking score
c-So	0.099464	173.146898	562.554055	−2846.40156	−1124.07459	−0.21253	−3.9046
t-DPPO	0.094524	185.147498	993.755053	−3255.50835	−2105.07149	−0.13708	−5.0465
JH-III	0.086421	199.147498	999.456503	−4101.15647	−2578.12564	0.05055	−6.3604

To prove further in in silico, the structural analyses were done with the deduced SoJHEH sequence.

All the three ligands were prepared, minimized and docked with the SoJHEH structure. The docking scores have come in the increasing order of −3.90465, −5.04655 and −6.36045 for c-So, t-DPPO and JH-III, respectively. The docking scores obtained signify that SoJHEH protein showed more binding affinity toward JH-III, indicating the juvenile hormone epoxide hydrolase in *Spilarctia obliqua* is highly JH specific and hence satisfies the name SoJHEH. However, the docking studies with several inhibitors have to be screened yet to further analyze the exact binding affinities of SoJHEH to check whether the derivatives of this binding site can be helpful in developing third-generation pesticides. The other side, Microsomal and Soluble epoxide hydrolases (mEH and sEH), is mentioned to harmonize one another in detoxifying various toxic, mutagenic, and xenobiotic residues cancerous in nature. Besides, it was also revealed that sEH is more related to the regulation of inflammation and blood pressure, whereas mEH detoxifies Xenobiotic deposits. Therefore, characterizing and fully understanding the EH mechanism in insects is a prerequisite for producing pesticides of the third-generation and exploring their function in bioremediation as many toxic substances are epoxy group-containing compounds.

References

1. Jones DT (1999) Protein secondary structure prediction based on position-specific scoring matrices. J Mol Biol 292:195–202
2. Laskowski RA, MacArthur MW, Moss DS, Thornton JM (1993) PROCHECK —a program to check the stereochemical quality of protein structures. J App Cryst 26:283–291
3. Singh KD, Muthusamy K (2013) Molecular modeling, quantum polarized ligand docking and structure-based 3D-QSAR analysis of the imidazole series as dual AT(1) and ET(A) receptor antagonists. Acta Pharmacol Sin 34:1592–1606
4. Wiederstein M, Sippl MJ (2007) ProSA-web: interactive web service for the recognition of errors in three-dimensional structures of proteins. Nucleic Acids Res 35:W407–W410
5. Bjellqvist B, Basse B, Olsen E, Celis JE (1994) Reference points for comparisons of two-dimensional maps of proteins from different human cell types defined in a pH scale where isoelectric points correlate with polypeptide compositions. Electrophoresis 15:529–539
6. Bjellqvist B, Hughes GJ, Pasquali Ch, Paquet N, Ravier F, Sanchez J-Ch, Frutiger S, Hochstrasser DF (1993) The focusing positions of polypeptides in immobilized pH gradients can be predicted from their amino acid sequences. Electrophoresis 14:1023–1031
7. Buchan DWA, Minneci F, Nugent TCO, Bryson K, Jones DT (2013) Scalable web services for the PSIPRED Protein Analysis Workbench. Nucl Acids Res 41 W1:W340–W348
8. Hofmann K, Stoffel W (1993) TMbase - A database of membrane spanning proteins segments. Biol Chem Hoppe-Seyler 374:166
9. Halgren TA, Murphy RB, Friesner RA, Beard HS, Frye LL, Pollard WT, Banks JL (2004) Glide: a new approach for rapid, accurate docking and scoring. 2. Enrichment factors in database screening. J Med Chem 47:1750–1759
10. Sali A, Blundell TL (1993) Comparative protein modelling by satisfaction of spatial restraints. J Mol Biol 234:779–815
11. Zhang Q-R, Wei-Hua X, Chen F-S, Li S (2005) Molecular and biochemical characterization of juvenile hormone epoxide hydrolase from the silkworm, Bombyx mori. Insect Biochem Mol Biol 35:153–164

12. Friesner RA, Murphy RB, Repasky MP, Frye LL, Greenwood JR, Halgren TA, Sanschagrin PC, Mainz DT (2006) Extra precision glide: docking and scoring incorporating a model of hydrophobic enclosure for protein-ligand complexes. J Med Chem 49:6177–6196
13. Laskowski RA, Rullmannn JA, MacArthur MW, Kaptein R, Thornton JM (1996) AQUA and PROCHECK-NMR: programs for checking the quality of protein structures solved by NMR. J Biomol NMR 8:477–486
14. Li S, Falabella P, Kuriachan I, Vinson SB, Borst DW, Malva C, Pennacchio F (2003) Juvenile hormone synthesis, metabolism, and resulting haemolymph titre in Heliothis virescens larvae parasitized by Toxoneuron nigriceps. J Insect Physiol 49:1021–1030
15. Gasteiger E, Hoogland C, Gattiker A, Duvaud S, Wilkins MR, Appel RD, Bairoch A (2005) Protein identification and analysis tools on the ExPASy server. Walker JM (ed) The proteomics protocols handbook. Humana Press, pp 571–607
16. Chubb AJ, Fitzgerald DJ, Nolan KB, Moman E (2006) The productive conformation of prostaglandin G2 at the peroxidase site of prostaglandin endoperoxide H synthase: docking, molecular dynamics and site-directed mutagenesis studies. Biochemistry 45:811–820
17. Suite S (2009) Induced Fit Docking protocol; Glide version 5, Schrodinger. LLC, New York
18. Hema M, Mamatha DM, Swetha Kumari K (2013) Cloning and Insilico characterization of juvenile hormone epoxide Hydrolase gene of Spilarctia obliqua. Int J Agric Sci Res, Transtellar Publications 3(2):275–288. ISSN 2250-0057
19. Berman HM, Westbrook J, Feng Z, Gilliland G, Bhat TN, Weissig H, Shindyalov IN, Bourne PE (2000) The Protein Data Bank Nucleic Acids Research 28:235–242
20. Swetha Kumari K, Mamatha DM, Kalpana S, Dadala B (2015) Expression and characterization of recombinant Juvenile Hormone Epoxide Hydrolase of Spilarctia obliqua, a major pest in Agri-Sericulture. Mol Entomol 6(1):1–11. ISSN: 1925-198X
21. Kamita SG, Yamamoto K, Dadala MM, Pha K, Samra AI, Morisseau C, Escaich A, Hammock BD (2013) Cloning and characterization of a microsomal epoxide hydrolase from Heliothis virescens. Insect Biochem Mol Biol 4:219–228

Optimisation of Forecast Error Through Combining the Forecast Results

M. Chandrasekhar Reddy and P. Balasubramanyam

Abstract The dynamics of forecasting are so complex that no single forecasting model will always give better accuracy for any single time series data. It usually changes from time to time and for data scientists, this is the biggest challenge to find the right forecast model for a given time series at a given time. In the current paper, we discussed an ensembling approach that will combine all the forecasted results from multiple models and give a single forecast value based on each fitted model's accuracy. This will remove the complexity of selecting the best model from time to time and improve the model's overall performance.

Keywords Time series forecasting · Ensembling approach · Accuracy improvement

1 Introduction to Time Series Analysis

Time series forecasting is one of the most complex and dynamic statistical procedure that has very wide applications in all most all domains starting from manufacturing, health care, banking and insurance etc. It is always giving a challenge to researchers in terms of getting the best accuracy and the best accurate models will give better decisions in the practical field. A lot of research has been done in the areas of time series forecasting and advanced models are developed to achieve the best accuracy [1]. The time series data has many more influential factors than its own past values. The key challenge here is to identify the best influential factors for the modelling purpose.

M. Chandrasekhar Reddy
Department of Statistics, College of Natural Sciences, Arba Minch University, P.O. Box no. 21, Arba Minch, Ethiopia
e-mail: m.chandrasekarreddy@gmail.com

P. Balasubramanyam (✉)
Goa Institute of Management, Goa, India
e-mail: balasubramanyam@gim.ac.in

S. Jyothi et al. (eds.), *Proceedings of the 2nd International Conference on Computational and Bio Engineering*, Lecture Notes in Networks and Systems 215,
https://doi.org/10.1007/978-981-16-1941-0_57

Time series analysis comprises methods for analysing time series data in order to extract meaningful statistics and other characteristics of the data like summary statistics and graphs. The primary target here is to find the meaningful relationships between the variables at different times and build a model based on those relationships. The decision on the correct form of the relationship purely depends on different data characteristics like how much the values are interrelated and the time interval between any two similar changes, etc. The irregular factors that are not covered in the model add to its error or deviation in the model. The regular practice of finding the trivial trends in the data is done through time series trend plots where the spikes away from regular data indicate the outliers and some unusual patterns.

2 Seasonality and Stationarity Test

Seasonality is a characteristic of time series that shows a repeated pattern at a regular time interval. The autocorrelation plots are used to identify there is a significant seasonality or not. If there is seasonality, the autocorrelation plot should show spikes at lags equal to the period. For example, for monthly data, if there is a seasonality effect, there is a significant spike at lag 12, 24, 36…… .

One of the important concepts in time series is stationarity also; it is usually seen as an assumption. It is the property that the mean and variance constant do not change over time and the autocovariance function between Xt_1 and Xt_2 only depends on the interval t_1 and t_2. Prior to the time series analysis. It is necessary to put out the process that makes up that series, whether constant over time or not [2]. To check whether the time series is stationary or not Augmented Dickey-Fuller test is used [3]. If the time series is not stationary, there are different methods to transform nonstationary data to stationary, one of the methods is normally differencing techniques that are used to transform a time series from a nonstationary to stationary by subtracting each datum in the series from its predecessor.

3 Modelling of Time Series

When working with forecasting of time series data, the underlying time series is assumed to be stationary. Assuming stationarity, there are several different approaches to construct forecasting models, for example, an autoregressive process, a moving average process, an autoregressive and moving average process, and an autoregressive integrated moving average process [4].

3.1 Autoregressive Process

An autoregressive process is used to forecast a time series. As mentioned earlier, a first-order autoregressive model is denoted AR(1) and Y_t regressed on Y_{t-1}. An autoregressive model of the *p*th order is denoted AR(*p*) and takes the form of

$$Y_t = \delta + \alpha_1 Y_{t-1} + \alpha_2 Y_{t-2} + \ldots\ldots\ldots\ldots + \alpha_p Y_{t-p} + u_t$$

where the constant is denoted by δ and u_t is the white noise stochastic error term [4].

3.2 A Moving Average (MA) Process

In a moving average process, the dependent variable is regressed on current and lagged error terms. It is therefore estimated through a constant and a moving average of the error terms [5]. If the dependent variable is regressed on the current and one lagged error term, it follows a first-order moving average process, denoted MA(1). Moreover, a model that includes q number of error terms follows a *q*th order moving average process, denoted MA(*q*). Moving average process is defined as

$$Y_t = \mu + \beta_0 u_t + \beta_1 u_{t-1} + \beta_2 u_{t-2} + \ldots\ldots\ldots\ldots + \beta_q u_{t-q}$$

where the error term u are assumed to be white noise and μ is the constant [4].

In a MA model, the error terms are usually scaled to make β_0 equal to one (Chatfield-2003).

3.3 An Autoregressive and Moving Average (ARMA) Process

It is possible to combine an autoregressive process (AR) and a moving average process (MA) since the dependent variable often possesses characteristics of both. This is known as an autoregressive and moving average process, or ARMA [6]. If both of the underlying AR and MA models are of the first first-order, the model is denoted ARMA(1, 0, 1) model can be generalised as

$$Y_t = \theta + \alpha_1 Y_{t-1} + \beta_o u_t + \beta_1 u_{t-1}$$

where θ is the constant. Suppose the underlying autoregressive model is of order *p* and the moving average model is of order *q*. In that case, the ARMA process is denoted by ARMA(*p*, *q*) [4].

3.4 An Autoregressive Integrated Moving Average (ARIMA) Process

Suppose the time series of an ARMA model has to be differenced a certain number of items to become stationary. In that case, the model becomes what is known as an autoregressive integrated moving average model, or an ARIMA model [7]. As mentioned earlier, a time series that has to be differenced d number of times in order to become stationary is integrated of order d, denoted $I(d)$. In its general form, the ARIMA model is denoted ARIMA(p, d, q), which means that the AR is of the pth order, the time series is integrated d number of times, and the moving average is of the qth-order. This further means that if the underlying AR and MA models are of the first-order, and the time series is stationary at the first difference, the ARIMA model is denoted by ARIMA (1, 1, 1). It is important to note that an ARIMA model is not derived from any economic theory, that is a theoretic model. The Box-Jenkins methodology can be followed to determine p, d, and q and estimate an ARIMA model [4].

4 What is Ensembling Approach

An ensembling approach combines the forecasted results from multiple models and gives out a single forecast result for any given time series data. So the researcher can build a library of models and each time, he can get results from all models in the library and combine them in such a way that the overall performance is good. The library set of models can be revisited at regular intervals of time and new models can be added to the library if needed.

4.1 Steps Involved in Esembling Approach

There are several approaches to ensemble the forecasting results from different models and some of the approaches are described below.

4.1.1 Enembling Through Simple Average

In this approach, we will get the forecasted results from multiple models and then a simple average of forecasting results is obtained as a single-point forecast for the given time series. This approach not very good since the bad model is taking away the advantage of a good model and hence we may not find good improvement in the accuracy of the model.

4.1.2 Ensembling by Weighted Average

In this approach, we will get the forecasted results from all the models along with their accuracy measure or error measure. In the current study, we used mean absolute percentage error (MAPE) [8] to find the accuracy of each model. Now, weights are assigned to each model based on the accuracy model. The highest accuracy model gets the highest weight, and the low accuracy model will get lower accuracy.

Model	Forecat results	Weights-based pon error
Model 1 (single exponenital model)	F1	W1
Model 2 (double exponential model)	F2	W2
Model 3 (ARIMA)	F3	W3

Now the ensemble forecast result is obtained as a weighted average as shown below

$$F = \frac{F1 \cdot W1 + F2 \cdot W2 + F2 \cdot W3}{W1 + W2 + W3}$$

4.1.3 Modified Weighted Average Approach

This is a modified version of calculating the weights for different models where an approach similar to the precision and recall concept is adopted. In the first step, we will calculate the forecasted results from multiple models and their error measure (MAPE) is obtained for all the models. For each model, we will obtain the average and standard deviation of the MAPE results. Now the weights are calculated by using the formula

$$W_i = \frac{\frac{1}{\text{MAPE}_i} + \frac{1}{\text{SD(MAPE)}_i}}{\sum_{i=1}^{8} \frac{1}{\text{MAPE}_i} + \frac{1}{\text{SD(MAPE)}_i}}$$

This method of calculating the weights will improve the overall performance of the model in many cases. There are some exceptions to this approach that if there is a huge difference between the best model and the next best, this may not give the required best results.

5 Empirical Analysis

We planned to test the above approach with an example time series data. We took global prices of gold in USD (https://datahub.io/core/gold-prices#data) [9] since the year 2000, and we took one year recent data aside for testing.

The time series plot of the data looks as follows, and we can see a clear trend of increasing patterns in the global price of gold.

We have tested the data for seasonality and stationarity and then applied the three popular models, namely single exponential, double exponential and ARIMA models. The forecasted values were obtained for one year period, namely from Jul 2019 to June 2020, and errors are calculated as per that.

The error rates for each individual models are obtained as follows:

Model name	MAPE value
Single exponential	2.3202
Double exponential	2.2587
ARIMA Model	2.2475

Now in the first approach, we combined the forecast through simple avaerage and then weighted average where weights are based MAPE values obtained above. The results of the two approaches were

MAPE for average forecast	MAPE for weighted average
2.2268	2.2263

Now in the final approach, we calculated the weights based on the formula described in 5.1.3, and then MAPE is compared with the above two approaches.

MAPE for the average forecast	MAPE for the weighted average	MAPE for modified weighted approach
2.2268	2.2263	2.2250

There is a very minute decrease in error rate we can observe, but when we increase the number of models in the library, it can be reduced further and give maximum benefit to the researcher.

6 Conclusions and Results

Forecasting is always a challenging area for data science persons, and there is a huge need to carry out the forecasting with utmost accuracy. Several time series forecasting models were developed, and each model will perform at its best only in few scenarios. So in the current paper, we demonstrated an approach to combine the forecasting results in the best way to get good forecast accuracy and gives more benefits to any organisation that adopts this approach.

Further to this, if the library of models can have more sophisticated models like TAR, IDM, ARFIMA, GARCH, which can improve the overall accuracy of the time series prediction. Also, for ensembling, instead of combining all models forecasted results, we can combine only top-performing three or four models to get the best results in the future studies.

References

1. Adhikari R (2003) An introductory study on time series modeling and forecasting, 2nd edn. Springer, pp 12–24
2. Douglas MK, Montomery C, Jennings CL (2008) Introduction to time series analysis and forecasting. Wiley, Hoboken, New Jersey
3. Obubu M, Nokolo PC (2016) Modeling and forecasting of armed robbery cases in nigeria using auto regressive integrated moving average (ARIMA) models. J Nat Sci Res
4. Gujarati DN (2006) Essentials of econometrics, 3d edn. McGraw-Hill, New York
5. Maddala GS (1977) Econometrics. McGraw-Hill, New York
6. Bari SH, Rahman MT, Hussain MM, Ray S (2015) Forecasting Monthly Precipitatio in Sylhet City Using ARIMA Model. Civ Environ Res 7(1):69–78
7. Agyemang B (2012) Autoregressive integrated moving average (ARIMA) interventio analysis model for the major crimes in ghana (the case of the eastern region), (Doctoral dissertation)
8. Galavi H, Mirzaei M, Shui LT, Valizadeh N (2013) Klang river-level forecasting using ARIMA and ANFIS models. J Am Water Works Assoc 105(9):81–82
9. https://datahub.io/core/gold-prices#data for gold price per gram in USD data from Jan 2000 to July 2019

A Comparative Performance Analysis of AOMDV and PAR Algorithms for Mobile Ad Hoc Networks

Salini Suresh, M. Manjunath, K. Ravikumar, and M. Hanumanthappa

Abstract Mobile Ad hoc Networks (MANET) is centred on wireless networks that enable users to communicate and exchange information in a simple yet and effective method. Multipath routing is of great interest to MANETs with the salient feature of dynamic topology and distributed infrastructure. Ad hoc On-Demand Multipath Distance Vector is the enhancement of AODV (Ad hoc On-demand Distance Vector) protocol that establishes optimal routes which are link disjoint and loop-free and based on hop count as the metric value from source to the destination. Although many routing protocols are proposed for routing in MANET, discovering an optimal path between source and destination is a key problem. The Ant Colony Optimisation (ACO) also has a greater possibility of choosing the right and shortest path between two nodes. Petal Ant Routing is a metaheuristics, nature-inspired, reactive routing algorithm for effective multipath routing. This research work intents to perform a comparative analysis of conventional AOMDV and nature-inspired based PAR algorithm for routing in MANET.

Keywords Mobile ad hoc networks · Ant colony optimisation · Petal ant routing · Multipath routing

S. Suresh (✉)
Dayananda Sagar College, Bangalore, India
e-mail: pnsalinisuresh@gmail.com

M. Manjunath
R V College of Engineering, Bangalore, India
e-mail: manjunathm@rvce.edu.in

K. Ravikumar
Kalinga Institute of Industrial Management, Bhuvaneshwar, India
e-mail: ravinaac@gmail.com

M. Hanumanthappa
Bangalore University, Bangalore, India
e-mail: hanu6572@gmail.com

S. Jyothi et al. (eds.), *Proceedings of the 2nd International Conference on Computational and Bio Engineering*, Lecture Notes in Networks and Systems 215,
https://doi.org/10.1007/978-981-16-1941-0_58

1 Introduction

MANET is among the most relevant and prevalent wireless network used in the mobile technology industry. Furthermore, the operation of wireless networks of this kind depends on routing algorithms. The Mobile Ad hoc Networks are wireless mobile node networks with adaptive network connectivity and excluding the need for any pre-defined wireless communications infrastructure. The key role in MANET transmission is errorless data broadcasting. Packet delivery ratio in a single path and multipath routing protocols is high regardless of network size, but average end-to-end delay increases in multipath routing due to route discovery requests. Owing to proactive routing, single-path routing protocols experience high throughput, whereas multipath routing experiences lower throughput as the network has more number of intermediate nodes.

The source node selects the optimal route from many routes available using a single routing discovery technique in multipath routing. This method deploys the route discovery process with an availability of a pool of routes that may be used if failure happens in the network, thus reducing end-to-end delay [1]. Concurrent multipath routing techniques makes the routing very effective in MANETs and are employed in applications to stream videos of top quality in heterogeneous wireless network [2]. Network congestion, end-to-end delay, network throughput and overhead depends on routing information. The routing protocols in MANET are classified as Proactive, Reactive and Hybrid.

One of the most familiar on-demand multipath routing techniques is AOMDV which is the enhancement of AODV protocol. The optimal routes, which are also loop-free and link disjoint from source to destination, are decided with hop count as the metric value. The hop count metric is mainly used in the AOMDV to establish the optimised routes. It uses advertised hop count rather than generally used hop count in AODV.

The Ant Colony Optimisation is a technique relying on swarm intelligence and imitates the behaviour of real ants is often deployed to find the optimal route in communication networks [3]. It is observed that, while the ants search and travel towards the food from the nest, they release a chemical named pheromone along its path. When returning back to the nest, the ant sticks to the same pattern travelled earlier, causing repeated pheromone deposits once more on the path. Every ant in the colony does the same pheromone deposition on the way to food to nest and back. In MANET, artificial ant agents like Forward ANT, Backward ANT replace the pheromone. The Petal Ant Routing consists of three segments that is route discovery phase, route maintenance and repair phase.

2 Literature Review

AOMDV depends on the routing information previously provided by the basic AODV algorithm to limit the overheads caused when multiple paths are discovered. There are no special control packets utilised. The added overheads incurred in AOMDV comparative to AODV are additional route reply (RREP) packets (RREP) and route request packets (RERQ) intended for multipath discovery [4, 5].

In PAR, a source that desires to communicate with a destination, calculates the petal region and broadcasts the artificial ant agents, Forward ANTS in the network. Upon receiving a Forward ANT, a node verifies whether it is located inside or outside the petal ant region. The mobile node that lies within the petal region accepts the Forward ANT and re-transmitted it to the next adjacent node in the network. This action continues until the Forward ANT reaches the destination node, then sends Backward ANT to the source. Source node initiates the data transfer through multiple intermediary nodes when it receives the Backward ANT. Thus multipath routing is made possible using PAR [6].

In [7] authors have put forth a nature-inspired Ant colony-based algorithm for multipath routing with an emphasis on Quality-of-service. The emphasis while selecting the best route is completely based on the reliability of the nodes, bandwidth, end-to-end delay and hop count so as to calculate disjoint multipath from the source node to the destination.

3 Methodology

The evaluation and comparison of the performance of AOMDV and PAR routing algorithms for multipath routing in MANETS is done by simulation using Network Simulator considering the network parameters packet sent and received, Packet Delivery Fraction (PDF), end-to-end delay, overheads, throughput and average energy consumed for varying network size.

As the hop count between sources to destination increases, the distance also tends to increase when the network grows. Therefore the overhead increases for the discovery of routes and the delay of end-to-end packets also rise. The size of the network is therefore directly proportional to end-to-end delay, overhead, energy consumed. Indirectly, the size of the network is also proportionate to the throughput and packet delivery ratio [8].

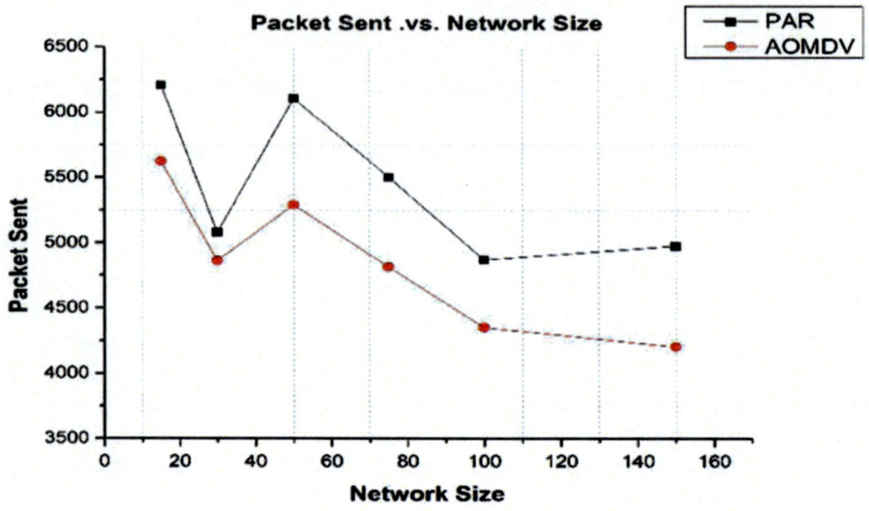

Fig. 1 Comparison of packet sent for PAR and AOMDV with respect to network size

3.1 Experimental Setup

The simulation experiments were carried out using NS2 version 2.35 and Linux Ubuntu. The simulation is performed for the multiple network sizes varying from 15 to 150 nodes for a time period of 160 s.

The throughput of a network is directly proportional to the number of packets that are sent from the source and the number of packets received at the destination. Figure 1 shows the comparison of the packet sent for PAR and AOMDV with respect to Network size. It shows that PAR algorithms can be sent 12.33% more packets than the AOMDV protocol. Hence meta-heuristic PAR algorithms can generate more packets and contribute to improved throughput.

The results of simulation of PAR and AOMDV routing protocols for a packet received by the destination with respect to network size are plotted in Fig. 2, which shows PAR receives 19.2% more packets than conventional AOMDV.

Packet delivery fraction corresponds to the ratio of data packets received at the destinations to data packets generated and sent from sources in a network. It is evident from Fig. 3 that the PAR algorithm has a 5.52% better PDF than AOMDV.

Figure 4 illustrates the end-to-end delay experienced by PAR and AOMDV with respect to Network size. The end-to-end delay experienced by PAR is 17.64% more than AOMDV because in ACO-based algorithms, due to more packets traversing the optimal path, the network congestion is higher and hence the delay across the network.

The throughput of the network is a measure of network efficiency in MANET. Figure 5 shows that PAR algorithms deliver a 15.39% better throughput than AOMDV.

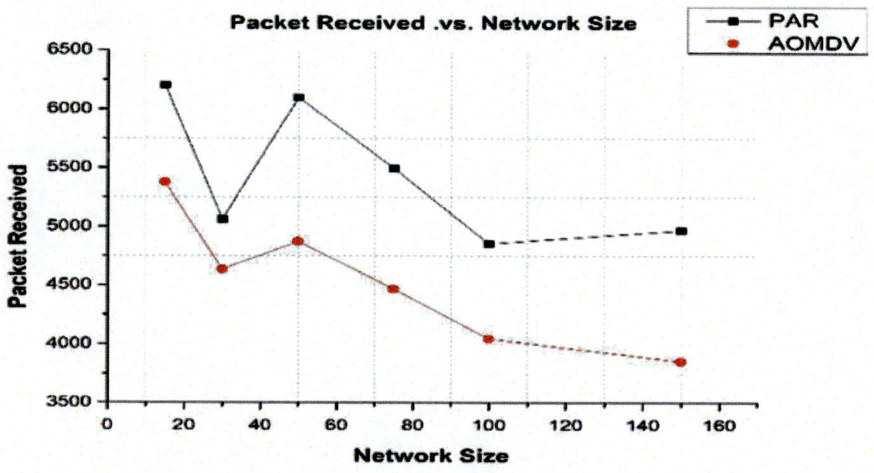

Fig. 2 Comparison of packet received for PAR and AOMDV with respect to network size

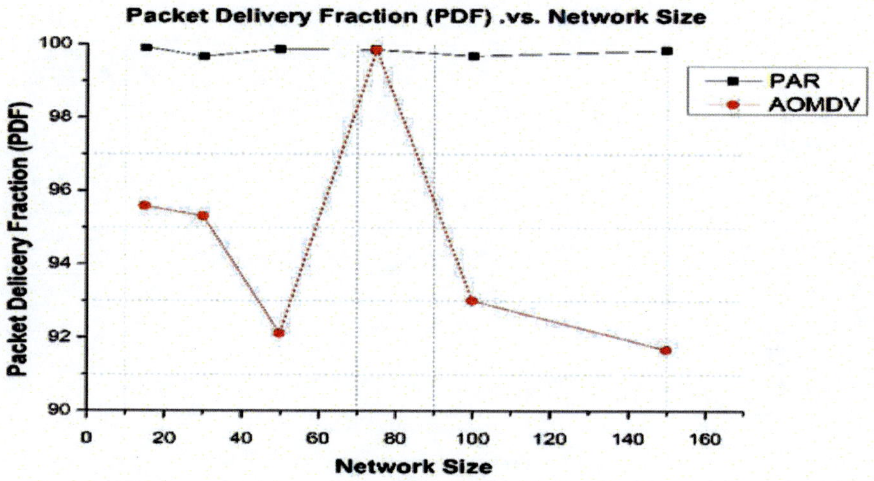

Fig. 3 Comparison of packet delivery fraction for PAR and AOMDV with respect to network size

Figure 6 shows that the overhead incurred in PAR is 99.4% less compared to AOMDV.PAR offers very lower routing overheads than the AOMDV routing protocol since the PAR algorithm requires less time to compute routes between end nodes.

The average energy consumed for routing packets in MANET during the simulation experiments is shown in Fig. 7. It shows that PAR consumes 11% lesser energy than the AOMDV routing protocol.

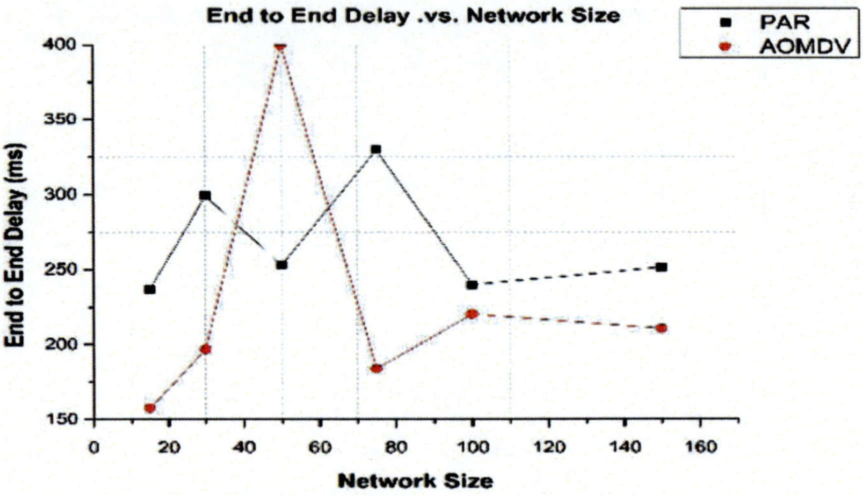

Fig. 4 End-to-end delay for PAR and AOMDV with respect to network size

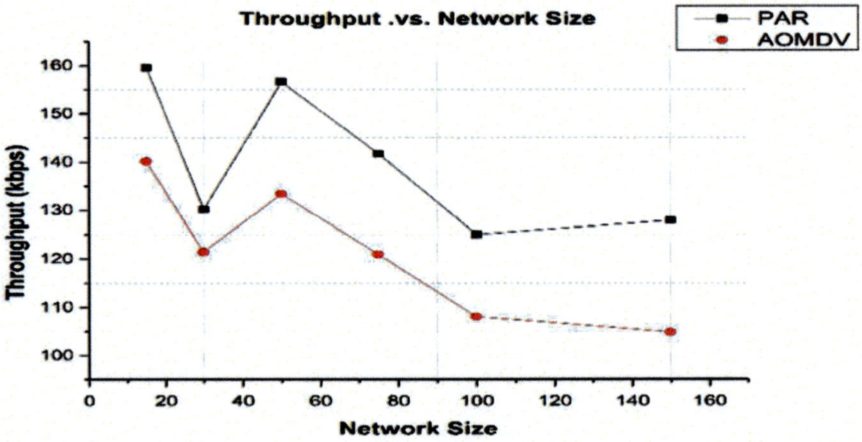

Fig. 5 Comparison of throughput for PAR and AOMDV with respect to network size

4 Conclusion

This research work emphasis the comparative performance analysis of multipath routing algorithms PAR with AOMDV for MANET. These routing algorithms were simulated using Network Simulator NS2 and assessed for various network parameters. As the PAR algorithm does optimisation in terms of the number of routes discovered and packet propagation, the routes establishment between source and destination is faster so that it aids in creating an extra count of packets across the

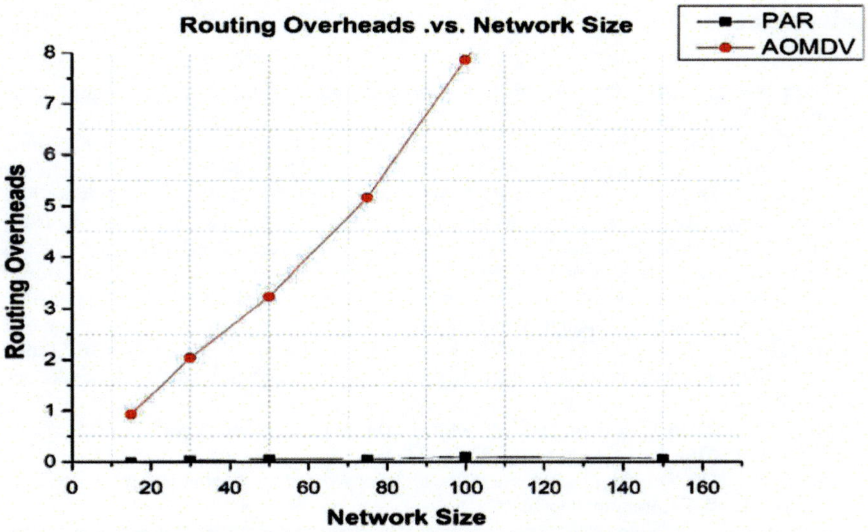

Fig. 6 Comparison of overhead for PAR and AOMDV with respect to network size

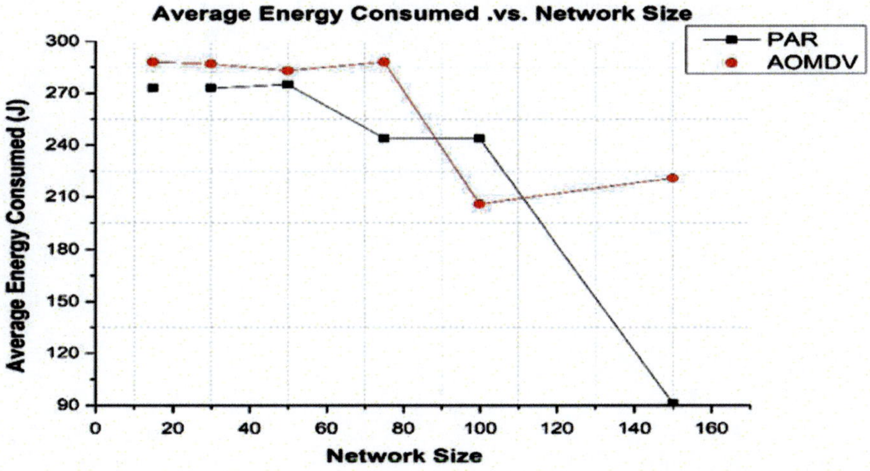

Fig. 7 Comparison of average energy consumed by PAR and AOMDV with respect to network size

network. PAR delivers better throughput through reduced delay, and thus the performance efficiency of PAR is improved than the AOMDV for multipath routing in MANETs.

References

1. Marina MK, Das SR (2006) Ad hoc on-demand multipath distance vector routing. Wirel Commun Mob Comput 6(7):969–988
2. Wu J, Tan R, Wang M (2019) Energy-efficient multipath TCP for quality-guaranteed video over heterogeneous wireless networks. IEEE Trans Multimed 21(6):1593–1608
3. Di Caro GA, Ducatelle F, Gambardella LM (2008) Ant colony optimization for routing in mobile ad hoc networks in urban environments. Technical Report No. IDSIA-05-08, Dalle Molle Institute for Artificial Intelligence Galleria 2, 6928 Manno, Switzerland
4. Nasipuri A, Castaneda R, Das SR (2001) Performance of multipath routing for on-demand protocols in mobile ad hoc networks. ACM/Kluwer Mob Netw Appl (MONET) 6(4):339–349
5. Ashoka S, Manjunath M, Hanumanthappa M (2018) Performance analysis of petal ant routing (PAR) and dynamic source routing (DSR) in MANET using network simulator (NS2). In: Second world conference on smart trends in systems, security and sustainability (WorldS4), London, pp 127–132
6. Kanani C, Sinhal A (2013) Ant colony optimization based modified AOMDV for multipath routing in MANET. Int J Comput Appl 82(10):14–19
7. Krishna PV (2012) Quality-of-service-enabledant colony-based multipath routing for mobile ad hoc networks. IET Commun 6(1):76–83
8. Bheemalingaiah M, Naidu MM, Sreenivasa Rao D, Vishvapathi P (2016) Energy aware on-demand multipath routing protocol in mobile ad hoc, networks. Int J Comput Netw Wirel Commun 6(5)

Effect of Chemoradiation on Haematological Parameters in Cervical Cancer

B. Sai Lalitha, M. Malini, and Mahendran Botlagunta

Abstract Cancer is the second leading cause of death and cervical cancer is one of the most common cancer types and ranks fourth largest for both incidence and mortality in females. Chemoradiation is a standard therapeutic model used to treat cervical cancer patients and often has shown to effect the haematopoietic system. Therefore, in this paper, we present the effect of chemoradiation on haematological parameters such as White Blood Cell (WBC), Platelets and Haemoglobin (Hgb). Eighteen subjects were enrolled from Basavatarakam Indo-American Cancer Hospital, Hyderabad, diagnosed with squamous cell carcinoma of the cervix. The study group received a combination of external beam radiation therapy (EBRT) (50 Gy in 25 fractions at 2 Gy per fraction), brachytherapy (21 Gy in 3 fractions at 30 gy, 40 gy and 50gy of EBRT), weekly cisplatin (40 mg/m^2) chemotherapy for 5 days a week for about 5 weeks. For estimating the changes in haematological parameters the clinical blood samples were collected from the patients before, and after the chemoradiation treatment regime. We found 17 patients in WBC, 17 patients in platelet, and 16 patients in Hgb in reference ranges. The reference range for WBC is 4–10 $\times$ 10^9 g/L, platelet is 150–410 $\times$ 10^9 g/L and Hbg is 120–150 g/L. Following therapy, patients were counselled to monitor the clinical complications. We found that the patients' low haemoglobin suffered stomach pain, heavy white vaginal discharge problems post-treatment, and patients under high haemoglobin group reported to be healthy with no post-treatment complications.

Keywords Cervical cancer · Chemoradiation · Haematological parameters

B. Sai Lalitha · M. Malini (✉)
Department of Biomedical Engineering, University College of Engineering, Osmania University, Hyderabad, India
e-mail: mudigonda.malini@gmail.com

M. Botlagunta
Department of Indo-American Cancer Research Foundation, Basavatarakam Indo-American Cancer Hospital and Research Centre, Hyderabad, Telangana, India

S. Jyothi et al. (eds.), *Proceedings of the 2nd International Conference on Computational and Bio Engineering*, Lecture Notes in Networks and Systems 215,
https://doi.org/10.1007/978-981-16-1941-0_59

1 Introduction

Cancer is the second leading cause of death and the most critical barrier to increasing life expectancy in every country of the world in the twenty-first century. Cervical cancer is one of the most common cancer types and ranks fourth largest for both incidence and mortality in females [1]. Cervical cancer occurs in the cervix cells, the lower part of the uterus that connects to the vagina, and is caused by Human papillomavirus (HPV). There are over 100 types of the HPV virus. Among them, HPV16 and HPV18 are high-risk oncogenic types [2]. Chemoradiation is one of the promising treatments in the present times and is used in about 40–50% of all cervical cancer cases [3].

The study of blood's physiology, anatomy, and blood aetiology is known as Haematology. Adult human contains average blood greater than 5 L. Blood is a precious fluid and has many functions. Blood delivers oxygen and nutrients to living cells, disposes of the waste products, and provides immune response cells. The main components of blood are red blood cells, haemoglobin (Hgb), white blood cells (WBC) and platelet. WBC, Platelets and haemoglobin are some of the fast-dividing cells produced by the bone marrow [4]. WBC fights infections from foreign particles or chemicals. Reduced WBC levels are shown to impair the immune system, and make cancer patients more vulnerable to many infections. For patients with very low WBC count, even mild infection may become life-threatening [5]. Platelets help in blood clotting during a cut or bruise by forming plugs in the blood vessels to stop the bleeding. The low platelet count is known as thrombocytopenia. If the count is very low, bleeding will become a significant problem. Therefore, platelet count needs to be monitored periodically during the chemoradiation for cancer patients.

Haemoglobin is a molecule present in red blood cells that carry oxygen to different organs/tissues in the body. RBC degradation due to chemoradiation in cancer patients contributes to decreased haemoglobin and iron concentration. Low haemoglobin in circulating blood or anaemia results in hypoxia. It is a common adverse effect in more than 60 percent of chemoradiation patients [6, 7]. Depending on the levels of haemoglobin, anaemia can vary from mild to life-threatening. If the patient's haemoglobin count is very low, there might be a need for blood transfusions, which can be crucial [8]. Chemoradiation is often shown to damage the bone marrow by reducing immune cells' production, slowing the mechanism and growth and division [3, 4]. As WBC, platelet, Hbg are some of the fast-growing cells produced by the bone marrow; they are effected [9, 10]. If the count is very low, the therapy dose should be reduced or stopped depending on the patient condition. This leads to a delayed procedure. Monitoring haematological parameters during chemoradiation can be crucial in assessing the patient's response to the treatment. Therefore, in this paper, we would like to study the effect of chemoradiation on haematological parameters such as white blood cells (WBC), platelet and Hgb.

2 Methodology

2.1 Subjects

Eighteen subjects under the mean age of 58 were enrolled from Basavatarakam Indo-American Cancer Hospital, Hyderabad. Subjects who are diagnosed with stage II and III squamous cell carcinoma of the cervix based on the American Joint Committee on Cancer (AJCC) were included in the study after signed informed consent.

Inclusion criteria: (i) patients' with histopathologically diagnosed cervical carcinoma, (ii) No prior treatment and (iii) tumour size was medium.

Exclusion criteria: (i) Not willing to sign the informed consent form, (ii) other than squamous cell carcinoma of the cervix and (iii) Patients' with HIV infection.

2.2 Design

For estimating the changes in haematological parameters the clinical blood samples were collected from the patients undergoing chemoradiation. The research group received a combination of cisplatin and radiation. External beam radiation (EBRT) was given for 5 days a week for about 5 weeks (50 Gy in 25 fractions at 2 Gy per fraction), High Dose Rate brachytheraphy (HDR) (21 Gy in 3 fractions at 30, 40 and 50 Gy of EBRT) and injection cisplatin at 40 mg/m^2 once per week. The clinical blood samples were collected before, during, and after the treatment. The first blood sample was collected before the treatment, and the second sample was collected at the 13th fraction and the third sample after the treatment (i.e., 25th fractions) Fig. 1. Immediately after collecting the samples, they were analysed for WBC, platelet and haemoglobin count using Sysmex 2000i automated blood cell counter machine.

3 Results and Discussion

To study the effect of chemoradiation we compared the change in the count of haematological parameters (such as haemoglobin, WBC, platelet count) between control (before radiation), reference haematological parameters values and experiment arm as shown in Table 1. Haematological counts may be an important prognostic factor and aim to minimise immune disturbances. We observed chemoradiation outcomes in 18 cervical cancer patients through complete blood count (CBC) and assessed the clinical significance of haematological parameters in chemoradiation. The before a sample was considered as the baseline value for two samples (during and after). The observation from the results is tabulated in Table 1.

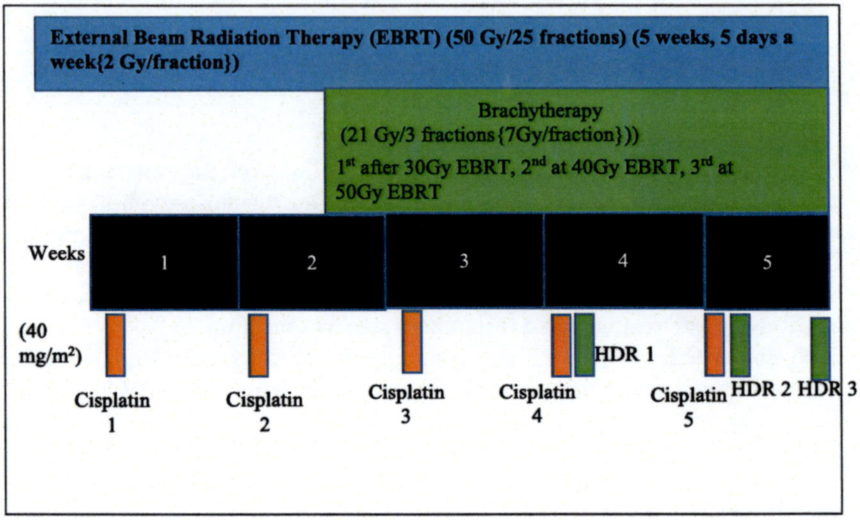

Fig. 1 Treatment cycle

(a) **White Blood cells:**

As seen from the above table (Table 1), when compared with WBC's reference value (4–10 × 10^9 g/L), We observed reduced WBC count during and after than before therapy. From Fig. 2, before therapy, all the patients had a higher value than during and after therapy. The decrease of the values was recorded in the 13th fraction of treatment and decreased even more in the 25th fraction (after). Initially, it was observed that 12 patients had WBC count in the normal range, and six patients had a higher value than the normal range. Eleven of 12 patients in the normal range and 6 of 6 patients in the higher than the normal range showed a decrease in the count during chemoradiation. After chemoradiation, all the patients showed decreased values than before and during the treatment/chemoradiation.

Seventeen patients were in the normal range but, 1 (patient no.12) patient had below standard range value.

The reduced value of WBC was because of the effect of therapy on fast-growing white blood cells. After 2 months of follow-up, we observed an increase in 17 patients and a decrease in one patient.

(b) **Platelets:**

The results (Table 1) were compared with platelets reference value (150–410 g/L). Platelet count decreased with time through the treatment course. We found a very drastic fall in all the patients during and after therapy compared with the baseline. From (Fig. 3), we observed induced platelet count for during and after than before therapy. The decrease in the level was recorded in the 13th fraction of treatment and decreased even more in the 25th fraction (after). Thirteen patients were under the

Table 1 Haematological parameters values before, during and after chemoradiation

Patient no	WBC				Platelets				Haemoglobin			
	Baseline	During	After	Follow-up	Baseline	During	After	Follow-up	Baseline	During	After	Follow-up
1	10.08	5.01	3.36	5.22	362	229	271	319	121	118	98	120
2	10.89	5.79	2.49	2.49	267	205	110	110	132	130	116	116
3	13.83	13.76	4.63	7.84	467	241	234	374	116	117	105	102
4	10.08	6.52	2.26	7.05	325	215	153	268	135	121	105	121
5	8.65	6.32	3.56	5.04	273	216	216	360	124	132	115	120
6	13.89	6.7	2.85	4.11	479	216	239	302	97	114	108	115
7	11.14	6.13	7.12	7.12	325	240	302	302	129	124	142	142
8	11.37	8.58	6.38	7.43	269	233	178	209	93	91	100	102
9	8.59	8.55	3.96	6.01	352	195	184	244	120	130	102	106
10	8.9	9.12	2.88	7.94	357	251	184	448	110	117	96	90
11	13.92	12.77	12.15	8.67	335	276	354	366	119	119	110	105
12	28.94	9.92	5.33	4.19	626	384	190	169	93	100	72	112
13	9.98	7.02	5.07	9.32	245	150	150	271	137	121	107	125
14	8.68	6.47	3.25	7.32	450	352	276	295	110	109	97	65
15	7.38	3.81	3.07	7.81	492	294	223	423	102	109	104	125
16	10.1	5.3	3.5	6.5	400	325	201	320	97	85	79	80
17	7.28	6.1	2.5	7.28	287	200	120	287	125	100	85	125
18	5.24	3.8	2.88	3.38	261	190	154	179	133	133	134	140

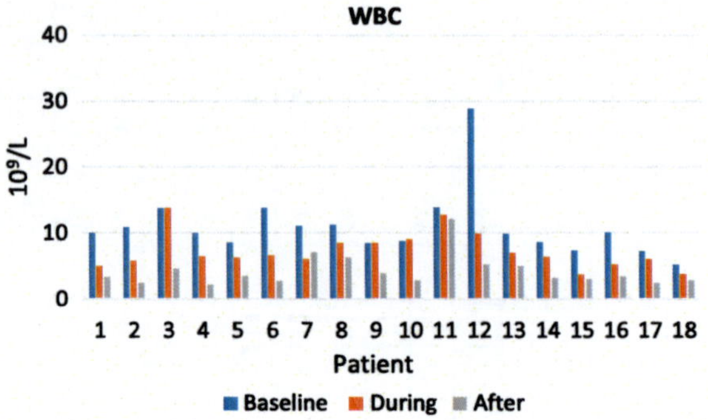

Fig. 2 The patient data of WBC count before, during and after the treatment

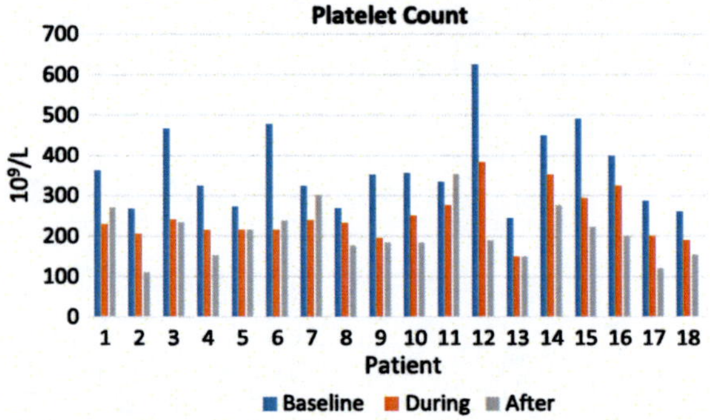

Fig. 3 The patient data of Platelet count before, during and after the treatment

reference range, and five patients had higher values than the reference range before radiation. The count was decreased in all the patients. Seventeen of 18 patients were under the normal range, and one patient showed below the normal range (Table 2). After two months of follow-up post-treatment, we observed that the bone marrow's platelet production was recovered as the value was increased. All the patients showed an increase in platelet production, but one patient (patient no. 12) was found with decreased platelet.

(c) **Haemoglobin:**

The results (Table 1) were compared with haemoglobin's reference value (120–150 g/L). From Fig. 3, before therapy 9 patients had haemogolobin in the reference range.

Table 2 Statistical values for haematological parameters

Parameter	P-Value	Mean of control	Mean of experiment	Standard deviation
WBC count	0.00001	116	104	16
Platelet count	0.000003	365	208	2
Haemoglobin	0.02	11	4	64

At the 13th fraction of treatment, 7 of 9 patients showed a decrease in values, and 2 of 9 patients showed an increase in values. At the 25th fraction of treatment, 7 of 9 patients showed a decrease in values, and 2 of 9 patients showed an increase in values. Before therapy 9 patients had below the reference range. Among the patients below the reference range, five patients had 100–120 g/L, and four patients had below 100 g/L. At 13th fraction, in patients ranging from 100 to 120 g/L 4 of 5 patients showed a decrease in values while one patient showed an increase in the value.

In patients below 100 g/L, 3 of 4 patients showed decrease, and 1 of 4 patients showed an increase in values. At 25th fraction patients ranging from 100 to 120 g/L, 4 of 5 patients showed a decrease in values, and 1of 5 patients showed an increase in value. In patients below 100 g/L 3 of 4 patients showed a decrease and 1 of 4 patients showed an increase in values.

In the statistical analysis, the p-value selected was <0.05, the obtained p-value for WBC was <0.00001, platelet was <0.000003, haemoglobin was 0.02 (Table 2). The obtained statistical analysis results showed the study was significant.

Our study compared the results of WBC, Platelet, and Hgb with the reference considering before radiation value as the baseline value. WBC, Platelet levels were decreased from the baseline value at 13th fraction and reaching the lowest levels at 25th fraction in all the subjects. Observing the results, the study group was divided into two groups low haemoglobin (LH) and high haemoglobin (HH). LH group indicates high haemoglobin early therapy and reduced after treatment. HH group has low haemoglobin early treatment and increased after treatment. Thirteen of 18 patients were included in the LH group, and 5 of 18 patients were included in the HH in this group. Seventeen of 18 patients suffered less haemoglobin count post-treatment (Fig. 4, Table 2). To correlate our findings with post-treatment complications, we conducted a telephonic survey. In the survey we found that patients under the LH group suffered stomach pain, heavy white vaginal discharge problems post-treatment and patients under HH group reported to be healthy with no post-treatment complications.

Chemoradiotherapy affects the production of total WBC, Platelet and Hbg count by suppressing the function of bone marrow. Our results are in agreement with

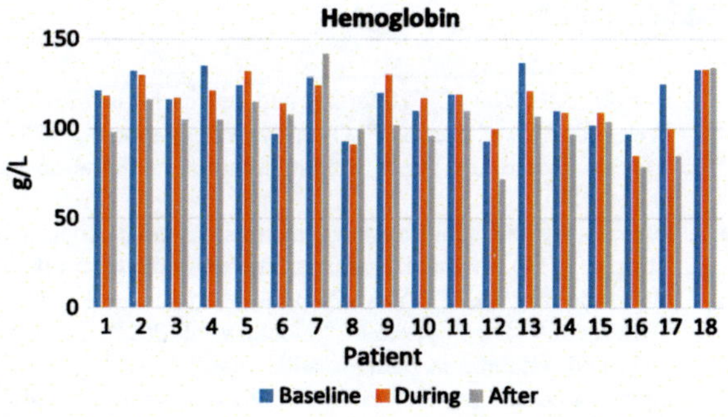

Fig. 4 The patient data of haemoglobin count before, during and after the Treatment

previously published clinical reports, which underlies the importance of haematological parameters. Haematological parameters have shown to get affected by either radiation, chemotherapy alone and or in combination with both (chemoradiation). For example Vitzthum et al., reported that a decrease in Neutrophil, lymphocyte, and platelet count in 90 cervical cancer patients.[11] On the other hand, chemotherapy has been shown to induce haematologic toxicity by decreasing the haemoglobin content, leukogram and plateletgram in 39 cervical cancer patients [12]. American cancer society showed that when chemotherapy is given with radiation, the blood counts tend to be lower when compared with radiotherapy alone. These side effects typically improve in the weeks after treatment is stopped [13]. Barbara Bachtiary et al., demonstrated that radiotherapy causes substantial immune suppression. In chemoradiation, an increased haematological toxicity was found in 34 cervical cancer patients [14].

The major limitation of this study is the small sample size. Also, it is a single-centre study, and thus, the generalisation of our findings is limited. Therefore, further large, prospective, and multi-centre clinical studies are warranted to determine the significance of response of WBC, Platelet, Hgb.

4 Conclusion

Chemoradiation is one of the common methods used for cervical cancer treatment. In our paper, we studied the effect of chemoradiation on haematological parameters in cervical cancer patients. Taken together from our results, we conclude that among 18 patients, 13 patients (72%) showed decreased levels of haemoglobin, all patients (100%) showed decreased levels in total WBC and platelet. To correlate haematological parameters with lifestyle index with the post chemoradiation complications, we conducted a telephonic survey. Our study offers unique data, where the post-treatment clinical complications correlated with haematological counts during chemoradiation treatment regime.

References

1. Murphy ME, The HSP70 family and cancer. https://doi.org/10.1093/carcin/bgt111
2. Ikwegbue PC, Masamba P, Oyinloye BE, Kappo AP (2018) Roles of heat shock proteins in apoptosis, oxidative stress, human inflammatory diseases, and cancer. https://doi.org/10.3390/ph11010002
3. https://www.texaschildrens.org/health/stem-cells-blood-and-immune-system
4. Etim N (2014) Haematological parameters and factors affecting their values. Agric Sci 2:37–47. https://doi.org/10.12735/as.v2i1p37
5. Gennigens C, De Cuypere M, Seidel L et al (2020) Correlation between hematological parameters and outcome in patients with locally advanced cervical cancer treated by concomitant chemoradiotherapy. Cancer Med 9:8432–8443. https://doi.org/10.1002/cam4.3465
6. F Bellati, M. Pernice, N. Manci et al Hemoglobin variation and blood transfusion rates in patients affected by locally advanced cervical cancer undergoing neo-adjuvant chemotherapy

followed by radical surgery: the role of erythropoietic growth factors. https://doi.org/10.1093/annonc/mdl491
7. Barbalato L, Pillarisetty LS (2020) Histology, "red blood cell", treasure Island. Stat Pearls Publishing, FL
8. Msdmanuals, blood disorders biology of blood formation of blood cells, https://www.msdmanuals.com/en-in/home/blood-disorders/biology-of-blood/formation-of-blood-cells#:~:text=Red%20blood%20cells%2C%20most%20white,mature%20in%20the%20thymus%20gland
9. van Meir H, Nout RA, Welters MJ, et al (2016) Impact of (chemo)radiotherapy on immune cell composition and function in cervical cancer patients. Oncoimmunology 6(2):e1267095. Published 2016 Dec 23. https://doi.org/10.1080/2162402X.2016.1267095
10. Chemocare.com. 2015. Cancer cells and chemotherapy: http://chemocare.com/chemotherapy/what-is-chemotherapy/cancer-cells-chemotherapy.aspx#:~:text=If%20the%20cancer%20cells%20are,self-death%20or%20apoptosis
11. Vitzthum Lucas K et al (2020) Comparison of hematologic toxicity and bone marrow compensatory response in head and neck vs. cervical cancer patients undergoing chemoradiotherap. Front Oncol. https://doi.org/10.3389/fonc.2020.01179
12. Araújo, David Felipe B et al (2020) Hematological and biochemical toxicity analysis of chemotherapy in women diagnosed with cervical cancer. J Bras Patol e Med Lab, 56, e1772020. Epub July 01, 2020.https://doi.org/10.5935/1676-2444.20200038
13. American Cancer Society: https://www.cancer.org/cancer/cervical-cancer/treating/radiation.html
14. Bachtiary B, Herbacek I, Zideck T, Knocke TH, Dimopoulos J, Poetter R, Micksche M (2005) Impact of radiotherapy with and without concurrent cisplatin on lymphocyte subpopulations in cervical cancer patients. Anticancer Res. Nov-Dec;25(6C):4673–4678. PMID: 16334159

A Novel Approach for Heart Disease Prediction Based on Risk Factors Using Machine Learning

K. Santhi and A. Rama Mohan Reddy

Abstract In the world, one of the most critical diseases that affect human life is heart disease. This disease occurs due to an insufficient amount of blood circulated to other parts of the body. The disease occurs due to many risk factors, such as age, diabetes, blood sugar, gender, cholesterol, and coronary heart disease, etc. Generally, this disease is diagnosed using medical history; in many aspects, this is not a reliable solution. So prevention of this disease using an accurate and efficient algorithm is needed. This paper developed a novel approach to prevent heart disease by considering two risk factors: coronary heart disease and diabetes. The performance of the proposed method has been validated on different machine learning algorithms. The proposed method can also assist the doctors in diagnosing patients suffering from heart attacks efficiently.

Keywords SVM · Naive bayesian · Random forest · Heart disease · Sampling

1 Introduction

In today's era, cardiovascular heart disease (CVD) is increasing rapidly, which is considered as one of the dangerous human disease, and it is a part of a heart attack. In the human body, one of the most important organs is the heart. According to the world health organisation every year, more than 10 a million people die due to CVD, particularly heart attacks and strokes in the United States. In this disease, insufficient blood and oxygen are circulated by the heart to other parts of the body, leading to heart failure. So if the blood is not circulated properly, the function of the heart reduces, which leads to death. Shortness of breath, weakness of the physical body, swollen feet, diabetes, tiredness, chest pain, and hypertension are symptoms of heart

K. Santhi (✉) · A. Rama Mohan Reddy
CSE Department, SV University College of Engineering, Tirupati, India
e-mail: santhi.kuraganti@gmail.com

A. Rama Mohan Reddy
e-mail: ramamohansvu@gmail.com

S. Jyothi et al. (eds.), *Proceedings of the 2nd International Conference on Computational and Bio Engineering*, Lecture Notes in Networks and Systems 215,
https://doi.org/10.1007/978-981-16-1941-0_60

disease. In United States countries, people are still unaware of risks and symptoms of heart disease, even advancement in technology and in health departments.

Medical problems like cancer, blood pressure, diabetes have been detected early, which leads to successful treatment. Biomedical diseases such as diabetes and heart have common symptoms in various bio medical databases like PUBMED [1]. Among all health-related problems, CVD is a very risky problem; therefore, prediction of such disease plays a major role today. In order to reduce the number of deaths from heart disease, there is a number of machine learning techniques used to identify the issues related to a heart attack. The development of plaque substance in coronary articles will block the vessel completely in the body, leading to a heart attack in the human. Every year more than two lakh open surgeries are done in India. Manually it is difficult to determine the reasons for getting heart disease. Many researchers use data mining techniques like machine learning techniques to predict heart disease from risk factors [2]. In this paper, a novel approach is developed and compared with machine learning techniques for identifying and predicting the risks related to heart disease and classifying the occurrence of disease. This method improves the accuracy of identifying the disease than previous approaches.

The rest of the paper is placed as follows. Section 2 describes a literature survey related to heart disease. Section 3 discusses the proposed work. Section 4 includes experimental analysis, which is discussed in detail. Section 5 is related to the conclusion of the paper.

2 Related Work

Health-related problems are solved by different researchers using machine learning and data mining techniques. They also used different approaches to predict and categorise CVD disease [3]. Considers the patient medical record and symptoms related to disease to analyse and predict heart disease. Different researcher's use the data set which is available in online, UCI machine repository for identifying different classification issues related to heart related disease [4].

Most of the data in the data sets are not normalised [5], which uses multilayer perception to find the effect of class imbalance in the data set [6]. Considers two data sets FHS and MHANES, for investigating issues related to heart disease. By applying KNN and Random forest, these data sets showed the tradeoff between the accuracy and execution time [7]. Developed a new approach that is used to monitor and detect coronary disease by using the UCI repository, which consists of 330 instances with 75 attributes. This data set is tested with three different machine learning algorithms: functional trees, SVM, and Naive Bayesian algorithms using the WEKA tool.

In [8], by applying these three algorithms using the WEKA tool on heart disease in diabetes patients. He considers data related to Chennai research institute with 500 patient's data. After the application of machine learning algorithms, 358 patients do not have heart disease and 142 patients have the disease. A new method to diagnose heart disease which consists of three phases and is based on an ANN machine learning algorithm [9]. Also uses the ANN approach to predict heart disease with statistical enterprise miner with improved accuracy.

In this paper, a novel approach is developed to predict heart disease on class imbalance data. The proposed approach is based on the attribute selection method with machine learning for diagnosing heart disease. This system considers the heart disease data set from the UCI repository. The contributions of this paper are following.

(i) Compares the performance of all classifiers on imbalanced heart disease data set on risk factors, such as CHD and diabetes.
(ii) Evaluate the performance of classifiers on balanced heart disease with risk factors CHD and diabetes data set.

3 Proposed Work

This section includes attribute selection methods and the application of various machine learning approaches used to predict heart disease. Figure 1 shows the pictorial representation of the proposed work. Firstly, describe the data set used in the proposed approach to classifying the disease followed by preprocessing method and then machine learning models used in experimentation.

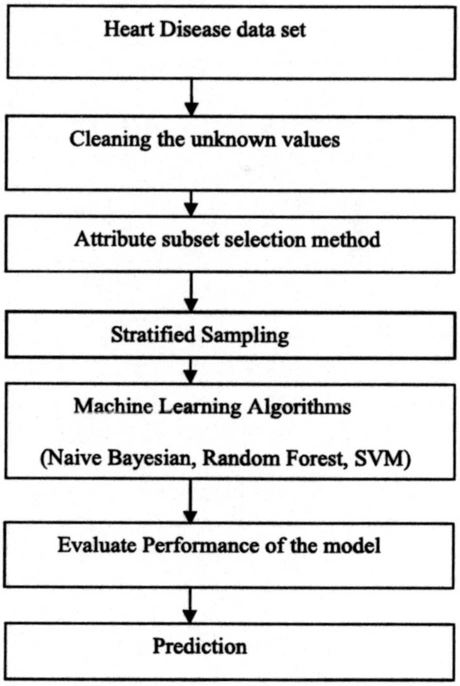

Fig. 1 Proposed methodology

3.1 Data Set Description

The heart disease data set is collected from the UCI machine learning repository. This repository has datasets related to different domains. We have used the CVD data set, which is extracted from the UCI machine learning repository. Generally, this data set is provided by different research institutions such as University hospital, Medical centre V.A., Cleveland clinical foundation and Hungarian institute of cardiology. Table 1 shows the attributes of the data set used for experimentation, interpretation and description.

The data set consists of many incomplete and inconsistent values. Preprocessing of this data set leads to in accurate results. So preprocessing is needed to get better results using machine learning algorithms and to take better decisions on the data. In the proposed work, we considered two preprocessing methods on the heart disease data set. By using the attribute subset selection method, we are eliminating irrelevant

Table 1 Attributes of data set

Attribute	Description	Interpretation
Cp	Chest pain type	1: Typical angina 2: Atypical angina 3: Non-angina pain 4: Asymptomatic
Chol	Cholesterol inside patients body in mg/dl	[126,564]
Thalach	Maximum heart rate achieved	[71,202]
Age	Age in years	[29,77]
Gender	Gender (male/female)	Male: 1 Female: 0
Smoker	Non-smoker/Smoker	Non-smoker: 0 Smoker: 1
Trestbps	Resting blood pressure in mm Hg	[94,200]
Class	Diagnosis of heart disease	1: Un healthy 2: Healthy
Fbs	Fasting blood sugar > 120 mg/dl	0: False 1: True
Restecg	Resting ECG results	0: Normal 1: ST-T wave abnormality 2: LV hypertrophy
Diabetes	Yes/No	0: No 1: Yes
Ca	Major vessels coloured by fluoroscopy	Range (0–3)
CHD	Risk of developing CHD	1: Yes 0: No

features and by using attribute mean filling the missing values. In many data mining applications, most of the data sets are suffering from class imbalance problem. To process the imbalanced classes, many machine learning algorithms failed. To overcome this, we used stratified sampling to balance it. This entire data set is divided into two classes or stratum. For the given data set, stratified sampling performs the following.

1. Identify the target class.
2. Label the categories for each class.
3. Assign a random number to each case.
4. Sort each case by number.
5. Take the decision rule to select the case.

This paper considered heart disease data set on two risk factors: coronary heart disease and diabetes. In this data set, 75% of individuals are not suffering from heart disease on risk factor whereas the remaining 25% of the patients are suffering from a disease.

3.2 Models Used in the Proposed Approach

3.2.1 Naive Bayesian

Naive Bayesian classifier is also called as probability-based classifier, and it is functioned based on the Bayesian theorem. To classify high-dimensional data sets, the Naive Bayesian algorithm is most widely used. In this algorithm, all features are mutually exclusive. Changes done in one feature is independent of another feature or attribute.

Let T be the training data which consists of set of tuples associated with class labels.

Each tuple consists of n attributes named as $A_1, A_2, A_3, \ldots, A_n$. Each tuple is labelled with a class. Assume there are m classes like cl, c2, ... , *cm*. For a given tuple *X*, the Bayesian classifier predicts the class as class C_i if $P(C_i|X) > P(C_j|X)$ for $1 \leq j \leq m, j \neq i$.

3.2.2 Support Vector Machine

A support vector machine is a machine learning classification algorithm. This algorithm is mainly used for solving a quadratic programming problem. Various applications use SVM because of high performance in classification. It solves the binary classification problem by using two coefficient vectors w and d for the hyperplane, $W^{T}x + b = 0$. To solve this type of problem, SVM uses lagrangian multipliers. The coefficient vectors w and d are normal variants to the hyperplane surface *b*.

3.2.3 Random Forest

Random forest algorithm is a supervised machine algorithm. It generates the decision trees randomly in the form of a forest. While generating decision tress it uses a bagging approach to compare and improve the performance of various machine learning models. To build an efficient model, it splits the node for finding the best feature and to refine the result. It is mainly used for detecting outliers. Based on impurity added to the features in the model, it gives ranking to the features.

4 Experimental Results

For the experimentation, we have collected the data set related to heart stroke, which is available in the UCI machine repository. This data set consists of 35,005 instances. These instances are related to different risk factors like diabetes, Hypertension, Nutrition, obesity, smoke, and stroke. The proposed model is implemented on this dataset for two risk factors: Diabetes, coronary heart disease. The total number of instances based on diabetes and coronary heart disease are 2704 and 1352, respectively.

The following measures are used to evaluate the performance of the proposed method. In this paper, to classify the occurrence of the number of members suffering heart disease, an efficient measure called confusion matrix is used. Generally, there are four terms associated with this matrix. These four terms are named True Positive, True Negative, False Positive, False Negative, respectively.

True Positive (TP): If the model correctly predicts the positive class.

True Negative (TN): If the model correctly predicts the negative class.

False Positive (FP): If the model incorrectly predicts the positive class

False Negative (FN): If the model incorrectly predicts the negative class.

Precision (*P*): It is the ratio of the number of people predicted as suffering from heart risk to the total number of people who had heart disease.

$$P = \frac{\mathrm{TP}}{\mathrm{TP} + \mathrm{FP}} \tag{1}$$

Recall (*R*): It is the ratio of the number of people suffering as heart disease predicted by the model to the individuals that are suffering from disease.

$$R = \frac{\mathrm{TP}}{\mathrm{TP} + \mathrm{FN}} \tag{2}$$

F_1 Score: It is the ratio of the harmonic mean of precision and recall.

$$F_1 \text{ Score} = 2 \times \frac{P \times R}{P + R} \tag{3}$$

Accuracy (A): It is the ratio of the total number of instances and the number of instances is correctly predicted by the model.

$$A = \frac{\mathrm{TP} + \mathrm{TN}}{T} \tag{4}$$

where $T = \mathrm{TP} + \mathrm{FP} + \mathrm{TN} + \mathrm{FN}$

In the heart disease data set on risk factor CHD, there are 26 unknown class instances and 1326 labelled instances and on risk factor diabetes 52 unknown class instances and 2652 labelled instances are used for experimentation. Tables 2 and 3 show the performance of classifiers for heart disease with risk factor coronary heart disease before balancing and after balancing the data.

Figures 2 and 3 show the performance of classifiers on heart disease data set and execution time to process the data set on risk factor coronary heart disease before balancing the data, and Figs. 4 and 5 show the performance of classifiers on heart disease data set and execution time to process the data set on risk factor coronary heart disease after balancing the data.

Tables 4 and 5 show the performance of classifiers for heart disease with risk factor diabetes heart disease before balancing and after balancing the data.

The Figs. 6 and 7 shows the performance of classifiers on heart disease data set and execution time to process the data set on risk factor diabetes disease before balancing the data and Figs. 8 and 9 shows the performance of classifiers on heart disease data set and execution time to process the data set on risk factor diabetes

Table 2 Performance of classifiers for heart disease with risk factor coronary heart disease before balancing data

Criteria	Naive bayesian		Random forest		SVM	
Folds	5	10	5	10	5	10
Time to build a model	0.02	0.04	0.34	0.77	0.024	0.35
Correctly classified Instances	1308	1320	1318	1323	1320	1324
Incorrectly classified Instances	18	6	8	3	6	2
Accuracy	93.5	95.5	94.8	96.2	95.5	96.8

Table 3 Performance of classifiers for heart disease with risk factor coronary heart disease after balancing data

Criteria	Naive bayesian		Random forest		SVM	
Folds	5	10	5	10	5	10
Time to build a model	0.01	0.02	0.15	0.52	0.02	0.2
Correctly classified Instances	1318	1324	1320	1325	1324	1325
Incorrectly classified Instances	8	2	6	1	2	1
Accuracy	94.5	98.5	95.8	98.8	98.5	99.2

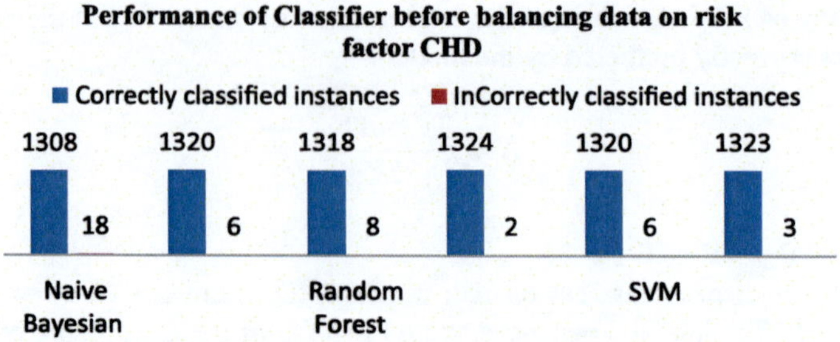

Fig. 2 Performance of classifier before balancing data on CHD

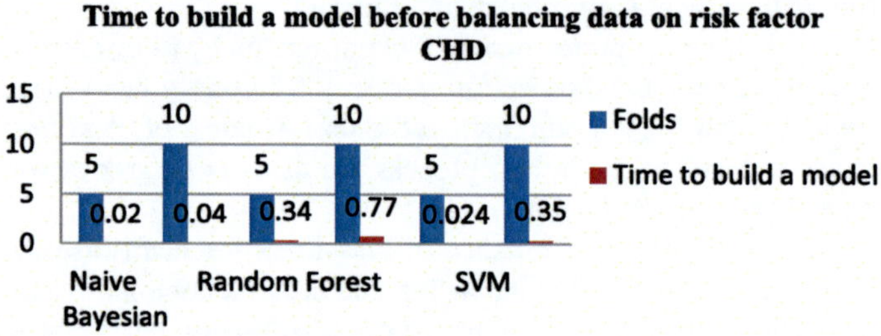

Fig. 3 Time to build a model before balancing data on CHD

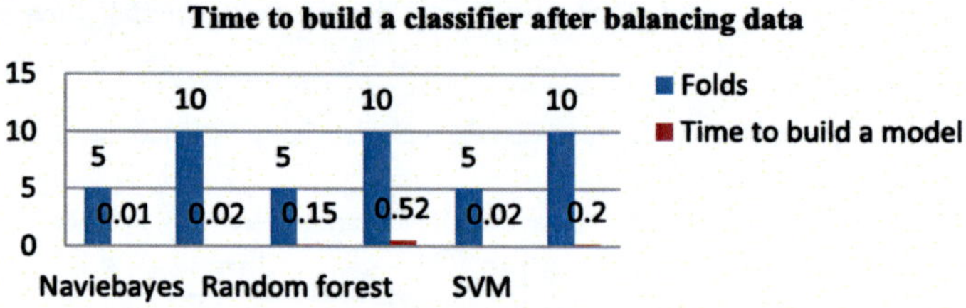

Fig. 4 Time to build a model after balancing data on CHD

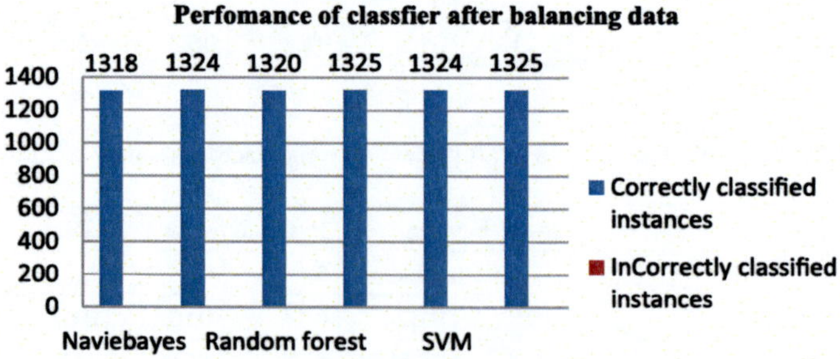

Fig. 5 Performance of classifier of balanced data on CHD

Table 4 Performance of classifiers for heart disease with risk factor diabetes before balancing data

Criteria	Naive bayesian		Random forest		SVM	
Folds	5	10	5	10	5	10
Time to build a model	0.02	0.01	0.32	0.65	0.024	0.35
Correctly classified Instances	2450	2558	2500	2552	2600	2640
InCorrectly classified Instances	202	106	152	100	52	12
Accuracy	90.8	94.2	92.4	95.5	94.3	96.8

Ignored Class Unknown Instances 26
Total Number of Instances 1326

Table 5 Performance of classifiers for heart disease with risk factor diabetes after balancing data

Criteria	Naive bayesian		Random forest		SVM	
Folds	5	10	5	10	5	10
Time to build a model	0.01	0.01	0.32	0.65	0.024	0.35
Correctly classified Instances	2552	2602	2600	2552	2626	2650
Incorrectly classified Instances	100	50	52	100	26	2
Accuracy	91.5	96.3	92.8	97.8	95.5	98.7

Ignored Class Unknown Instances 52
Total Number of Instances 2652

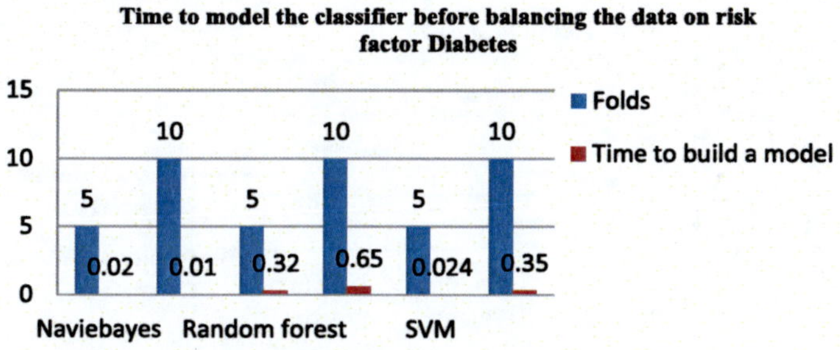

Fig. 6 Time to build a model before balancing data on diabetes

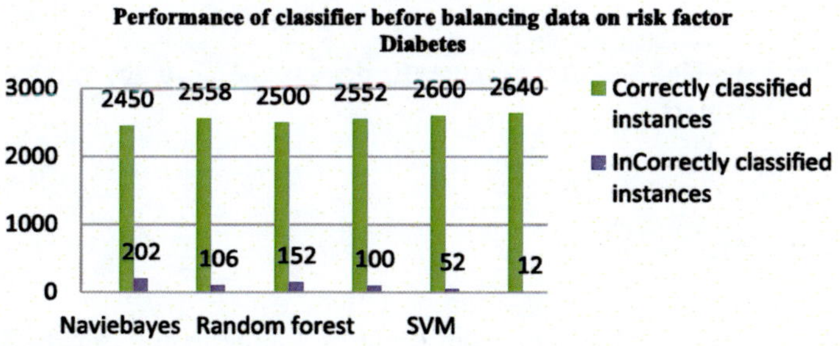

Fig. 7 Performance of classifier before balancing data on diabetes

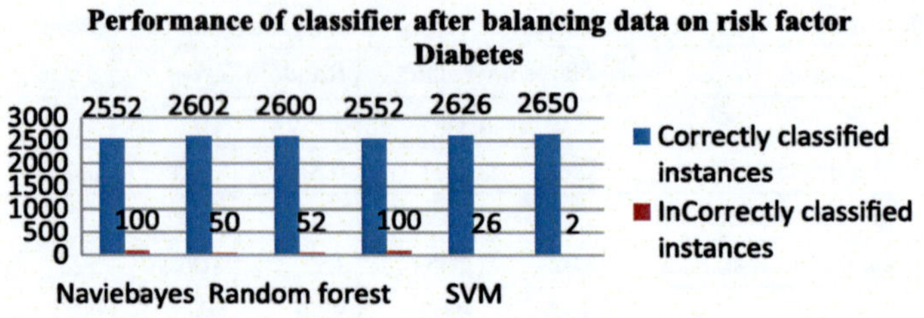

Fig. 8 Performance of classifier after balancing data on diabetes

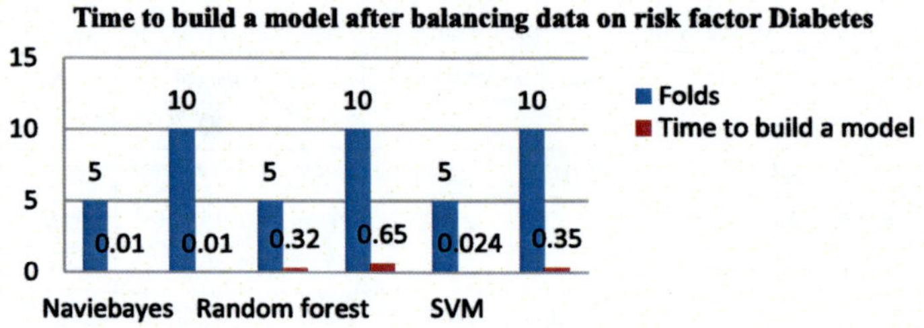

Fig. 9 Time to build a model after balancing data on diabetes

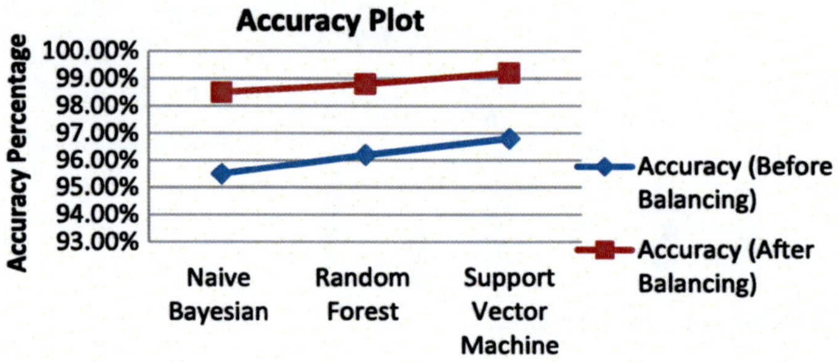

Fig. 10 Accuracy plot of classifiers

disease after balancing the data. Figure 10 depicts the accuracy plot comparison between the classifiers.

5 Conclusion

Every day patients suffering from heart diseases are increasing rapidly. To resolve this, there is a need for an efficient approach to classify the disease. In this paper, we proposed a novel approach to predict the disease by using efficient preprocessing techniques followed by machine learning algorithms. We also checked the performance of different machine learning algorithms like Naive Bayesian, Random forest and support vector machine on heart disease data set with two risk factors coronary heart disease and diabetes. In this paper, we performed experimentation and on different machine learning algorithms with 5 and 10 cross-validation. The proposed approach reduces the execution time after balancing the data on heart disease data. The performance of Naive Bayesian, Random forest and SVM algorithms before balancing data on CHD are 95.5, 96.2, 96.8. The performance of Naive Bayesian, Random forest and SVM algorithms after balancing data are 98.5,98.8 and 99.2. The performance of Naive Bayesian, Random forest and SVM algorithms before balancing data on diabetes are 94.2, 95.5, 96.8. The performance of Naive Bayesian, Random Forest and SVM algorithms after balancing data on diabetes are 96.3, 97.8, 98.7.

References

1. Dhanalakshmi P, Ramani K, Eswara Reddy B (2017) An improved rank based disease prediction using web navigation patterns on bio-medical databases. Futur Comput Inform J 2:133–147
2. Khazaee A (2013) Heart beat classification using particle swarm optimization. In tell Syst Appl, 25–33
3. Vanisree K, Singaraju J (2011) Decision support system for congenital heart disease diagnosis based on signs and symptoms using neural networks. Int J Comput Appl 19(6):6–12
4. Nazir S, Shahzad S, Mahfooz S, Nazir M (2015) Fuzzy logic based decision support system for component security evaluation. Int Arab J Inf Technol 15:1–9
5. Yildirim P (2017) Chronic kidney disease prediction on imbalanced data by multilayer perceptron. In: IEEE 41st annual computer software and applications conference, pp 193–198
6. Rajliwall NS, Davey R, Chetty G (2018) Machine learning based models for cardiovascular risk prediction. In: IEEE International conference on machine learning and data engineering (ICMLDE)
7. Otoom AF, Abdallah EE, Kilani Y, Kefaye A, Ashour M (2015) Effective diagnosis and monitoring of heart disease. Int J Softw Eng Its Appl 9(1):143–156
8. Parthiban G, Srivatsa SK (2012) Applying machine learning methods in diagnosing heart disease for diabetic patients. Int J Appl Inf Syst 3(7):2249–0868
9. Olaniyi EO, Oyedotun OK (2015) Heart diseases diagnosis using neural networks arbitration. Int J Intell Syst Appl 7(12):75–82

Big Data-Based Smart Grid Data Management Model for Improved Monitoring of Power Consumption

Doosetty Padmaja Kalala and P. Bhargavi

Abstract Smart grid is treated as the future of the power grid, practices two-way flows of power and data to create an automated, extensively distributed power conveyance network. This composition explores how a big data environment can be utilised for data management in the smart grid and presented a model called a big data-based smart grid data management (BSGDM). We research the advantages and prospects of both smart grid and cloud computing environment. For security purpose, we designed a security management organisation and also defined applications of BSGDM.

Keywords The smart grid · Cloud computing · Security · Trading · Services

1 Introduction

In recent years, the consumption of electricity has changed in nature and in practice. A smart grid is an advancement to the electric grid in the 21st century. The smart grid is associated with computation, information and communication technologies which are associated with power generation, transmission and computation [1]. The development of the smart grid combines both electricity usage and information technology. The advancements in applications and services make the grid smarter. But, the large data produced by the smart grid impacts the tracking and managing of the environment. We require a cost-effective model for data storage, analysis, and management [6, 7]. Therefore, the introduction of information technology to the smart grid must handle large volumes of data. Cloud computing and big data will serve the purpose of handling large volumes of power grid data [10]. The cloud data centres are associated with huge infrastructure to provide computing services to the

D. P. Kalala (✉)
Department of Computer Science and Engineering, Sri Padmavati Mahila University, Tirupati, AP, India
e-mail: padmajakalala@gmail.com

P. Bhargavi
Department of Computer Science, Sri Padmavati Mahila University, Tirupati, AP, India

S. Jyothi et al. (eds.), *Proceedings of the 2nd International Conference on Computational and Bio Engineering*, Lecture Notes in Networks and Systems 215,
https://doi.org/10.1007/978-981-16-1941-0_61

smart grid environment. The big data-based smart grid data management model is required to handle the smart grid data (BSGDM).

This paper is organised as follows. Section 2 discusses the advantages and prospects of BSGDM. Section 3 describes the new model called as big data-based smart grid data management. Section 4 presents the relation between the cloud and the smart grid. Section 5 deals with the applications of BSGDM, and the conclusion are drawn in Sect. 6.

2 Advantages and Prospects

This section explains the advantages of smart grid data management in big data and the opportunities to integrate the big data and smart grid.

A. The Smart Grid Environment

Smart grid is treated as the intelligent network model based on trending technologies, equipment, and sensors to optimise energy resources and improve efficiency, reliability, and security. The major aim of the smart grid environment is to integrate the different energy generation platforms into the system and to monitor energy production and consumption.

BSGDM can produce possibilities for the success of the environment of the smart grid. The technology allowing individuals and small companies the ability to produce usable energy lowers entry barriers for new players and liberalises the marketplaces in the smart grid sector since the energy produced from the end-users can be traded. BSGDM further decreases the access limitations since it would be much more cost-effective for end-users on-demand and to use pay-as-you-go on cloud computing to recognise data control, in contrast to the case where they design and implementation their own data model. Furthermore, at the moment, cloud computing is one of the most popular places of research where new interesting solutions and applications keep growing. Various programmes designed and implemented by companies provide a range of value-added services to users. This pattern may also motivate new interesting data models in the smart grid sector, which further enhance the success of the environment of the smart grid [2].

B. Cloud Computing Environment

Smart grid is one of the trending market which utilises the cloud computing. The smart grid is treated as the alternative solution to the future power systems, which concentrates on higher profits. Smart grid is a combination of power and communication that allows the IT industry to be involved in the new market. Cloud computing is an environment that serves different areas like e-commerce, mobile industries and data management. Microsoft offers Hohm [3], an online application used by the customers to monitor and evaluate power utilisation and provide power-saving suggestions. Google offers PowerMeter [4] for monitoring the power utilisation and option to discuss with other customers to reduce the power consumption.

3 Big Data-Based Smart Grid Data Management

Figure 1 shows the functionality of the power grid environment, which consists of the generation, transmission and distribution phase. The power grid environment is operated with cloud computing and big data domain to operate smartly. The smart grid environment is composed of different participants like power generation, power transmission, power distribution. Local customers, markets and cloud service providers. Generally, electricity is produced at different power plants by different mechanical generators. These power generators are located in less populated areas and contain huge infrastructure. The generated electricity in the power plants is distributed through the power cables for huge distances to connect with the substations. As a next step, the substations convert the high voltage power into distribution voltage and transmitted to the customer location [5]. Figure 2 shows the

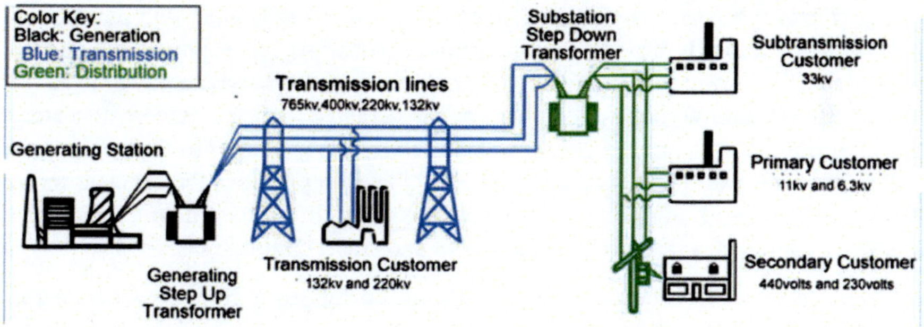

Fig. 1 Functionality of the power grid

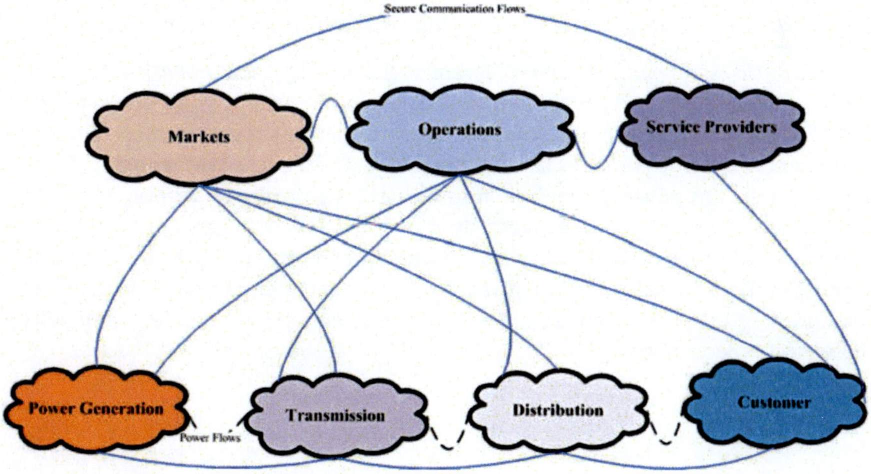

Fig. 2 Power grid conceptual model [5]

Table 1 Power grid participants

Domains	Participants
Power generator	The generation of power in huge amounts, may also store power for the future purposes
Transmission	The carriers of electricity through power cables for longer distances
Distribution	The distribution of power to and from the customers
Customers	The users of electricity, may also manage, store and generate electricity
Markets	The managers and participants in the electricity market
Operations	The organizers of movement of electricity
Service providers	The companies providing services to end users and applications

complete process of power generation and distribution process. Table 1 shows the explanation of smart grid participants.

Smart grid infrastructure: smart grid environment is associated with power, data and communication. It provides two-way communication, one is power and another one is data. The power grid operates in a unidirectional mode where power is generated, transmitted through the power lines and distributed to the customers. In a smart grid environment, the unused power is returned by the customer to the power grid. For instance, the customers who have installed the solar panels in their houses return the power to the grid. In this research work, the smart grid infrastructure is operated with three modules: smart power, smart data and smart communication.

- *Smart power*: This module is responsible for power generation, power transmission and power distribution and consumption.
- *Smart Data*: This module is responsible for metering, tracking, and managing data with respect to smart grid.
- *Smart Communication*: This module is responsible for making communication between the systems, devices and applications with respect to smart grid.

Smart grid management system: The smart grid management provides the maintenance of the controls and functionalities [5]. The development of advanced applications and services makes the use of technology to operate the grid smarter. The smart management module performs smarter operations to operate smarter goals [17]. The smarter goals are power efficiency improvement, demand and supply, operational cost reduction and exhaust management.

Smart grid protection: It delivers reliability, security and privacy and failure protection. With the help of smart infrastructure, the smart grid provides smart protection for the grid environment and provides smart management to effectively deliver the security mechanisms.

A. Enhanced algorithm for BSGDM

Algorithm 1: Data Management

Begin

1. Initialize the power grid
2. After activating the power grid, initialize the time T
3. Compute the power distribution based on the data available at the cloud environment
4. Calculate the cost function from Eq. 1

$$_p \quad M \times \frac{U}{CP} \qquad (1)$$

Where M represents the amount of power consumption
U represents the utilization of the power by the device
CP represents the device capacity

5. Forward the C_p details to the cloud
6. Cloud analyzes the details of the cost function
7. Compute the grid power utilization
    For each device in the grid compute power consumption do
        Track power utilization
            Retrieve the power utilization data (Instance)
        Store the power utilization data

End

In algorithm 1, the power utilisation of each device is monitored and analysed, and this information is forwarded to the cloud to compute the cost function for usage billing purpose. As a next step, compare the instance power utilisation data with the previously stored data in the cloud. The billing information is stored in the cloud for future purposes.

4 Security Management in BSGDM

Guaranteeing privacy, security and QoS would be important to the BSGDM. Privacy and security may be the major concern in outsourcing big data management to the cloud [8]. Moreover, ensuring the quality of service is vital for grid operations. For instance, the diagnosis of system failure is not worked properly, it leads to unexpected results. So here we are proposing a component called security management (SM), shown in Fig. 3. SM resides in the smart grid environment but acts as an intermediate between participants in both the smart grid and cloud computing environments.

The SM functionalities are predefined or defined on the go. The SM selects suitable strategies according to the security requirements defined by the smart grid participants. If there is no strategy it meets the requirements identified, the SM then reminds the participants and lets them decide to check the security levels. Here we are focused on three strategies. It is first choosing the appropriate cloud computing service provider. For instance, for storing and processing the regulated data, the SM selects the private cloud to fulfil the needs of the data management system [8]. In this method, the information can be stored, maintained and updated. Second, an encryption process is needed for securing data in the data management system. For storing

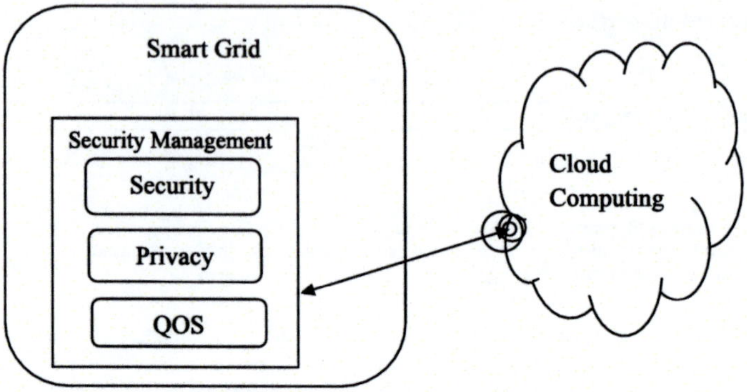

Fig. 3 Security management

sensitive information, the SM adds an extra layer of data encryption according to the requirements of the participants in the smart grid [9]. Third, let us look at the technique of improving the redundancy of data storing and calculation. For storing the data, the SM can use data storage virtualisation. It improves the redundancy of data items stores different parts of this data into different cloud computing service providers. By careful design of redundancy, if any part of data loss occurs, we may still regain all the data for some data computations [16]. The SM creates the duplicates of data and runs in different cloud computing service providers and results will be finally delivered back to the smart grid participants.

5 Applications of BSGDM

In this section, we are going to address the four applications of BSGDM. The applications are related to both the smart grid and big data environment.

A. Data and service management in BSGDM

First, the user data is usually stored by, for example, the electricity utility for billing data [11]. However, user actions will not always be related to the electricity utility in the evolving smart grid. For instance, the smart grid highlights a grid pattern, microgrid, which is a local group of power generations, storage and maintenance [5]. A microgrid can operate individually without obtaining power from the utility. The user data and operations in a microgrid may not be stored in the utility because it works individually. Using cloud computing to store and study this user data would be a good option. The user data storage and analysis are provided by the cloud computing service providers as a "pay-as-you-go" theme [12]. The end-users in the smart grid environment transfer the user data to the cloud environment and request the related services when needed.

The important feature of the smart grid is that the use of two-way flows of data and power enables many progressive applications to control the functionalities of devices in the smart grid. One key part of recognising this is the smart metre, which can acquire data from user devices and equipment while also managing the actions of the devices. Many essential features, such as real-time end-user behaviour analysis and demand response, are noticed depending on this system. With the progression of the smart grid, more and more features may be developed and implemented.

B. High-level Trading

In the advancing smart grid, a high-level trading service is needed to support more flexible power grid assets for dealing functions. As mentioned before, distributed generation of power decreases the power barriers for new players and liberalises the marketplaces in the smart grid, since individual customers could business their generated power. Therefore, the infrastructure of the smart grid leads to too many potential marketplaces [13]. Developing and maintaining a high-level trading service is probably costly, especially for small electric resources or individual customers who just want to do deals without the inspiration to create a trading plan [19]. Providing trading support as a "pay-as-you-go" utility is hence powerful. Cloud computing service providers offer data storage, analysis, computation, trading strategy, and even advertising as services [20]. The customers provide the requirement of what they want to buy or sell, and ask the service providers for a trading support. This is similar to the online stock brokerage process. The trading support service will find the match for the customers.

C. Grid Operation data Analysis

The advancing smart grid will generate a lot of data due to the wide-scale metering, detecting, and tracking functions [21]. Storage space, real-time analysis, and promotion of such a lot of data is a non-trivial procedure for traditional power sources. Incorporating this procedure into cloud computing may be a powerful option, since the cloud service providers are usually equipped with a large number of massive data processing and storage centres [14]. Therefore, energy sources can transfer the data to the cloud computing environment, and data storage, analysis, or optimisation of the data as services [17]. Furthermore, once such information from multiple power sources is incorporated by cloud computing service providers with privacy rules required, the service provider may be able to provide improved grid operation support based on more ample data.

D. Endorsing, Checking and Marketing Service

The customer power usage information is packed with customer behaviour details [15], which can be excavated to offer power endorsements or expert assistance for customers to help them minimise electric bills or help the grid function well. This data can also be used to promote appropriate power-saving appliances [18]. From the viewpoint of product sellers, this data can improve the hit rate of a marketing. The

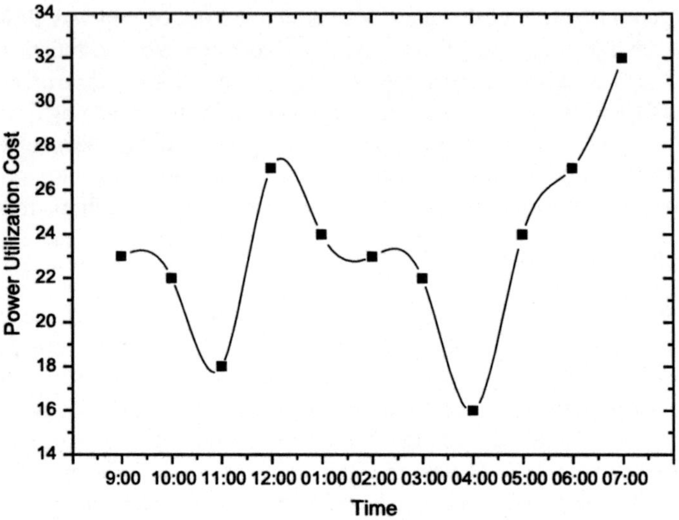

Fig. 4 Power utilisation cost versus time

customer can use such suggesting, checking, and marketing services if they allow the cloud computing service providers to access the customer data with privacy rules required and then ask the service providers to deliver these services. This support can become wiser if more information is incorporated, with privacy required.

6 Experimental Analysis

To evaluate the proposed model, the Amazon cloud platform is hired and accessed from windows PC. Figure 4 explains the power utilisation cost based on data provided by the APSPDCL electricity board. The presented data is recorded on 10th September 2020 from 9:00 am to 7:00 pm.

Figure 5 shows the power grid overall utilisation. The power utilisation is more at the grid from 5:00 pm to 7:00 pm. If the power utilisation of the customer is more than the threshold, the penalty price imposed on the customer and the corresponding financial value for that particular period increases automatically [22].

7 Conclusion

This paper discussed the advantages and prospects of using cloud data management in smart grid environments and developed a model called BSGDM to interconnect the big data and smart grid and presented the applications. Though there is more

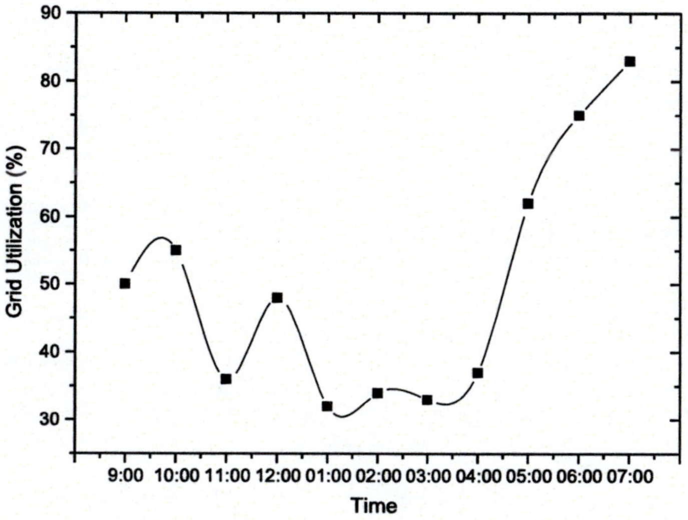

Fig. 5 Power grid utilisation versus time

scope of using data management of smart grid in the cloud; still there is a lot of issues to solve. For instance, it would be more significant to preserve the privacy and security of the grid, then the participants in both cloud computing and smart grid environment have to communicate with the intermediate communication provider domain to achieve the common goal. Therefore, a systematic study is needed to solve optimisation problems, implementation cost and some relevant parameters in both environments.

References

1. Fang X et al. Smart grid—the new and improved power grid: a survey. IEEE Commun Surv Tutor
2. Simmhan Y et al (2011) An analysis of security and privacy issues in smart grid software architectures on clouds. In: IEEE Int'l. Conf. Cloud Computing
3. Microsoft hohm, http://www.microsoft-hohm.com
4. Google Powermeter, http://www.google.com/powermeter
5. National Institute of Standards and Technology (2010) NIST framework and roadmap for smart grid interoperability standards, release 1.0, http://www.nist.gov/public_affairs/releases/upload/smartgrid_interop-erability_final.pdf, Jan. 2010
6. Endo P et al (2011) Resource allocation for distributed cloud: concepts and research challenges. IEEE Netw, 42–46
7. Buyyaa R et al (2009) Cloud computing and emerging it platforms: vision, hype, and reality for delivering computing as the 5th utility. Futur Gener Comp Sys, 599–616
8. Simmhan Y et al (2011) An informatics approach to demand response optimization in smart grids. Tech. rep., Comp. Sci. Dept., Univ. Southern CA

9. Wang C, Ren K, Wang J (2011) Secure and practical outsourcing of linear programming in cloud computing. In: INFOCOM 2011 Proc. IEEE, pp 820–28
10. Wang K, Wang Y, Xiaoxuan H, Sun Y, Deng D-J, Vinel A, Zhang Y (2017) Wireless big data computing in smart grid. IEEE Wirel Commun 24(2):58–64
11. Bhattarai BP, Paudyal S, Luo Y, Mohanpurkar M, Cheung K, Tonkoski R, Hovsapian R, Myers KS, Zhang R, Zhao P, Manic M (2019) Big data analytics in smart grids: state-of-the-art, challenges, opportunities, and future directions. IET Smart Grid 2(2):141–154
12. Jaradat M, Jarrah M, Bousselham A, Jararweh Y, Al-Ayyoub M (2015) The internet of energy: smart sensor networks and big data management for smart grid
13. Chin W-L, Li W, Chen H-H (2017) Energy big data security threats in IoT-based smart grid communications. IEEE Commun Mag 55(10):70–75
14. Diamantoulakis PD, Kapinas VM, Karagiannidis GK (2015) Big data analytics for dynamic energy management in smart grids. Big Data Res 2(3):94–101
15. Wang B, Fang B, Wang Y, Liu H, Liu Y (2016) Power system transient stability assessment based on big data and the core vector machine. IEEE Trans Smart Grid 7(5):2561–2570
16. Mayilvaganan M, Sabitha M (2013) A cloud-based architecture for Big-Data analytics in smart grid: a proposal. In: 2013 IEEE international conference on computational intelligence and computing research, pp 1–4. IEEE
17. Kezunovic M, Xie L, Grijalva S (2013) The role of big data in improving power system operation and protection. In: 2013 IREP symposium bulk power system dynamics and control-IX optimization, security and control of the emerging power grid, pp 1–9. IEEE
18. Li L, Bo Z (2009) High-speed power grid data acquisition system based on embedded TCP/IP protocol. Chin J Sci Instrum 2(9):35–41
19. Ghorbanian M, Dolatabadi SH, Siano P (2019) Big data issues in smart grids: a survey. IEEE Syst J 13(4):4158–4168
20. Ahsan U, Bais A (2017) Distributed big data management in smart grid. In: 2017 26th Wireless and Optical Communication Conference (WOCC), pp 1–6. IEEE
21. Huang Z, Nieplocha J (2008) Transforming power grid operations via high performance computing. In: 2008 IEEE power and energy society general meeting-conversion and delivery of electrical energy in the 21st century, pp 1–8. IEEE
22. Raju DN, Saritha V (2016) Cost optimisation approach for cloud-based smart grid data management. Int J Smart Grid Green Commun 1(2):143–155

Metaheuristic Optimization Using Hybrid Algorithm in Cloud-Based Big Data Analytics

Amitkumar Manekar and G. Pradeepini

Abstract Task Scheduling is a prominent research topic in cloud computing. There are several objectives associated with Optimise Task Scheduling and Resource allocation as cloud computing systems are more complex than the traditional distributed system. There are several challenges such as resolving the task mapped to the node on which task to be executed. Modelling deadline or make spam parameter, task reliability, various task allocation strategies, etc. A simplified but near-optimal proposed nature-inspired algorithms are focus in this paper. In this paper, a basic idea about optimisation, reliability, and complexity is considered while designing a modern BDA solution (Big Data Application). In this paper, we focused on Dragonfly algorithm and Sea lion algorithms which are nature-inspired algorithms. These algorithms are efficient for optimisation purposes for solving task scheduling and resource allocation problem. Finally, the performance of the DA algorithm and Sea lion is compared with the Genetic Algorithm (GA) and Particle Swarm Optimisation (PSO) for modern BDA such as Hadoop Map reduce. Simulation results prove the efficacy of the suggested algorithms

Keywords Resource allocation · Cloud · Big data · Deadline · Utilisation cost · Migration · CDSLnO

1 Introduction

In the modern era of computing such as bioinformatics, astronomy, physics, smart computing, weather data analysis, and modelling, and for any data-driven scientific applications, cloud computing with Big data application is used. All those technologies are dependent on fair share policy of task scheduling and resource

A. Manekar (✉) · G. Pradeepini
Department of CSE, Koneru Lakshmaiah Education Foundation, Green Fields, Vaddeswaram, Andhra Pradesh, India
e-mail: asmanekar24@gmail.com

G. Pradeepini
e-mail: pardeepini.gera@gmail.com

S. Jyothi et al. (eds.), *Proceedings of the 2nd International Conference on Computational and Bio Engineering*, Lecture Notes in Networks and Systems 215,
https://doi.org/10.1007/978-981-16-1941-0_62

sharing. Execution of all data-driven and data-intensive workflow depend of fast and parallel executions of BDA. BDA uses extensive use of the reasonable amount of time demands a high-performance computing environment. Cloud computing needs computing resources on demand and virtualisation is a prominent solution to this [1]. It is necessary to schedule workflows on cloud in a way that reduces the cost of leasing resources. Scheduling tasks on resources is a NP hard problem and using meta-heuristic algorithms is an obvious choice for the same. In this paper, work is stimulated to Nature-inspired algorithms like particle swarm optimisation such as Dragonfly algorithm DA and Sea lion algorithm [9, 10].

2 Related Works

Table 1 shows the reviews on task scheduling in cloud computing. At first, MR

Table 1 Features and challenges of task scheduling in cloud computing using various techniques

Author [citation]	Adopted methodology	Features	Challenges
Lee et al. [1]	MR algorithm	• High speed • Reliability and scalability	• Failure rate noise has to be focused more
Chen et al. [2]	MUMTO approach	• Minimal cost • Optimal run time	• Have to analyse the user mobility
Simic et al. [3]	ANN framework	• Speedy delivery of results • Minimised infrastructure cost	• No consideration on QoS requirements
Najme et al. [4]	PSO model	• Minimised implementation time • High efficiency	• Fault tolerance constraints are not taken into account
Fredy et al. [4]	MHRA scheme	• Better scheduling solution • Minimises the cost function	• Need consideration on energetic models to evaluate the cost
Zhang et al. [5]	PSO	• Minimised execution cost • Optimal convergence speed	• Dynamic slack time could not be exploited here
Srichandan et al. [6]	BF algorithm	• Minimises the make span • Reduced energy consumption	• Similarity function has to be focused more
Zhang et al. [7]	Dynamic game theory	• Minimised make span • Reduced workload of machines	• The real time generation of scheduling is not considered

algorithm was introduced in [1] that offers high speed, and it also offers better reliability and scalability. However, failure rate noise has to be focused more. MUMTO approach was exploited in [2] that offers minimal cost and it also offers optimal run time, but it has to analyse the user mobility. "ANN framework was used in [3] that reduce the infrastructure cost and it also offers speedy delivery of results. However, there was no consideration on QoS requirements. In addition, PSO was implemented in [4] that attain minimised implementation time along with high efficiency, nevertheless, fault tolerance constraints are not taken into account." MHRA theory was presented in [4] that offers minimised the cost function with better scheduling solution, but the energetic model is not considered. Moreover, PSO was implemented in [5], which minimised execution cost and optimal convergence speed. Nevertheless, it poses low searching efficiency. In addition, BF algorithm was suggested in [6], which reduced make span and minimal energy consumption. However, the similarity function has to be focused more. Dynamic game theory was introduced in [7], which minimises the make span with a reduced workload. These limitations have to be considered for improving the performance of cost optimisation in the cloud effectively in the current research work.

3 Problem Definition

In cloud computing, especially workflows in IaaS clouds using nature-inspired algorithms are considered in this work. Metaheuristic optimisation algorithms are simpler to implement and best suited for cloud computing [8]. Above all, they can minimise local optima and they can be applied in a wide range of issues covering various disciplines [5, 6]. In ultra large scale and high scalability such as cloud computing, large resource pools and idle resources are available. Dynamic allocation of application and mapping resources to these applications are treated with fair resource sharing policy in all BDA.

These natures-inspired metaheuristic EA are designed on multi-objective functions. In this work, we have considered Deadline, Utilisation Cost, and Migration Cost as objective for solving large-scale real-world optimisation problems over heterogeneous computing resources. The main objectives of the research are

(1) "To make a clear review on different research works related to cost optimisation-resource allocation and task scheduling in cloud and also to define the clear problem statement on this aspect."
(2) "To design a new model for cost optimisation, particularly resource allocation in cloud under big data analytics with Hadoop or Spark, etc."
(3) "To introduce a new hybrid algorithm for solving the cost optimisation problem that obviously enhances with respect to better convergence rate and speed, respectively."

(4) "To make a clear comparison of proposed model with other state-of-the-art models with respect to different analysis under cost saving, deadline and task completion."

4 Methodology

This model insists on optimising resource allocation and migration with respect to certain constraints or parameters such as Deadline, cost saving and utilisation cost (resource saving). Moreover, this proposal aims to define a new objective function that combines fine-grained resource allocation and migration process execution. For this, a new hybrid algorithm that hybrids the concept of Dragonfly (DA) and Sea Lion Algorithm (SLnO) is introduced. DA [4] is a new meta-heuristic optimisation approach that solves the single-objective, discrete and multi-objective problems. The "SLnO algorithm [4] imitates the hunting behaviour of sea lions in nature. Moreover, it is inspired by sea lions' whiskers that are used in order to detect the prey."

A. **Proposed CD-SLnO Algorithm**

"Minimal internal memory, and slow convergence" is a drawbacks of existing DA [4–6] though it performs precisely. To overcome this we can use the concept of SLnO is merged with it to introduce a new hybrid model. Hybrid optimisation algorithms have been reported to be promising for certain search problems [5]. The working computation procedure of proposed hybrid CD-SLnO model is as follows: "DA includes two significant stages: (i) Exploration and (ii) Exploitation." The proposed logic which we are introducing as follows is as follows: "Conventionally, the modelling for separation and cohesion are computed based on S_l and S." As per the proposed logic, the separation and cohesion formula are computed as per "Eqs. (1) and (2), where $\bar{B}$ denotes a random vector obtained from SLnO algorithm [4], S_l indicates the lth nearer individual position,S signifies current individual position and U reveals the count of the nearby individuals."

$$C_i = -\sum_{l=1}^{U} \left(2\bar{B}S - S_l\right) \tag{1}$$

$$O_i = \frac{\sum_{l=1}^{U_b} S_l}{U_b} - 2\bar{B}S \tag{2}$$

B. **Computational Analysis by varying No. of Tasks**

The computational time analysis of the suggested CD-SLnO model is summarised in this section by varying the counts of VM from 20 and 40. Table 2 "reveal the time (represented in minutes) utilised by the presented CD-SLnO approach over the conventional models with respect to varied number of tasks from 200, 400, 600, 800,

Table 2 Analysis on proposed model over existing models by varying the number of virtual machines

No. of tasks	MTA-S [1]	PSO + GA [6]	MSDE [9]	CD-SLnO
VM = 30				
200	7.300469	1.585228	2.834687	0.944865
400	9.634184	3.848147	5.559878	3.711675
600	21.8097	16.27314	21.47189	3.744312
800	23.61543	20.57337	23.11756	7.027397
1000	29.73418	25.47303	29.22491	9.739561
1200	38.1482	33.61701	37.94493	12.98874
VM = 40				
200	7.11672	1.558116	2.934807	1.033701
400	9.420927	3.901276	6.140882	2.043776
600	12.29334	7.368898	9.946115	3.32982
800	22.0944	17.67865	20.88112	6.864799
1000	30.24161	25.93365	29.28711	9.751759
1200	38.53181	34.13123	37.9366	12.69217

1000 and 1200. For a better performance of the system, the computing time has to be minimal, which is attained by the presented model. Predominantly, the computing time of the presented method for VM = 20 is 87.05%, 40.39% and 66.67% better than existing MTA-S, PSO + GA and MSDE models when the number of tasks is 200." In addition, the computing time of offered scheme for "VM = 40 has attained a minimal value of 3.32982, while the existing MTA-S, PSO + GA and MSDE schemes have achieved the higher computing time of 12.29334, 16.27314 and 21.47189 when the number of tasks is 600." Thus, the improvement of the presented CD-SLnO method has been established from the analysis outcomes.

5 Conclusion

This paper has introduced a novel resource allocation model, which combined resource allocation and the migration process. Moreover, an optimisation concept named CD-SLnO was developed for resource allocation and migration by considering the parameters such as deadline, utilisation cost and migration cost. In the end, comparison was made for validating the betterment of the presented scheme over existing models. On observing the outcomes, the presented CD-SLnO model has accomplished optimal values when computed over the compared models. Primarily, the computing time of the suggested method for "task = 250 was 85.1%, 43.35% and 66.35% superior to traditional MTA-S, PSO + GA and MSDE models when the number of VM was 50. Furthermore, the computing time of the implemented scheme for task = 500 has obtained a minimal value of 2.524837, whereas the evaluated MTA-S, PSO + GA and MSDE schemes have obtained higher computing time values 11.09581, 5.413038 and 7.63648 when the number of VM was 50. Thus, the enhancement of the presented CD-SLnO model was proved in an effectual manner."

References

1. Lee JW, Jang G, Jung H, Lee JG, Lee U (2019) Maximizing MapReduce job speed and reliability in the mobile cloud by optimizing task allocation. Pervasive Mob Comput 60, Article 101082
2. Chen M, Liang B, Dong M (2018) Multi-user multi-task offloading and resource allocation in mobile cloud systems. IEEE Trans Wireless Commun 17(10):6790–6805
3. Simic Visnja, Stojanovic Boban, Ivanovic Milos (2019) Optimizing the performance of optimization in the cloud environment–an intelligent auto-scaling approach. Futur Gener Comput Syst 101:909–920
4. Koti P, Dhavachelvan P, Amudhavel J (2017) Particle swarm optimization based profiling and scheduling of big data applications. J Adv Res Dyn Control Syst 9(Special Issue 12):2144–2155
5. Kiran KVD (2018) Authorization of data in Hadoop using apache sentry. Int J Eng Technol (UAE) 7(3.6 Special Issue 6):234–236, ISSN: 2227524X
6. Phani Sheetal A, Ravindranath K (2018) Software metric evaluation on cloud based applications. Int J Eng Technol (UAE) 7(1.5 Special Issue 5):13–18
7. Balaji P, Nagaraju O, Haritha D (2017) CommTrust: Reputation based trust evaluation in E-commerce applications. Paper presented at the Proceedings of the 2017 International Conference on Big Data Analytics and Computational Intelligence, ICBDACI 2017, 318–323. https://doi.org/10.1109/icbdaci.2017.8070856
8. Jena SR, Lavanya DR, Gadde SS (2017) Minimization of execution time over cloud computing environment using fuzzy technique. J Adv Res Dyn Control Syst 9(Special Issue 18)
9. Prasad CMVS, Rao KR, Satish Kumar D, Prabhu AV (2018) Performance evaluation of power optimization in wireless sensor networks using particle swarm optimization. Int J Eng Technol (UAE) 7:404–408
10. Srichandan S, Kumar TA, Bibhudatta S (2018) Task scheduling for cloud computing using multi-objective hybrid bacteria foraging algorithm. Futur Comput Inform J 3(2):210–230, December 2018

An Enhanced Autism Spectrum Disorder Detection Model Using Convolutional Neural Networks and Machine Learning Algorithms

T. Lakshmi Praveena and N. V. Muthu Lakshmi

Abstract Neurodevelopmental disorders affect the nervous system and functioning of the brain, which affects expressing emotions, learning ability, and memorising things. ASD—Autism Spectrum Disorder is a type of by-birth neurodevelopmental disorder and is difficult to diagnose because there are no clinical tests such as blood test to diagnose ASD. Manual detection model is time-consuming process and requires a lot of specialists to analyse individual condition. In recent years, most of the research is carried out on automated ASD detection by employing advanced technologies. The proposed model automates ASD detection and overcomes limitations of existing methods. In this model both convolutional neural networks and supervised system studying methods are used jointly to improve accuracy in ASD detection compared to other models. The proposed ASD detection model has two phases, in the first phase, features are extracted utilising the convolutional neural network model and in the second phase extracted features are given as input to the detection process employed with supervised machine learning algorithms. Findings of the suggested paradigm are contrasted with different supervised machine learning algorithms.

Keywords Neurodevelopmental disorder · Autism spectrum disorder · Convolutional neural networks · Supervised machine learning

1 Introduction

Neurodevelopmental disorders affect the nervous system and functioning of the brain, which affects expressing emotions, learning ability, memorising things, and self-controlling of behaviour. ASD is a type of neurological syndrome that influences

T. Lakshmi Praveena (✉) · N. V. Muthu Lakshmi
Sri Padmavathi Mahila Visvavidyalayam, Tirupati, AP, India
e-mail: praveenalaxmi1@gmail.com

N. V. Muthu Lakshmi
e-mail: nvmuthulakshmi@gmail.com

S. Jyothi et al. (eds.), *Proceedings of the 2nd International Conference on Computational and Bio Engineering*, Lecture Notes in Networks and Systems 215,
https://doi.org/10.1007/978-981-16-1941-0_63

at birth. In general, neurodevelopmental disorders result in more anxiety, the deficit in social communication, weakness in learning, expressing emotions and language proficiency. In the paper [1], America's children and the environment (ACE) wrote an essay on neurodevelopmental maladies with a comprehensive analysis of ASD and discussed about symptoms of ASD. ASD has more impact on toddlers, children, young and adults compared to the children of age less than 2 years. The major objective of ASD research is detecting ASD accurately at early stage and with less cost in less time. ASD Diagnosis is difficult because there is no single test to diagnose neural disorder immediately. As ASD diagnosis dependent on wide range of symptoms, severity, and behavioural changes in ASD children, similar to non-autistic children [2]. The prematurely recognition of ASD can improve the living standards of ASD individuals and improves communication ability [3]. Standard detection model of ASD uses manual analysis tools like observation analysis, questionnaires with parents or caregivers. This is a time-consuming process and requires many specialists to analyse data to decide ASD level and severity.

In recent years, most research is made on automated ASD detection by employing advanced technologies like statistical tools, Data mining algorithms, artificial neural network algorithms, and machine learning (ML) algorithms in detecting ASD with less computation cost and less detection time. This paper proposes a model which automates ASD detection easily and also overcomes the limitations of existing methods. Proposed autism detection model adopts convolutional neural networks (CNN) and ML algorithms to improve the detection accuracy compared to existing models. CNN are part of artificial neural network and its functionality is similar to biological neural networks in brain. CNN is a computing system that is designed based on the human brain system. CNN are efficient and accurate in classification, prediction, and detection. Supervised machine learning is the ML method that learns based on the existing input, output samples. The supervised ML (SML) methods are suitable for taxonomy models and gives accurate results. These are individually used in different applications to get accurate results. The proposed model uses both artificial neural networks and SML methods to generate most accurate results in ASD detection compared to individual models.

The proposed ASD detection model has two phases, in the first phase features are extracted using CNN model and in the second phase extracted features are given as input to detection process employed with supervised machine learning algorithms. Results of the anticipated model are related with different SML algorithms like SVM, Decision trees, K-Nearest Neighbour, Random Forest. This enhanced model improves ASD detection accuracy compared to existing models.

The paper is organised as three sections. Section 2 discusses work related to ASD detection methods like manual methods and automated methods. Section 3 explains about process and architecture of the proposed model. Section 4 analyses and compares the performance of the proposed model with existing models.

2 Related Work

The common person's basic ability is social communication, expressing emotions, controlling behaviour, and learning [4]. These all depends on nerves system functionality of brain. The deficiency of these activities is known neural disorders. Neural disorders are classified into different types based on the symptoms in person [5]. One of the types of neurodevelopmental disorder is ASD, that affects at birth but is generally diagnosed after 3 years of age because ASD has deficiency in behaviour which exhibits at 3 years age. The detection methods play main role in ASD diagnosis and analysis. Centre for disease control (CDC) provides details of different ASD diagnosis tools like Autism Diagnosis Interview-Revised (ADI-R) [6], Autism Diagnostic Observation Schedule-Generic (ADOS-G) [6], Childhood Autism Rating Scale (CARS) [6] and Gilliam Autism Rating Scale-Second Edition (GARS-2) [6]. Manual diagnosis process takes around 3–5 days' time to report by different specialists and psychiatrists. Diagnosis after 3 years age can delay in developmental growth. Automated diagnosis methods/tools of ASD are necessary for better developmental growth, for better treatment and to diagnose in less time with less manpower. Automated methods are developed by advanced technologies like statistical methods, datamining algorithms, machine learning algorithms and artificial neural networks algorithms.

Recent research for automated diagnosis and detection of ASD is increased. Machine learning methods are accurately detecting ASD from structured dataset of different ages using convolutional neural networks(CNN) compared to other machine learning algorithms [7]. Automated ASD detection with the analysis of brain imaging dataset ABIDE using CNN gives accurate results and is used for early detection of ASD in less than 3 years age [8]. ASD and Attention Deficit Hyperactivity Disorder (ADHD) are automatically detected by using dynamic deep learning methods by visual analysis of behaviour [9]. The visual analysis and questionnaire analysis are combined to detect ASD using machine learning algorithms for better results in ASD detection [10]. In my previous work relative assessment of SML algorithms in the detection of ASD is performed [11] and a modified regression algorithm is proposed to improve the accuracy of ASD detection [12]. Research on ASD detection methods is increasing and improving every year for better detection accuracy. This research is useful for doctors, special educators, caretakers and parents to impart better treatment and therapies to ASD individuals at an early stage.

3 Proposed ASD Detection Model

An enhanced ASD detection model is proposed in this paper. This model integrates both SML algorithms and CNN of artificial neural networks framework for automated ASD detection. CNN has multiple hidden layers with the combination of convolutional 2D layers, batch normalisation layer, dropout layer and max pooling layer. Hidden layers are connected to fully connected layers, which extract features from given data. The output layer or SoftMax layer is replaced with a binary SML layer whose result is 0 (No ASD) or 1 (ASD).

Supervised machine learning algorithms used in this paper are SVM, RF, Decision trees and XGBoost. Support vector machine (SVM) reduces the generalisation error but not suited for noise data. The good qualities of SVM are used efficiently by combining with CNN [13]. Random forest (RF) reduces the overfitting that is generally observed in decision trees but becomes slow when dataset size increases because of increased trees. These limitations are overcome by combining RF with CNN [14]. Decision trees require less pre-processing of data than other algorithms, but the limitation is its complexity in computing features. Integrating decision trees with CNN reduces the computational time and increases the accuracy [15].

3.1 Architecture of Proposed Model

In this paper to analyse the ASD we used dataset provided by Fadi tabah et al. provided structured ASD dataset in UCI machine learning repository [16] in four variations based on the different age groups [17]. The ASD dataset has four sets of data with different age ranges which are divided as toddler, child, Adolcent and adult. Figure 1 describes about flow of proposed model. Table 1 describes the attributes of ASD dataset.

The steps of proposed model are given below:

Step 1: This dataset is extracted from CSV file and loaded into data frame. ASD dataset has A1–A10 binary attributes.

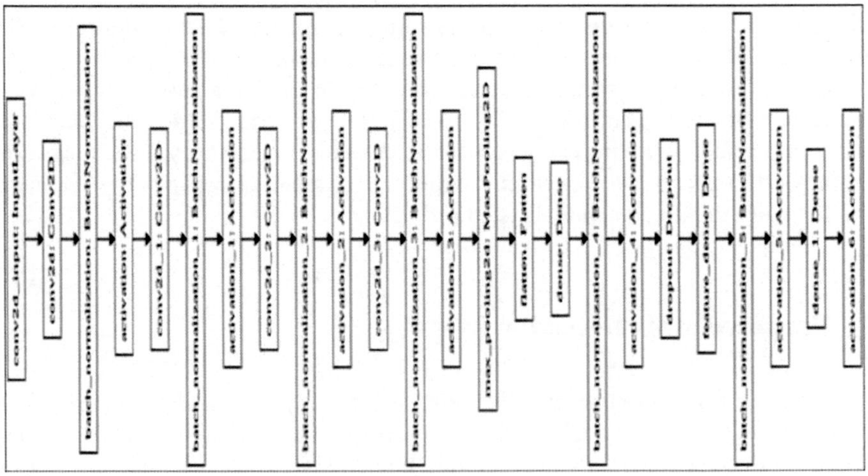

Fig. 1 Phase-1 CNN model of proposed model

Table 1 Dataset description of ASD data amassed from UCI ML source

Sl. no.	Attribute name	Description and possible values
1–10	Al... A10	Y-l, N-0
11	Age	min value: 1 year and max value: 80 years
12	Sex	M-Male, F-Female
13	Ethnicity	White, Black, Hispanic, Ab original Latino Middle Eastern, Asian, South Asia, Others
14	Jaundice	Y-l, N-0
15	Family ASD	Y-l, N-0
16	Residence	States and countries in Asia, South Asia and others
17	Used App Before	Y-l, N-0
18	Score	Min value: 0 and Max Value: 10
19	Screening Type	1–3, 4–11, 12–16, 17 and above
20	User	Seif, Parents, Others, Relative
21	ASD	Yes and No

Step 2: Missing values are imputed by using substitution method for categorical attributes and mean value substitution for numerical attributes.
Step 3: Label encoder is used to transform categorical attributes into numerical attributes by using substitution method applied on categorical attribute.
Step 4: This step performs preparation of data to process structured data with CNN. Structured data to be converted in the form of image matrix, so the data is transformed as $n \times m \times p \times q$ dimension matrix where n is total of annotations, m is amount of channels, each channel is with separate category of attributes. ASD dataset has ten behaviour related questionnaire data and another ten attributes are additional information of person, these are created as two channels. And p is the count of attributes in each channel, q is number of planes whose value is 1. Table 2 gives the dimensional values of transformed dataset.
Step 5: The preparation step data is divided into train dataset, test dataset and validation dataset. Train dataset is used to train the model, test dataset is used to test the model and validation dataset is used to evaluate the model. The split is performed with the ratio of (70:20:10).

Table 2 ASD dataset dimensions for processing with proposed model

Dataset	Number of observations (n)	Amount of channels (m)	Quantity of features per channel (p)	Total planes (q)
Toddler	1054	2	10	1
Child	510	2	10	1
Adolescent	248	2	10	1
Adult	1118	2	10	1

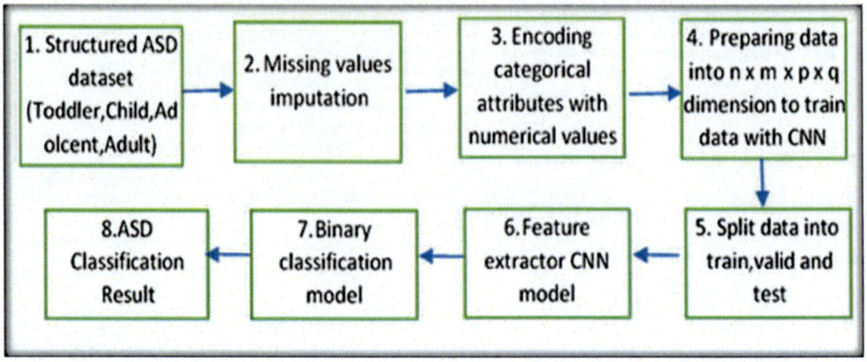

Fig. 2 Flow chart of proposed model

Step 6: The suggested CNN model is intended as automated feature extractor model. The CNN model comprises four convolution layers and two fully connected (FC) layers as shown in Fig. 1. The FC layers output is the features extracted from dataset give as input. These extracted features are given as response to SML algorithm.
Step 7: SML algorithm is used as binary classification model. The extracted features in step 6 are given as input or training data to the model. Model is fit on train data and evaluated on validation data and test data. Different algorithms like SVM, Random forest, XG Boost, and decision tree algorithms are used to find accurate models to detect ASD with high accuracy.
Step 8: Predicted result of ASD is given as output from the model. Figure 2 illustrates the flow graph of the recommended model with all steps in model.

4 Performance Analysis

The performance of CNN model with supervised machine learning algorithms is evaluated and presented in Table 2. Table 3 shows the accuracy and loss of different algorithms individually and with the combination of CNN on toddler dataset, child dataset, adolescent dataset, adult dataset and on a complete dataset of ASD are given. Figure 3 is the graphical representation of performance comparison analysis.

5 Conclusion and Future Work

ASD is a neuro-developing disorder, its diagnosis and detection are a lengthy process that may delay a person development and delay in treatment. Automated ASD detection-related research is required to overcome limitations in ASD diagnosis and

Table 3 Evaluation of SML algorithms on structured ASD data

Model	Toddler		Child		Adolscent		Adult		Complete	
	Accuracy (%)	Loss (%)	Accuracy (%)	Loss (%)	Accuracy (%)	Loss (%)	Accuracy (%) (%)	Loss (%)	Accuracy (%)	Loss (%)
Decision Ttee(J48)	99	31.25	99	31.25	99	31.25	99	31.25	99	31.25
Random Forest	100	30.52	100	30.52	100	30.52	100	30.52	100	30.52
SVM Model	100	30.59	100	30.59	100	30.59	100	30.59	100	30.59
XGBoost	100	30.91	100	30.91	100	30.91	100	30.91	100	30.91
Convolutional Neural Networks	100	32.33	100	32.33	100	32.33	100	32.33	100	32.33
Linear Regression	98	32.05	98	32.05	98	32.05	98	32.05	98	32.05
Modified Linear Regression	99	31.05	99	31.05	99	31.05	99	31.05	99	31.05
CNN-SVM	100	30.91	100	30.91	100	30.91	100	30.91	100	30.91
CNN-RF	100	31.23	100	31.23	100	31.23	100	31.23	100	31.23
CNN-Decision Trees	100	30.51	100	30.51	100	30.51	100	30.51	100	30.51
CNN+XGB	100	30.59	100	30.59	100	30.59	100	30.59	100	30.59

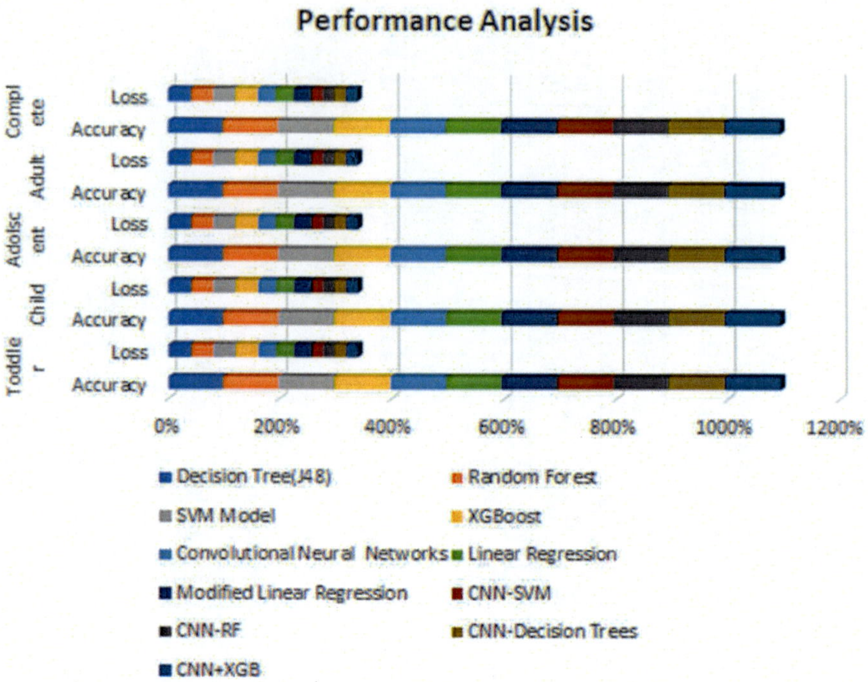

Fig. 3 Performance analysis of proposed model with other algorithms

treatment. In this paper proposes CNN model with a combination of supervised machine learning algorithms. This model shows an improvement in accuracy and reduces the loss value. The advanced algorithms of supervised machine learning algorithm can also be implemented with this model. This model can also be used in other medical applications to improve detection accuracy.

Acknowledgements Fundamental information was gathered by F. F. Thabtah, he developed an upwardly mobile product for identifying ASD. F. F. Thabtah sited dataset in UCI ML depository along with open gain access and benefitted to research scholars those are working on ASD. Thankful to F. F. Thabtah for encouraging investigation upon ASD.

References

1. Article on "Neurodevelopmental Disorders" America's Children and the Environment | Third Edition, Updated October 2015
2. Charman T (2014) Variability in neurodevelopmental disorders: evidence from autism spectrum disorders. In: Neurodevelopmental disorders
3. Becerra TA, Massolo ML, Yau VM, Owen-Smith AA, Lynch FL, Crawford PM, Pearson KA, Pomichowski ME, Quinn VP, Yoshida CK, Croen LA (2017) A survey of parents with children

on the autism spectrum: experience with services and treatments. https://doi.org/10.7812/TPP/16-009
4. Cattaneo L, Fabbri-Destro M, Boria S et al (2007) Impairment of actions chains in autism and its possible role in intention understanding. Proc Natl Acad Sci 104(45):17825–17830
5. Elsabbagh M, Divan G, Koh YJ, Kim YS, Kauchali S, Marcin C et al (2012) Global prevalence of autism and other pervasive developmental disorders. Autism Res. 5:160–79
6. https://www.cdc.gov/ncbddd/autism/hcp-screening.html
7. Raj S, Masood S (2020) Analysis and detection of autism spectrum disorder using machine learning techniques. Procedia Comput Sci 167:994–1004. ISSN 1877-0509. https://doi.org/10.1016/j.procs.2020.03.399
8. Zeinab S, Mohammadsadegh A, Soorena S, Mariam Z-M, Moloud A, Rajendra AU, Reza K, Vahid S (2020) Automated detection of autism spectrum disorder using a convolutional neural network. Front Neurosci 13:1325. Https://Www.Frontiersin.Org/Article/10.3389/Fnins.2019.01325, https://doi.org/10.3389/fnins.2019.01325, Issn = 1662-453x
9. Jaiswal S, Valstar MF, Gillott A, Daley D (2017) Automatic detection of ADHD and ASD from expressive behaviour in RGBD data. In: 2017 12th IEEE international conference on automatic face & gesture recognition (FG 2017), Washington, DC, pp 762–769. https://doi.org/10.1109/fg.2017.95
10. Abbas H, Garberson F, Glover E, Wall DP (2018) Machine learning approach for early detection of autism by combining questionnaire and home video screening. J Am Med Inform Assoc 25(8):1000–1007. https://doi.org/10.1093/jamia/ocy039 PMID: 29741630
11. Lakshmi Praveena T, Muthu Lakshmi NV (2019) Prediction of autism spectrum disorder using supervised machine learning algorithms. Asian J Comput Sci Technol 8(S3):142–145. ISSN: 2249-0701. © The Research Publication. www.trp.org.in
12. Omar KS, Mondal P, Khan NS, Rizvi MRK, Islam MN (2019) A machine learning approach to predict autism spectrum disorder. In: 2019 international conference on electrical, computer and communication engineering (ECCE) Cox's. IEEE, Bazar, Bangladesh, pp. 1–6
13. Ahlawat S, Choudhary A (2020) Hybrid CNN-SVM classifier for handwritten digit recognition. Procedia Comput Sci 167:2554–2560. ISSN 1877-0509. https://doi.org/10.1016/j.procs.2020.03.309
14. Cao G, Wang S, Wei B, Yin Y, Yang G (2013) A hybrid CNN-RF method for electron microscopy images segmentation. J Biomim Biomater Tissue Eng 18:114. https://doi.org/10.4172/1662-100X.1000114
15. Yin S, Duan J, Ouyang P, Liu L, Wei S (2017) Multi-CNN and decision tree-based driving behavior evaluation. In: Proceedings of the symposium on applied computing (SAC '17). Association for Computing Machinery, New York, NY, USA, pp 1424–1429. DOI:https://doi.org/10.1145/3019612.3019649
16. UCI Machine Learning Repository. http://archive.ics.uci.edu/ml
17. Thabtah F (2018) Machine learning in autistic spectrum disorder behavioral research: a review and ways forward. Inform. Health Soc Care: 1–20

Building an Efficient Feature Selection for Intrusion Detection System on UNSW-NB15

P. G. V. Suresh Kumar and Shaheda Akthar

Abstract Network interruption location is one of the urgent security parts and guard instruments of organisation security frameworks. Feature selection should be treated as a vital pre-handling step to improve the general framework execution while mining on gigantic datasets. In this specific situation, in this paper, we centre around a two-venture approach of feature determination dependent on Recursive Feature Elimination (RFE) calculation. We present another methodology comprises with converging of feature selection and characterisation for UNSW-NB15 interruption location dataset utilising decision tree. The goal is to improve the capability of interruption order with an altogether decreased arrangement of information features from the preparation information. The initial step chooses the features with a higher variable significance score and aides the introduction of quest measure for the second step, which yields the last element subset for arrangement and translation. In the test, we have applied decision tree classifier on a few info include subsets of preparing dataset of UNSW-NB15 dataset. The test results indicated the proposed strategy effectively brings 97.34% grouping accuracy utilising just 20 features, while every 42 preparing features accomplished 94.32% characterisation accuracy.

Keywords IDS · Decision tree · RFE and UNSW-NB15

1 Introduction

Security has been an urgent factor in this advanced computerised period because of the quick improvement of data innovation, trailed by genuine computer violations that prompted the development of Intrusion Detection Systems (IDSs). The web's developing need made us rely upon computer network so the network security

P. G. V. Suresh Kumar (✉)
Department of CSE, Acharya Nagarjuna University, Guntur, AP, India
e-mail: pendemsuresh@gmail.com

S. Akthar
Department of Computer Science, Government College for Women, Guntur, AP, India
e-mail: shahedaakthar76@gmail.com

S. Jyothi et al. (eds.), *Proceedings of the 2nd International Conference on Computational and Bio Engineering*, Lecture Notes in Networks and Systems 215,
https://doi.org/10.1007/978-981-16-1941-0_64

assumes a significant job [1]. Accordingly, we have to locate a most ideal approach to ensure our frameworks. Unapproved admittance to classified assets, spontaneous duplication and modification of records, wholesale fraud, and some other activities focusing on the obliteration of the data framework and organisation foundation are among security penetrates. To react to these issues, network interruption-based security apparatuses are utilised to review each arrange parcel, and dependent on the past organisation traffic information, it chooses whether such traffic is trustful or noxious [2].

In such manner, given the criticalness of feature selection and the significance of actualising new strategies for distinguishing interruptions in computer networks, this paper expects to utilise a component subset of the notable ongoing IDS dataset UNSWNB15 [3] to assess and analyse the prescient precision of a decision tree learning classifier to decide whether decision tree model are in every case better while identifying interruption in computer networks [4, 5].

2 Interruption Detection System (IDS)

IDS is one of the significant examination territories in field of data and organisation security to shield data or information from unapproved access. IDS is a classifier that can characterise the information as typical or assault [6]. The IDS assume a fundamental function in distinguishing peculiarities and assaults on the organisation that have increased and are more unavoidable. A few IDS distinguish by searching for explicit marks of the known dangers like the way antivirus programming ordinarily identifies and secures against malware [7]. There is likewise an alternate sort of IDS that distinguish assaults dependent on contrasting traffic designs against a pattern and searching for oddities. IDS basically screens and alarms while some different IDS play out an activity or activities because of a recognised danger [5].

The essential point of the organisation assault was to bargain the trustworthiness, accessibility or secrecy of the organisation framework that is done through the information stream on a PC network by a gatecrasher. Hence, IDS is expected to identify noxious or unapproved exercises on the organisation and square the gatecrasher traffic association with keep the framework from additional harm. IDS by and large fall into two principle types: abnormality location frameworks and Misuse identification frameworks [8, 9]. In peculiarity-based procedures, the characterisation depends on decides where any endeavour that drops out of the ordinary conduct is treated as an assault, not at all like abuse discovery methods where a rundown of marks of realised assaults is kept in the framework and contrasted and caught information, which suggests that novel assaults can't be detected [3, 10].

3 Proposed Methodology

In this proposed system, feature decrease strategy utilising RFE is directed as an underlying advance towards diminishing the quantity of qualities without losing the principle reason and target data from the first information. The subsequent stage is building up an indicator with an improved exactness to arrange informational collection. There are various stages in the proposed engineering for an efficient IDS. We are proposing another model for productive component choice and interruption identification. This methodology is of the accompanying strides as follows:

Stage 1: Read the UNSW-NB15 dataset.
Stage 2: Preprocess the dataset.
Stage 3: Select the huge highlights utilising RFE calculation.
Stage 4: Perform Classification utilising decision tree calculation on the dataset to choose the best features.
Stage 5: Evaluate execution of the classifier.

3.1 Decision Tree

Decision tree is likely the most well-known information digging procedure ordinarily utilised for arrangement [11]. The standard thought of a choice tree is to part our information recursively into subsets so every subset contains pretty much homogeneous conditions of our objective variable. All info credits are assessed at each split in the tree for their effect on the anticipated characteristic. At the point when this recursive cycle is finished, a choice tree is shaped which can be changed over in basic If–Then standards.

Decision trees can examine information and recognise huge attributes in the organisation that show vindictive exercises. It can increase the value of some ongoing security frameworks by examining huge arrangement of interruption discovery information. It can constantly perceive patterns that help further examination, the advancement of assault marks, and different checking exercises. The fundamental preferred position of utilising decision trees rather than other characterisation strategies is that they give a rich arrangement of decides that are straightforward. Decision tree tends in some cases to over-fitting, so to locate the best boundaries to fit the model, a comprehensive matrix search boundaries tuning was figured and data gain is utilised to choose features. Henceforth, working from the preparation information, a tree was gotten with its leaves being class names. When fabricating a decision tree, just each element is utilised in turn to part the hub and segment the information. Consequently, features are utilised in a univariate way.

3.2 Feature Selection

The organisation information has numerous features and to execute machine learning proficiently, just productive features must be utilised. Feature choice is a vital cycle wasteful interruption location. Utilising just the significant features gives more precise consistency. Information handling is utilised to improve the nature of informational collection. To make IDS more effective disentanglement of feature is finished by diminishing the information measurement and unpredictability. Features choice is a strategy which is utilised to eliminate superfluous features from the first informational collection [12, 13]. Feature selection is otherwise called variable subset choice of pertinent features for the utilisation of model development. Feature Selection is help full at the hour of examination of model. Picking a subset of good features eliminates the unessential information, expands the learning exactness and improves the understandability [4]. Great component subset determination calculations are essential for AI applications. In the wake of acquiring the sufficient number of features during the univariate determination measure, a recursive component end (RFE) was worked with the quantity of features passed as boundary to distinguish the features chose.

3.2.1 Recursive Feature Elimination (RFE)

During the RFE cycle, first, the classifier is prepared on the first arrangement of features and loads are ascribed to each feature. At that point, includes whose supreme loads are the littlest are pruned from the current set features [3]. That cycle is recursively rehashed on the pruned set until the ideal number of features to pick is at last reached.

A recursive component end technique which consistently fabricate a model, setting the element aside and afterward rehashing the cycle with the remained features until all features in the dataset are depleted [10]. The goal of the recursive component is to recover includes by recursively keeping more modest and more modest gathering of features. A decent element positioning basis doesn't really create a decent component subset age. The a few models gauge the impact of eliminating each component in turn dependent on the objective to accomplish [14]. They become extremely imperfect with regards to eliminating a few highlights all at once, which is important to get a little element subset.

4 Experimental Results

The UNSW-NB15 dataset was utilised to assess the presentation of the proposed strategy [15]. The execution of the proposed strategy was composed primarily with

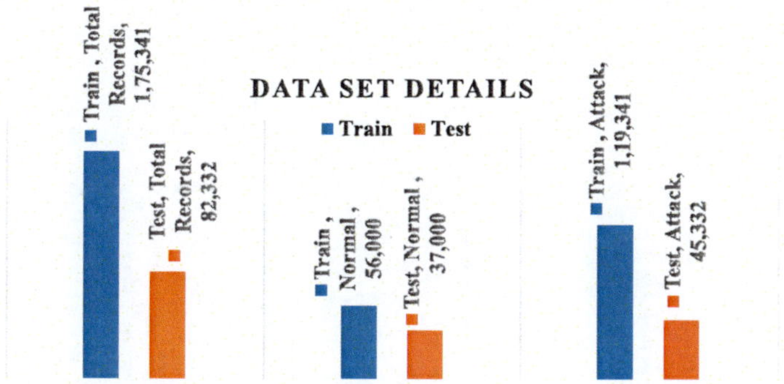

Fig. 1 Training and Testing details of UNSW-NB15

Python 3.6 programming language and a Machine Learning Library known as Scikit-Learn which has a few learning calculations. The climate on which the application is manufactured is called Jupyter, which is a workbench given by an Anaconda stage. As for the equipment utilised, a PC having an arbitrary access memory of 4 GB RAM, stockpiling size 1 TB with Processor type Intel(R) Core (TM) i3 CPU @2.53 GHz speed was utilised.

4.1 Dataset Description

The UNSW-NB15 dataset created utilising IXIA Perfect Storm, is a complete network-based dataset that reflects present day network traffic situations. In 2015, the UNSW-NB15 dataset [3] was made at the Australian Center for Cyber Security. Dataset contains the two records about network associations and the traffic itself (around 100 GB) from the test PC organisation (3 servers). The accessible UNSW-NB15 dataset contains 42 features (barring the names) and ten classes: Normal, Fuzzers, Analysis, Backdoors, DoS, Exploits, Generic, Reconnaissance Shellcode, and Worms. Its training set includes 175,341 records, while the testing set has 82,332 records as shown in Fig. 1. The point-by-point circulation of these classes regarding number of records per assault just as preparing/testing set appropriation are represented in the Fig. 2.

4.2 Experimental Measurements

This paper breaks down the impacts of feature determination for decision tree arrangement for UNSW-NB15 dataset. Accuracy, Detection rate and False alarm rate are utilised to assess and analyse the model.

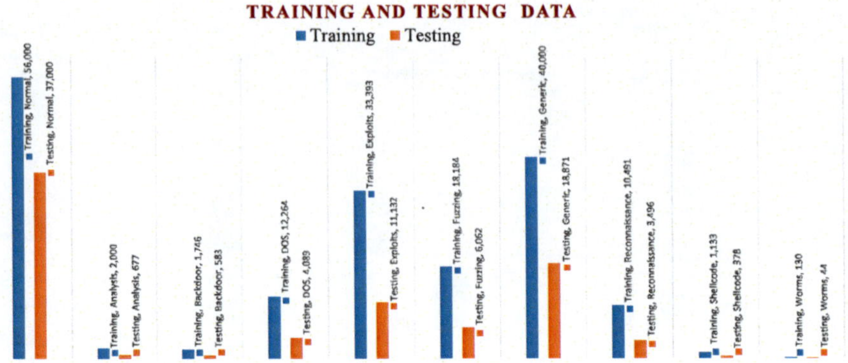

Fig. 2 Training and Testing data distribution

Accuracy

Accuracy is the measure of accurately ordered occurrences of the complete cases, characterised as the proportion of the quantity of right forecasts to the all-out number of expectations.

$$\text{Accuracy} = \frac{\text{TP} + \text{TN}}{\text{TP} + \text{TN} + \text{FP} + \text{FN}} \tag{1}$$

Detection Rate (DR)

DR is the proportion of effectively recognised positive occurrences from all the real sure cases, characterised as the proportion of right certain forecasts to the all-out number of positive expectations. The higher it's worth the better.

$$\text{DR} = \frac{TP}{TP + FN} \tag{2}$$

False Alert Rate (FAR)

FPR, is the proportion of the inaccurately ordered negative (ordinary) occurrence as a peculiarity from all the genuine negative occasions, or characterised as the extent of negative expectation this is erroneously considered as certain (abnormality) for every negative forecast. The lower its worth the better [11].

$$\text{FAR} = \frac{FP}{FP + TN} \tag{3}$$

where TP = True positive, TN = True negative, FP = False positive, FN = False negative.

4.3 Results and Discussion

The primary stage decision tree learning calculation is prepared on the first arrangement of features utilised in the examination. In the second stage, we actualise a RFE calculation to get the sufficient number of features to distinguish the features chosen. The outcomes that we got for decision tree without feature determination and include choice appear beneath in Fig. 3 with their comparing esteems. The information in UNSW-NB15 dataset is either marked as would be expected or as one of the nine various types of assault. We have separated from 42 trait we have separated to 20 element vectors by utilising RFE procedure to get an ideal choice from complete dataset for preparing just as for testing tests. Figure 3 shows the presentation of arranging UNSW-NB15 assaults by utilising the decision tree calculation for the full measurement information and furthermore after the component decrease with RFE strategy.

From Fig. 3, we notice the presentation of decision tree without RFE dependent on exactness, DR, FAR has are 94.32, 97.54 and 13.74%. Though the decision tree presentation with RFE includes determination dependent on precision, DR, FAR has accomplished are 95.42, 98.58 and 10.96%. The location pace of the decision tree model-based generally speaking features prior to applying the RFE include determination procedure is 97.54%, yet the recognition rate subsequent to applying RFE features selection strategy is 98.58% shows an improvement of roughly 1.04% in the wake of applying the proposed feature selection. In any case, there had all the earmarks of being in progress in the FAR with include choice. The FAR rate is dropped from 13.74 to 10.96%. In our trial result the decision tree with RFE calculation shows the most noteworthy precision contrasted and decision tree without RFE. With the improvement the exactness, the proposed model exhibited that it performs well in the wake of choosing important features. This outcome gave new understanding utilising a characterisation learning calculation and decrease method to choose applicable and significant components to improve the framework's exactness and recognise potential features that may add to this improvement. The greater part

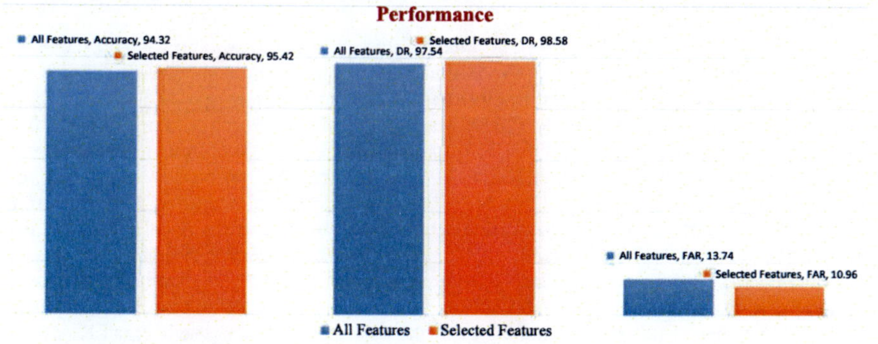

Fig. 3 Performance of the model with and without features

of the proposed research framework could viably use the choice cycle to improve the discovery pace of their framework and significantly limit the false alert rate.

5 Conclusion

Network interruption discovery is one of the significant security parts and guard components of organisation security frameworks. In this paper, the importance of utilising a bunch of applicable features with the decision tree grouping learning calculation for displaying an IDS has been illustrated. The proposed technique can assist with identifying interruption in computer networks with high location precision, and we accept that this work will fill in as a rule for future exploration in the area of organisation interruption recognition frameworks. The assessment of the adequacy of the technique utilising diverse arrangement metric estimation has been made. It has been demonstrated that by lessening the quantity of features, the exactness of the model was improved. To distinguish interruption from huge traffic information, recognition calculation, and feature determination technique need to more proficient.

References

1. Liao H-J, Lin C-HR, Lin Y-C, Tung K-Y (2013) 'Intrusion detection system: a comprehensive review. J Netw Comput Appl 36(1):16–24
2. Thaseen IS, Kumar CA (2016) Intrusion detection model using fusion of chi-square feature selection and multi class SVM. J King Saud Univ Comput Inf Sci (in press)
3. Moustafa N, Slay J (2015) Unsw-nb15: a comprehensive data set for network intrusion detection systems (unsw-nb15 network data set). In: 2015 military communications and information systems conference (MilCIS). IEEE, pp 1–6
4. Yan J, Zhang B, Liu N, Yan S, Chen Z (2006) Effective and efficient dimensionality reduction for large scale and streaming data processing. IEEE Trans Knowl Data Eng 18(3):320–333
5. Lakhina S, Joseph S, Verma B (2010) Feature reduction using principal component analysis for effective anomaly–based intrusion detection on NSL-KDD. Int J Eng Sci Technol 2(6):1790–1799
6. Yu K-M, Wu M-F, Wong W-T (2008) Protocol-based classification for intrusion detection. In: WSEAS international conference. Proceedings. Mathematics and computers in science and engineering. World Scientific and Engineering Academy and Society, pp 29–34
7. Ahmed T, Oreshkin B, Coates M (2007) Machine learning approaches to network anomaly detection. In: Proceedings of USENIX workshop tackling computer system problems with machine learning techniques, Cambridge, MA, USA, April 2007, pp 1–6
8. Ma T, Wang F, Cheng J, Yu Y, Chen X (2016) A hybrid spectral clustering and deep neural network ensemble algorithm for intrusion detection in sensor networks. Sensors 16(10):1701
9. Bace R, Mell P (2001) NIST special publication on intrusion detection systems
10. Mukherjee S, Sharma N (2012) Intrusion detection using Naive Bayes Classifier with feature reduction. Procedia Technol 4:119–128
11. Han J, Kamber M (2006) Data mining concepts and techniques. In: The Morgan Kaufmann series in data management systems, 2nd edn. Morgan Kaufmann, San Mateo, CA

12. Ravi Kumar G, Kongara VS, Ramachandra GA (2013) An efficient ensemble based classification techniques for medical diagnosis. Int J Latest Technol Eng Manag Appl Sci II(VIII):5–9. ISSN-2278-2540
13. Nkiama H, Said SZM, Saidu M (2016) A subset feature elimination mechanism for intrusion detection system. Int J Adv Comput Sci Appl 7(4):148–157
14. Pang Y, Yuan Y, Li X (2008) Effective feature extraction in high dimensional space. IEEE Trans, Syst
15. UNSW-NB15 Dataset Description (2015) https://www.unsw.adfa.edu.au/australian-centre-for-cyber-security/cybersecurity/ADFA-NB15-Datasets/. Accessed May 2015

An Adaptive Channel Assignment Strategy for Multi-radio Multi-channel Wireless Mesh Networks with Mobile Nodes Based on Internet Traffic

V. Sujatha, C. Shoba Bindu, G. Vijay Kumar, and P. Dileep Kumar Reddy

Abstract Wireless Mesh Networks are considered to be the best among the wireless networks to achieve last-mile connectivity. The conventional Mesh Routers with single radio limit the entire network to a single channel so that the total capacity of multiple channels available is not attained. Mesh Routers have multi-radio interfaces that are allocated to various partially overlapping/orthogonal channels to enhance WMN efficiency. Users need to connect with one of the available Mesh Routers to access the Internet via Wireless Mesh Network. Many previous techniques considered traffic only in the downlink direction but not in the uplink direction. Nowadays, most users share their data in uplink direction also, that is, from user to gateway router. This project proposes a new strategy of channel allocation to enhance the user throughput in both uplink and downlink directions of internet traffic. The proposed strategy will be a distributed solution to the channel allocation problem in WMN. This work also focuses on the mobility of user nodes and their effect on the channel allocation problem. The results of the simulation indicate a major improvement in user performance.

Keywords Multi-radio multi-channel wireless mesh networks · Channel assignment · Internet traffic

---

V. Sujatha (✉) · C. Shoba Bindu
Department of CSE, JNTUA, Ananthapuramu, Andhra Pradesh 515001, India
e-mail: sujatha.varada52@gmail.com

C. Shoba Bindu
e-mail: shobabindhu.cse@jntua.ac.in

G. Vijay Kumar
Department of CSE, G.Pulla Reddy Engineering College, Kurnool, Andhra Pradesh 518007, India
e-mail: gvjykumar@gmail.com

P. Dileep Kumar Reddy
Department of CSE, SV College of Engineering, Tirupathi, Andhra Pradesh 517507, India
e-mail: dileepreddy503@gmail.com

S. Jyothi et al. (eds.), *Proceedings of the 2nd International Conference on Computational and Bio Engineering*, Lecture Notes in Networks and Systems 215,
https://doi.org/10.1007/978-981-16-1941-0_65

1 Introduction

The Wireless Mesh Network has the potential to dynamically afford internet connectivity to everyone else in the world by establishing network nodes. WMN provides wide-scale association at a lower price than the wired network. Wireless Mesh Networks (WMNs) are robust, provide reliable service coverage, and easy to maintain in addition to the cost advantage [1]. Wireless mesh networks (WMN) are multi-hop networks with the ability to transmit packets among wireless mesh routers. The IEEE 802.11 MAC protocol is used for Wireless Mesh Networks.

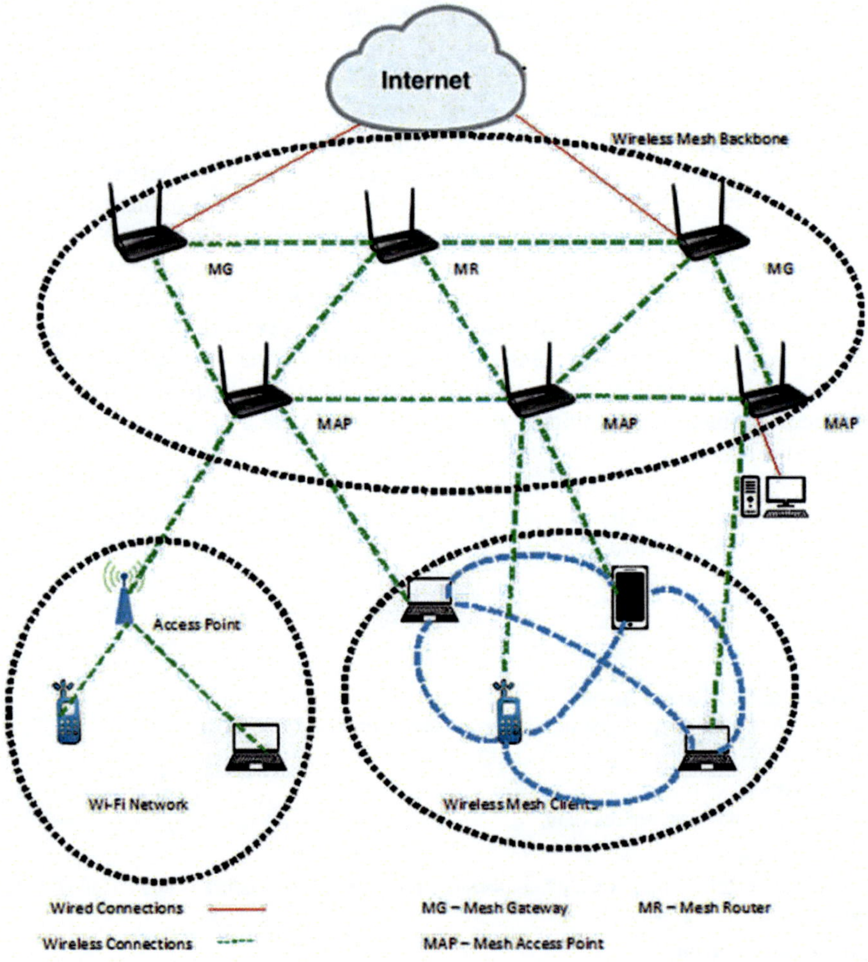

Fig. 1 Wireless mesh network architecture

As shown in Fig. 1, the architecture of Wireless Mesh Network comprises of mesh routers, gateway nodes, and clients. Mesh clients (MC) can act as a client transferring its packets to the network. Mobile/stationary end-user devices are considered as Mesh Clients. Mesh routers (MR) interconnect with others in a multi-hop fashion so that packets are often transferred over many wireless connections (hops). Routers that are associated with the external world are known as gateway nodes (GWNs). These GWNs transfer traffic to and fro of the mesh network. The structure of an infrastructure-based Wireless Mesh Network (also called mesh backbone) is the group of routers and gateway nodes linked together in a multi-hop fashion. WMNs can be used as extensions to ad hoc networks (MANET), IEEE 802.16 based broadband wireless (WiMAX) networks, IEEE 802.11 WLANs (Wi-Fi), sensor and vehicular networks, and cellular networks [1]. A mesh router can have one or more functionalities of a gateway, router, and access point.

Usually, a mesh router has two or more radio interfaces to provide other routers with simultaneous transmissions and receptions and thus to increase the performance of WMN. Suppose the radios are allocated to the non-overlapping channels available in the spectrum in such a way that no interference exists with the communication of adjacent routers. In that case, the user throughput will be maximum. In MRMC WMN, the problem of channel selection is therefore important to consider in improving user performance [2].

The Internet traffic was considered only in the downlink direction by most approaches or researchers [2]. The proposed method allocates channels to multi-radio mesh routers in WMN such that the uploading speed of Internet users connected to WMN improves. Simulation results show a significant increase in the performance of the proposed work.

2 Related Work

The majority of the current research on Channel Allocation algorithms in Wireless Mesh Networks has been the emphasis on the allocation of non-overlapping channels [3–5] to the neighbouring node to reduce network interference. Since nodes operating on non-overlapping channels do not interfere, it may be possible to increase overall network throughput through simultaneous transmissions. As the number of radio interfaces in WMN is less than the number of orthogonal channels, channel assignment schemes have a significant need for the proper use of a restricted number of channels. The research on partially overlapping channels (POCs) for channel assignment has recently drawn tremendous attention to this realisation [6].

Previously, a channel allocation algorithm based on non-overlapping channels had to deal with only co-channel interference. The adjacent channel interference, which is interference between two neighbours configured on adjacent (partially overlapping) channels, is one of the key issues in the design of efficient channel assignment schemes using POCs. The main problem lies in need to consider interference between adjacent channels. Therefore, recently some attention has been given to POCs for channel allocation in wireless networks [6–11].

In multi-radio and multi-channel WMN, there exists a trade-off between network connectivity and channel interference. To minimise co-channel interference and to retain backhaul connectivity, all mesh routers need to maintain a common channel. It is a major challenge for the researchers to develop a channel allocation algorithm that compromises connectivity and interference issues. A wide range of methods is proposed for channel allocation of multi-radio and multi-channel in the WMN literature [12]. Different channel assignment schemes have different criteria with various parameters. Interference-aware channel selection is presented in [13, 14], channel selection scheme for maximising throughput is given in [15].

This paper offers an adaptive algorithm for effectively using multi-channel multi-channel multi-radio mesh routers to attain improved performance through partially overlapping channels.

3 Proposed Model

This paper designs a new strategy to improve the throughput of the Internet users connected to WMN in the uplink direction along with the downlink direction.

3.1 Assumptions

The hierarchical structure of WMN is assumed as shown in Fig. 1. Mesh routers have at least three interfaces. Among them, two interfaces are access interfaces and the remaining interface is for routing purpose. The routing interface is used for linking routers and used for sending the packets in a mesh backbone. One of the access interfaces is used to provide users with network access in the uplink direction and another for downlink direction.

The end-to-end user throughput is measured in two stages between the gateway node and the user which is associated with the Internet. The user's end-to-end throughput is the combination of access link throughput and the mesh backhaul throughput. The detailed formulation of the channel selection metric is described as follows:

3.2 Formulation of Channel Selection Metric

Consider a WMN based on IEEE 802.11 as shown in Fig. 1 with 'm' number of Mesh Routers and 'n' number of users. The MRs are denoted as $R_1, R_2, \ldots, R_m$, and the users are denoted as $u_1, u_2, \ldots, u_n$. One user can associate with only one MR and an MR can associate with multiple users. Therefore the user ui associated with an MR 'R_j' is then defined as u_i^j and the set of all users associated with an MR R_j is defined as U_j.

According to the channel access mechanism as defined by IEEE 802.11 CSMA/CA (Carrier Sense Multiple Access with Collision Avoidance), if a user has data to transmit, the user must listen to the channel to find out if the channel is free or busy. The user needs to wait until the channel becomes free if the channel is busy. The user has to wait for the DIFS (DCF Inter Frame Space) period if the channel becomes free. Also after DIFS time, If the channel is idle, the user has to wait for a SIFS time and then transfer a DATA packet. The receiver waits for a SIFS time after receiving the DATA packet and responds with an ACK frame. All the wireless nodes repeat this simulation within the network for each packet transmission. When a collision occurs during the transmission, all the nodes involved in the collision have to wait for a certain random amount of time and restart the procedure.

Accordingly, the transmission time T_t for a packet of size 's' bits with data rate 'r' bits per second is given by

$$T_t = T_{DCF} = T_{DIFS} + T_B + T_{DATA} + T_{SIFS} + T_{ACK} \quad (1)$$

where T_{DATA} = Length of data frame/bit rate = s/r.

Throughput of the active user is given by

$$Th\left(u_i^j\right) = T_{DATA}/T_t \quad (2)$$

The end-to-end throughput of a user is the aggregation of access link throughput and backhaul link throughput. The access link throughput of user u_i associated with a router r_j is calculated as shown in Eq. (2) and is given as

$$Th_{access\text{-}link} = Th\left(u_i^j\right) \quad (3)$$

It is also assumed that each mesh router r_j calculates throughput on its routing interface as

$$Th(r_j) = (DATA * L)/T \quad (4)$$

where *DATA* is the number of data packets received with size L in a reference time-period T.

The average backhaul link throughput of user u_i connected to router r_j is given as:

$$Th_{avg}b = \frac{Th(rj)}{|Uj|} \quad (5)$$

where U_j is the number of active users connected with router r_j.

Now the end-to-end throughput of user u_i^j is given as

$$Th_{E2E}\left(u_i^j\right) = w\,(Th_{access\text{-}link}) + (1 - w)Th_{avg}^b \tag{6}$$

where *w* is the access link weighing factor and its value ranges between 0 and 1. The value of *w* can be determined based on network conditions and application requirements. Each mesh router r_j calculates average throughput Th $_{E2E}$ (u_i^j) of active user u_i connected with it and if the throughput falls below the predefined threshold value $Th_{threshold}$ then the router r_j recalculates Th $_{E2E}$ (u_i^j) for all channels (except routing channel and current channel) as in Eq. (6) and selects a channel in which Th $_{E2E}$ (u_i^j) is maximum and the new channel number is conveyed through beacon/probe response frames to all its users. Threshold value $Th_{threshold}$ is determined based on network conditions and application requirements.

3.3 Proposed Algorithm

Input:
R = {r_1, r_2, r_3,…., r_m} given set of 'm' mesh routers
U = {u_1, u_2, u_3,…, u_n} given set of 'n' users
Step-1: User u_i checks for availability of free channel by sending probe request frame.
If free channel is available then
Go to step 2
Else
Wait for some fixed time till free channel is available
End if
Step 2: for each mesh router $r_j \in R$ in WMN do
begin
for each current active user $u_i \in U_j$ associated with mesh router r_j do
begin
Calculate transmission time taken for every active user
$T_t = T_{DCF} = T_{DIFS} + T_B + T_{DATA} + T_{SIFS} + T_{ACK}$
$Th(u_i^j) = T_{DATA} / T_t$
Where T_{DATA} = Length of data frame/ bit rate
Step 3: Calculate access link throughput of all active users associated with router r_j
$Th_{access\text{-}link}(u_i^j) = Th(u_i^j)$
Step 4: Calculates backhaul link throughput of its active users u_i^j as
$Th_{avg}^{b} = \frac{Th(r_j)}{|U_j|}$
Step 5: Calculates end-to-end throughput for each of its active users u_i^j as
$Th_{E2E}(u_i^j) = w\,(Th_{access\text{-}link}) + (1-w)\,Th_{avg}^{b}$
Step 6: if ($Th_{E2E}(u_i^j)$) <= Th threshold then
Repeat steps 2 to 5 for all channels except current channel and routing channel.
End for
Step 7: Select channel with maximum Th $_{E2E}$(u_i^j) and inform the same to all users through beacon/probe frames.
Step 8: User can upload data through selected channel.
End for

4 Performance Analysis

The proposed method is analysed using the NS3 simulator, an open-source and free simulation environment apt for networking. The simulation scenario with users and routers distributed randomly in the WMN is considered as shown in Fig. 2. The IEEE 802.11b WMN with 35 nodes connected to the Internet with high speed (1 Gbps) wired link has been considered for simulation. Each mesh router utilises the "Open Shortest Path First (OSPF)" routing protocol to route the packets. The simulation depends on the standard IEEE 802.11b that has three Orthogonal Channels out of 11 available channels, and the physical layer bit rate is 1 Mbps. Each node is configured with three radios. For network routing one radio and the other two radios are for data transmission. The "User Datagram Protocol (UDP)" is considered under the saturated internet traffic through gateway to all users with standard packet size 1024 bytes. The transfer rate of traffic flow is set to 1Mbps, and the simulation runs for 100 s.

By measuring the system throughput, the performance of the proposed approach is analysed by gradually increasing the number of users accessing the Internet through WMN and compared with the conventional WLAN technology. In Table 1, the concrete simulation settings are described.

Fig. 2 Wireless mesh network simulation

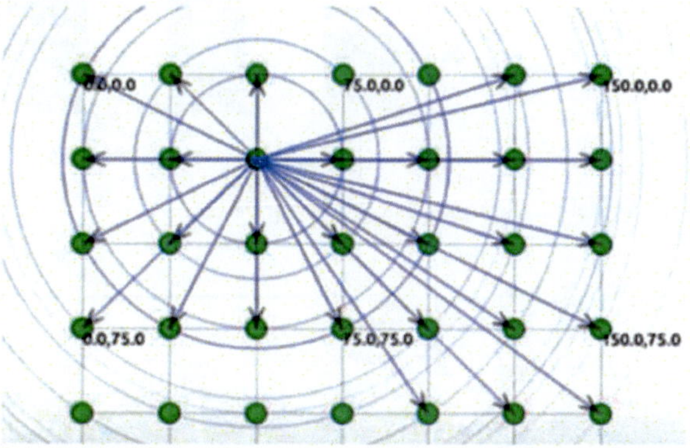

Table 1 Simulation parameters

Simulation parameters	Values
Number of nodes in the mesh topology	35 (5 × 7 grid)
Traffic type	UDP
Packet size	1024 bytes
Physical standard	IEEE 802.11b
Transmission rate	1 Mbps
Simulation time	100 s
Number of radios	3

The results of the simulation are compared with uplink throughput in IEEE 802.11 WLANs for validation of the proposed system as in [16]. Simulation results assess the throughput of the designed strategy. In the proposed method, the router calculates the end-to-end throughput for all users on both access link and backhaul link by taking into account different parameters such as channel conditions, access interface load, interference, etc., and selects the channel that provides the highest throughput and thereby providing required end-to-end Quality of Service. The results are averaged in various simulation runs as in Fig. 3.

As the proposed method considers the routing interface load and access interface load into channel selection decision, both downlink throughput and uplink throughput can be analysed based on internet traffic. The simulation results are as shown in Fig. 4.

This proposed system also focuses on the mobility of user nodes and their effect on the channel allocation problem. So, by changing the user speed in each simulation,

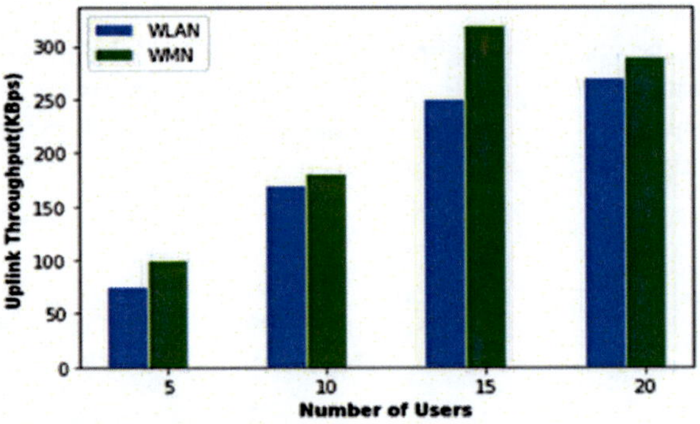

Fig. 3 Uplink throughput in WMN and WLAN

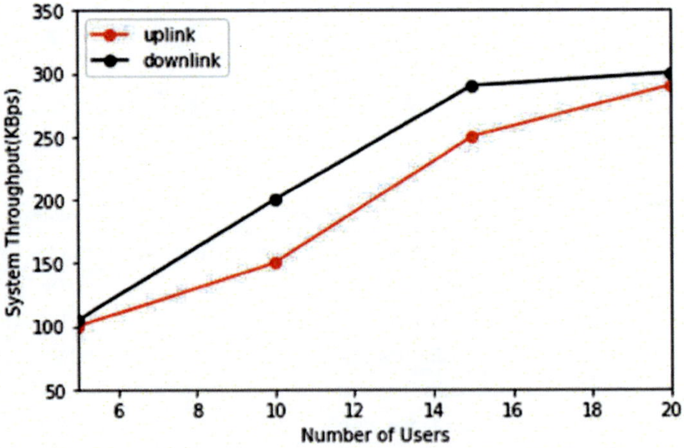

Fig. 4 Downlink and uplink throughput of WMN

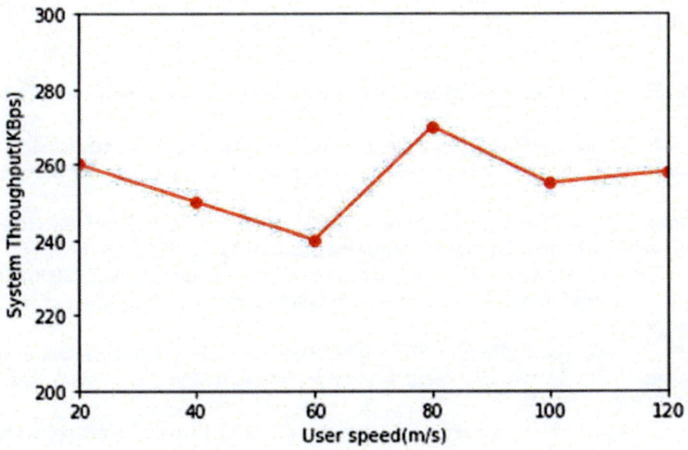

Fig. 5 Impact of user speed on throughput

the results are observed as shown in Fig. 5. There is no much impact on the throughput of WMN even the user is moving at a particular speed. Based on the feasible end-to-end throughput, the proposed method selects the channel and outperforms the WLAN technology.

5 Conclusion and Future Work

Wireless Mesh Network is a powerful technology for facilitating mobile users with wireless broadband Internet access. To increase the efficiency of Internet users, proper channels need to be allocated to various radio interfaces of a mesh router. This paper proposes a new method of channel selection to the access interface of the mesh router retaining a static channel for the routing interface of the mesh router, assuming that the mesh router has three radios. The results of the simulation show a significant improvement in WMN throughput in the uplink direction of data transfer. The proposed strategy can be generalised to mesh router with multiple interfaces. Further enhancement of the proposed model may involve better optimising techniques for improvement of user throughput, multicast channel allocation issue, etc.

References

1. Akyildiz IF, Wang X, Wang W (2005) Wireless mesh networks: a survey. Comput Netw 47:445–487
2. Lavanya M, Shoba Bindu C, Vijay Kumar G (2019) Internet traffic based channel selection in multi-radio multi-channel wireless mesh networks. Int J Commun Netw Inf Secur 11(2):262–269
3. Ramachandran KN, Belding EM, Almeroth KC, Buddhikot MM (2006) Interference aware channel assignment in multi-radio wireless mesh networks. In: IEEE INFOCOM
4. Barrett CL, Istrate G, Kumar VSA, Marathe MV, Thite S, Thulasidasan S (2006) Strong edge coloring for channel assignment in wireless radio networks. In: Proceedings of IEEE PerCom workshops
5. Das AK, Vijayakumar R, Roy S (2006) WLC30-4: static channel assignment in multi-radio multi-channel 802.11 wireless mesh networks: issues, metrics, and algorithms. In: IEEE GLOBECOM
6. Mishra A, Shrivastava V, Banerjee S, Arbaugh W (2006) Partially overlapped channels not considered harmful. In: ACM SIGMetrics/performance, pp 63–74(2
7. Cui Y, Li W, Cheng X (2011) Partially overlapping channel assignment based on "Node Orthogonality" for 802.11 wireless networks. In: IEEE INFOCOM, Shanghai
8. Bukkapatanam V, Franklin A, Murthy C (2009) Using partially overlapped channels for end-to-end flow allocation and channel assignment in wireless mesh networks. In: IEEE ICC, Dresden, Germany
9. Wang D, Lv P, Chen Y, Xu M (2011) POCAM: partially overlapped channel assignment in multi-radio multi-channel wireless mesh networks. In: IEEE ISCIT '11, Hangzhou, China
10. Hoque M, Hong X, Afroz F (2009) Multiple radio channel assignment utilizing partially overlapped channels. In: IEEE GLOBECOM, Honolulu, US
11. Duarte PBF, Fadlullah ZM, Hashimoto K, Kato N (2010) Partially overlapped channel assignment on wireless mesh network backbone. In: IEEE GLOBECOM, Miami
12. Lavanya BM, Bindu CS, Kumar GV (2017) A Study on channel assignment algorithms for multi-channel multi-radio wireless mesh networks. In: International conference on intelligent sustainable systems (ICISS), pp 568–575
13. Subramanian AP, Gupta H, Das SR, Cao J (2008) Minimum interference channel assignment in multi radio wireless mesh networks. IEEE Trans Mob Comput 7(12):1459–1473
14. Yiltas-Kaplan D, Durukan-Odabasi S, Kirci P (2016) An interference-aware channel assignment scheme for wireless mesh networks. Ad-Hoc Sens Wirel Netw 30(3–4);183–199
15. Chieochan S, Hossain E (2013) Channel assignment for throughput optimization in multichannel multi radio wireless mesh networks using network coding. IEEE Trans Mob Comput 12(1):118–135
16. Choi N, Seok Y, Kwon T (2013) Exploiting multi-user diversity for uplink throughput In IEEE 802.11 Wlans. IEEE Commun Lett 17(9):1710–1713

Multi-standard Schema-Based Classification of Geospatial Metadata in Spatial Data Infrastructures Using Feature Weight Induced Probabilistic Learning Scheme

Gangothri Rajaram and K. R. Manjula

Abstract Its metadata depicts the 'fitness-of-use' of the geospatial data. The Geospatial metadata contains information related to quality, lineage, bounding coordinates, etc., that is basic for users to select datasets suitable for them. Validating the metadata resource against a standard schema improves the visibility and visualises the fitness of the dataset. We propose a multi-standard perspective approach that aims to help users classify metadata that could support repositories in schema-based metadata validation process in Spatial Data Infrastructures and crosswalk between metadata standards. In this paper, a delineating study on the impact of a number of classifiers on the geospatial metadata set is performed. The metadata set is accumulated using metadata from NOAA, GEOSS, ISRO-Bhuvan clearing houses and Geonetwork, narrowing the choice of metadata towards its compliance to FGDC, INSPIRE and NSDI metadata standards for ease of study. A Weight induced Probabilistic Neural Network (W-PNN) Classifier with Gaussian kernel was proposed for the same and for the particular problem at hand, it is found to perform well in accuracy for large metadata set. With a small amount of metadata, the proposed W-PNN classifier and all other canonical classifiers had the same impact on metadata in terms of performance. But, in the case of an increased metadata size, W-PNN would be the better choice for metadata classification in Spatial Data Infrastructures (SDI).

Keywords Geospatial metadata · Spatial data infrastructures · Clearinghouse · Crosswalk · Fitness-of-use · Neogeographers · Neural network · Metadata validation

G. Rajaram (✉) · K. R. Manjula
Jain University, School of Computing, SASTRA Deemed University, Thanjavur, Tamil Nadu, India
e-mail: gangaprabhu1329@gmail.com

K. R. Manjula
e-mail: manjula@cse.sastra.edu

S. Jyothi et al. (eds.), *Proceedings of the 2nd International Conference on Computational and Bio Engineering*, Lecture Notes in Networks and Systems 215,
https://doi.org/10.1007/978-981-16-1941-0_66

1 Introduction

The usage and sharing of geographic information have become enormous and mandatory over the years. Spatial Data Infrastructures act as a platform to exchange and distribute spatial data between partners and stakeholders in the geospatial community [1, 2]. The current state of development in SDIs, geoportals and data standards [3] provides ease of access to geospatial information through a single repository containing an enormous volume of data from various providers. The availability of spatial data and its increasing amount offers users wide more options on spatial datasets. Thereby, more choices of datasets lead to more concern on retrieval of quality data and rise problems in decision-making for users. Researchers postulate that usually, geospatial data users have the least knowledge on the geo-information field, and subsequently, they are not aware of the risks and prospects of misused spatial data [4, 5]. This could end up with inappropriate conclusions for problems in a spatial context. Certainly, discussions in the literature on spatial data quality indicators [6] proved that metadata plays a vital role in the spatial data selection process. Besides, [6] the ample existence of numerous recommendations and the metadata standards data quality information is frequently not communicated in a consistent way. The study has identified that geospatial data users are highly interested in good quality metadata. Hence, there is a necessity to evaluate the quality of the geospatial metadata from various standards in the clearinghouses before storing, reproducing and populating them.

The clearinghouses worldwide have adapted their suitable metadata standards with an intimate architecture that links users to data providers through web-based Geoportals [7]. The most significant approach found in today's geoportals of National clearinghouses is to permit and encourage geospatial data users to contribute metadata in response to the increase in neogeographers. They provide tools such as a metadata editor for users to create proper descriptions about datasets in compliance with standard elements. Similarly, they have metadata validator that inspects and validates the metadata generated by users against standards. Being a vital resource, these tools become vague in the case of neogeographers as they are ambiguous about the various available metadata standards, elements, and usage. Therefore, when geospatial users tend to use/contribute spatial data from/to different (country's) geoportals it is incoherent to be aware of all available standards that various organisations incorporate. In fact, all the clearinghouses inhere to traditional standards such as ISO 19115, FGDC, INSPIRE and NSDI. These standards have common elements, but not all elements are identical to each other as they have their own uniqueness. Furthermore, [8–10] revealed the importance of metadata visualisation, validation and assessment.

Also, studies publicised the nexus between metadata validation and quality data sharing strategies [11]. Hence we hypothesise that users need an affable metadata validation process that perpends them to focus on description documentation than searching standard metadata templates in simple words, a schema-based validation tool that could support multi-standards. The various metadata validator tool available on the internet are narrowed to a specific standard viz, the European Inspire

Metadata Validator, which makes it tough for neogeographers to use it. To overcome this difficulty, this paper postulates the concept of a multi-standard metadata validation process, which accepts metadata of any traditional standard and checks its compliance with that particular standard. Hence, it relieves the stress of users to study and understand a specific standard that the portal supports, thus paving the way for them to concentrate only on descriptions of data. We propose a probabilistic learning scheme for classifying geospatial metadata based on the schema of different metadata-standard classes. For ease of study, this paper takes three widely used standards such as FGDC, INSPIRE and ISRO's NSDI ver 2.0 for study.

This work aims to improve the schema-based metadata validation process of portals and repositories with a central and common methodology (conceptually falling on legacy standards such as ISO 19115, FGDC, INSPIRE) means of pattern recognition and machine learning technique. We demonstrate the effectiveness of the methodology by testing it to diverse metadata sets from numerous clearing houses facilitating traditional standards. Experiments were carried out on a customised dataset formed as a combination of hundreds of metadata extracted from various geoportals using various standards and synthetic data. And the performance of W-PNN over existing traditional classifiers such as Multi-layer Perceptron Neural Network (MPNN), Back Propagation Neural Network (BPNN), Support Vector Machine (SVM), Naïve Bayes (NB), Random Forest (RF), Logistic Regression (LR), K-nearest neighbour (KNN) is evaluated with respect to various performance metrics (classifiers accuracy, precision, recall). The rest of the paper is organised as follows: Materials and Methods that initiated the need of classification in Geospatial Metadata; followed by an overview of Probabilistic Neural Network; the proposed methodology and architecture; the experiments carried out from metadata acquisition, preprocessing and the results observed with validation; Then finally, it concludes with future works.

2 Materials and Methods

2.1 Geospatial Metadata, Standards and Crosswalks

Once created, geospatial information is widely distributed and used by multiple software systems for various purposes. Governments have recognised Spatial Data Infrastructures (SDI) as a vital infrastructure for the development of the country. The tremendous development in technology has helped the accelerated distribution of digital resources that could aid as the foundation for the development of SDI's processes and methods. One such vital piece of an essential component for the development of SDI is the appropriate documentation of the data and services. This documentation is termed as 'metadata'. Metadata describes the content, quality, authorship and condition and other essential information about the data resources. Metadata files each describing specific resources are published through catalogue systems/directories/registries. Metadata and Catalogs are the key components of SDI that could facilitate ease of accessibility and interoperability of data resources and services. Furthermore, the performance and quality of services offered by the SDIs are directly proportional to the enhancement in the use of metadata. Many organisations have created and published geospatial metadata standards that suggest the mandatory and additional elements that could better describe the spatial data resources. The important standards are Content Standard for Digital Geospatial Metadata by Federal Geographic Data Committee (FGDC), INSPIRE Metadata Standard with reference to ISO 19115 and Open Geospatial Consortium (OGC) and much more. There exist vast diversity and heterogeneity in metadata standards and content. Also, there occurs heterogeneity in the schemas used for metadata. There is a Metadata knowledge base in the catalogue system that supports various metadata standards for the data resources in the SDI. The crosswalk is the mechanism that enables the transformation of metadata in accordance with the source standard to metadata in accordance with the target standard. Crosswalks have paved the way for the possibility of developing discovery services that could search effectively across heterogeneous metadata holdings. Thus, enabling metadata interoperability. Due to an increase in the number, size and complexity of metadata standards, the task of facilitating metadata for different standards is quite complex [12]. To maintain the interoperability of metadata standards, it is necessary to create software systems to convey several metadata dialects, that is, systems providing crosswalks between metadata standards. Crosswalks help to promote interoperability, supporting several search engines to search across heterogeneous databases effectively.

Presumably, crosswalks' construction is a difficult and error-prone task requiring profound knowledge and vast experience with metadata standards. In [13], the author has proposed some semi-formal methods for the crosswalk, but still, they are error-prone yet, it proposes the future need for automating crosswalk in SDIs, which has been established in [14]. Thus, we hypothesise: schema-based effective quantified classification of a stream of input/available metadata in a clearinghouse/repository prior to any crosswalk procedure in automatic crosswalk systems is imperative and reduces complexity. And the learning scheme for multi-standard schema-based classification in this paper would address this issue that could lend atomic support to promote automating crosswalk on a large scale.

2.2 *Quantification of Geospatial Metadata in Hypergraph Based Geospatial Metadata Framework*

The current work is the perpetuation of the preliminary study published by the same authors in [15]; development of a computational framework for Geospatial Metadata corresponding to the same standards considered in this study. The framework combines TopicMaps and Hypergraphs (HXTM) to exploit the interdependency of the metadata elements. It also employs a Hypergraph-Helly Property-based Weight-Assignment Algorithm (HHWA) that combines with Rank Order Centroid (ROC) method to provide standard weights for metadata elements that could be applicable to all standards. Rajaram et al. [15] provides a clear investigation of the need for the implementation of intelligent techniques on geospatial metadata. Also, it has been clearly indicated that the quantification outcome will constitute a significant feature for researchers to develop algorithms and techniques towards envisioning metadata services.

The technical content of the methodologies found in [15] are detailed in precise as follows: The metadata structure is bounded to a hypergraph structure, with the schema bounded to essential quality metrics (Completeness, Consistency, Conformance, Currency, Readability, Linkage, Accuracy and Richness of Information) and the nature of contents in the metadata elements (Thematic, Spatial, and Temporal) as hypergraph edges and metadata elements itself as hypergraph vertices. The hypergraph representation is treated against the recursive use of helly property harvesting dominance to pass through rank order centroid to determine deserving standard weights for all metadata elements. Therefore, the weights obtained are reliable to be taken forth for further value-adding services for metadata. The framework has also been exploited as a prototype web service tool "MetDEVViz—MetaData Editor Validator & Vizualization" that analyses the quality of metadata using the weight values. The quality assessment is done for individual standards that demand a prerequisite from the user to understand/know the standard and schema of the given metadata. The automation of this tool and making it useful for neogeographers would be possible if there is a learning mechanism for identifying which standard the particular input metadata belongs to and thereby classifying it to proceed further quality assessment techniques carried out in the tool that has been predefined for each standard. This has highly motivated the authors to propose a classification learning scheme with a comparison to existing state-of-art techniques that could be tailormade to suit the characteristics of metadata classification problem.

2.3 *Overview of Probabilistic Neural Network*

Probabilistic Neural Network (PNN) is developed by Specht in 1990 [16] as a simple neural network structure, that relies on Parzen Windows classifier by obtaining multivariate probability density estimates with a non-parametric estimator. Probabilistic Neural Networks belong to the Radial Basis Networks (RBN) family with parallel algorithms based on Bayesian Classification criteria and are closely related to conventional estimators for probability density function. It learns more quickly than major neural network models such as Multilayer Perceptron (MLP), radial basis function (RBF) and self-organising map (SOM) with a faster training process [17, 18]. The significance of PNN that makes it suitable for a wide variety of applications is the one-pass network [19] training phase where the network's initial weight not needed to set with no iterations for weight adjustments.

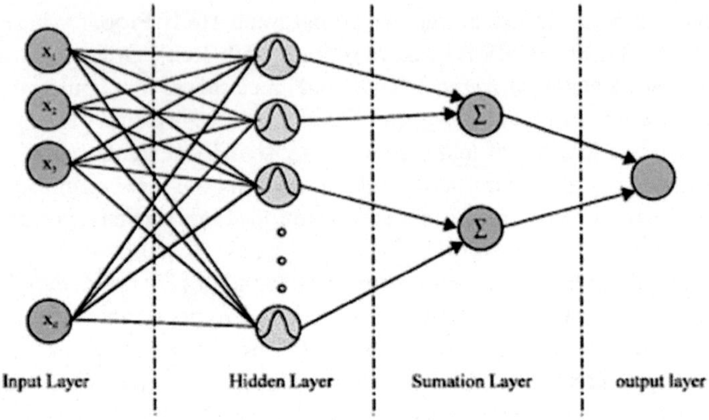

Fig. 1 Standard structure of probabilistic neural network

Thereby, does not suffer local maxima issues. The PNN architecture (Fig. 1) is composed of successive layers of interconnected neurons. PNN is a dense multi-layer network including an input layer, pattern or exemplar layer, summation or accumulation layer, and an output layer.

- The *Input Layer* is consists of n input vector to be identified and responsible for the distribution of patterns in input parameters to the next layer. The input layer does not perform any computation tasks.
- The *Pattern/Exemplar Layer* is composed of nodes called pattern units with exponential functions fully connected with the input nodes. There are many pattern units as training samples. The pattern unit is associated with feature values that act as connection weights to the input nodes.
- The *Summation/Accumulation Layer* contains the summation units that is responsible for probability estimation. The number of summation units depends on the classes. The summation unit receives data input from the corresponding pattern units of that class.
- The *Output Layer* acts as the decision layer with the maximum of the summation node outputs. With a simple comparison, the unique unit corresponding to the class that has the highest summation unit value resulting in the output one, while others generate zero to indicate the classification decision.

The input layer calculates the distance between the provided input vector and the learning vectors and the resultant vector point out the distance between the input and the training vector. This vector is multiplied element by element and sent to the radial-based transfer function. The second layer calculates the sum of all vectors for each class of entries to produce an output vector of probability. Finally, the activation function receives the output from the second layer, identifies the maximum of the subsequent probabilities and produces 1 for that class and zero for other classes. So, the networks vector classes' entry into a class that contributes to high probability determines the correct design parameter of the probabilistic neural network. Probabilistic Neural Network is suitable for pattern recognition, non-linear mapping, and classification problems. It is quicker in terms of magnitude when compared to back propagation as the training set size increases; it guarantees to represent an optimal solution with an elegant training method. The advantages of PNN are notably rapid training speed, learning capacity, guaranteed convergence, incremental training with generalisation and robust to noise. The prominent use of PNN are simplified Model update and fluctuating the training data does not need long-term training, however, offers faster learning speed. In this paper, we induce weights into the PNN model and evaluate the effectiveness of classification using existing performance metrics [20].

3 Proposed Methodology

Ever since its introduction, the probabilistic neural network was predominantly a classifier with the non-iterative training process. In the case of Geospatial Metadata, this unsupervised learning scheme is used because it is adaptive to structural changes where increasing the training size guarantees an optimal solution for the given classification problem. It is relevant to note that the betterment of the PNN falls on tuning the smoothing parameter to make the learning more robust. Since there exists an apparent variation in a number of elements found in each metadata standard the learning system is limited to features for classification. Also, some elements (features) are found to be identical for two or more standards such as Bounding Coordinates, Geographical Coverage, and Name of the Dataset etc. Thereby, a small perturbation in the input data (relative error) leads to large variation of results by amplifying the relative error marking the metadata's nature to be ill-conditioned statistically. Due to this ill-conditioned nature of the metadata, we prefer a probabilistic neural network using Gaussian activation function for classification based on standards since there is no iteration nor computation of weights and for a large number of Gaussians in the sum there is a significant error buildup.

As we know, computation of weights is not required in the probabilistic neural network whereas, in our context, the computation of metadata feature weights is performed earlier by Hypergraph Helly property. These weights are significant for distinguishing the metadata based on their standards which in turn is thereby helpful for classification. The proposed 'Weight induced Probabilistic Neural Network for Hypergraph based Geospatial Metadata Framework (W-PNN)' model propounds to induce these weights into the probabilistic neural network Fig. 2 illustrates the architecture of the proposed Probabilistic Neural Network Model for classifying the Geospatial Metadata based on various standards.

In this context, the traditional network is adapted by introducing multiple metadata recordset (M) as weight vector (ω) constituting 'n' weights (W) (vector of vectors), replacing the feature values with attribute weights.

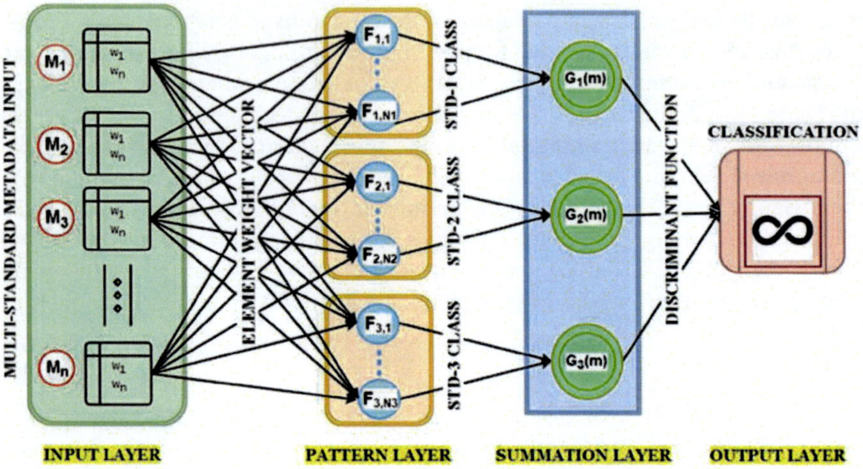

Fig. 2 Architecture of the proposed W-PNN model for improved metadata classification

$$M = \{1, m_2, \ldots .m\} \tag{1}$$

$$\omega = \{1, w_2, \ldots .w\} \tag{2}$$

The input to the proposed model is the multi-standard geospatial metadata list (vector array) that are considered as topographies for identifying patterns during classification. The format of each input metadata is a vector of weight values of the elements that are present in that metadata. There is one pattern node (P) for every standard's training sample. The weight vectors form a product with feature values associated with the pattern layer that adopts a pattern based on the training set. It passes through a Gaussian activation function (3); where ω^T is the training weight vectors, W_k is the networks weight

$$e^{(\omega^T W_{ki} - 1)/\sigma^2} \tag{3}$$

Each summation node receives the output from pattern nodes associated with a given standard class. The Gaussian functional values corresponding to all groups are feed into the output layer. At each class output node, all inputs are summed and multiplied with a constant (4).

$$\sum_{i=1}^{NK} e^{(\omega^T W_{ki} - 1)/\sigma^2} \tag{4}$$

The maximum of summed functional value at the output node procedure the classification decision (5).

$$\sum_{i=1}^{NK} e^{(\omega^T W_{ki}-1)/\sigma^2} > \sum_{i=1}^{Nj} e^{(\omega^T W_{ki}-1)/\sigma^2} \quad (5)$$

The most important factor to concentrate is the selection of smoothing factor, that is, the deviation of the activation (Gaussian) function for training the data, for small smoothing factors cause jagged approximations that cannot generalise well, whereas large deviations could smooth out details.

The flowchart in Fig. 3 picturises the procedure of the proposed algorithm, and Table 1 lists down the pseudocode of the proposed algorithm depicting the necessary inputs and resultant output. The algorithm requires a subroutine call to Hypergraph-Helly Property-based Weight-Assignment Algorithm (HHWA) [15] for metadata preprocessing. Then it passes through the training and testing phases of the neural network. The classification of metadata based on standards is achieved through W-PNN mainly in two stages: recognising patterns from feature weight vectors during

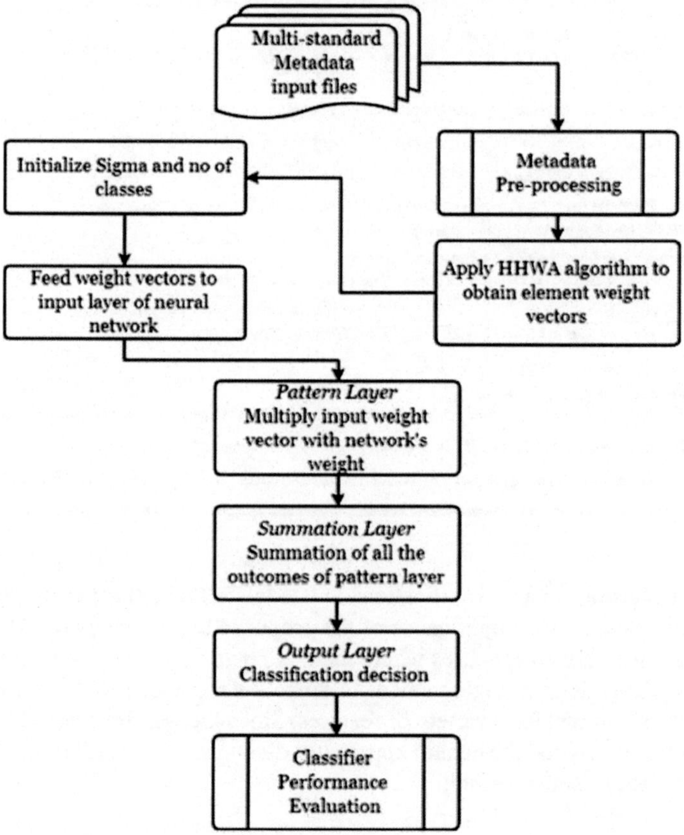

Fig. 3 Flowchart of the proposed algorithm

Table 1 Algorithm for the proposed classification methodology

Weight induced Probabilistic Neural Network for Hypergraph based Geospatial Metadata (W-PNN)

INPUT:

$M = \{m_1, m_2, m_3, \ldots, m_N\}$

// N – MultiStandard Metadata Samples, comprising training & testing samples//

$\omega = \{w_1, w_2, w_3, \ldots, w_n\}$ *// n – Weight Vectors//*

$S = \{S_1, S_2, S_3, \ldots, S_k\}$ *// k – classes/Standards for the given metadata set //*

$\sigma \leftarrow$ *smoothing parameter*

OUTPUT:

$C(M) \leftarrow$ *Class of the testing Metadata*

ALGORITHM:

- *In the MultiStandard Metadata Samples Dataset M,*
  Parse each Metadata files to extract elements in the metadata(m).
- *Apply for all metadata samples →*
  Hypergraph – Helly property based Weight Assignment Algorithm(HHWA) [15]
- ***Training phase:***
  1. *The individual metadata samples in training input pattern* $M = \{m_1, m_2, m_3, \ldots, m_N\}$ *each carry weight vectors* $\omega = \{w_1, w_2, w_3, \ldots, w_n\}$.
  2. *Create Pattern unit* P_M *that considers features of each class i.e, standards.*
  3. *Each Pattern unit* P_M *forms a dot product with the input weight vector* ω *and network's weight value* W_{ki}; $Z_i = \omega_i . W_{ki}$
  4. *Connect Pattern unit to Summation unit* G_M;
     to learn the pattern of all classes for classification decision.
- ***Testing Phase:***
  1. *For all Testing Samples M with weight vectors* ω*; repeat step 2 & 3*
  2. ***Pattern unit:*** *Compute* $Z_i = \omega^T W_i$
  3. ***Summation unit:*** *Compute* $\sum_i = 1^{Nk} e^{(\omega^T W_{ki} - 1)/\sigma^2}$;
     to calculate class membership for maximum probability on a discriminant function
  4. *Classification Decision* $C(M) = \arg\max(\sum_i(M))\,; i = 1, 2, \ldots, N$
- ***Return Classification decision for each metadata sample with Confusion Matrix***

the training phase and class identification of metadata in the testing phase based on the learning carried out. The significance of the proposed methodology over traditional PNN is; it is well-suited for cases where there is no quantification of feature values. The system depends on a single-valued classification parameter (in this case, it is the weight of the features) for a variety of features. Since the weights obtained are Rank based Centroid weights, they could apparently distinguish the predominant features that contribute class membership.

4 Experimental Evaluation

This section describes the dataset used for supporting the experiments, presents the experimental setup applied, and discusses the obtained results. We aimed to express a comparative study of performance metrics of our proposed model with other state-of-art classification techniques through experiments. For all classification techniques to be implemented, we have given the feature weights as inputs. In the case of PNN; the approach of giving weighted inputs has produced better accuracy. There exist numerous metadata standards affiliated to various geospatial agencies all over the world; though the proposed algorithm is capable of incorporating any number of standards for classification, due to ease of illustration, we consider three main metadata standards: FGDC, INSPIRE and ISRO_NSDI for experimentation point of view.

4.1 Metadata Acquisition and Preprocessing

In general, metadata are available in XML formats adapting various standards in various metadata catalogues. We have downloaded ample metadata sets from numerous earth observation centres' catalogue such as GEOSS, NEODA, INSPIRE Directive, Copernicus and ISRO-NRSC Open EO Data Archive comprising of ISO 19115/FGDC, INSPIRE and NSDI standards. A synthetic dataset suitable for multi-standard classification problem is customised containing the downloaded metadata and manually edited metadata for efficient testing of the proposed model. The created metadata set represents a mixed metadata collection that incorporates some standardised structure for the representation of elements and descriptions. The accumulated metadata are now considered to be raw collection of metadata elements that may neither be complete nor incomplete and even sometimes missing. Each metadata file is parsed to extract the presence of values for elements that are available with it. The quantified values of metadata elements in [15] can be used, when treating metadata over machine learning techniques. We can thus leverage this information to build our dataset. The entire synthetic metadata set is then converted to a comma-separated values (CSV) file by filling the standard weights as values for the metadata element that are present in the particular metadata file with the missing elements carrying null weight; here the standard weights are those calculated and published in the HXTM Framework for Geospatial Metadata Standards [15]. This file contains 5000 metadata records with 155 elements acting as features for predicting the standards and these are the input for training and testing of in all models.

Table 2 Characteristics of metadata standards under study

Geospatial metadata standards	No of metadata elements (No. of features)	No of elements similar to other standards
Fedderal Geographic Data Committee[FGDC]	85	16
Infrasturcture for Spatial Information in Europe [INSPIRE]	39	10
National Spatial Data Infrastructure [NSDI] ver 2.0-ISRO	31	11

4.2 Discussion

Our experiment relied on the dataset created using 5000 metadata with a mixed proportion of FGDC, INSPIRE and ISRO_NSDI standard metadata. The experiments were carried out on a PC with Intel Core i3 (2.4 GHz) processor with 6 GB RAM running on Windows 8 operating system, and the proposed algorithm is implemented in Python.

Table 2 represents the characterisation statistics for the dataset. Notice that most of the instances in our dataset have decisively added incomplete metadata for testing the efficiency of the proposed algorithm. A variety of experimental methodologies exists to evaluate the effectiveness of methods for classification [20]. In this work, we use precision and recall to evaluate the proposed methodology's quality, perform a comparison with other canonical classifiers proven in literature; and complement these values with a detailed analysis of the scenarios where the proposed algorithm produced a fluctuating performance. Precision and recall are per-class metrics to evaluate the classification algorithm on different aspects. Precision is the ratio of the number of data instances correctly classified to a class divided by the total number of instances assigned to that class. In a metadata classification problem; with three possible classes, that is, three different metadata standards, precision for each standard class corresponds to the number of metadata instances that are correctly identified by the classifier, over the total number of metadata (both correctly and incorrectly) identified by the classifier. Whereas, recall is the ratio of the number of instances correctly assigned to a class divided by a total number of items in the class. Recall for the metadata standard identification problem it is the number of matches that were correctly identified by the total number of metadata in the dataset. Since precision can raise at the expense of a recall, we also compute F1 measure that could equally weigh precision and recall through their harmonic mean. We have also measured the performance of the results through the accuracy metric, which is the proportion of correct decisions produced by the proposed method under evaluation. To leverage the given dataset, we used a five-fold and ten-fold cross-validation methodology. This means that the available metadata is split into 5/10 distinct subsets, with an equal number of metadata from the three standard classes. Since this work has made a primitive attempt to classify metadata based on its standards, there is no evidence or reference for the performance of standard classifiers over geospatial metadata. Therefore, analysis and comparison befit this paper to be liable for readers to conceive the impact of various classifiers over geospatial metadata including the proposed technique. Hence, we introduce the ranking statistics for comparison of W-PNN and the state-of-art classification algorithms. This shows in turn, which model performs best among all others. Finally, we also perform the Friedman test to present that the utilised comparison method is appropriate.

5 Results

The feature weights obtained from Hypergraph-Helly property-based Weight Assignment technique were represented in the form of a vector, and these weight vectors are used to train the PNN classifier. Table 3 presents the obtained results (highlighted in Table 3) and effectively compares the performance of various classifiers on the metadata set. However, in order to verify the scalability of the proposed algorithm and other classifier models, we have observed its performance on metadata sets with increasing dataset size; viz, 500, 1000, 2000 and 5000 metadata files. The algorithms marginally outperform each other, clouding the identification of prominent classifiers. Therefore, to have a keen notice on its significant outcome, the experiments are handled twice in each case with five and ten-fold cross-validation. The results of the proposed W-PNN was evaluated using the following metrics:

- *Sensitivity (True Positive Rate): Measures the proportion of sample metadata correctly identified with the appropriate standard.*
- *Specificity (True Negative Rate): Measures the proportion of sample metadata that are correctly identified to not belong to a standard.*
- *Stability: Rate of a number of successful training with total number of training.*

In the comparative study, canonical classifiers proven in literature and implemented in Matlab, R, or WEKA systems have been evaluated, and all of them worked on the metadata set. The experiments, in the beginning, was tried by converting the element numbers in the metadata as id values (ranging from 1 to155) and given as input to classifier algorithms. In many cases, it has failed to produce good results as id values could not project significant differences among similar elements in various standards. After the development of Hypergraph-Helly property used Weight Assignment technique, we have got weight vectors that could be possibly fed as input to any machine learning technique for better results. This is because of the weighted input is given to the machine learning techniques; that was formed as a fine-tuned values obtained from Hypergraph Helly Property and Rank Order Centroid.

From the experimental analysis in Table 3, it was clear that, the performance of the W-PNN and existing classifiers were found to be similar in the case of less number of metadata samples. Although, weighted inputs have been given to other classifiers. Also, the probabilistic learning nature of the PNN classifier has gone hand-in-hand with the deterministic weights generated through hypergraphs and

Table 3 Experimental results of classifiers with reference to performance metrics

Dataset Size 'k'-fold cross validation	Metrics	MPNN	BPNN	SVM	NB	RF	LR	KNN	W-PNN
500	*Presicion*	94.21	80.22	89.25	74.65	89.32	85.24	82.11	94.13
	Recall	94.58	83.95	88.63	74.52	89.24	84.58	82.02	93.11
	Accuracy	**95.12**	**89.24**	**92.13**	**88.56**	**90.85**	**90.45**	**89.92**	**94.54**
5-fold	*Rank*	1	7	3	8	4	5	6	2
500	*Presicion*	93.85	87.25	90.58	82.64	84.28	79.64	78.25	94.15
	Recall	93.98	87.10	90.55	82.57	78.54	80.54	88.21	94.08
	Accuracy	**95.12**	**88.95**	**92.25**	**88.94**	**90.88**	**90.54**	**89.88**	**94.62**
10-fold	*Rank*	1	7	3	8	4	5	6	2
1000	*Presicion*	92.25	84.62	94.12	80.54	87.54	86.32	80.81	93.67
	Recall	91.06	79.21	94.35	80.21	87.62	86.44	80.94	92.95
	Accuracy	**95.14**	**90.13**	**94.96**	**82.36**	**89.12**	**89.13**	**82.17**	**96.41**
5-fold	*Rank*	2	4	3	7	6	5	8	1
1000	*Presicion*	95.24	87.21	93.51	81.18	84.98	86.35	71.25	91.48
	Recall	94.95	86.37	93.89	81.88	84.66	86.24	73.54	91.14
	Accuracy	**96.23**	**90.15**	**94.82**	**82.66**	**88.59**	**88.50**	**81.11**	**96.56**
10-fold	*Rank*	2	4	3	7	5	6	8	1
2000	*Presicion*	94.62	79.69	94.58	75.34	79.98	76.54	69.34	96.35
	Recall	94.22	79.45	94.10	75.29	79.20	76.77	69.87	95.95
	Accuracy	**97.89**	**91.94**	**95.10**	**81.17**	**81.20**	**79.81**	**71.07**	**97.98**
5-fold	*Rank*	2	4	3	6	5	7	8	1
2000	*Presicion*	94.02	88.56	93.86	78.64	78.64	77.54	65.59	95.64
	Recall	93.79	86.41	93.75	78.79	77.95	77.48	66.95	94.68
	Accuracy	**97.88**	**91.56**	**94.58**	**80.35**	**80.98**	**79.11**	**71.95**	**97.92**
10-fold	*Rank*	2	4	3	6	5	7	8	1
5000	*Presicion*	95.54	82.15	94.56	74.24	78.24	75.18	69.86	96.21
	Recall	94.21	81.45	94.17	73.16	78.55	71.14	70.89	96.01
	Accuracy	**97.94**	**91.15**	**94.89**	**78.21**	**80.44**	**76.37**	**70.78**	**98.68**
5-fold	*Rank*	2	4	3	6	5	7	8	1
5000	*Presicion*	94.87	80.54	94.89	70.11	79.68	70.58	68.96	96.47
	Recall	94.71	81.24	94.52	69.79	79.77	69.54	69.82	95.32
	Accuracy	**97.32**	**92.56**	**95.38**	**79.89**	**80.34**	**76.23**	**70.11**	**98.54**
10-fold	*Rank*	2	4	3	6	5	7	8	1
Total Ranking Points:		*14*	*38*	*24*	*54*	*39*	*49*	*60*	*10*

centroid algorithms. This proves PNN to be the best-suited classifier for metadata, in particular for the ill-conditioned low-dimensional semistructured datasets. In addition, the performance of W-PNN was also evaluated with the existing classifiers with respect to stability and training time. Table 3 clearly states that the stability of the proposed algorithm is better than existing neural network-based classifiers. Although the training time of the proposed algorithm is quite higher than existing canonical

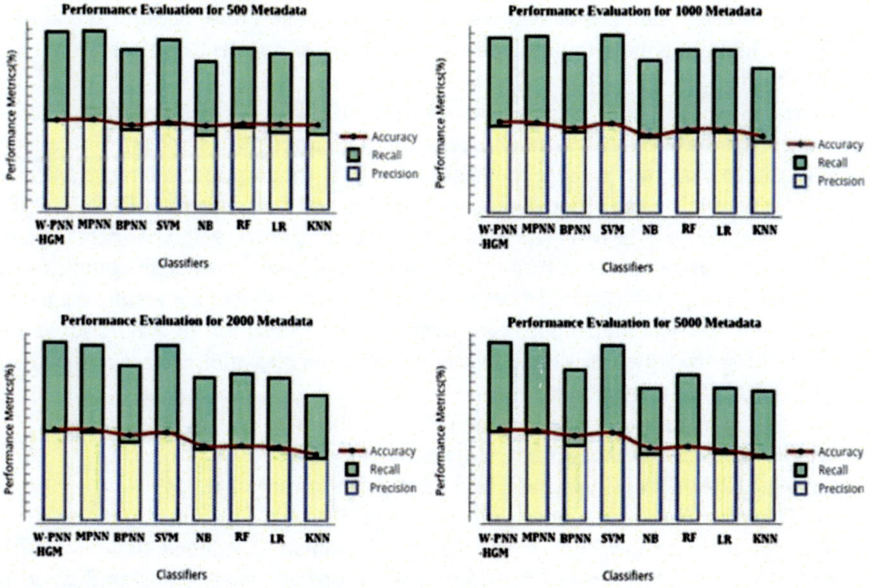

Fig. 4 Performance evaluation of the proposed algorithm

classifiers such as Random Forest, Naïve Bayes, Logistic Regression, K-nearest Neighbour and Support

Vector Machine, it is lesser than the existing neural network-based classifiers. However, with the increasing data size, W-PNN outperforms the existing classifiers through high precision, recall and accuracy (Fig. 4). Additionally, the ranking points (RP) is provided as a comparative factor that informs about the order of the achieved results for all analysed classification algorithms. = 1 indicates the model with highest accuracy and RP = 8 point the model with the lowest accuracy. The last row of the Table 3 describes the Total Ranking Points (TRP) the cumulative rank of any technique in all data size cases. The minimum the TRP value the most significant the algorithm; and the increasing TRP value gives proceeding better techniques. TRP of the W-PNN is 10 indicating it to have the best predicting ability with high accuracy in all cases of datasize; while for MPNN is 14, it is the second-best classifier with respect to all cases. It is worth collating the result of W-PNN, MPNN and BPNN as well. The W-PNN model outperforms the proposed technique in cases of increased data sizes. However, if we look more carefully at the results, we can certainly realise that the gain in accuracy is more substantial. In spite of achieving reduced RP during lesser metadata experiments, the W-PNN model is the best among all tested models with TRP = 10. Still, with reference to the results discussed in Table 3, we could see the deprived performance of W-PNN during smaller data size. Also, in those cases, MPNN has produced better results. In the case of large-sized metadata sets, W-PNN has given a better performance than other classifiers. It is also necessary to note down the performance of SVM throughout all the cases has given the same proportion during comparison with other classifiers. Therefore, none of the above classifiers, including the proposed one could be a best suited generic classifier for all cases of tests.

There are producing different results when the data size parameter differs. Therefore, the users have to make the choice of the classifier with a focus to data size parameters. In case of a smaller metadata set it is a good option to choose MPNN than W-PNN; if it is a large metadata set, it is the best choice to go with W-PNN. Still, when the user does not have a prior idea on the size of metadata set to be analysed and also could not afford a biased result than SVM could give a moderate approximate performance for classification. In order to statistically estimate the results acquired by all models, we accomplish the Non-parametric Freidman test in our work. Based on the ranking strategies presented in Table 3 for each classifier, we assume NULL and ALTERNATIVE hypothesis.

The NULL hypothesis of the test is framed as follows:

h0: All the classifier models have identical effects.

The ALTERNATIVE hypothesis is framed as follows:

h1: The classifier models have differnt effects.

With reference to the TRP in Table 3, the Freidman Q-value statistic is 49.70, while the number of degrees of freedom is 49. Supposing statistical significance at 0.99 level, the critical test value (CTV) is 27.21. The calculated Q statistic value exceeds CTV; thereby the NULL hypothesis can be rejected. Thus, the classifier models discussed in the experimental evaluation have different effects on the metadata classification problems with respect to data size. The conclusion of this test infers the reliability of the comparative study handled using ranking results in this paper.

6 Conclusion

The aim of the paper was to study and propose classification techniques that evaluate pictorial differences between metadata standards through feature weights. This has been accomplished with the proposed Weight induced Probabilistic Neural Network along with a detailed study on seven other traditional canonical classifiers. The results of the comparative study evaluated with various performance metrics are presented. The significant merits of the proposed work are:

- *The HXTM framework of Geospatial Metadata [12] is exploited for quantification of metadata elements for performing classification of metadata.*
- *A Weight induced Probabilistic Neural Network is proposed for classifying Geospatial Metadata on a Hypergraph-based framework (W-PNN) based on standards using metadata element weights.*
- *The proposed algorithm and other state-of-art classification techniques have experimented with different sized metadata sets belonging to various standards.*

- *The performance of all canonical classifiers and the proposed algorithm is evaluated with respect to precision, recall, accuracy and validated through ranking statistics.*

The results reveal that the effect of the classification algorithms varies with an increase in a number of metadata samples considered for the study. This has been validated through a non-parametric Friedman's Statistical test. This paper aids the metadata research community develop visions on exploiting machine learning techniques to better metadata services. In future, the algorithm can be embedded with a metadata validator for automatic multi-standard quality evaluation in geospatial metadata. Also, it can be combined with automatic crosswalk techniques in Spatial Data Infrastructures.

Appendix: Glossary

NOAA	National Oceanic and Atmospheric Administration
NEODA	NRSC Open EO Data Archive
GEOSS	Global Earth Observation System of Systems
ISRO	Indian Space Research Organization
FGDC	Federal Geographic Data Committee
INSPIRE	Infrastructure for Spatial Information in Europe
NSDI	National Spatial Data Infrastructure
W-PNN	Weight induced Probabilistic Neural Network
SDI	Spatial Data Infrastructure
ISO	International Organization for Standardization
MPNN	Multi-layer Perceptron Neural Network
BPNN	Back Propogation Neural Network
SVM	Support Vector Machine
NB	Naïve Bayes
RF	Random Forest
LR	Logistic Regression
KNN	K-nearest neighbour
GC	Open Geospatial Consortium
HXTM	Hypergraph-based TopicMap framework for Geospatial Metadata
HHWA	Hypergraph-Helly Property-based Weight Assignment Algorithm
ROC	Rank Order Centroid
MetDEVViz	MetaData Editor Validator & Vizualization
PNN	Probabilistic Neural Network
RBN	Radial Basis Networks
SOM	Self-organising map
RP	Ranking Points
TRP	Total Ranking Points

References

1. Rajabifard A, Feeney, Williamson IP (2002) Future directions for SDI development. Int J Appl Earth Observ Geoinf 4(1):11–22
2. Crompvoets J, Bregt A, Rajabifard A, Williamson I (2004) Assessing the worldwide developments of national spatial data clearinghouses. Int J Geogr Inf Sci 18(7):665–689
3. Percivall G (2010) The application of open standards to enhance the interoperability of geoscience information. Int J Digit Earth S1:14–30
4. Goodchild MF (1995) Sharing imperfect data. Shar Geogr Inf: 413–425
5. Grira J (2014) Improving knowledge about the risks of inappropriate uses of geospatial data by introducing a collaborative approach in the design of geospatial databases. Dissertation, University of Laval
6. Devillers R, Bédard Y, Jeansoulin R (2005) Multidimensional management of geospatial data quality information for its dynamic use within GIS. Photogramm Eng Remote Sens 71(2):205–215
7. Crompvoets J (2016) Geoportals. In: Richardson D (ed) International Encyclopedia of geography: people, the Earth, environment and technology, 1st edn. Wiley, pp 1–6
8. Zabala A, Riverola A, Serral I, Díaz P, Lush V, Masó J, Habermann T (2013) Rubric-Q: Adding quality-related elements to the GEOSS clearinghouse datasets. IEEE J Select Top Appl Earth Observ Remote Sens 6(3):1676–1687
9. Hillmann DI, Phipps J (2007) Application profiles: exposing and enforcing metadata quality. In: International conference on dublin core and metadata applications, pp 52–62
10. Assaf A, Senart A, Troncy R (2015) Roomba: automatic validation, correction and generation of dataset metadata. In: Proceedings of the 24th international conference on World Wide Web, pp 159–162
11. Tolosana-Calasanz R, Álvarez-Robles JA, Lacasta J, Nogueras-Iso J, Muro-Medrano PR, Zarazaga-Soria FJ (2006) On the problem of identifying the quality of geographic metadata. In: International conference on theory and practice of digital libraries, pp 232–243
12. Giuliani G, Guigoz Y, Lacroix P, Ray N, Lehmann A (2016) Facilitating the production of ISO-compliant metadata of geospatial datasets. Int J Appl Earth Obs Geoinf 44:239–243
13. Nogueras-Iso J, Zarazaga-Soria FJ, Muro-Medrano PR (2005) Geographic information metadata for spatial data infrastructures. In: Resources, interoperability and information retrieval. Springer, Berlin, Heidelberg
14. Yang H, Feng G (2012) Automatic creation of crosswalk for geospatial metadata standard interoperability. In: International conference on geographic information science, pp 311–324
15. Rajaram G, Karnatak HC, Venkatraman S, Manjula KR, Krithivasan K (2018) A novel computational knowledge-base framework for visualization and quantification of geospatial metadata in spatial data infrastructures. GeoInformatica 22(2):269–305
16. Pecht DF (1990) Probabilistic neural networks. Neural Netw 3(1):109–118
17. Wrobel K, Doroz R, Porwik P, Naruniec J, Kowalski M (2017) Using a probabilistic neural network for lip-based biometric verification. Eng Appl Artif Intell 64:112–127
18. Nayak DR, Dash R, Majhi B, Prasad V (2017) Automated pathological brain detection system: a fast discrete curvelet transform and probabilistic neural network based approach. Expert Syst Appl 88:152–164
19. Lotfi A, Benyettou A (2011) Using probabilistic neural networks for handwritten digit recognition. J Artif Intell 4(4):288–294
20. Seliya N, Khoshgoftaar TM, Van Hulse J (2009) A study on the relationships of classifier perform metrics. In: ICTAI'09 21st international conference on tools with artificial intelligence, pp 59–66

Solar Irradiation Prediction Using Weather Forecast Data

S. Gayathri Devi, K. R. Manjula, P. P. Sasisharan, V. S. Rohith, and N. Ramya

Abstract Solar radiation is an important parameter for the planning and operation of solar energy systems. Black-out occurs when the proportion of electricity that has been provided is greater or lesser than the proportion of electricity devoured. This has become an extra compound due to the hike in renewable energy, largely due to weather. Conventional power plants need to reach up to speed with these consistent variations as it is difficult to store power in enormous amounts throughout a prolonged time frame. The major objective of this work is to discover a machine learning model which accurately forecasts solar radiation. In this work, Linear Regression, ElasticNet, Gradient Boosting Regressor, Support Vector Regression, Random Forest Regressor, Extra Trees Regressor, Sequential and MultiLayer Perceptron models are used. The evaluation measures which are used in this work to identify the best model are MAE and RMSE.

Keywords Solar irradiation prediction · Solar energy · Weather forecast

S. Gayathri Devi (✉) · K. R. Manjula · P. P. Sasisharan · V. S. Rohith · N. Ramya
Computer Science & Engineering, School of Computing, SASTRA Deemed to be University, Thanjavur, India
e-mail: gayathridevi@cse.sastra.edu

K. R. Manjula
e-mail: manjula@cse.sastra.edu

P. P. Sasisharan
e-mail: sasisharan14@gmail.com

V. S. Rohith
e-mail: rohith.vs7@gmail.com

N. Ramya
e-mail: ramya.selvi1118@gmail.com

S. Jyothi et al. (eds.), *Proceedings of the 2nd International Conference on Computational and Bio Engineering*, Lecture Notes in Networks and Systems 215,
https://doi.org/10.1007/978-981-16-1941-0_67

1 Introduction

Solar energy is eco-friendly. During the production of this energy, no harmful gases are evolved, which will affect the atmosphere. Photovoltaic (PV) cell is the main source of solar energy. The PV cell captures radiation from the sun and turns it into Direct Current, which will be stored on to a battery. The solar cells need a little less continuance in their life ahead of installation so that the cost in forthcoming years is much reduced. So these solar-powered machineries are observed to be completely eco-friendly, structured, noiseless, and extremely dependable. For an uninterrupted energy supply, the sun is the most pre-eminent source which has been in power for long years that have gone. The total energy which can be drawn from renewable sources is predicted to be 11,100 TWh in the year 2035 [1]. The best out of those would surely be the solar energy which would give hands in most of our needs. The comprehensive worldwide installed solar power had a capacity of 1.5 GW in 2000 which has now exceeded its limits and has reached 40 GW in 2010 (Fig. 1).

1.1 Photovoltaic Cells

The photovoltaics, also known as PV, has developed from a small-scale application to become a mainstream electricity source. The PV converts the light into electricity using the photoelectric effect. It reduces pollution, gives healthier air,

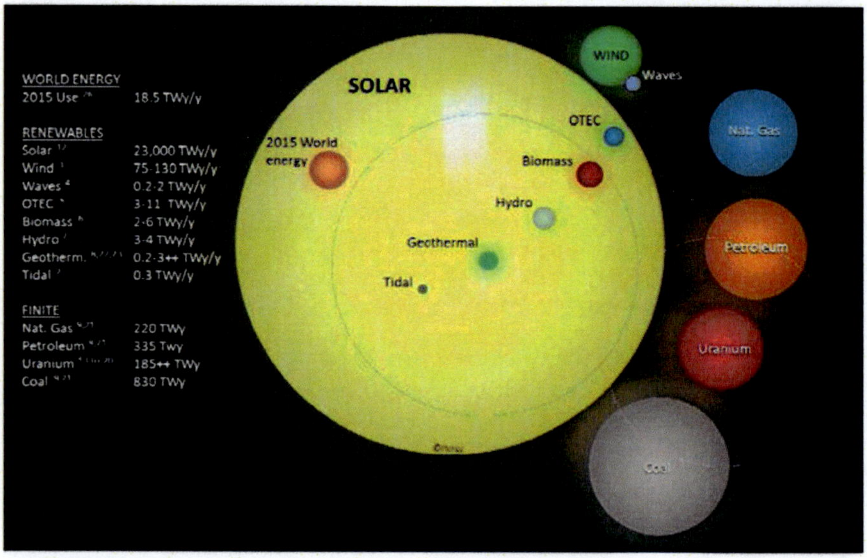

Fig. 1 Comparison of potential of energy sources

reduces dangerous greenhouse gases from cars, increases countries energy security by depending on an indigenous and inexhaustible resource [2, 3].

2 Materials and Methods

2.1 *Dataset*

This dataset consists of meteorological data from the Hawaii Space Exploration Analog and Simulation weather station for four months from September 2016 to December 2016. The dataset comprises attributes such as wind direction, wind speed, humidity, temperature and Solar radiation. It contains measurements for the past 4 months and the level of solar radiation is predicted. The total dataset resulted in 32,686 observations for all 11 attributes. It is divided into a training dataset and test dataset. The training dataset is used to develop a prediction model whereas the test dataset is used to evaluate the developed model.

2.2 *Performance Metrics*

Root Mean Square Error (RMSE) and Mean Absolute Error (MAE) are used as performance metrics assuming n as the number of test data, yi as the observed solar radiation and $\hat{y}$ as the predicted solar radiation.

$$\mathbf{MAE} = \frac{1}{n}\sum_{i=1}^{n}|\hat{\mathbf{y}}\mathbf{i} - \mathbf{yi}| \quad \mathbf{RMSE} = \sqrt{\frac{1}{n}\sum_{i=1}^{n}(y - \hat{\mathbf{y}}\mathbf{i})^2}$$

3 Methodology

In any prediction of the radiation, an understanding of its characteristics and phases is of fundamental importance [4] (Fig. 2).

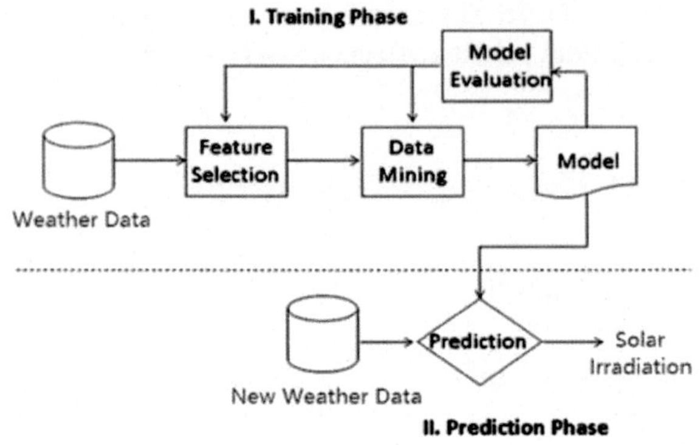

Fig. 2 Workflow model

3.1 Normalization and Scaling

Basically, normalization is a technique as often applied as a portion of data preparation for ML. The point of the Normalization strategy is to change over the estimations of numeric sections in the dataset to utilize a typical scale, without contorting contrasts in the scopes of values or losing data. Feature scaling can be used to normalize the features or the range of independent variables. The range of all features should be normalized to contribute an approximately corresponding proportion to the final distance.

Similarly, standardizing the dataset is the fundamental need in case of machine learning estimators while implementing through scikit-learn; otherwise, they can behave in an unexpected manner if the individual features are not in the form of expected standard; that is, Gaussian having zero mean and unit variance.

Most elements present in the objective function of the machine learning algorithm are assumed to have features with zero centrality and variance of a similar order. The disadvantage of having a feature with an unusually large variance in comparison with others is that it can dominate the function and prevent the estimator from learning the other features fairly.

3.2 Correlation

Using heatmap (Fig. 3), the exploratory data analysis is done for visualizing the data. Correlation between the parameters in this work is found using the Pearson Correlation Coefficient and it is given by:

$$r = \frac{\sum(x_i - \bar{x})(y_i - \bar{y})}{\sqrt{\sum(x_i - \bar{x})^2 \sum(y_i - \bar{y})^2}}$$

r correlation coefficient

x_i values of the x-variable in a sample
$\bar{x}$ mean of the values of the x-variable
y_i values of the y-variable in a sample
$\bar{y}$ mean of the values of y-variable.

The dataset has 80% as training data and 20% as testing data. The dataset is trained with LinearRegression, ElasticNet, GradientBoostingRegressor, SVR, RandomForestRegressor, ExtraTreesRegressor, MLP and Sequential model.

3.3 Linear Regression and Support Vector Machine

Linear Regression models the connection between two factors by fitting a direct condition to observed data (Training Data). One variable is viewed as an illustrative variable(Value to be predicted), and the others expect there exists a straight relationship and are viewed as a reliant variable. The graph is plotted for dependent variables and independent and a suitable linear equation is written for the observed data points.

Support Vector Machine is a supervised machine learning algorithm that can be used for classification or regression purposes. A hyperplane is being generated using which objects are being classified based on their classes. This is termed SVM as it uses extremes of cases known as support vectors, which support generating the hyperplane.

3.4 Bagging Ensemble Methods

Random forest incorporate various discrete decision trees that work as a group. The main idea behind random forest is a straightforward yet incredible one. It joins numerous choice trees to decrease the danger of overfitting. Combining the forecasts from each tree lessens the variance of the expectations, overhauling the performance on test data. The extra trees are another ensemble method that creates numerous decision trees, and a mean is taken for them. Random forest method takes optimal split point while extra trees take random split point.

3.5 Boosting Ensesmble Methods

Gradient Boosted Trees (GBTs) are also a bunch of decision trees. They repetitively train decision trees to limit the loss function. Each iteration uses the current ensemble to foresee the label of every training occurrence and afterwards it compares the prediction with the genuine label. The dataset is re-named to give more significance

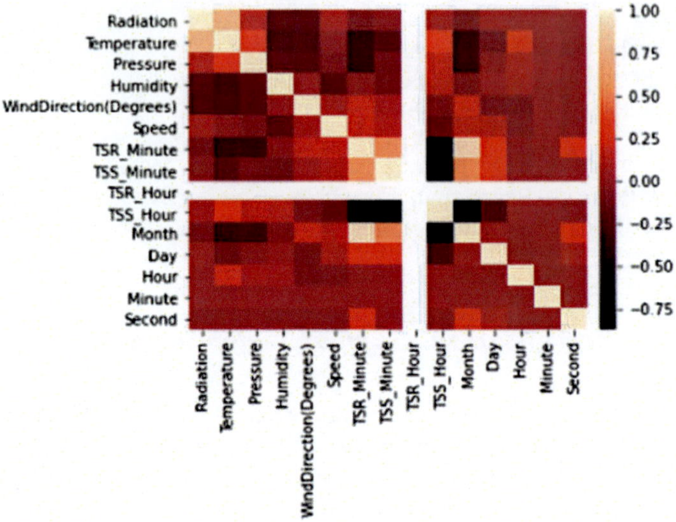

Fig. 3 Correlation Heatmap

to preparing cases with poor expectations. In this manner, in the following cycle, the decision tree can encourage rectification for past mix-ups.

4 Results

For the training data, the Sequential model's MAE value is 109.81334769662789, whereas the MLP model has 110.11306324234464 as the MAE value (Table 1).

The plot for the training dataset are given in (Fig. 4):

For testing data, the sequential model's MAE value is 2589.060891871356, whereas the MLP model's MAE is 3723.320752962702 (Table 2). The plot for the testing dataset are given in (Fig. 5).

5 Conclusion

Solar power is a much more optimal resource than other conventional energy sources, but due to the fact that there is no proper mechanism to predict the solar energy produced on a day-to-day basis, it cannot be used as the primary source of energy. Through this paper, leveraging of weather parameters has been done to predict solar irradiation. From the experimental results, it is clear that the Gradient Boosting Regressor model outperforms all the other models with a very high margin. This model can predict solar energy production to a certain extent. This process is based

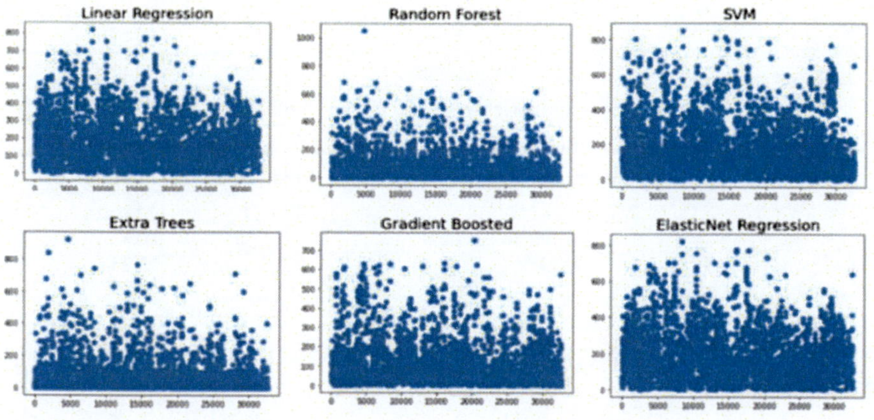

Fig. 4 Absolute error difference for various models in the training set

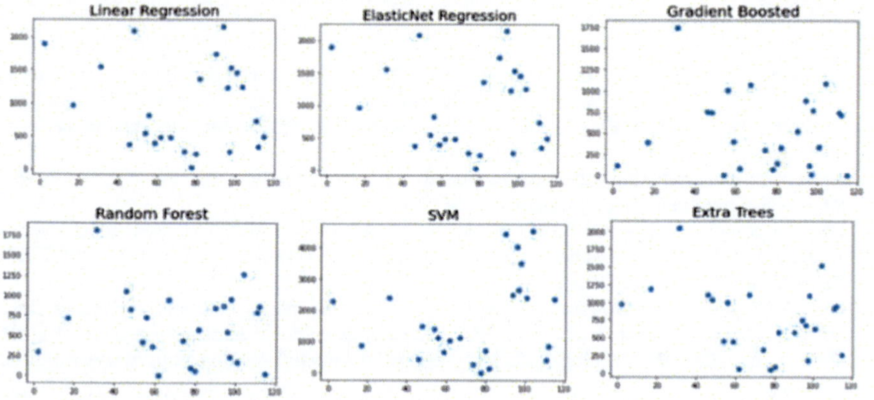

Fig. 5 Absolute error difference for various models in the testing set

Table 1 Comparison of MAEs and RMSEs of various models (training data)

Model	MAE	RMSE
Linear Regression	178.336	224.542
ElasticNet Regression	178.358	224.553
Random Forest	71.073	121.562
Extra Trees	68.01	122.658
SVM	176.645	231.631
Gradient Boosted	111.889	160.292
MLP	110.113	109.819

Table 2 Comparison of MAEs and RMSEs of various models (testing data)

Model	MAE	RMSE
Linear Regression	945.282	1140.2
ElasticNet Regression	945.127	1139.87
Random Forest	620.666	745.919
Extra Trees	784.904	889.18
SVM	1736.9	2186.14
Gradient Boosted	513.937	673.213
MLP	3723	3723

on a novel approach. Further research is required to make the prediction feasible for practical applications.

References

1. A solar radiation prediction model using weather forecast data and regional atmospheric data. In: 2018 IEEE 7th world conference on photovoltaic energy conversion (WCPEC) (A joint conference of 45th IEEE PVSC, 28th PVSEC & 34th EU PVSEC). https://ieeexplore.ieee.org/document/8547750
2. Sauer FJ (2013) Predicting solar radiation based on available weather indicators. Masters theses. https://scholarsmine.mst.edu/masters_theses/7361
3. Detyniecki M, Marsala C, Krishnan A, Siegel M (2012) Weather-based solar energy prediction. In: IEEE international conference on fuzzy systems, pp 1–7. 10.1109/FUZZ-IEEE.2012.6251145
4. Park S-W (2014) An irradiation prediction model for photovoltaic power generations under limited weather information. In: FAC proceedings volumes. https://doi.org/10.3182/20140824-6-ZA-1003.02142National. Center for Biotechnology Information. http://www.ncbi.nlm.nih.gov

EEG-Based Brain-Electric Activity Detection During Meditation Using Spectral Estimation Techniques

Padmavathi Kora, A. Rajani, M. C. Chinnaiah, K. R. Madhavi, K. Swaraja, and K. Meenakshi

Abstract The meditation consists of various stages of concentrating on the feeling of peaceful realization of becoming a person and communion of the human soul with the supreme soul brain electric source position in the frequency domain used on multichannel EEG recordings to create activation differences between meditation and open eyed, task-free resting. EEG signals are collected after 3rd, 6th, 9th and 12th week of training with three types meditation Trans dental, Raja yoga and Mindfulness meditation. Then EEG signal are classified using transfer learning methods (VGG-16, VGG-19, ResNet-18 and GoogleNet). EEG showed reduced activity in delta and increased activity in low alpha frequencies. The percentage of the alpha activity in the total power was a better indicator of the state of meditation. With the opening of the eyes the total power and the percentage of alpha activity came down. The percentage of alpha activity was higher which signifies perfect meditation. Alpha and beta are highest in the midline central area (Cz) during the relaxed meditation state, and theta is higher in C3 and C4. After experimental evaluation, we observed that the outcomes of these models gives 99.4% accuracy with the VGG-16 transfer learning model.

Keywords Meditation · EEG · Spectral analysis · Transfer learning

P. Kora (✉) · K. Swaraja · K. Meenakshi
GRIET, Hyderabad 500090, Telangana, India

A. Rajani
JNTUH, Hyderabad 500085, Telangana, India

M. C. Chinnaiah
BVRIT, Hyderabad 500090, Telangana, India

K. R. Madhavi
Sree Vaidyanathan Engineering College, Tirupati 517102, AP, India

S. Jyothi et al. (eds.), *Proceedings of the 2nd International Conference on Computational and Bio Engineering*, Lecture Notes in Networks and Systems 215,
https://doi.org/10.1007/978-981-16-1941-0_68

1 Introduction

EEG is a simple way to research functional features of Brain and brain-body link condition. The aim of this research is during meditation, the brain activity is to be examined; the brain activity that is represented as a state in a relaxed and calm state. The goal is to investigate and 0bserve EEG during meditation, functions present in the EEG signal through the use of FFT and DWT. The strategy taken to extract meditation-related EEG features are carried out by study of time-series, using the normalized voltage of the signal as the basic trait, say. The EEG signal is registered at the level of the point F3 and point F4, before and after meditation. In recent meditation studies, a lot of attention has been focused on the comparison between the meditation-group and the non- meditation group (control groups), with studies that show certain good qualities, such as stronger mental function, being in the meditation community, more clearly shown. However, questions were raised whether the positive traits were a result of meditation activities, or people with certain traits are naturally drawn to engage in meditation activity. For the purpose of clearer study, the effects of meditation, a physiological marker of meditation must first be determined. Then, using the physiological forecaster of meditation, objective analysis, as an independent measure, more of meditation's advantages are accomplished with more simplicity. Since meditation is mostly a mental exercise, the physiological inquiry, of course, starts with the brain. For frequency domain root localization, we used identify variations in intra-cortical electrical activity occurring When meditating. We used frequency domain source localization to detect intra-cortical electrical activity changes occurring during meditation. Centered on the above description of the required cognitive and effective details of this activity to be areas in the brain that are also involved in focus control as encoding of emotion and memory to demonstrate alterations via electrical operation. Several efforts have been made in order to categorize the methods of meditation. The very popular systems of classification differentiate between oriented structures center, open monitoring, and automated self-transcendence professions.

After the 3rd, 6th, 9th and 12th EEG signals are collected, after the training of three kinds meditations like Trans Dental Meditation, Raja-yoga and Meditation on Mindfulness meditation. EEG features are collected using techniques of spectral estimation (DWT and FFT). The derived features are then categorized using methods of transfer learning (VGG-16, VGG-19,ResNet-18 and GoogleNet), as seen in Fig. 1, respectively. Decreased delta activity and elevated activity have been shown with medium frequencies of alpha.

2 Related Work

Several advantages have been correlated with [1–4] meditation in previous trials, such as a drop in stress level and decreased inflammatory responsiveness; Increased

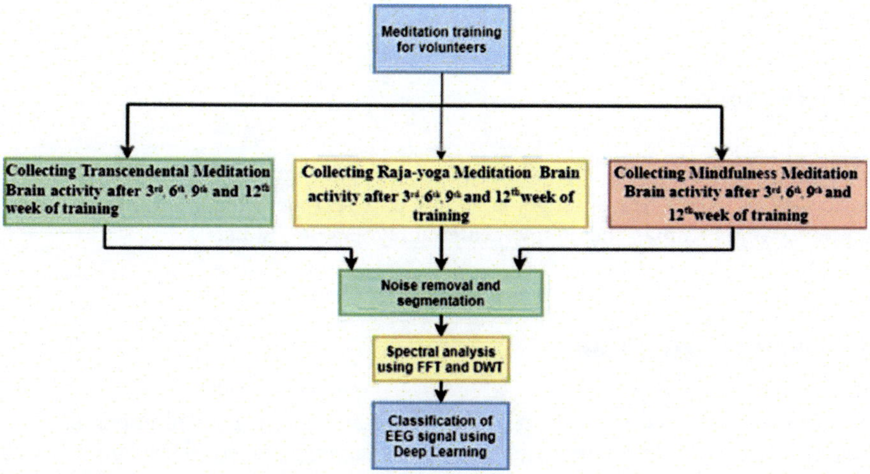

Fig. 1 EEG feature extraction and classification

production of dopamine in limbic brain regions correlated with inspiration and sensation of joy; increased amounts of melatonin and serotonin in the blood plasma, both neurochemicals that play a major role in mood regulation, tension and prevention of ageing [5, 6]; Superior regulation in selective and sustained concentration and inhibitory control [7–9]; lower scoring of fear, neuroticism, psychoticism, and depression personality traits, and higher scoring in emotion detection and speech [10], which indicated improved expertise in emotion management; and reduced fatigue, anxiety, and ruminatatism [11]. Meditation as a formal procedure, however, was not generally developed as a viable therapeutic application. While several experiments have been able to display positive meditation-related characteristics, findings taken from a comparison of long-term practitioners vs. control groups raised concerns as to whether the positive characteristics found were direct results of meditation practice or were naturally innate between different population groups. Before meditation may be implemented as a prescribed technique for a medicinal reason, a bet- ter definition of meditation behavior and its correlation with benefits must be checked. According to a study by Cahn and Polich, sixty-four meditation EEG studies from 1957 to 2005, according to a review by Cahn and Polich [12], processed EEG data using Power Spectrum, sixty-four meditation EEG studies spectral analysis EEG data. The conclusion reached by Cahn and Polich was that although meditation tended to influence theta and alpha PSD operations, no other inferences were evident.

3 Methods

3.1 Database

EEG from 52 meditators (age: 42.0 ± 10.1 years, range: 25-59; meditation experience: 17.5 ± 10.8 years, range: 4-43; 14 females) were recorded during winter season Brahma Kumari meditation center at Hyderabad, India.

4 Feature Extraction

An EEG signal is recorded from 52 subjects. Each record generated important findings (F3 and F4). Data from each channel is sampled twice; the length of each sampling segment is one second. A total of 52 data from meditation and 52 data from post meditation were reported. The instantaneous statistics of the signal, such as the signal average, were investigated in order to measure this qualitative finding. The fast Fourier transform (FFT) was added to the resting state EEG data with the rejected first and last 1 min data to comply with the portion of the study. The EEG signals were first re-referenced to the mean potential electrode reference on the scalp and then filtered between 0.5 and 30 Hz by bandpass. The analysis period was then divided by an epoch of 1 s, and epochs where the voltage exceeded 150 micro volts were refused. The FFT was introduced after these preprocessing steps. In steps of 0.1 Hz, the log transformed power spectrum was measured and was averaged over frequencies within the frequencies of theta (4-7 Hz), alpha (8-12 Hz) and beta (13-18 Hz).

5 Results

We created a CNN in Keras (Python 3.7), which takes a 2D matrix input (like a grayscale image) and outputs a one-dimensional vector. CNN fed with a raw 2D array also has height, width and depth=1, like an image. The original input size is 1, 32*512 and the input is given as .csv file to the convolution layer. The time duration 512 samples(1 s) are resampled as 128. With the number of time samples as the width and the number of electrodes as the height, we represent the input as a 2D series. Compared with the "EEG-as-an-image" approach, this approach also greatly reduced the input dimensionality. In the proposed research, the pre-defined transfer learning architectures are trained and validated for binary classification. For the classification task, four different transfer learning architectures (VGG-16, VGG-19,ResNet-18 and GoogleNet) are examined. The training and validation loss, confusion matrix and ROC and AUC of the binary- class classification are shown in Fig. 2 and Table 1. Initially, the network tuned every model for 50-epochs with a learning rate of 0.0001.

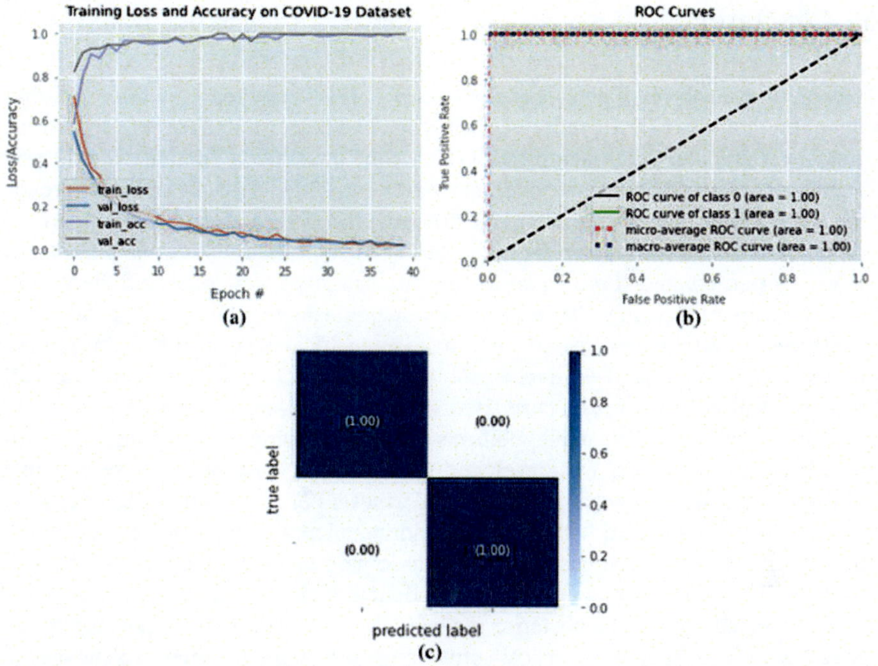

Fig. 2 **a** Accuracy curve of VGG-16 **b** ROC curve of VGG-16 **c** Confusion matrix of VGG-16

Table 1 EEG based Brain classification

	Sensitivity (%)	Specificity(%)	Precision (%)	F1-score (%)	Accuracy (%)
VGG-16	100	98.6	99	99.5	99.4
VGG-19	99	97.2	97.9	98.9	98.8
ResNet-18	97.9	95.8	96.9	97.4	97
GoogleNet	95.8	94.4	95.8	95.7	95.2

For this sort of data using SVMs and KNNs, our findings using deep networks are complementary to previous work. For different purposes, though, we consider this transfer learning approach to be especially important. Second, it largely mitigates the need to pick features (spectral bands and channels). The works on smaller data-sets, after experimental evaluation, we observed that the outcomes of these models gives 99.4% accuracy with the VGG-16 transfer learning model.

6 Conclusion

This method analyses electroencephalogram (EEG) knowledge obtained during intense meditations (Transcendental, Raja-yoga and Mindfulness) over four months. First, novel tools for preprocessing EEG data were developed. These devices helped to semi-automatically eliminate ocular artifacts, muscular artifacts, and interference from power lines. Secondly, using spectral analysis, longitudinal shifts in cortical function have been assessed in order to classify the cortical causes of meditation. In the retreat participants, three major longitudinal changes were observed: decreased individual alpha frequency after exercise, comparable decrease was consistently seen in seasoned meditators; decreased alpha-band intensity in the frontal mid-line region, correlated with increased vigilance efficiency; and decreased beta-band intensity in the parietal-occipital areas, correlated with daily vigilance efficiency.In order to provide a concrete and testable hypothesis of the underlying mechanisms, a systematic computational model was developed that showed (1) decreased intra-thalamic gain after exercise, suggesting increased alertness; (2) increased cortico-thalamic latency, closely associated with decreased individual alpha frequency (found during spectral analysis); (3) decreased intra-thalamic gain to the brain; and (4) anterior-posterior division in the modeled thalamus reticular nucleus layer, and improved post-training connectivity in the posterior modeled thalamus reticular nucleus region. In order to base the changes in cortical activation and model parameters on changes in behavior and self-reported psychological functions, correlation analysis was carried out. A concrete theory was built of the processes behind concentrated concentration meditation. Behind the improvements seen during meditation preparation, this principle offers both mechanistic and teleological reasoning. The hypothesis also contributes to other predictions, including the idea that people suffering with neurodevelopmental conditions and seizures will be handled with personalized meditation strategies. This study aims to relate the hypothesis to the long-held views that perception, concentration, stability, and psychological well-being are strengthened by meditation.

References

1. Rosenkranz MA, Lutz A, Perlman DM, Bachhuber DR, Schuyler BS, MacCoon DG, Davidson RJ (2016) Reduced stress and inflammatory respon- siveness in experienced meditators compared to a matched healthy control group. Psychoneuroendocrinology 68:117–125
2. Kjaer TW, Bertelsen C, Piccini P, Brooks D, Alving J, Lou HC (2002) Increased dopamine tone during meditation-induced change of consciousness. Cognitive Brain Res 13(2):255–259
3. Rubia K (2009) The neurobiology of meditation and its clinical effectiveness in psychiatric disorders. Biol psychol 82(1):1–11
4. Meenakshi K, Swaraja K, Kora P (2019) A robust DCT-SVD based video watermarking using zigzag scanning. In: Soft computing and signal processing. Springer, pp 477–485
5. Harinath K, Malhotra AS, Pal K, Prasad R, Kumar R, Kain TC, Rai L, Sawhney RC (2004) Effects of hatha yoga and omkar meditation on cardiorespiratory performance, psychologic profile, and melatonin secretion. J Altern Complement Med 10(2):261–268

6. Solberg EE, Holen A, Ekeberg Ø, Østerud B, Halvorsen R, Sandvik L (2004) The effects of long meditation on plasma melatonin and blood serotonin. Med Sci Monit 10(3):CR96–CR101
7. Slagter HA, Lutz A, Greischar LL, Francis AD, Nieuwenhuis S, Davis JM, Davidson RJ (2007) Mental training affects distribution of limited brain resources. PLoS Biol 5(6):e138
8. Jha AP, Krompinger J, Baime MJ (2007) Mindfulness training modifies subsystems of attention. Cogn Affect Behav Neurosci 7(2):109–119
9. Padmavathi K, Ramakrishna KS (2015) Detection of atrial fibrillation using au-toregressive modeling. Int J Electr Comput Eng (IJECE) 5(1):64–70
10. Aftanas L, Golosheykin S (2005) Impact of regular meditation practice on eeg activity at rest and during evoked negative emotions. Int J Neurosci 115(6):893–909
11. Hofmann SG, Sawyer AT, Witt AA, Oh D (2010) The effect of mindfulness-based therapy on anxiety and depression: a meta-analytic review. J Consult Clin Psychol 78(2):169
12. Cahn BR, Polich J (2006) Meditation states and traits: Eeg, erp, and neuroimaging studies. Psychol Bull 132(2):180

Design and Implementation of Opportunistic Routing in WSN Using Multi-hop River Formation for IOT-Based Indoor Patient Smart Monitoring

Venkateswari Pichaimani and K. R. Manjula

Abstract The Internet of Things (IOT) is aiding to cause well-formed cities, hospitals, homes and the economic by preparing the private and public associations to monitor remotely with secured features. The passive infrared motion (PIR) sensors are helpful and connected to a camera to fetch information about anybody crossing the premises' entrance or exit. The extracted information (images) are shared to the sink server for future processing and maintained in a storage scheme. Aside from the security angle, the system is supersmart enough to excellence energy consumption by switch ON/OFF fan, lights and camera. These features will be more helpful for disabled or Alzheimer patients. A classical graph algorithm - Dijkstra algorithm is compared with meta-heuristic Enhanced River Formation Dynamics for Multi-node Routing Protocol (ERFD-MRP) and the performance compared to the state-of-the-art solutions in expressions of energy absorption system, network outturn, hand-to-hand detention and packet release routing efficiency.

Keywords Indoor patient monitoring · Wireless sensor network · Internet of things · Dijkstra · River formation dynamics

1 Introduction

The Internet of Things aids to cause well-formed cities, hospitals, homes and economic by preparing the private and public associations to monitor remotely with secured features using Wireless Sensor Networks (WSN). WSN is an amazing technology that is to materialize monitoring chore in various platforms and to take necessary actions with respect to captured data or information. WSN is useful in numerous IOT applications (smart city, smart hospital, smart maintaining of agriculture, etc.)

V. Pichaimani (✉)
CSE, SRC, SASTRA Deemed University, Thanjavur 613401, TN, India
e-mail: Venkat_eswar2000@src.sastra.edu

K. R. Manjula
SAP/CSE, SoC, SASTRA Deemed University, Thanjavur 613401, TN, India
e-mail: manjula@cse.sastra.edu

S. Jyothi et al. (eds.), *Proceedings of the 2nd International Conference on Computational and Bio Engineering*, Lecture Notes in Networks and Systems 215,
https://doi.org/10.1007/978-981-16-1941-0_69

and has gained attention as well as importance during the last decades. The power consumption of indoor appliances is a merit attention in the energy-saving systems. With respect to the World Energy Council, the indoor power absorption rate is much more. Electricity consumption in India has tripled since 2000. The percentage of indoor consumption has increased by electricity used from 55% in 2001 to more than 80% in 2017. According to the survey of economic reports, India is with one-third of the global average energy consumption, which is about to reflect on its GDP by increasing prices, and it is a heavy burden or stress put on the upper-middle-income cluster of people. Thus, saving power is the main concern.

1.1 Wsn

Wireless Sensor Network is of a sensor node that senses our surroundings and transmits data to the sink node. The sensor nodes are probably microcomputers with options to sense data and process extracted data that can even be transmitted to the storage sink or to the main base station of the information. The sensing unit extracts the necessary information and moves to the calculating section that integrates a variety of data together and then maintains as packets in storage units. After some time, the sink receives the transmitted data and then the application users make access of WSN data to process. When WSN is activated, various tasks are performed that establish the necessary infrastructure to the network, such as the distribution of sensor nodes and routing of transmitted data, which will allow the sensor node to performing the applications normally. In particular, when a node transfers data to other nodes, it must ensure the connectivity, direct communication with other nodes, and its radio power. In the WSN, an appropriate energy-efficient algorithm must be adopted to less the consumption of transmitted data and significantly less the energy in routing.

1.2 Clustering

Clustering is an activity of splitting networks into sub-groups. In WSN, nodes are sub-grouped into clusters. Each and every cluster has a head and is termed as cluster head. In a cluster, the nodes are not directly communicating with the sink (base station); instead, they transmit the data to cluster head and then which in turn sends accumulated data to the sink.

The sensor nodes carry forward the data to the central base station in the data collection process by following direct communication or multi-hop communication. The direct communication scheme from sensor node to sink (server) is an expensively energy-consuming one due to the distance between sensor nodes and sink. This affects the lifespan of the network. On the other hand, for a better network lifetime, the multi-hop communication schemes are advisable. In multi-hop communication, every sensor node is forwarding the data in a busy manner to the neighbor one as nearest intermediate nodes or by the sink node using multi-hop routing paths. Multi-hop techniques give better network energy conservation and the lifetime.

An energy-efficient routing protocol emerges and is used in a clustered environment with hierarchical routing between nodes and shortest paths between the nodes. The clustering protocol used here is (LEACH) Low Energy Adaptive Clustering Hierarchy.

1.3 Network Simulation

The evaluation or performance measuring is a typical component of research systems to assess new thoughts or ideas, problem identification, and optimize the traditional systems. The generic approaches that are to be followed are as below:

- Prototyping is to build and observe works.
- Analytical modeling is a mathematical scheme (model) that is used to examine.
- Simulation is to build a model using a software program as a system.

Prototyping is a faster mechanism and also provides limited control and observes over large-scale systems. Likewise, analytical modeling is not able to capture systems that are highly complicated. Then, simulation is an attractive solution that is flexible available to perform performance evaluation of computer mechanisms. The network simulators named as qualnet, opnet, SSFNET, NS-2 are commonly utilized. In this project, the Ns2 simulator is used.

2 Literature Survey

Wireless sensor network (WSN) is an essential technology to practically apply in the area of IoT-related applications. Automation and monitoring applications are made much easier almost in all kinds of infrastructure using IoT systems. In the paper of home monitoring [1], passive infrared (PIR) motion sensors are implemented by river formation dynamics-based protocol with multi-hop routing mechanism. But in our proposed model, it is enriched with meta-heuristic enhanced technique. Vector primarily depends on the shortest path analysis in the geographic system [2] and is often applied to road pattern networks. Despite the algorithms that relate to network may be applicable to watercourse sections (layers), they do not typically go through the flow direction. The explicit interests in the river layer network lie as utilization of Peruvian Amazon Rural Livelihoods & Poorness (PARLAP) work in evaluating the significance of river layer structure. Then it markets access to communities within the Peruvian Amazon and its influence on diverse. The distance matrix is required to identify the shortest path in this present endeavor.

GIS generally aims to follow the oversimplified method, by concentrating on shortest path identification algorithms for road networks. It follows dijkstra's identification of shortest path expressions, by listing the supply and destination nodes, and come back to the route pure mathematics, then the full distance basis and the total upstream and downstream of to identify every shortest path. The formula may be used for performing numerous routes. However, the load (or distance) worth value must be non-negative values. In the paper on better routing protocols for IoT [3], this work contributes to simulate the standard routing protocols by using adhoc on-demand distance vector, Destination sequence distance vector, and Dynamic Supply routing by changing network traffic type, density of the node, and node speed. Their estimation on protocols is predicted by packet release or delivery relative amount, the quantitative relation of dropped packets, and therefore the outturn of network by protocol usage.

Smart electrical observation system [4] wherever users will operate wirelessly their beam, fan, TV, freeze, etc., within the house even anyplace within the world. In our sensible monitoring system, the whole electrical part is controlled through machine applications. Here, the Bluetooth HC-06 module is employed to transfer information from the mobile application. Additionally, a PIR motion detector is employed to sense any kind of human presence. So, we have the tendency to use this detector to gap the door mechanically within the house.

Smart agriculture system using a hominoid application and GSM system that additionally operated by the blessing of IoT. Once the wet of the garden's soil becomes low, it produces a proof transmitted to Arduino mega and ensured that wet level of the garden soil is low. Moreover, receiving the signal from wetness sensor, Arduino starts the pump. However, this project is enforced by lacking internet communication and no server is employed to store all the records of a sensible house. In the paper of mixed transmission strategy [5], authors estimate the issues of energy balance in data collection of WSN, getting to equalize the energy usage of all sensor nodes when the information propagation is processed. With our extensive simulation results, let demonstrate that the energy balancing, lifespan, and decrease the delay of the proposed protocol and compare the hop-by-hop transmission and another routing protocol cluster-based.

3 Problem Statement

The control mechanism of power in electrical appliances is a highly demandable task in smart city schemes. To monitor home or hospitals, cameras are targeted at the entrance of the patient room to record videos of anyone supposed to enter or leaving the room for surveillance. Videos require bulk storage unit when it is compared with photos. The captured photos are communicated to the server (base station).

In WSN, routing is the mainly following task for data communication between cluster head to sink. Due to data communication the energy is depleted. Some routing algorithm does not work with the networks that are implemented in the large scale region due to the usage of single hop routing. Here, each and every node can transmit directly to the cluster head and sink or base station. No categorize propagation models are followed for a different environment to minimize the path loss, which is identified as the main weakness.

In WSN, the limitation of the sensor nodes merged with the specific location based rules or conditions to make efficient energy and the scalability metric factors more crucial. The routing protocols have the following issues while forwarding data:

- Dynamic network topology
- Redundant data is often forwarded
- Data storm is an issue or network congestion that is near to the base station
- Failure prone

A Graph G of (V, E) represents WSN, in which V is a set of n number of nodes (sensor nodes and a base station) and E is a set of wireless links between nodes. In each round, all sensor nodes in the network forward its data to the base station. The problem is to increase the number of rounds that before the network dies. This is achieved by identifying the energy-efficient multi-hop paths from each sensor node to the base station. The main aim is to select the next hop node finding the energy-efficient multiple hop paths from each sensor node to the base station. In this paper, the RFD scheme is utilized to find the next successive hop node to establish an energy-efficient path.

3.1 Motivation

The major contributions of this paper are cost-effective, energy-efficient, and computationally intelligent wireless sensor network (WSNs) platform over the multi-node remote monitoring system. This smart patient monitoring system (SPMS) consists of N number of WSN nodes with a single server node. The server node is capable of storing and routing the motion image. All sensor nodes can capture a moving image and transmit the image to the sink node via multi-hops. ERFD-MRP is an enhanced river formation dynamic-based multi-node algorithm for collecting data and communicating from sensors and forwarding to the sink [6]. The transmission of data is done, which reduces energy consumption in routing.

3.2 Objectives

The device designed in this project is installed at the main entrance of the disabled or Alzheimer patient room. The passive infrared motion (PIR) sensors are helpful and connected to a camera to fetch information about anybody crossing the premises' entrance or exit. The extracted information (images) are shared to the sink server for future processing and maintained in a storage scheme. The smart patient monitoring system starts with the detection of motion and notification is forward to the application user with the help of SMS or e-mail. The proposed scheme's deployment of hardware is carried out by Arduino, PIR sensors, and GSM functionality module for SMS alert or e-mail notification. In addition to the security perspective, the system is supersmart enough to excellence energy consumption by switch ON/OFF fan, lights and camera.

The smart patient monitoring system (SPMS) consists of N number of WSN nodes with a single server node. The server node is capable of storing and routing the motion image. All sensor nodes can capture a motion image and transmit the image to the sink node via multi-hops. This paper presents a comparison of two optimization algorithms implemented to identify the better shortest path in dynamic environments. A classical graph algorithm - Dijkstra algorithm is compared with metaheuristic Enhanced River Formation Dynamics for Multi-node Routing Protocol (ERFD-MRP) which is used to send the source node image periodically to sink node (server). The performance compared to the state-of-the-art solutions in expressions of energy absorption system, network outturn, hand-to-hand detention and packet release routing efficiency.

3.3 Methodology

3.3.1 Node Initialization

Initialize the wireless network nodes with a multi-hop network by randomly deploying 40 nodes in an area of 1000 × 1000 sq m. Nodes are assigned with initial energy, transmitting energy, and receiving energy. It randomly selects the sender's node and receiver's nodes. Then, the communication is initiated.

3.3.2 Wsn Deployment

All the nodes are configured to exchange the location and initial energy information among all the nodes. Base Station is configured with the highest communication range. Establish a data transmission between the nodes using agent of UDP and traffic of CBR. A LEACH is implemented where based on cluster head the nodes are divide from the cluster. The CH is selected randomly in a distributed manner. The cluster head or zone head is opted, and packets collected and sent by the better path have been selected for cluster head or zone head selection process.

3.3.3 Dijkstra's Algorithm

Dijkstra's algorithm is helpful for finding the better shortest paths from the given source node to all the other nodes in a graph. It chooses a single node as the "source" node and identifies the shortest paths from the source to all other nodes in the graph based on the distance between them. It computes the single source shortest path issue for a graph with non-negative edge path costs and to produce a shortest better path. Dijkstra's algorithm is performed to solve the single-source shortest problem which employs the greedy approach [7].

Dijkstra algorithm steps:

- Assume starting node to the value of zero and assign all other nodes to infinity
- Starting node will send to be the queue of minimum priority depends on the distance and vertex.
- The minimum distance vertex will be moved from the priority queue and changed in the queue.
- Check the current node distance is less than the next node distance and then send the next node distance to the priority queue.
- Repeat the mentioned steps ii to iv until the priority queue becomes empty.

3.3.4 ERFD-MRP Algorithm

River Formation Dynamics is involved in the proposed mechanism and it is one of the heuristic optimization methods and subgroup topics of swarm-intelligence. RFD is based on replicating how rainwater makes rivers by eroding the ground level and deposit of sediments. These rivers merged into the sea by selecting the shortest better path by considering altitudes as a parameter of the land. Likewise, the sensors are to sense the data from the environment and forward it to the base station with the routers arranged in the intermediate. This RFD technique is considered for multiple hop communication. A mechanism is proposed to route protocol called Enhanced River Formation Dynamics-based Multi-node Routing Protocol (ERFD MRP) to collect the data from the surroundings and moving to base station (Fig. 1).

procedure RFD-MRP Algorithm()

begin
/Initialization of WMN parameters/
Define Source Node, Number of Nodes, location of nodes, Initial energy
nodeDeployment()
while (not all nodes are dead) do repeat
    forward Node Selection()
    forward Data()
    update Energy() until
data reaches to server **end**
end procedure

In the beginning, every node is placed in the given location. After that, every node of the graph has initialized and this basically consists of two functions. The destination altitude or terminal selected node is set to the value zero. In expressions of RFD, this node is known as the sea, i.e., the last ambition of every drop, and the

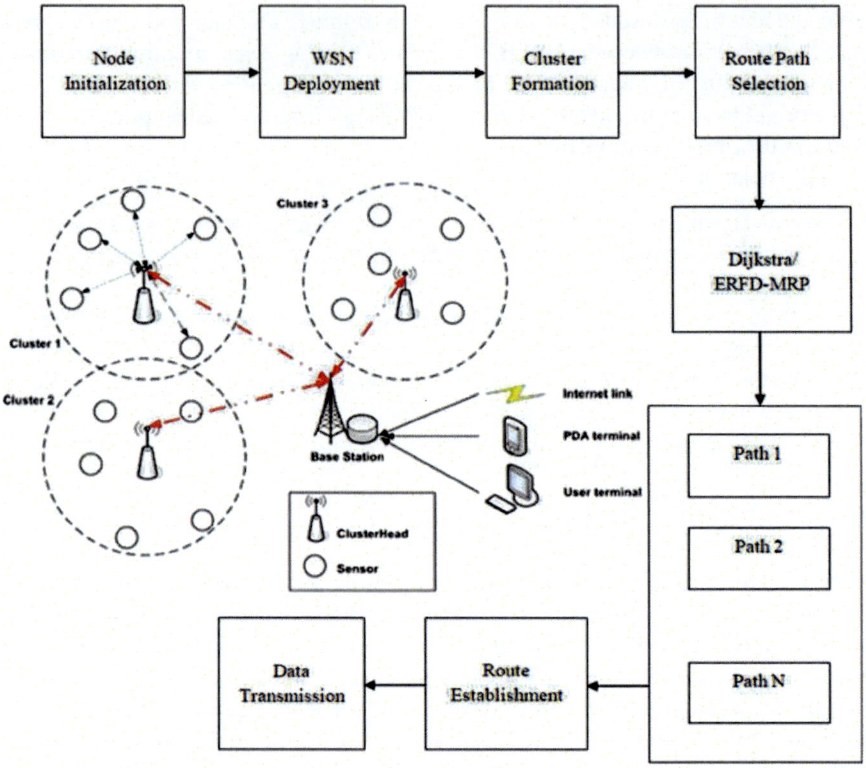

Fig. 1 Architecture of ERFD-MRP

altitude of the remaining nodes is placed to several identical values. All the nodes are configured to exchange the location and initial energy information among the nodes. Base Station is configured with the highest communication range. The while loop for this algorithm is carried out until either every drop finds a similar response, that is, every drop passes through the identical series of nodes, or one more another optional ending condition is satisfied. The initial action of the loop body consists of traveling the data across the nodes in a random manner. The node 'i' is having the probability to select node 'j' as a forward node is denoted by P(i,j) and is calculated as follows

$$P(i, j) = \begin{matrix} \frac{\mathrm{H(i,j)}}{\sum \mathrm{H(i,l)}} \; i(j \in NN_i \& H(i, j) > 0) \\ 0 \qquad\qquad \text{otherwise} \end{matrix} \tag{1}$$

where, ∈ and (,) > 0

$$H(i, j) = \left(\frac{(\mathrm{HC(i, BS)} - \mathrm{HC(j, BS)})}{\mathrm{distance}(i, j)}\right)\mathrm{RE(j)} \tag{2}$$

where, HC is Hop count

distance(i,j) is the Euclidean distance between node i to node j (i.e., $\sqrt{(x_i - x_j)^2 + (y_i - y_j)^2}$)

NN_i is Neighbor node of node i

RE(j) is the residual energy of node j

The nodes are forward to the next node until it reaches the base station. For forwarding or data receiving, the residual energy of these nodes ‘i’ and ‘j’ must be higher than the threshold value (T_E). The threshold value is equal to 20% of the initial energy.

According to the first order radio model, the energy consumed for transmitting (E_{Tx})and receiving (E_{Rx})b-bits of data can be calculated as,

$$E_T(b) = b\left(E_{elec} + \varepsilon_{amp} d^2\right) \tag{3}$$

$$E_{Rx} b = b \cdot E_{elec} \tag{4}$$

where, d is the distance from sender to a receiver node

E_{elec} is energy dissipation in the transmitter or receiver circuit

is the energy required per bit meter square by the transmitter amplifier.

The total energy consumed (E_{TE}) can be calculated as,

$$E_{TE}(b) = E_{Tx} + E_{Rx} = b\left(2E_{elec} + \varepsilon_{amp} d^2\right) \tag{5}$$

The residual energy (RE) for node i can be calculated when transmitting or receiving data packets,

$$R(i) = RE(i) - (E_{Tx}(b) + E_{Rx}(b)) \tag{6}$$

The data will reach the sink node (base station). The energy of all nodes is updated by the difference of current node energy value with the next node energy value. The process will repeat until the source nodes reach the server.

4 Comparison and Result

In this section, the performance of the Dijkstra and river-formation dynamics-based multiple hop algorithm has experimented. As of the observed results, X_graphs are nicely plotted. The energy consumption, throughput, delay and packet delivery are the generic parameters to be treated and X_graphs are plotted for the mentioned parameters (Table 1).

Table 1 Simulation Parameters

Parameter	Value
Simulator	NS2
Simulation time	10 s
Area (m^2)	1000X1000
Number of nodes (nn)	40
Position of BS (X,Y)	600,300
Routing Protocol	AODV
Radio Type	802.11a/g
Transmission rate	10 packets/s
Transmitting Power	20
Packet size	1000

4.1 Performance Metrics

The performance metrics such as Throughput, Delay, Energy consumption, and Packet delivery ratio were used.

Throughput

It is the ratio of the total number of bits received (RecievedBytes) to unit time, i.e. the difference of end time (t_{end}) and start time (t_{start}).

$$\text{Throughput} = (\text{RecievedBytes}) * 8/\,(t_{end} - t_{start})\ \text{bps}$$

Average delay from end-to-end

It is the average difference between the time at which the number of mentioned packets sent by sender and the time which the number of mentioned packets received by the receiver.

Packet delivery ratio

It is defined as the ratio of the number of mentioned packets received at the destination and the number of mentioned packets sent by the source.

$$\text{Packet delivery ratio} = (\text{Received packet}/\ \text{Generated packet}) * 100$$

Energy Consumption

A particular amount of energy loses when a node transmits or receives packets. As a result, energy gets decreased. The energy consumption level can be determined by finding the difference of the total energy and the total energy-consumed at a particular instance of time.

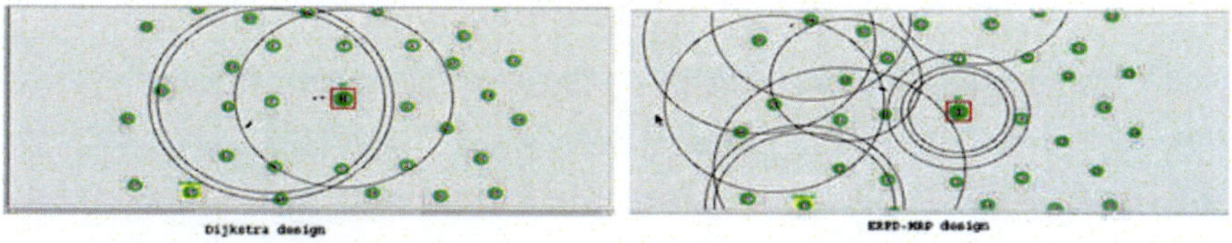

Fig. 2 Simulation evaluation

5 Experimental Evaluation

5.1 Network Simulator 2 (NS2)

NS-2 is an occasion-driven parcel-level system test, and a system created as a piece of the VINT (Virtual Internet Testbed) extend. In 1995 Version 1 of NS was created and with variant 2 out of 1996. The NS2 with TCL/OTCL is a combination include. Adaptations 2 of NS2 have been encased in a dialect which is alluded to as Object orientated TCL (OTCL). Scripting is an open supply bundle accessible for each Win32 and Linux stages (Fig. 2).

5.2 Network Animator (NAM)

NAM is a visual demonstration of NS output. Then, TCL/TK-based animation-tool or a graphical visually interpreted for visualizing network-simulation traces and real-world traces of packet. This tool supports topology layout, protocol graph, and different data inspection tools.

5.3 Xgraph

Xgraph is a plotting software program that is for creating graphic representations of simulated results. A graph is drawn on an X-display with given data that is read by data files or standard input where there are no files to mention.

5.4 Results

See Fig. 3.

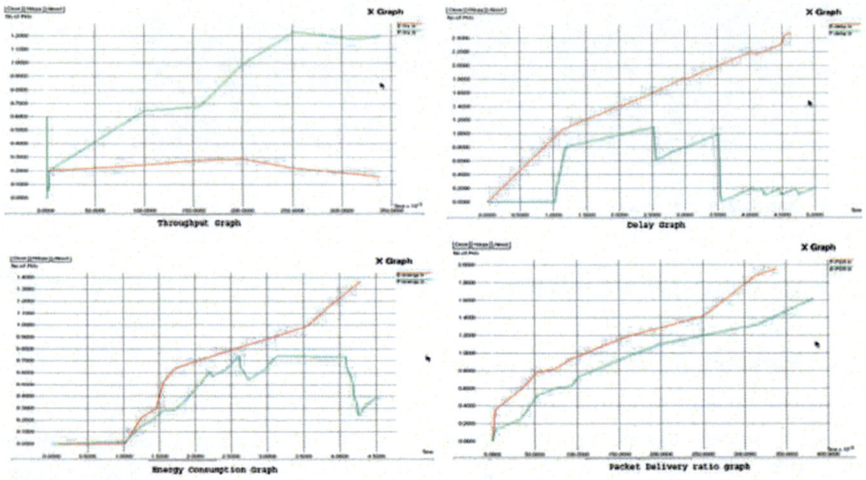

Fig. 3 Experimental results

6 Conclusion and Future Enhancements

In multi-hop algorithm routing, the selection process of a forward node for the transfer of the data plays a major role. The ERFD-MRP is a river-formation dynamic-based multi-hop protocol in WSN to save energy and enhance the performance of the network. The comparison shows that the ERFD MRP algorithm illustrates better than the Dijkstra algorithm according to energy consumption, throughput, average hand-to-hand delay and packet release ratio. In the future, the proposed approach will be implemented in a real testbed with varying traffic to measure the actual performance and computation overhead. Moreover, it will extend to the 5G emerging smart indoor-based patient monitoring system and secure route optimization protocol in smart indoor monitoring is expected to be implemented.

References

1. Guravaiah K, Velusamy RL (2019) Prototype of home monitoring device using internet of things and river formation dynamics-based multi-hop routing protocol (RFDHM). IEEE Trans Consum Electron 65(3)
2. Kalacska M, Tanaka M, Cadieux N, Coomes OT, Takasaki Y (2020) A python algorithm for shortest-path river network distance calculations considering river flow direction. MDPI J 5(8)
3. Ahmed A, Raza M, Ali S (2019) Towards better routing protocols for IoT. In: International conference on computing, mathematics and engineering technologies (iCoMET 2019)
4. Rahman MM, Rashid HA, Rahman W, Islam R (2018) Embodiment of IOT based smart home security system. Int J Res Appl Sci Eng Technol (IJRASET) 6:2321–9653
5. Jin N, Gu T, Liu T, Zhu Y(2017) A mixed transmission strategy to achieve energy balancing in wireless sensor networks. IEEE Trans Wireless Commun 16(4)

6. Sharma S, Gupta N (2016) Routing in wireless mesh network using river formation dynamics. Int J Res Appl Sci Eng Technol (IJRASET), 2(7)
7. Rajakumari KE, Mahalaxmi G (2017) Multi-agent technology to improve the internet of things routing algorithm using ant colony optimization. Ind J Sci Technol 10(31)

A Cross-Cultural Assessment of the Influence of Knowledge Management Practices and Acceptance of Massive Open Online Courses (Moocs) Among Technical Students in India

Preethika Basava, Chetan Reddy Bayana, Priyanka Battula, and K. R. Manjula

Abstract Massive Open Online Courses (MOOCs) render great learning content which includes videos and quizzes with the help of social communication tools on several online platforms that are accessible worldwide and allow participants to learn at their own pace. Many contextual factors can be considered by engineering students while making a decision for acknowledgment of MOOCs of which the cultural dimension plays a vital role. The Technology Acceptance Model (TAM) is integrated with Knowledge Management (KM) practices and is used to scrutinize a student's behaviour towards MOOCs' acceptance. Structural Equation Modelling (SEM) is utilized to test the hypothesized relationships and to conduct the cross-cultural analysis. Finally, a structural equation-based multi-group analysis is performed to get the patterns of cultural differences in assessing the effects of the Knowledge Management practices for undertaking MOOCs by engineering students.

Keywords MOOCs · Technology acceptance model · Knowledge management · Structural equation modelling · Cross-cultural analysis

1 Introduction

The term MOOC first came into existence to refer to a course offered by Stephen Downes and George Siemens in 2008 [1, 2]. With the advancements in technologies, a

P. Basava (✉) · C. R. Bayana · P. Battula · K. R. Manjula
Department of Computer Science Engineering, SASTRA University, Thanjavur, India
e-mail: preethikabasava6@gmail.com

C. R. Bayana
e-mail: chetanreddy4137@gmail.com

P. Battula
e-mail: 121003043@sastra.ac.in

K. R. Manjula
e-mail: manjula@cse.sastra.edu

S. Jyothi et al. (eds.), *Proceedings of the 2nd International Conference on Computational and Bio Engineering*, Lecture Notes in Networks and Systems 215,
https://doi.org/10.1007/978-981-16-1941-0_70

rich learning environment is created, and low-cost and effective distance education is made available to students by many top universities in the form of MOOCs. In recent times, MOOCs have been gaining popularity by promoting the democratization of education as they enable students to have access to learning resources, materials at any time and from anywhere across the world. MOOCs accommodate unlimited learners of similar interest at free of cost. However, a small amount is charged for providing certificates of accomplishment to the learners for a few courses. Based on several theoretical paradigms, there are four variants of MOOCs. They are cMOOCs, xMOOCs, hMOOCs, and ahMOOCs [2].

Students mostly enroll in the course, but only a few of them will complete the course. The main challenges to be considered in this regard are the absence of a student's participation and lack of proper interaction as these are offered web-based. Students' decision for undertaking MOOCs depends on several contextual factors. Apart from them, the cultural backgrounds of the students also play a vital role. So, a cross-cultural analysis is performed with two different cultures to understand the effect of Knowledge Management practices on the engineering student's adoption rate of MOOCs.

Several studies and researches began for MOOCs in 2009. From Web of Science databases, there are around 3000 articles that included MOOCs in their title. Figure 1 depicts the bibliometric network [3, 4] of frequently used keywords by authors in the articles obtained from the VOS viewer, and Fig. 2 shows a bibliometric network of the frequently used words in the abstracts [4] of the above articles from the VOS viewer. All the results obtained from the bibliometric analysis tells that there are

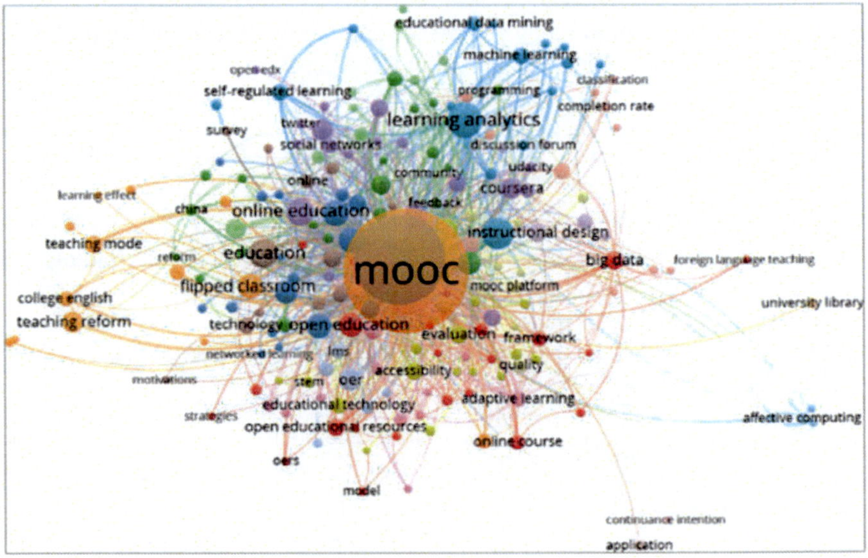

Fig. 1 Bibliometric network of frequently used keywords by authors

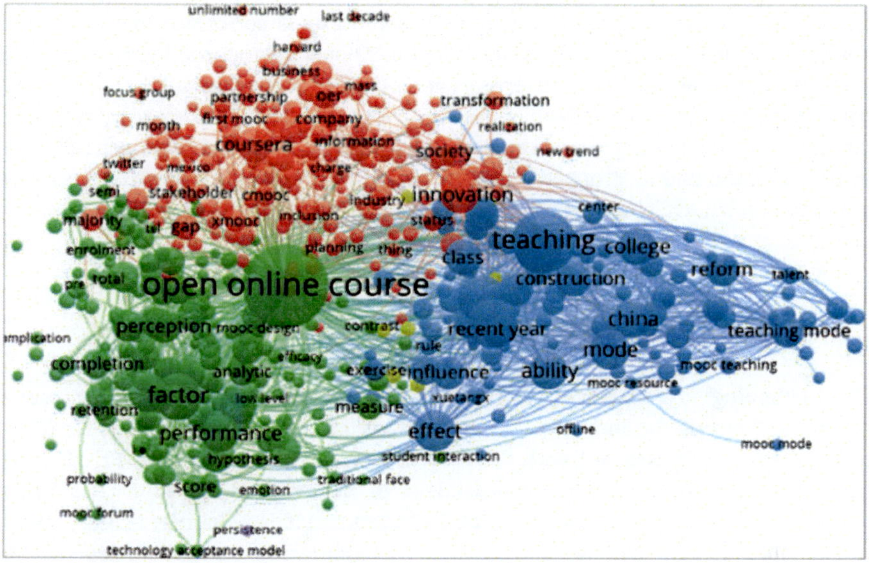

Fig. 2 Bibliometric network of frequently used words in the abstract

many studies that concentrate on the antecedents and consequences of MOOCs. But they did not focus on the cultural backgrounds of students and their variations in relationship with Knowledge Management practices and MOOCs.

2 Problem Statement

The fundamental aim of this paper is to discover the effect of cultural variations and the KM practices on the acknowledgment of MOOCs using extended TAM. Here the data is grouped into north and south divisions of India and identified the variations of the culture on acquiring MOOCs.

2.1 Advantages of Proposed Method

The cross-cultural analysis provides more accurate results on the MOOCs' acceptance by considering the effect of KM practices [5]. Usually, the role of KM practices is visualized on information systems and business models. Here KM practices are integrated with the TAM to analyze the user behaviour towards the MOOC platform. Previously the TAM was used to predict the acceptance of Google applications for collaborative technologies. Here, it is applied to identify the user behaviour on the MOOC acceptance. SEM-based hypothesis testing is performed to evaluate the implications of the KM practices and TAM attributes. These implications provide the accessibility, usage, and transferring capability of course materials of MOOCs by engineering students. The SEM-based multi-group analysis is performed by grouping the data based on any criteria and then analyzing each group's hypothesized relationships. Here, the data is grouped based on cultural differences to identify the effect of culture on MOOCs' acceptance.

3 Proposed Methodology

The Technology Acceptance Model is popular to efficiently explain the user's behaviour towards the intake of any technology [6]. The Actual Behaviour (AB) of the user can be assessed with the help of Perceived Ease of Use (PEOU) and Perceived Usefulness (PU) in accordance with Behavioural Intention (BI). TAM is integrated with KM-practices, which include knowledge sharing, access, application, and storage, and is used to explain engineering student's intention for acknowledgment of MOOCs.

3.1 Data Description

The data is gathered from engineering students across the two divisions of India (north and south) through an online survey. The valid sample is taken from the collected data. The participants consist of engineering students with a mean age group of 21. They are majoring in Computer Science, Electronics and Communication, Mechanical, Civil, Electricals and Electronics, Bio-Technology, Electronics and Instrumentation, Information Technology, Mechatronics, and Aerospace. They are doing their courses on various platforms such as Udemy, Swayam, NPTEL, Udacity, Edx, Khan Academy, and others.

A 26 component scale with estimation values from one to five is taken to relate the knowledge management practices for MOOCs' acceptance. The scale has four dimensions of knowledge management practices, including four elements of knowledge access, three elements of knowledge storage, three elements of knowledge sharing, and three elements of knowledge application. The components adapted from TAM include four elements of perceived usefulness, three elements of perceived ease of use, three elements of behavioural intention, and three elements of actual behaviour.

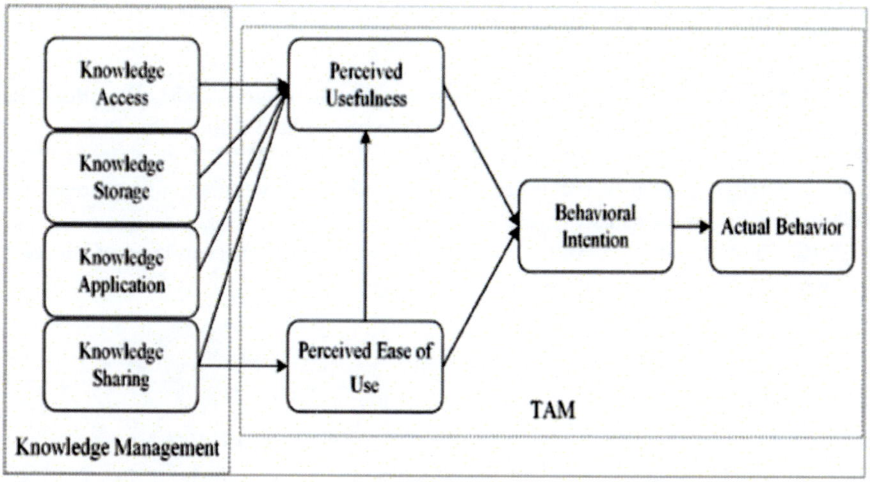

Fig. 3 Research model

3.2 *Knowledge Management Practices*

Knowledge Access. "Acquiring and transferring information and knowledge from a particular system" [7].

Knowledge Storage. "Storing and retrieving all kinds of data, information, and knowledge" [7].

Knowledge Application. "Accessing knowledge with the help of storage and retrieval techniques" [8].

Knowledge Sharing. "Distributing various resources among individuals as a part of various activities" (Fig. 3) [8].

3.3 *Technology Acceptance Model*

Perceived Ease of Use (PEOU). "The degree to which a person believes that using a particular system would be free of effort" [9].

Perceived Usefulness (PU). "The degree to which person believes that using a particular system would enhance his/her job performance" [9–11].

Behavioral Intention (BI). "The individual's overall reaction of a user towards a particular system".

The following null hypotheses are formulated considering the above dimensions in relationship with MOOCs.

3.4 Cross-Cultural Variations

By integrating the Knowledge Management practices with TAM, the proposed model's application in MOOCs is with respect to context and cultural dimensions [12] across the two divisions of our country, north and south. Considering the fact that the Knowledge Management practices in MOOCs' context are unexplored, the below-stated hypothesis is formed.

H10. The relationship between KM practices and MOOC acceptance would vary between the north and south divisions of India.

4 Results

4.1 Reliability and Validity Measures

The reliability of the data is found by calculating the mean, standard deviation, skewness, and kurtosis. Tables 1 shows the null hypotheses [11] and Table 2 show the descriptive statistics, normality, and reliability coefficients of the data [5]. All the scales show the internal consistency ranging from 0.81 to 0.86, and the overall Cronbach's alpha is 0.84. Table 3 shows the correlation between all the components, i.e., components of Knowledge Management practices and components of TAM. It can be inferred from the results that each item is remarkably correlated to its corresponding component.

Table 1 Null hypotheses

H. No.	Hypotheses
H1	Knowledge access would have a positive and significant impact on the PU of MOOCs
H2	Knowledge storage would have a positive and significant impact on the PU of MOOCs
H3	Knowledge application would have a positive and significant impact on the PU of MOOCs
H4	Knowledge sharing would have a positive and significant impact on the PU of MOOCs
H5	Knowledge sharing would have a positive and significant impact on the PEOU of MOOCs
H6	PEOU would have a positive and significant impact on the PU of MOOCs
H7	PEOU would have a positive and significant impact on the BI to use MOOCs
H8	PU would have a positive and significant impact on the BI to use MOOCs
H9	BI would have a positive and significant impact on the actual behavior

Table 2 Descriptive statistical measurements and reliability

Constructs	Items	Mean	SD	Skewness	Kurtosis	Alpha (α)
Knowledge Access (KAC)	KAC1. "MOOCs enable me to access video lectures anytime and anywhere." KAC2. "MOOCs facilitate my access to video lectures." KAC3. "MOOCs enable me to quickly access video lectures and learning materials." KAC4. "MOOCs enable me to have ubiquitous access to learning materials and video lectures."	16.65	3.13	−1.096	1.718	0.869
Knowledge Storage (KST)	KST1. "MOOCs enable me an always accessible storage of video lectures." KST2. "MOOCs enable me an easily accessible storage of learning materials." KST3. "MOOCs enable me a ubiquitous storage of learning materials and video lectures."	11.99	2.57	−0.705	0.327	0.821
Knowledge Sharing (KSH)	KSH1. "MOOCs enable me to exchange lecture notes with classmates." KSH2. "MOOCs enable me to share learning materials with classmates." KSH3. "MOOCs enable me to share video lectures with others."	10.59	3.19	−0.381	0.454	0.840

(continued)

Table 2 (continued)

Constructs	Items	Mean	SD	Skewness	Kurtosis	Alpha (α)
Knowledge Application (KAP)	KAP1. "MOOCs help me to learn a new concept." KAP2. "MOOCs help me to expand my knowledge." KAP3. "MOOCs can be employed in problem-solving."	12.54	2.30	−1.078	1.611	0.833
Perceived Usefulness (PU)	PU1. "Using MOOCs would improve my academic performance." PU2. "Using MOOCs would improve my effectiveness." PU3. "Using MOOCs would improve my skills." PU4. "Using MOOCs would improve my efficiency."	16.62	3.03	−0.746	0.463	0.859
Perceived Ease of Use (PEOU)	PEOU1. "Using MOOCs would be easy for me." PEOU2. "Using MOOCs, I can easily watch a video lecture." PEOU3. "Using MOOCs, I can easily share learning materials."	12.04	2.51	−0.619	0.093	0.810
Behavioral Intention (BI)	BI1. "I intend to continue to use MOOCs for learning new concepts in the future." BI2. "I predict that I would use MOOCs for sharing learning materials in the future." BI3. "I predict that I would frequently use MOOCs to watch video lectures in the future."	11.89	2.61	−0.766	0.512	0.826

(continued)

Table 2 (continued)

Constructs	Items	Mean	SD	Skewness	Kurtosis	Alpha (α)
Actual Behavior (AB)	AB1. "I use MOOCs to watch video lectures." AB2. "I use MOOCs to learn a new concept." AB3. "I use MOOCs for sharing learning materials."	12.14	2.47	−0.802	0.711	0.843

Table 3 Correlation matrix

S. No.	KAc	KSt	KSh	KAp	PU	PEOU	BI	AB
KAc	1							
KSt	0.715	1						
KSh	0.280	0.367	1					
KAp	0.580	0.582	0.359	1				
PU	0.612	0.592	0.479	0.794	1			
PEOU	0.529	0.591	0.501	0.647	0.713	1		
BI	0.439	0.523	0.460	0.620	0.697	0.704	1	
AB	0.542	0.610	0.552	0.657	0.729	0.781	0.724	1

4.2 Hypothesis Testing and SEM Analysis

SEM analysis is done on the formulated null hypotheses with the help of SPSS-AMOS26. Figure 4 shows the results of hypothesis testing with structural models along with coefficients for each path. Knowledge access, sharing, and application had a remarkable positive effect on PU. Hence, H1, H3, and H4 are accepted. Knowledge storage does not have a notable impact on PU. So, H2 is rejected. Similarly, from the other implications H5, H6, H7, H8, and H9 are supported.

The confirmatory factor analysis is performed to find the impact of each item on the corresponding theoretical construct. It is clear from the results that each item has a remarkable impact on the corresponding components of extended TAM. Table 4 shows the hypothesis testing results. Estimates, standard error (S.E.), critical ratio (C.R.), and P value are identified for each implication.

4.3 Cross-Cultural Analysis

The cultural variations are observed between the north and south divisions of India by performing the SEM-based multi-group analysis. Table 5 shows the SEM analysis results for NORTH (left) and SOUTH (right). The results show that in the north,

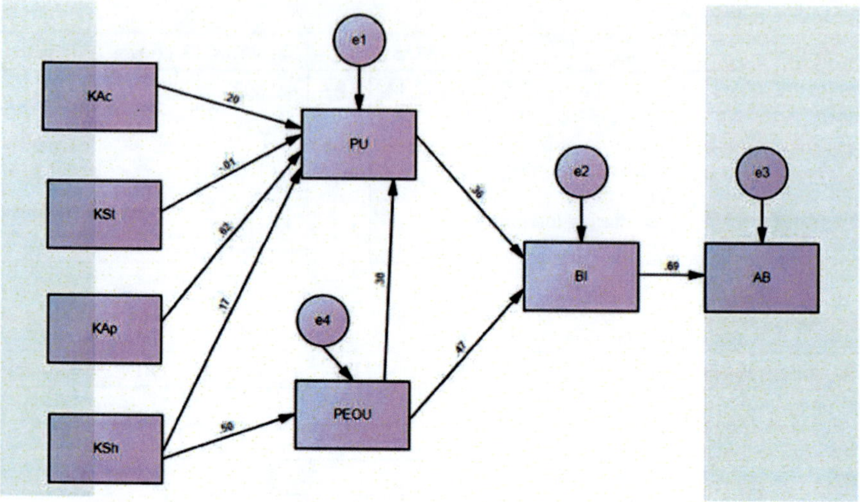

Fig. 4 Structural model with hypothesis testing results

Table 4 Hypothesis testing results

	Estimate	S.E.	C.R.	PLabel
PEOU <— KSh	0.393	0.050	7.891	***
PU <— KAc	0.156	0.037	4.203	***
PU <— KSt	−0.006	0.045	−0.125	0.900
PU <— KAp	0.657	0.050	13.042	***
PU <— KSh	1.129	0.042	3.071	0.002
PU <— PEOU	0.291	0.053	5.543	***
BI <— PU	0.342	0.055	6.206	***
BI <— PEOU	0.439	0.054	8.136	***
AB <— BI	0.683	0.053	12.823	***

Table 5 SEM analysis results for NORTH and SOUTH

	Estimate	S.E.	C.R.	P	Estimate	S.E.	C.R.	P
PEOU <— KSh	0.296	0.072	4.117	***	0.497	0.067	7.441	***
PU <— KAc	0.144	0.058	2.477	0.013	0.125	0.047	2.689	0.007
PU <— KSt	0.018	0.065	0.282	0.778	0.003	0.059	0.042	0.966
PU <— KAp	0.580	0.069	0.282	0.778	0.003	0.059	0.042	0.966
PU <— KSh	0.120	0.055	2.204	0.028	0.176	0.062	2.823	0.005
PU <— PEOU	0.244	0.073	3.318	***	0.308	0.076	4.046	***
BI <— PU	0.431	0.077	5.631	***	0.232	0.083	2.802	0.005
BI <— PEOU	0.268	0.076	3.511	***	0.609	0.077	7.942	***
AB <— BI	0.661	0.091	7.264	***	0.695	0.063	11.111	***

Table 6 Multi-group analysis results

S. No.	Implication	North	South
1	PEOU → KSH	Positive and significant	Positive and significant
2	PU → KAC	Not Significant	Not Significant
3	PU → KST	Not Significant	Not Significant
4	PU → KAP	Positive and significant	Positive and significant
5	PU → KSH	Not Significant	Positive and significant
6	PU → PEOU	Positive and significant	Positive and significant
7	BI → PU	Positive and significant	Not Significant
8	BI → PEOU	Positive and significant	Positive and significant
9	AB → BI	Positive and significant	Positive and significant

their knowledge access, storage, and sharing do not have a notable effect on PU. The results of the south show that knowledge access, storage, and sharing do not have a remarkable effect on PU. Also, PU does not have a notable effect on BI.

From the multi-group analysis results, it is visible that the implications have certain changes in the north and south divisions of India. Table 6 shows the multi-group analysis results. Based on the results, some important differences in terms of acceptance of MOOCs and cultural variations are achieved. So, H10 is supported.

5 Conclusions and Future Directions

It is clearly observed from the findings that there exist some remarkable cultural variations between the north and south divisions of India having a notable impact on the acceptance of MOOCs by engineering students. Knowledge sharing does not have a remarkable effect on PU in North India. Logically, if students are not equipped or fully aware of accessing resources and retrieving the information correctly, they are unable to share them with their peers. So, this shows that students from North India are unable to access MOOCs correctly. PU does not have a remarkable positive effect on PU in South India. Even though students from South India are capable of accessing, storing, and sharing, they are not utilizing MOOCs in an effective manner. However, students from both divisions did not explore MOOCs to the fullest or they might be unaware of the functionalities correctly. Educators are urged to prepare the students about the MOOCs' capacities and their uses. Students must know about the fact that knowledge sharing assumes a vital role in the selection of any online platform for learning activities.

Regardless of the effects of cultural differences identified, the findings obtained are even useful for improving the acceptance rate of MOOCs by students across India. Since the data is collected through online surveys, obtaining neutral samples from both the division of the country is required. Further improvements can be done for data collection. The study here is limited to engineering students across India. This can be extended to students pursuing other degrees and also to other countries across the world. There are facts that education is a two-way process between students and educators. This study mainly focused on the student's point of view. So, education impact can also be considered in further studies.

References

1. Fang J, Tang L, Yang J, Peng M (2019) Social interaction in MOOCs: the mediating effects of immersive experience and psychological needs satisfaction. Telemat Inf 39:75–91
2. García-Peñalvo FJ, Fidalgo-Blanco Á, Sein-Echaluce ML (2018) An adaptive hybrid MOOC model: Disrupting the MOOC concept in higher education. Telemat Inf 35(4):1018–1030. https://doi.org/10.1016/j.tele.2017.09.012
3. Hew KF, Cheung WS (2014) Students' and instructors' use of massive open online courses (MOOCs): motivations and challenges. Educ Res Rev 12:45–58.https://www.sciencedirect.com/science/article/abs/pii/S1747938X14000128
4. Jọrdan K (2014) Initial trends in enrolment and completion of massive open online courses. Int Rev Res Open Distrib Learn 15(1):133–160.https://doi.org/10.19173/irrodl.v15i1.1651
5. Arpaci I, Al-Emran M, Al-Sharafi MA (2020) The impact of knowledge management practices on the acceptance of Massive Open Online Courses (MOOCs) by engineering students: a cross-cultural comparison. Telematics and Informatics. https://doi.org/10.1016/j.tele.2020.101468
6. Cheung R, Vogel D (2013) Predicting user acceptance of collaborative technologies: an extension of the technology acceptance model for e-learning. Comput Educ 63:160–175
7. Arpaci I (2019) A hybrid modeling approach for predicting the educational use of mobile cloud computing services in higher education. Comput Hum Behav 90:181–187. https://doi.org/10.1016/j.chb.2018.09.005
8. Al-Emran M, Mezhuyev V, Kamaludin A, Shaalan K (2018) The impact of knowledge management processes on information systems: a systematic review. Int J Inf Manag 43:173–187
9. Wu B, Chen X (2017) Continuance intention to use MOOCs: Integrating the technology acceptance model (TAM) and task technology fit (TTF) model. Comput Hum Behav 67:221–232
10. Tao D, Fu P, Wang Y, Zhang T, Qu X (2019) Key characteristics in designing massive open online courses (MOOCs) for user acceptance: an application of the extended technology acceptance model. Interact Learn Environ: 1–14
11. Teo T, Dai HM (2019) The role of time in the acceptance of MOOCs among Chinese university students. Interact. Learn Environ: 1–14
12. Tarhini A, Hone K, Liu X (2015) A cross-cultural examination of the impact of social, organisational and individual factors on educational technology acceptance between British and Lebanese university students. Br J Educ Technol 46(4):739–755

Review Paper on Anomaly Detection in Data Streams

G. Sandhya Madhuri, Yamuna, and M. Usha Rani

Abstract Anomaly is in general defined as deviation or diversion from the normal. The word anomaly came from the Greek word *anomalia* which means "uneven" or "irregular". In our day-to-day lives, we have seen many such irregularities or deviations from normalcy. For example, a condition monitoring system beeps an alarm when it detects any value or parameter of the machine away from the minimum value limit to the maximum value limit, or a credit card fraud alerts the bank and the customer immediately. Now, the crucial task here is how we detect anomalies in data streams. When there is streaming data that is continuously generated from any source, it is called a data stream. The task of finding anomalies from such a stream of data will be a challenging job. In this paper, we will discuss elaborately about data streams and anomaly detection in data streams by reviewing several papers and articles written on this topic.

Keywords Anomaly · Data streams · Outliers · Anomaly detection algorithms

1 Introduction

What is a data stream? A continuous stream of data generated by different sources or a singular source is known as a data stream. Now, to apply any algorithm and to draw knowledge or obtain information about anomalies is a difficult task.

In many areas which are growing with a lot of application importance, data stream mining is one among them. This is developing with novel methods and discoveries in several applications associated with medical world, computers, stock exchange,

G. Sandhya Madhuri (✉) · Yamuna · M. Usha Rani
Department of Computer Science, SPMVV, Tirupati, India
e-mail: sandhyamadhuri@gmail.com

Yamuna
e-mail: ryamunaspmvv@gmail.com

M. Usha Rani
e-mail: musha_rohan@yahoo.com

S. Jyothi et al. (eds.), *Proceedings of the 2nd International Conference on Computational and Bio Engineering*, Lecture Notes in Networks and Systems 215,
https://doi.org/10.1007/978-981-16-1941-0_72

bio-informatics and also predictions like weather, text, share market, etc. Among all other concerns in this data stream mining, the important one is 'data'. Why? Because with the massive generation of data from various sensors, chats, Facebook, online banking and Twitter, the concept of vigorously changing data is the core difficulty [10].

We need algorithms that can be used on such rapidly changing data to mine the data like predict, detect anomalies, identify frequent patterns, etc. Because of the ever-evolving features of streaming data, we need forecasting models to become accustomed to concept drifts [6].

In such scenarios, detecting anomalies is challenging. Initial anomaly detection is important and should be done even though it is difficult in practice. There must be systems to process real-time data and not batch-wise data. Algorithms should learn from the evolving data [1].

1.1 Challenges and Techniques to Mine Data Streams

Existing approaches may be insufficient when they have to handle voluminous and boundless data. Also, when trying to perform stream operations on real-time data, you should wish to obtain accurate results.

The existing query operators of the database will not serve the purpose of finding the output when we are trying to implement some of them like sort, join, etc.

Therefore, alternate methods or extensions to the existing methods are required to analyse the data streams. The methods like 'Data reduction', 'Manipulating semantics of data streams', 'Approximation', 'Punctuation to integrate knowledge', and 'Windowed versions of classical operators' are some of the techniques surveyed so that can be viewed as nonstop user query [8].

In such scenarios where continuous data is streaming, applying techniques to approaches like personalization and recommendation is a challenging task, for example, OTT platform recommendations, YouTube, e-commerce websites like Amazon, e-bay, etc.

Mining large streams of data for personal recommendations:

Figure 1 classifies the information systems and how to create personal recommendations using this classification [2].

2 Anomaly Detection

What is an anomaly? Any data object that does not belong to a group (cluster) or behaves differently than other data objects, we call such data points or objects as anomalies.

Anomalies (outliers) are of different types:

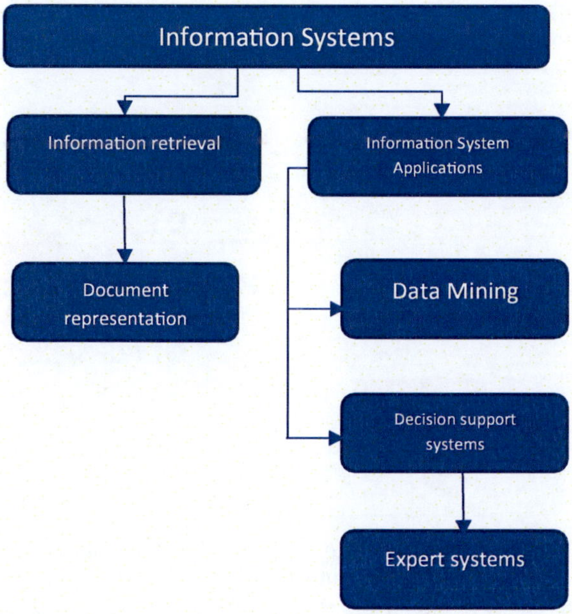

Fig. 1 Personal recommendations

1. Point outliers.
2. Contextual outliers.
3. Collective outliers.

1. Point outliers: A data object or point is considered as a point anomaly or global outlier when it is far away from the other data points in the data set.
2. Contextual outliers: The value of any data point that diverges from those of the values of all other data in a particular context. If it is giving the same value in a different context, then it might not be an anomaly, for example, time-series data.
3. Collective outliers: These are similar to the above anomalies, but the only difference is these are a group of data points whose values are deviated from the rest of the data points. But if we check any singular anomaly, it would look like its value is closer to the neighbouring data points.

2.1 Anomalies in Data Streams

Anomalies or outliers in data streams can be found from the real world like fraud prevention, energy demand, cybersecurity, medical diagnosis, etc. The IoT data is obtained by collecting the sensor data from all varieties of systems. We cannot predict

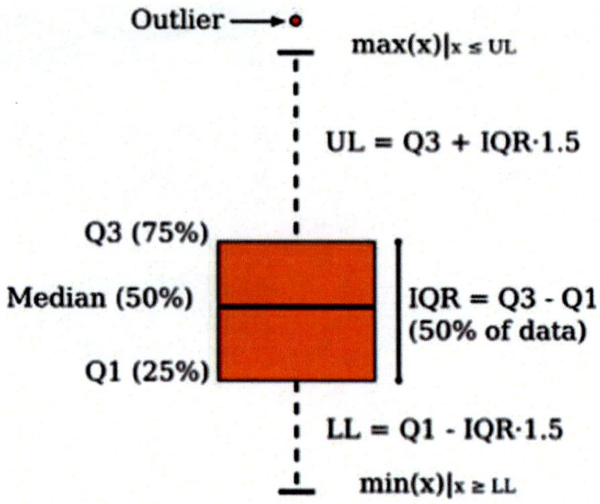

Fig. 2 Quartiles based solution

the behaviour of the systems. Some algorithms, like Welford's, quartile-based, a z-score metric-based solution and half-space trees in machine learning, offer solutions to detect anomalies in streams. Given below (Fig. 2) is the quartiles-based solution.

2.2 *Different Techniques to Detect Anomalies*

There are several techniques to find anomalies. Traditional anomaly detection techniques analyse each data instance separately like univariate and multivariate records. Usually, anomaly detection is carried out by grouping the instances together in data streams like time series. Symbolic instances and time series create a new problem to identify anomalies. Chandola et al. [4] have formulated a new technique called RBA (reference-based analysis) which can map data set to a reference data set and find out anomalies in the continuous data stream.

Even though there are multiple anomaly detection techniques identified so far, there are limitations such as all the data points should be labeled for train data and map with the testing data in techniques like statistical or machine learning techniques. In context with the network anomalies, a new framework called GEE is developed by Nguyen et al. [9], which detects the anomalies in network traffic.

3 IoT and Anomaly Detection

In our daily life, we see various scenarios where IoT is being used regularly. A person uses a smartwatch to monitor heartbeat, a vehicle uses a sensor for parking or experts monitor wild animals with a biochip transponder. Nowadays IoT is widely used in various industries to monitor production and customer service. Connectivity is the fundamental feature of IoT. Multiple IoT devices interact with each other using the internet which in turn collects lot of data and continuous data.

Let us take the example of monitoring the indoor atmosphere to safeguard human health. In a scenario, vertical plant wall systems are installed in the walls of a room and sensors are connected to monitor the CO_2 level in the room. Now 'machine learning outlier detection technique' is used to enhance the quality of the atmosphere in the room and also predict the future level of Co_2 levels in the room. In this condition, two types of anomalies are categorized, point anomalies and contextual anomalies. After applying 'prediction based and pattern recognition methods', it is finally decided that neural network-based models like autoencoder (AE) and long short-term memory encoder-decoder (LSTM-ED) have outperformed others [7]. Figure 3 shows the setup of IoT with vertical plant wall and sensors installed.

The important factor that has to be considered when IoT is established is security. Cybersecurity is a vital aspect that has to be monitored in important and crucial infrastructures like electricity grids, power plants, etc. Usually, these systems are controlled by SCADA systems called supervisory control and data acquisition. Misusing the logins, delay in the time intervals and unauthorized entries are monitored in such infrastructure. Using temporal pattern recognition methods, hidden Markov model and artificial neural networks are the two algorithms that give perfect detection of anomalies in SCADA systems [5].

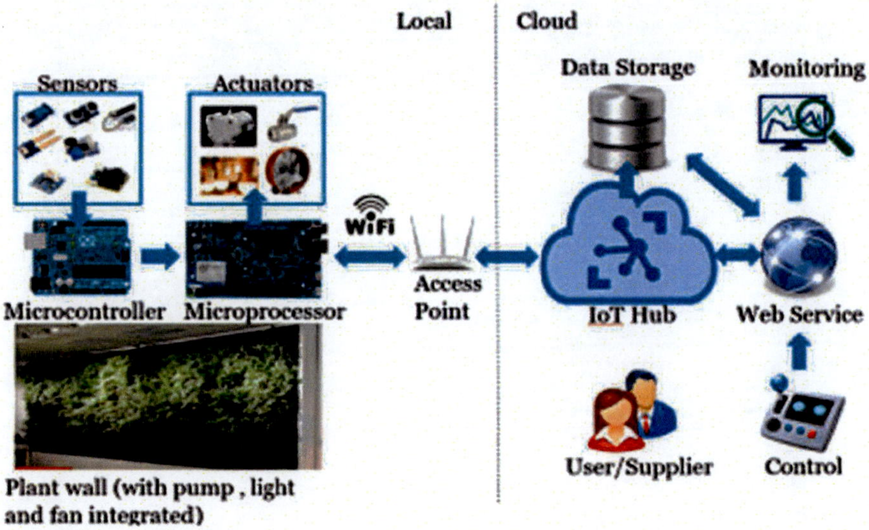

Fig. 3 IoT setup to observe the Co_2 level in a room using sensors

4 Massive Online Analysis (MOA)

Furthermore, for detecting outliers, techniques like clustering, load shedding, frequent pattern mining, etc. are used, which are not useful for detecting anomalies in the data streams. Therefore, we need a new methodology or technique to detect anomalies or outliers in data streams. One such tool that is useful in outlier detection is the MOA tool which is called the massive online analysis tool. This tool consists of algorithms like MCOD, COD, EXACT storm, Apprx STORM, Any Out, Abstract C, etc. This tool provides a setup where we can compare two algorithms on any given data set to check the performance analysis. We can also generate a synthetic data set using this MOA tool (Fig. 4).

There are three parameters for every algorithm to be set up to check the performance. The user can change the set of parameters and check the performance of them as a comparison between the two methods.

We can load any data stream or data set and set the visualization speed to check the outliers pointed out in the form of red dots, whereas the inliers, i.e., normal data points, are pointed out in grey.

If we pass IoT data like animal health monitoring data or sensor-generated data, we can observe if there are any deviations in the data points generated by such data streams.

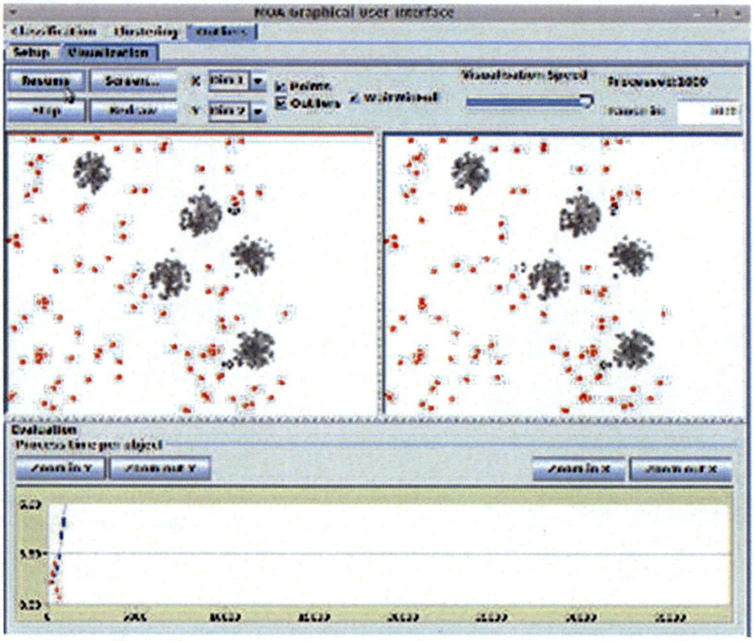

Fig. 4 Visualization in MOA

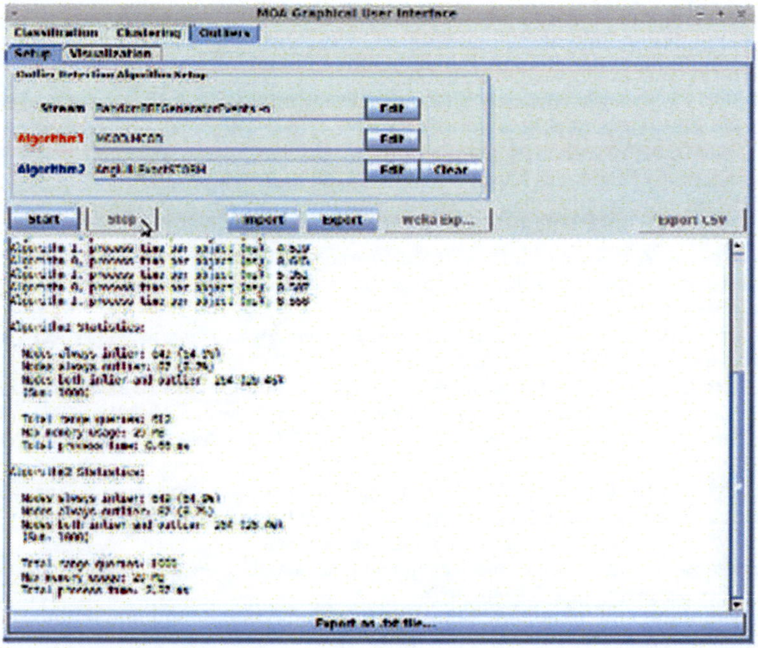

Fig. 5 Performance analysis in MOA

The results of the algorithms are measured based on the CPU time, memory used and total range queries generated. Below is a screenshot of such an outcome from the MOA tool [3] (Fig. 5).

5 Conclusion

Finally, we conclude that we have tried to review all the possible anomaly detections in data streams here. Anomalies are a threat to evolving data, be it IoT data, monitoring data, time-series data or data generated from SCADA systems. Therefore, we need a standardized system in place to detect such threats.

References

1. Ahmad S, Lavin A, Purdy S, Agha Z (2017) Unsupervised real-time anomaly detection for streaming data. Neurocomputing. https://doi.org/10.1016/j.neucom.2017.04.070
2. Amatriain X (2013) Mining large streams of user data for personalized recommendations. ACM SIGKDD Explor Newslett 10(1145/2481244):2481250

3. Bifet A, Holmes G, Kirkby R, Pfahringer B (2010) MOA: massive online analysis. J Mach Learn Res 1601–1604. Submitted 11/09; Published 4/10, Department of Computer Science University of Waikato Hamilton, New Zealand
4. Chandola V, Banerjee A, Kumar V (2009) Anomaly detection: a survey. ACM Comput Surv 10(1145/1541880):1541882
5. Kalech M (2019) Cyber-attack detection in SCADA systems using temporal pattern recognition techniques. Comput Secur. https://doi.org/10.1016/j.cose.2019.03.007
6. Krawczyk B, Minku LL, Gama J, Stefanowski J, Woźniak M (2017) Ensemble learning for data stream analysis: a survey. Inf Fusion. https://doi.org/10.1016/j.inffus.2017.02.004
7. Liu Y, Pang Z, Karlsson M, Gong S (2020) Anomaly detection based on machine learning in IoT-based vertical plant wall for indoor climate control. Build Environ. https://doi.org/10.1016/j.buildenv.2020.107212
8. Maier D, Tucker PA, Garofalakis M (2005) Filtering, punctuation, windows and synopses. https://doi.org/10.1007/0-387-25229-0_3
9. Nguyen QP, Lim KW, Divakaran DM, Low KH, Chan MC (2019) GEE: a gradient-based explainable variational autoencoder for network anomaly detection. In: 2019 IEEE conference on communications and network security, CNS 2019. https://doi.org/10.1109/CNS.2019.8802833
10. PhridviRaj MSB, GuruRao CV (2014) Data mining—past, present and future—a typical survey on data streams. Procedia Technol. https://doi.org/10.1016/j.protcy.2013.12.483
11. https://moa.cms.waikato.ac.nz/details/outlier-detection/
12. https://towardsdatascience.com/detecting-real-time-and-unsupervised-anomalies-in-streaming-data-a-starting-point-760a4bacbdf8
13. https://towardsdatascience.com/outliers-analysis-a-quick-guide-to-the-different-types-of-outliers-e41de37e6bf6

Significance of Primary Baseline Databases for Conservation of Biodiversity

Suvarna Latha Anchapakula, Bharathi Depuru, Lakshmi Padmavathi Pydipati, and Hemavathi Buddareddy

Abstract As our earth is in the middle of unprecedented climate change due to overpopulation, urbanization, overexploitation, industrialization and habitat degradation which has led to biodiversity decline, documentation of indigenous knowledge needs to be focused. This historical extinction decline in the biodiversity resources is to be preserved for integrating previously disconnected data resources for more effective and efficient conservation and management. Considerable efforts are made by the scientists in various organizations for preserving the resources as online databases. It is analysed that only 10% of biological resources or biocollections are available in digital form. Digitalization of bioresources at global levels is in the initial stages of data compilation, so integrative and novel applications are very much needed for the data retrieval of quality records for scientific studies in the future. Data digitalization, integration, analysis and publication are essential to make biodiversity informatics more efficient and relevant in our fast-changing environment. The significance of primary baseline databases and biodiversity conservation are discussed.

Keywords Biodiversity conservation · Primary baseline databases · Digitalization · Bioresources

1 Introduction

Biological diversity is defined as the variety and variability of various living forms present on the earth. It is the valuable treasure house on the earth, which provides all the resources necessary for the survival of human beings from food to shelter. It has been seen that biodiversity is getting degraded day by day due to anthropogenic activities, like urbanization, industrialization, overexploitation and habitat fragmentation, which needs to be focused on preserving our nature's gift for future sustenance. Conservation of biodiversity in the form of databases is very useful for

S. L. Anchapakula · B. Depuru (✉) · L. P. Pydipati · H. Buddareddy
Department of Biosciences and Sericulture, Sri Padmavati Mahila Visvavidyalayam (Woman's University), Tirupati, Andhra Pradesh, India
e-mail: depurubharathi1960@gmail.com

S. Jyothi et al. (eds.), *Proceedings of the 2nd International Conference on Computational and Bio Engineering*, Lecture Notes in Networks and Systems 215,
https://doi.org/10.1007/978-981-16-1941-0_73

the retrieval, access and analysis for scientific studies by the present and near future. Ongoing technological advancements in information technology towards the incorporation, investigation and representation of biodiversity data are prompting towards the advanced field of research called biodiversity informatics.

It is the application of information technology to provide data about the conservation status of all living organisms. Even before the term "biodiversity informatics" was coined, applications of information technology in biodiversity were in practice in the field of biological sciences without the use of the definite term to bring these activities under a single umbrella leading to the status of a recognized discipline within biological sciences. The term "biodiversity informatics" was first used by John Whiting in 1992 while establishing the Canadian Biodiversity Informatics Consortium (CBIC) [1]. However, it was used in its broader sense during the discussions on Organisation for Economic Co-operation and Development (OCED) in 1996, which later recommended the formation of the Global Biodiversity Information Facility (GBIF) an international network established in 2001 to promote the mobilization and utilization of biodiversity data in Asia, also the most prime objective is to provide online open-access data to everyone about origin and occurrence of species are recorded. GBIF derives this information from many global sources, authentic scientific data from conservationists, researchers, also includes museum specimens data collected from the past centuries.

About 5–50 million organisms inhabit this planet of which nearly 1.8 million organisms have been named and classified so far [2]. However, there is no single repository either offline or online that could provide access to baseline data about these 1.8 million known organisms. Several national, regional and taxon specific efforts have been initiated in the last decade or so to collate baseline data about these known organisms. Some of the significant ones which deal with the development of Indian Fauna, an electronic catalogue of known organisms, are enlisted in catalogue of Indian Biota [3]. However, it must be mentioned that GBIF focuses on recording the data of 98% organisms of 1.8 million known organisms [4]. GBIF signed a memorandum of cooperation with Species 2000, Integrated Taxonomic Information System (ITIS) and Catalogue of Life (CoL) Partnership to expedite the process of cataloguing all known species [5]. About 1,636,434,865 species occurrence records belonging to diverse living forms like archaea, viruses, bacteria, fungi, chromista, bryophytes, pteridophytes, gymnosperms, protozoa, Animalia and Plantae are available, and 55,059 dataset information is available to anyone from anywhere.

The Species 2000, ITIS and CoL provide comprehensive and authentic information on indexing of species currently present, their status, species synonyms, common names, integration with other species and distribution.

Thus, environmental informatics and molecular bioinformatics are the two main connecting subjects to biodiversity informatics which provide the generalized scientific information infrastructure in biology.

2 Significance

Biological informatics is an unexplored field which provides tremendous information on the distribution and critical condition of organisms present throughout the world. This field of science has significant importance in various educational, scientific and governmental sectors at global, national and regional levels. This emerging field provides biodiversity data for its investigation and comprehension in the near future.

Biological informatics has been following potential diverse applications for the protection of biodiversity:

- Information of various species (known and unknown) can be accessed
- Information on geographic parameters
- Infectious disease vectors distribution in various ecological conditions
- Assessment of invasive species
- Climate change assessment and its impact on biodiversity.

3 Global and National Databases

Global and national databases of all Biota provide data for conservation and sustainable development research.

Online databases providing information on various valuable biological resources are enlisted below [6]:

- Global Biodiversity Information Facility [GBIF] [www.gbif.org]
- Biodiversity Information System of Europe [BISE] [https://biodiversity.europa.eu/]
- Inter-American Biodiversity Information Network [IABIN] [http://www.iabin-us.org]
- Australian Biodiversity Information Facility [ABIF] [www.deh.gov.au/biodiversity/digir/]
- All Taxa Biodiversity Inventory [ATBI] [www.dlia.org/atbi/]
- European Natural History Specimen Information Network [ENHSIN] [www.nhm.ac.uk/science/rco/enhsin]
- Integrated Taxonomic Information System [ITIS] [www.eti.uva.nl/]
- National Biological Information Infrastructure [NBII] [www.nbii.gov/]
- World Biodiversity Information Network [REMIB] [www.conabio.gob.mx/remib_ingles/doctors/remibing.html]
- World Biodiversity Database [www.eti.uva.nl/ Database/WBD.html]
- Species 2000 [www.sp2000.org]
- Nature Serve: An Online Encyclopedia of Life [www.natureserve.org]
- Environmental Information System [ENVIS] [http://www.bsienvis.nic.in/]
- International Union for Conservation of Nature-Species Information System [IUCN SIS] [www.iucn.org/themes/ssc/our_work/sis.htm]
- IUCN Redlist [www.iucnredlist.org]

- Indian Biodiversity Information System [IBIS] [http://www.ibin.gov.in/]
- Invasive Species Specialist Group [IISC] [http://www.issg.org/]
- Global Invasive Species Programme [GISP] [https://www.gisp.org/]
- **Bacteria** [www.bacterio.cict.fr/]
- **Fungi**: National Fungus Collection, USDA [nt.ars-grin.gov/sbmlweb/collections/fungusCollection/Index.cfm]
- Oregon State University Mycological Collection [ocid.nacse.org/research/herbarium/myco/]
- University of Michigan Fungus Collection [http://www.lsa.umich.edu/ herb/]
- **Viruses**: The Universal Virus Database [www.ncbi.nlm.gov/ ICTVdb/]
- Genetics National Microbial Germplasm and Invertebrate Genetic Resources Program [www.nscalliance.org/bioinformatics/database.asp].

4 Digitalized Inventories on Bioresources of India

During the year 1999, the Ministry of Science and Technology established the National Bioresources Development Board (NBDB) under the aegis of the Department of Biotechnology with a mandate on research and development, policy framework and sustainable utilization of the country's bioresources by developing databases. The databases on all these bioresources are compiled on to a common platform named as Jeeva Sampada meaning Wealth of Bioresources at the Biodiversity Documentation Unit, UAS Bangalore centre. The database is now ready in its first version and is currently being made available through the website (www.ibin.gov.in) and also through CDs (Jeeva Sampada) for public use.

Department of Biotechnology (DBT), Government of India, New Delhi, has been promoting the development of a national level database on vegetation cover, biological resources and biological diversity of the country. As a first step towards compilation, three sets of databases are constructed.

a. **Biodiversity Characterization at Landscape Level in India Database**—A database covering the entire India has been prepared during the past 12 years. Comprehensive information on geospatial data of vegetation pattern, land cover, habitat destruction, species distribution, species abundance, species density and information are added biodiversity values.
b. **Jeeva Sampada**—This database was compiled by over 400 scientists from about 200 organizations which provide information on various bioresources' wealth of the country.
c. **Jeeva Manachitra**—A spatial database based on satellite images and ground inventory done over the last two decades. This data also highlights the conservation value of the species.

The three data sets are the largest data sets ever created on the country's biological resources and biodiversity. It was realized that there are a large number of more diverse data sets available in India, which could be brought together along with these core data sets. In this context, DBT is planning to develop a national level portal that can network diverse data sets of the country such that

a. The country shall have one window for the entire range of databases on bio-resources and the biodiversity.
b. Each of the data provider offers and gains information by sharing.
c. The end-user can retrieve a range of information through one portal.

4.1 Establishment of Bio-Resource Information Centers (BRICs)

Indian Bioresource Information Network (IBIN) serves to analysis critical information on diverse scope of problems of bioresources of our country. The primary aim of IBIN is to integrate the biological databases of the country. It is involved in designing relevant information on biological resources utilization, its access, bio-prospecting and protecting the indigenous knowledge from biopiracy of plantae, nnimalia and microbial diversity.

The portal encourages the scientists to host their findings with ownership authority, which is the main objective to create, promote, create a database network for integration on biodiversity in various languages. This information is an added value to the researchers, conservationists, common man and policymakers. IBIN invites the data from identified laboratories to publish their databases as BRICS.

5 Opportunities and Challenges of Biological Informatics

Considerable expansions in database structuring by various organizations are used for research purposes to explore the biocollections on the globe. This field enables exchange, integration and promotion of information on biodiversity conservation for better sustainability, optimizes the usage of bioresources and enhances the productivity worldwide. The areas which are prone to damage by the climatic effects can be identified and can be reclaimed. It facilitates scientific progress in data collection and expands the horizons for the training of researchers to do innovative research in biodiversity informatics.

6 Biodiversity Informatics—Future Perspectives

Biodiversity informatics will bring new strategies in predicting and mitigating the damage caused to bioresources, and visualization of biodiversity information through new web services and software tools can be analysed. Online open source web services will provide statistics of the distributed species and organisms thereby biodiversity conservation at regional, national and global levels can be achieved. Also through GIS, geospatial data of vegetation pattern, land use, green cover, habitat destruction, species distribution, species abundance and species density can be valuable information for sustainable utilization of bioresources. A spatial database based on satellite images and ground inventory done over the last two decades highlights the conservation of endangered species.

7 Conclusion

Biodiversity informatics will address the needs of researchers, policy-makers, environmental managers and the wider public for a consistent and up-to-date listing of all the world's known species. Also supports those who need to manage their own taxonomic information and species lists. Our knowledge of the world's species is still incomplete, but hundreds of thousands of researchers have worked over centuries to build this knowledge and assign names to the species they have recorded. Every day, taxonomists continue to publish new scientific names and refine our understanding of the world's species. Biodiversity informatics contain the accessible list of all species and show which species is referenced by any scientific name, but the tools and services offered also enable taxonomists and other stakeholders to publish and revise species lists for any purpose.

References

1. Berendsohn WG (2001) Biodiversity Informatics. In: Proceedings of 2nd national colloquium on global change research, Bad Honnef Jan 26–27. http://www.bgbm.org/BioDivInf/def-e.htm
2. Edwards JL, Lane MA, Nielsen ES (2000) Interoperability of biodiversity databases: biodiversity information on every desktop. Science 289(5488):2312–2314
3. Chavan V, Watvc AV, Londhe MS, Rane NS, Pandit AT, Krishnan S (2004) Cataloguing Indian biota: the electronic catalogue of known Indian fauna. Curr Sci 87(6):749–763
4. GBIF Plans 2007–2011: from prototype towards full operation. In: Lane M (eds) Global biodiversity information facility, Copenhagen, Denmark, 85 pp, (2006)
5. GBIF and Catalogue of Life Partnership sign memorandum of cooperation. GBIF press release dated 07, January (2004). http://www.gbif.org/Stories/STORY1073492201
6. Padmanabh D, Kumar P (2012) Biodiversity informatics and digitalizing the biodiversity of India. J Funct Environ Bot 2(1):12–21

Effectiveness of Plant-Based Biomaterials in the Field of Biomedical Engineering

Lakshmi Padmavathi Pydipati, Bharathi Depuru, Suvarna Latha Anchapakula, and Hemavathi Buddareddy

Abstract Naturally derived biomaterials are capable of promoting metabolic transformation processes, suggesting their potential usage in drug delivery and in making of therapeutic supportive devices. In particular, plant-based biosubstances can boost cell attachment and immigration when implanted in defective areas, thereby initiating extracellular matriculation and promoting tissue repair. These biomaterials are therefore used to repair or replace damaged tissues and organs in individuals. Furthermore, they are readily adaptable and inexpensive—making them the ideal material to be used as "scaffolds" for both tissue engineering and cellular processing. Since their structure and function mimic human tissues, plant vasculature tissues are a seductive alternative to scaffolding. Therefore, we are addressing the efficacy of biomaterials based on plants in the field of biomedical engineering in this article.

Keywords Synthetic biomaterials · Biosubstances · Immigration · Photosynthetic · Scaffolds

1 Introduction

Human civilizations have been dependent since ancient times on plant products for their fundamental needs, e.g. for food, fuel, clothing and health needs [1, 2]. Indeed, the ancient Egyptians and Romans were able to shape limb prostheses from wood and used fibres of vegetables to stitch skin around lesions. Some of the oldest applications of biomaterial were from old Phoenicia (The inhabitants of old Lebanon, Syria and Israel), where golden wires were used to bind loose teeth with artificial ones. Bone plates were successfully used in the early 1900s to support fractured bone and to speed up the healing process. Blood vessel repair was being developed in clinical trials during the 1950s and 1960s as well as artificial heart valves and hip joints [3]. Conventional tissue repair methods such as autograph and allograft have

L. P. Pydipati · B. Depuru (✉) · S. L. Anchapakula · H. Buddareddy
Faculty, Department of Biosciences and Sericulture, Sri Padmavati Mahila Visvavidyalayam (Women's University), Tirupati 517502, AP, India
e-mail: depurubharathi1960@gmail.com

S. Jyothi et al. (eds.), *Proceedings of the 2nd International Conference on Computational and Bio Engineering*, Lecture Notes in Networks and Systems 215,
https://doi.org/10.1007/978-981-16-1941-0_75

many drawbacks, including several operations, high risk of infection, painful procedures and donor shortages. Demand for 'biorepair needs' is gradually increasing as a consequence of growing population and urbanization. The industrial revolution has allowed a range of synthetic biomaterials (first metallic and then polymers) to be produced with features that are highly suitable for developing medical devices. The composition and structure of synthetic biomaterials, however, do not fit native tissue/organisms, are of minimal bioconsistency and have limited capacity to induce tissue remodelling. In recent times, materials of natural origin and synthetic biomaterials have become an integral element of the strategies for regenerative therapeutics and tissue engineering. Out of 3 lakh known plant species worldwide, only five thousand have been explored for their therapeutic applications to this date. Enormous amounts of plant-based natural materials are available for medicinal use and most of these have yet to be explored.

The creation of life-changing and life-saving concepts and products such as artificial organs, pacemakers, artificial hips, surgical robot systems, advanced prosthetic systems, kidney dialysis, MRIs, EKGs, ECGs, pharmaceutical and therapeutics has been significantly contributed by biological understanding integrated with engineering technology principles in order to meet medical needs. In this regard, the demand for alternative strategies for successful regeneration of the damaged tissues is constantly growing [4].

Tissue engineering with biomimetic 3D cell-seeded scaffolding appears as a promising method for the repair of tissue [5–7] oriented innovations like stem cell engineering and 3D printing of bioorganisms. Though the skills of bioengineers are focused on engineering and biology, they include multiple fields of knowledge. Computer software is often intended to control complex structures, such as 3D X-ray machines. Many of these technicians often use their chemical and biological expertise to develop new medical therapies, mathematics and statistics to construct models to describe tramsmission of brain and heart signals.

2 Classification of Biomaterials Based on Nature and Tissue Responses

Biomaterials which are derived naturally can be categorized into many groups based on their nature [8]. For example, biomaterials based on protein (silk, gelatin, collagen, etc.); biomaterials based on polysaccharide (chitin, cellulose, glucose, pectin, etc.); and biomaterials derived from decellularized tissue (decellularized blood vessels, heart valves, liver). In general, there are three terms for defining a biomaterial through tissue responses. They are biologically inert biomaterials, biologically active biomaterials and biologically absorbable biomaterials.

a. **Biologically Inert Biomaterials**: Any material that has minimal interaction with its surrounding tissue, after it is placed in the human body, is said to be 'bioinert'. Examples of these are titanium, stainless steel, partially stabilized zirconia, alumina and ultrahigh molecular weight polyethylene. A fibrous capsule may typically grow around bioinert implants, and thus the functionality of the implant relies on the integration of tissue around it.

b. **Biologically Active Biomaterials**: Bioactive refers to a substance that interacts with the surrounding tissue and often even soft tissue after being inserted in the human body; it is also referred as 'biologically active material'. This happens over time—a kinetic alteration of the surface that is induced based on its implantation within the living bone. An ion-exchange reaction between a 'bioactive' implant and surrounding body fluids leads to the formation of 'carbonate apatite' layer similar to mineral phase in the bone. Synthetic hydroxyapatite, A-W glass ceramic and bioglass are the main examples of these products.
c. **Biologically Absorbable Biomaterials**: Biologically absorbable means a substance that begins to dissolve (resorbed) once it is inserted inside the human body and is slowly replaced with progressive tissue (like bone cartilage). Polylactic–polyglycolic acid copolymers and tri calcium phosphate are typical examples of bioabsorbable content. Other common materials that have been used over the past three decades are calcium oxide, calcium carbonate and gypsum.

(Source: https://www.azom.com/article.aspx?ArticleID=2630) [26]

3 Three Dimensional Bioprinting

The area of tissue technology obtained new incitation only with the advent of 3D cell culture technologies. Primary part of the twenty-first century showed that cells cultivated in 3D represent better human physiology than cells cultivated in regular 2D monolayers. Further research broadened the spectrum of usage from regenerative to the delivery of drug, testing of substance and medicine for personalized use and to provide 3D environment to the cell biomaterials used as scaffolds in 3D cell cultures. Cellulose, alginate, Matrigel, Collagen I, fibrin, hyaluronic acid, gelatin or mixtures of natural biomaterials are widely used in the production of 3D cell cultures. Polyurethane, poly ethylene glycol and poly vinyl alcohol are the synthetic polymers used in 3D cell cultures.

During the bioprinting process, a biomaterial solution or multiple biomaterial mixtures in the form of hydrogel, encasing the required types of cells (bioink) is used for structuring the tissue models. Alginate and nanocellulose are two of the most commonly used biomaterials derived from marine and land forms (Fig. 1). Tissue engineering can be revolutionized by bioprinting since the regulated deposition of cells and biologically active molecules in 3D space enables tissue generation by way of additive means. Tissue engineering that uses cell-seeded biomimetic 3D scaffolds appears to be a promising strategy and successful alternative in the future to encourage tissue repair, especially in the orthopaedic sector. But their success strictly depends on an ever deeper knowledge about the characteristics of the biomaterials and the potentialities of their combinations.

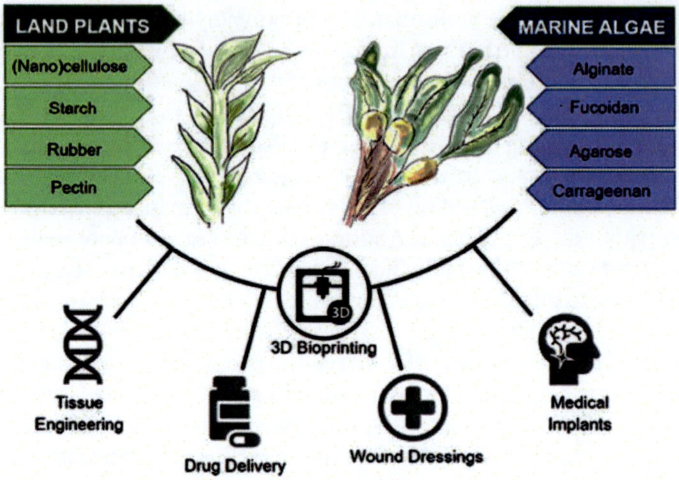

Fig. 1 Plant derived biomaterials from land and marine sources and their potential in 3D bioprinting. *Source* https://www.frontiersin.org/articles/446057 [25]

Source: https://www.innovationnewsnetwork.com/why-innovation-in-biomaterials-is-key-to-the-success-of-bioprinting/5763/ [24]

4 Plant Based Biomaterials for Advanced Biomedical Applications

The regeneration of tissues and structuring of artificial organs can serve as a means of resolving the disparity between the number of donating organs and people's needs. The bone implant is a dangerous procedure such as the hip replacement: an infection may cause rejection reaction. Biomaterial implant coatings make osseointegration more easy (integrate into the surrounding bone), promoting bone development and avoiding reaction of rejection. Kokkenen and colleagues (2010) [9] separated sugar subgroups from their principal carbon column from enzyme-treated apples and coated apple pectin on titanium dental implants at thickness of nanometers. Their experiment showed positive results by promoting osseointegration in rat bone tissue culture with reduced rejection responses. Thus photosynthetic creation of the universe are diversified, adaptable and inexpensive—making them the ideal material for the development of novel biomaterials.

5 Hydrogels and Their Use in Tissue Engineering

Hydrogels are water-swollen and cross-linked hydrophilic polymeric networks derived from plant cell wall materials. Due to their structural similarity with bio-extra cellular matrix (ECM), hydrogels are defined as smart and sophisticated materials over the past two decades with many applications. For cell existence and enlargement, hydrogels imitate a real environment of 3D cell culture. The biomedical field of research has shown a particular interest in hydrogels as they can be made into therapeutic structures from various polymers due to their tunable properties.

Hydrogels from biorenewable polymers are common in biomedical applications, as they have low toxicity, environmental friendliness, bio-compatibility, easy access and low cost in parallel. Nevertheless, the multifaceted challenge is to find in tissue engineering an ideal plant-based green hydrogel that can imitate fundamental structural, functional and performing properties of human tissues [10]. Moreover, these natural polymers are often optimized so that their chemistry and physical actions can be controlled for the delivery of drugs and the regeneration of stem cell guided tissues.

Ngoenkam et al. [11] documented heat sensitive hydrogel which is injectable for the delivery of chondrocytes in tissue engineering of cartilage. Ciolacu et al. [12] prepared hydrogels of less density with cellulose/lignin to release high drug content. The encapsulation of bioactive compounds into pectin hydrogels that are intended for delivery of drug inside colon is a reliable technique to protect them against deterioration due to high acidity or degradation caused by gastric and intestinal enzymes by stomach. Some pectin displayed a promising and beneficial characteristics of in vitro anti-inflammative properties in the tuning of healthcare products [13]. A new class of nanoscale cellulose-based particulate matter and polymers are hairy nanocelluloses, which can overcome traditional nanocellulose structural limitations as well as provide extra mechanical tissue engineering materials resistance. In vivo experiments of Hadebe et al. (2014) [14] on type 1 diabetic animal models have showed a lower blood glucose level with the use of insulin-filled hydrogel patches made of amidated pectin.

6 Biomaterials as Scaffolds

Recently, considerable attention was given to the advancement of 3D printing technology for the development of cellulose-based hydrogels and tissue engineering scaffoldings. Plant tissue vasculature provides an enticing alternative scaffolding as it imitates human tissues in its form and functions. The cellulose/hydroxylapatite nanocomposite scaffolding for bone tissue engineering has been developed by Zadegan et al. [15] which showed increased cell viability and chondrocyte cells proliferation. Apple cellulose scaffolds were developed by Andrew Pelling et al. [16] from the University of Ottawa. They provide a complementary approach to

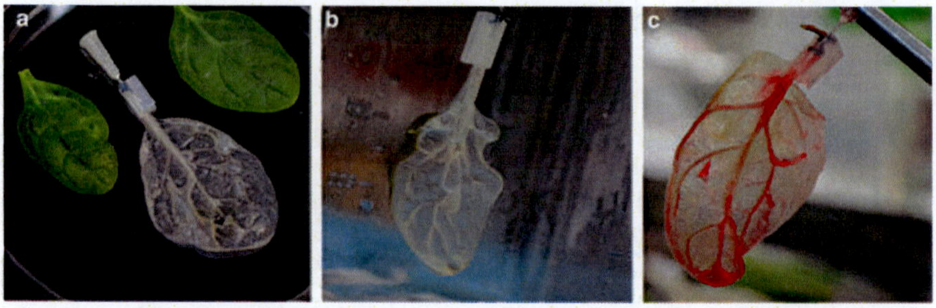

Fig. 2 The Gaudette Group's decellularized spinach leaf-heart. **a** decellularized spinach leaf and intact spinach leaf, **b** shows the catheter apparatus used to perfuse the tissue, **c** shows red dye perfused through spinach scaffold recellularized with cardiomyocytes. *Source:* https://thebiochemistblog.com/2018/03/07/plant-based-biomaterials-engineering-the-future/ [27]

present techniques for in vitro culture of mammalian cells in a 3D environment, making them simple to generate, cost efficient and from a renewable source.

The creation of Gaudette group (2017) [17] decellularized plants for scaffolding for the production of tissue perfusion has opened up the possibility of a new research wing in scientific field which investigates the imitation between kingdoms (plant and animal). Plant vasculature tissue imitates mammalian vasculature. The central xylem and phloem allow water and nutrients, like blood vessels in humans, to be transported to distant cells. They decellularized a spinach leaf (from their local market) and tested leaf with red dye which showed low fluid leakage. They went on to recellularization with human cells; in the lumen of plant vasculature, human endothelial cells were placed, while mesenchyma cells were placed into the outer tissue. These adhered and grew quickly. In addition, cardiomyocytes (cardiomyocytes) have been seeded onto the leaf surfaces. Cardiomyocytes have the ability to beat on their own which is essential to the heart function. The cardiomyocytes seeded on the spinach could unbelievably beat and develop by themselves (Fig. 2).

7 Hybridized Implants in Bioengineering

Composite scaffolds have been developed to combine various biomaterials [18]. They are important in terms of technology and industrial application as they combine bio-compatibility, biodegrading and a significant mechanical power. In addition, these scaffolding forms can be used to regenerate both hard and soft tissues and to imitate the tissue architecture consisting of cells and extracellular matrices. Several experiments demonstrated the success of tissue engineering techniques with composite supportives (synthetic/natural polymers) [19, 20]. Other researchers showed that the addition of metallic implants to polymer or metal/ceramically grounded coatings could serve as an other interesting solution [21]. Generally speaking, natural hydrogels were less toxic than synthetic ones, for example, with natural cellulose-hydroxyapathetic hydrogel created for bone tissue engineering.

Other researchers demonstrated that Pasqui et al. developed a natural cellulose-hydroxyapatite hybrid hydrogel for bone tissue engineering. Another interesting solution may be the combination of metallic implants with polymer coating or metal/ceramic scaffolds [22]. Generally, natural hydrogels are less toxic and more tolerated than synthetic ones, and Gilabert Chirivella et al. [23] developed coating material for the coating of titanium dental implants, mixed with silver and chitosan. The prepared particles showed to have an inhibitory effect on dental pathogens (*S. mutans* and *P. gingivalis*). It also prevented the formation and eventual adhesion of biofilm.

8 Conclusions

Plant derived products represent novel and interesting candidates for biomaterial applications, especially for 3D bioprinting. Enormous appeal lies in their gelation capacity, hydrophilicity and natural rigidity which appear to emulate the extracellular matrix and render them ideal biological and mechanical candidates for bioprinting tissues, organs and biomedical adjuncts. The achievement and use of plant products does not pose ethical concerns because of their botanical roots. Typically, they are readily accessed and cost-effective, most of them with low immunogenicity and are not harmful by themselves, but still bioactive at low concentrations. The polymers are obviously beneficial, used effectively in wound dressings, implantable health devices and drug delivery applications, which are enabled by integrating bioprinting technologies for genuinely personalized and biomedical therapies with non-antigenicity, non-toxicity, convenience to chemical and physical properties, as well as bio-degradability. The properties of the bioinks have to be further optimized and improved by the combination with other materials to satisfy the biological and mechanical requirements of different kinds of tissues. Before in vivo implementation, a rigorous analysis of the toxicity of the material is further required. The combination of polymers (hybridization) with different characteristics has been especially advantageous and interesting to improve functionality and overcome polymer limitations in biomaterials formation. This sustainable, natural and plentiful collection of flexible plant-based biomaterials provides tremendous potential for the future of bioprinting, regardless of the existing challenges to be addressed in the clinical practice. Enhanced knowledge of polymers and subsequent applications for the future of plant-based hydrogels is important for research to be encouraged with modern and feasible applications because plant-based compounds are becoming hot topics.

References

1. Karunamoorthi K, Jegajeevanram K, Vijayalakshmi J, Mengistie E (2013) Traditional medicinal plants: a source of phytotherapeutic modality in Resource-Constrained Health care settings. Evid Based Complement Altern Med 18:67–74
2. Sen R, Chatterjee M (2011) Plant derived therapeutics for the treatment of Leishmaniasis. Phytomedicine 18:1056–1069
3. Heness G, Ben-Nissan B (2004) Innovative bioceramics. In: Materials forum, vol 27
4. Walker BW, Lara RP, Mogadam E, Yu CH, Kimball W, Annabi N (2019) Rational design of microfabricated electroconductive hydrogels for biomedical applications. Prog Polym Sci 92:135–157
5. Bedian L, Rodríguez AMV, Vargas GH, Parra-Saldivar R, Iqbal HM (2017) Bio-based materials with novel characteristics for tissue engineering applications–A review. Int J Biolog Macromolecules 98:837–846
6. Nasajpour A, Ansari S, Rinoldi C, Rad AS, Aghaloo T, Shin SR, Mishra YK, Adelung R, Swieszkowski W, Annabi N, Khademhosseini A, Moshaverinia A, Tamayol A (2018) A multifunctional polymeric periodontal membrane with osteogenic and antibacterial characteristics. Adv Func Mater 28:1703437
7. Taale M, Schütt F, Carey T, Marx J, Mishra YK, Stock N, Fiedler B, Torrisi F, Adelung R, Selhuber-Unkel C (2019) Biomimetic carbon fiber systems engineering: A modular design strategy to generate biofunctional composites from graphene and carbon nanofibers. ACS Appl Mater Interfaces 11:5325–5335
8. Bao Ha TL, Minh Quan T, Nguyen Vu D, Minh Si D (2017) Naturally derived biomaterials: preparation and application. https://doi.org/10.5772/55668
9. Kokkonen H, Niiranen H, Schols HA, Morra M, Stenbäck F, Tuukkanen J (2010) Pectin-coated titanium implants are well-tolerated in vivo. J Biomed Mater Res A 93:1404–1409
10. Mohammadinejad R, Maleki H, Larrañeta E, Fajardo AR, Nik AB, Shavandi A, Sheikhi A, Ghorbanpour M, Farokhi M, Govindh P, Cabane E, Azizi S, Aref AR, Mozafari M, Mehrali M, Thomas S, Mano JF, Mishra YK, Thakur VK (2019) Status and future scope of plant-based green hydrogels in biomedical engineering. Appl Mater Today 16:213–246. ISSN 2352-9407
11. Ngoenkam J, Faikrua A, Yasothornsrikul S, Viyoch J (2010) Potential of an injectable chitosan/starch/β-glycerol phosphate hydrogel for sustaining normal chondrocyte function. Int J Pharm 391:115–24 [Crossref], [PubMed], [Web of Science ®], [Google Scholar]
12. Ciolacu DE, Nicu R, Ciolacu F. Cellulose-based hydrogels as sustained drug-delivery systems. Materials (Basel) 13(22):5270. https://doi.org/10.3390/ma13225270
13. Sakurai MH, Matsumoto T, Kiyohara H, Yamada H (1999) B-cell proliferation activity of pectic polysaccharide from a medicinal herb, the roots of Bupleurum falcatum L. and its structural requirement. Immunology 97:540–547
14. Silindile SI, Phikelelani PS, Serumula MR, Musabayane C (2014) Transdermal delivery of insulin by amidated pectin hydrogel matrix patch in streptozotocin -induced diabetic rats: effects on some selected metabolic parameters. PloS one 9:e101461. https://doi.org/10.1371/journal.pone.0101461
15. Zadegan S, Hosainalipour M, Rezaie H, Hossein G, Mohammad S (2011) Synthesis and biocompatibility evaluation of cellulose/hydroxyapatite nanocomposite scaffold in 1- n-allyl-3-methylimidazolium chloride. Mater Sci Eng 31:954–961. https://doi.org/10.1016/j.msec.2011.02.021
16. Modulevsky DJ, Lefebvre C, Haase K, Al-Rekabi Z (2014) Pelling AE apple derived cellulose scaffolds for 3d mammalian cell culture. PLoS ONE 9(5): https://doi.org/10.1371/journal.pone.0097835
17. Gershlak JR, Hernandez S, Fontana G, Perreault LR, Hansen KJ, Larson SA, Binder BYK, Dolivo DM, Yang T, Dominko T, Rolle MW, Weathers PJ, Medina-Bolivar F, Cramer CL, Murphy WL, Gaudette GR (2017) Crossing kingdoms: using decellularized plants as perfusable tissue engineering scaffolds. Biomaterials 125:13–22. ISSN 0142-9612, https://doi.org/10.1016/j.biomaterials

18. Dolcimascolo A, Calabrese G, Conoci S, Parenti R (2018) Innovative biomaterials for tissue engineering. https://doi.org/10.5772/intechopen.83839
19. Qian J, Xu W, Yong X, Jin X, Zhang W (2014) Fabrication and in vitro biocompatibility of biomorphic PLGA/nHA composite scaffolds for bone tissue engineering. Mater Sci Eng C Mater Biol Appl 1(36):95–101. 10.1016/j.msec.2013.11.047
20. Sekiya N, Ichioka S, Terada D, Tsuchiya S, Kobayashi H (2013) Efficacy of a poly glycolic acid [PGA]/collagen composite nanofibre scaffold on cell migration and neovascularization in vivo skin defect model. J Plast Surg Hand Surg 47(6):498–502
21. Oughlis S, Lessim S, Changotade S, Bollotte F, Poirier F, Helary G et al (2011) Development of proteomic tools to study protein adsorption on a biomaterial titanium grafted with poly(sodium styrene sulfonate). J Chromatogr B 879:3681–3687. https://doi.org/10.1016/j.jchromb.2011.10.006
22. Wu C, Zhou Y, Xu M, Han P, Chen L, Chang J et al (2013) Copper-containing mesoporous bioactive glass scaffolds with multifunctional properties of angiogenesis capacity, osteostimulation and antibacterial activity. Biomaterials 34(2):422–433. https://doi.org/10.1016/j.biomaterials.2012.09.066
23. Gilabert-chirivella E, Pérez-feito R, Ribeiro C et al (2017) Chitosan patterning on titanium implants. Prog Organ Coat 111:23–28
24. https://www.innovationnewsnetwork.com/why-innovation-in-biomaterials-is-key-to-the-success-of-bioprinting/5763/
25. https://www.frontiersin.org/articles/446057
26. https://www.azom.com/article.aspx?ArticleID=2630
27. https://thebiochemistblog.com/2018/03/07/plant-based-biomaterials-engineering-the-future/

Real-Time Video Tracking Framework With Moving Object Segmentation in Stream Data

K. R. Manjula and S. Anand Kumar Varma

Abstract Object tracking is an active application-oriented research topic. Analysis of video requires human operators to monitor human activities. This framework should be capable of detecting and tracking moving objects accurately. The unsupervised segmentation of moving objects affects the overall performance of an object tracking framework. The ideal aim of unsupervised segmentation method is to split the image into meaningful objects. In this proposed video tracking framework, initially, segmentation of the moving object with displacement vectors in an unsupervised manner is done. Using the displacement vector as a feature, the input frame segmentation is achieved with the Expectation–Maximization (E–M) technique. In this work, the E–M algorithm initialized with the number of objects present in the frame is obtained by clustering objects using the K-means clustering technique. The resulting objects obtained may contain shadows of the objects. Hence, shadow removal approach is used to remove shadows surrounding objects. Finally, the objects tracked using displacement vectors and the behaviour of objects are analyzed. The experiments on various datasets such as standard videos, surveillance videos, and nonstandard videos demonstrated, outperforms in single and multiple cases.

1 Introduction

The principal objective of video tracking framework [1] is to comprehend the spatio-temporal changes of objects with motion. Surveillance has gained a lot of importance. Their adaptation to various environments leads to specific research areas like wireless, mobile, aerial, traffic surveillance, human object identification, detection of anomalous behaviour, interactive surveillance using multiple cameras, etc [1]. The goal of the tracking framework is to draw the trajectory of an object in every frame

K. R. Manjula (✉)
SAP/CSE, SoC, SASTRA Deemed University, Thanjavur 613401, TN, India
e-mail: manjula@cse.sastra.edu; manju_sakv@yahoo.co.in

S. Anand Kumar Varma
E.G.S. Pillay College, Namakkal 611002, Nagapattinam, India

S. Jyothi et al. (eds.), *Proceedings of the 2nd International Conference on Computational and Bio Engineering*, Lecture Notes in Networks and Systems 215,
https://doi.org/10.1007/978-981-16-1941-0_76

over time by finding its location [2]. The initial step for this proposed video tracking framework is to extract the objects associated with a displacement vector, from the video or foreground segmentation. So the segmentation of moving object affects the overall performance of an object tracking system [3]. If any object missed segmentation, it could not be tracked because the object edges are given as the input to the tracking module.

The segmentation objective is to distinguish the semantically significant pixels of the frame and group the pixels fitting in such segments [4]. Foreground segmentation is extricating fascinating patterns or objects from background in the video sequence [5]. These intriguing patterns or objects are called foreground or forefront objects. The foreground segmentation is a vital pre-processing task in image and video processing like video and image coding, compression, moving object tracking and recovery [6, 7].

1.1 The Proposed Tracking Framework Overview

Monitoring and tracking frameworks require the segmentation of every single moving object in a video. The performance of the tracking framework is extraordinarily dependent on segmentation modules [8]. It assumes a key part as it diminishes the amount of data to be prepared and used by object tracking. The overview of the proposed tracking framework is shown in Fig. 1.

The data obtained from the sources are converted into frames and fed into the pre-processing module. DCT (Discrete Cousin Transform) is applied for obtaining the intensity values as DC coefficient in the pre-processing block. D.C. coefficients of the frequency domain values are used for the estimation of displacement vectors (DV) [9, 10]. Using this displacement vector, we can find whether the block is still

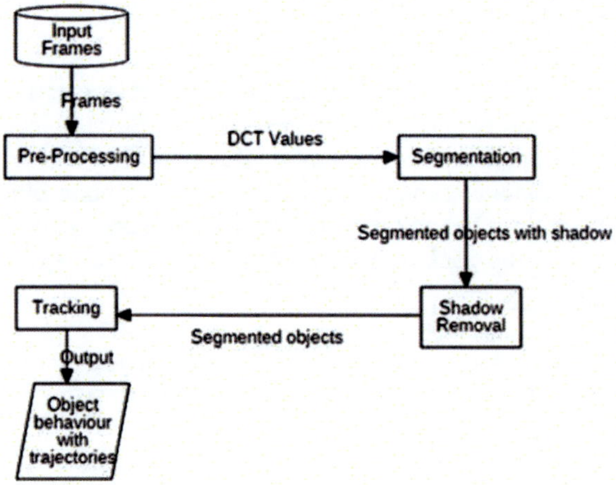

Fig. 1 Overview of the proposed tracking framework

or moving. K-means clustering is used to find the number of objects with DV in the frame. Then Expectation and Maximization (E–M) algorithm is applied to get the segmented objects. The object shadows are wrongly identified as objects along with objects in the segmentation step [11].

In shadow removal block, cast shadows are identified and removed. To remove these shadows initially, the background image is constructed, used by the shadow removal algorithm [12]. The final block in this proposed tracking system is tracking block, in which the trajectory of objects are drawn, and finally, the behaviour of the objects are analyzed.

2 Related Work

A popular and at the same time, a simple approach is a background subtraction technique. A background subtraction approach can target appearance variations such as illuminations, background clutter proposed by Chien et al. [13]. But handling the fuzzy boundary will be a significant issue in the background subtraction technique. Lijun Guo et al. [14] proposed an unsupervised video segmentation model, based on the spatio-temporal graphical model. This model extends the multi-object segmentation from the binary segmentation using Conditional Random Framework (CRF). Texture-based approach for shadow removal algorithms are also gaining importance and used in few of research application. Photometric properties of shadows and its textural characteristics are combined to detect them efficiently as presented in Jun Tao Xue [12]. Recently, the GAME theory-based techniques in which each object is a player and competes for maximum utilities gain more research attention. It can track the object efficiently under mutual occlusion [9]. The individual estimations inside the impediment region are viewed as the players in the developed game competing for the greatest utilization of certain systems.

A technique reported by Nascimento et al. [8] concentrates only on pedestrian trajectories. The methods aimed to model the directions through a little arrangement of vector or movement fields because they tend to follow a set of typical trajectories. A combined shape- and feature-based video analysis which uses size, point and region but does not rely on motion vectors capable of handling non-rigid motion are all discussed in [2]. The tracker framework is initiated by evaluating the region of an object in every frame. This type of object tracking is called silhoutte tracking. Silhouettes tracking is done by shape evolution or shape coordinating the given object model, which utilizes object's information [15, 16].

Table 1 Input data details

Video source	Video type	Number of frames (pers)	Size in M.B. (approximately)
Hall_Monitoring	.mpeg	29	953
PETS	.cif	28	967
Surveillance	.dav	25	876

3 The Proposed Tracking Framework

3.1 *Input Data*

The standard videos such as PETS and surveillance videos (which are taken from university surveillance database) are converted into frames and stored into the local database for further processing. Details of the video data used in this project are given in the Table 1.

3.2 *Pre-processing*

Pre-processing is the initial step in the proposed tracking framework in which the frequency domain conversion is carried out using a transforming function (1),

$$Y_{x_1,x_2} = \sum_{p_1=0}^{N_1-1} \sum_{p_1=0}^{N_2-1} Y_{p_1,p_2} \cos\left[\frac{\pi}{N_1}\left(p_1 + \frac{1}{2}\right)x_1\right] \cos\left[\frac{\pi}{N_2}\left(p_2 + \frac{1}{2}\right)x_2\right] \quad (1)$$

where the (p_1, p_2) is the spatial location or coordinates of each pixel.

These DC coefficients of the frequency domain values fed as input for the next module are called unsupervised segmentation method.

3.3 *Unsupervised Segmentation Method*

The unsupervised video segmentation, which separates the shape data of moving object from the video, is the key operation for object tracking framework. The proposed video segmentation method is shown in Fig. 2. The A.C. coefficient obtained from the pre-processing module is fed as input to the segmentation module. Initially, the Displacement Vector (DV) is estimated from the estimation of DV block. If the DV is equalled to zero, then it is eliminated else it is fed as input to the number of moving objects estimation block. The total number of objects presented in the

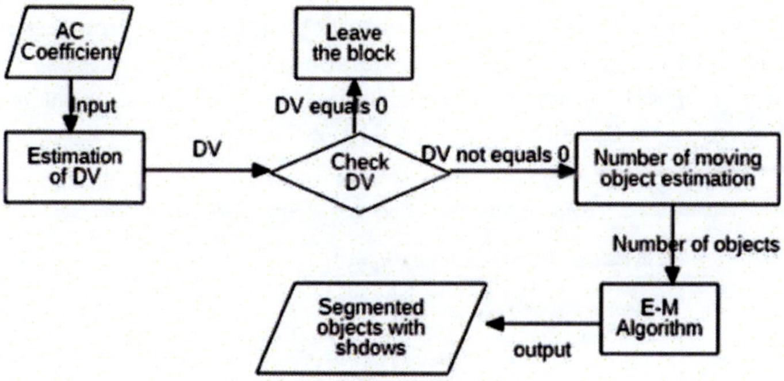

Fig. 2 Architecture of the proposed segmentation method

Fig. 3 Displacement of Vector estimation

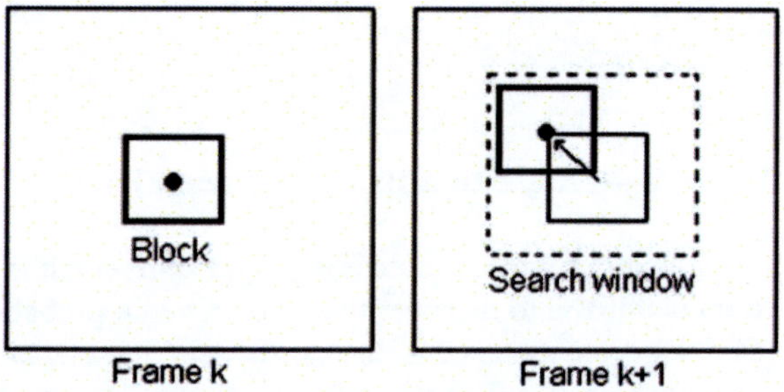

frame is found out by this module. And it is fed as input to the E–M algorithm block. The detailed explanation of each block is explained below [14, 17].

3.3.1 Find the Displacement Vector (DV)

The difference between the block's position in the current frame and the best matching block's position in the reference frame is known as DV. The Fast Three Step Search (F3SS) is the method for estimating the DV at eight locations $\pm 2^{N-1}$ (for a search window of $\pm 2^{N-1}$samples). The DC coefficient is the measure to compare and estimate the DV in F3SS. This method includes only three steps of searching [3, 18]. The sample search window of the proposed approach is shown in Fig. 3.

3.3.2 Motion Model Estimation

The segmentation is exceptionally delicate to the number of movement models, so estimating the number of movement models (layers) is essential. If the number of motion models is less, then the objects are merged, resulting in under segmentation. But, if the quantity of moving object is vast, it results in splitting the objects, leading to over-segmentation. Thus, it is essential to determine the appropriate number of motion models before starting the segmentation process. This process is achieved by algorithm 1. The resultant number of objects $|c_i|$ is sent as input to the E–M algorithm to fragment the moving objects [19, 20].

***Algorithm** 1: To find the number of moving objects (K-Means Clustering)*

***Initialize** the cluster centre* $k_i = 1, \ldots, k$

***Repeat** until no change in clusters*

{ *1. Add each pixel to cluster (Nearest)*

$$c_i = \{j : d(x_j, k_i) \leq d(x_j, k_l), l \neq i, j = 1, \ldots, n\}$$

2. Set the cluster mean to each cluster

$$k_i = \frac{1}{|c_i|} \sum_{j \in c_i} x_j, \forall i$$

} ***Return*** $|c_i|$

3.3.3 E–M Algorithm-Based Segmentation

An E–M algorithm is a clustering algorithm which is iterative in nature. This algorithm is utilized to discover the most extreme probability or maximum a posteriori (M.A.P.) evaluations of statistical models' parameters. This model relies on secret idle variables. The EM iteration substitutes between performing a desire or expectation (E) step and maximization (M) step. The E-step makes a capacity for the log-probability desire, assessed utilizing the current appraisal for the parameters. The M-step creates the parameters maximizing the expected log-likelihood found on the E-step. These parameter-evaluations are then used to decide the dissemination of the inactive variables in the following E step.

To start with, the static object is segmented by allocating the pixels with zero. This is done to decrease the weight of an object by the E.M. algorithm. The remaining pixels with DV are fragmented into different layers by applying the E.M. algorithm, as shown in algorithm 2.

***Algorithm** 2: To cluster the pixels*

***Repeat** until convergence*

{ 1. *For each* c_i, *(E-Step)*

$$Set\ E_{c_i}(x^i) = p(x^i | 2^i; \theta)$$

2. *Set (M-Step)*

$$\theta := arg\, max \sum_{c_i} \sum_{x^i} E_{c_i}(x^i)\ log \frac{p(2^i, x^i, \theta)}{E_{c_i}(x^i)}$$ }

3.4 Shadow Removal

Object segmentation is affected by the presence of shadows. The detection and removal of shadows are the primary tasks for processing the outdoor image or video. These shadows are removed by using morphological operations such as erosion and dilation. Before that, shadows are detected using shadow detection algorithm as shown in algorithm 3.

***Algorithm** 3: To detect the shadows in the objects*

***For** each segmented pixel, p compute*

$$\mu_R(p) = \frac{\Sigma p}{N}, \text{and } \sigma_R(p) = \sqrt{\frac{1}{N}\Sigma_{j=1}^{N}(p_j - \mu_R(p))^2}$$

Find

$$D(x) = p(x) - \mu_R(p) \sim N(0, \sigma_D(x)^2)$$
$$\sigma_D(x)^2 = \frac{1}{N^2}((N-1)^2\sigma_R^2(p) + \Sigma_{p \neq x}(p)\sigma_R^2(u))$$

***If** $\{|D(x)| > K_s(\beta)\sigma_D(x)\} < \beta$ then*

Associated with shadows, so examine the neighbourhood pixel

If yes then $S(x)$ is incremented

***Check** $\mu_R(p)$ and $S(x)$ with a threshold to determine shadows*

The erosion (morphological operation) of a binary image b by an organizing component creates another similar image with binary values m = bθ s with ones in all spatial location s (x,y) as shown in Fig. 4a. The dilation of an image b by an organizing component s creates another binary image n = b ⊕ s with ones in all axes (x,y) which is shown in Fig. 4b.

4 Moving ObjectsTracking

Object tracking is the issue of deciding (assessing) the positions and other vital data of moving objects in image groupings. For a given segmented region, the object tracker tracks the moving object at each timestamp. In the primary case, object regions in the frame are acquired by an object detection algorithm, and then the tracking framework

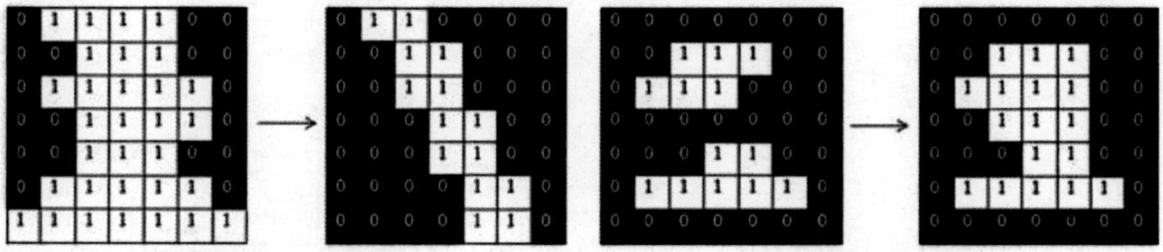

(a) Erosion operation *(b) Dilation operation*

Fig. 4 Morphological operations

finds the corresponding objects. In the last case, the region is assessed by location and region information of the objects, which is iteratively acquired from previous frames. The proposed tracking algorithm is explained in algorithm 4.

***Algorithm 4:** To track the objects*

***Find** O_t^n in frame t*
***Find** O_{t+1}^n in frame t+1*
***Check** $O_t^n > O_{t+1}^n$ then*
    Object may be occluded or end.
***Find** $D_{i,j} = \sqrt{(x_j - x_i)^2 + (y_j - y_i)^2}$*
    Matching $D_{i,j}$ in t and (t+1)
***Check** $DV_i = DV_{i+1}$*
    Find $p_{i,j}$ then
    Draw trajectory between $p_{i,j}$, $p_{i+1,j+1}$

In the initial step, for the first frame find the number of objects (O_t^n). Then, find the number of the objects (O_{t+1}^n) in the upcoming frames, which is compared with the previous number of objects; if it is decreased then the objects are occluded.

The (t−1) frame's identified objects are compared to find the matching object in t frame using the shortest Euclidean distance (E.D.) measure or value. Then draw a trajectory between the centres of the bounding box of the objects. If the number of objects in the next frame decreases compared with the previous frame, then the objects are merged. We may look in some cases, and it may not be able to get the best matching block. In that case, mark the one bounding box that is nearer to the object as O_{t-1}^n in the previous frame. If the number of objects in the next frame increases compared to the previous frame, then the occluded objects may be seen in the frame. Check whether the objects are new by checking with the previously found objects. If a match is found, then track it. Else start a new path. The proposed tracking model explained in Fig. 5 is for easy implementation of the algorithm.

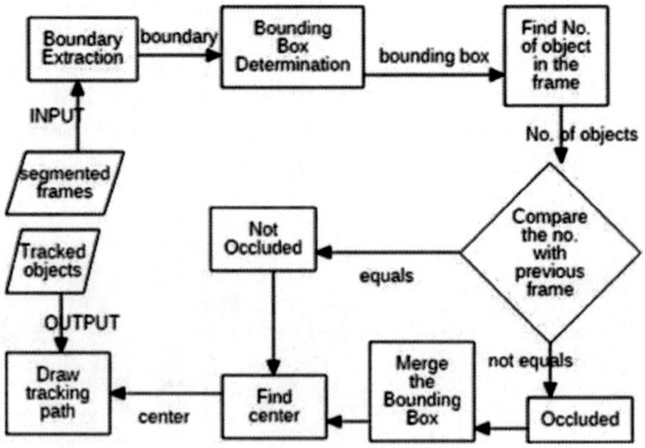

Fig. 5 Block diagram of the tracking model

5 Result, Analysis and Discussion

The proposed video surveillance segmentation and tracking framework using motion vectors are experimented with different video frames, such as standard videos and live surveillance video, which is shown in Figs. 6 and 7. The hall monitoring video sequence, the corresponding segmentation results and intermediate results are shown in Fig. 6. Figure 7 shows the segmentation results of both PETS video sequence and surveillance video sequence, and the multiple objects are segmented using the proposed video segmentation method shown in Fig. 7.

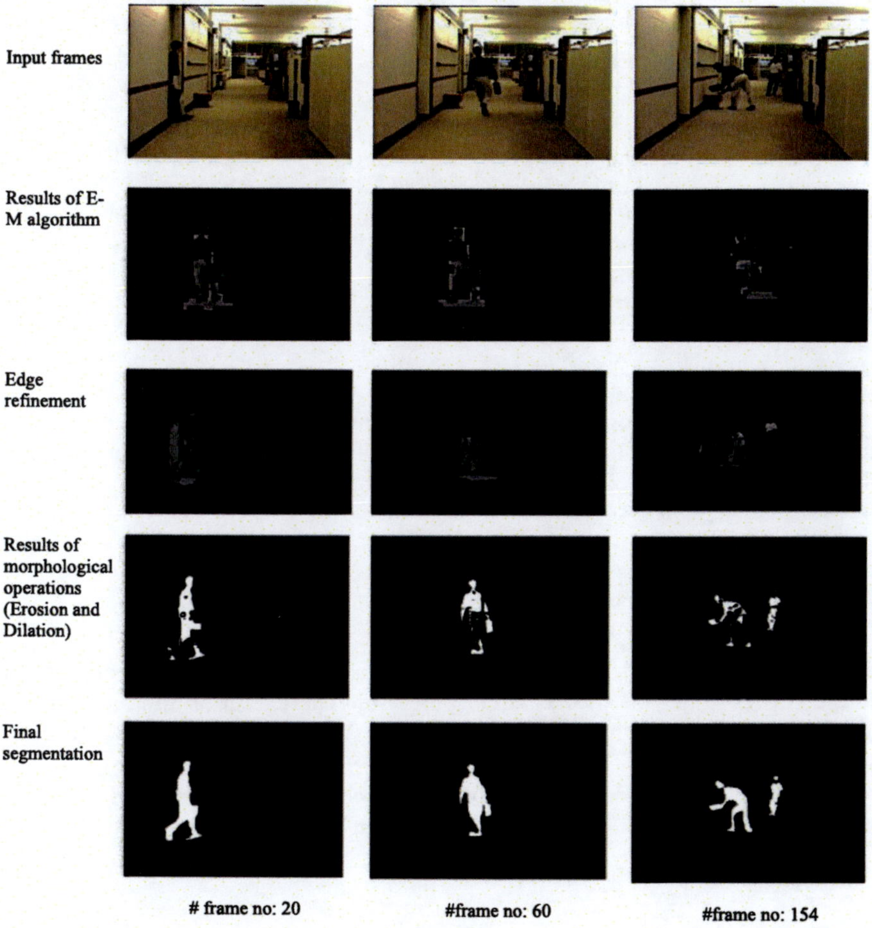

Fig. 6 Results of the proposed segmentation approach with an input video sequence and intermediate results

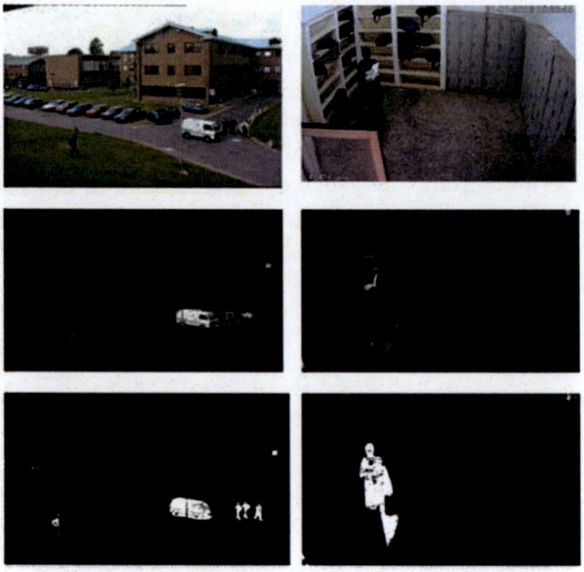

Fig. 7 Results of the proposed video segmentation method for standard PETS video and surveillance video

The result of the proposed tracking framework is shown in Fig. 8. In Fig. 8, the intermediate results of the framework such as estimating the bounding box and finding the centre point of each object are shown.

5.1 Performance Analysis

The proposed framework's performance analysis is done by executing the algorithms in the H.C.L. notebook with Pentium processor. The execution time is given in Table 2, and it is found that the processing time (s) is proportional to the number of frames per sec.

The following chart depicts the processing time versus the number of frames as shown in the Fig. 9 corresponding to the values in Table 2.

6 Conclusion

For identifying the moving objects in the surveillance video, a tracking system is proposed. The proposed system segmented the moving objects in the video using

(a) Bounding box determination of each objects

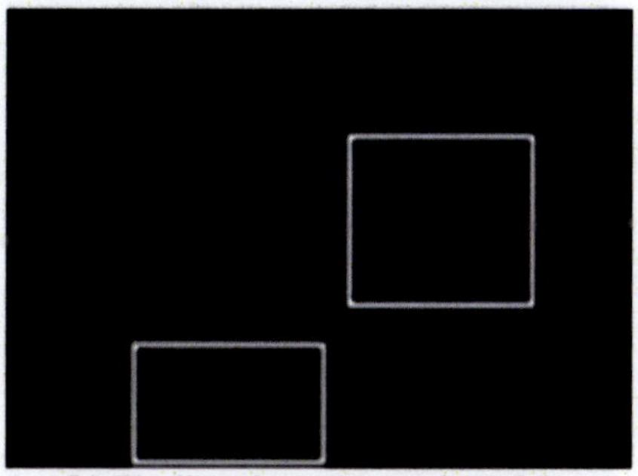

(b) Finding the centre point of the objects

(c) Tracking results (with objects trajectories)

Fig. 8 The result of the proposed video tracking framework

Table 2 Execution time versus Number of frames

No of frames (pers)	Execution time (s)
15	35
35	74
50	105
100	213
250	511
360	730
569	1152
767	1530
1053	2102

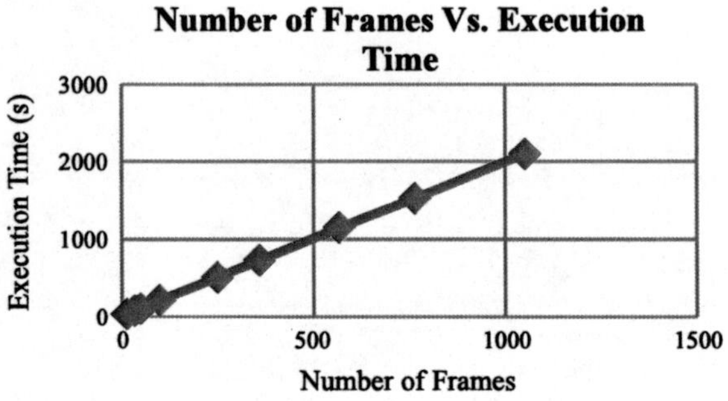

Fig. 9 The graphical representation of processing time(s) with the number of frames(pers)

motion vectors estimated from the frequency domain by applying the Expectation–Maximization Algorithm. The system can eliminate the shadows that are segmented as objects along with the objects; hence, the system used shadow removal approach. The proposed method can robustly segment the moving objects even in the presence of shadows. Then the moving objects are tracked using motion vectors. The results of the proposed tracking system have been experimented with a good number of videos, and it performs well. This proposed work can be extended to address the tracking of dynamic background-based objects.

Acknowledgements The authors would like to thank the Department of Science and Technology, India, for their financial support extended through Fund for Improvement of S and T Infrastructure (FIST) programme (SR/FST/ETI-349/2013).

References

1. Zhou X, Li Y, He B (2012) Tracking humans in mutual occlusion based on game theory. In: IEEE international conference on information and automation (ICIA), pp 156–161
2. Ghosh A, Subudhi BN, Ghosh S (2012) Object detection from videos capturing by moving camera by fuzzy edge incorporated markov random field and local histogram matching. IEEE Trans Circuits Syst Video Technol 22(8):1127–1135
3. Manjula KR, Dinesh Kumar E (2014) Segmentation of multispectral images using enhanced watershed transform. IEEE Xplore 6:978–93
4. Dinesh Kumar E, Manjula KR (2014) Segmentation of multispectral images using enhanced watershed transform. Int J Appl Eng Res (IJAER) 9(15):2967–2974
5. Afonso MV, Nascimento JC, Marques JS (2014) Automatic estimation of multiple motion fields from video sequences using a region matching based approach. IEEE Trans Multimedia 16(1):1–14
6. Shoban Raj G, Manjula KR (2015) Enhancement of security in block-based visual secret sharing. Int J Appl Eng Res (IJAER) 109(10):25909–25916
7. Zhou Xiaowei, Yang Can, Weichuan Yu (2013) Moving obejct detection by detecting contiguous outliers in the low-rank representation. IEEE Trans Pattern Anal Mach Intell 35(3):597–610
8. Nascimento JC, Figueiredo MAT, Marques JS (2013) Activity recognition using a mixture of vector fields. IEEE Trans Image Process 22(5):1712–1725

9. Roy A, Chattopadhyay P, Sural S et al (2015) Modelling, synthesis and characterization of occlusion in videos. IET Comput Vis 9:943–949
10. Anandhakumar P, Priyadarshini J, Rajeswari L, Srividhya S, Niveditha CK (2011) Occlusion detection and object tracking using filter banks. In: 2011 International conference on recent trends in information technology (ICRTIT), June 2011, pp 992–998
11. Jianyong B, Runfen Y, Yang Y (2012) A novel vehicle's shadow detection and removal algorithm. In: 2nd International conference on consumer electronics, communications and networks (CECNet), Apr 2012, pp 822–826
12. Xue JT, Hui LY, Xing SF (2012) Research on shadow elimination in intelligent traffic monitoring. In: International conference on machine learning and cybernetics, vol 4, pp 1350–1355
13. Chien SY, Chan WK, Tseng YH, Chen HY (2013) Video object segmentation and tracking framework with improved threshold decision and diffusion distance 23(6):921–934
14. Guo L, Cheng T, Huang Y, Zhao J, Zhang R (2015) Unsupervised video object segmentation by spatiotemporal graphical model. Int J Multimedia Tools Appl:1–17
15. Ahmadi P, Tabandeh M, Gholampour I (2016) Abnormal event detection and localization in traffic videos based on group sparse topical coding. IET Image Process 10:235–246
16. Porikli F, Bashir F, Sun H (2010) Compressed domain video object segmentation. IEEE Trans Circ Syst Video Technol 20(1):2–14
17. Kong W, Hussain A, Saad MHM, Tahir NM (2012) Hand detection from silhouette for video surveillance application. In: IEEE 8th International colloquium on signal processing and its applications (CSPA), pp 514–518
18. Manjula KR, Keshari AK, Pahalzani A (2015) An approach to perform uncertainty analysis on the spatial dataset using clustering and distance-based outlier detection technique. Ind J Sci Technol 8
19. Kim T, Lee S, Paik J (2011) Combined shape and feature-based video analysis and its application to non-rigid object tracking. IET Image Process 5(2):87–100
20. Huang S, Hong J (2011) Moving object tracking system based on camshaft and kalman filter. In: International conference on consumer electronics, communication and networks (CECNet), pp 1423–1426
21. Chiranjeevi P, Sengupta S (2014) Neighbourhood supported model level fuzzy aggregation for moving object segmentation. IEEE Trans Image Process 23(2):645–657

Short Circuit Fault Analysis of Three Phase Transmission System Using Matlab

D. Srilatha and R. V. S. Lakshmi Kumari

Abstract Electric power sector is broadly categorised into four sections as electric power generation, transmission, distribution and utilisation. Generated power has to be transmitted from generating station to load centre through transmission system with higher efficiency. It is necessary to know various types of faults and their impact on power system reliability. Power system faults are broadly classified into open circuit and short circuit faults. In this paper, effect of short circuit faults on a three phase transmission system has been analysed. Entire work has been done using MATLAB simulink environment. The complete modelling and proposed simulation work have been done, and the analysis of transient faults was carried out with the MATLAB simulink software. In this work, MATLAB software is used for creating various types of faults for shorter duration on transmission lines such as L-G fault, LL-G fault, LL fault and three phase short circuit and it also analysed the effect of faults on voltage and current through different phases.

Keywords Electric power generation · L-G fault · LL-G fault · Voltage

1 Introduction

As various faults occur in the system, the fault analysis helps to assess the bus voltage and the line current. The control framework forms of gear associated with it are affected by these faults. Today, due to the continuous extension of the Power System Network, it is unavoidable to command and regulate power systems. The words 'bus voltage' and 'line current' are very important when dealing with the power system. There are primarily two types of faults in the power system [1, 2]. Accurate and fast

D. Srilatha (✉)
Department of EEE, Vasireddy Venkatadri Institute of Technology, Nambur, AP, India
e-mail: Srilatha.dande@gmail.com

R. V. S. Lakshmi Kumari
Department of EEE, Gayatri Vidya Parishad College of Engineering for Women, Vishakapatnam, AP, India
e-mail: sharmalaks@gmail.com

S. Jyothi et al. (eds.), *Proceedings of the 2nd International Conference on Computational and Bio Engineering*, Lecture Notes in Networks and Systems 215,
https://doi.org/10.1007/978-981-16-1941-0_77

detection of the path and location of faults under different types of fault conditions are essential requirements from the fault point of service restoration and versatility. Study of transmission line fault analysis is to secure the power system. We use the circuit breaker and relay mechanism as switch gear units to protect the power system [3]. The identification and analysis of high voltage transmission line faults helps to pick and establish a better safety objective and transmission line protection. The safe device is a circuit breaker and the L-L-L fault depends entirely on its rating. As compared to other flaws, the triple line fault current is much larger. Simulation is carried out on the computer using MATLAB simulation, and it is easy to identify and analyse over-voltage transmission line faults [4]. This is because the single line to ground fault is more common, and the three phase fault is very high relative to the other transmission system faults. If there are various faults in the system, then the fault analysis helps to assess the voltage of the bus and the bus present line. The term is used when dealing with the power grid bus voltage, and line current is very critical. Short circuit faults are mainly classified into two types: the first one is the symmetrical fault and the other is the unsymmetrical fault.

All three phase faults are symmetrical, and the remaining, line to ground (L-G), double line to ground (LL-G) faults, and line to line (L-L), are unsymmetrical faults. And, line to ground fault (L-G) is the most common fault, and 65–70% of faults are of this type. This makes the conductor to contact with the ground. Three phase fault is the most severe fault. Occurrence of these faults in the transmission system causes rapid increase of line or phase current magnitude and decrease of voltage magnitude. These electrical faults cause over-current flow through lines when the system has electrical faults, the low impedance direction results in the flow of the fault current. A large amount of current drawn from the power supply allows the relay to trigger and destroy the electrical equipment. The main objective of this work is to study the general faults occurring in a transmission system, such as both symmetrical and unsymmetrical faults on the transmission line of power system.

2 Modelling of Three Phase Transmission System

The simulation of various faults related to the simulation model made in the MATLAB programme using the Sim Power System is shown in Figs. 1 and 2. The circuit consists of transmission line section, three phase voltage source, three phase load and three phase fault blocks. The rating of different components, i.e. three-phase sources, three-phase transformers, three-phase loads and different transmission line lengths, should be carefully selected for the simulation analysis. Under different fault conditions, a discrete powergui with sampling time Ts = 3e−05 s is used for the simulation analysis.

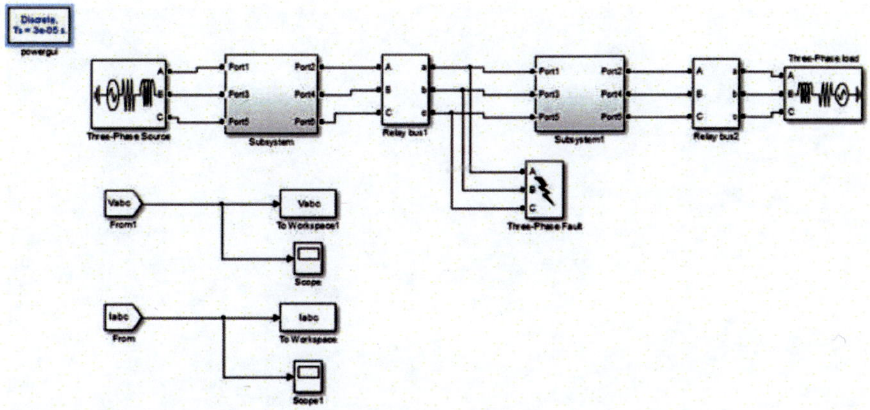

Fig. 1 Simulink model of three phase transmission system

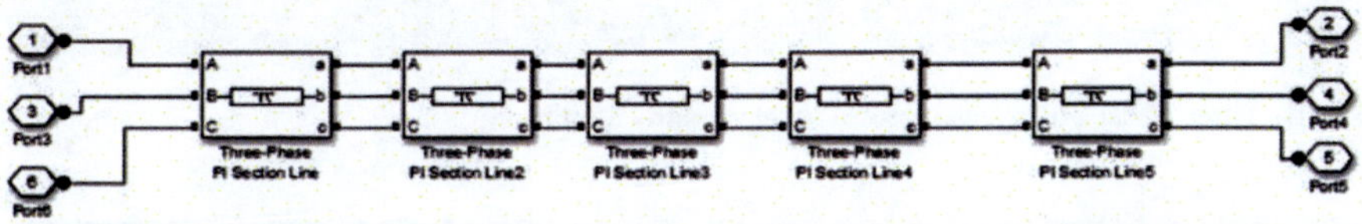

Fig. 2 Subsystem model

2.1 *Simulation Results and Analysis*

a. A three phase transmission system has been simulated under healthy conditions without any fault, and voltage and current wave forms of healthy system is shown in Figs. 3 and 4. All the three phase voltage and current waveforms are having phase difference of 120 degrees with each other with equal magnitudes.

b. When a triple line to ground fault is created over 0.15 s, all the three phase voltage magnitude became zero, and current rapidly increases. Variation of voltage and current waveforms is shown in Figs. 5 and 6.

c. When a single line to ground fault is created over 0.15 s, the single phase voltage magnitude became zero, and the phase current rapidly increases. Variation of voltage and current waveforms are shown in Figs. 7 and 8.

d. When a double line to ground fault is created over 0.15 s, the two phase voltage magnitude became zero, and their phase current rapidly increases. The variation of voltage and current waveforms is shown in Figs. 9 and 10.

e. When a line to line fault is created over 0.15 s, three phase voltage magnitude and their phase current variations are shown in Fig. 11 and 12.

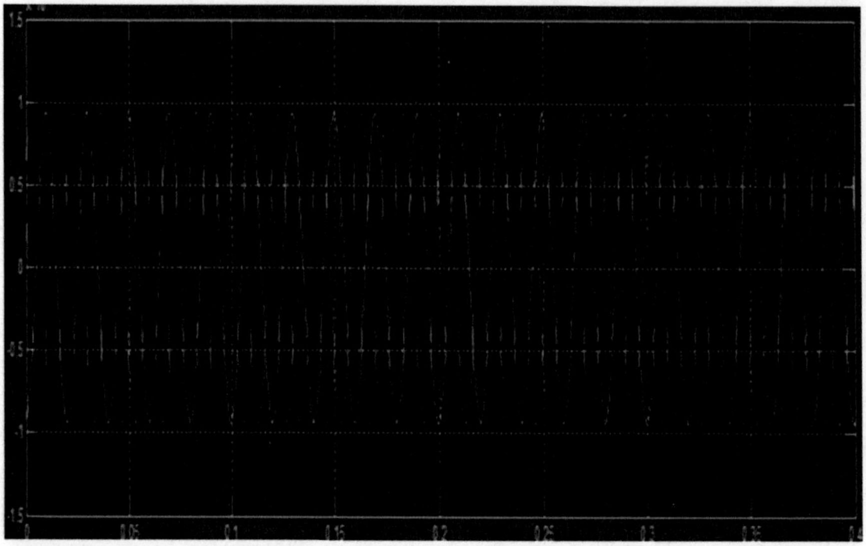

Fig. 3 Three phase voltage waveform under healthy condition

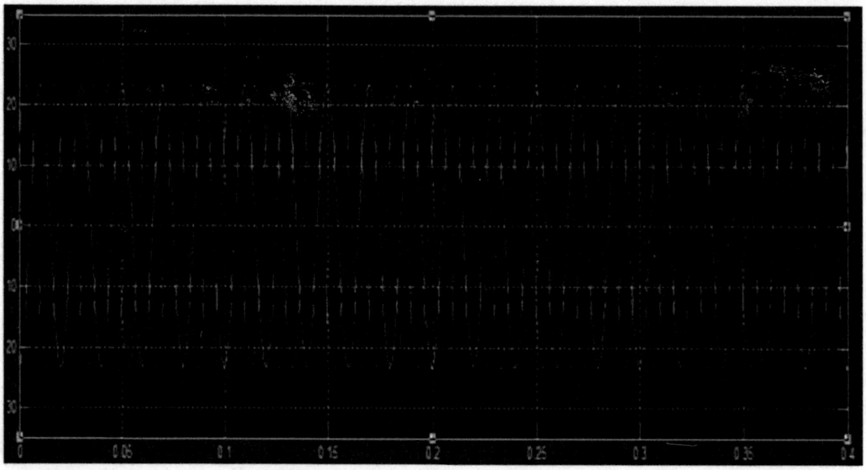

Fig. 4 Three phase current waveform under healthy condition

3 Conclusion

A simulation analysis based on MATLAB simulink is conducted on the different forms of symmetrical and unsymmetrical faults that occurred in the transmission line. Circuit breaker is introduced to eliminate the impact of failure in the healthy system due to faults. The effect of both symmetrical and unsymmetrical faults on

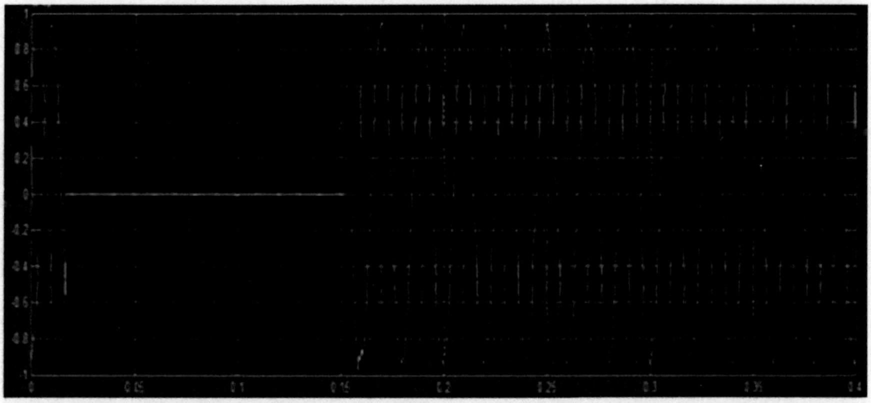

Fig. 5 Three phase voltage waveform for LLL-G fault

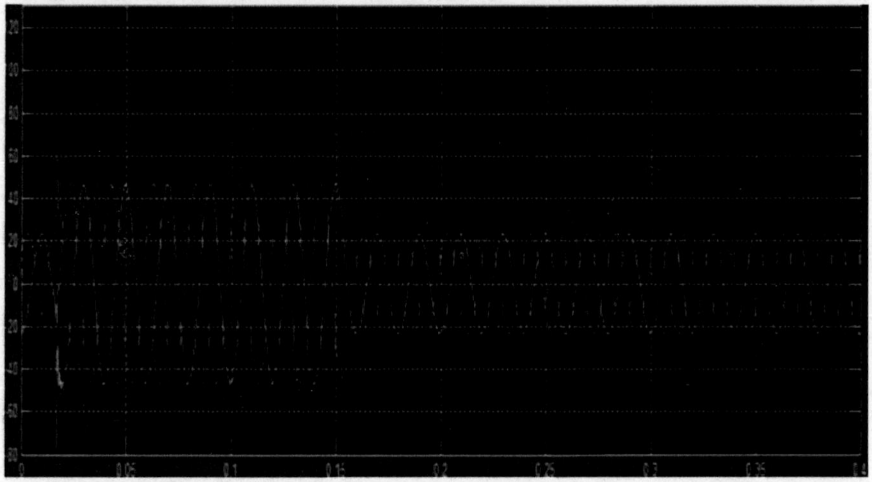

Fig. 6 Three phase current waveform for LLL-G fault

transmission system has been created, and its impact on line voltage and current has been analysed during prefault and post fault conditions.

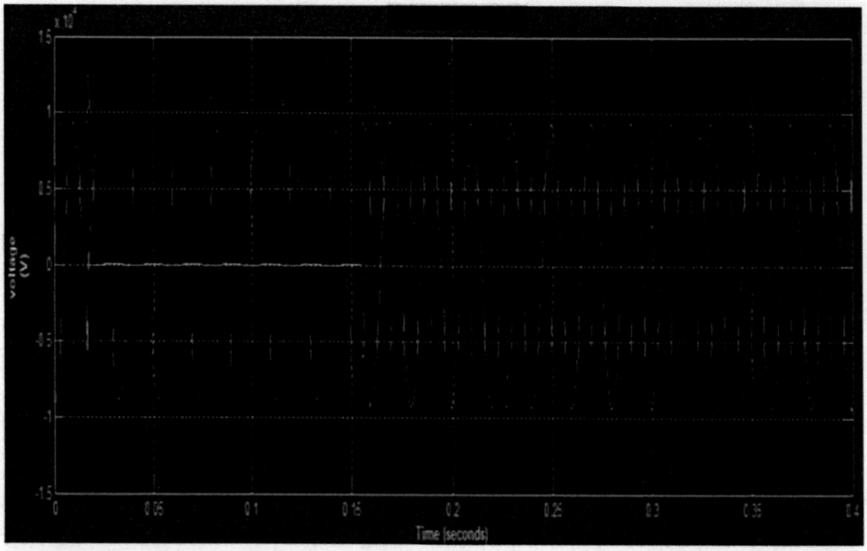

Fig. 7 Three phase voltage wave form for L-G fault

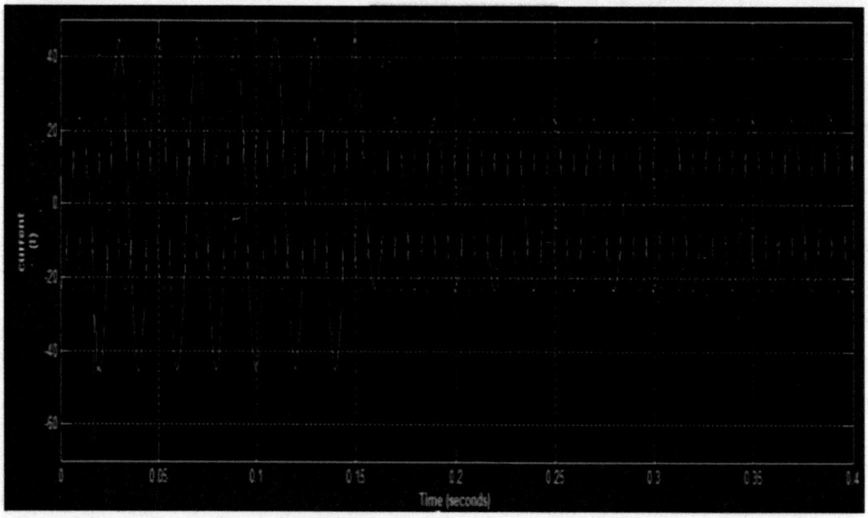

Fig. 8 Three phase current wave form for L-G fault

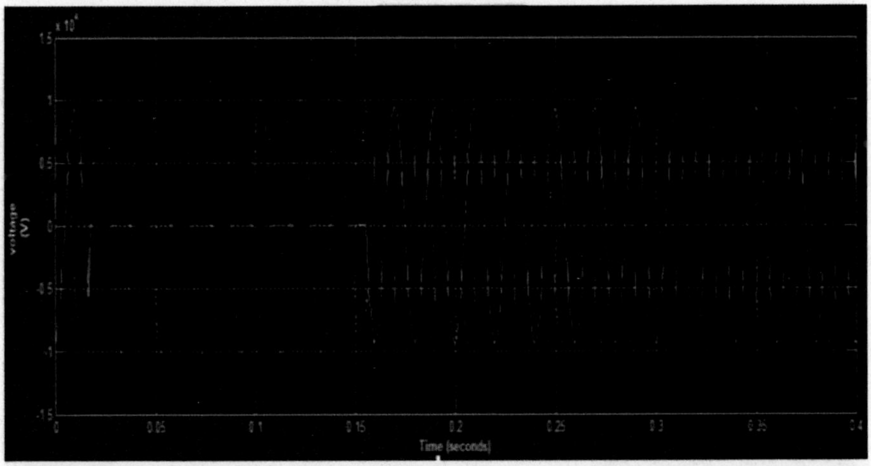

Fig. 9 Three phase voltage wave form for LL-G fault

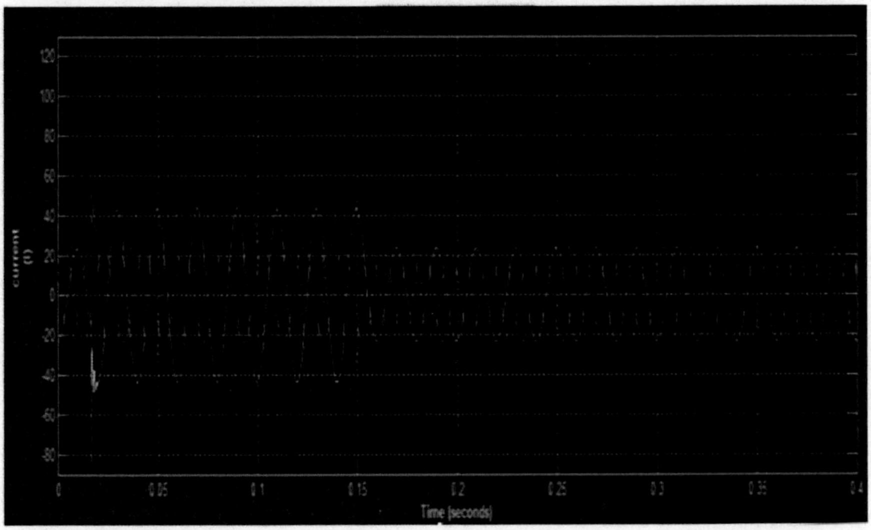

Fig. 10 Three phase current wave form for LL-G fault

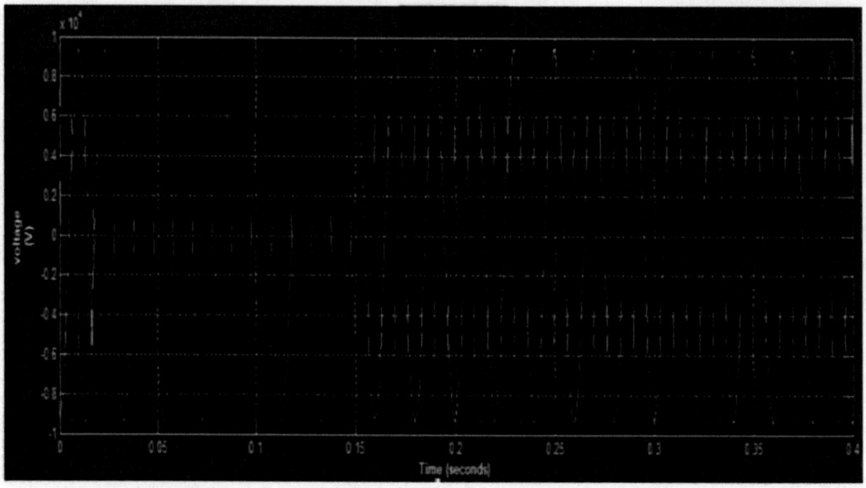

Fig. 11 Three phase voltage wave form for LL fault

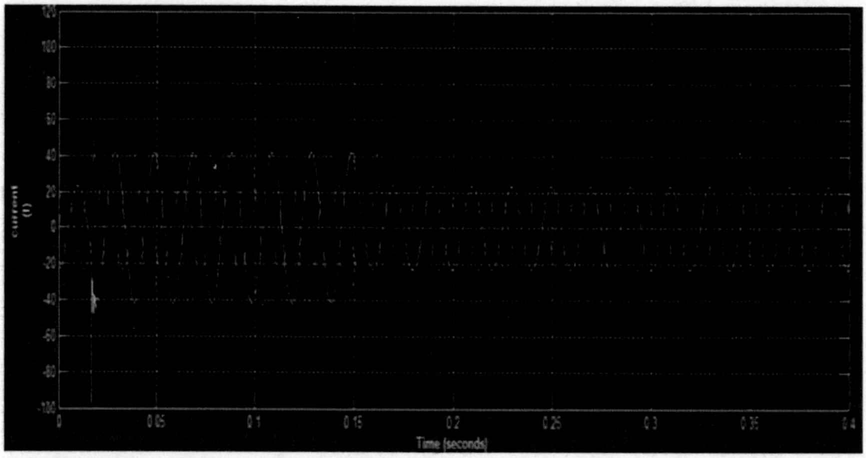

Fig. 12 Three phase current wave form for LL fault

References

1. Karekar S, Barik T (2016) A modelling of 440 KV EHV transmission line faults identified and analysis by using MATLAB. IJAREEIE 5(3):1242–1249
2. Naghate Swapnil C, Dhuldhar Saurab M, Nagdewate Ashvani B (2015) Transmission line fault analysis by using MATLAB simulation. Int J Eng Sci Res Technol 4(2):330–333
3. Mahapatra S, Singh M (2016) Analysis of symmetrical fault in IEEE 14 bus system for enhancing over current protection scheme. Int J Future Gener Commun Network 9(4):51–62

4. Manju, Maharana S, Sharma C (2014) Fault analysis of transmission line approach to MATLAB simulation. Taraksh J Web Serv (1)1
5. Venkateswara Rao K, Srilatha D (2020) Disease prediction and diagnosis implementing fuzzy neural classifier based on IOT and Cloud. Int J Adv Sci Technol 29 (5):737–745
6. Jose P, Bindu VR (2014) Wavelet-based transmission line fault analysis. Int J Eng Innov Technol (IJEIT) 3(8):55–60
7. Kesharwani S, Singh DK (2014) Simulation of fault detection for protection of transmission line using neural network. Int J Sci Eng Technol Res (IJSETR) 3(5)
8. Srilatha D, Lakshmi Kumari RVS (2020) Analysis of active power load dispatch in economic and environmental aspects with pseudo power flow. Int J Future Gener Commun Network 13(3):3665–3675
9. Srilatha D, Lakshmi Kumari RVS, Jain B (2018) Analysis of multi area load frequency control system with hybrid fruit fly algorithm. J Adv Res Dynam Control Syst 10, 10-Special Issue:2669–2678

Text Summarization of Multiple Documents Using Binary Fruit Fly Optimization Algorithm

Kishore Kumar Mamidala and Suresh Kumar Sanampudi

Abstract Due to the availability of huge data on the Internet, it becomes a time-consuming process for the user to find the most relevant text related to their interesting topic. This process can be simplified by text summarization. Initially, a set of multiple documents are preprocessed separately. Later, word frequency and the similarity of the sentences from every preprocessed document are calculated and then most redundant sentences are eliminated. Next, the content from these multiple preprocessed document is merged into a single input document. Finally, the merged document is given as input to the proposed Binary Fruit Fly Optimization Algorithm to generate a text summary. We used the ROUGE system to evaluate the results.

Keywords Text summarization · Fruit fly optimization algorithm · ROUGE · Abstractive summary · Natural language processing · Term frequency

1 Introduction

Digital information on the Internet is growing at an exponential rate day by day. Due to this, users need to spend a lot of time to get interesting information from it. Therefore, creating summaries from such huge data is essential. When such work is done by humans, it can be very intense. So, new tools, methods and designs are needed to create text summaries automatically. A text summary is a set of phrases or sentences that best summarize the relevant concepts of the document [1]. It aim is to generate a fluent and consistent summary that contains only the key points mentioned in the document. It serves as a perfect guide to the users to study, research

K. K. Mamidala (✉)
Department of Computer Science & Engineering, Vivekananda Institute of Technology & Science, Karimnagar, India
e-mail: kishore.mamidala@gmail.com

S. K. Sanampudi
Department of Information Technology, JNTUH College of Engineering Jagtial, Telangana, India
e-mail: sureshsanampudi@jntuh.ac.in

S. Jyothi et al. (eds.), *Proceedings of the 2nd International Conference on Computational and Bio Engineering*, Lecture Notes in Networks and Systems 215,
https://doi.org/10.1007/978-981-16-1941-0_78

and make further decisions on their favourite topic. This also saves users' time to make a decision on whether to read a particular document or not.

The different approaches of automating text summarization can be broadly categorized into extractive and abstractive [2, 3]. In an Extraction-based summarization, the key phrases are identified from the source document and combined to create a summary without making any changes to the texts. So, it may contain grammatical errors in essence. In an Abstraction-based summarization, the summary is created as entirely new phrases and sentences that capture the meaning of the original lengthy document. It can overcome grammatical inconsistencies in the sentences of the summary. This approach is similar to humans preparing abstract, but it is a more challenging task to a machine due to the lack of complex linguistic information.

In natural language processing (NLP), text summarization is one of the fastest-growing fields in recent years with a lot of attention from researchers. The task of text summarization can be done through a machine learning model by considering whether or not to include a sentence in the generated summary as a classification problem [4, 5]. Machine learning models are usually trained to understand the document and gather useful information. Swarm intelligence can be introduced in the development of machine learning models to prepare better a text summary [4]. We propose the use of the Binary Fruit Fly Optimization Algorithm to prepare a more accurate text summary.

The basic organization of the paper is arranged in five sections as follows: Sect. 2 presents related work on text summarization and use of the Fruit Fly Optimization Algorithm (FFOA) to solve various optimization problems, Sect. 3 describes the proposed model to generate a summary, Sect. 4 presents the experiment and results and the conclusions are given in Sect. 5.

2 Related Work

The main objective of extractive summarization is to identify the most significant sentences in the source document. Many researchers have made various efforts to advance text summarization using different techniques, tools and approaches. The quality of a good extractive summary often depends on two important factors such as relevance (include the most important sentences of the source text) and non-repetition (excluding duplicate sentences) [f].

Many researchers used TF-IDF to measure the importance of sentences for preparing a summary [6]. TF-IDF is a numerical-based criterion that specifies the importance of a word in a document. TF is a word frequency in the document and if a word appears many times in a document, then its TF value is high. IDF reduces the weight of the more frequently appearing words and increases the weight of the infrequently appearing words in the corpus. So, the words with high TF-IDF values indicate that they are highly correlated with the document in which they found.

Some authors have changed the summarization process as optimization or a search problem. They used nature-inspired optimization approaches to extract a summary.

Sanchez- Gomez et al. [5] generated a summary using Artificial Bee Colony (ABC) by simultaneously increasing content coverage and removing redundancy from multiple documents. Liu Dexi, et al. [7] used the genetic algorithm (GA), a well-known search and optimization technique for generating maximum quality summaries by identifying sets of very important sentences from input documents. GA initiates the summary process by randomly selecting sentences as the initial population. MS Binwahlan, et al. [4] used the swarm intelligence-based Particle Swarm Optimization model to generate the final summary based on sentence scores by emphasizing textual features.

Fruit Fly Optimization Algorithm (FFOA) [8] is one of the swarm intelligence-based novel heuristic search algorithms proposed in 2011. It is inspired by the natural behaviour of fruit flies to find the optimal solution of the objective function. This algorithm has a simple design, less parameters and is easy to implement. Choubey, Nitin S. et al. [9] used Fruit Fly Optimization Algorithm (FFOA) for solving the Travelling Salesperson Problem. Shen, Liming, et al. [10] used FFOA to tune the parameters of support vector machines (SVM) to classify the medical data more accurately. Wang Ling, et al. [11] used a modified version of FFOA called binary FFOA (BFFOA) to solve the multi-dimensional knapsack problem using three different types of evolutionary searches: smell-based search, local vision-based search and global vision-based search.

3 Fruit Fly Optimization Algorithm

The Fruit Fly Optimization Algorithm (FFOA) [8] is a novel evolutionary method for finding optimal solutions inspired by the food-finding behaviour of the fruit fly. A fruit fly is superior to other species in sensing and perception, especially in osphresis and vision. The osphresis organs of fruit flies can find all kinds of food scents floating in the air and fly towards the corresponding location. After getting near to the food locations, they use their sensitive vision to find the best food location and fly towards that direction. The procedure of the FOA is summarized as follows.

Step 1: Initialization phase: Initialize FFOA parameters such as the maximum number of generations and population size.
Step 2: Initialize the starting position of the fruit fly group.
Step 3: Osphresis foraging phase: Every individual Fruit Fly searches in a random direction for food around the current location of the fruit fly group using osphresis.
Step 4: Vision foraging Phase: Evaluate all the new food locations found by each Fruit Fly based on the smell concentration values (fitness value).
Step 5: Find the best Fruit Fly based on the highest smell concentration value, and then the Fruit Fly group moves towards the best one utilizing vision.
Step 6: Repetition phase: End the algorithm if the maximum number of iterations is reached; otherwise, go back to Step 3.

4 Proposed Model

Before implementing the proposed Binary Fruit Fly Optimization Algorithm (*BFFOA*) on a text document to generate a text summary, preprocessing should be performed to convert the data into a "ready-to-use" form. Initially, preprocessing is applied separately on each document of a set of documents and modify them accordingly. Later, calculate sentence informative weight (*sw*) for each sentence in every document separately and also calculate inter-sentence similarity weight to eliminate redundant sentences. Finally, all the contents from multiple documents are merged into a single document. The merged document is given as input to the *BFFOA* to generate text summarization.

4.1 Preprocessing

The following steps are performed to preprocess the text documents [12]:

- **Sentence Tokenization:** It is the process of recognizing where the sentences start and end to make further processing of the text relatively easy.
- **Noise Removal:** This step purely depends on the text domain and goal of the analysis. It removes unwanted information, such as special characters, numeric digits, punctuation, accents and html tags from the text.
- **Word tokenization:** It is the process of splitting the sentence into words. This step is essential to calculate the feature scores of individual words of a sentence for finding important sentences in a document.
- **Removing Stop words:** It is the process of removing words which do not convey any information, such as "and", "the" and "it" which are insignificant in feature score calculation to simplify the summarization task.
- **Stemming:** It is the process of transforming the linguistically derived word to its root word. This step further simplifies the task deducing the crux of the document.

4.2 Calculate Sentence Score

We used the Term Frequency–Inverse Document Frequency (TF-IDF) [6] method to assign the weights to the words in the document. The weighting process is done by applying the below-given equations:

$$\mathrm{tf(w,doc)} = \frac{\mathrm{freq_{t,\,doc}}}{\mathrm{\|D\|}} \tag{1}$$

$$\mathrm{idf(w)} = \frac{\log \mathrm{T}}{\mathrm{df(w)}} \tag{2}$$

$$sw_{ik} = tf(w, doc)*idf(w) \quad (3)$$

where $freq_{w,doc}$ represents the number of times a word w occurs in a document *doc* and ||D|| represents the total number of unique words in that document, T is the total number of sentences in the document, $df(w)$ is the total number of documents containing the word w and sw_{ik} is the ith sentence informative weight with respect to word w.

4.3 Preparing Final Input Document

After calculating each sentence weight (*sw*), calculate inter-sentence similarities using Eq. (4)

$$\mathrm{sim}(s_i, s_j) = \frac{\sum_{k=1}^{m} sw_{ik} * sw_{jk}}{\sqrt{\sum_{k=1}^{m} sw_{ik}^2} \cdot \sqrt{\sum_{k=1}^{m} sw_{ik}^2}} \quad (4)$$

From each document, select the least similar sentences according to predefined threshold similarity value. Merge all the selected least similar sentences from each document into a single document D_{input}. Once again apply Eq. (4) on D_{input} to eliminate redundant sentences. Now, $D_{input} = \{St_1, St_2, \ldots, St_N\}$ contains N sentences which act as an input to the proposed *BFFOA* to produce the text summary as output.

4.4 Binary Fruit Fly Optimization Algorithm

Initialization phase: The Binary Fruit Fly Optimization Algorithm (*BFFOA*) starts with initializing the fruit fly swarm location. In this phase, we start with n solutions (Fruit Flies) represented by $Sol_{(g)i}$, (i = 1, 2, ..., n), and g is the generation number initialized as $g = 0$. Each $S_{(g)i}$ is a N-bit binary vector such that $Sol_{(g)i} = [s_{(g)i1}, s_{(g)i2}, \ldots, s_{(g)iN}]$ where $s_{(g)ij}$ is either 0 or 1. If $s_{(g)ij} = 1$, then *jth* sentence from input document is included in the *ith* solution. Otherwise, it is discarded. The solution in each generation is updated by using a probability vector $Prob_{(g)} = [P_{(g)1}, P_{(g)2}, \ldots, P_{(g)N}]$, where the probability value $P_{(g)j} \in [0,1]$. Each solution is updated by generating N random decimal numbers between 0 and 1. If *jth* random number, $rand_j < P_{(g)j}$, then $s_{(g)ij} = 1$; otherwise $s_{(g)ij} = 0$. The initial probabilty vector $Prob_{(0)} = [0.5, 0.5, \ldots, 0.5]$ is considered to provide uniform chances for every sentence in the input search space to be included in the solution space.

Osphresis foraging phase: In this phase, each Fruit Fly performs a smell-based search to explore food locations around it. This is implemented by generating Q number of neighbour solutions around each n solution. We did this by performing a mutation operation by inverting randomly selected L number of bits. To restrict to a neighbour search, L is chosen such that $\frac{L}{N}<0.1$. To control each generated new solution, not to include almost all sentences in the summary, a repair operation is performed. In this operation, if the number of 1 s in each solution is greater than $\frac{N}{\sqrt[3]{N}}$, then excess 1 s are converted to 0 s by random selection.

Vision foraging Phase: In this phase, each Fruit Fly moves to its best neighbour food location. It is implemented by replacing each parent solution with their best neighbour solution by evaluating its *Q* neighbour solutions using the below defined fitness function.

The main objective of the good summary is to maximize the cohesion among the sentences, relevant to the query and preserving readability [13]. Therefore, the fitness function should include three factors: cohesion factor, topic relationship factor and readability factor. We considered the weighted average of these three factors as a fitness function $f(Sol_{(g)k})$ to generate the summary. It is given as follows:

$$f\left(Sol_{(g)k}\right) = \frac{\alpha f_{relv}\left(Sol_{(g)k}\right) + \beta f_{coh}\left(Sol_{(g)k}\right) + \gamma f_{read}\left(Sol_{(g)k}\right)}{\alpha + \beta + \gamma} \tag{5}$$

$$f_{relv}\left(Sol_{(g)k}\right) = \mathrm{sim}\left(\mathrm{s_p}, \mathrm{C}\right) p = 1, 2, \ldots, N \tag{6}$$

$$f_{coh}\left(Sol_{(g)k}\right) = 1 - \mathrm{sim}\left(\mathrm{s}_p, \mathrm{s_q}\right) p \neq q = 1, 2, \ldots, N \tag{7}$$

$$f_{read}(Sol_{(g)k}) = sim(s_p, s_q) \; p \neq q = 1, 2, \ldots, N \tag{8}$$

where α, β and γ are adjustable parameters that specify the relative importance of the three factors. $f_{relv}(Sol_{(g)k})$ is the relevance factor that indicates the coverage of source document content in the summary. *C* is the weighted average of sentences in the source document. s_p and s_q are the *p*th and *q*th sentences in the input document that is present in the *kth* solution in the *gth* generation. $f_{coh}(Sol_{(g)k})$ is the cohesion factor that indicates the connection between the sentences. The higher the cohesion factor, the summary quality is high and vice versa. $f_{read}(Sol_{(g)k})$ is the readability factor that refers to how the sentences in the summary are interrelated.

After calculating the fitness value of each *Q* neighbour solution of a parent solution, the best neighbour solution (maximum fitness) replaces its parent solution if any. To further enhance and to avoid converge of solution in local minima, the group probability vector $Prob_{(g)}$ is updated by selecting the best two solutions $s_{(g)best1}$ and $s_{(g)best2}$ using the below equation:

$$\mathrm{P}_{(g)i} = \frac{1}{1+e^{-b\left(s_{(g)best1i} - s_{(g)best2i}\right)*\mathrm{P}_{(g)i}}} \tag{9}$$

where b is a vision search parameter. Now update each *n* solution using the updated group probability vector $Prob_{(g)}$, if its solution at a global location is better than the current location. This step allows the Fruit Fly to move to a new global position.

Repeat the Osphresis foraging phase and Vision foraging phase one after the other desired result is achived or the maximum number of iterations (generations) is reached.

5 Results and Experiment

The widely used evaluation method in the text summary is to compare the results of the proposed algorithm on the standard corpus published by the Document Understanding Collection (DUC). The DUC has many sets, but mostly many researchers primarily use DUC01 and DUC02 for experimentation. DUC01 and DUC02 contain 309 and 567 news articles in English, respectively, each with a source document and Golden Standard summaries created by two different human experts. We compared our results with the below-given methods.

The Fuzzy Evolutionary Optimization Model (FEOM) [14] generates a summary using a fuzzy-based controller from optimized text clusters formed based on sentence content. UnifiedRank [15] is a graph-based representation formed using sentence ranking. It uses a mutual relationship between extraction of the summary from both single documents and proposes a novel unified approach to the simultaneous generation of the summary. Manifold-Ranking [16] provides a ranking score to sentences based on information richness and uses a greedy algorithm to impose a diversity penalty on each sentence. The summary is generated by selecting the sentences with both high information richness and high information novelty. Conditional Random Fields (CRF) [17] treat the summarization process as a sequence labelling problem by considering interactions between sentences. They use heuristic rules to select the most informative sentences into a summary by treating each sentence as either 0 or 1. Querying-Clustering-Summarizing (QCS) [18] is a tool used to generate a summary from the document by removing redundant information.

The summary generated by BFFOA is evaluated by comparing it with predefined golden standard summaries using F-measure from ROUGE [19].

$$\mathrm{FM}(Sol) = \frac{2{*}\mathrm{Rc}(Sol){*}\mathrm{Pr}(Sol)}{\mathrm{Rc}(Sol) + \mathrm{Pr}(Sol)} \tag{10}$$

$$Rc(Sol) = \frac{\sum_{S \in S_{ref}} \sum_{gram_n \in S} Count_{match}(gram_n)}{\sum_{S \in S_{ref}} \sum_{gram_n \in S} Count(gram_n)} \tag{11}$$

$$\mathrm{Pr}(Sol) = \frac{\sum_{S \in} S_{ref} \sum_{gram_n \in S} Count_{match}(gram_n)}{S \in Sol \sum_{gram_n \in S} Count(gram_n)}. \tag{12}$$

where *Rc* is the recall, *Pr* is precision and *FM* is the F-measure metric of the best final solution (summary) *Sol* using ROUGE. *n-gram* is the contiguous sequence of *n* words, S_{ref} is a set of human-produced summaries usually known a gold-standard summaries set and *Countmatch*(*gramn*) is the number of *n-grams* that co-occur between the BFFOA-generated summary and the human-produced reference summaries.

From the above table, it is clear that the BFFOA shows better results when compared with other existing methods. Unlike abstractive text summarization methods, the proposed method produces a summary with grammatical errors and less proficiency. But it shows superior results when compared to other methods in the Extractive Text Summarization category. BFFOA is one of the swarm intelligence evolutionary algorithms, so its comprehensive search capability increases and gives better results with increasing computing power and time.

6 Conclusion

This paper focuses on the modified version of Fruity Fly-based search to generate a better extraction summary from multiple documents. The sentence feature analysis is performed using TF-IDF to find important sentences. After studying all the sentence scores, the sentences with similar content are removed to minimize the redundancy and to improve the performance of the proposed algorithm. The proposed method generated a summary using inter-sentence similarity and readability metric by removing redundancy, preserving cohesiveness and readability of the summary. All experiments are evaluated in the DUC01 and DUC02 datasets using ROUGE-1 and ROUGE-2. The BFFOA summary is compared with other existing methods in the extractive text summarization category, and it is observed that the proposed model shows a better ROUGE score.

Table 1 Comparison of F-measure score of the proposed model with other related models on DUC01 and DUC02 collections using ROUGE-1 and ROUGE-2

Document set	Method	ROUGE-1	ROUGE-2
DUC01	Proposed BFFOA	0.49864	0.20026
	FEOM	0.47728	0.18549
	UnifiedRank	0.45377	0.17649
	Manifold Ranking	0.43359	0.16635
	CRF	0.45512	0.17327
	QCS	0.44852	0.18523
DUC02	Proposed BFFOA	0.49236	0.23592
	FEOM	0.46575	0.12490
	UnifiedRank	0.48478	0.21462
	Manifold Ranking	0.42325	0.10677
	CRF	0.44006	0.10924
	QCS	0.44865	0.18766

References

1. Gillick Daniel et al. (2009) The icsi/utd summarization system at tac 2009. Tac.
2. Gupta Vishal, Gurpreet Singh Lehal (2010) A survey of text summarization extractive techniques. J Emerg Technol Web İntell 2.3(2010):258–268
3. Khan A, Salim N (2014) A review on abstractive summarization methods. J Theor Appl Inf Technol 59(1):64–72
4. Binwahlan Mohammed Salem, Naomie Salim, Ladda Suanmali (2009) Swarm based text summarization. 2009 International Association of Computer Science and Information Technology-Spring Conference. IEEE
5. Sanchez-Gomez JM, Vega-Rodríguez MA, Pérez CJ (2018) Extractive multi-document text summarization using a multi-objective artificial bee colony optimization approach. Knowl-Based Syst 159:1–8
6. Ledeneva Yulia, Alexander Gelbukh, René Arnulfo García-Hernández (2008) Terms derived from frequent sequences for extractive text summarization. International Conference on Intelligent Text Processing and Computational Linguistics. Springer, Berlin, Heidelberg
7. Liu Dexi et al. (2006) Genetic algorithm based multi-document summarization. Pacific rim ınternational conference on artificial ıntelligence. Springer, Berlin, Heidelberg
8. Pan Wen-Tsao (2011) A new evolutionary computation approach: fruit fly optimization algorithm. 2011 Conference of Digital Technology and Innovation Management
9. Choubey NS (2014) Fruit fly optimization algorithm for travelling salesperson problem. Int J Comp Appl 107(18):22–27
10. Shen Liming et al. (2016) Evolving support vector machines using fruit fly optimization for medical data classification. Knowledge-Based Systems 96:61–75
11. Wang L, Zheng X-L, Wang S-Y (2013) A novel binary fruit fly optimization algorithm for solving the multidimensional knapsack problem. Knowl-Based Syst 48:17–23
12. Manne Suneetha, Sameen Fatima S (2012) An extensive empirical study of feature terms selection for text summarization and categorization. Proceedings of the Second International Conference on Computational Science, Engineering and Information Technology
13. Chatterjee Niladri, Amol Mittal, Shubham Goyal. "Single document extractive text summarization using genetic algorithms." 2012 Third International Conference on Emerging Applications of Information Technology. IEEE, 2012.
14. Song Wei et al. (2011) Fuzzy evolutionary optimization modeling and its applications to unsupervised categorization and extractive summarization. Expert Syst Appl 38.8:9112–9121

15. Wan Xiaojun (2010) Towards a unified approach to simultaneous single-document and multi-document summarizations. Proceedings of the 23rd international conference on computational linguistics (Coling 2010)
16. Wan Xiaojun, Jianwu Yang, Jianguo Xiao (2007) Manifold-ranking based topic-focused multi-document summarization. IJCAI, vol. 7
17. Shen Dou et al. (2007) Document summarization using conditional random fields. IJCAI, vol. 7
18. Dunlavy Daniel M, John Conroy, Dianne P O'leary (2003) QCS: a tool for querying, clustering, and summarizing documents. Companion Volume of the Proceedings of HLT-NAACL 2003-Demonstrations
19. Lin Chin-Yew (2004) Rouge: a package for automatic evaluation of summaries. Text summarization branches out

Correction to: Classification of Normal/Abnormal Heart Sound Recording Through Convolution Neural Network Through the Integration of Baseline and AdaBoost Classifier

P. Jyothi and G. Pradeepini

Correction to:
Chapter "Classification of Normal/Abnormal Heart Sound Recording Through Convolution Neural Network Through the Integration of Baseline and AdaBoost Classifier" in:
S. Jyothi et al. (eds.),
***Proceedings of the 2nd International Conference on Computational and Bio Engineering*,**
Lecture Notes in Networks and Systems 215,
https://doi.org/10.1007/978-981-16-1941-0_44

In the original version of the book, the following belated corrections have been incorporated:

In chapter "Classification of Normal/Abnormal Heart Sound Recording Through Convolution Neural Network Through the Integration of Baseline and AdaBoost Classifier", the affiliation from "CSE Department, VNRVJIET, Hyderabad, India and CSE, and Department, KL University, Vijayawada, India" of author "P. Jyothi and G. Pradeepini" respectively is to be changed to:

P. Jyothi Department of CSE, Koneru Lakshmaiah Education Foundation, Vaddeswaram, India.

G. Pradeepini Department of CSE, Koneru Lakshmaiah Education Foundation, Vaddeswaram, India.

The chapter and book have been updated with the changes.

The updated version of this chapter can be found at
https://doi.org/10.1007/978-981-16-1941-0_44

S. Jyothi et al. (eds.), *Proceedings of the 2nd International Conference on Computational and Bio Engineering*, Lecture Notes in Networks and Systems 215,
https://doi.org/10.1007/978-981-16-1941-0_79

Author Index

A
Akthar, Shaheda, 641
Ambedkar, Y. R., 299
Amul Mary, F., 115
Anand Kumar Varma, S., 745
Anchapakula, Suvarna Latha, 729, 735
Anitha, Raju, 395
Aparna, Gira, 239
Aruna Mastani, S., 33

B
Bai, Vadetay Saraswathi, 251
Balasubramanyam, P., 355, 577
Bandi, Sai Vennela, 267
Basava, Preethika, 709
Battula, Priyanka, 709
Bayana, Chetan Reddy, 709
Bhagyamma, C. T., 177
Bhargavi, M., 333
Bhargavi, P., 615
Bhavani, B. G., 85
Bhogapathi, Raveendra Babu, 497
Bolla, Sai Jyothi, 163
Botlagunta, Mahendran, 593
Buddareddy, Hemavathi, 729, 735

C
Chandrasekhar Reddy, M., 577
Chinnaiah, M. C., 687

D
Deepika, V., 51
Depuru, Bharathi, 729, 735
Dileep Kumar Reddy, P., 533, 651
Divyaja, P., 1
Divya, Recharla, 477

E
Eluri, Ramesh, 423

G
Gayathri Devi, S., 679
Gnaneswari, K., 209
Gowthami, K., 219

H
Hanumanthappa, M., 585
Hari Priya, K., 507
Hema, K., 51
Hemalatha, K., 51
Himabindu, K., 11, 17

I
Indhravathi, Ch., 381

J
Jaya Madhuri, R., 195, 219
Jyothi, P., 441
Jyothi, S., 17, 43, 93, 115, 163, 403, 507

K
Kakarla, Naga Madhavi Latha, 459

S. Jyothi et al. (eds.), *Proceedings of the 2nd International Conference on Computational and Bio Engineering*, Lecture Notes in Networks and Systems 215,
https://doi.org/10.1007/978-981-16-1941-0

Kalala, Doosetty Padmaja, 615
Kallepalli, Suseela, 59
Karanam, Swetha Rani, 59
Koduru, Swetha kumari, 565
Kora, Padmavathi, 687
Kurni, Muralidhar, 449

L

Lakshmi, C., 187
Lakshmi Kumari, R. V. S., 759
Lakshmi Praveena, T., 631
Lakshmi Thanuja, B., 209
Lalasa, M., 25, 129
Lal, Pallavaram Venkateswar, 135
Lavanya, Korada, 477

M

Maddu, Saraswathi, 343
Madhavi, K., 449
Madhavi, K. R., 687
Madhuri, R. Jaya, 333
Malarvizhi, N., 417
Malini, M., 593
Mamatha, D. M., 17, 403, 507, 553, 565
Mamidala, Kishore Kumar, 769
Manekar, Amitkumar, 625
Manjula, K. R., 661, 679, 695, 709, 745
Manjunath, M., 585
Maravajhala, Vidyavathi, 279
Mary, Gade Anitha, 239
Mathakala, Vani, 229
Meda, Srikanth, 497
Meenakshi, K., 687
Monica, T., 365
Morusu, Keerthana, 279
Mounika Raj, K., 487
Mukku, Samykya, 129
Muthu Lakshmi, N. V., 631
Muzib, Y. Indira, 299

N

Naga Deepthi, P., 459
Nagadevi, D., 449
Nagalakshmamma, K., 25, 129
Nagalakshmi, G., 291
Nagendra Prasad, Y., 533
Nageshwar Rao, P., 25
Nalini Durga, S., 151
Nalla, Sarath, 423
Nasina, Sai Sindhuri, 323
Nithya, S., 25

P

Palemaplli, Uma Maheswari Devi, 229
Patra, Raj Kumar, 433
Peddiboyana, Vijaya Lakshmi, 267
Pendyala, Vijetha, 423
Pichaimani, Venkateswari, 695
Pradeepini, G., 441, 625
Priyadarshini, R., 417
Priya, Narasimhappagari Chandana, 311
Pydipati, Lakshmi Padmavathi, 729, 735

R

Rajani, A., 687
Rajaram, Gangothri, 661
Raju, P. J., 403
Rama Mohan Reddy, A., 603
Ramya, N., 679
Rani, M. Usha, 311
Ravikumar, K., 585
Ravuri, Jaya Madhuri, 343
Reddy, A. Rama Mohan, 323
Revathi, A., 43, 93
Rohith, V. S., 679

S

Sai Lalitha, B., 593
Sai Mounika, Avvaru, 467
Sai Sruthi, K., 543
Sai Susheel, K., 259
Sanampudi, Suresh Kumar, 769
Sandhya Madhuri, G., 721
Sanober, Sumaya, 523
Santhi, K., 603
Sarada, V., 467
Sarath, T., 291
Saritha, K., 449
Sasisharan, P. P., 679
Savitri, G., 129
Seetharamulu, J., 403
Sekar, K., 11
Shanthi Sree, K. S., 129
Sharmila, S., 507
Shoba Bindu, C., 533, 651
Shobha, S., 543
Somasekhar, G., 433
Sreedevi, A., 543
Sreenivasa Murthy, K., 355
Srilakshmi, Uppalapati, 135
Srilatha, D., 759
Sripriya, Chilakapati, 467
Sriramadasu, Kalpana, 553

Srujan Raju, K., 433
Sucharita, V., 259
Sudha, T., 251
Sudheer Kumar, E., 533
Sujata, P., 129
Sujatha, M., 195
Sujatha, V., 651
Sujitha, Y. Sarah, 299
Sukanya, K., 75, 107
Sukeerthi, T., 75, 107
Sultana, Sufia, 403
Sumana, Gollapalli, 239
Suneel, A. Sai, 365
Suneetha, K., 487
Suneetha, T., 75
Suresh Kumar, P. G. V., 641
Suresh, Salini, 585
Suvarnalatha, A., 25
Swaraja, K., 687
Swathi, K., 459
Swathi, P., 43, 93

T
Tata, Ravi Kumar, 395

U
Usha, R., 381
Usharani, Bhimavarapu, 395
Usha Rani, K., 151, 187, 523
Usha Rani, M., 1, 721

V
Vandana Rao, K., 11, 107
Vani, A., 75
Vani, M., 209
Venkateswara Rao, P., 259
Venkateswarlu, D., 135
Vijaya Kumari, N., 177
Vijay Kumar, G., 651

Y
Yamuna, 721
Yerragolla, Sathvik, 267

Z
Zaheer Ahamed, M., 33